T0234799

GEOMETRY OF QUANTUM STATES
An Introduction to Quantum Entanglement

Quantum information theory is a branch of science at the frontiers of physics, mathematics and information science, and offers a variety of solutions that are impossible using classical theory. This book provides a detailed introduction to the key concepts used in processing quantum information and reveals that quantum mechanics is a generalisation of classical probability theory.

The second edition contains new sections and entirely new chapters: the hot topic of multipartite entanglement; in-depth discussion of the discrete structures in finite dimensional Hilbert space, including unitary operator bases, mutually unbiased bases, symmetric informationally complete generalised measurements, discrete Wigner functions and unitary designs; the Gleason and Kochen–Specker theorems; the proof of the Lieb conjecture; the measure concentration phenomenon; and the Hastings' non-additivity theorem.

This richly-illustrated book will be useful to a broad audience of graduates and researchers interested in quantum information theory. Exercises follow each chapter, with hints and answers supplied.

INGEMAR BENGTSSON is a professor of physics at Stockholm University. After gaining a Ph.D. in theoretical physics from the University of Göteborg (1984), he held post-doctoral positions at CERN, Geneva, and Imperial College, London. He returned to Göteborg in 1988 as a research assistant at Chalmers University of Technology, before taking up a position as Lecturer in Physics at Stockholm University in 1993. He was appointed Professor of Physics in 2000. Professor Bengtsson is a member of the Swedish Physical Society and a former board member of its Divisions for Particle Physics and for Gravitation. His research interests are related to geometry, in the forms of classical general relativity and quantum theory.

KAROL ŻYCZKOWSKI is a professor at the Institute of Physics, Jagiellonian University, Kraków, Poland and also at the Center for Theoretical Physics, Polish Academy of Sciences, Warsaw. He gained his Ph.D. and habilitation in theoretical physics at Jagiellonian University, and has followed this with a Humboldt Fellowship in Essen, a Fulbright Fellowship at the University of Maryland, College Park, and a visiting research position at the Perimeter Institute, Waterloo, Ontario. He has been a docent at the Academy of Sciences since 1999 and a full professor at Jagiellonian University since 2004. Professor Życzkowski is a member of the Polish Physical Society and Academia Europaea. He works on quantum information, dynamical systems and chaos, quantum and statistical physics, applied mathematics and the theory of voting.

GEOMETRY OF QUANTUM STATES

An Introduction to Quantum Entanglement

SECOND EDITION

INGEMAR BENGTSSON

Stockholm University, Sweden

KAROL ŻYCZKOWSKI

Jagiellonian University, Poland

CAMBRIDGE
UNIVERSITY PRESS

CAMBRIDGE
UNIVERSITY PRESS

University Printing House, Cambridge CB2 8BS, United Kingdom

One Liberty Plaza, 20th Floor, New York, NY 10006, USA

477 Williamstown Road, Port Melbourne, VIC 3207, Australia

314-321, 3rd Floor, Plot 3, Splendor Forum, Jasola District Centre, New Delhi - 110025, India

79 Anson Road, #06-04/06, Singapore 079906

Cambridge University Press is part of the University of Cambridge.

It furthers the University's mission by disseminating knowledge in the pursuit of education, learning and research at the highest international levels of excellence.

www.cambridge.org
Information on this title: www.cambridge.org/9781107656147
DOI: 10.1017/9781139207010

© Ingemar Bengtsson and Karol Życzkowski 2017

First published 2006
Second edition 2017
First paperback edition 2020

A catalogue record for this publication is available from the British Library

ISBN 978-1-107-02625-4 Hardback
ISBN 978-1-107-65614-7 Paperback

Contents

Preface

Preface to the first edition

The geometry of quantum states is a highly interesting subject in itself. It is also relevant in view of possible applications in the rapidly developing fields of quantum information and quantum computing.

But what is it? In physics words like 'states' and 'system' are often used. Skipping lightly past the question of what these words mean – it will be made clear by practice – it is natural to ask for the properties of the space of all possible states of a given system. The simplest state space occurs in computer science: a 'bit' has a space of states that consists simply of two points, representing on and off. In probability theory the state space of a bit is really a line segment, since the bit may be 'on' with some probability between zero and one. In general the state spaces used in probability theory are 'convex hulls' of a discrete or continuous set of points. The geometry of these simple state spaces is surprisingly subtle – especially since different ways of distinguishing probability distributions give rise to different notions of distance, each with their own distinct operational meaning. There is an old idea saying that a geometry can be understood once it is understood what linear transformations are acting on it, and we will see that this is true here as well.

The state spaces of classical mechanics are – at least from the point of view that we adopt – just instances of the state spaces of classical probability theory, with the added requirement that the sample spaces (whose 'convex hull' we study) are large enough, and structured enough, so that the transformations acting on them include canonical transformations generated by Hamiltonian functions.

In quantum theory the distinction between probability theory and mechanics goes away. The simplest quantum state space is these days known as a 'qubit'. There are many physical realisations of a qubit, from silver atoms of spin 1/2 (assuming that we agree to measure only their spin) to the qubits that are literally designed in today's laboratories. As a state space a qubit is a three–dimensional ball; each diameter of the ball is the state space of some classical bit, and there are so many bits that their sample spaces conspire to form a space – namely the surface of the ball – large enough to carry the canonical transformations that are characteristic of mechanics. Hence the word quantum mechanics.

It is not particularly difficult to understand a three–dimensional ball, or to see how this description emerges from the usual description of a qubit in terms of a complex two-dimensional Hilbert space. In this case we can take the word *geometry* literally – there will exist a one-to-one correspondence between pure states of the qubit and the points of the surface of the Earth. Moreover, at least as far as the surface is concerned, its geometry has a statistical meaning when transcribed to the qubit (although we will see some strange things happening in the interior).

As the dimension of the Hilbert space goes up, the geometry of the state spaces becomes very intricate, and qualitatively new features arise – such as the subtle way in which composite quantum systems are represented. Our purpose is to describe this geometry. We believe it is worth doing. Quantum state spaces are more wonderful than classical state spaces, and in the end composite systems of qubits may turn out to have more practical applications than the bits themselves already have.

A few words about the contents of our book. As a glance at the Contents will show, there are 17 chapters, culminating in a long chapter on 'entanglement'. Along the way, we take our time to explore many curious byways of geometry. We expect that you – the reader – are familiar with the principles of quantum mechanics at the advanced undergraduate level. We do not really expect more than that, and should you be unfamiliar with quantum mechanics we hope that you will find some sections of the book profitable anyway. You may start reading any chapter – if you find it incomprehensible we hope that the cross-references and the index will enable you to see what parts of the earlier chapters may be helpful to you. In the unlikely event that you are not even interested in quantum mechanics, you may perhaps enjoy our explanations of some of the geometrical ideas that we come across.

Of course there are limits to how independent the chapters can be of each other. Convex set theory (Chapter 1) pervades all statistical theories, and hence all our chapters. The ideas behind the classical Shannon entropy and the Fisher–Rao geometry (Chapter 2) must be brought in to explain quantum mechanical entropies (Chapter 12) and quantum statistical geometry (Chapters 9 and 13). Sometimes we have to assume a little extra knowledge on the part of the reader, but since no chapter in our book assumes that all the previous chapters have been understood, this should not pose any great difficulties.

We have made a special effort to illustrate the geometry of quantum mechanics. This is not always easy, since the spaces that we encounter more often than not have a dimension higher than three. We have simply done the best we could. To facilitate self-study each chapter concludes with problems for the reader, while some additional geometrical exercises are presented in the final appendix.

Once and for all, let us say that we limit ourselves to finite-dimensional state spaces. We do this for two reasons. One of them is that it simplifies the story very much, and the other is that finite-dimensional systems are of great independent interest in real experiments.

The entire book may be considered as an introduction to quantum entanglement. This very non-classical feature provides a key resource for several modern applications of quantum mechanics including quantum cryptography, quantum computing and quantum communication. We hope that our book may be useful for graduate and postgraduate students of physics. It is written first of all for readers who do not read the mathematical literature everyday, but we hope that students of mathematics and of the information sciences will find it useful as well, since they also may wish to learn about quantum entanglement.

We have been working on the book for about five years. Throughout this time we enjoyed the support of Stockholm University, the Jagiellonian University in Cracow, and the Center for Theoretical Physics of the Polish Academy of Sciences in Warsaw. The book was completed in Waterloo during our stay at the Perimeter Institute for Theoretical Physics. The motto at its main entrance – ΑΣΠΟΥΔΑΣΤΟΣ ΠΕΡΙ ΓΕΩΜΕΤΡΙΑΣ ΜΗΔΕΙΣ ΕΙΣΙΤΩ – proved to be a lucky omen indeed, and we are pleased to thank the Institute for creating optimal working conditions for us, and to thank all the knowledgeable colleagues working there for their help, advice, and support.

We are grateful to Erik Aurell for his commitment to Polish–Swedish collaboration; without him the book would never have been started. It is a pleasure to thank our colleagues with whom we have worked on related projects: Johan Brännlund, Åsa Ericsson, Sven Gnutzmann, Marek Kuś, Florian Mintert, Magdalena Sinołęcka, Hans-Jürgen Sommers and Wojciech Słomczyński. We are grateful to them, and to many others who helped us to improve the manuscript. If it never reached perfection, it was our fault, not theirs. Let us mention some of the others: Robert Alicki, Anders Bengtsson, Iwo Białynicki-Birula, Rafał Demkowicz-Dobrzański, Johan Grundberg, Sören Holst, Göran Lindblad, and Marcin Musz. We have also enjoyed constructive interactions with Matthias Christandl, Jens Eisert, Peter Harremoës, Michał, Paweł and Ryszard Horodeccy, Vivien Kendon, Artur Łoziński, Christian Schaffner, Paul Slater, and William Wootters.

Five other people provided indispensable support: Martha and Jonas in Stockholm, and Jolanta, Jaś, and Marysia in Cracow.

Waterloo, 12 March 2005   *Ingemar Bengtsson*
                                                            Karol Życzkowski

Preface to the second edition

More than a decade has passed since we completed the first edition of the book. Much has happened during it. We have not tried to take all recent valuable contributions into account, but some of them can be found here. The Lieb conjecture for spin coherent states has been proved by Lieb and Solovej, and an important additivity conjecture for quantum channels has been disproved. Since these conjectures were discussed at some length in the first edition we felt we had to say more about them. However, our main concern with this second edition has been to improve explanations and to remove mistakes (while trying not to introduce new ones).

There are two new chapters: one of them is centred around the finite Weyl–Heisenberg group, and the discrete structures it gives rise to. The other tries to survey the vast field of multipartite entanglement, which was relegated to a footnote or two in the first edition. There are also a few new sections: They concern Gleason's theorem and quantum contextuality, cross–sections and projections of the body of mixed quantum states, and the concentration of measure in high dimensions (which is needed to understand what is now Hastings' non-additivity theorem).

We are grateful to several people who helped us with the work on this edition. It is a pleasure to thank Felix Huber, Pedro Lamberti and Marcus Müller for their comments on the first edition. We are indebted to Radosław Adamczak, Ole Andersson, Marcus Appleby, Adán Cabello, Runyao Duan, Shmuel Friedland, Dardo Goyeneche, David Gross, Michał and Paweł Horodeccy, Ted Jacobson, David Jennings, Marek Kuś, Ion Nechita, Zbigniew Puchała, Wojciech Roga, Adam Sawicki, Stanisław Szarek, Stephan Weis, Andreas Winter, Iwona Wintrowicz, and Huangjun Zhu, for reading some fragments of the text and providing us with valuable remarks. Had they read all of it, we are sure that it would have been perfect! We thank Kate Blanchfield, Piotr Gawron, Lia Pugliese, Konrad Szymański, and Maria Życzkowska, for preparing for us two-dimensional figures and three-dimensional printouts, models and photos for the book.

We thank Simon Capelin and his team at Cambridge University Press for the long and fruitful work with us. Most of all we thank Martha and Jolanta for their understanding and infinite patience.

Toruń, 11 June 2016

Ingemar Bengtsson
Karol Życzkowski

Acknowledgements

Figures 4.10–13 and 16.1–3 have appeared already in our reference [112], which was published in the *International Journal of Physics* **A17** Issue 31 (2002), coauthored with Johan Brännlund. They appear here with the kind permission of the copyright holder, World Scientific.

Figure 5.2 and the Figure accompanying the Hint to Problem 5.2 are reprinted from Dr. Kate Blanchfield's PhD thesis, with the kind permission of its copyright holder and author.

1

Convexity, colours and statistics

1.1 Convex sets

What picture does one see, looking at a physical theory from a distance, so that the details disappear? Since quantum mechanics is a statistical theory, the most universal picture which remains after the details are forgotten is that of a convex set.

—*Bogdan Mielnik*[1]

Our object is to understand the geometry of the set of all possible states of a quantum system that can occur in nature. This is a very general question; especially since we are not trying to define 'state' or 'system' very precisely. Indeed we will not even discuss whether the state is a property of a thing, or of the preparation of a thing, or of a belief about a thing. Nevertheless we can ask what kind of restrictions are needed on a set if it is going to serve as a space of states in the first place. There is a restriction that arises naturally both in quantum mechanics and in classical statistics: the set must be a *convex set*. The idea is that a convex set is a set such that one can form 'mixtures' of any pair of points in the set. This is, as we will see, how probability enters (although we are not trying to define 'probability' either).

From a geometrical point of view a *mixture* of two states can be defined as a point on the segment of the straight line between the two points that represent the states that we want to mix. We insist that given two points belonging to the set of states, the straight line segment between them must belong to the set too. This is certainly not true of any set. But before we can see how this idea restricts the set of states we must have a definition of 'straight lines' available. One way to proceed is to regard a convex set as a special kind of subset of a flat Euclidean space $\mathbf{E}^n$. Actually we can get by with somewhat less. It is enough to regard a convex set as a subset of an affine space. An *affine space* is just like a vector space, except that no

---

[1] Reproduced from [659].

special choice of origin is assumed. The *straight line* through the two points $\mathbf{x}_1$ and $\mathbf{x}_2$ is defined as the set of points

$$\mathbf{x} = \mu_1 \mathbf{x}_1 + \mu_2 \mathbf{x}_2, \qquad \mu_1 + \mu_2 = 1. \tag{1.1}$$

If we choose a particular point $\mathbf{x}_0$ to serve as the origin, we see that this is a one-parameter family of vectors $\mathbf{x} - \mathbf{x}_0$ in the plane spanned by the vectors $\mathbf{x}_1 - \mathbf{x}_0$ and $\mathbf{x}_2 - \mathbf{x}_0$. Taking three different points instead of two in Eq. (1.1) we define a *plane*, provided the three points do not belong to a single line. A k-dimensional k–*plane* is obtained by taking $k + 1$ generic points, where $k < n$. For $k = n$ we describe the entire space $\mathbf{E}^n$. In this way we may introduce *barycentric coordinates* into an n-dimensional affine space. We select $n + 1$ points $\mathbf{x}_i$, so that an arbitrary point $\mathbf{x}$ can be written as

$$\mathbf{x} = \mu_0 \mathbf{x}_0 + \mu_1 \mathbf{x}_1 + \ldots + \mu_n \mathbf{x}_n, \qquad \mu_0 + \mu_1 + \ldots + \mu_n = 1. \tag{1.2}$$

The requirement that the barycentric coordinates μ_i add up to one ensures that they are uniquely defined by the point $\mathbf{x}$. (It also means that the barycentric coordinates are not coordinates in the ordinary sense of the word, but if we solve for μ_0 in terms of the others then the remaining independent set is a set of n ordinary coordinates for the n-dimensional space.) An *affine map* is a transformation that takes lines to lines and preserves the relative length of line segments lying on parallel lines. In equations an affine map is a combination of a linear transformation described by a matrix $\mathbf{A}$ with a translation along a constant vector $\mathbf{b}$, so $\mathbf{x}' = \mathbf{A}\mathbf{x} + \mathbf{b}$, where $\mathbf{A}$ is an invertible matrix.

By definition a subset S of an affine space is a *convex set* if for any pair of points $\mathbf{x}_1$ and $\mathbf{x}_2$ belonging to the set it is true that the *mixture* $\mathbf{x}$ also belongs to the set, where

$$\mathbf{x} = \lambda_1 \mathbf{x}_1 + \lambda_2 \mathbf{x}_2, \qquad \lambda_1 + \lambda_2 = 1, \quad \lambda_1, \lambda_2 \geq 0. \tag{1.3}$$

Here λ_1 and λ_2 are barycentric coordinates on the line through the given pair of points; the extra requirement that they be positive restricts $\mathbf{x}$ to belong to the segment of the line lying between the pair of points.

It is natural to use an affine space as the 'container' for the convex sets since convexity properties are preserved by general affine transformations. On the other hand it does no harm to introduce a flat metric on the affine space, turning it into an Euclidean space. There may be no special significance attached to this notion of distance, but it helps in visualizing what is going on. See Figures 1.1 and 1.2. From now on, we will assume that our convex sets sit in Euclidean space, whenever it is convenient to do so.

Intuitively a convex set is a set such that one can always see the entire set from whatever point in the set one happens to be sitting at. Still they can come in a variety of interesting shapes. We will need a few definitions. First, given any subset of the

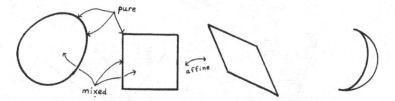

Figure 1.1 Three convex sets, two of which are affine transformations of each other. The new moon is not convex. An observer in Singapore will find the new moon tilted but still not convex, since convexity is preserved by rotations.

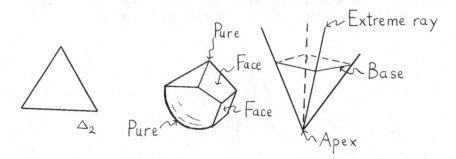

Figure 1.2 The convex sets we will consider are either convex bodies (like the simplex on the left, or the more involved example in the centre) or convex cones with compact bases (an example is shown on the right).

affine space we define the *convex hull* of this subset as the smallest convex set that contains the set. The convex hull of a finite set of points is called a *convex polytope*. If we start with $p + 1$ points that are not confined to any $(p - 1)$-dimensional subspace then the convex polytope is called a *p-simplex*. The *p*-simplex consists of all points of the form

$$\mathbf{x} = \lambda_0 \mathbf{x}_0 + \lambda_1 \mathbf{x}_1 + \ldots + \lambda_p \mathbf{x}_p, \quad \lambda_0 + \lambda_1 + \ldots + \lambda_p = 1, \quad \lambda_i \geq 0. \quad (1.4)$$

(The barycentric coordinates are all non–negative.) The *dimension* of a convex set is the largest number n such that the set contains an n-simplex. When discussing a convex set of dimension n we usually assume that the underlying affine space also has dimension n, to ensure that the convex set possesses interior points (in the sense of point set topology). A closed and bounded convex set that has an interior is known as a *convex body*.

The intersection of a convex set with some lower dimensional subspace of the affine space is again a convex set. Given an n-dimensional convex set S there is also a natural way to increase its dimension with one: choose a point $\mathbf{y}$ not belonging to the n-dimensional affine subspace containing S. Form the union of all the *rays* (in this chapter a ray means a half line), starting from $\mathbf{y}$ and passing through S.

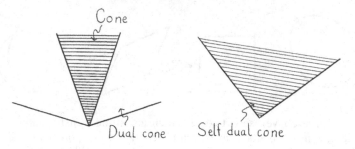

Figure 1.3 Left: a convex cone and its dual, both regarded as belonging to Euclidean 2-space. Right: a self dual cone, for which the dual cone coincides with the original. For an application of this construction see Figure 11.6.

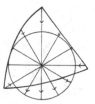

Figure 1.4 A convex body is homeomorphic to a sphere.

The result is called a *convex cone* and $\mathbf{y}$ is called its *apex*, while S is its *base*. A ray is in fact a one dimensional convex cone. A more interesting example is obtained by first choosing a p-simplex and then interpreting the points of the simplex as vectors starting from an origin O not lying in the simplex. Then the $p + 1$ dimensional set of points

$$\mathbf{x} = \lambda_0 \mathbf{x}_0 + \lambda_1 \mathbf{x}_1 + \ldots + \lambda_p \mathbf{x}_p, \qquad \lambda_i \geq 0 \tag{1.5}$$

is a convex cone. Convex cones have many nice properties, including an inbuilt partial order among its points: $\mathbf{x} \leq \mathbf{y}$ if and only if $\mathbf{y} - \mathbf{x}$ belongs to the cone. Linear maps to $\mathbb{R}$ that take positive values on vectors belonging to a convex cone form a dual convex cone in the dual vector space. Since we are in the Euclidean vector space $\mathbf{E}^n$, we can identify the dual vector space with $\mathbf{E}^n$ itself. If the two cones agree the convex cone is said to be *self dual*. See Figure 1.3. One self dual convex cone that will appear now and again is the *positive orthant* or *hyperoctant* of $\mathbf{E}^n$, defined as the set of all points whose Cartesian coordinates are non-negative. We use the notation $\mathbf{x} \geq 0$ to denote the fact that $\mathbf{x}$ belongs to the positive orthant.

From a purely topological point of view all convex bodies are equivalent to an n–dimensional ball. To see this choose any point $\mathbf{x}_0$ in the interior and then for every point in the boundary draw a ray starting from $\mathbf{x}_0$ and passing through the boundary point (as in Figure 1.4). It is clear that we can make a continuous transformation of the convex body into a ball with radius one and its centre at $\mathbf{x}_0$ by moving the points of the container space along the rays.

Convex bodies and convex cones with compact bases are the only convex sets that we will consider. Convex bodies always contain some special points that cannot be obtained as mixtures of other points – whereas a half space does not! These points are called *extreme points* by mathematicians and *pure* points by physicists (actually, originally by Weyl), while non-pure points are called *mixed*. In a convex cone the rays from the apex through the pure points of the base are called *extreme rays*; a point $\mathbf{x}$ lies on an extreme ray if and only if $\mathbf{y} \le \mathbf{x} \Rightarrow \mathbf{y} = \lambda\mathbf{x}$ with λ between zero and one. A subset F of a convex set that is stable under mixing and purification is called a *face* of the convex set. What the phrase means is that if

$$\mathbf{x} = \lambda\mathbf{x}_1 + (1 - \lambda)\mathbf{x}_2, \qquad 0 < \lambda < 1 \tag{1.6}$$

then $\mathbf{x}$ lies in F if and only if $\mathbf{x}_1$ and $\mathbf{x}_2$ lie in F. The 'only if' part of the definition forces $\mathbf{x}$ to lie on the boundary of the set. A face of dimension k is a k-face. A 0-face is an extreme point, and an $(n - 1)$-face is also known as a *facet*. It is interesting to observe that the set of all faces on a convex body form a partially ordered set; we say that $F_1 \le F_2$ if the face F_1 is contained in the face F_2. It is a partially ordered set of the special kind known as a *lattice*, which means that a given pair of faces always have a greatest lower bound (perhaps the empty set) and a lowest greater bound (perhaps the convex body itself).

To stem the tide of definitions let us quote two theorems that have an 'obvious' ring to them when they are stated abstractly but which are surprisingly useful in practice:

Minkowski's theorem. *Any convex body is the convex hull of its pure points.*

Carathéodory's theorem. *Any point in an n-dimensional convex set X can be expressed as a convex combination of at most $n + 1$ pure points in X.*

Thus any point $\mathbf{x}$ of a convex body S may be expressed as a *convex combination* of pure points:

$$\mathbf{x} = \sum_{i=1}^{p} \lambda_i\mathbf{x}_i, \qquad \lambda_i \ge 0, \qquad p \le n + 1, \qquad \sum_i \lambda_i = 1. \tag{1.7}$$

This equation is quite different from Eq. (1.2) that defined the barycentric coordinates of $\mathbf{x}$ in terms of a fixed set of points $\mathbf{x}_i$, because – with the restriction that all the coefficients be non-negative – it may be impossible to find a finite set of $\mathbf{x}_i$ so that every $\mathbf{x}$ in the set can be written in this new form. An obvious example is a circular disk. Given $\mathbf{x}$ one can always find a finite set of pure points $\mathbf{x}_i$ so that the

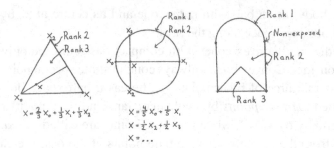

Figure 1.5 In a simplex a point can be written as a mixture in one and only one way. In general the rank of a point is the minimal number of pure points needed in the mixture; the rank may change in the interior of the set as shown in the rightmost example. The set on the right has two non-exposed points which form faces of their own. Note also that its insphere is not unique.

equation holds, but that is a different thing. The exact number of pure points one needs is related to the face structure of the body, as one can see from the proof of Carathéodory's theorem (which we give as Problem 1.1.)

It is evident that the pure points always lie in the boundary of the convex set, but the boundary often contains mixed points as well. The simplex enjoys a very special property, which is that any point in the simplex can be written as a mixture of pure points in one and only one way (as in Figure 1.5). This is because for the simplex the coefficients in Eq. (1.7) are barycentric coordinates and the result follows from the uniqueness of the barycentric coordinates of a point. No other convex set has this property. The *rank* of a point $\mathbf{x}$ is the minimal number p needed in the convex combination (1.7). By definition the pure points have rank one. In a simplex the edges have rank two, the faces have rank three, and so on, while all the points in the interior have maximal rank. From Eq. (1.7) we see that the maximal rank of any point in a convex body in $\mathbb{R}^n$ does not exceed $n + 1$. In a ball all interior points have rank two and all points on the boundary are pure, regardless of the dimension of the ball. It is not hard to find examples of convex sets where the rank changes as we move around in the interior of the set (see Figure 1.5).

The simplex has another quite special property, namely that its lattice of faces is *self dual*. We observe that the number of k-faces in an n dimensional simplex is

$$\binom{n+1}{k+1} = \binom{n+1}{n-k}. \tag{1.8}$$

Hence the set of $n-k-1$ dimensional faces can be put in one-to-one correspondence with the set of k-faces. In particular, the pure points ($k = 0$) can be put in one-to-one correspondence with the set of *facets* (by definition, the $n - 1$ dimensional faces).

Figure 1.6 Support hyperplanes of a convex set.

For this, and other, reasons its lattice of subspaces will have some exceptional properties, turning it into what is technically known as a *Boolean* lattice.[2]

There is a useful dual description of convex sets in terms of supporting hyperplanes. A *support hyperplane* of S is a hyperplane that intersects the set and which is such that the entire set lies in one of the closed half spaces formed by the hyperplane (see Figure 1.6). Hence a support hyperplane just touches the boundary of S, and one can prove that there is a support hyperplane passing through every point of the boundary of a convex body. By definition a *regular point* is a point on the boundary that lies on only one support hyperplane, a *regular support hyperplane* meets the set in only one point, and the entire convex set is regular if all its boundary points as well as all its support hyperplanes are regular. So a ball is regular, while a convex polytope or a convex cone is not – indeed all the support hyperplanes of a convex cone pass through its apex. A face is said to be *exposed* if it equals the intersection of the convex set and some support hyperplane. Convex polytopes arise as the intersection of a finite number of closed half-spaces in $\mathbb{R}^n$, and any pure point of a convex polytope saturates n of the inequalities that define the half-spaces; again a statement with an 'obvious' ring that is useful in practice.

In a flat Euclidean space a linear function to the real numbers takes the form $\mathbf{x} \to \mathbf{a} \cdot \mathbf{x}$, where $\mathbf{a}$ is some constant vector. Geometrically, this defines a family of parallel hyperplanes. We have the important

Hahn–Banach separation theorem. *Given a convex body and a point $\mathbf{x_0}$ that does not belong to it. Then one can find a linear function f and a constant k such that $f(\mathbf{x}) > k$ for all points belonging to the convex body, while $f(\mathbf{x_0}) < k$.*

This is again almost obvious if one thinks in terms of hyperplanes.

It is useful to know a bit more about dual convex sets. For definiteness let us start out with a three dimensional vector space in which a point is represented by a vector $\mathbf{y}$. Then its dual plane is the set of vectors $\mathbf{x}$ such that

$$\mathbf{x} \cdot \mathbf{y} = -1. \tag{1.9}$$

---

[2] Because it is related to what George Boole thought were the laws of thought; see Varadarajan's book [916] on quantum logic for these things.

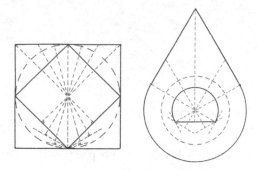

Figure 1.7 A square is dual to another square; on the left we see how the points on the edge at the top define a corner at the bottom of the dual square. To the right we see a more complicated convex set with non-exposed faces (that are points). Its dual has non-polyhedral corners. In both cases the unit circle is shown dashed.

The constant on the right hand side was set to -1 for convenience. The dual of a line is the intersection of a one-parameter family of planes dual to the points on the line. This is in itself a line. The dual of a plane is a point, while the dual of a curved surface is another curved surface – the envelope of the planes that are dual to the points on the original surface. To define the dual of a convex body with a given boundary we change the definition slightly, and include all points on one side of the dual planes in the dual. Thus the *dual* X° of a convex body X is defined to be

$$X^\circ = \{\mathbf{x} \mid c + \mathbf{x} \cdot \mathbf{y} \geq 0 \ \forall \mathbf{y} \in X\}, \tag{1.10}$$

where c is a number that was set equal to 1 above (and also when drawing Figure 1.7). The dual of a convex body including the origin is the intersection of the half-spaces defined by the pure points $\mathbf{y}$ of X. The dual of the dual of a body that includes the origin is equal to the convex hull of the original body. If we enlarge a convex body the conditions on the dual become more stringent, and hence the dual shrinks. The dual of a sphere centred at the origin is again a sphere, so a sphere (of suitable radius) is self dual. The dual of a cube is an octahedron. The dual of a regular tetrahedron is another copy of the original tetrahedron, possibly of a different size. The copy can be made to coincide with the original by means of an affine transformation, hence the tetrahedron is a self dual body.[3]

We will find much use for the concept of *convex functions*. A real function $f(\mathbf{x})$ defined on a closed convex subset X of $\mathbb{R}^n$ is called *convex*, if for any $\mathbf{x}, \mathbf{y} \in X$ and $\lambda \in [0, 1]$ it satisfies

$$f(\lambda\mathbf{x} + (1-\lambda)\mathbf{y}) \leq \lambda f(\mathbf{x}) + (1-\lambda)f(\mathbf{y}). \tag{1.11}$$

---

[3] To readers who wish to learn more about convex sets – or who wish to see proofs of the various assertions that we left unproved – we recommend the book by Eggleston [283].

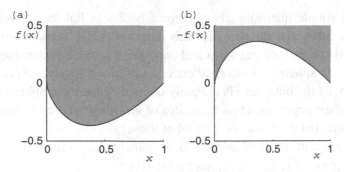

Figure 1.8 (a): the convex function $f(x) = x \ln x$ (b): the concave function $g(x) = -x \ln x$. The names stem from the shaded epigraphs of the functions which are convex and concave, respectively.

The name refers to the fact that the *epigraph* of a convex function, that is the region lying above the curve $f(\mathbf{x})$ in the graph, is convex. Applying the inequality $k - 1$ times we see that

$$f\left(\sum_{j=1}^{k} \lambda_j \mathbf{x}_j\right) \leq \sum_{j=1}^{k} \lambda_j f(\mathbf{x}_j), \tag{1.12}$$

where $\mathbf{x}_j \in X$ and the nonnegative weights sum to unity, $\sum_{j=1}^{k} \lambda_j = 1$. If a function f from $\mathbb{R}$ to $\mathbb{R}$ is differentiable, it is convex if and only if

$$f(y) - f(x) \geq (y - x) f'(x). \tag{1.13}$$

If f is twice differentiable it is convex if and only if its second derivative is nonnegative. For a function of several variables to be convex, the matrix of second derivatives must be positive definite. In practice, this is a very useful criterion. A function f is called *concave* if $-f$ is convex.

One of the main advantages of convex functions is that it is (comparatively) easy to study their minima and maxima. A minimum of a convex function is always a global minimum, and it is attained on some convex subset of the domain of definition X. If X is not only convex but also compact, then the global maximum sits at an extreme point of X.

1.2 High dimensional geometry

In quantum mechanics the spaces we encounter are often of very high dimension; even if the dimension of Hilbert space is small the dimension of the space of density matrices will be high. Our intuition on the other hand is based on two and three dimensional spaces, and frequently leads us astray. We can improve ourselves by asking some simple questions about convex bodies in flat space. We choose to look at balls, cubes and simplices for this purpose. A flat metric is assumed. Our questions will concern the *inspheres* and *outspheres* of these bodies (defined as the largest inscribed sphere and the smallest circumscribed sphere, respectively). For any convex body the outsphere is uniquely defined, while the insphere is not – one can show that the upper bound on the radius of inscribed spheres is always attained by some sphere, but there may be several of those.

Let us begin with the surface of a ball, namely the *n*-dimensional *sphere*. In equations a sphere of radius *r* is given by the set

$$X_0^2 + X_1^2 + \cdots + X_n^2 = r^2 \tag{1.14}$$

in an $n + 1$ dimensional flat space $\mathbf{E}^{n+1}$. A sphere of radius one is denoted $\mathbf{S}^n$. The sphere can be parametrized by the angles $\phi, \theta_1, \ldots, \theta_{n-1}$ according to

$$\begin{cases} X_0 = r\cos\phi\sin\theta_1\sin\theta_2\ldots\sin\theta_{n-1} \\ X_1 = r\sin\phi\sin\theta_1\sin\theta_2\ldots\sin\theta_{n-1} \\ X_2 = \quad r\cos\theta_1\sin\theta_2\ldots\sin\theta_{n-1} \\ \ldots \qquad\qquad\qquad \ldots \\ X_n = \qquad\qquad\qquad r\cos\theta_{n-1} \end{cases} \quad \begin{matrix} 0 < \theta_i < \pi \\ 0 \le \phi < 2\pi \end{matrix} . \tag{1.15}$$

The volume element d*A* on the unit sphere then becomes

$$dA = d\phi d\theta_1 \ldots d\theta_{n-1} \sin\theta_1 \sin^2\theta_2 \ldots \sin^{n-1}\theta_{n-1}. \tag{1.16}$$

We want to compute the volume vol($\mathbf{S}^n$) of the *n*-sphere, that is to say its 'hyperarea' – meaning that vol($\mathbf{S}^2$) is measured in square metres, vol($\mathbf{S}^3$) in cubic metres, and so on. A clever trick simplifies the calculation: Consider the well known Gaussian integral

$$I = \int e^{-X_0^2 - X_1^2 - \cdots - X_n^2} dX_0 dX_1 \ldots dX_n = (\sqrt{\pi})^{n+1}. \tag{1.17}$$

Using the spherical polar coordinates introduced above our integral splits into two, one of which is related to the integral representation of the Euler Gamma function, $\Gamma(x) = \int_0^\infty e^{-t} t^{x-1} dt$, and the other is the one we want to do:

$$I = \int_0^\infty dr \int_{S^n} dA e^{-r^2} r^n = \frac{1}{2}\Gamma\left(\frac{n+1}{2}\right) \text{vol}(\mathbf{S}^n). \tag{1.18}$$

We do not have to do the integral over the angles. We simply compare these results and obtain (recalling the properties of the Gamma function)

$$\text{vol}(S^n) = 2\frac{\pi^{\frac{n+1}{2}}}{\Gamma(\frac{n+1}{2})} = \begin{cases} \dfrac{2(2\pi)^p}{(2p-1)!!} & \text{if} \quad n = 2p \\[3mm] \dfrac{(2\pi)^{p+1}}{(2p)!!} & \text{if} \quad n = 2p+1 \end{cases}, \tag{1.19}$$

where double factorial is the product of every other number, $5!! = 5 \cdot 3 \cdot 1$ and $6!! = 6 \cdot 4 \cdot 2$. An alarming thing happens as the dimension grows. For large x we can approximate the Gamma function using Stirling's formula

$$\Gamma(x) = \sqrt{2\pi}\, e^{-x} x^{x-\frac{1}{2}}\left(1 + \frac{1}{12x} + o\left(\frac{1}{x^2}\right)\right). \tag{1.20}$$

Hence for large n we obtain

$$\text{vol}(S^n) \sim \sqrt{2}\left(\frac{2\pi e}{n}\right)^{\frac{n}{2}}. \tag{1.21}$$

This is small if n is large! In fact the 'biggest' unit sphere – in the sense that it has the largest hyperarea – is S^6, which has

$$\text{vol}(S^6) = \frac{16}{15}\pi^3 \approx 33.1. \tag{1.22}$$

Incidentally Stirling's formula gives 31.6, which is already rather good. We hasten to add that vol(S^2) is measured in square metres and vol(S^6) in (metre)^6, so that the direct comparison makes no sense.

There is another funny thing to be noticed. If we compute the volume of the n-sphere without any clever tricks, simply by integrating the volume element dA using angular coordinates, then we find that

$$\text{vol}(S^n) = 2\pi \int_0^\pi d\theta \sin\theta \int_0^\pi d\theta \sin^2\theta \ldots \int_0^\pi d\theta \sin^{n-1}\theta = \tag{1.23}$$

$$= \text{vol}(S^{n-1}) \int_0^\pi d\theta \sin^{n-1}\theta.$$

As n grows the integrand of the final integral has an increasingly sharp peak close to the equator $\theta = \pi/2$. Hence we conclude that when n is high most of the hyperarea of the sphere is confined to a 'band' close to the equator. What about the volume of an n-dimensional unit ball $\mathbf{B}^n$? By definition it has unit radius and its boundary is S^{n-1}. Its volume, using the radial integral $\int_0^1 r^{n-1} dr = 1/n$ and the fact that $\Gamma(x+1) = x\Gamma(x)$, is

$$\text{vol}(\mathbf{B}^n) = \frac{\text{vol}(S^{n-1})}{n} = \frac{\pi^{\frac{n}{2}}}{\Gamma(\frac{n}{2}+1)} \sim \frac{1}{\sqrt{2\pi}}\left(\frac{2\pi e}{n}\right)^{\frac{n}{2}}. \tag{1.24}$$

Again, as the dimension grows the denominator grows faster than the numerator and therefore the volume of a unit ball is small when the dimension is high. We can turn this around if we like: a ball of unit volume has a large radius if the dimension is high. Indeed since the volume is proportional to r^n, where r is the radius, it follows that the radius of a ball of unit volume grows like $\sqrt{n}$ when Stirling's formula applies.

The fraction of the volume of a unit ball that lies inside a radius r is r^n. We assume $r < 1$, so this is a quickly shrinking fraction as n grows. The curious conclusion of this is that when the dimension is high almost all of the volume of a ball lies very close to its surface. In fact this is a crucial observation in statistical mechanics. It is also the key property of n-dimensional geometry: when n is large the 'amount of space' available grows very fast with distance from the origin.

In some ways it is easier to see what is going on if we consider hypercubes $\square_n$ rather than balls. Take a cube of unit volume. In n dimensions it has 2^n corners, and the longest straight line that we can draw inside the hypercube connects two opposite corners. It has length $L = \sqrt{1^2 + \ldots + 1^2} = \sqrt{n}$. Or expressed in another way, a straight line of any length fits into a hypercube of unit volume if the dimension is large enough. The reason why the longest line segment fitting into the cube is large is clearly that we normalised the volume to one. If we normalise $L = 1$ instead we find that the volume goes to zero like $(1/\sqrt{n})^n$. Concerning the insphere – the largest inscribed sphere, with *inradius* r_n – and the outsphere – the smallest circumscribed sphere, with *outradius* R_n – we observe that

$$R_n = \frac{\sqrt{n}}{2} = \sqrt{n} r_n. \tag{1.25}$$

The ratio between the two grows with the dimension, $\zeta_n \equiv R_n/r_n = \sqrt{n}$. Incidentally, the somewhat odd statement that the volume of a sphere goes to zero when the dimension n goes to infinity can now be interpreted: since $\mathrm{vol}(\square_n) = 1$ the real statement is that $\mathrm{vol}(\mathbf{S}^n)/\mathrm{vol}(\square_n)$ goes to zero when n goes to infinity.

Now we turn to simplices, whose properties will be of some importance later on. We concentrate on *regular simplices* Δ_n, for which the distance between any pair of corners is one. For $n = 1$ this is the unit interval, for $n = 2$ a regular triangle, for $n = 3$ a regular tetrahedron, and so on. Again we are interested in the volume, the radius r_n of the insphere, and the radius R_n of the outsphere. We will also compute χ_n, the angle between the lines from the 'centre of mass' to a pair of corners. For a triangle it is $\arccos(-1/2) = 2\pi/3 = 120$ degrees, but it drops to $\arccos(-1/3) \approx 110°$ for the tetrahedron. A practical way to go about all this is to think of Δ_n as a *pyramid* (a part of a cone, namely the convex

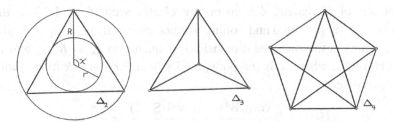

Figure 1.9 Regular simplices in two, three and four dimensions. For Δ_2 we also show the insphere, the circumsphere, and the angle discussed in the text.

hull of its apex and its base) having Δ_{n-1} as its base. It is then not difficult to show that

$$R_n = nr_n = \sqrt{\frac{n}{2(n+1)}} \quad \text{and} \quad r_n = \sqrt{\frac{1}{2(n+1)n}}, \tag{1.26}$$

so their ratio grows linearly, $\zeta_n = R_n/r_n = n$. The volume of a pyramid is $V = Bh/n$, where B is the area of the base, h is the height of the pyramid and n is the dimension. For the simplex we obtain

$$\text{vol}(\Delta_n) = \frac{1}{n!}\sqrt{\frac{n+1}{2^n}} \tag{1.27}$$

We can check that the ratio of the volume of the largest inscribed sphere to the volume of the simplex goes to zero. Hence most of the volume of the simplex sits in its corners, as expected. The angle χ_n subtended by an edge as viewed from the centre is given by

$$\sin\frac{\chi_n}{2} = \frac{1}{2R_n} = \sqrt{\frac{n+1}{2n}} \quad \Leftrightarrow \quad \cos\chi_n = -\frac{1}{n}. \tag{1.28}$$

When n is large we see that χ_n tends to a right angle. This is as it should be. The corners sit on the outsphere, and for large n almost all the volume of the outsphere lies close to the equator – hence, if we pick one corner and let it play the role of the north pole, all the other corners are likely to lie close to the equator. Finally it is interesting to observe that it is known for convex bodies in general that the radius of the outsphere is bounded by

$$R_n \leq L\sqrt{\frac{n}{2(n+1)}}, \tag{1.29}$$

where L is the length of the longest line segment L contained in the body. The regular simplex saturates this bound.

The effects of increasing dimension are clearly seen if we look at the ratio between surface (hyper) area and volume for bodies of various shapes. Rather than fixing the scale, let us study the dimensionless quantities $\zeta_n = R_n/r_n$ and $\eta(X) \equiv R\,\text{vol}(\partial X)/\text{vol}(X)$, where X is the body, ∂X its boundary, and R its outradius. For n-balls we receive

$$\eta_n(\mathbf{B}^n) = R\,\frac{\text{vol}(\partial \mathbf{B}^n)}{\text{vol}(\mathbf{B}^n)} = R\,\frac{\text{vol}(\mathbf{S}^{n-1})}{\text{vol}(\mathbf{B}^n)} = \frac{Rn}{R} = n. \tag{1.30}$$

Next consider a hypercube of edge length L. Its boundary consists of $2n$ facets, that are themselves hypercubes of dimension $n - 1$. This gives

$$\eta_n(\square_n) = R\,\frac{\text{vol}(\partial \square_n)}{\text{vol}(\square_n)} = \frac{\sqrt{n}L}{2}\,\frac{2n\,\text{vol}(\square_{n-1})}{\text{vol}(\square_n)} = \frac{n^{3/2}L}{L} = n^{3/2}. \tag{1.31}$$

A regular simplex of edge length L has a boundary consisting of $n + 1$ regular simplices of dimension $n - 1$. We obtain the ratio

$$\eta_n(\Delta_n) = R\,\frac{\text{vol}(\partial \Delta_n)}{\text{vol}(\Delta_n)} = L\sqrt{\frac{n}{2(n+1)}}\,\frac{(n+1)\text{vol}(\Delta_{n-1})}{\text{vol}(\Delta_n)} = n^2. \tag{1.32}$$

In this case the ratio η_n grows quadratically with n, reflecting the fact that simplices have sharper corners than those of the cube.

The reader may know about the five regular Platonic solids in three dimensions. When $n > 4$ there are only three kinds of regular solids, namely the simplex, the hypercube, and the *cross-polytope*. The latter is the generalisation to arbitrary dimension of the octahedron. It is dual to the cube; while the cube has 2^n corners and $2n$ facets, the cross-polytope has $2n$ corners and 2^n facets. The two polytopes have the same values of ζ_n and η_n.

These results are collected in Table 15.2, which includes a comparison to another very interesting convex body. We observe that $\eta_n = n\zeta_n$ for all these bodies. There is a reason for this. When Archimedes computed volumes, he did so by breaking them up into pyramids and using the formula $V = Bh/n$, where V is the volume of the pyramid B is the area of its base. Then we get

$$\eta_n = R\,\frac{\sum_{\text{pyramids}} B}{\left(\sum_{\text{pyramids}} B\right)h/n} = \frac{nR}{h}. \tag{1.33}$$

If the height h of the pyramids is equal to the inradius of the body, the result follows.[4]

---

[4] For more information on the subject of this section, consult Ball [87]. For a discussion of rotations in higher dimensions consult Section 8.3.

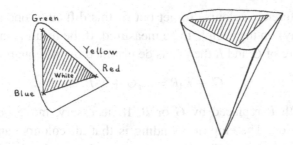

Figure 1.10 Left: the chromaticity diagram, and the part of it that can be obtained by mixing red, green and blue. Right: when the total illumination is taken into account, colour space becomes a convex cone.

1.3 Colour theory

How do convex sets arise? An instructive example occurs in colour theory, and more particularly in the psychophysical theory of colour. (This means that we will leave out the interesting questions about how our brain actually processes the visual information until it becomes a percept.) In a way tradition suggests that colour theory should be studied before quantum mechanics, because this is what Schrödinger was doing before inventing his wave equation.[5] The object of our attention is *colour space*, whose points are the colours. Naturally one might worry that the space of colours may differ from person to person but in fact it does not. The perception of colour is remarkably universal for human beings, colour blind persons not included. What has been done experimentally is to shine mixtures of light of different colours on white screens; say that three reference colours consisting of red, green and blue light are chosen. Then what one finds is that after adjusting the mixture of these colours, the observer will be unable to distinguish the resulting mixture from a given colour C. To simplify matters, suppose that the overall brightness has been normalised in some way. Then a colour C is a point on a two dimensional *chromaticity diagram*. Its position is determined by the equation

$$C = \lambda_0 R + \lambda_1 G + \lambda_2 B. \tag{1.34}$$

The barycentric coordinates λ_i will naturally take positive values only in this experiment. This means that we only get colours inside the triangle spanned by the reference colours R, G and B. Note that the 'feeling of redness' does not enter into the experiment at all.

But colour space is not a simplex, as designers of TV screens learn to their chagrin. There will always be colours C' that cannot be reproduced as a mixture

---

[5] Schrödinger [802] wrote a splendid review of the subject (1926). Readers who want a more recent discussion may enjoy the book by Williamson and Cummins [970].

of three given reference colours. To get out of this difficulty one shines a certain amount of red (say) on the sample to be measured. If the result is indistinguishable from some mixture of G and B then C' is determined by the equation

$$C' + \lambda_0 R = \lambda_1 G + \lambda_2 B. \tag{1.35}$$

If not, repeat with R replaced by G or B. If necessary, move one more colour to the left hand side. The empirical finding is that all colours can be assigned a position on the chromaticity diagram in this way. If we take the overall intensity into account we find that the full colour space is a three dimensional convex cone with the chromaticity diagram as its base and complete darkness as its apex (of course this is to the extent that we ignore the fact that very intense light will cause the eyes to boil rather than make them see a colour). The pure colours are those that cannot be obtained as a mixture of different colours; they form the curved part of the boundary. The boundary also has a planar part made of purple.

How can we begin to explain all this? We know that light can be characterised by its spectral distribution, which is some positive function I of the wave length λ. It is therefore immediately apparent that the space of spectral distributions is a convex cone, and in fact an infinite dimensional convex cone since a general spectral distribution $I(\lambda)$ can be defined as a convex combination

$$I(\lambda) = \int d\lambda' I(\lambda') \delta(\lambda - \lambda'), \quad I(\lambda') \geq 0. \tag{1.36}$$

The delta functions are the pure states. But colour space is only three dimensional. The reason is that the eye will assign the same colour to many different spectral distributions. A given colour corresponds to an equivalence class of spectral distributions, and the dimension of colour space will be given by the dimension of the space of equivalence classes. Let us denote the equivalence classes by $[I(\lambda)]$, and the space of equivalence classes as colour space. Since we know that colours can be mixed to produce a third quite definite colour, the equivalence classes must be formed in such a way that the equation

$$[I(\lambda)] = [I_1(\lambda)] + [I_2(\lambda)] \tag{1.37}$$

is well defined. The point here is that whatever representatives of $[I_1(\lambda)]$ and $[I_2(\lambda)]$ we choose we always obtain a spectral distribution belonging to the same equivalence class $[I(\lambda)]$. We would like to understand how this can be so.

In order to proceed it will be necessary to have an idea about how the eye detects light (especially so since the perception of sound is known to work in a quite different way). It is reasonable – and indeed true – to expect that there are chemical substances in the eye with different sensitivities. Suppose for the sake of the argument that there are three such 'detectors'. Each has an adsorption

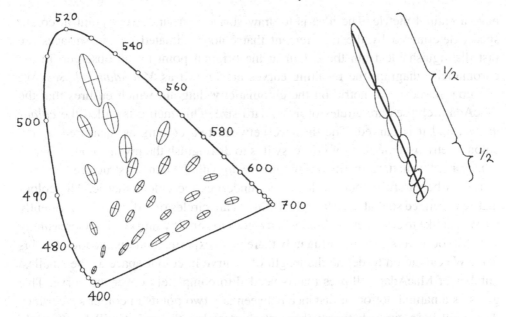

Figure 1.11 To the left, we see the MacAdam ellipses [632]. They show the points where the colour is just distinguishable from the colour at the centre of the ellipse. Their size is exaggerated by a factor of ten. To the right, we see how these ellipses can be used to define the length of curves on the chromaticity diagram – notice the position of the half way mark.

curve $A_i(\lambda)$. These curves are allowed to overlap; in fact they do. Given a spectral distribution each detector then gives an output

$$c_i = \int d\lambda \, I(\lambda)A_i(\lambda). \tag{1.38}$$

Our three detectors will give us only three real numbers to parametrize the space of colours. Equation (1.37) can now be derived. According to this theory, colour space will inherit the property of being a convex cone from the space of spectral distributions. The pure states will be those equivalence classes that contain the pure spectral distributions. On the other hand the dimension of colour space will be determined by the number of detectors, and not by the nature of the pure states. This is where colour blind persons come in; they are missing one or two detectors and their experiences can be predicted by the theory. By the way, many birds (such as hens) enjoy a colour space of four dimensions, dogs have two, while seals make do with one.

Like any convex set, colour space is a subset of an affine space and the convex structure does not single out any natural metric. Nevertheless colour space does have a natural metric. The idea is to draw surfaces around every point in colour space, determined by the requirement that colours situated on the surfaces are just distinguishable from the colour at the original point by an observer. In the chromaticity diagram the resulting curves are known as *MacAdam ellipses*. We can now introduce a metric on the chromaticity diagram which ensures that the MacAdam ellipses are circles of a standard size. This metric is called the *colour metric*, and it is curved. The distance between two colours as measured by the colour metric is a measure of how easy it is to distinguish the given colours. On the other hand this natural metric has nothing to do with the convex structure *per se*.

Let us be careful about the logic that underlies the colour metric. The colour metric is defined so that the MacAdam ellipses are circles of radius ϵ, say. Evidently we would like to consider the limit when ϵ goes to zero (by introducing increasingly sensitive observers), but unfortunately there is no experimental justification for this here. We can go on to define the length of a curve in colour space as the smallest number of MacAdam ellipses that is needed to completely cover the curve. This gives us a natural notion of distance between any two points in colour space since there will be a curve between them of shortest length (and it will be uniquely defined, at least if the distance is not too large). Such a curve is called a *geodesic*. The *geodesic distance* between two points is then the length of the geodesic that connects them. This is how distances are defined in Riemannian geometry, but it is worthwhile to observe that only the 'local' distance as defined by the metric has a clear operational significance here. There are many lessons from colour theory that are of interest in quantum mechanics, not least that the structure of the convex set is determined by the nature of the detectors.

1.4 What is 'distance'?

In colour space distances are used to quantify distinguishability. Although our use of distances will mostly be in a similar vein, they have many other uses too – for instance, to prove convergence for iterative algorithms. But what are they? Though we expect the reader to have her share of inborn intuition about the nature of geometry, a few indications of how this can be made more precise are in order. Let us begin by defining a *distance* $D(\mathbf{x}, \mathbf{y})$ between two points in a vector space (or more generally, in an affine space). This is a function of the two points that obeys three axioms:

  i) The distance between two points is a non-negative number $D(\mathbf{x}, \mathbf{y})$ that equals zero if and only if the points coincide.
 ii) It is symmetric in the sense that $D(\mathbf{x}, \mathbf{y}) = D(\mathbf{y}, \mathbf{x})$.
iii) It satisfies the triangle inequality $D(\mathbf{x}, \mathbf{y}) \leq D(\mathbf{x}, \mathbf{z}) + D(\mathbf{z}, \mathbf{y})$.

Actually both axiom ii) and axiom iii) can be relaxed – we will see what can be done without them in Section 2.3 – but as is often the case it is even more interesting to try to restrict the definition further, and this is the direction that we are heading in now. We want a notion of distance that meshes naturally with convex sets, and for this purpose we add a fourth axiom:

iv) It obeys $D(\lambda\mathbf{x}, \lambda\mathbf{y}) = \lambda D(\mathbf{x}, \mathbf{y})$ for non-negative numbers λ.

A distance function obeying this property is known as a *Minkowski distance*. Two important consequences follow, neither of them difficult to prove. First, any convex combination of two vectors becomes a metric straight line in the sense that

$$\mathbf{z} = \lambda\mathbf{x} + (1 - \lambda)\mathbf{y} \quad \Rightarrow \quad D(\mathbf{x}, \mathbf{y}) = D(\mathbf{x}, \mathbf{z}) + D(\mathbf{z}, \mathbf{y}), \quad 0 \leq \lambda \leq 1. \quad (1.39)$$

Second, if we define a unit ball with respect to a Minkowski distance we find that such a ball is always a convex set.

Let us discuss the last point in a little more detail. A Minkowski metric is naturally defined in terms of a *norm* on a vector space, that is a real valued function $||\mathbf{x}||$ that obeys

$$i_n) \quad ||\mathbf{x}|| \geq 0, \text{ and } ||\mathbf{x}|| = 0 \Leftrightarrow \mathbf{x} = 0.$$

$$ii_n) \quad ||\mathbf{x} + \mathbf{y}|| \leq ||\mathbf{x}|| + ||\mathbf{y}||. \quad (1.40)$$

$$iii_n) \quad ||\lambda\mathbf{x}|| = |\lambda| \, ||\mathbf{x}||, \quad \lambda \in \mathbf{R}.$$

The distance between two points $\mathbf{x}$ and $\mathbf{y}$ is now defined as $D(\mathbf{x}, \mathbf{y}) \equiv ||\mathbf{x} - \mathbf{y}||$, and indeed it has the properties i-iv. The unit ball is the set of vectors $\mathbf{x}$ such that $||\mathbf{x}|| \leq 1$, and it is easy to see that

$$||\mathbf{x}||, \, ||\mathbf{y}|| \leq 1 \quad \Rightarrow \quad ||\lambda\mathbf{x} + (1 - \lambda)\mathbf{y}|| \leq 1. \quad (1.41)$$

So the unit ball is convex. In fact the story can be turned around at this point – any centrally symmetric convex body can serve as the unit ball for a norm, and hence it defines a distance. (A centrally symmetric convex body K has the property that, for some choice of origin, $\mathbf{x} \in K \Rightarrow -\mathbf{x} \in K$.) Thus the opinion that balls are round is revealed as an unfounded prejudice. It may be helpful to recall that water droplets are spherical because they minimise their surface energy. If we want to understand the growth of crystals in the same terms, we must use a notion of distance that takes into account that the surface energy depends on direction.

We need a set of norms to play with, so we define the l_p-*norm* of a vector by

$$||\mathbf{x}||_p \equiv (|x_1|^p + |x_2|^p + \ldots + |x_n|^p)^{\frac{1}{p}}, \quad p \geq 1. \quad (1.42)$$

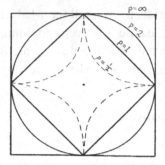
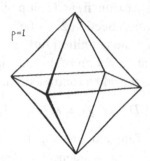

Figure 1.12 Left: points at distance 1 from the origin, using the l_1-norm for the vectors (the inner square), the l_2-norm (the circle) and the l_∞-norm (the outer square). The $l_{\frac{1}{2}}$-case is shown dashed – the corresponding ball is not convex because the triangle inequality fails, so it is not a norm. Right: in three dimensions one obtains respectively an octahedron, a sphere and a cube. We illustrate the $p = 1$ case.

In the limit we obtain the *Chebyshev norm* $||\mathbf{x}||_\infty = \max_i |x_i|$. The proof of the triangle inequality is non-trivial and uses *Hölder's inequality*

$$\sum_{i=1}^{N} |x_i y_i| \leq ||\mathbf{x}||_p ||\mathbf{y}||_q, \qquad \frac{1}{p} + \frac{1}{q} = 1, \qquad (1.43)$$

where $p, q \geq 1$. For $p = 2$ this is the *Cauchy–Schwarz inequality*. If $p < 1$ there is no Hölder inequality, and the triangle inequality fails. We can easily draw a picture (namely Figure 1.12) of the unit balls $\mathbf{B}_p$ for a few values of p, and we see that they interpolate beween a cube (for $p \to \infty$) and a cross-polytope (for $p = 1$), and that they fail to be convex for $p < 1$. We also see that in general these balls are not invariant under rotations, as expected because the components of the vector in a special basis were used in the definition. The topology induced by the l_p-norms is the same, regardless of p. The corresponding distances $D_p(\mathbf{x}, \mathbf{y}) \equiv ||\mathbf{x} - \mathbf{y}||_p$ are known as the l_p-*distances*.

Depending on circumstances, different choices of p may be particularly relevant. The case $p = 1$ is relevant if motion is confined to a rectangular grid (say, if you are a taxi driver in Manhattan). As we will see (in Section 14.1) it is also of particular relevance to us. It has the slightly awkward property that the shortest path between two points is not uniquely defined. Taxi drivers know this, but may not be aware of the fact that it happens only because the unit ball is a polytope, i.e., it is convex but not strictly convex. The l_1-distance goes under many names: *taxi cab, Kolmogorov,* or *variational distance.*

The case $p = 2$ is consistent with Pythagoras' theorem and is the most useful choice in everyday life; it was singled out for special attention by Riemann when he made the foundations for differential geometry. Indeed we used a $p = 2$ norm when we defined the colour metric at the end of Section 1.3. The idea is that once we have some coordinates to describe colour space then the MacAdam ellipse surrounding a point is given by a quadratic form in the coordinates. The interesting thing – that did not escape Riemann – is the ease with which this 'infinitesimal' notion of distance can be converted into the notion of geodesic distance between arbitrary points. (A similar generalisation based on other l_p-distances exists and is called *Finslerian geometry*, as opposed to the *Riemannian geometry* based on $p = 2$.)

Riemann began by defining what we now call differentiable manifolds of arbitrary dimension;[6] for our purposes here let us just say that this is something that locally looks like $\mathbb{R}^n$ in the sense that it has open sets, continuous functions, and differentiable functions; one can set up a one-to-one correspondence between the points in some open set and n numbers x^i, called *coordinates*, that belong to some open set in $\mathbb{R}^n$. There exists a *tangent space* $\mathbf{T}_q$ at every point q in the manifold; intuitively we can think of the manifold as some curved surface in space and of a tangent space as a flat plane touching the surface in some point. By definition the tangent space $\mathbf{T}_q$ is the vector space whose elements are *tangent vectors* at q, and a tangent vector at a point of a differentiable manifold is defined as the tangent vector of a smooth curve passing through the point. Intuitively, it is a little arrow sitting at the point. Formally, it is a *contravariant* vector (with index upstairs). Each tangent vector V^i gives rise to a directional derivative $\sum_i V^i \partial_i$ acting on the functions on the space; in differential geometry it has therefore become customary to think of a tangent vector as a derivative operator. In particular we can take the derivatives in the directions of the coordinate lines, and any directional derivative can be expressed as a linear combination of these. Therefore, given any coordinate system x^i the derivatives ∂_i with respect to the coordinates form a basis for the tangent space – not necessarily the most convenient basis one can think of, but one that certainly exists. To sum up a tangent vector is written as

$$\mathbf{V} = \sum_i V^i \partial_i, \qquad (1.44)$$

where $\mathbf{V}$ is the vector itself and V^i are the components of the vector in the coordinate basis spanned by the basis vectors ∂_i.

[6] Riemann lectured on the hypotheses which lie at the foundations of geometry in 1854, in order to be admitted as a Dozent at Göttingen. As Riemann says, only two instances of continuous manifolds were known from everyday life at the time: the space of locations of physical objects, and the space of colours. In spite of this he gave an essentially complete sketch of the foundations of modern geometry. For a more detailed account see (for instance) Murray and Rice [682]. A very readable albeit old-fashioned account is by our Founding Father: Schrödinger [806]. For beginners the definitions in this section can become bewildering; if so our advice is to ignore them, and look at some examples of curved spaces first.

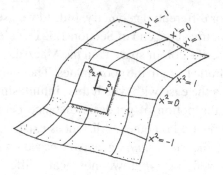

Figure 1.13 The tangent space at the origin of some coordinate system. Note that there is a tangent space at every point.

It is perhaps as well to emphasise that the tangent space $\mathbf{T}_q$ at a point q bears no a priori relation to the tangent space $\mathbf{T}_{q'}$ at a different point q', so that tangent vectors at different points cannot be compared unless additional structure is introduced. Such an additional structure is known as 'parallel transport' or 'covariant derivatives', and will be discussed in Section 3.2.

At every point q of the manifold there is also a *cotangent space* $\mathbf{T}_q^*$, the vector space of linear maps from $\mathbf{T}_q$ to the real numbers. Its elements are called *covariant vectors* and they have indices downstairs. Given a coordinate basis for $\mathbf{T}_q$ there is a natural basis for the cotangent space consisting of n covariant vectors dx^i defined by

$$dx^i(\partial_j) = \delta^i_j, \tag{1.45}$$

with the Kronecker delta appearing on the right hand side. The tangent vector ∂_i points in the coordinate direction, while dx^i gives the level curves of the coordinate function. A general element of the cotangent space is also known as a *one-form*. It can be expanded as $U = U_i dx^i$, so that covariant vectors have indices downstairs. The linear map of a tangent vector $\mathbf{V}$ is given by

$$U(\mathbf{V}) = U_i dx^i(V^j \partial_j) = U_i V^j dx^i(\partial_j) = U_i V^i. \tag{1.46}$$

From now on the *Einstein summation convention* is in force, which means that if an index appears twice in the same term then summation over that index is implied. A natural next step is to introduce a scalar product in the tangent space, and indeed in every tangent space. (One at each point of the manifold.) We can do this by specifying the scalar products of the basis vectors ∂_i. When this is done we have in fact defined a *Riemannian metric tensor* on the manifold, whose components in the coordinate basis are given by

$$g_{ij} = \langle \partial_i, \partial_j \rangle. \tag{1.47}$$

It is understood that this has been done at every point q, so the components of the metric tensor are really functions of the coordinates. The metric g_{ij} is assumed to have an inverse g^{ij}. Once we have the metric it can be used to raise and lower indices in a standard way ($V_i = g_{ij}V^j$). Otherwise expressed it provides a canonical isomorphism between the tangent and cotangent spaces.

Riemann went on to show that one can always define coordinates on the manifold in such a way that the metric at any given point is diagonal and has vanishing first derivatives. In effect – provided that the metric tensor is a positive definite matrix, which we assume – the metric gives a 2-norm on the tangent space at that special point. Riemann also showed that in general it is not possible to find coordinates so that the metric takes this form everywhere; the obstruction that may make this impossible is measured by a quantity called the *Riemann curvature tensor*. It is a linear function of the second derivatives of the metric (and will make its appearance in Section 3.2). The space is said to be flat if and only if the Riemann tensor vanishes, which is if and only if coordinates can be found so that the metric takes the same diagonal form everywhere. The 2-norm was singled out by Riemann precisely because his grandiose generalisation of geometry to the case of arbitrary differentiable manifolds works much better if $p = 2$.

With a metric tensor at hand we can define the length of an arbitrary curve $x^i = x^i(t)$ in the manifold as the integral

$$\int ds = \int \sqrt{g_{ij}\frac{dx^i}{dt}\frac{dx^j}{dt}}dt \qquad (1.48)$$

along the curve. The shortest curve between two points is called a geodesic, and we are in a position to define the geodesic distance between the two points just as we did at the end of Section 1.3. The geodesic distance obeys the axioms that we laid down for distance functions, so in this sense the metric tensor defines a distance. Moreover, at least as long as two points are reasonably close, the shortest path between them is unique.

One of the hallmarks of differential geometry is the ease with which the tensor formalism handles coordinate changes. Suppose we change to new coordinates $x^{i'} = x^{i'}(x)$. Provided that these functions are invertible the new coordinates are just as good as the old ones. More generally, the functions may be invertible only for some values of the original coordinates, in which case we have a pair of partially overlapping *coordinate patches*. It is elementary that

$$\partial_{i'} = \frac{\partial x^j}{\partial x^{i'}}\partial_j. \qquad (1.49)$$

Since the vector **V** itself is not affected by the coordinate change – which is after all just some equivalent new description – Eq. (1.44) implies that its components must change according to

$$V^{i'} \partial_{i'} = V^i \partial_i \quad \Rightarrow \quad V^{i'}(x') = \frac{\partial x^{i'}}{\partial x^j} V^j(x). \tag{1.50}$$

In the same way we can derive how the components of the metric change when the coordinate system changes, using the fact that the scalar product of two vectors is a scalar quantity that does not depend on the coordinates:

$$g_{i'j'} U^{i'} V^{j'} = g_{ij} U^i V^j \quad \Rightarrow \quad g_{i'j'} = \frac{\partial x^k}{\partial x^{i'}} \frac{\partial x^l}{\partial x^{j'}} g_{kl}. \tag{1.51}$$

We see that the components of a tensor, in some basis, depend on that particular and arbitrary basis. This is why they are often regarded with feelings bordering on contempt by professionals, who insist on using 'coordinate free methods' and think that 'coordinate systems do not matter'. But in practice few things are more useful than a well chosen coordinate system. And the tensor formalism is tailor made to construct scalar quantities invariant under coordinate changes.

In particular the formalism provides invariant measures that can be used to define lengths, areas, volumes, and so on, in a way that is independent of the choice of coordinate system. This is because the square root of the determinant of the metric tensor, $\sqrt{g}$, transforms in a special way under coordinate transformations:

$$\sqrt{g'}(x') = \left(\det \frac{\partial x'}{\partial x} \right)^{-1} \sqrt{g}(x). \tag{1.52}$$

The integral of a scalar function $f'(x') = f(x)$, over some manifold **M**, then behaves as

$$I = \int_{\mathbf{M}} f'(x') \sqrt{g'}(x') \, \mathrm{d}^n x' = \int_{\mathbf{M}} f(x) \sqrt{g}(x) \, \mathrm{d}^n x \tag{1.53}$$

– the transformation of $\sqrt{g}$ compensates for the transformation of $\mathrm{d}^n x$, so that the measure $\sqrt{g} \mathrm{d}^n x$ is invariant. A submanifold can always be locally defined via equations of the general form $x = x(x')$, where x' are intrinsic coordinates on the submanifold and x are coordinates on the *embedding space* in which it sits. In this way Eq. (1.51) can be used to define an *induced metric* on the submanifold, and hence an invariant measure as well. Equation (1.48) is in fact an example of this construction – and it is good to know that the geodesic distance between two points is independent of the coordinate system.

Since this is not a textbook on differential geometry we leave these matters here, except that we want to draw attention to some possible ambiguities. First there is an ambiguity of notation. The metric is often presented in terms of the squared *line element*,

$$\mathrm{d}s^2 = g_{ij} \mathrm{d}x^i \mathrm{d}x^j. \tag{1.54}$$

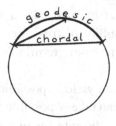

Figure 1.14 Here is how to measure the geodesic and the chordal distances between two points on the sphere. When the points are close these distances are also close; they are consistent with the same metric.

The ambiguity is this: in modern notation dx^i denotes a basis vector in cotangent space, and ds^2 is a linear operator acting on the (symmetrised) tensor product $\mathbf{T} \otimes \mathbf{T}$. There is also an old fashioned way of reading the formula, which regards ds^2 as the length squared of that tangent vector whose components (at the point with coordinates x^i) are dx^i. A modern mathematician would be appalled by this, rewrite it as $g_x(ds, ds)$, and change the label ds for the tangent vector to, say, A. But a liberal reader will be able to read Eq. (1.54) in both ways. The old fashioned notation has the advantage that we can regard ds as the distance between two 'nearby' points given by the coordinates x and $x + dx$; their distance is equal to ds plus terms of order higher than two in the coordinate differences. We then see that there are ambiguities present in the notion of distance too. To take the sphere as an example, we can define a distance function by means of geodesic distance. But we can also define the distance between two points as the length of a chord connecting the two points, and the latter definition is consistent with our axioms for distance functions. Moreover both definitions are consistent with the metric, in the sense that the distances between two nearby points will agree to second order. However, in this book we will usually regard it as understood that once we have a metric we are going to use the geodesic distance to measure the distance between two arbitrary points.

1.5 Probability and statistics

The reader has probably surmised that our interest in convex sets has to do with their use in statistics. It is not our intention to explain the notion of probability, not even to the extent that we tried to explain colour. We are quite happy with the Kolmogorov axioms, that define probability as a suitably normalised positive measure on some set Ω. If the set of points is finite, this is simply a finite set of positive numbers adding up to one. Now there are many viewpoints on what the meaning of it all may be, in terms of frequencies, propensities, and degrees of reasonable beliefs. We do not have to take a position on these matters here because the geometry of probability distributions is invariant under changes of interpretation.[7] We do need to fix some terminology however, and will proceed to do so.

Consider an experiment that can yield N possible outcomes, or in mathematical terms a *random variable* X that can take N possible values x_i belonging to a *sample space* Ω, which in this case is a discrete set of points. The probabilities for the respective outcomes are

$$P(X = x_i) = p^i. \tag{1.55}$$

For many purposes the actual outcomes can be ignored. The interest centres on the probability distribution $P(X)$ considered as the set of N real numbers p^i such that

$$p^i \geq 0, \qquad \sum_{i=1}^{N} p^i = 1. \tag{1.56}$$

(We will sometimes be a little ambiguous about whether the index should be up or down – although they should be upstairs according to the rules of differential geometry.) Now look at the space of all possible probability distributions for the given random variable. This is a simplex with the p^i playing the role of barycentric coordinates; a convex set of the simplest possible kind. The pure states are those for which the outcome is certain, so that one of the p^i is equal to one. The pure states sit at the corners of the simplex and hence they form a zero dimensional subset of its boundary. In fact the space of pure states is isomorphic to the sample space. As long as we keep to the case of a finite number of outcomes – the *multinomial probability distribution* as it is known in probability theory – nothing could be simpler.

Except that, as a subset of an n dimensional vector space, an n dimensional simplex is a bit awkward to describe using Cartesian coordinates. Frequently it is more convenient to regard it as a subset of an $N = n + 1$ dimensional vector space instead, and use the unrestricted p^i to label the axes. Then we can use the l_p-norms to define distances. The meaning of this will be discussed in Chapter 2; meanwhile we observe that the probability simplex lies somewhat askew in the vector space, and we find it convenient to adjust the definition a little. From now on we set

$$D_p(P, Q) \equiv ||P - Q||_p \equiv \left(\frac{1}{2} \sum_{i=1}^{N} |p_i - q_i|^p \right)^{\frac{1}{p}}, \qquad 1 \leq p. \tag{1.57}$$

---

[7] The reader may consult the book by von Mises [939] for one position, and the book by Jaynes [499] for another. Jaynes regards probability as quantifying the degree to which a proposition is plausible, and finds that $\sqrt{p_i}$ has a status equally fundamental as that of p_i.

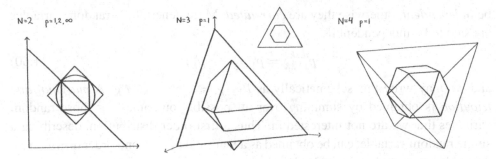

Figure 1.15 For $N = 2$ we show why all the l_p–distances agree when the definition (1.57) is used. For $N = 3$ the l_1–distance gives hexagonal 'spheres', arising as the intersection of the simplex with an octahedron. For $N = 4$ the same construction gives an Archimedean solid known as the cub-octahedron.

The extra factor of $1/2$ ensures that the edge lengths of the simplex equal 1, and also has the pleasant consequence that all the l_p–distances agree when $N = 2$. However, it is a little tricky to see what the l_p–balls look like inside the probability simplex. The case $p = 1$, which is actually important to us, is illustrated in Figure 1.15; we are looking at the intersection of a cross–polytope with the probability simplex. The result is a convex body with $N(N - 1)$ corners. For $N = 2$ it is a hexagon, for $N = 3$ a cuboctahedron, and so on.

The l_1–distance has the interesting property that probability distributions with orthogonal *supports* – meaning that the product $p_i q_i$ vanishes for each value of i – are at maximal distance from each other. One can use this observation to show, without too much effort, that the ratio of the radii of the in- and outspheres for the l_1–ball obeys

$$\frac{r_{\text{in}}}{R_{\text{out}}} = \sqrt{\frac{2}{N}} \text{ if } N \text{ is even,} \qquad \frac{r_{\text{in}}}{R_{\text{out}}} = \sqrt{\frac{2N}{N^2 - 1}} \text{ if } N \text{ is odd.} \tag{1.58}$$

Hence, although some corners have been 'chopped off', the body is only marginally more spherical than is the cross-polytope. Another way to say the same thing is that, with our normalisation, $\|\mathbf{p}\|_1 \leq \|\mathbf{p}\|_2 \leq R_{\text{out}} \|\mathbf{p}\|_1 / r_{\text{in}}$.

We end with some further definitions, that will put more strain on the notation. Suppose we have two random variables X and Y with N and M outcomes and described by the distributions P_1 and P_2, respectively. Then there is a *joint probability distribution* P_{12} of the joint probabilities,

$$P_{12}(X = x_i, Y = y_j) = p_{12}^{ij}. \tag{1.59}$$

This is a set of NM non–negative numbers summing to one. Note that it is not implied that $p_{12}^{ij} = p_1^i p_2^j$; if this does happen the two random variables are said to be *independent*, otherwise they are *correlated*. More generally K random variables are said to be independent if

$$p_{12\ldots K}^{ij\ldots k} = p_1^i \, p_2^j \, \cdots \, p_K^k, \tag{1.60}$$

and we may write this schematically as $P_{12\ldots K} = P_1 P_2 \, \ldots \, P_K$. A *marginal distribution* is obtained by summing over all possible outcomes for those random variables that we are not interested in. Thus a first order distribution, describing a single random variable, can be obtained as a marginal of a second order distribution, describing two random variables jointly, by

$$p_1^i = \sum_j p_{12}^{ij}. \tag{1.61}$$

There are also special probability distributions that deserve special names. Thus the *uniform* distribution for a random variable with N outcomes is denoted by $Q_{(N)}$ and the distributions where one outcome is certain are collectively denoted by $Q_{(1)}$. The notation can be extended to include

$$Q_{(M)} = \left(\frac{1}{M}, \frac{1}{M}, \ldots, \frac{1}{M}, 0, \ldots, 0 \right), \tag{1.62}$$

with $M \leq N$ and possibly with the components permuted.

With these preliminaries out of the way, we will devote Chapter 2 to the study of the convex sets that arise in classical statistics, and the geometries that can be defined on them – in itself, a preparation for the quantum case.

Problems

1.1 Prove Carathéodory's theorem.
1.2 Compute the inradius and the outradius of a simplex, that is prove Eq. (1.26).
1.3 Prove that the unit balls of the l_p and l_q norms are dual convex sets whenever $1/p + 1/q = 1$.

2
Geometry of probability distributions

Some people hate the very name of statistics, but I find them full of beauty and interest.
—*Sir Francis Galton*[1]

In quantum mechanics one often encounters sets of non–negative numbers that sum to unity, having a more or less direct interpretation as probabilities. This includes the squared moduli of the coefficients when a pure state is expanded in an orthonormal basis, the eigenvalues of density matrices, and more. Continuous distributions also play a role, even when the Hilbert space is finite dimensional. From a purely mathematical point of view a probability distribution is simply a *measure* on a sample space, constrained so that the total measure is one. Whatever the point of view one takes on this, the space of states will turn into a convex set when we allow probabilistic mixtures of its pure states. In classical mechanics the sample space is phase space, which is typically a continuous space. This leads to technical complications but the space of states in classical mechanics does share a major simplifying feature with the discrete case, namely that every state can be expressed as a mixture of pure states in a unique way. This was not so in the case of colour space, nor will it be true for the convex set of all states in quantum mechanics.

2.1 Majorisation and partial order

Our first aim is to find ways of describing probability distributions; we want to be able to tell when a probability distribution is 'more chaotic' or 'more uniform' than another. One way of doing this is provided by the theory of *majorisation*.[2] We will

---

[1] Reproduced from Francis Galton. *Natural Inheritance*. Macmillan, 1889.
[2] This is a large research area in linear algebra. Major landmarks include the books by Hardy, Littlewood and Pólya [401], Marshall and Olkin [645], and Alberti and Uhlmann [16]. See also Ando's review [37]; all unproved assertions in this section can be found there.

regard a probability distribution as a vector $\vec{x}$ belonging to the positive hyperoctant in $\mathbb{R}^N$, and normalised so that the sum of its components is unity.

The set of all normalised vectors forms an $N-1$ dimensional simplex Δ_{N-1}. We are interested in transformations that take probability distributions into each other, that is transformations that preserve both positivity and the l_1–norm of positive vectors.

Now consider two positive vectors, $\vec{x}$ and $\vec{y}$. We order their components in decreasing order, $x_1 \geq x_2 \geq, .. \geq x_N$. When this has been done we may write $x_i^{\downarrow}$. We say that $\vec{x}$ is *majorised* by $\vec{y}$, written

$$\vec{x} \prec \vec{y} \quad \text{if and only if} \quad \begin{cases} \text{(i):} \quad \sum_{i=1}^{k} x_i^{\downarrow} \leq \sum_{i=1}^{k} y_i^{\downarrow} \quad \text{for} \quad k = 1, \ldots, N \\[2mm] \text{(ii):} \quad \sum_{i=1}^{N} x_i = \sum_{i=1}^{N} y_i. \end{cases} \tag{2.1}$$

We assume that all our vectors are normalised in such a way that their components sum to unity, so condition (ii) is automatic. It is evident that $\vec{x} \prec \vec{x}$ (majorisation is reflexive) and that $\vec{x} \prec \vec{y}$ and $\vec{y} \prec \vec{z}$ implies $\vec{x} \prec \vec{z}$ (majorisation is transitive) but it is not true that $\vec{x} \prec \vec{y}$ and $\vec{y} \prec \vec{x}$ implies $\vec{x} = \vec{y}$, because one of these vectors may be obtained by a permutation of the components of the other. But if we arrange the components of all vectors in decreasing order then indeed $\vec{x} \prec \vec{y}$ and $\vec{y} \prec \vec{x}$ does imply $\vec{x} = \vec{y}$; majorisation does provide a *partial order* on such vectors. The ordering is only partial because given two vectors it may happen that none of them majorise the other. Moreover there is a *smallest element*. Indeed, for every vector $\vec{x}$ it is true that

$$\vec{x}_{(N)} \equiv (1/N, 1/N, \ldots, 1/N) \prec \vec{x} \prec (1, 0, \ldots, 0) \equiv \vec{x}_{(1)}. \tag{2.2}$$

Note also that

$$\vec{x}_1 \prec \vec{y} \quad \text{and} \quad \vec{x}_2 \prec \vec{y} \quad \Rightarrow \quad (a\vec{x}_1 + (1-a)\vec{x}_2) \prec \vec{y} \tag{2.3}$$

for any real $a \in [0, 1]$. Hence the set of vectors majorised by a given vector is a convex set. In fact this set is the convex hull of all vectors that can be obtained by permuting the components of the given vector.

Vaguely speaking it is clear that majorisation captures the idea that one vector may be more 'uniform' or 'mixed' than another, as seen in Figure 2.1. We can display all positive vectors of unit l_1-norm as a probability simplex; for $N = 3$ the convex set of all vectors that are majorised by a given vector is easily recognised (Figure 2.2). For special choices of the majorizing vector we get an equilateral triangle or a regular tetrahedron; for $N = 4$ a number of Platonic and Archimedean solids appear in this way (an Archimedean solid has regular but not equal faces [238]). See Figure 2.3.

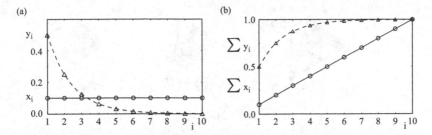

Figure 2.1 Idea of majorisation: in panel (a) the vector $\vec{x} = \{x_1, \ldots, x_{10}\}$ (○) is *majorised* by $\vec{y} = \{y_1, \ldots, y_{10}\}$ (△). In panel (b) we plot the distribution functions and show that Eq. (2.1) is obeyed.

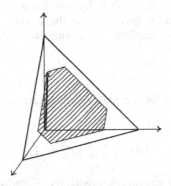

Figure 2.2 The probability simplex for $N = 3$ and the shaded convex set that is formed by all vectors that are majorised by a given vector; its pure points are obtained by permuting the components of the given vector.

Many processes in physics occur in the direction of the majorisation arrow (because the passage of time tends to make things more uniform). Economists are also concerned with majorisation. When Robin Hood robs the rich and helps the poor he aims for an income distribution that is majorised by the original one (provided that he acts like an isometry with respect to the l_1–norm, that is that he does not keep anything in his own pocket). We will need some information about such processes and we begin by identifying a suitable class of transformations. A *stochastic* matrix is a matrix B with N rows, whose matrix elements obey

$$\text{(i):} \quad B_{ij} \geq 0$$

$$\text{(ii):} \quad \sum_{i=1}^{N} B_{ij} = 1. \tag{2.4}$$

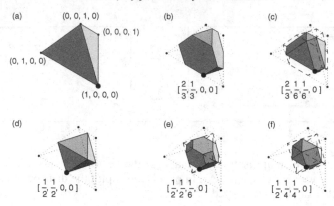

Figure 2.3 Panel (a) shows the probability simplex for $N = 4$. The set of vectors majorised by a given vector gives rise to the convex bodies shown in (b–f); these bodies include an octahedron (d), a truncated octahedron (e), and a cuboctahedron (f).

A *bistochastic* or *doubly stochastic* matrix is a square stochastic matrix obeying the additional condition[3]

$$\text{(iii):} \quad \sum_{j=1}^{N} B_{ij} = 1. \tag{2.5}$$

Condition (i) means that B preserves positivity. Condition (ii) says that the sum of all the elements in a given column equals one, and it means that B preserves the l_1–norm when acting on positive vectors, or in general that B preserves the sum $\sum_i x_i$ of all the components of the vector. Condition (iii) means that B is *unital*, that is that it leaves the 'smallest element' $\vec{x}_{(N)}$ invariant. Hence it causes some kind of contraction of the probability simplex towards its centre, and the classical result by Hardy, Littlewood and Pólya [401] does not come as a complete surprise:

HLP lemma. *$\vec{x} \prec \vec{y}$ if and only if there exists a bistochastic matrix B such that $\vec{x} = B\vec{y}$.*

For a proof see Problem 2.3. The product of two bistochastic matrices is again bistochastic; they are closed under multiplication but they do not form a group.

  A general two by two bistochastic matrix is of the form

$$T = \begin{pmatrix} t & 1-t \\ 1-t & t \end{pmatrix}, \qquad t \in [0,1]. \tag{2.6}$$

---

[3] Bistochastic matrices were first studied by Schur (1923) [814].

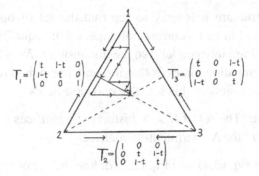

Figure 2.4 How T–transforms, and sequences of T–transforms, act on the probability simplex. The distribution $(3/4, 1/4, 0)$ is transformed to the uniform ensemble with an infinite sequence of T–transforms, while for the distribution $1/24(14, 7, 3)$ we use a finite sequence $(T_2 T_1)$.

In higher dimensions there will be many bistochastic matrices that connect two given vectors. Of a particularly simple kind are T–*transforms* ('T' as in transfer), that is matrices that act non-trivially only on two components of a vector. It is geometrically evident from Figure 2.4 that if $\vec{x} \prec \vec{y}$ then it is always possible to find a sequence of not more than $N - 1$ T–transforms such that $\vec{x} = T_{N-1} T_{N-2} \ldots T_1 \vec{y}$. (Robin Hood can achieve his aim using T–transforms to transfer income.) On the other hand (except for the two by two case) there exist bistochastic matrices that cannot be written as sequences of T–transforms at all.

A matrix B is called *unistochastic* if there exists a unitary matrix U such that $B_{ij} = |U_{ij}|^2$. (No sum is involved – the equality concerns individual matrix elements.) In the special case that there exists an orthogonal matrix O such that $B_{ij} = (O_{ij})^2$ the matrix B is called *orthostochastic*. Due to the unitarity condition every unistochastic matrix is bistochastic, but the converse does not hold, except when $N = 2$. On the other hand we have [449]:

Horn's lemma. $\vec{x} \prec \vec{y}$ *if and only if there exists an orthostochastic matrix B such that $\vec{x} = B\vec{y}$.*

There is an easy to follow algorithm for how to construct such an orthostochastic matrix [130], which may be written as a product of $(N - 1)$ T–transforms acting in different subspaces. In general, however, a product of an arbitrary number of T–transforms needs not be unistochastic [746].

A theme that will recur is to think of a set of transformations as a space in its own right. The space of linear maps of a vector space to itself is a linear space of its own in a natural way; to superpose two linear maps we define

$$\left(a_1 T_1 + a_2 T_2\right)\vec{x} \equiv a_1 T_1 \vec{x} + a_2 T_2 \vec{x}. \tag{2.7}$$

Given this linear structure it is easy to see that the set of bistochastic matrices forms a convex set and in fact a convex polytope. Of the equalities in Eq. (2.4) and Eq. (2.5) only $2N-1$ are independent, so the dimension is $N^2 - 2N + 1 = (N-1)^2$. We also see that permutation matrices (having only one non-zero entry in each row and column) are pure points of this set. The converse holds:

Birkhoff's theorem. *The set of $N \times N$ bistochastic matrices is a convex polytope whose pure points are the $N!$ permutation matrices.*

To see this note that Eq. (2.4) and Eq. (2.5) define the set of bistochastic matrices as the intersection of a finite number of closed half-spaces in $\mathbb{R}^{(N-1)^2}$. (An equality counts as the intersection of a pair of closed half-spaces.) According to Section 1.1 the pure points must saturate $(N-1)^2 = N^2 - 2N + 1$ of the inequalities in condition (i). Hence at most $2N - 1$ matrix elements can be non-zero; therefore at least one row (and by implication one column) contains one unit and all other entries zero. Effectively we have reduced the dimension one step, and we can now use induction to show that the only non-vanishing matrix elements for a pure point in the set of bistochastic matrices must equal 1, which means that it is a permutation matrix. Note also that using Carathéodory's theorem (again from Section 1.1) we see that every $N \times N$ bistochastic matrix can be written as a convex combination of $(N-1)^2 + 1$ permutation matrices, and it is worth adding that there exist easy-to-follow algorithms for how to actually do this.

Functions which preserve the majorisation order are called *Schur convex*;

$$\vec{x} \prec \vec{y} \quad \text{implies} \quad f(\vec{x}) \leq f(\vec{y}). \tag{2.8}$$

If $\vec{x} \prec \vec{y}$ implies $f(\vec{x}) \geq f(\vec{y})$ the function is called *Schur concave*. Clearly $-f(\vec{x})$ is Schur concave if $f(\vec{x})$ is Schur convex, and conversely. The key theorem here is:

Schur's theorem. *A differentiable function $F(x_1, \ldots, x_N)$ is Schur convex if and only if F is permutation invariant and if, for all $\vec{x}$,*

$$(x_1 - x_2) \left(\frac{\partial F}{\partial x_1} - \frac{\partial F}{\partial x_2} \right) \geq 0. \tag{2.9}$$

Permutation invariance is needed because permutation matrices are (the only) bistochastic matrices that have bistochastic inverses. The full proof is not difficult when one uses T–transforms [37]. Using Schur's theorem, and assuming that $\vec{x}$ belongs to the positive orthant, we can easily write down a supply of Schur convex functions. For this purpose we can also rely on *Karamata's inequality*: Any function of the form

$$F(\vec{x}) = \sum_{i=1}^{N} f(x_i) \tag{2.10}$$

is Schur convex, provided that $f(x)$ is a convex function on $\mathbb{R}$ (in the sense of Section 1.1).[4] In particular, the l_p–norm of a vector is Schur convex. Schur concave functions include the *elementary symmetric functions*

$$s_2(\vec{x}) = \sum_{i<j} x_i x_j, \qquad s_3(\vec{x}) = \sum_{i<j<k} x_i x_j x_k, \qquad (2.11)$$

and so on up to $s_N(\vec{x}) = \prod_i x_i$.

This concludes our tour of the majorisation order and the transformations that go with it. Since we laid so much stress on bistochastic matrices, let us end with an interesting theorem that applies to stochastic matrices in general. We quote it in an abbreviated form (see [130] for more):

The Frobenius–Perron theorem. *An irreducible $N \times N$ matrix whose matrix elements are non–negative has a real, positive and simple eigenvalue and a corresponding positive eigenvector. The modulus of any other eigenvalue never exceeds this.*

For a stochastic matrix the leading eigenvalue is 1 and all others lie in the unit disk of the complex plane – all stochastic matrices have an invariant eigenvector, although it need not sit in the centre of the probability simplex.

Stochastic maps are also called *Markov maps*. A *Markov chain* is a sequence of Markov maps and may be used to provide a discrete time evolution of probability distributions. It is not easy to motivate why time evolution should be given by linear maps, but as a matter of fact Markov maps are useful in a wide range of physical problems. In classical mechanics it is frequently assumed that time evolution is governed by a Hamiltonian. This forces the space of pure states to be infinite dimensional (unless one adopts finite number fields [979]). Markovian evolution does not require this, so we will see a lot of Markov maps in this book. Another context in which stochastic matrices appear is the process of *coarse graining* (or *randomisation*, as it is known in the statistical literature): suppose that we do not distinguish between two of three outcomes described by the probability distribution $\vec{p}$. Then we may map (p_0, p_1, p_2) to $(q_0, q_1) \equiv (p_0, p_1 + p_2)$ and this map is effected by a stochastic matrix. In this book we will be much concerned with functions that are monotonely increasing, or decreasing, under stochastic maps.

2.2 Shannon entropy

Let P be a probability distribution for a finite number N of possible outcomes, that is we have a vector $\vec{p}$ whose N components obey $p_i \geq 0$ and $\sum_i p_i = 1$. We ask for functions of the variables p_i that can tell us something interesting about the

---

[4] For a proof of this (and of many other inequalities), see Beckenbach and Bellman [102].

distribution. Perhaps surprisingly there is a single choice of such a function that (at least arguably) is more interesting than any other. This is the *Shannon entropy*

$$S(P) = -k \sum_{i=1}^{N} p_i \ln p_i \tag{2.12}$$

where k is a positive number that we usually set equal to 1. Note that the entropy is associated to a definite random variable X and can be written as $S(X)$, but the only property of the random variable that matters is its probabilitity distribution P. The reason why it is called entropy was explained by Shannon in a disarming way:

My greatest concern was what to call it. I thought of calling it 'information', but the word was overly used, so I decided to call it 'uncertainty'. When I discussed it with John von Neumann, he had a better idea. Von Neumann told me, "You should call it entropy, for two reasons. In the first place your uncertainty function has been used in statistical mechanics under that name, so it already has a name. In the second place, and more important, nobody knows what entropy really is, so in a debate you will always have an advantage."[5]

Entropy is a concept that has evolved a long way from its thermodynamic origins – and as so often happens a return to the origin may be difficult, but this will not concern us much in this book. The Shannon entropy can be interpreted as a measure of the uncertainty about the outcome of an experiment that is known to occur according to the probability distribution P, or as the amount of information needed to specify the outcome that actually occurs, or very precisely as the expected length of the communication needed for this specification. It takes the value zero if and only if one outcome is certain (that is for a pure state, when one component of $\vec{p}$ equals one) and its maximum value $k \ln N$ when all outcomes are equally likely (the uniform distribution). In this section only we set $k = 1/\ln 2$, which in effect means that we use logarithms to the base 2 in the definition of S:

$$S(P) = -\sum_{i=1}^{N} p_i \log_2 p_i. \tag{2.13}$$

With this choice the entropy is said to be measured in units of *bits*. If we have $N = 2^a$ possible outcomes we see that the maximum value of S is $\log_2 N = a$ bits, which is the length of the string of binary digits one can use to label the outcomes. If the outcomes occur with unequal probabilities it will pay, on the average, to label the more likely outcomes with shorter strings, and indeed the entropy goes down.

---

[5] Quoted by Tribus and McIrvine [898]; Shannon's original work (1948) [828] is available in book form as Shannon and Weaver [829]. A version of Eq. (2.12), with all the p_i equal to $1/W$, is engraved on Boltzmann's tombstone.

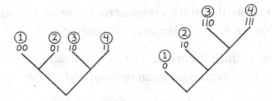

Figure 2.5 'Code trees' for four possible outcomes. The one on the right gives a shorter expected length if $p_1 > p_3 + p_4$.

The context in which the Shannon entropy has an absolutely precise interpretation along these lines is when we have a *source* that produces outcomes of an infinite sequence of *independent and identically distributed* random variables. Suppose we want to code the outcomes of a set of such *i.i.d.* events as a string of binary digits (zeros and ones). We define a *code* as a map of a sequence of outcomes into strings of binary numbers. The coding has to be done in such a way that a given string has an unambiguous interpretation as a sequence of outcomes. Some examples will make this clear: if there are four possible outcomes then we can code them as 00, 01, 10 and 11 respectively. These are *code words*. Without ambiguity, the string 0100010100 then means $(01, 00, 01, 01, 00)$. Moreover it is clear that the length of string needed to code one outcome is always equal to 2 bits. But other codes may do better, in the sense that the average number of bits needed to encode an outcome is less than 2; we are interested in codes that minimise the *expected length*

$$L = \sum_i p_i l_i, \qquad (2.14)$$

where l_i is the length of the individual code words (in bits). In particular, suppose that $p_1 \geq p_2 \geq p_3 \geq p_4$, and label the first outcome as 0, the second as 10, the third as 110 and the fourth as 111. Then the string we had above is replaced by the shorter string 10010100, and again this can be broken down in only one way, as $(10, 0, 10, 10, 0)$. That the new string is shorter was expected because we used a short code word to encode the most likely outcome, and we can read our string in an unambiguous way because the code has the *prefix property*: no code word is the prefix of any other. If we use both 0 and 1 as code words the prefix property is lost and we cannot code for more than two outcomes. See Figure 2.5 for the structure of the codes we use.

We are now faced with the problem of finding an *optimal code*, given the probability distribution. The expected length L_* of the code words used in an optimal code obeys $L_* \leq L$ where L is the expected length for an arbitrary code. We will not describe the construction of optimal codes here, but once it is admitted that such a code can be found then we can understand the statement of

Shannon's noiseless coding theorem. *Given a source distribution P let L_* be the expected length of a codeword used in an optimal code. Then*

$$S(P) \leq L_* \leq S(P) + 1. \tag{2.15}$$

For the proof of this interesting theorem – and how to find an optimal code – we must refer to the literature [237]. Our point is that the noiseless coding theorem provides a precise statement about the sense in which the Shannon entropy is a measure of information. (It may seem as if L_* would be a better measure. It is not for two reasons. First, L_* is not described by any nice analytical formula. Second, as we will see below, when many codewords are being sent the expected length per codeword can be made equal to S.)

The Shannon entropy has many appealing properties. Let us make a list:

- **Positivity**. Clearly $S(P) \geq 0$ for all discrete probability distributions.
- **Continuity**. $S(P)$ is a continuous function of the distribution.
- **Expansibility**. We adopt the convention that $0 \ln 0 = 0$. Then it will be true that $S(p_1, \ldots, p_N) = S(p_1, \ldots, p_N, 0)$.
- **Schur concavity**. As explained in Section 2.1, the Shannon entropy is Schur concave, so it tells us something about the majorisation order that we imposed on the set of all probability distributions.
- **Concavity**. It is also a concave function in the sense of convex set theory; mixing of probability distributions increases the entropy.
- **Additivity**. If we have a joint probability distribution P_{12} for two random variables and if they are independent, so that the joint probabilities are products of the individual probabilities, then

$$S(P_{12}) = S(P_1) + S(P_2). \tag{2.16}$$

In words, the information needed to describe two independent random variables is the sum of the information needed to describe them separately.[6]

- **Subadditivity**. If the two random variables are not independent then

$$S(P_{12}) \leq S(P_1) + S(P_2), \tag{2.17}$$

with equality if and only if the random variables are independent. Any correlation between them means that once we know the result of the first trial the amount of information needed to specify the outcome of the second decreases.

---

[6] In the statistical mechanics community additivity is usually referred to as *extensitivity*.

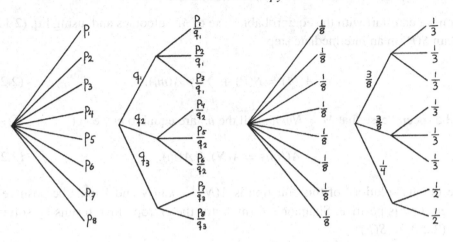

Figure 2.6 The recursion property illustrated; to the right it is used to determine $S(\frac{3}{8}, \frac{3}{8}, \frac{1}{4})$ in terms of the Shannon entropy for uniform distributions.

- **The recursion property**. Suppose that we coarse grain our description in the sense that we do not distinguish between all the outcomes. Then we are dealing with a new probability distribution Q with components

$$q_1 = \sum_{i=1}^{k_1} p_i, \quad q_2 = \sum_{i=k_1+1}^{k_2} p_i, \quad \ldots, \quad q_r = \sum_{i=N-k_r+1}^{N} p_i, \qquad (2.18)$$

for some partition $N = k_1 + k_2 + \ldots + k_r$. It is easy to show that

$$S(P) = S(Q) + q_1 S\left(\frac{p_1}{q_1}, \ldots, \frac{p_{k_1}}{q_1}\right) + \ldots + q_r S\left(\frac{p_{k_{N-k_r+1}}}{q_r}, \ldots, \frac{p_N}{q_r}\right). \qquad (2.19)$$

This is the recursion property and it is illustrated in Figure 2.6; it tells us how a choice can be broken down into successive choices.

The recursion property can be used as an axiom that singles out the Shannon entropy as unique.[7] Apart from the recursion property we assume that the entropy is a positive and continuous function of P, so that it is enough to consider rational values of $p_i = m_i/M$, where $M = \sum_i m_i$. We also define

$$A(N) = S(1/N, 1/N, \ldots, 1/N). \qquad (2.20)$$

---

[7] Increasingly elegant sets of axioms for the Shannon entropy were given by Shannon [829], Khinchin [524] and Faddejew [307]. Here we only give the key points of the argument.

Then we can start with the equiprobable case of M outcomes and, using Eq. (2.19), obtain $S(P)$ in an intermediate step:

$$A(M) = S(P) + \sum_{i=1}^{N} p_i A(m_i).$$ (2.21)

In the special case that $M = Nm$ and all the m_i are equal to m we get

$$A(Nm) = A(N) + A(m).$$ (2.22)

The unique solution[8] of this equation is $A(N) = k \ln N$ and k must be positive if the entropy is positive. Shannon's formula for the entropy then results by solving Eq. (2.21) for $S(P)$.

The argument behind Shannon's coding theorem can also be used to define the Shannon entropy, provided that we rely on additivity as well. Let the source produce the outcomes of $\mathcal{N}$ independent events. These can be regarded as a single outcome for a random variable described by a joint probability distribution and must be provided by a codeword. Let $L_*^{\mathcal{N}}$ be the expected length for the codeword used by an optimal code. Because the Shannon entropy is additive for independent distributions we get

$$\mathcal{N}S(P) \le L_*^{\mathcal{N}} \le \mathcal{N}S(P) + 1.$$ (2.23)

If we divide through by $\mathcal{N}$ we get the expected length per outcome of the original random variable – and the point is that this converges to $S(P)$, exactly. By following this line of reasoning we can give a formulation of the noiseless coding theorem that is more in line with that used later (in Section 13.2). But here we confine ourselves to the observation that, in this specific sense, the Shannon entropy is exactly what we want a measure of information to be!

Now what does the function look like? First we switch back to using the natural logarithm in the definition (we set $k = 1$), then we take a look at the Shannon entropy (2.12) for $N = 2$. Figure 2.7 also shows that

$$2 \ln 2 \min\{x, 1 - x\} \le S(x, 1 - x) \le 2 \ln 2 \sqrt{x(1 - x)}.$$ (2.24)

Much sharper power (and logarithmic) bounds have been provided by Topsøe [897]. Bounds valid for $N = 2$ only are more interesting than one might think, as we will see when we prove the Pinsker inequality in Section 14.1.

---

[8] This is the tricky point – see Rényi [773] for a comparatively simple proof.

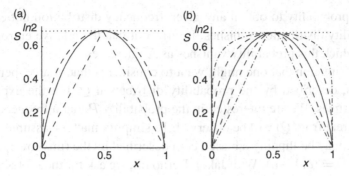

Figure 2.7 The Shannon entropy $S(x, 1 - x)$ for $N = 2$ (solid curve). In panel (a) we show the simple power bounds (2.24), and in panel (b) some bounds provided by the Rényi entropies from Section 2.7, upper bounds with $q = 1/5$ (dotted) and $1/2$ (dashed) and lower bounds with $q = 2$ (dashed, concave) and 5 (not concave).

2.3 Relative entropy

The Shannon entropy is a function that tells us something about a probability distribution. We now consider two probability distributions P and Q, and we look for a measure of how different they are – a distance between them, or at least a *distinguishability measure*. Information theory offers a favoured candidate: the *relative entropy*

$$S(P||Q) = \sum_{i=1}^{N} p_i \ln \frac{p_i}{q_i}. \tag{2.25}$$

It is also known as the *Kullback–Leibler entropy* or the *information divergence*.[9] In terms of lengths of messages, the relative entropy $S(P||Q)$ measures how much the expected length of a codeword grows, if the coding is optimal but made under the erroneous assumption that the random variable is described by the distribution Q. Relative entropy is not a distance in the sense of Section 1.4 since it is not symmetric under interchange of P and Q. Nevertheless it does provide a measure of how different the two distributions are from each other. We should be clear about the operational context in which a given distinguishability measure is supposed to work; here we assume that we have a source of independent and identically distributed random variables, and we try to estimate the probability distribution by observing the frequency distribution of the various outcomes. The Law of Large Numbers guarantees that in the limit when the number $\mathcal{N}$ of samplings goes to

---

[9] The relative entropy was introduced by Kullback and Leibler in 1951 [563], and even earlier than that by Jeffreys in 1939 [500]. Sanov's theorem [790] appeared in 1957. In mathematical statistics an asymmetric distance measure is referred to as a divergence.

infinity the probability to obtain any given frequency distribution not equal to the true probability distribution vanishes. It turns out that the relative entropy governs the rate at which this probability vanishes as $\mathcal{N}$ increases.

A typical situation that one might want to consider is that of an experiment with N outcomes, described by the probability distribution Q. Let the experiment be repeated $\mathcal{N}$ times. We are interested in the probability $\mathcal{P}$ that a frequency distribution P (different from Q) will be observed. To simplify matters, assume $N = 2$. For instance, we may be flipping a biased coin described by the (unknown) probability distribution $Q = (q, 1 - q)$. With Jakob Bernoulli, we ask for the probability $\mathcal{P}$ that the two outcomes occur with frequencies $P = (m/\mathcal{N}, 1 - m/\mathcal{N})$. This is a matter of counting how many strings of outcomes there are with each given frequency, and the exact answer is

$$\mathcal{P}\left(\frac{m}{\mathcal{N}}\right) = \left(\begin{array}{c} \mathcal{N} \\ m \end{array}\right) q^m (1 - q)^{\mathcal{N}-m}. \tag{2.26}$$

For large $\mathcal{N}$ we can approximate this using Stirling's formula, in the simplified form $\ln \mathcal{N}! \approx \mathcal{N} \ln \mathcal{N} - \mathcal{N}$. For $\mathcal{N} \approx 100$ this is accurate within a percent or so. The result is

$$\ln \left(\mathcal{P}(m/\mathcal{N})\right) \approx -\mathcal{N}\left[\frac{m}{\mathcal{N}}\left(\ln \frac{m}{\mathcal{N}} - \ln q\right) + \left(1 - \frac{m}{\mathcal{N}}\right)\left(\ln \left(1 - \frac{m}{\mathcal{N}}\right) - \ln (1 - q)\right)\right]$$
$$\tag{2.27}$$

and we recognise the relative entropy on the right hand side. Indeed the probability to obtain the frequency distribution P is

$$\mathcal{P}(P) \approx e^{-\mathcal{N}S(P\|Q)}. \tag{2.28}$$

This simple exercise should be enough to advertise the fact that relative entropy means something.

There are more precise ways of formulating this conclusion. We have

Sanov's Theorem. *Let an experiment with N outcomes described by the probability distribution Q be repeated $\mathcal{N}$ times, and let E be a set of probability distributions for N outcomes such that E is the closure of its interior. Then, for $\mathcal{N}$ large, the probability $\mathcal{P}$ that a frequency distribution belonging to E will be obtained is*

$$\mathcal{P}(E) \sim e^{-\mathcal{N}S(P_*\|Q)}, \tag{2.29}$$

where P_* is that distribution in E that has the smallest value of $S(P\|Q)$.

That E is the closure of its interior means that it is a 'nice' set, without isolated points, 'spikes', and so on. An important ingredient in the proof is that the number of frequency distributions that may be realised is bounded from above by $(\mathcal{N}+1)^N$, since any one of the N outcomes can occur at most $\mathcal{N}+1$ times. On the other hand the number of possible ordered strings of outcomes grows exponentially with $\mathcal{N}$.

But this is all we have to say about the proof; our main concern is the geometrical significance of relative entropy, not to prove the theorems of information theory.[10]

The relative entropy is always a positive quantity. In fact

$$S(P||Q) = \sum_{i=1}^{N} p_i \ln \frac{p_i}{q_i} \geq \frac{1}{2} \sum_{i=1}^{N} (p_i - q_i)^2 = D_2^2(P, Q). \qquad (2.30)$$

To prove this, note that any smooth function f obeys

$$f(x) = f(y) + (x - y)f'(y) + \frac{1}{2}(x - y)^2 f''(\xi), \qquad \xi \in (x, y). \qquad (2.31)$$

Now set $f(x) = -x \ln x$, in which case $f''(x) \leq -1$ when $0 \leq x \leq 1$. A little rearrangement proves that the inequality (2.30) holds term by term, so that this is not a very sharp bound.

The relative entropy is asymmetric. That this is desirable can be seen if we consider two coins, one fair and one with heads on both sides. Start flipping one of them to find out which is which. A moment's thought shows that if the coin we picked is the fair one, this will most likely be quickly discovered, whereas if we picked the other coin we can never be completely sure – after all the fair coin can give a sequence of a thousand heads, too. If statistical distance is a measure of how difficult it will be to distinguish two probability distributions in a repeated experiment then this argument shows that such a measure must have some asymmetry built into it. Indeed, using notation from Eq. (1.62),

$$S(Q_{(N)}||Q_{(1)}) = \infty \quad \text{and} \quad S(Q_{(1)}||Q_{(N)}) = \ln N. \qquad (2.32)$$

The asymmetry is pronounced. For further intuition about this, consult Section 14.1. The asymmetry can also be studied in Figure 2.8c and d, for $N = 2$ and in Figure 2.9 for $N = 3$.

There is a kind of 'Pythagorean theorem' that supports the identification of relative entropy as a (directed) distance squared:

Pythagorean theorem. *The distribution P lies in a convex set E and Q lies outside E. Choose a distribution P_* on the boundary of the convex set such that $S(P_*||Q)$ assumes its minimum value for P_* belonging to E and fixed Q. Then*

$$S(P||Q) \geq S(P||P_*) + S(P_*||Q). \qquad (2.33)$$

To prove this, consider distributions on the straight line $P_\lambda = \lambda P + (1 - \lambda)P_*$, which lies inside E because E is convex. A minor calculation shows that

$$\lambda = 0 \quad \Rightarrow \quad \frac{d}{d\lambda} S(P_\lambda||Q) = S(P||Q) - S(P||P_*) - S(P_*||Q). \qquad (2.34)$$

---

[10] All unproved assertions in this section are proved in the book by Cover and Thomas [237]. A number of further results can be found in their chapter 10.

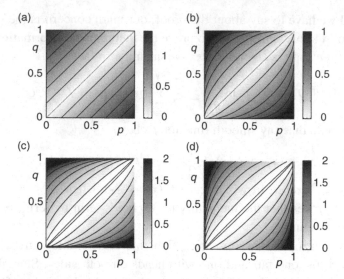

Figure 2.8 Measures of distinguishability between probability distributions $P = (p, 1 - p)$ and $Q = (q, 1 - q)$: a) Euclidean distance, b) Bhattacharyya distance (see Section 2.5), c) and d) relative entropies $S(P||Q)$ and $S(Q||P)$.

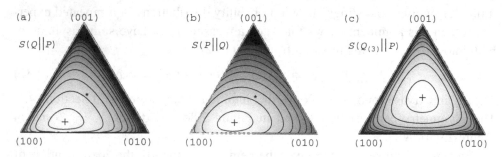

Figure 2.9 Contour lines of the relative entropy $S(Q||P)$ (a) and $S(P||Q)$ for $Q = (0.6, 0.4, 0.1)$ ($+$) as a function of P. Panel (c) shows an analogous picture for $S(Q_{(3)}||P)$, while the contours of $S(P||Q_{(3)})$ consisting of points of the same Shannon entropy are shown in Figure 2.14c.

But the derivative cannot be negative at $\lambda = 0$, by the assumption we made about P^*, and the result follows. It is called a Pythagorean theorem because, under the same conditions, if we draw Euclidean straight lines between the points and measure the Euclidean angle at P_* then that angle is necessarily obtuse once P_* has been chosen to minimise the Euclidean distance from the convex set to Q. If D_{PQ} denotes the Euclidean distance between the points then Pythagoras' theorem in its usual formulation states that $D_{PQ}^2 \geq D_{PP_*}^2 + D_{P_*Q}^2$. In this sense the relative entropy really behaves like a distance squared.

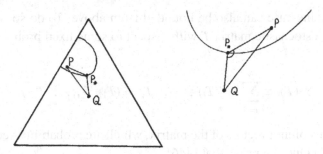

Figure 2.10 A Pythagorean property of relative entropy (left), and the same property in Euclidean geometry where $D_{PQ}^2 \geq D_{PP_*}^2 + D_{P_*Q}^2$. The angle is obtuse because the set to which P belongs is convex (right).

The convexity properties of the relative entropy are interesting. It is *jointly convex*. This means that when $\lambda \in [0, 1]$ it obeys

$$S(\lambda P_1 + (1 - \lambda)P_2 || \lambda Q_1 + (1 - \lambda)Q_2) \leq \lambda S(P_1||Q_1) + (1 - \lambda)S(P_2||Q_2). \quad (2.35)$$

Convexity in each argument separately is an easy consequence (set $Q_1 = Q_2$). Moreover relative entropy behaves in a characteristic way under Markov maps. Let T be a stochastic map (that is, a matrix that takes a probability vector to another probability vector). Then we have the following property:

- **Monotonicity under stochastic maps**.

$$S(TP||TQ) \leq S(P||Q). \quad (2.36)$$

In words, the relative entropy between two probability distributions always decreases in a Markov chain, so that the distinguishability of two distributions decreases with time. It is really this property that makes relative entropy such a useful concept; functions of pairs of probability distributions that have this property are known as *monotone* functions. The increase of entropy follows if T is also bistochastic, that is to say that it leaves the uniform probability distribution invariant. The uniform distribution is denoted $Q_{(N)}$. Clearly

$$S(P||Q_{(N)}) = \sum_i p_i \ln (Np_i) = \ln N - S(P). \quad (2.37)$$

Since $TQ_{(N)} = Q_{(N)}$ the decrease of the relative entropy implies that the Shannon entropy of the distribution P is increasing with time; we are getting closer to the uniform distribution. (And we have found an interesting way to show that $S(P)$ is a Schur concave function.)

The entropy increase can also be bounded from above. To do so we first define the entropy of a stochastic matrix T with respect to some fixed probability distribution P as

$$S_P(T) \equiv \sum_{i=1}^{N} p_i S(\vec{T}_i) \; ; \qquad \vec{T}_i = (T_{1i}, T_{2i}, \ldots, T_{Ni}). \tag{2.38}$$

(Thus $\vec{T}_i$ are the column vectors of the matrix, which are probability vectors because T is stochastic.) One can prove that [846]

$$S_P(T) \leq S(TP) \leq S_P(T) + S(P). \tag{2.39}$$

Again, this is a result that holds for all stochastic matrices.

2.4 Continuous distributions and measures

The transition from discrete to continuous probability distributions appears at first sight to be trivial – just replace the sum with an integral and we arrive at the *Boltzmann entropy*

$$S_B = - \int_{-\infty}^{\infty} \mathrm{d}x \, p(x) \ln p(x). \tag{2.40}$$

This is a natural generalisation of Shannon's entropy to continuous probability distributions, but depending on the behaviour of the distribution function $p(x)$ it may be ill-defined, and it is certainly not bounded from below in general, since $p(x)$ need not bounded from above. To see this let $p(x)$ be a step function taking the value t^{-1} for $x \in [0, t]$ and zero elsewhere. The entropy S_B is then equal to $\ln t$ and goes to negative infinity as $t \to 0$. In particular a pure classical state corresponds to a delta function distribution and has $S_B = -\infty$. In a way this is as it should be – it takes an infinite amount of information to specify such a state exactly.

The next problem is that there is no particular reason why the entropy should be defined in just this way. It is always a delicate matter to take the continuum limit of a discrete sum. In fact the formula as written will change if we change the coordinate that we are using to label the points on the real line. In the continuous case random variables are scalars, but the probability density transforms like a density. What this means is that

$$\langle A \rangle = \int \mathrm{d}x \, p(x) A(x) = \int \mathrm{d}x' \, p'(x') A'(x') \tag{2.41}$$

where $A'(x') = A(x)$. Hence $p'(x') = J^{-1} p(x)$, where J is the Jacobian of the coordinate transformation. This means that the logarithm in our definition of the Boltzmann entropy will misbehave under coordinate transformations. There is a simple solution to this problem, consisting in the observation that the relative entropy with respect to some *prior* probability density $m(x)$ does behave itself. In equations,

$$S_B = - \int_{-\infty}^{\infty} dx\, p(x) \ln \frac{p(x)}{m(x)} \qquad (2.42)$$

is well behaved. The point is that the quotient of two densities transforms like a scalar under coordinate changes. Now, depending on how we take the continuum limit of the Shannon entropy, different densities $m(x)$ will arise. Finding an appropriate $m(x)$ may be tricky, but this should not deter us from using continuous distributions when necessary. We cannot avoid them in this book, because continuous distributions occur also in finite dimensional quantum mechanics. In most cases we will simply use the 'obvious' translation from sums to integrals along the lines of Eq. (2.40), but it should be kept in mind that this really implies a special choice of prior density $m(x)$.

At this point we begin to see why mathematicians regard probability theory as a branch of measure theory. We will refer to expressions such as $d\mu(x) \equiv p(x)dx$ as *measures*. Mathematicians tend to use a slightly more sophisticated terminology at this point. Given a suitable subset A of the space coordinatised by x, they define the measure as a real valued function μ such that

$$\mu(A) \equiv \int_A d\mu(x). \qquad (2.43)$$

The measure measures the volume of the subset. This terminology can be very helpful, but we will not use it much.

When the sample space is continuous the space of probability distributions is infinite dimensional. But in many situations we may be interested in a finite dimensional subset. For instance, we may be interested in the two dimensional submanifold of *normal distributions*

$$p(x; \mu, \sigma) = \frac{1}{\sqrt{2\pi}\sigma} e^{-\frac{(x-\mu)^2}{2\sigma^2}}, \qquad (2.44)$$

with the mean μ and the standard deviation σ serving as coordinates in the submanifold. In general a finite dimensional submanifold of the space of probability distributions is defined by the function $p(x; \theta^1, \ldots, \theta^n)$, where θ^a are coordinates in the submanifold. When we think of this function as a function of θ^a it is known as the *likelihood function*. This is a setup encountered in the theory of *statistical inference*, that is to say the art of guessing what probability distribution that governs an experiment, given only the results of a finite number of samplings, and perhaps some kind of prior knowledge. (The word 'likelihood' is used because, to people of resolutely frequentist persuasions, there can be no probability distribution for θ^a.)

The likelihood function is a hypothesis about the form that the probability distribution takes. This hypothesis is to be tested, and we want to design experiments that allow us to make a 'best guess' for the values of the parameters θ^a. As a first step we want to design a statistical geometry that is helpful for this kind of question.

A somewhat idiosyncratic notation is used in the statistical literature to represent tangent spaces. Consider the *log–likelihood function*

$$l(x; \theta) = \ln p(x; \theta). \tag{2.45}$$

Here x denotes coordinates on the sample space Ω and θ are coordinates on a subspace of probability distributions. The idea is to describe tangent space through the natural isomorphism between its basis vectors ∂_a (used in Section 1.4) and the *score vectors* l_a, which are the derivatives with respect to the coordinates θ^a of the log–likelihood function. That is to say

$$\partial_a \equiv \frac{\partial}{\partial \theta^a} \quad \leftrightarrow \quad l_a \equiv \frac{\partial l}{\partial \theta^a}. \tag{2.46}$$

For this to make any sense we must assume that the score vectors form a set of n linearly independent functions of x for any given θ. (It is not supposed to be self-evident why one should use precisely the log-likelihood function here, but it is an interesting function to consider – the Shannon entropy is nothing but its expectation value.) The expectation values of the score vectors vanish. To see this recall that the expectation value is a sum over the sample space which we denote schematically as an integral:

$$\langle l_a \rangle \equiv \int_\Omega dx \, p(x; \theta) l_a(x; \theta) = \int_\Omega dx \, \partial_a p(x; \theta) = \frac{\partial}{\partial \theta^a} \int_\Omega dx \, p(x; \theta) = 0. \tag{2.47}$$

It is assumed that the interchange of the order of integration with respect to x and differentiation with respect to θ is allowed. A general tangent vector $A(x) = A^a l_a(x)$ is a linear combination of score vectors; as such it is a random variable with expectation value zero. These at first sight peculiar definitions actually achieve something: sets of probability distributions are turned into manifolds, and random variables are turned into tangent vectors. In the next section we will see that scalar products between tangent vectors have a statistical meaning, too.

2.5 Statistical geometry and the Fisher–Rao metric

One way of introducing a metric on the space of probability distribution is to consider the situation where a large number $\mathcal{N}$ of samplings are made from a given probability distribution P, at first assumed to be discrete. This case is sufficiently simple so that we can do all the details. We ask if we can find out what P actually

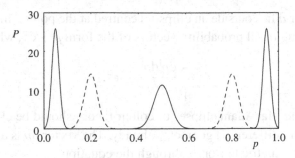

Figure 2.11 Bernoulli's result: the shapes of the Gaussians will determine how easily points can be distinguished.

is, given only the result of the $\mathcal{N}$ samplings. We considered this problem in Section 2.3; when there are only two possible outcomes the probability to obtain the frequencies $(f_1, f_2) = (m/\mathcal{N}, 1 - m/\mathcal{N})$ is

$$P\left(\frac{m}{\mathcal{N}}\right) = \binom{\mathcal{N}}{m} p^m (1-p)^{\mathcal{N}-m}. \tag{2.48}$$

We give the answer again because this time we denote the true probability distribution by $P = (p, 1 - p)$. When $\mathcal{N}$ is large, we can use Stirling's formula to approximate the result; in textbooks on probability it is usually given as

$$P\left(\frac{m}{\mathcal{N}}\right) = \frac{1}{\sqrt{2\pi \mathcal{N} p(1-p)}} e^{-\frac{\mathcal{N}}{2}\frac{(\frac{m}{\mathcal{N}}-p)^2}{p(1-p)}} = \frac{1}{\sqrt{2\pi \mathcal{N} p_1 p_2}} e^{-\frac{\mathcal{N}}{2}\sum_{i=1}^{2}\frac{(f_i-p_i)^2}{p_i}}. \tag{2.49}$$

This is consistent with Eq. (2.28) when the observed frequencies are close to P. The error that we have committed is smaller than $k/\mathcal{N}$, where k is some real number independent of $\mathcal{N}$; we assume that $\mathcal{N}$ is large enough so that this does not matter. Figure 2.11 shows that the Gaussians are strongly peaked when the outcomes of the experiment are nearly certain, while they are much broader in the middle of the probability simplex. Statistical fluctuations are small when we are close to pure states; in some sense the visibility goes up there.

The result is easily generalised to the case of N outcomes; we obtain the normal distribution

$$\mathcal{P}(F) \propto e^{-\frac{\mathcal{N}}{2}\sum_{i=1}^{N}\frac{(f^i-p^i)^2}{p^i}}. \tag{2.50}$$

We raised all indices, because we will be doing differential geometry soon! Our question is: given only the frequencies f^i, do we dare to claim that the probabilities take the values $p^i = f^i$, rather than some other values q^i? This is clearly a matter of taste. We may choose some number ϵ and decide that we dare to do it if the probability vector $\vec{q}$ lies outside an ellipsoid centred at the point $\vec{p}$ in the probability simplex, consisting of all probability vectors of the form $\vec{p} + d\vec{p}$, where

$$\sum_{i=1}^{N} \frac{dp^i dp^i}{p^i} \leq \epsilon^2. \tag{2.51}$$

The analogy to the MacAdam ellipses of colour theory should be clear – and if ϵ is small we are doing differential geometry already. The vector $d\vec{p}$ is a tangent vector, and we introduce a metric tensor g_{ij} through the equation

$$ds^2 = \sum_{i,j} g_{ij} dp^i dp^j = \frac{1}{4} \sum_{i=1}^{N} \frac{dp^i dp^i}{p^i} \quad \Leftrightarrow \quad g_{ij} = \frac{1}{4} \frac{\delta_{ij}}{p^i}. \tag{2.52}$$

(For clarity, we do not use Einstein's summation convention here.) This Riemannian metric is known as the *Fisher–Rao metric*, or as the *Fisher information matrix*.[11]

What is it? To see this, we will try to deform the probability simplex into some curved space, in such a way that all the little ellipsoids become spheres of equal size. A rubber model of the two dimensional case would be instructive to have, but since we do not have one let us do it with equations. We introduce new coordinates X^i, all of them obeying $X^i \geq 0$, through

$$X^i = \sqrt{p^i} \quad \Rightarrow \quad dX^i = \frac{dp^i}{2\sqrt{p^i}}. \tag{2.53}$$

Then the Fisher–Rao metric takes the very simple form

$$ds^2 = \sum_{i=1}^{N} dX^i dX^i. \tag{2.54}$$

All the little error ellipsoids have become spheres. If we remember the constraint

$$\sum_{i=1}^{N} p^i = \sum_{i=1}^{N} X^i X^i = 1, \tag{2.55}$$

then we see that the geometry is that of a unit sphere. Because of the restricted coordinate ranges we are in fact confined to the positive hyperoctant of $\mathbf{S}^{N-1}$. Figure 2.12 illustrates the transformation that we made.

Section 3.1 will provide more details about spheres if this is needed. But we can right away write down the geodesic distance D_{Bhatt} between two arbitary probability distributions P and Q, now regarded as the length of the great circle between them.

---

[11] The factor of $1/4$ is needed only for agreement with conventions that we use later on. The Fisher–Rao metric was introduced by Rao (1945) [766], based on earlier work by Fisher (1925) [312] and Bhattacharyya (1943) [132].

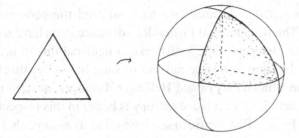

Figure 2.12 The convex (flat) and statistical (round) geometry of a simplex when $N = 3$.

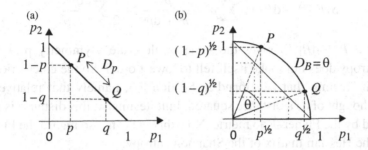

Figure 2.13 For $N = 2$ the space of discrete probabilities reduces to (a) a straight line segment along which all the l_p distances agree, or (b) a quarter of a circle equipped with the Bhattacharyya distance, measured by the angle θ.

This is simply the angle between two vectors, with components $X^i = \sqrt{p^i}$ and $Y^i = \sqrt{q^i}$. Hence

$$\cos D_{\text{Bhatt}} = \sum_{i=1}^{N} \sqrt{p^i q^i} = B(P, Q). \tag{2.56}$$

In statistics this distance is known as the *Bhattacharyya distance*. See Figure 2.13. The right hand side of the equation is known as the *Bhattacharyya coefficient*, and it looks intriguingly similar to the scalar product in quantum mechanics. Its square is known as the *classical fidelity*. Alternatively we can define the chordal distance D_H as the distance between the points in the flat embedding space where the round hyperoctant sits, that is

$$D_H = \left(\sum_{i=1}^{N} \left(\sqrt{p^i} - \sqrt{q^i} \right)^2 \right)^{\frac{1}{2}}. \tag{2.57}$$

In statistics this is known as the *Hellinger distance*. It is clearly a monotone function of the Bhattacharyya distance.

In writing down these distances we have allowed the geometry to run ahead of the statistics. The infinitesimal Fisher–Rao distance does have a clear cut operational significance, in terms of the statistical distinguishability of nearby probability distributions, in the limit of a large number of samplings. But this is not really the case for the finite Bhattacharyya and Hellinger distances, at least not as far as we know. The Kullback–Leibler relative entropy is better in this respect, and should be connected to the Fisher–Rao metric in some way. Let us assume that two probability distributions are close, and let us expand the relative entropy (2.25) to lowest order. We find

$$S(P||P + \mathrm{d}P) = \sum_i p^i \ln \frac{p^i}{p^i + \mathrm{d}p^i} \approx \frac{1}{2} \sum_i \frac{\mathrm{d}p^i \mathrm{d}p^i}{p^i}. \qquad (2.58)$$

Expanding $S(P + \mathrm{d}P||P)$ gives the same result – the asymmetry present in the relative entropy does not make itself felt to lowest order. So the calculation simply confirms the feeling that developed in Section 2.3, namely that relative entropy should be thought of as a distance squared. Infinitesimally, this distance is precisely that defined by the Fisher–Rao metric. Note that we can also regard the Fisher–Rao metric as the Hessian matrix of the Shannon entropy:

$$g_{ij} = -\frac{1}{4} \partial_i \partial_j S(p) = \frac{1}{4} \partial_i \partial_j \sum_{k=1}^{N} p^k \ln p^k. \qquad (2.59)$$

The fact that Shannon entropy is concave implies that the Fisher–Rao metric is positive definite. In Section 3.2 we will tell the story backwards and explain that relative entropy exists precisely because there exists a 'potential' whose second derivatives form the components of the Riemannian metric.

As we made clear in Section 1.4 there are many ways to introduce the concept of distance on a given space, and we must ask in what sense – if any – our concept of statistical distance is unique? There is a clear cut answer to this question. We will consider maps between two probability distributions P and Q of the form $Q = TP$, where T is a stochastic matrix. We define a *monotone* function as a function of pairs of probability distributions such that

$$f(TP, TQ) \leq f(P, Q). \qquad (2.60)$$

Then there is a comfortable result:

Čencov's theorem. *For multinomial distributions the Fisher–Rao metric is (up to normalisation) the only metric whose geodesic distance is a monotone function.*

In this desirable sense the Fisher–Rao metric is unique. It is not hard to see that the Fisher–Rao metric has the desired property. We must show that the length of an arbitrary tangent vector dp^i at the point p^i decreases under a stochastic map, that is to say that

$$||Tdp||^2 \leq ||dp||^2 \quad \Leftrightarrow \quad \sum_{i=1}^{N} \frac{(Tdp)^i (Tdp)^i}{(Tp)^i} \leq \sum_{i=1}^{N} \frac{dp^i dp^i}{p^i}. \qquad (2.61)$$

To do so, we first prove that

$$\left(\sum_j T_{ij} dp_j \right)^2 \leq \left(\sum_j T_{ij} \frac{dp_j dp_j}{p_j} \right) \left(\sum_j T_{ij} p_j \right). \qquad (2.62)$$

Looking closely at this expression (take square roots!) we see that it is simply the Cauchy–Schwarz inequality in disguise. Dividing through by the rightmost factor and summing over i we find – precisely because $\sum_i T_{ij} = 1$, that is because the map is stochastic – that Eq. (2.61) follows.[12]

It often happens that one is interested in some submanifold of the probability simplex, coordinatised by a set of coordinates θ^a. The Fisher–Rao metric will induce a metric on such submanifolds, given by

$$g_{ab} = \sum_{ij} \frac{\partial p^i}{\partial \theta^a} \frac{\partial p^j}{\partial \theta^b} g_{ij} = \frac{1}{4} \sum_i \frac{\partial_a p^i \partial_b p^i}{p^i}. \qquad (2.63)$$

As long as the submanifold – the *statistical model*, as it may be referred to here – is finite dimensional, this equation is easily generalised to the case of continuous probability distributions $p(x)$, where the probability simplex itself is infinite dimensional:

$$g_{ab} = \frac{1}{4} \int_\Omega dx \frac{\partial_a p \partial_b p}{p}. \qquad (2.64)$$

This metric has the nice property that it is unaffected by reparametrizations of the sample space, that is to say by changes to new coordinates $x'(x)$ – and indeed obviously so since the derivatives are with respect to θ.

The odd-looking notation using score vectors comes into its own here. Using them we can rewrite Eq. (2.64) as

$$g_{ab} = \frac{1}{4} \int_\Omega dx \, p l_a l_b = \frac{1}{4} \langle l_a l_b \rangle. \qquad (2.65)$$

---

[12] We do not prove uniqueness here, but in section 14.1 we show that the flat metric is not monotone. Unfortunately Čencov's (1972) [199] proof is difficult; for an accessible proof of his key point see Campbell [192]. By the way, Čencov will also appear in our book under the name of Chentsov.

The components of the metric tensor are the scalar products of the basis vectors in tangent space. A general tangent vector is, in the language we now use, a random variable with vanishing mean, and the scalar product of two arbitrary tangent vectors becomes

$$\frac{1}{4}\langle A(x)B(x)\rangle = \frac{1}{4}\int_{\Omega} dx\, p(x)A(x)B(x). \tag{2.66}$$

Except for the annoying factor of $1/4$, this is the *covariance* of the two random variables.

In order to get used to these things let us compute the Fisher–Rao metric on the two dimensional family of normal distributions, with the mean μ and variance σ as coordinates. See Eq. (2.44). We have two score vectors

$$l_{\mu} = \partial_{\mu}\ln p(x;\mu,\sigma) = \frac{x-\mu}{\sigma^2} \qquad l_{\sigma} = \partial_{\sigma}\ln p(x;\mu,\sigma) = \frac{(x-\mu)^2}{\sigma^3} - \frac{1}{\sigma}. \tag{2.67}$$

Taking the appropriate expectation values we find (after a short calculation) that the statistical metric on the normal family is

$$ds^2 = \frac{1}{4\sigma^2}\left(d\mu^2 + 2d\sigma^2\right). \tag{2.68}$$

This is a famous metric of constant negative curvature, known as the *Poincaré metric* on the upper half plane. The μ–axis itself consists of the pure states, and points there are infinitely far away from any point in the interior of the upper half plane; if the outcome is certain (so that the standard deviation σ is zero) then it is trivial to distinguish this distribution from any other one.

As our parting shot in this section, consider a statistical inference problem where we want to find the values of the parameters θ^a from samplings of some random variable. A random variable ξ^a is said to be an *unbiased estimator* of the parameter θ^a if $\langle \xi^a \rangle = \theta^a$. So, there are no systematic errors in the experiment. Still, there is a limit on how well the unbiased estimators can perform their task:

The Cramér–Rao theorem. *The variance of an unbiased estimator obeys*

$$\langle \xi^a \xi^b \rangle - \langle \xi^a \rangle \langle \xi^b \rangle \geq \frac{1}{4}g^{ab}. \tag{2.69}$$

The inequality means that the left hand side minus the right hand side is a positive semi–definite matrix.[13]

---

[13] We have left much unsaid here. Amari [30] contains a highly recommended book length discussion of statistical geometry. We also recommend the review by Ingarden [488].

2.6 Classical ensembles

Let us define an *ensemble* as a set equipped with a probability measure. Choosing a probability simplex as our set, we obtain an ensemble of probability distributions, or states, characterised by a distribution $P(\vec{p})$. The physical situation may be that of a machine producing an unlimited set of independent and identically distributed copies of a physical system each characterised by the same unknown probability distribution $\vec{p}$, in which case $P(\vec{p})$ tells us something about the machine, or about our knowledge of the machine.

We seem to be on the verge of breaking our promise not to commit ourselves to any particular interpretation of probability, since we appear to imply that the original probability $\vec{p}$ is an inherent property of the individual physical systems that are being produced. We have not introduced any agent whose degree of reasonable belief can reasonably be said to be quantified by $\vec{p}$. Closer inspection reveals that the subjective interpretation of probability can be saved. To do so one observes that there is really only one thing that we are ignorant of, namely the full sequence of observations. But beyond that we make use of some prior assumptions that specify exactly what we mean by 'independent and identically distributed copies'. Then *de Finetti's theorem* tells us that this situation can be treated as if it was described by the distribution $P(\vec{p})$ on the probability simplex. This is a subtle point, and we refer to the literature for a precise account of our ignorance.[14]

We have now explained one setting in which a distribution $P(\vec{p})$ arises. Others can be imagined. From the subjective point of view, the prior should summarise whatever information we have before the data have been analysed, and it should be systematically updated as the observations are coming in. The prior should obey one important requirement: it should lead to manageable calculations in practice. What we want to do here is much simpler. We ask what we should expect to observe if the state of a physical system is picked 'at random'. A *random state* is a state picked from the probability simplex according to some measure, and that measure is now to be chosen in a way that captures our idea of randomness. Hence we ask for the *uniform prior* corresponding to complete ignorance. But we are using an alarmingly vague language. What is complete ignorance?

Presumably, the uniform prior is such that the system is equally likely to be found anywhere on the simplex, but there are at least two natural candidates for what this should mean. We could say that the simplex is flat. Then we use the measure

$$dP_\Delta = (N-1)! \; \delta \left(\sum_{i=1}^{N} p_i - 1 \right) dp_1 dp_2 \ldots dp_N. \qquad (2.70)$$

---

[14] For an engaging account both backwards (with references to earlier literature) and forwards (to quantum information theory) see Caves, Fuchs and Schack [197]. But note that from now on we will use whatever language we find convenient, rather than get involved in issues like this.

(At the cost of a slight inconsistency, we have lowered the index on p_i again. We find this more congenial!) This probability measure is correctly normalised, because we get one when we integrate over the positive cone:

$$\int_{\mathbb{R}_+^N} dP_\Delta = (N-1)! \int_0^1 dp_1 \int_0^{1-p_1} dp_2 \cdots \int_0^{1-p_1-\cdots-p_{N-2}} dp_{N-1}$$

$$= \int_0^1 dp_1 \int_0^1 dp_2 \cdots \int_0^1 dp_{N-1} = 1. \qquad (2.71)$$

The volume of the simplex is $1/(N-1)!$ times the volume of the parallelepiped spanned by its edges.

But if we use the Fisher–Rao rather than the flat metric, then the simplex is round rather than flat, and the appropriate density to integrate is proportional to the square root of the determinant of the Fisher–Rao metric. The measure becomes

$$dP_{\text{FR}} = \frac{\Gamma\left(\frac{N}{2}\right)}{\pi^{N/2}} \delta\left(\sum_{i=1}^N p_i - 1\right) \frac{dp_1 dp_2 \dots dp_N}{\sqrt{p_1 p_2 \cdots p_N}}. \qquad (2.72)$$

To check the normalisation we first change coordinates to $X^i = \sqrt{p_i}$, and then use the fact that we already know the volume of a round hyperoctant – namely $1/2^N$ times the volume of $\mathbf{S}^{N-1}$, given in Eq. (1.19). This choice of prior is known as *Jeffreys' prior* and is generally agreed to be the best choice. Jeffreys arrived at his prior through the observation that for continuous sample spaces the corresponding expression has the desirable property of being invariant under reparametrizations of the sample space. Note the implication: somehow Jeffreys is claiming that 'most' of the probabilities that we encounter in the world around us are likely to be close to one or zero.

Further choices are possible (and are sometimes made in mathematical statistics). Thus we have the *Dirichlet distribution*

$$dP_s \propto \delta\left(\sum_{i=1}^N p_i - 1\right) (p_1 p_2 \dots p_N)^{s-1} dp_1 dp_2 \dots dp_N, \qquad (2.73)$$

which includes the flat and round measures for $s = 1$ and $s = 1/2$, respectively. To find a simple way to generate probability vectors according to these distributions study Problem 2.4.

2.7 Generalised entropies

The Shannon entropy is arguably the most interesting function of $\vec{p}$ that one can find. But in many situations the only property that one really requires is that the function be Schur concave, that is consistent with the majorisation order. For this reason we are willing to call any Schur concave function a *generalised entropy*. Similarly a *generalised relative entropy* must be monotone under stochastic maps. Equation 2.10, when adjusted with a sign, provides us with a liberal supply of Schur concave functions. To obtain a supply of generalised monotone entropies, let g be a convex function defined on $(0, \infty)$ such that $g(1) = 0$. Then the expression

$$S_g(P||Q) = \sum_i p_i \, g(q_i/p_i) \tag{2.74}$$

is monotone under stochastic maps. The choice $g(t) = -\ln t$ gives the Kullback–Leibler relative entropy that we have already studied. All of these generalised relative entropies are consistent with the Fisher–Rao metric, in the sense that

$$S_g(P||P + dP) = \sum_i p_i g \left(1 + \frac{dp_i}{p_i} \right) \approx \frac{g''(1)}{2} \sum_i \frac{dp_i dp_i}{p_i}. \tag{2.75}$$

As long as we stay with classical probability theory the Fisher–Rao geometry itself remains unique.

We will study generalised entropies in some detail. We define the *moments* of a probability distribution as

$$f_q(\vec{p}) = \sum_{i=1}^{N} p_i^q. \tag{2.76}$$

They are Schur concave for $q \leq 1$, and Schur convex for $q \geq 1$. The set of moments $f_q(\vec{p})$ for $q = 2, \ldots, N$ determines the vector $\vec{p}$ up to a permutation of its components, just as knowing the traces, $\mathrm{Tr}A^k$, $k = 1, \ldots, N$, of a Hermitian matrix A one can find its spectrum. The analogy is exact: in Chapter 13 we will think of classical probability distributions as diagonal matrices, and face the problem of generalizing the definitions of the present chapter to general Hermitian matrices. Instead of the Shannon entropy, we will have the von Neumann entropy that one can calculate from the spectrum of the matrix. But it is much easier to raise a matrix A to some integer power, than to diagonalise it. Therefore it is easier to compute the moments than to compute the von Neumann entropy.

When $q = 2$ the moment f_2 is also known as the *purity* (because it equals one if and only if the state is pure) or as the *index of coincidence* (because it gives the probability of getting identical outcomes from two independent and equally distributed events). The *linear entropy* is defined as $S_L = 1 - f_2$ and the *participation number* as $R = 1/f_2$ (it varies between 1 and N and is interpreted as the 'effective number of events' that contribute to it).

Table 2.1 *Properties of generalised entropies.*

Entropy	Shannon	Rényi	Havrda-Charvat
Formula	$-\sum_{i=1}^{N} p_i \ln p_i$	$\frac{1}{1-q} \ln\left(\sum_{i=1}^{N} p_i^q\right)$	$\frac{1}{1-q}\left(\sum_{i=1}^{N} p_i^q - 1\right)$
Recursivity	yes	no	no
Additivity	yes	yes	no
Concavity	yes	for $0 < q \leq 1$	for $q > 0$

In order to bring the moments closer to the Shannon entropy we can define the *Havrda–Charvát entropies* [15] as

$$S_q^{HC}(P) \equiv \frac{1}{1-q}\left[\sum_{i=1}^{N} p_i^q - 1\right]. \tag{2.77}$$

We now get the Shannon entropy in the limit $q \to 1$, and the linear entropy when $q = 2$. These entropies are Schur concave but not recursive. They are not additive for independent random variables; when $P_{12}^{ij} = p_1^i p_2^j$ we have

$$S_q^{HC}(P_{12}) = S_q^{HC}(P_1) + S_q^{HC}(P_2) + (1-q)S_q^{HC}(P_1)S_q^{HC}(P_2). \tag{2.78}$$

To ensure additivity a logarithm can be used in the definition. A one parameter family of Schur concave and additive entropies are the *Rényi entropies* [16]

$$S_q(P) \equiv \frac{1}{1-q} \ln\left[\sum_{i=1}^{N} p_i^q\right]. \tag{2.79}$$

We assume that $q \geq 0$. An added advantage of the Rényi entropies is that they are normalised in a uniform way, in the sense that they vanish for pure states and attain their maximal value $\ln N$ at the uniform distribution.

We summarise some properties of generalised entropies in Table 2.1. There we see a possible disadvantage of the Rényi entropies for $q > 1$, which is that we cannot guarantee that they are concave in the ordinary sense. In fact concavity is lost for $q > q_* > 1$, where q_* is N dependent.[17]

Special cases of the Rényi entropies include $q = 0$, which is the logarithm of the number of non-zero components of the distribution and is known as the *Hartley entropy.*[18] When $q \to 1$, we have the Shannon entropy (sometimes denoted S_1),

---

[15] They were first studied by Havrda and Charvát (1967) [414]; in statistical physics they go under the name of *Tsallis entropies*. For a review of their uses, see Tsallis [899] or Kapur [515].

[16] Rényi (1961) [773] introduced them as examples of functions that are additive, and obey all of Khinchin's axioms for an entropy except the recursion property. They have an elegant thermodynamical interpretation [84].

[17] Peter Harremoës has informed us of the bound $q_* \leq 1 + \ln(4)/\ln(N-1)$.

[18] The idea of measuring information regardless of its contents originated with Hartley (1928) [408].

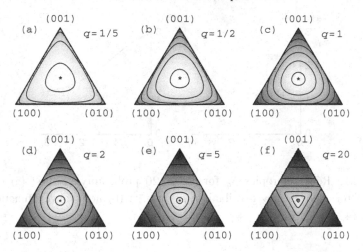

Figure 2.14 The Rényi entropy is constant along the curves plotted for (a) $q = 1/5$, (b) $q = 1/2$, (c) $q = 1$, (d) $q = 2$, (e) $q = 5$ and (f) $q = 20$.

and when $q \to \infty$ the *Chebyshev entropy* $S_\infty = -\ln p_{\max}$, a function of the largest component $p_{\max}$. Figure 2.14 shows some isoentropy curves in the $N = 3$ probability simplex; equivalently we see curves of constant $f_q(\vec{p})$. The special cases $q = 1/2$ and $q = 2$ are of interest because their isoentropy curves form circles, with respect to the Bhattacharyya and Euclidean distances respectively. For $q = 20$ we are already rather close to the limiting case $q \to \infty$, for which we would see the intersection of the simplex with a cube centred at the origin in the space where the vector $\vec{p}$ lives – compare the discussion of l_p–norms in Chapter 1. For $q = 1/5$ the maximum is already rather flat. This resembles the limiting case S_0, for which the entropy reflects the number of events which may occur: it vanishes at the corners of the triangle, is equal to $\ln 2$ at its sides and equals $\ln 3$ for any point inside it.

For any given probability vector P the Rényi entropy is a continuous, non-increasing function of its parameter,

$$S_t(P) \leq S_q(P) \qquad \text{for any} \qquad t > q. \tag{2.80}$$

To show this, introduce the auxiliary probability vector $r_i \equiv p_i^q / \sum_i p_i^q$. Observe that the derivative $\partial S_q / \partial q$ may be written as $-S(P||R)/(1-q)^2$. Since the relative entropy $S(P||R)$ is non–negative, the Rényi entropy S_q is a non-increasing function of q. In Figure 2.7b we show how this fact can be used to bound the Shannon entropy S_1.

In a similar way one proves [101] analogous inequalities valid for $q \geq 0$:

$$\frac{d}{dq}\left[\frac{q-1}{q}S_q\right] \geq 0, \qquad \frac{d^2}{dq^2}\left[(1-q)S_q\right] \geq 0. \tag{2.81}$$

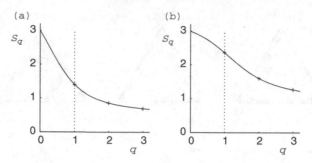

Figure 2.15 Rényi entropies S_q for $N = 20$ probability vectors: (a) convex function for a power–law distribution, $p_j \sim j^{-2}$; (b) non–convex function for $P = \frac{1}{100}(43, 3, \ldots, 3)$.

The first inequality reflects the fact that the l_q–norm is a non–increasing function. It allows one to obtain useful bounds on Rényi entropies,

$$\frac{q-1}{q}S_q(P) \leq \frac{s-1}{s}S_s(P) \qquad \text{for any} \qquad q \leq s. \tag{2.82}$$

Due to the second inequality the function $(1 - q)S_q$ is convex. However, this does not imply that the Rényi entropy itself is a convex function[19] of q; it is non-convex for probability vectors P with one element dominating. See Figure 2.15.

The Rényi entropies are correlated. For $N = 3$ we can see this if we superpose the various panels of Figure 2.14. Consider the superposition of the isoentropy curves for $q = 1$ and 5. Compared to the circle for $q = 2$ the isoentropy curves for $q < 2$ and $q > 2$ are deformed (with a three-fold symmetry) in the opposite way: together they form a kind of David's star with rounded corners. Thus if we move along a circle of constant S_2 in the direction of decreasing S_5 the Shannon entropy S_1 increases, and conversely.

The problem, what values the entropy S_q may admit, provided S_t is given, has been solved by Harremoës and Topsøe [405]. They proved a simple but not very sharp upper bound on S_1 by S_2, valid for any distribution $P \in \mathbb{R}^N$

$$S_2(P) \leq S_1(P) \leq \ln N + 1/N - \exp\left(-S_2(P)\right). \tag{2.83}$$

The lower bound provided by a special case of (2.80) is not tight. Optimal bounds are obtained[20] by studying both entropies along families of interpolating probability distributions

$$Q_{(k,l)}(a) \equiv aQ_{(k)} + (1 - a)Q_{(l)} \qquad \text{with} \qquad a \in [0, 1]. \tag{2.84}$$

---

[19] As erroneously claimed in [1009].
[20] This result [405] was later generalised [127] for other entropy functions.

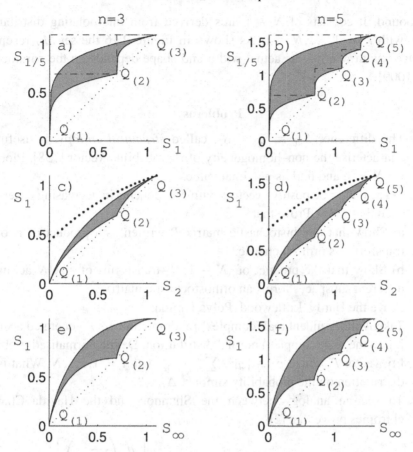

Figure 2.16 The set M_N of all possible discrete distributions for $N = 3$ and $N = 5$ in the Rényi entropies plane S_q and S_t: $S_{1/5}$ and S_1 (a,b); S_1 and S_2 (c,d), and S_1 and S_∞ (e,f). Thin dotted lines in each panel stand for the lower bounds (2.80), dashed–dotted lines in panels (a) and (b) represent bounds between S_0 and S_1, while bold dotted curves in panel (c) and (d) are the upper bounds (2.83).

For instance, the upper bound for S_q as a function of S_t with $t > q$ can be derived from the distribution $Q_{(1,N)}(a)$. For any value of a we compute S_t, invert this relation to obtain $a(S_t)$ and arrive at the desired bound by plotting $S_q[a(S_t)]$. In this way we may bound the Shannon entropy by a function of S_2,

$$S_1(P) \leq (1 - N)\frac{1 - a}{N} \ln \frac{1 - a}{N} - \frac{1 + a(N - 1)}{N} \ln \frac{1 + a(N - 1)}{N}. \quad (2.85)$$

where $a = [(N \exp[-S_2(P)]/(N - 1)]^{1/2}$. This bound is shown in Figure 2.16c and d for $N = 3$ and $N = 5$. Interestingly, the set M_N of possible distributions plotted in the plane S_q versus S_t is not convex. All Rényi entropies agree at the distributions $Q_{(k)}$, $k = 1, \ldots, N$. These points located at the diagonal, $S_q = S_t$, belong to the lower bound. It consists of $N - 1$ arcs derived from interpolating distributions $Q_{(k,k+1)}$ with $k = 1, \ldots, N - 1$. As shown in Figure 2.16 the set M_N resembles a *medusa*[21] with N arms. Its actual width and shape depends on the parameters t and q [1009].

Problems

2.1   The difference $S_{str} \equiv S_1 - S_2$, called *structural entropy*, is useful to characterise the non–homogeneity of a probability vector [738]. Plot S_{str} for $N = 3$, and find its maximal value.

2.2   Let $\vec{x}, \vec{y}$ and $\vec{z}$ be positive vectors with components in decreasing order and such that $\vec{z} \prec \vec{y}$. Prove that $\vec{x} \cdot \vec{z} \le \vec{x} \cdot \vec{y}$.

2.3   a) Show that any bistochastic matrix B written as a product of two T–transforms is orthostochastic.

b) Show that the product of $(N - 1)$ T–transforms of size N acting in different subspaces forms an orthostochastic matrix.

2.4   Prove the Hardy–Littlewood–Pólya lemma.

2.5   Take N independent real (complex) random numbers z_i generated according to the real (complex) normal distribution. Define normalised probability vector P, where $p_i \equiv |z_i|^2 / \sum_{i=1}^{N} |z_i|^2$ with $i = 1, \ldots, N$. What is its distribution on the probability simplex Δ_{N-1}?

2.6   To see an analogy between the Shannon and the Havrda–Charvat entropies prove that [2]

$$S(P) \equiv - \sum_i p_i \ln p_i = - \left[\frac{d}{dx} \left(\sum_i p_i^x \right) \right] \bigg|_{x=1} \quad (2.86)$$

$$S_q^{HC}(P) \equiv \frac{1}{1-q} \left(\sum_i p_i^q - 1 \right) = - \left[D_q \left(\sum_i p_i^x \right) \right] \bigg|_{x=1} \quad (2.87)$$

where the 'multiplicative' *Jackson q–derivative* reads

$$D_q \left(f(x) \right) \equiv \frac{f(qx) - f(x)}{qx - x}, \quad \text{so that} \quad \lim_{q \to 1} D_q(f(x)) = \frac{df(x)}{dx}. \quad (2.88)$$

2.7   For what values of q are the Rényi entropies concave when $N = 2$?

---

[21] Polish or Swedish readers will know that a medusa is a (kind of) jellyfish.

3

Much ado about spheres

Geometry has always struck me as a kind of express lane to the truth.

—Shing-Tung Yau[1]

In this chapter we will study spheres, mostly two- and three-dimensional spheres, for two reasons: because spheres are important, and because they serve as a vehicle for introducing many geometric concepts (such as symplectic, complex and Kähler spaces, fibre bundles, and group manifolds) that we will need later on. It may look like a long detour, but it leads into the heart of quantum mechanics.

3.1 Spheres

We expect that the reader knows how to define a round n-dimensional sphere through an embedding in a flat ($N = n + 1$)-dimensional space. Using Cartesian coordinates in the latter, the n-sphere $\mathbf{S}^n$ is the surface

$$X \cdot X = \sum_{I=0}^{n} X^I X^I = (X^0)^2 + (X^1)^2 + \cdots + (X^n)^2 = 1, \qquad (3.1)$$

where we gave a certain standard size (unit radius) to our sphere and also introduced the standard scalar product in $\mathbb{R}^N$. The Cartesian coordinates $(X^0, X^1, \ldots, X^n) = (X^0, X^i) = X^I$, where $1 \leq i \leq n$ and $0 \leq I \leq n$, are known as *embedding coordinates*. They are not intrinsic to the sphere but useful anyway.

Our first task is to introduce a few more intrinsic coordinate systems on $\mathbf{S}^n$, in addition to the polar angles used in Section 1.2. Eventually this should lead to the insight that coordinates are not important, only the underlying space itself counts. We will use a set of coordinate systems that are obtained by projecting the sphere

---

[1] Reproduced from Shing-Tung Yau. *The Shape of Inner Space*. Basic Books, 2010.

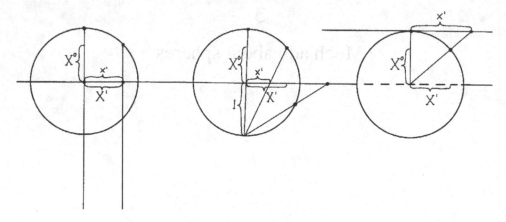

Figure 3.1 Three coordinate system that we use. To the left, orthographic projection from infinity of the northern hemisphere into the equatorial plane. In the middle, stereographic projection from the south pole of the entire sphere (except the south pole itself) onto the equatorial plane. To the right, gnomonic projection from the centre of the northern hemisphere onto the tangent plane at the north pole.

from a point on the axis between the north and south poles to a plane parallel to the equatorial plane. Our first choice is perpendicular projection to the equatorial plane, known as *orthographic projection* among mapmakers. See Figure 3.1. The point of projection is infinitely far away. We set

$$X^i = x^i \qquad X^0 = \sqrt{1 - r^2}, \qquad r^2 \equiv \sum_{i=1}^{n} x^i x^i < 1. \tag{3.2}$$

This coordinate patch covers the region where $X^0 > 1$; we need several coordinate patches of this kind to cover the entire sphere. The metric when expressed in these coordinates is

$$ds^2 = dX^0 dX^0 + \sum_{i=1}^{n} dX^i dX^i = \frac{1}{1 - r^2} \left[(1 - r^2) dx \cdot dx + (x \cdot dx)^2 \right] \tag{3.3}$$

where

$$x \cdot dx \equiv \sum_{i=1}^{n} x^i dx^i \qquad \text{and} \qquad dx \cdot dx \equiv \sum_{i=1}^{n} dx^i dx^i. \tag{3.4}$$

A nice feature of this coordinate system is that – as a short calculation shows – the measure defined by the metric becomes simply $\sqrt{g} = 1/X^0$.

An alternative choice of intrinsic coordinates – perhaps the most useful one – is given by *stereographic projection* from the south pole to the equatorial plane, so that

$$\frac{x^i}{X^i} = \frac{1}{1+X^0} \quad \Leftrightarrow \quad X^i = \frac{2x^i}{1+r^2} \qquad X^0 = \frac{1-r^2}{1+r^2}. \tag{3.5}$$

A minor calculation shows that the metric now becomes manifestly *conformally flat*, that is to say that it is given by a conformal factor Ω^2 times a flat metric:

$$ds^2 = \Omega^2 \delta_{ij} dx^i dx^j = \frac{4}{(1+r^2)^2} dx \cdot dx. \tag{3.6}$$

This coordinate patch covers the region $X^0 > -1$, that is the entire sphere except the south pole itself. To cover the entire sphere we need at least one more coordinate patch, say the one that is obtained by stereographic projection from the north pole. In the particular case of $\mathbf{S}^2$ one may collect the two stereographic coordinates into one complex coordinate z; the relation between this coordinate and the familiar polar angles is

$$z = x^1 + ix^2 = \tan\frac{\theta}{2} e^{i\phi}. \tag{3.7}$$

We will use this formula quite frequently.

A third choice is *gnomonic* or *central projection*. (The reader may want to know that the gnomon being referred to is the vertical rod on a primitive sundial.) We now project one half of the sphere from its centre to the tangent plane touching the north pole. In equations

$$x^i = \frac{X^i}{X^0} \quad \Leftrightarrow \quad X^i = \frac{x^i}{\sqrt{1+r^2}} \qquad X^0 = \frac{1}{\sqrt{1+r^2}}. \tag{3.8}$$

This leads to the metric

$$ds^2 = \frac{1}{(1+r^2)^2} \left[(1+r^2)dx \cdot dx - (x \cdot dx)^2 \right]. \tag{3.9}$$

One hemisphere only is covered by gnomonic coordinates. (The formalism presented in Section 1.4 can be used to transform between the three coordinate systems that we presented, but it was easier to derive each from scratch.)

All coordinate systems have their special advantages. Let us sing the praise of stereographic coordinates right away. The topology of coordinate space is $\mathbb{R}^n$, and when stereographic coordinates are used the sphere has only one further point not

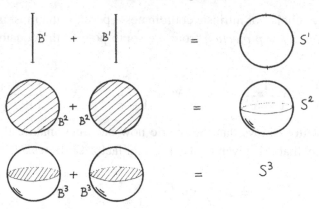

Figure 3.2 A circle is the sum of two intervals, a 2-sphere is the sum of two discs glued together along the boundaries, and the 3-sphere is the sum of two balls again with the boundaries identified. In the latter case the glueing cannot be done in three dimensions. See Appendix C for a different picture in the same vein.

covered by these coordinates, so the topology of $\mathbf{S}^n$ is the topology of $\mathbb{R}^n$ with one extra point attached 'at infinity'. The conformal factor ensures that the round metric is smooth at the added point, so that 'infinity' in coordinate space lies at finite distance on the sphere. One advantage of these coordinates is that all angles come out correctly if we draw a picture in a flat coordinate space, although distances far from the origin are badly distorted. We say that the map between the sphere and the coordinate space is *conformal*, that is, it preserves angles. The stereographic picture makes it easy to visualise $\mathbf{S}^3$, which is conformally mapped to ordinary flat space in such a way that the north pole is at the origin, the equator is the unit sphere, and the south pole is at infinity. With a little training one can get used to this picture, and learn to disregard the way in which it distorts distances. If the reader prefers a compact picture of the 3-sphere this is easily provided: use the stereographic projection from the south pole to draw a picture of the northern hemisphere only. This gives the picture of a solid ball whose surface is the equator of the 3-sphere. Then project from the north pole to get a picture of the southern hemisphere. The net result is a picture consisting of two solid balls whose surfaces must be mentally identified with each other.

When we encounter a new space, we first ask what symmetries it has, and what its geodesics are. Here the embedding coordinates are very useful. An infinitesimal *isometry* (a transformation that preserves distances) is described by a *Killing vector field* pointing in the direction that points are transformed. We ask for the flow lines of the isometry. A sphere has exactly $n(n+1)/2$ linearly independent Killing vectors at each point, namely

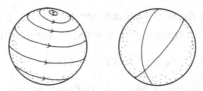

Figure 3.3 Killing flows and geodesics on the 2-sphere.

$$J_{IJ} = X_I \partial_J - X_J \partial_I. \tag{3.10}$$

(Here we used the trick from Section 1.4, to represent a tangent vector as a differential operator.) On the 2-sphere the flow lines are always circles at constant distance from a pair of antipodal *fixed points* where the flow vanishes. See Figure 3.3. The situation gets more interesting on the 3-sphere, as we will see.

A geodesic is the shortest curve between any pair of nearby points on itself. On the sphere a geodesic is a great circle, that is the intersection of the sphere with a two-dimensional plane through the origin in the embedding space. Such a curve is 'obviously' as straight as it can be, given that it must be confined to the sphere. (Incidentally this means that gnomonic coordinates are nice, because the geodesics will appear as straight lines in coordinate space.) The geodesics can also be obtained as the solutions of the Euler–Lagrange equations coming from the constrained Lagrangian

$$L = \frac{1}{2}\dot{X} \cdot \dot{X} + \Lambda(X \cdot X - 1), \tag{3.11}$$

where Λ is a Lagrange multiplier and the overdot denotes differentiation with respect to the *affine parameter* along the curve. We rescale the affine parameter so that $\dot{X} \cdot \dot{X} = 1$, and then the general solution for a geodesic takes the form

$$X^I(\tau) = k^I \cos\tau + l^I \sin\tau, \qquad k \cdot k = l \cdot l = 1, \ k \cdot l = 0. \tag{3.12}$$

The vectors k^I and l^I span a plane through the origin in $\mathbb{R}^N$. Since $X^I(0) = k^I$ and $\dot{X}^I(0) = l^I$, the conditions on these vectors say that we start on the sphere, with unit velocity, in a direction tangent to the sphere. The entire curve is determined by these data.

Let us now choose two points along the geodesic, with different values of the affine parameter τ, and compute

$$X(\tau_1) \cdot X(\tau_2) = \cos(\tau_1 - \tau_2). \tag{3.13}$$

With the normalisation of the affine parameter that we are using $|\tau_1 - \tau_2|$ is precisely the length of the curve between the two points, so we get the useful formula

$$\cos d = X(\tau_1) \cdot X(\tau_2), \tag{3.14}$$

where d is the geodesic distance between the two points. It is equal to the angle between the unit vectors $X^I(\tau_1)$ and $X^I(\tau_2)$ – and we encountered this formula before in Eq. (2.56).

3.2 Parallel transport and statistical geometry

Let us focus on the positive octant (or hyperoctant) of the sphere. In the previous chapter its points were shown to be in one-to-one correspondence to the set of probability distributions over a finite sample space, and this set was sometimes thought of as round (equipped with the Fisher metric) and sometimes as flat (with convex mixtures represented as straight lines). See Figure 3.4. How can we reconcile these two ways of looking at the octant? To answer this question we will play a little formal game with connections and curvatures.[2]

Curvature is not a very immediate property of a space. It has to do with how one can compare vectors sitting at different points with each other, and we must begin with a definite prescription for how to *parallel transport* vectors from one point to another. For this we require a *connection* and a *covariant derivative*, such

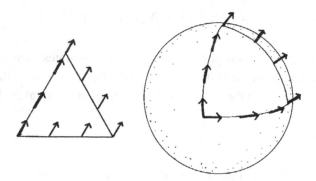

Figure 3.4 A paradox? On the left we parallel transport a vector around the edge of the flat probability simplex. On the right the same simplex is regarded as a round octant, and the obvious (Levi-Civita) notion of parallel transport gives a different result. It is the same space but two different affine connections!.

---

[2] In this section we assume that the reader knows some Riemannian geometry. Readers who have forgotten this can refresh their memory with Appendix A.2. Readers who never knew about it may consult, say, Murray and Rice [682] or Schrödinger [806] – or take comfort in the fact that Riemannian geometry is used in only a few sections of our book.

as the *Levi-Civita connection* that is defined (using the metric) in Appendix A.2. Then we can transport a vector V^i along any given curve with tangent vector X^i by solving the ordinary differential equation $X^j \nabla_j V^i = 0$ along the curve. But there is no guarantee that the vector will return to itself if it is transported around a closed curve. Indeed it will not if the *curvature tensor* is non-zero. It must also be noted that the prescription for parallel transport can be changed by changing the connection. Assume that in addition to the metric tensor g_{ij} we are given a totally symmetric *skewness tensor* T_{ijk}. Then we can construct the one-parameter family of affine connections

$$\overset{(\alpha)}{\Gamma}_{ijk} = \Gamma_{ijk} + \frac{\alpha}{2} T_{ijk}. \tag{3.15}$$

Here Γ_{ijk} is the ordinary Levi-Civita connection (with one index lowered using the metric). Since T_{ijk} transforms like a tensor all α-connections transform as connections should. The covariant derivative and the curvature tensor will be given by Eqs. (A.5) and (A.10), respectively, but with the Levi-Civita connection replaced by the new α-connection. We can also define α-geodesics, using Eq. (A.8) but with the new connection. This is *affine differential geometry*; a subject that at first appears somewhat odd, because there are 'straight lines' (geodesics) and distances along them (given by the affine parameter of the geodesics), but these distances do not fit together in the way they would do if they were consistent with a metric tensor.[3]

In statistical geometry the metric tensor has a potential, that is to say there is a convex function Φ such that the Fisher–Rao metric is

$$g_{ij}(p) = \frac{\partial^2 \Phi}{\partial p^i \partial p^j} \equiv \partial_i \partial_j \Phi(p). \tag{3.16}$$

In Eq. (2.59) this function is given as minus the Shannon entropy, but for the moment we want to play a game and keep things general. Actually the definition is rather strange from the point of view of differential geometry, because it uses ordinary rather than covariant derivatives. The equation will therefore be valid only with respect to some preferred *affine coordinates* that we call p^i here, anticipating their meaning; we use an affine connection defined by the requirement that it vanishes in this special coordinate system. But if we have done one strange thing we can do another. Therefore we can define a totally symmetric third-rank tensor using the same preferred affine coordinate system, namely

$$T_{ijk}(p) = \partial_i \partial_j \partial_k \Phi(p). \tag{3.17}$$

Now we can start the game.

---

[3] The statistical geometry of α-connections is the work of Čencov [199] and Amari [30].

Using the definitions in Appendix A.2 it is easy to see that

$$\Gamma_{ijk} = \frac{1}{2}\partial_i\partial_j\partial_k\Phi(p).\tag{3.18}$$

But we also have an α-connection, namely

$$\overset{(\alpha)}{\Gamma}_{ijk} = \frac{1+\alpha}{2}\partial_i\partial_j\partial_k\Phi(p) = (1+\alpha)\Gamma_{ijk}.\tag{3.19}$$

A small calculation is now enough to relate the α-curvature to the usual one:

$$\overset{(\alpha)}{R}_{ijkl} = (1-\alpha^2)R_{ijkl}.\tag{3.20}$$

Equation (3.19) says that the α-connection vanishes when $\alpha = -1$, so that the space is (-1)-flat. Our coordinate system is preferred in the sense that a line that looks straight in this system is indeed a geodesic with respect to the (-1)-connection. The surprise is that Eq. (3.20) shows that the space is also $(+1)$-flat, even though this is not obvious just by looking at the $(+1)$-connection in this coordinate system.

We therefore start looking for a coordinate system in which the $(+1)$-connection vanishes. We try the functions

$$\eta^i = \frac{\partial\Phi}{\partial p^i}.\tag{3.21}$$

Then we define a new function $\Psi(\eta)$ through a *Legendre transformation*, a trick familiar from thermodynamics:

$$\Psi(\eta) + \Phi(p) - \sum_i p^i\eta^i = 0.\tag{3.22}$$

Although we express the new function as a function of its 'natural' coordinates η^i, it is first and foremost a function of the points in our space. By definition

$$d\Psi = \sum_i p^i d\eta^i \quad\Rightarrow\quad p^i = \frac{\partial\Psi}{\partial\eta^i}.\tag{3.23}$$

Our coordinate transformation is an honest one in the sense that it can be inverted to give the functions $p^i = p^i(\eta)$. Now let us see what the tensors g_{ij} and T_{ijk} look like in the new coordinate system. An exercise in the use of the chain rule shows that

$$g_{ij}(\eta) = \frac{\partial p^k}{\partial\eta^i}\frac{\partial p^l}{\partial\eta^j}g_{kl}(p(\eta)) = \frac{\partial p^j}{\partial\eta^i} = \frac{\partial^2\Psi}{\partial\eta^i\partial\eta^j} \equiv \partial_i\partial_j\Psi(\eta).\tag{3.24}$$

For T_{ijk} a slightly more involved exercise shows that

$$T_{ijk}(\eta) = -\partial_i\partial_j\partial_k\Psi(\eta).\tag{3.25}$$

To show this, first derive the matrix equation

$$\sum_{k=1}^{n} g_{ik}(\eta)\, g_{kj}(p) = \delta_{ij}. \tag{3.26}$$

The sign in (3.25) is crucial since it implies that $\overset{(+1)}{\Gamma}{}_{ijk} = 0$; in the new coordinate system the space is indeed manifestly $(+1)$-flat.

We now have two different notions of affine straight lines. Using the (-1)-connection they are

$$\dot{p}^j \overset{(-1)}{\nabla}{}_j \dot{p}^i = \ddot{p}^i = 0 \quad \Rightarrow \quad p^i(t) = p_0^i + t p^i. \tag{3.27}$$

Using the $(+1)$-connection, and working in the coordinate system that comes naturally with it, we get instead

$$\dot{\eta}^j \overset{(+1)}{\nabla}{}_j \dot{\eta}^i = \ddot{\eta}^i = 0 \quad \Rightarrow \quad \eta^i(t) = \eta_0^i + t \eta^i. \tag{3.28}$$

We will see presently what this means.

There is a final manoeuvre that we can do. We go back to Eq. (3.22), look at it, and realise that it can be modified so that it defines a function of pairs of points P and P' on our space, labelled by the coordinates p and η' respectively. This is the function

$$S(P||P') = \Phi\left(p(P)\right) + \Psi\left(\eta'(P')\right) - \sum_i p^i(P)\, \eta'^i(P'). \tag{3.29}$$

It vanishes when the two points coincide, and since this is an extremum the function is always positive. It is an asymmetric function of its arguments. To lowest non-trivial order the asymmetry is given by the skewness tensor; indeed

$$S(p||p + dp) = \sum_{i,j} g_{ij}\, dp^i dp^j - \sum_{i,j,k} T_{ijk}\, dp^i dp^j dp^k + \ldots \tag{3.30}$$

$$S(p + dp||p) = \sum_{i,j} g_{ij}\, dp^i dp^j + \sum_{i,j,k} T_{ijk}\, dp^i dp^j dp^k + \ldots \tag{3.31}$$

according to Eqs (3.17) and (3.25). (We are of course entitled to use the same coordinates for both arguments, as we just did, provided we do the appropriate coordinate transformations.)

To bring life to this construction it remains to seed it with some interesting function Φ and see what interpretation we can give to the objects that come with it. We already have one interesting choice, namely minus the Shannon entropy. First we play the game on the positive orthant $\mathbb{R}_+^N$, that is to say we use the index i ranging from 1 to N, and assume that all the $p^i > 0$ but leave them otherwise unconstrained. Our input is

$$\Phi(p) = \sum_{i=1}^{N} p^i \ln p^i. \qquad (3.32)$$

Our output becomes

$$ds^2 = \sum_{i=1}^{N} \frac{dp^i dp^i}{p^i} = \sum_{i=1}^{N} dx^i dx^i; \qquad 4p^i = (x^i)^2, \quad (3.33)$$

$$\eta^i = \ln p^i + 1, \qquad (3.34)$$

$$\Psi(\eta) = \sum_{i=1}^{N} p^i, \qquad (3.35)$$

$$S(p||p') = \sum_{i=1}^{N} p^i \ln \frac{p^i}{p'^i} + \sum_{i=1}^{N} (p'^i - p^i), \qquad (3.36)$$

$$(+1) \text{ geodesic}: \quad p^i(t) = p^i(0) e^{t[\ln p^i(1) - \ln p^i(0)]}. \qquad (3.37)$$

In Eq. (3.33) the coordinate transformation $p^i = 4(x^i)^2$ shows that the Fisher–Rao metric on $\mathbb{R}^N_+$ is flat. To describe the situation on the probability simplex we impose the constraint

$$\sum_{i=1}^{N} p^i = 1 \qquad \Leftrightarrow \qquad \sum_{i=1}^{N} x^i x^i = 4. \qquad (3.38)$$

Taking this into account we see that the Fisher–Rao metric on the probability simplex is the metric on the positive octant of a round sphere with radius 2. For maximum elegance, we have chosen a different normalisation of the metric, compared to what we used in Section 2.5. The other entries in our list of output have some definite statistical meaning, too. We are familiar with the relative entropy $S(p||p')$ from Section 2.3. The geodesics defined by the (-1)-connection, that is to say the lines that look straight in our original coordinate system, are convex mixtures of probability distributions. The (-1)-connection is therefore known as the *mixture connection* and its geodesics are (one–dimensional) *mixture families*. The space is flat with respect to the mixture connection, but (in a different way) also with respect to the $(+1)$-connection. The coordinates in which this connection vanishes are the η^i. The $(+1)$-geodesics, that is the lines that look straight when we use the coordinates η^i, are known as (one–dimensional) *exponential families* and the $(+1)$-connection as the *exponential connection*. Exponential families are important in the theory of statistical inference; it is particularly easy to pin down (using samplings) precisely where you are along a given curve in the space of probability

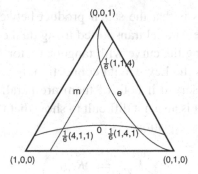

Figure 3.5 Here we show three different kinds of geodesics – mixture (m), exponential (e) and metric (0) – on the simplex; since the mixture coordinates p^i are used, only the mixture geodesic appears straight in the picture.

distributions, if that curve is an exponential family. We notice one interesting thing: it looks reasonable to define the mean value of $p^i(0)$ and $p^i(1)$ as $p^i(1/2)$, where the parameter is the affine parameter along a geodesic. If our geodesic is a mixture family, this is the arithmetic mean, while it will be the geometric mean $p^i(1/2) = \sqrt{p^i(0)p^i(1)}$ if the geodesic is an exponential family.

Since we have shown that there are three different kinds of straight lines on the probability simplex – mixture families, exponential families, and geodesics with respect to the round metric – we should also show what they look like. Figure 3.5 is intended to make this clear. The probability simplex is *complete* with respect to the exponential connection, meaning that the affine parameter along the exponential geodesics goes from minus to plus infinity – whereas the mixture and metric geodesics cross its boundary at some finite value of the affine parameter.

Our story is almost over. The main point is that a number of very relevant concepts – Shannon entropy, Fisher–Rao metric, mixture families, exponential families, and the relative entropy – have appeared in natural succession. But the story can be told in different ways. Every statistical manifold has a totally symmetric skewness tensor built in. Indeed, following Section 2.4, we can use the score vectors l_i and define, using expectation values,

$$g_{ij} = \langle l_i l_j \rangle \quad \text{and} \quad T_{ijk} = \langle l_i l_j l_k \rangle. \tag{3.39}$$

In particular, a skewness tensor is naturally available on the space of normal distributions. This space turns out to be (± 1)-flat, although the coordinates that make this property manifest are not those used in Section 2.5. Whenever a totally symmetric tensor is used to define a family of α-connections one can show that

$$X^k \partial_k (g_{ij} Y^i Z^j) = g_{ij} X^k \overset{(\alpha)}{\nabla}_k Y^i Z^j + g_{ij} Y^i X^k \overset{(-\alpha)}{\nabla}_k Z^j. \tag{3.40}$$

What this equation states is that the scalar product between two vectors remains constant if the vectors are parallel transported using dual connections (so that their covariant derivatives along the curve with tangent vector X^i vanish by definition). The 0-connection, that is the Levi-Civita connection, is self dual in the sense that the scalar product is preserved if both of them are parallel transported using the Levi-Civita connection. It is also not difficult to show that if the α-connection is flat for any value of α then the equation

$$\overset{(\alpha)}{R}_{ijkl} = \overset{(-\alpha)}{R}_{ijkl} \qquad (3.41)$$

will hold for all values of α. Furthermore it is possible to show that if the space is α-flat then it will be true, in that preferred coordinate system for which the α-connection vanishes, that there exists a potential for the metric in the sense of Eq. (3.16). The point is that Eq. (3.40) then implies that

$$\overset{(-\alpha)}{\Gamma}_{jki} = \partial_k g_{ij} \quad \text{and} \quad \overset{(-\alpha)}{\Gamma}_{[jk]i} = 0 \quad \Rightarrow \quad \partial_j g_{ki} - \partial_k g_{ji} = 0. \qquad (3.42)$$

The existence of a potential for the metric follows:

$$g_{ij} = \partial_i V_j \quad \text{and} \quad g_{[ij]} = 0 \quad \Rightarrow \quad V_j = \partial_j \Phi. \qquad (3.43)$$

At the same time it is fair to warn the reader that if a space is compact, it is often impossible to make it globally α-flat [79].

3.3 Complex, Hermitian, and Kähler manifolds

We now return to the study of spheres in the global manner and forget about statistics for the time being. It turns out to matter a lot whether the dimension of the sphere is even or odd. Let us study the even dimensional case $n = 2m$, and decompose the intrinsic coordinates according to

$$x^i = \left(x^a, x^{m+a}\right), \qquad (3.44)$$

where the range of a goes from 1 to m. Then we can, if we wish, introduce the complex coordinates

$$z^a = x^a + i x^{m+a}, \qquad \bar{z}^{\bar{a}} = x^a - i x^{m+a}. \qquad (3.45)$$

We deliberately use two kinds of indices here – barred and unbarred – because we will never contract indices of different kinds. The new coordinates come in pairs connected by the complex conjugation,

$$(z^a)^* = \bar{z}^{\bar{a}}. \qquad (3.46)$$

This equation ensures that the original coordinates take real values. Only the z^a count as coordinates, and once they are given the $\bar{z}^{\bar{a}}$ are fully determined. If we choose stereographic coordinates to start with, we find that the round metric becomes

$$ds^2 = g_{ab}dz^a dz^b + 2g_{a\bar{b}}dz^a d\bar{z}^{\bar{b}} + 2g_{\bar{a}\bar{b}}d\bar{z}^{\bar{a}}d\bar{z}^{\bar{b}} = \frac{4}{(1+r^2)^2}\delta_{a\bar{b}}\,dz^a d\bar{z}^{\bar{b}}. \tag{3.47}$$

Note that we do not complexify the manifold. We would obtain the *complexified* sphere by allowing the coordinates x^i to take complex values, in which case the real dimension would be multiplied by two and we would no longer have a real sphere. What we actually did may seem like a cheap trick in comparison, but for the 2-sphere it is anything but cheap, as we will see.

To see if the introduction of complex coordinates is more than a trick we must study what happens when we try to cover the entire space with overlapping coordinate patches. We choose stereographic coordinates, but this time we also include a patch obtained by projection from the north pole:

$$x'^a = \frac{X^a}{1 - X^0}, \qquad x'^{m+a} = \frac{-X^{m+a}}{1 - X^0}. \tag{3.48}$$

Now the whole sphere is covered by two coordinate systems. Introducing complex coordinates in both patches, we observe that

$$z'^a = x'^a + ix'^{m+a} = \frac{X^a - iX^{m+a}}{1 - X^0} = \frac{2(x^a - ix^{m+a})}{1 + r^2 - 1 + r^2} = \frac{\bar{z}^a}{r^2}. \tag{3.49}$$

These are called the *transition functions* between the two coordinate systems. In the special case of $\mathbf{S}^2$ we can conclude that

$$z'(z) = \frac{1}{z}. \tag{3.50}$$

Remarkably, the transition functions between the two patches covering the 2-sphere are holomorphic (that is complex analytic) functions of the complex coordinates. In higher dimensions this simple manoeuvre fails.

There is another peculiar thing that happens for $\mathbf{S}^2$, but not in higher dimensions. Look closely at the metric:

$$g_{z\bar{z}} = \frac{2}{(1 + |z|^2)^2} = \frac{2}{1 + |z|^2}\left(1 - \frac{|z|^2}{1 + |z|^2}\right) = 2\partial_z\partial_{\bar{z}}\ln(1 + |z|^2). \tag{3.51}$$

Again a 'potential' exists for the metric of the 2-sphere. Although superficially similar to Eq. (3.16) there is a difference – the latter is true in very special coordinate systems only, while Eq. (3.51) is true in every complex coordinate system connected to the original one with holomorphic coordinate transformations of the form $z' = z'(z)$. Complex spaces for which all this is true are called *Kähler manifolds*.

The time has come to formalise things. A differentiable manifold is a space which can be covered by coordinate patches in such a way that the transition functions are differentiable. A *complex manifold* is a space which can be covered by coordinate patches in such a way that the coordinates are complex and the transition functions are holomorphic.[4]

Any even-dimensional manifold can be covered by complex coordinates in such a way that, when the coordinate patches overlap,

$$z' = z'(z, \bar{z}), \qquad\qquad \bar{z}' = \bar{z}'(z, \bar{z}). \qquad\qquad (3.52)$$

The manifold is complex if and only if it can be covered by coordinate patches such that

$$z' = z'(z), \qquad\qquad \bar{z}' = \bar{z}'(\bar{z}). \qquad\qquad (3.53)$$

Since we are using complex coordinates to describe a real manifold a point in the manifold is specified by the n independent coordinates z^a – we always require that

$$\bar{z}^{\bar{a}} \equiv (z^a)^*. \qquad\qquad (3.54)$$

A complex manifold is therefore a real manifold that can be described in a particular way. Naturally one could introduce coordinate systems whose transition functions are non-holomorphic, but the idea is to restrict oneself to holomorphic coordinate transformations (just as, on a flat space, it is convenient to restrict oneself to Cartesian coordinate systems).

Complex manifolds have some rather peculiar properties caused ultimately by the 'rigidity properties' of analytic functions. By no means all even dimensional manifolds are complex, and for those that are there may be several inequivalent ways to turn them into complex manifolds. Examples of complex manifolds are $\mathbb{C}^n = \mathbb{R}^{2n}$ and all orientable two-dimensional spaces, including $\mathbf{S}^2$ as we have seen. An example of a manifold that is not complex is $\mathbf{S}^4$. It may be difficult to decide whether a given manifold is complex or not; an example of a manifold for which this question is open is the 6-sphere.[5]

An example of a manifold that can be turned into a complex one in several inequivalent ways is the *torus* $\mathbf{T}^2 = \mathbb{C}/\Gamma$. Here Γ is some discrete isometry group generated by two (commuting) translations, and $\mathbf{T}^2$ will inherit the property of being a complex manifold from the complex plane $\mathbb{C}$. See Figure 3.6. A better way

---

[4] A standard reference on complex manifolds is Chern [212].
[5] As we were preparing the first edition, a rumour spread that Chern had proved, just before his untimely death at the age of 93, that $\mathbf{S}^6$ does not admit a complex structure. This time there is a claim that it does [302], and another that it does not [65]. We have no opinion.

Figure 3.6 A flat torus is a parallelogram with sides identified; it is also defined by a pair of vectors, or by the lattice of points that can be reached by translations with these two vectors.

to say this is that a flat torus is made from a flat parallelogram by gluing opposite sides together. It means that there is one flat torus for every choice of a pair of vectors. The set of all possible tori can be parametrized by the relative length and the angle between the vectors, and by the total area. Since holomorphic transformations cannot change relative lengths or angles this means that there is a two-parameter family of tori that are inequivalent as complex manifolds. In other words the 'shape space' of flat tori (technically known as *Teichmüller space*) is two dimensional. Note though that just because two parallelograms look different we cannot conclude right away that they represent inequivalent tori – if one torus is represented by the vectors **u** and **v**, then the torus represented by **u** and **v** + **u** is intrinsically the same.

Tensors on complex manifolds are naturally either real or complex. Consider vectors: since an n complex dimensional complex manifold is a real manifold as well, it has a *real tangent space* **V** of dimension $2n$. A real vector (at a point) is an element of **V** and can be written as

$$V = V^a \partial_a + \bar{V}^{\bar{a}} \partial_{\bar{a}}, \qquad (3.55)$$

where $\bar{V}^{\bar{a}}$ is the complex conjugate of V^a. A complex vector is an element of the *complexified tangent space* $\mathbf{V}^{\mathbb{C}}$, and can be written in the same way but with the understanding that $\bar{V}^{\bar{a}}$ is independent of V^a. By definition we say that a real vector space has a *complex structure* if its complexification splits into a direct sum of two complex vector spaces that are related by complex conjugation. This is clearly the case here. We have the direct sum

$$\mathbf{V}^{\mathbb{C}} = \mathbf{V}^{(1,0)} \oplus \mathbf{V}^{(0,1)}, \qquad (3.56)$$

where

$$V^a \partial_a \in \mathbf{V}^{(1,0)} \qquad \bar{V}^{\bar{a}} \partial_{\bar{a}} \in \mathbf{V}^{(0,1)}. \qquad (3.57)$$

If **V** is the real tangent space of a complex manifold, the space $\mathbf{V}^{(1,0)}$ is known as the *holomorphic tangent space*. This extra structure means that we can talk of vectors of type $(1, 0)$ and $(0, 1)$ respectively; more generally we can define both tensors and differential forms of *type* (p, q). This is well defined because analytic coordinate transformations will not change the type of a tensor, and it is an important part of the theory of complex manifolds.

We still have to understand what happened to the metric of the 2-sphere. We define an *Hermitian manifold* as a complex manifold with a metric tensor of type $(1, 1)$. In complex coordinates it takes the form

$$ds^2 = 2g_{a\bar{b}}\, dz^a dz^{\bar{b}}. \tag{3.58}$$

The metric is a symmetric tensor, and this implies

$$g_{\bar{b}a} = g_{a\bar{b}}. \tag{3.59}$$

The reality of the line element will be ensured if the matrix $g_{a\bar{b}}$ (if we think of it as a matrix) is also Hermitian,

$$(g_{a\bar{b}})^* = g_{b\bar{a}}. \tag{3.60}$$

This is assumed as well.

Just to make sure that you understand what these conditions are, think of the metric as an explicit matrix. Let the real dimension $2n = 4$ in order to be fully explicit; then the metric is

$$\begin{pmatrix} 0 & g_{a\bar{b}} \\ g_{\bar{a}b} & 0 \end{pmatrix} = \begin{pmatrix} 0 & 0 & g_{1\bar{1}} & g_{1\bar{2}} \\ 0 & 0 & g_{2\bar{1}} & g_{2\bar{2}} \\ g_{\bar{1}1} & g_{\bar{1}2} & 0 & 0 \\ g_{\bar{2}1} & g_{\bar{2}2} & 0 & 0 \end{pmatrix}. \tag{3.61}$$

It is now easy to see what the conditions on the Hermitian metric really are. By the way $g_{a\bar{b}}$ is not the metric tensor, it is only one block of it.

An Hermitian metric will preserve its form under analytic coordinate transformations, hence the definition is meaningful because the manifold is complex. If the metric is given in advance the property of being Hermitian is non-trivial, but as we have seen $\mathbf{S}^2$ equipped with the round metric provides an example. So does $\mathbb{C}^n$ equipped with its flat metric.

Given an Hermitian metric we can construct a differential form

$$J = 2ig_{a\bar{b}}\, dz^a \wedge dz^{\bar{b}}. \tag{3.62}$$

This trick – to use an n by n matrix to define both a symmetric and an anti-symmetric tensor – works only because the real manifold has even dimension equal to $2n$. The imaginary factor in front of the form J is needed to ensure that the form is a real 2-form. The manifold is *Kähler* – and *J* is said to be a *Kähler form* – if *J* is closed, that is to say if

$$\mathrm{d}J = 2ig_{a\bar{b},c}\mathrm{d}z^c \wedge \mathrm{m}\mathrm{d}z^a \wedge \mathrm{d}\bar{z}^{\bar{b}} + 2ig_{a\bar{b},\bar{c}}\mathrm{d}\bar{z}^{\bar{c}} \wedge \mathrm{d}z^a \wedge \mathrm{d}\bar{z}^{\bar{b}} = 0, \qquad (3.63)$$

where the comma stands for differentiation with respect to the appropriate coordinate. Now this will be true if and only if

$$g_{a\bar{b},\bar{c}} = g_{a\bar{c},\bar{b}}, \qquad g_{a\bar{b},c} = g_{c\bar{b},a}. \qquad (3.64)$$

This implies that in the local coordinate system that we are employing there exists a scalar function $K(z,\bar{z})$ such that the metric can be written as

$$g_{a\bar{b}} = \partial_a\partial_{\bar{b}}K. \qquad (3.65)$$

This is a highly non-trivial property because it will be true in all allowed coordinate systems, that is in all coordinate systems related to the present one by an equation of the form $z' = z'(z)$. In this sense it is a more striking statement than the superficially similar Eq. (3.16), which holds in a very restricted class of coordinate systems only. The function $K(z,\bar{z})$ is known as the *Kähler potential* and determines both the metric and the Kähler form.

We have seen that $\mathbf{S}^2$ is a Kähler manifold. This happened because any 2-form on a two-dimensional manifold is closed by default (there are no 3-forms), so that every Hermitian two dimensional manifold has to be Kähler.

The Levi-Civita connection is constructed in such a way that the length of a vector is preserved by parallel transport. On a complex manifold we have more structure worth preserving, namely the complex structure: we would like the type (p,q) of a tensor to be preserved by parallel transport. We must ask if these two requirements can be imposed at the same time. For Kähler manifolds the answer is 'yes'. On a Kähler manifold the only non-vanishing components of the Christoffel symbols are

$$\Gamma_{ab}^{\ c} = g^{c\bar{d}}g_{\bar{d}a,b}, \qquad \Gamma_{\bar{a}\bar{b}}^{\ \bar{c}} = g^{\bar{a}d}g_{d\bar{a},\bar{b}}. \qquad (3.66)$$

Now take a holomorphic tangent vector, that is a vector of type $(1,0)$ (such as $V = V^a\partial_a$). The equation for parallel transport becomes

$$\nabla_X V^a = \dot{V}^a + X^b\Gamma_{bc}^{\ a}V^c = 0, \qquad (3.67)$$

together with its complex conjugate. If we start with a vector whose components $\bar{V}^{\bar{a}}$ vanish and parallel transport it along some curve, then the components $\bar{V}^{\bar{a}}$ will stay zero since certain components of the Christoffel symbols are zero. In other words a vector of type $(1,0)$ will preserve its type when parallel transported. Hence the complex structures on the tangent spaces at two different points are compatible, and it follows that we can define vector fields of type $(1,0)$ and $(0,1)$ respectively, and similarly for tensor fields.

All formulae become simple on a Kähler manifold. Up to index permutations, the only non-vanishing components of the Riemann tensor are

$$R_{a\bar{b}c\bar{d}} = g_{\bar{d}d} \, \Gamma_{ac}^{\ \ d}{}_{,\bar{b}}. \tag{3.68}$$

Finally, a useful concept is that of *holomorphic sectional curvature*. Choose a 2-plane in the complexified tangent space at the point z such that it is left invariant by complex conjugation. This means that we can choose coordinates such that the plane is spanned by the tangent vectors dz^a and $d\bar{z}^{\bar{a}}$ (and here we use the old fashioned notation according to which dz^a are the components of a tangent vector rather than a basis element in the cotangent space). Then the holomorphic sectional curvature is defined by

$$R(z, dz) = \frac{R_{a\bar{b}c\bar{d}} \, dz^a d\bar{z}^{\bar{b}} dz^c d\bar{z}^{\bar{d}}}{(ds^2)^2}. \tag{3.69}$$

Holomorphic sectional curvature is clearly analogous to ordinary scalar curvature on real manifolds, and (unsurprisingly to those who are familiar with ordinary Riemannian geometry) one can show that if the holomorphic sectional curvature is everywhere independent of the choice of the 2-plane then it is independent of the point z as well. Then the space is said to have constant holomorphic sectional curvature. Since there was a restriction on the choice of the 2-planes, constant holomorphic sectional curvature does not imply constant curvature.

3.4 Symplectic manifolds

Kähler manifolds have two kinds of geometry: Riemannian and symplectic. The former concerns itself with a non-degenerate symmetric tensor field, and the latter with a non-degenerate anti-symmetric tensor field that has to be a closed 2-form as well. This is to say that a manifold is *symplectic* only if there exist two tensor fields Ω_{ij} and Ω^{ij} (not related by raising indices with a metric – indeed no metric is assumed) such that

$$\Omega_{ij} = -\Omega_{ji}, \qquad \Omega^{ik}\Omega_{kj} = \delta^i_j. \tag{3.70}$$

This is a *symplectic 2-form* Ω if it is also closed,

$$d\Omega = 0 \quad \Leftrightarrow \quad \Omega_{[ij,k]} = 0. \tag{3.71}$$

These requirements are non-trivial: it may well be that a (compact) manifold does not admit a symplectic structure, although it always admits a metric. Indeed $\mathbf{S}^2$ is the only sphere that is also a symplectic manifold.

A manifold may have a symplectic structure even if it is not Kähler. Phase spaces of Hamiltonian systems are symplectic manifolds, so that – at least in the guise of Poisson brackets – symplectic geometry is quite familiar to physicists.[6] The point is that the symplectic form can be used to associate a vector field with any function $H(x)$ on the manifold through the equation

$$V_H^i = \Omega^{ij}\partial_j H. \tag{3.72}$$

This is known as a *Hamiltonian vector field*, and it generates *canonical transformations*. These transformations preserve the symplectic form, just as isometries generated by Killing vectors preserve the metric. But the space of canonical transformations is always infinite dimensional (since the function H is at our disposal), while the number of linearly independent Killing vector fields is always rather small – symplectic geometry and metric geometry are analogous but different. The *Poisson bracket* of two arbitrary functions F and G is defined by

$$\{F, G\} = \partial_i F \Omega^{ij} \partial_j G. \tag{3.73}$$

It is bilinear, anti-symmetric and obeys the Jacobi identity precisely because the symplectic form is closed. From a geometrical point of view the role of the symplectic form is to associate an area with each pair of tangent vectors. There is also an interesting interpretation of the fact that the symplectic form is closed, namely that the total area assigned by the symplectic form to a closed surface that can be contracted to a point (within the manifold itself) is zero. Every submanifold of a symplectic manifold inherits a 2-form from the manifold in which it sits, but there is no guarantee that the inherited 2-form is non-degenerate. In fact it may vanish. If this happens to a submanifold of dimension equal to one-half of the dimension of the symplectic manifold itself, the submanifold is *Lagrangian*. The standard example is the subspace spanned by the coordinates q in a symplectic vector space spanned by the coordinates q and p, in the way familiar from analytical mechanics.

A symplectic form gives rise to a natural notion of volume, invariant under canonical transformations; if the dimension is $2n$ then the volume element is

$$dV = \frac{1}{n!} \left(\frac{1}{2}\Omega\right) \wedge \left(\frac{1}{2}\Omega\right) \wedge \ldots \left(\frac{1}{2}\Omega\right). \tag{3.74}$$

The numerical factor can be chosen at will – unless we are on a Kähler manifold where the choice just made is the only natural one. The point is that a Kähler manifold has both a metric and a symplectic form, so that there will be two notions of volume that we want to be consistent with each other. The symplectic form is precisely the Kähler form from Eq. (3.62), $\Omega = J$. The special feature of Kähler

---

[6] For a survey of symplectic geometry by an expert, see Arnold [53].

manifolds is that the two kinds of geometry are interwoven with each other and with the complex structure. On the 2-sphere

$$dV = \frac{1}{2}\Omega = \frac{2i}{(1+|z|^2)^2}dz \wedge d\bar{z} = \frac{4}{(1+r^2)^2}dx \wedge dy = \sin\theta d\theta \wedge d\phi. \quad (3.75)$$

This agrees with the volume form as computed using the metric.

3.5 The Hopf fibration of the 3-sphere

The 3-sphere, being odd-dimensional, is neither complex nor symplectic, but like all the odd-dimensional spheres it is a fibre bundle. Unlike all other spheres (except S^1) it is also a Lie group. The theory of fibre bundles was in fact created in response to the treatment that the 3-sphere was given in 1931 by Hopf and Dirac, and we begin with this part of the story. (By the way, Dirac's concern was with magnetic monopoles.)

The 3-sphere can be defined as the hypersurface

$$X^2 + Y^2 + Z^2 + U^2 = 1 \quad (3.76)$$

embedded in a flat four-dimensional space with (X, Y, Z, U) as its Cartesian coordinates. Using stereographic coordinates (Section 3.1) we can visualise the 3-sphere as $\mathbb{R}^3$ with the south pole ($U = -1$) added as a point 'at infinity'. The equator of the 3-sphere ($U = 0$) will appear in the picture as a unit sphere surrounding the origin. To get used to it we look at geodesics and Killing vectors.

Using Eq. (3.12) it is easy to prove that geodesics appear in our picture either as circles or as straight lines through the origin. Either way – unless they are great circles on the equator – they meet the equator in two antipodal points. Now rotate the sphere in the X–Y plane. The appropriate Killing vector field is

$$J_{XY} = X\partial_Y - Y\partial_X = x\partial_y - y\partial_x, \quad (3.77)$$

where x, y (and z) are the stereographic coordinates in our picture. This looks like the flow lines of a rotation in flat space. There is a geodesic line of fixed points along the z-axis. The flow lines are circles around this geodesic, but with one exception they are not themselves geodesics because they do not meet the equator in antipodal points. The Killing vector J_{ZU} behaves intrinsically just like J_{XY}, but it looks quite different in our picture (because we singled out the coordinate U for special treatment). It has fixed points at

$$J_{ZU} = Z\partial_U - U\partial_Z = 0 \quad \Leftrightarrow \quad Z = U = 0. \quad (3.78)$$

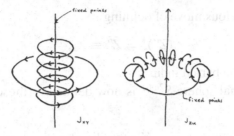

Figure 3.7 Flow lines and fixed points of J_{XY} and J_{ZU}.

This is a geodesic (as it must be), namely a great circle on the equator. By analogy with J_{XY} the flow lines must lie on tori surrounding the line of fixed points. A little calculation confirms this; the flow of J_{ZU} leaves the tori of revolution

$$(\rho - a)^2 + z^2 = a^2 - 1 > 0, \qquad \rho^2 \equiv x^2 + y^2, \tag{3.79}$$

invariant for any $a > 1$. So we can draw pictures of these two Killing vector fields, the instructive point of the exercise being that intrinsically these Killing vector fields are really 'the same'. See Figure 3.7.

A striking thing about the 3-sphere is that there are also Killing vector fields that are everywhere non-vanishing. This is in contrast to the 2-sphere; a well known theorem in topology states that 'you can't comb a sphere', meaning to say that every vector field on $\mathbf{S}^2$ has to have a fixed point somewhere. An example of a Killing field without fixed points on $\mathbf{S}^3$ is clearly

$$\xi = J_{XY} + J_{ZU}; \qquad ||\xi||^2 = X^2 + Y^2 + Z^2 + U^2 = 1. \tag{3.80}$$

Given our pictures of Killing vector fields it is clear that this combination must have flowlines that lie on the tori that we drew, but which wind once around the z-axis each time they wind around the circle $\rho = 1$. This will be our key to understanding the 3-sphere as a fibre bundle.

Remarkably all the flow lines of the Killing vector field ξ are geodesics as well. We will prove this in a way that brings complex manifolds back in. The point is that the embedding space $\mathbb{R}^4$ is also the complex vector space $\mathbb{C}^2$. Therefore we can introduce the complex embedding coordinates

$$\begin{bmatrix} Z^1 \\ Z^2 \end{bmatrix} = \begin{bmatrix} X + iY \\ Z + iU \end{bmatrix}. \tag{3.81}$$

The generalisation to n complex dimensions is immediate. Let us use P and Q to denote vectors in complex vector spaces. The scalar product in $\mathbb{R}^{2n}$ is the real part of an *Hermitian form* on $\mathbb{C}^n$, namely

$$P \cdot \bar{Q} = \delta_{\alpha\bar{\alpha}} P^\alpha \bar{Q}^{\bar{\alpha}} = P^\alpha \bar{Q}_\alpha. \tag{3.82}$$

Here we made the obvious move of defining

$$(Z^\alpha)^* = \bar{Z}^{\bar{\alpha}} \equiv \bar{Z}_\alpha, \tag{3.83}$$

so that we get rid of the barred indices.

The odd-dimensional sphere $\mathbf{S}^{2n+1}$ is now defined as those points in $\mathbb{C}^{n+1}$ that obey

$$Z \cdot \bar{Z} = 1. \tag{3.84}$$

Translating the formula (3.12) for a general geodesic from the real formulation given in Section 3.1 to the complex one that we are using now we find

$$Z^\alpha(\sigma) = m^\alpha \cos\sigma + n^\alpha \sin\sigma, \quad m \cdot \bar{m} = n \cdot \bar{n} = 1, \; m \cdot \bar{n} + n \cdot \bar{m} = 0, \quad (3.85)$$

where the affine parameter σ measures distance d along the geodesic,

$$d = |\sigma_2 - \sigma_1|. \tag{3.86}$$

If we pick two points on the geodesic, say

$$Z_1^\alpha \equiv Z^\alpha(\sigma_1) \qquad Z_2^\alpha \equiv Z^\alpha(\sigma_2), \tag{3.87}$$

then a short calculation reveals that the distance between them is given by

$$\cos d = \frac{1}{2}\left(Z_1 \cdot \bar{Z}_2 + Z_2 \cdot \bar{Z}_1\right). \tag{3.88}$$

This is a useful formula to have.

Now consider the family of geodesics given by

$$n^\alpha = im^\alpha \quad \Rightarrow \quad Z^\alpha(\sigma) = e^{i\sigma} m^\alpha. \tag{3.89}$$

Through any point on $\mathbf{S}^{2n+1}$ there will go a geodesic belonging to this family since we are free to let the vector m^α vary. Evidently the equation

$$\dot{Z}^\alpha = iZ^\alpha \tag{3.90}$$

holds for every geodesic of this kind. Its tangent vector is therefore given by

$$\partial_\sigma = \dot{Z}^\alpha \partial_\alpha + \dot{\bar{Z}}^{\bar{\alpha}} \partial_{\bar{\alpha}} = i(Z^\alpha \partial_\alpha - \bar{Z}^{\bar{\alpha}} \partial_{\bar{\alpha}}) = J_{XY} + J_{ZU} = \xi. \tag{3.91}$$

But this is precisely the everywhere non-vanishing Killing vector field that we found before. So we have found that on $\mathbf{S}^{2n+1}$ there exists a *congruence* – a space-filling family – of curves that are at once geodesics and Killing flow lines. This is a quite remarkable property; flat space has it, but very few curved spaces do.

In flat space the distance between parallel lines remains fixed as we move along the lines, whereas two skew lines eventually diverge from each other. In a positively curved space – like the sphere – parallel geodesics converge, and we ask if it is possible to twist them relative to each other in such a way that this convergence is cancelled by the divergence caused by the fact that they are skew. Then we would have a congruence of geodesics that always stay at the same distance from each other. We will call the geodesics *Clifford parallels* provided that this can be done. A more stringent definition requires a notion of parallel transport that takes tangent vectors of the Clifford parallels into each other; we will touch on this in Section 3.7.

Now the congruence of geodesics given by the vector field ξ are Clifford parallels on the 3-sphere. Two points belonging to different geodesics in the congruence must preserve their relative distance as they move along the geodesics, precisely because the geodesics are Killing flow lines as well. It is instructive to prove this directly though. Consider two geodesics defined by

$$P^\alpha = e^{i\sigma} P_0^\alpha, \qquad Q^\alpha = e^{i(\sigma+\sigma_0)} Q_0^\alpha. \tag{3.92}$$

We will soon exercise our right to choose the constant σ_0. The scalar product of the constant vectors will be some complex number

$$P_0 \cdot \bar{Q}_0 = re^{i\phi}. \tag{3.93}$$

The geodesic distance between two arbitrary points, one on each geodesic, is therefore given by

$$\cos d = \frac{1}{2} \left(P \cdot \bar{Q} + Q \cdot \bar{P} \right) = r \cos (\phi - \sigma_0). \tag{3.94}$$

The point is that this is independent of the affine parameter σ, so that the distance does not change as we move along the geodesics (provided of course that we move with the same speed on both). This shows that our congruence of geodesics consists of Clifford parallels.

The perpendicular distance d_0 between a pair of Clifford parallels is obtained by adjusting the zero point σ_0 so that $\cos d_0 = r$, that is, so that the distance attains its minimum value. A concise way to express d_0 is by means of the equation

$$\cos^2 d_0 = r^2 = P_0 \cdot \bar{Q}_0 Q_0 \cdot \bar{P}_0 = P \cdot \bar{Q} Q \cdot \bar{P}. \tag{3.95}$$

Before we are done we will see that this formula plays an important role in quantum mechanics.

It is time to draw a picture – Figure 3.8 – of the congruence of Clifford parallels. Since we have the pictures of J_{XY} and J_{ZU} already this is straightforward. We draw a family of tori of revolution surrounding the unit circle in the $z = 0$ plane, and eventually merging into the z-axis. These are precisely the tori defined in Eq. (3.79). Each torus is itself foliated by a one-parameter family of circles and they are twisting relative to each other as we expected; indeed any two circles in

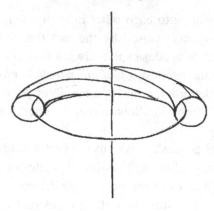

Figure 3.8 The Hopf fibration of the 3-sphere. Ref. [723] has a better version.

the congruence are linked (their linking number is one). The whole construction is known as the *Hopf fibration* of the 3-sphere, and the geodesics in the congruence are called *Hopf circles*. It is clear that there exists another Hopf fibration with the opposite twist that we would arrive at through a simple sign change in Eq. (3.80). By the way the metric induced on the tori by the metric on the 3-sphere is flat (as you can easily check from Eq. (3.98) below, where a torus is given by $\theta =$ constant); you can think of the 3-sphere as a one-parameter family of flat tori if you please, or as two solid tori glued together.

Now the interesting question is: how many Hopf circles are there altogether, or more precisely what is the space whose points consists of the Hopf circles? A little thinking gives the answer directly. On each torus there is a one-parameter set of Hopf circles, labelled by some periodic coordinate $\phi \in [0, 2\pi[$. There is a one-parameter family of tori, labelled by $\theta \in]0, \pi[$. In this way we account for every geodesic in the congruence except the circle in the $z = 0$ plane and the one along the z-axis. These have to be added at the endpoints of the θ-interval, at $\theta = 0$ and $\theta = \pi$ respectively. Evidently what we are describing is a 2-sphere in polar coordinates. So the conclusion is that the space of Hopf circles is a 2-sphere.

It is important to realise that this 2-sphere is not 'sitting inside the 3-sphere' in any natural manner. To find such an embedding of the 2-sphere would entail choosing one point from each Hopf circle in some smooth manner. Equivalently, we want to choose the zero point of the coordinate τ along all the circles in some coherent way. But this is precisely what we cannot do; if we could we would effectively have shown that the topology of $\mathbf{S}^3$ is $\mathbf{S}^2 \times \mathbf{S}^1$ and this is not true (because in $\mathbf{S}^2 \times \mathbf{S}^1$ there are closed curves that cannot be contracted to a point, while there are no such curves in $\mathbf{S}^3$). We can almost do it though. For instance, we can select those points where the geodesics are moving down through the $z = 0$ plane. This works fine except for the single geodesic that lies in this plane; we have mapped all of the 2-sphere except one point onto an open unit disc in the picture. It is instructive to make a few more attempts in this vein and see how they always fail to work globally.

The next question is whether the 2-sphere of Hopf circles is a round 2-sphere or not, or indeed whether it has any natural metric at all. The answer turns out to be 'yes'. To see this it is convenient to introduce the *Euler angles*, which are intrinsic coordinates on $\mathbf{S}^3$ adapted to the Hopf fibration. They are defined by

$$\begin{bmatrix} Z^1 \\ Z^2 \end{bmatrix} = \begin{bmatrix} X + iY \\ Z + iU \end{bmatrix} = \begin{bmatrix} e^{\frac{i}{2}(\tau+\phi)} \cos\frac{\theta}{2} \\ e^{\frac{i}{2}(\tau-\phi)} \sin\frac{\theta}{2} \end{bmatrix}, \qquad (3.96)$$

where

$$0 \le \tau < 4\pi, \qquad 0 \le \phi < 2\pi, \qquad 0 < \theta < \pi. \qquad (3.97)$$

We have seen that the periodic coordinate τ goes along the Hopf circles, in fact $\tau = 2\sigma$ in the congruence (3.89). (Making τ periodic with period 2π gives the Hopf fibration of real projective 3-space – checking this statement is a good way of making sure that one understands the Hopf fibration.) The coordinate ϕ runs along Hopf circles of the opposite twist. Finally a little calculation verifies that the coordinate θ labels the tori in Eq. (3.79) with $\cos(\theta/2) = 1/a$. The intrinsic metric on the 3-sphere becomes

$$ds^2 = |dZ^1|^2 + |dZ^2|^2 = \frac{1}{4}(d\tau^2 + d\theta^2 + d\phi^2 + 2\cos\theta d\tau d\phi). \qquad (3.98)$$

Here it is easy to see that the tori at $\theta =$ constant are flat.

To continue our argument: since the coordinates θ and ϕ label the 2-sphere's worth of geodesics in the congruence we can try to map this $\mathbf{S}^2$ into $\mathbf{S}^3$ through an equation of the form

$$\tau = \tau(\theta, \phi). \qquad (3.99)$$

But we have already shown that no such map can exist globally, and therefore this is not the way to define a natural metric on our 2-sphere. We still do not know if it is a round 2-sphere in any natural way.

Nevertheless the space of Clifford parallels is naturally a round 2-sphere. We simply define the distance between two arbitrary Clifford parallels as the perpendicular distance d_0 between them. This we have computed already, and it only remains to rewrite Eq. (3.95) in terms of the Euler angles. If the coordinates of the two points on the 2-sphere are (θ_1, ϕ_1) and (θ_2, ϕ_2) we obtain

$$\cos^2 d_0 = \frac{1}{2}\left(1 + \cos\theta_1\cos\theta_2 + \cos(\phi_1 - \phi_2)\sin\theta_1\sin\theta_2\right). \qquad (3.100)$$

This formula should be familiar from spherical trigonometry. If the two points are infinitesimally close to each other we can expand the left-hand side as

$$\cos^2 d_0 \approx 1 - d_0^2 \equiv 1 - ds^2. \tag{3.101}$$

A short calculation then verifies that the metric is

$$ds^2 = \frac{1}{4}(d\theta^2 + \sin^2\theta d\phi^2). \tag{3.102}$$

Precisely one-quarter of the usual round metric. From now on, when we talk about the Hopf fibration it will be understood that the 2- and 3-spheres are equipped with the above metrics.

3.6 Fibre bundles and their connections

Let us now bring in the general theory of fibre bundles.[7] By definition a *fibre* of a map $P \to M$ between two spaces is the set of points in P that are mapped to a given point in M. The situation gets more interesting if all fibres are isomorphic. The definition of a fibre bundle also requires that there is a group acting on the fibres: a *fibre bundle* over a *base manifold M* consists of a *bundle space P* and a *structure group G* acting on the bundle space in such a way that the base manifold is equal to the quotient P/G (a point in M is an orbit of G in P; the orbits are assumed to be isomorphic to each other). In this way we get a canonical projection $\Pi : P \to M$. The set of points that project to a particular point p on M is known as the fibre F over p. It is also required that the bundle space is locally equal to a Cartesian product, that is to say that P can be covered by open sets of the form $U \times F$, where U is an open set in M. A *principal fibre bundle* is a fibre bundle such that the fibres are copies of the group manifold of G.

A *Cartesian product* $M \times F$ (as in $\mathbb{R}^2 \equiv \mathbb{R} \times \mathbb{R}$) is a trivial example of a fibre bundle. The 3-sphere on the other hand is non-trivial because it is not just a Cartesian product of $\mathbf{S}^2$ and $\mathbf{S}^1$, although locally it behaves like that. It should be clear from the previous discussion that the 3-sphere really is a principal fibre bundle with structure group $U(1)$, whose group manifold is indeed a circle. The fact that the structure group is abelian is a simplifying feature; in general the fibres can be many-dimensional and the structure group can be non-abelian.

---

[7] A standard reference for connections and fibre bundles (and for a taste of the precision used in modern differential geometry) is Kobayashi and Nomizu [550]. There are many readable introductions for physicists; examples can be found in Shapere and Wilczek [830].

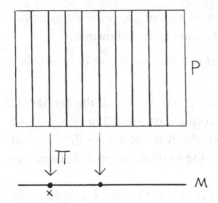
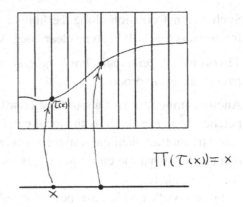

Figure 3.9 Left: the definition of a fibre bundle, including a fibred bundle space *P*, a base manifold *M* and a projection Π. Right: a section of the bundle that is an embedding of the base manifold in the bundle space (which may fail to exist globally).

For many purposes it is convenient to introduce a local coordinate system on the bundle that is adapted to the bundle structure. Schematically, let x^a be some coordinates on the base manifold and τ a coordinate along the fibres, assumed one-dimensional for simplicity. A full set of coordinates is then given by $x^i = (\tau, x^a)$. On S^3 the coordinates θ and ϕ play the part of x^a while τ is the fibre coordinate. Now the idea is to restrict oneself to coordinates that can be reached from x^a, τ through coordinate transformations of the general form

$$x'^a = x'^a(x), \qquad \tau' = \tau'(x, \tau). \qquad (3.103)$$

Such 'triangular' coordinate transformations appear because there is no natural way of identifying the fibres with each other.

To take a *section* of the bundle means to specify the fibre coordinate τ as a function of x^a,

$$\tau = \tau(x). \qquad (3.104)$$

See Figure 3.9. Locally this defines an embedding of the base manifold into the bundle. In coordinate independent terms, a section is defined as an embedding of the base manifold into the bundle such that if we follow it up with the projection down to the base we get back to the point we started out from. For a non-trivial bundle such as the 3-sphere no global section exists, but it is possible to take local sections on the coordinate patches U. In the overlap regions where two different local sections are defined one can go from one to the other provided that one moves along the fibres according to

$$\tau = \tau(x) \quad \rightarrow \quad \tau' = \tau'(x). \qquad (3.105)$$

Such a transformation along the fibres is known as a *local gauge transformation*, for reasons that will become clear later. A key theorem is the following.

Theorem. *A principal fibre bundle admits a global section if and only if it is a trivial Cartesian product.*

Another important fact about fibre bundles is that if one knows all the coordinate patches $U \times F$ as well as the local gauge transformations needed to go from one patch to another, then one can reconstruct the bundle. It is not always the case that one can see what the entire bundle looks like, in the intimate manner that one can see the 3-sphere.

Fibre bundles which are not principal are also of interest. An example is the *vector bundle*, where the fibres consist of vector spaces on which the group G acts, rather than of G itself. Note that the theorem that a fibre bundle has the trivial product topology if and only if it admits a global section holds for principal bundles, and only for principal bundles. A famous example of a non-trivial vector bundle that admits a global section is the Möbius strip; the group G that acts on the fibre is the finite group $\mathbb{Z}_2$ and indeed the principal bundle does not admit a global section. It is so well known how to construct a Möbius strip made of paper that we simply recommend the reader to make one and check these statements. The *tangent bundle* over a space **M** is also a vector bundle. The fibre over a point q is the tangent space $\mathbf{T}_q$ at this point. The group of general linear transformations acts on the fibres of the tangent bundle. A section of the tangent bundle – one element of the fibre at each point in the base space – is just a vector field.

One often wants to *lift* a curve in the base manifold M to the bundle P. Since many curves in the bundle project down to the same curve in M, this requires a further structure, namely a *connection* on the bundle. The idea is to decompose the tangent space of the bundle into *vertical* and *horizontal* directions, that is directions along the fibres and 'perpendicular' to the fibres, respectively. Then the idea is that the *horizontal lift* of a curve threads its way in the horizontal directions only. The problem is that there may be no notion of 'perpendicular' available, so this must be supplied. To a mathematician, a connection is a structure that defines a split of the tangent space into vertical and horizontal. To a physicist, a connection is like the vector potential in electrodynamics, and it takes a little effort to see that the same structure is being referred to. The key fact that must be kept in mind throughout the discussion is that if there is an embedding of one manifold into another, then a differential form on the latter automatically defines a differential form on the former (see Appendix A.3).

We simplify matters by assuming that the fibres are one dimensional, and let the coordinate τ run along the fibres. We want to decompose tangent space so that

$$\mathbf{T}_p = \mathbf{V}_p \oplus \mathbf{H}_p. \tag{3.106}$$

Any tangent vector at p can then be written as the sum of one vector belonging to the vertical subspace $\mathbf{V}_p$ and one belonging to the horizontal subspace $\mathbf{H}_p$. Since a vertical vector points along the fibres it must be proportional to

$$\partial_\tau \in \mathbf{V}_p. \tag{3.107}$$

But it is not so clear what a horizontal vector should look like. The trick is to choose a special one-form ω (that is a covariant vector ω_i) and to declare that a horizontal vector h^i is any vector that obeys

$$h^i \in \mathbf{H}_p \quad \Leftrightarrow \quad \omega_i h^i = 0. \tag{3.108}$$

Such a statement does not require any metric. But how do we choose ω? The obvious guess $d\tau$ has a problem since we are permitting coordinate transformations under which

$$d\tau = \frac{\partial \tau}{\partial \tau'} d\tau' + \frac{\partial \tau}{\partial x'^a} dx'^a. \tag{3.109}$$

Hence the naive definition of the horizontal subspace is tied to a particular set of coordinates. There will be many ways to perform the split. A priori none of them is better than any other. This is something that we will have to accept, and we therefore simply choose an ω of the general form

$$\omega = d\tau + U = d\tau + U_a dx^a. \tag{3.110}$$

In this way a connection, that is a decomposition of the tangent space, is equivalent to the specification of a special one-form on the bundle.

With a connection in hand, we define *parallel transport* as follows. Fix a curve $x^a(\sigma)$ in the base manifold. Lift this curve to a curve in the bundle by insisting that the 1-form induced on the lifted curve by the connection 1-form ω vanishes. Thus

$$\omega = d\tau + U = \left(\frac{d\tau}{d\sigma} + U_a \frac{dx^a}{d\sigma} \right) d\sigma = 0. \tag{3.111}$$

This leads to an ordinary differential equation that always admits a solution (at least for some range of the parameter σ), and it follows that we can associate a unique family of curves $(\tau(\sigma), x^a(\sigma))$ in the bundle to any curve in the base manifold. Through any point in the fibre there passes a unique curve. This is the horizontal lift of the curve in the base manifold. See Figure 3.10.

There are two things to notice. First, that this notion of parallel transport is somewhat different than that discussed in Section 3.2. Then we were transporting vectors, now we are transporting the position along the fibres. (Actually, the parallel transport defined in Section 3.2 is a special case, taking place in the tangent bundle.) Second, it may seem that we now have the means to identify the fibres with each other. But this is false in general, because parallel transport may depend on the

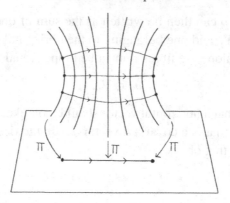

Figure 3.10 Horizontal lifts of a curve in the base manifold, using the notion of 'horizontal subspaces'.

path – and it may do so even if the bundle is topologically trivial. Indeed the condition that parallel transport be independent of the path, or in other words that the horizontal lift of a closed curve is itself closed, is (by Stokes' theorem, see Appendix A.1)

$$\oint_C \omega = \int_S d\omega = 0, \tag{3.112}$$

where S is any surface enclosed by a curve C that projects to a given closed curve in the base manifold. The 2-form

$$\Omega \equiv d\omega = \mathrm{d}U = \frac{1}{2} \left(\partial_a U_b - \partial_b U_a \right) \mathrm{d}x^a \wedge \mathrm{d}x^b \tag{3.113}$$

is known as the *curvature 2-form*. The point is that when the curvature 2–form is non-zero, and when Stokes' theorem applies, Eq. (3.112) will not hold. Then parallel transport along a closed curve will give rise to a shift along the fibre, known to mathematicians as a *holonomy*. See Figure 3.11.

Throughout, the connection is a 1-form that is defined on the bundle space – not on the base manifold. If we take a section of the bundle we have an embedding of the base manifold into the bundle, and the connection on the bundle will induce a definite 1 form on the base manifold. This is what happens in electrodynamics; in this way the physicist's connection is recovered from that of the mathematician. There is a catch: if the section is not globally defined, the connection on the base manifold will be ill-defined in some places. On the other hand there is no such problem with the curvature 2-form – Ω does exist globally on the base manifold.

Formula (3.113) is valid in this simple form for the case of one-dimensional fibres only. But in physics we often need many-dimensional fibres – and a non-abelian group G. The main novelty is that the connection takes values in the Lie

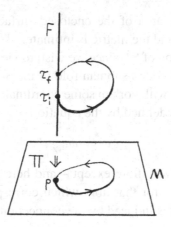

Figure 3.11 Holonomy. For the 3-sphere, the holonomy $\tau_f - \tau_i$ is proportional to the area surrounded by the loop in the base manifold (i.e. the 2-sphere).

algebra of G; since the dimension of the Lie algebra is equal to the dimension of the fibres this will be enough to single out a unique horizontal subspace. One difficulty that arises is that it is hard to get explicit about holonomies – Stokes' theorem is no longer helpful because the line integrals must be *path ordered* due to the fact that the connections at different points along the curve do not commute with each other.

Coming back to the 3-sphere, we observe that the fibres are the Hopf circles and the base manifold is a 2-sphere. Now the 3-sphere is not just any fibre bundle. Its bundle space is equipped with a metric, which means that there is a *preferred connection* singled out by the requirement that the vertical and the horizontal subspaces be orthogonal in the sense of that metric. If the horizontal lift of a curve in the base manifold is parametrized by σ its tangent vector is

$$\partial_\sigma = \dot{\tau} \partial_\tau + \dot{\theta} \partial_\theta + \dot{\phi} \partial_\phi \equiv h^i \partial_i. \qquad (3.114)$$

The tangent vector of a fibre is $\partial_\tau \equiv v^i \partial_i$. We require

$$g_{ij} v^i h^j = g_{\tau\tau} \dot{\tau} + g_{\tau\theta} \dot{\theta} + g_{\tau\phi} \dot{\phi} = \dot{\tau} + \cos\theta \dot{\phi} = 0, \qquad (3.115)$$

where we used Eq. (3.98) for the metric. Hence the metrically preferred connection is

$$\omega = d\tau + \cos\theta \, d\phi. \qquad (3.116)$$

The curvature of this connection is non-vanishing; indeed if parallel transport occurs along a closed curve ∂S, bounding a surface S in the base manifold, then the holonomy is

$$\tau_{\text{final}} - \tau_{\text{initial}} = \int d\tau = -\oint_{\partial S} \cos\theta \, d\phi = \int_S \sin\theta \, d\theta \, d\phi. \qquad (3.117)$$

This is proportional to the area of the enclosed surface on S^2, so the interplay between parallel transport and the metric is intimate.

We will now take a section of S^3, since we wish to view – as far as possible – the base manifold S^2 as an embedded submanifold of the 3-sphere so that ω induces a 1-form on the 2-sphere. We will work in some coordinate patch, say using the Euler angles. One local section is defined by the equation

$$\tau = \phi. \tag{3.118}$$

It picks out one point on every fibre except – and here we have to recall how the Euler angles were defined – for that fibre which corresponds to the north pole of the 2-sphere. The form that is induced by the connection on the 2-sphere is

$$\omega^+ = (1 + \cos\theta)\, d\phi. \tag{3.119}$$

When looking at this expression we must remember that we are using a coordinate system on S^2 that is ill-defined at the poles. But we are particularly interested in the behaviour at the north pole, where we expect this form to be singular somehow. If – with a little effort – we transform from θ and ϕ to our standard complex coordinate system on S^2 we find that

$$\omega^+ = \frac{i}{1 + |z|^2}\left(-\frac{dz}{z} + \frac{d\bar{z}}{\bar{z}}\right). \tag{3.120}$$

This is indeed ill-defined at the north pole ($z = 0$). On the other hand there is no problem at the south pole; transforming to complex coordinates that cover this point ($z \to z' = 1/z$) we find that

$$\omega^+ = \frac{i}{1 + |z'|^2}\left(\bar{z}'dz' - z'd\bar{z}'\right), \tag{3.121}$$

and there is no problem at the south pole ($z' = 0$).

By taking a different section (say $\tau = -\phi$) another form is induced on S^2. If we choose $\tau = -\phi$ we obtain

$$\omega^- = (\cos\theta - 1)\, d\phi. \tag{3.122}$$

This turns out to be well defined everywhere except at the south pole. In the region of the 2-sphere where both sections are well defined – which happens to be the region where the polar coordinates are well defined – the forms are related by

$$\omega^+ - \omega^- = 2d\phi. \tag{3.123}$$

This is a local gauge transformation in the ordinary sense of the word as used in electrodynamics.

In spite of the little difficulties there is nothing wrong with the connection; we have only confirmed the conclusion that no global section of the bundle exists. The curvature 2-form

$$\Omega_{\theta\phi} = \sin\theta \tag{3.124}$$

is an everywhere regular 2-form on the base manifold, and in fact equal to the symplectic form on $\mathbf{S}^2$. Quite incidentally, if we regard the 2-sphere as a sphere embedded in Euclidean 3-space our curvature tensor corresponds to a radially directed magnetic field. There is a magnetic monopole sitting at the origin – and this is what Dirac was doing in 1931, while Hopf was fibring the 3-sphere.

3.7 The 3-sphere as a group

The 3-sphere is not only a fibre bundle, it is also a group. More precisely it is a *group manifold*, that is to say a space whose points are the elements of a group. Only two spheres are group manifolds – the other case is $\mathbf{S}^1$, the group manifold of $U(1)$. What we are going to show for the 3-sphere is therefore exceptional as far as spheres are concerned, but it is typical of all *compact semi-simple* Lie groups. Compact means that the group manifold is compact; semi-simple is a technical condition which is obeyed by the *classical groups* $SO(N)$, $SU(N)$ and $Sp(N)$, with the exception of $SO(2) = U(1)$. What we tell below about $SU(2)$ will be typical of all the classical groups, and this is good enough for us.[8]

The classical groups are *matrix groups* consisting of matrices obeying certain conditions. Thus the group $SU(2)$ is by definition the group of unitary two-by-two matrices of determinant one. An arbitrary group element g in this group can be parametrized by two complex numbers obeying one condition, namely as

$$g = \begin{bmatrix} Z^1 & Z^2 \\ -\bar{Z}^2 & \bar{Z}^1 \end{bmatrix}, \qquad |Z^1|^2 + |Z^2|^2 = 1. \tag{3.125}$$

This means that there is a one-to-one correspondence between $SU(2)$ group elements on the one hand and points on the 3-sphere on the other, that is to say that whenever $X^2 + Y^2 + Z^2 + U^2 = 1$ we have

$$(X, Y, Z, U) \quad \leftrightarrow \quad g = \begin{bmatrix} X + iY & Z + iU \\ -Z + iU & X - iY \end{bmatrix}. \tag{3.126}$$

We begin to see the 3-sphere in a new light.

---

[8] There are many good books on group theory available where the reader will find a more complete story; an example that leans towards quantum mechanics is Gilmore [334].

Given two points in a group the group law determines a third point, namely $g_3 = g_1 g_2$. For a matrix group it is evident that the coordinates of g_3 will be given by real analytic functions of the coordinates of g_1 and g_2. This property defines a *Lie group*. Now a unitary matrix g of unit determinant can always be obtained by exponentiating a traceless anti-Hermitian matrix m. We can insert a real parameter τ in the exponent and then we get

$$g(\tau) = e^{\tau m}. \tag{3.127}$$

From one point of view this is a *one-parameter subgroup* of $SU(2)$, from another it is a curve in the group manifold $\mathbf{S}^3$. When $\tau = 0$ we are sitting at a point that is rather special from the first viewpoint, namely at the unit element $\mathbb{1}$ of the group. The tangent vector of the curve at this point is

$$\dot{g}(\tau)_{|\tau=0} = m. \tag{3.128}$$

In this way we see that the vector space of traceless anti-Hermitian matrices make up the tangent space at the unit element of the group. (Physicists usually prefer to work with Hermitian matrices, but here anti-Hermitian matrices are more natural – to convert between these preferences just multiply with i. Incidentally the at first sight confusing habit to view a matrix as an element of a vector space will grow upon us.) This tangent space is called the *Lie algebra* of the group, and it is part of the magic of Lie groups that many of their properties can be reliably deduced from a study of their Lie algebras. A reminder about Lie algebras can be found in Appendix B.

A group always acts on itself. A group element g_1 transforms an arbitrary group element g by *left translation* $g \to g_1 g$, by *right translation* $g \to g g_1^{-1}$ and by the *adjoint action* $g \to g_1 g g_1^{-1}$. (The inverse occurs on the right-hand side to ensure that $g \to g g_1^{-1} \to g g_1^{-1} g_2^{-1} = g(g_2 g_1)^{-1}$.) Fix a one-parameter subgroup $g_1(\sigma)$ and let it act by left translation; $g \to g_1(\sigma) g$. This defines a curve through every point of $\mathbf{S}^3$, and moreover the group laws ensure that these curves cannot cross each other, so that they form a congruence. The congruence of Clifford parallels that we studied in Section 3.5 can be viewed in this light. To do so consider the one-parameter subgroup obtained by exponentiating $m = i\sigma_3$, where σ_i denotes a Pauli matrix. It gives rise to an infinitesimal transformation $\delta g = mg$ at every point g. If we use the coordinates (X, Y, Z, U) to describe g we find that this equation becomes

$$\begin{bmatrix} \delta(X+iY) & \delta(Z+iU) \\ -\delta(Z+iU) & \delta(X-iY) \end{bmatrix} = \begin{bmatrix} i & 0 \\ 0 & -i \end{bmatrix} \begin{bmatrix} X+iY & Z+iU \\ -Z+iU & X-iY \end{bmatrix}. \tag{3.129}$$

Working this out we find it to be the same as

$$(\delta X, \delta Y, \delta Z, \delta U) = (J_{XY} + J_{ZU})(X, Y, Z, U). \tag{3.130}$$

This is precisely the Killing vector field that points along a congruence of Hopf circles. Choosing $m = i\sigma_1$ and $m = i\sigma_2$ will give us a total of three nowhere-vanishing linearly independent vector fields. They are

$$J_1 = J_{XU} + J_{YZ}, \qquad J_2 = J_{XZ} + J_{UY}, \qquad J_3 = J_{XY} + J_{ZU} \qquad (3.131)$$

and they form a representation of the Lie algebra of $SU(2)$. Their number is the same as the dimension of the manifold. We say that the 3-sphere is *parallelizable*. All group manifolds are parallelizable, for the same reason, but among the spheres only $\mathbf{S}^1$, $\mathbf{S}^3$ and $\mathbf{S}^7$ are parallelizable – so this is an exceptional property shared by all group manifolds. Such everywhere-non-vanishing vector fields can be used to provide a canonical identification of tangent spaces at different points of the group manifold. One can then modify the rules for parallel transport of vectors so that it respects this identification – in fact we can use Eq. (3.15) but with the totally anti-symmetric structure constants of the Lie algebra in the role of the tensor T_{ijk}. For the 3-sphere, this allows us to give a more stringent definition of Clifford parallels than the one we gave in Section 3.5.

So much for left translation. Right translations of the group also give rise, in an analogous way, to a set of three linearly independent vector fields

$$\tilde{J}_1 = J_{XU} - J_{YZ}, \qquad \tilde{J}_2 = J_{XZ} - J_{UY}, \qquad \tilde{J}_3 = J_{XY} - J_{ZU} \qquad (3.132)$$

These three represent another $SU(2)$ Lie algebra and they commute with the previous three, that is those in Eq. (3.131).

A useful fact about group manifolds is that one can always find a Lie algebra valued 1-form that is invariant under (say) right translation. This is the *Maurer–Cartan form* $dg g^{-1}$. That it takes values in the Lie algebra is clear if we write g as an exponential of a Lie algebra element as in Eq. (3.127). That it is invariant under right translation by a fixed group element g_1 is shown by the calculation

$$\mathrm{d}g \, g^{-1} \; \rightarrow \; \mathrm{d}(gg_1) \, (gg_1)^{-1} \; = \; \mathrm{d}g \, g_1 \, g_1^{-1} g^{-1} \; = \; \mathrm{d}g \, g^{-1}. \qquad (3.133)$$

Given a basis for the Lie algebra the Maurer–Cartan form can be expanded. For $SU(2)$ we use the Pauli matrices multiplied with i for this purpose, and get

$$\mathrm{d}g \, g^{-1} \; = \; i\sigma_1 \Theta_1 + i\sigma_2 \Theta_2 + i\sigma_3 \Theta_3. \qquad (3.134)$$

Doing the explicit calculation, that is multiplying two two-by-two matrices, we can read off that

$$\Theta_1 = -U\mathrm{d}X - Z\mathrm{d}Y + Y\mathrm{d}Z + X\mathrm{d}U = \tfrac{1}{2}\left(\sin\tau \, \mathrm{d}\theta - \cos\tau \, \sin\theta \, \mathrm{d}\phi\right)$$

$$\Theta_2 = -Z\mathrm{d}X + U\mathrm{d}Y + X\mathrm{d}Z - Y\mathrm{d}U = \tfrac{1}{2}\left(\cos\tau \, \mathrm{d}\theta + \sin\tau \, \sin\theta \, \mathrm{d}\phi\right) \qquad (3.135)$$

$$\Theta_3 = -Y\mathrm{d}X + X\mathrm{d}Y - U\mathrm{d}Z + Z\mathrm{d}U = \tfrac{1}{2}\left(\mathrm{d}\tau + \cos\theta \, \mathrm{d}\phi\right).$$

In the last step we use Euler angles as coordinates; Θ_3 is evidently our familiar friend, the connection from Eq. (3.116). There exists a left invariant Maurer–Cartan form $g^{-1}\, dg$ as well.

It is now natural to declare that the group manifold of $SU(2)$ is equipped with its standard round metric; the round metric is singled out by the requirement that left and right translations correspond to isometries. Since left and right translations can be performed independently, this means that the isometry group of any Lie group G equipped with its natural metric is $G \times G$, with a discrete factor divided out. For the 3-sphere the isometry group is $SO(4)$, and it obeys the isomorphism

$$SO(4) \;=\; SU(2) \times SU(2)/Z_2. \tag{3.136}$$

The Z_2 in the denominator arises because left and right translation with $-\mathbb{1}$ cancel each other. It is this peculiar isomorphism that explains why the 3-sphere manages to serve as the manifold of a group; no other $SO(N)$ group splits into a product in this way. In general (for the classical groups) the invariant metric is uniquely, up to a constant, given by

$$ds^2 = -\frac{1}{2}\,\mathrm{Tr}(dg\,g^{-1}dg\,g^{-1}) = -\frac{1}{2}\mathrm{Tr}(g^{-1}dgg^{-1}dg) = \Theta_1^2 + \Theta_2^2 + \Theta_3^2. \tag{3.137}$$

In the second step we rewrote the metric in terms of the left invariant Maurer–Cartan form, so it is indeed invariant under $G \times G$. The square root of the determinant of this metric is the *Haar measure* on the group manifold, and will figure prominently in Chapter 15.

Entirely by the way, but not uninterestingly, we observe that we can insert a real parameter α into the 3-sphere metric:

$$ds^2 \;=\; \Theta_1^2 + \Theta_2^2 + \alpha\Theta_3^2. \tag{3.138}$$

For $\alpha = 1$ this is the metric on the group and for $\alpha = 0$ it is the metric on the 2-sphere since it does not depend on the fibre coordinate τ. Because it is made from right invariant forms it is an $SO(3)$ invariant metric on $\mathbf{S}^3$ for any intermediate value of α. It is said to be the metric on a *squashed 3-sphere*; 3-spheres that are squashed in this particular way are called *Berger spheres*.

The three everywhere-non-vanishing Killing vector fields in Eq. (3.131) do not commute and therefore it is impossible to introduce a coordinate system such that all three coordinate lines coincide with the Killing flow lines. But they are linearly independent, they form a perfectly good basis for the tangent space at every point, and any vector can be written as a linear combination of the J_I, where I runs from one to three. We can introduce a basis Θ_I in the dual cotangent space through

$$\Theta_I(J_J) \;=\; \delta_{IJ}. \tag{3.139}$$

The Θ_I are precisely the three forms given in Eq. (3.135). There is no globally defined function whose differential is our connection Θ_3 because we are no longer using a coordinate basis.

We do have a number of coordinate systems especially adapted to the group structure at our disposal. The coordinate lines will arise from exponentiating elements in the Lie algebra. First we try the *canonical parametrization*

$$g = e^{i(x\sigma_1 + y\sigma_2 + z\sigma_3)}. \tag{3.140}$$

If we switch to spherical polars after performing the exponentiation we obtain

$$g = \begin{bmatrix} X + iY & Z + iU \\ -Z + iU & X - iY \end{bmatrix} = \begin{bmatrix} \cos r + i\sin r \cos\theta & i\sin r \sin\theta e^{-i\phi} \\ i\sin r \sin\theta e^{i\phi} & \cos r - i\sin r \cos\theta \end{bmatrix}. \tag{3.141}$$

The metric when expressed in these coordinates becomes

$$ds^2 = dr^2 + \sin^2 r(d\theta^2 + \sin^2\theta \, d\phi^2). \tag{3.142}$$

So these are geodesic polar coordinates – the parameter r is the arc length along geodesics from the origin. The Euler angles from Section 3.5 is another choice; following in Euler's footsteps we write an arbitrary group element as

$$g = e^{\frac{i\tau}{2}\sigma_3} e^{\frac{i\theta}{2}\sigma_2} e^{\frac{i\phi}{2}\sigma_3} = \begin{bmatrix} e^{\frac{i}{2}(\tau+\phi)}\cos\frac{\theta}{2} & e^{\frac{i}{2}(\tau-\phi)}\sin\frac{\theta}{2} \\ -e^{-\frac{i}{2}(\tau-\phi)}\sin\frac{\theta}{2} & e^{-\frac{i}{2}(\tau+\phi)}\cos\frac{\theta}{2} \end{bmatrix}. \tag{3.143}$$

Equation (3.96) results. It seems fair to warn the reader that, although similar coordinate systems can be erected on all classical group manifolds, the actual calculations involved in expressing, say, the metric tensor in terms of coordinates tend to become very long, except for $SU(2)$.

3.8 Cosets and all that

We have already used the notion of *quotient spaces* such as M/H quite freely. The idea is that we start with a space M and a group H of transformations of M; the quotient space is then a space whose points consists of the *orbits* of H, that is to say that by definition a point in M/H is an equivalence class of points in M that can be transformed into each other by means of transformations belonging to H. We have offered no guarantees that the quotient space is a 'nice' space however. In general no such guarantee can be given. We are in a better position when the space M is a Lie group G and H is some subgroup of G, and we assume that H is a closed subset of G. By definition a *left coset* is a set of elements of a group G that can be written in the form gh, where h is any element of a fixed subgroup H. This gives a partition of the group into disjoint cosets, and the space of cosets is called the *coset space* and denoted by G/H. The points of the coset space are written schematically as gH; they are in fact the orbits of H in G under right action. For coset spaces there is a comfortable theorem saying that G/H is always a manifold of dimension

$$\dim(G/H) = \dim(G) - \dim(H). \tag{3.144}$$

Right cosets and right coset spaces can be defined in an analogous way.

A description of a space as a coset space often arises as follows. We start with a space M that is a *homogeneous space*, which means that there is a group G of isometries such that every point in M can be reached from any given point by means of a transformation belonging to G. Hence there is only one G orbit, and – since G is a symmetry group – the space looks the same at each point. The technical term here is that the group G acts *transitively* on M. Now fix a point p and consider the *isotropy* or *little group* at this point; by definition this is the subgroup $H \subset G$ of transformations h that leave p invariant. In this situation it follows that $M = G/H$. At first sight it may seem that this recipe depends on the choice of the point where the isotropy group was identified but in fact it does not; if we start from another point $p' = gp$ where $g \in G$ then the new little group H' is *conjugated* to H in the sense that $H' = gHg^{-1}$. The coset spaces G/H and G/H' are then identical, it is only the description that has changed a little.

A warning: the notation G/H is ambiguous unless it is specified which particular subgroup H (up to conjugation) that is meant. To see how an ambiguity can arise consider the group $SU(3)$. It has an abelian subgroup consisting of diagonal matrices

$$h = \begin{bmatrix} e^{i\alpha} & 0 & 0 \\ 0 & e^{i\beta} & 0 \\ 0 & 0 & e^{-i(\alpha+\beta)} \end{bmatrix}. \tag{3.145}$$

In the group manifold this is a two-dimensional surface known as the *Cartan torus*. Now consider the coset space $SU(3)/U(1)$, where the subgroup $U(1)$ forms a circle on the Cartan torus. But there are infinitely many different ways in which a circle can wind around on a 2-torus, and the coset space will get dramatically different properties depending on the choice one makes. On the other hand the space $SU(3)/U(1) \times U(1)$ suffers from no such ambiguities – any two–dimensional abelian subgroup of $SU(3)$ is related by conjugation to the subgroup of diagonal matrices.

It is interesting to ask again why the 2-sphere of Hopf fibres is naturally a round 2-sphere, this time adopting our new coset space view of things. We can write a point in the coset space as Ωh, where Ω is the *coset representative* and h is an element in H. The group $SU(2)/Z_2 = SO(3)$ acts naturally on this coset space through left action;

$$g \to g_1 g \quad \Rightarrow \quad \Omega h \to g_1 \Omega h. \tag{3.146}$$

It is therefore natural to select a coset space metric that is invariant under $SO(3)$, and for $\mathbf{S}^2$ the round metric is the answer. In general this construction gives rise also to a uniquely defined measure on the coset space, induced by the Haar measure on the group manifold.

If we take stock of the situation we see that we now have a rich supply of principal fibre bundles to play with, since any coset space G/H is the base manifold of a principal bundle with bundle space G and fibre H. This construction will occur again and again. To see how common it is we show that every sphere is a coset space. We start from an orthogonal matrix belonging to $SO(N)$. Its first column is a normalised vector in $\mathbb{R}^N$ and we can choose coordinates so that it takes the form $(1, 0, \ldots, 0)$. The group $SO(N)$ clearly acts transitively on the space of all normalised vectors (that is, on the sphere $\mathbf{S}^{N-1}$), and the isotropy group of the chosen point consists of all matrices of the form

$$h = \begin{bmatrix} 1 & 0 & \cdots & 0 \\ 0 & & & \\ \vdots & & SO(N-1) & \\ 0 & & & \end{bmatrix}. \tag{3.147}$$

It follows that

$$\mathbf{S}^{N-1} = SO(N)/SO(N-1) = O(N)/O(N-1), \tag{3.148}$$

with the subgroup chosen according to Eq. (3.147). The round metric on $\mathbf{S}^{N-1}$ arises naturally because it is invariant under the group $SO(N)$ that acts on the coset space. A similar argument in $\mathbb{C}^N$, where a column of a unitary matrix is in effect a normalised vector in $\mathbb{R}^{2N}$, shows that

$$\mathbf{S}^{2N-1} = SU(N)/SU(N-1) = U(N)/U(N-1). \tag{3.149}$$

And with this observation we finally know enough about spheres.

Problems

3.1　Derive the intrinsic metric of the sphere using stereographic projection. For the sake of variety, project from the south pole to the tangent plane at the north pole rather than to the equatorial plane.

3.2   A coordinate patch covering part of the sphere – a map from the sphere
       to the plane – can be obtained by projecting from an arbitrary point in
       space. Show geometrically (by drawing a figure) that the only conformal
       projection – a projection that preserves angles – is the stereographic
       projection from a point on the sphere.

3.3   On the complex plane, identify points connected by $x \rightarrow x + 1$ and also
       $y \rightarrow y + 1$. Do the same for $x \rightarrow x + 1$ and $y \rightarrow y + 2$. Show that the two
       quotient spaces are inequivalent as complex manifolds.

3.4   Show that the Poisson brackets obey the Jacobi identity because the
       symplectic form is closed.

3.5   Verify that the geodesics on S^3 meet the equator in two antipodal points.

3.6   The coordinates τ and ϕ run along Hopf circles, and they appear
       symmetrically in the metric (3.98). But $0 \leq \tau < 4\pi$ and $0 \leq \phi < 2\pi$.
       Why this difference?

3.7   Take sections of the Hopf bundle by choosing $\tau = -\phi$, $\tau = \phi$ and $\tau = \phi + \pi$ respectively, and work out what they look like in the stereographic
       picture.

3.8   Everybody knows how to make a Möbius strip by means of a piece of
       paper and some glue. Convince yourself that the Möbius strip is a fibre
       bundle and show that it admits a global section. Identify the group acting
       on its fibres and construct its principal bundle. Show that the principal
       bundle does not admit a global section, and hence that the bundle is non-
       trivial.

4

Complex projective spaces

In the house of mathematics there are many mansions and of these the most elegant is projective geometry.

—Morris Kline[1]

An attentive reader of Chapter 3 will have noticed that there must exist a Hopf fibration of any odd–dimensional sphere; it is just that we studiously avoided to mention what the resulting base space is. In this chapter it will be revealed that

$$\mathbb{C}\mathbf{P}^n = \mathbf{S}^{2n+1}/\mathbf{S}^1. \tag{4.1}$$

The space on the left-hand side is called *complex projective space* or – when it is used as a space of pure states in quantum mechanics – *projective Hilbert space*. It is not a sphere unless $n = 1$. The study of projective spaces was actually begun by artists during the Renaissance, and this is how we begin our story.

4.1 From art to mathematics

Painters, engaged in projecting the ground (conveniently approximated as a flat plane) onto a flat canvas, discovered that parallel lines on the ground intersect somewhere on a *vanishing line* or *line at infinity* that exists on the canvas but not on the ground, where parallel lines do not meet. Conversely, there exists a line on the ground (right below the artist) that does not appear on the canvas at all. When the subject was taken over by mathematicians this led to the theory of the *projective plane* or *real projective 2-space* $\mathbb{R}\mathbf{P}^2$. The idea is to consider a space whose points consists of the one-dimensional subspaces of a three-dimensional vector space $\mathbb{R}^3$. Call them *rays*. The origin is placed at the eye of the artist, and the rays are his lines

---

[1] Reproduced from Morris Kline. *Mathematics in Western Culture*. George Allen & Unwin, 1954.

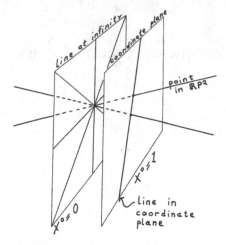

Figure 4.1 Affine coordinates for (almost all of) the projective plane.

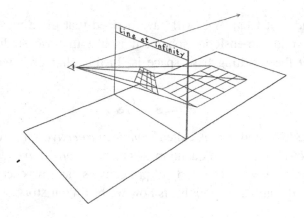

Figure 4.2 Renaissance painter, engaged in changing affine coordinates.

of sight. The ground and the canvas serve as two *affine coordinate planes* for $\mathbb{R}\mathbf{P}^2$. A *projective line* is a certain one-parameter family of points in the projective plane, and is defined as the space of rays belonging to some two-dimensional subspace $\mathbb{R}^2$. In itself, this is the real projective 1-space $\mathbb{R}\mathbf{P}^1$. To see how all this works out in formulae, we observe that any triple of real numbers defines a one-dimensional subspace of $\mathbb{R}^3$. But the latter does not determine the triple of numbers uniquely. Hence a point of $\mathbb{R}\mathbf{P}^2$ is given by an equivalence class

$$(X^0, X^1, X^2) \sim k(X^0, X^1, X^2) ; \qquad k \in \mathbb{R} \quad k \neq 0. \qquad (4.2)$$

The numbers X^α are known as *homogeneous coordinates* on $\mathbb{R}\mathbf{P}^2$. See Figures 4.1 and 4.2. If we want to use true coordinates, one possibility is to choose a plane in

$\mathbb{R}^3$ such as $X^0 = 1$, representing an infinite canvas, say. We think of this plane as an affine plane (a vector space except that the choice of origin and scalar product is left open). We can then label a ray with the coordinates of the point where it intersects the affine plane. This is an affine coordinate system on the projective plane; it does not cover the entire projective plane since the rays lying in the plane $X^0 = 0$ are missing. But this plane is a vector space $\mathbb{R}^2$, and the space of rays lying in it is the projective line $\mathbb{R}\mathbf{P}^1$. Hence the affine coordinates display the projective plane as an infinite plane to which we must add an extra projective line 'at infinity'.

Why did we define our rays as one-dimensional subspaces, and not as directed lines from the origin? The answer does not matter to the practising artist, but for the mathematician our definition results in simple *incidence properties* for points and lines. The projective plane will now obey the simple axioms that:

1. Any two different points lie on a unique line;
2. Any two different lines intersect in a unique point.

There are no exceptional cases (such as parallel lines) to worry about. That axiom (2) holds is simply the observation that two planes through the origin in $\mathbb{R}^3$ always intersect, in a unique line.

There is nothing special about the line at infinity; the name is just an artefact of a special choice of affine coordinates. Every projective line is the space of rays in some two-dimensional subspace of $\mathbb{R}^3$, and it has the topology of a circle – to its image in the affine plane we must add one point 'at infinity'.

It is interesting to observe that the space of all lines in $\mathbb{R}\mathbf{P}^2$ is another $\mathbb{R}\mathbf{P}^2$ since one can set up a one-to-one correspondence between rays and planes through the origin in $\mathbb{R}^3$. In fact there is a complete *duality* between the space of points and the space of lines in the projective plane. The natural way to select a specific one-to-one correspondence between $\mathbb{R}\mathbf{P}^2$ and the dual $\mathbb{R}\mathbf{P}^2$ is to introduce a metric in the underlying vector space, and to declare that a ray there is dual to that plane through the origin to which it is orthogonal. But once we think of the vector space as a metric space as well as a linear one we will be led to the round metric as the natural metric to use on $\mathbb{R}\mathbf{P}^2$. Indeed a moment's reflection shows that $\mathbb{R}\mathbf{P}^2$ can be thought of as the sphere $\mathbf{S}^2$ with antipodal points identified,

$$\mathbb{R}\mathbf{P}^2 = \mathbf{S}^2/Z_2. \tag{4.3}$$

See Figure 4.3. This is one of the reasons why the real projective plane is so famous: it is a space of constant curvature where every pair of geodesics intersect once (as opposed to twice, as happens on a sphere). In fact all of Euclid's axioms for points and straight lines are valid, except the fifth, which shows that the parallel axiom is independent of the others.

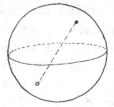

Figure 4.3 Real projective space is the sphere with antipodal points identified.

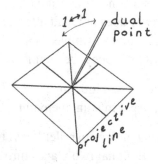

Figure 4.4 Polarity: a metric for the projective plane.

In general a specific one-to-one map between a projective space and its dual is known as a *polarity*, and choosing a polarity singles out a particular metric on the projective space as being the most natural one. See Figure 4.4.

There are other geometrical figures apart from points and lines. A famous example is that of the *conic section*. Consider a (circular, say) cone of rays with its vertex at the origin of $\mathbb{R}^3$. In one affine coordinate system the points of intersection of the cone with the affine plane will appear as a circle, in another as a hyperbola. Hence circles, ellipses and hyperbolae are projectively equivalent. In homogeneous coordinates such a conic section appears as

$$(X^1)^2 + (X^2)^2 = (X^0)^2. \tag{4.4}$$

This equation defines a submanifold of $\mathbb{R}P^2$ because it is homogeneous in the X^α, and therefore unaffected by scaling. If we use affine coordinates for which $X^0 = 1$ it appears as a circle, but if we use affine coordinates for which $X^1 = 1$ it appears as a hyperbola. We can use this fact to give us some further insight into the somewhat difficult topology of $\mathbb{R}P^2$: if we draw our space as a disc with antipodal points on the boundary identified – this would be the picture of the topology implied by Figure 4.3 – then we see that the hyperbola becomes a topological circle because

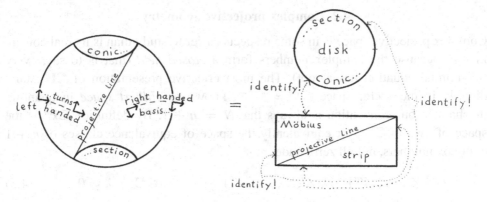

Figure 4.5 The topology of the real projective plane. For a different view, see Figure 4.12.

we are adding two points at infinity. Its interior is an ordinary disc while its exterior is a *Möbius strip* because of the identifications at the boundary – if you glue a disk to the boundary of a Möbius strip you obtain a projective plane. By the way $\mathbb{RP}^2$, unlike $\mathbb{RP}^1$, is a *non-orientable* space; this means that a 'right-handed' basis in the tangent space can be turned into a 'left-handed' one by moving it around so that in fact the distinction between right and left handedness cannot be upheld. See Figure 4.5.

If you think a little bit further about the $\mathbb{RP}^2$ topology you will also notice a – possibly reassuring – fact, namely that there is a topological difference between projective lines and conic sections. Although they are both circles intrinsically, the conic sections have the property that they can be continually deformed to a point within the projective plane, while the lines cannot. Indeed if you cut the projective plane open along a projective line it remains connected, while it splits into two pieces if you cut along a conic section.

We could go on to discuss n-dimensional real projective spaces $\mathbb{RP}^n$, but we confine ourselves to two topological remarks: that $\mathbb{RP}^n = S^n/Z_2$ (obvious), and that $\mathbb{RP}^{2n+1}$ is orientable while $\mathbb{RP}^{2n}$ is not (easy to prove).

We can change our *field* from the real numbers $\mathbb{R}$ to rational, complex or quaternionic numbers. In fact one can also use finite fields, such as the integers modulo p, where p is some prime number. If we repeat the construction of the projective plane starting from a three-dimensional vector space over the field of integers modulo p then we will obtain a projective plane with exactly $p + 1$ points on each line. In Chapter 12 we will try to convince the reader that this is an interesting construction, but like real and quaternionic projective spaces it is of only secondary importance in quantum mechanics, where complex projective space occupies the centre of the stage.

4.2 Complex projective geometry

Complex projective space is in some respects easier to study than is its real cousin, mainly because the complex numbers form a *closed field* (that is to say, every polynomial equation has a root). The most effective presentation of $\mathbb{CP}^n$ starts directly in the vector space $\mathbb{C}^N = \mathbb{C}^{n+1}$, known as *Hilbert space* in quantum mechanics; our convention is always that $N = n + 1$. By definition $\mathbb{CP}^n$ is the space of rays in $\mathbb{C}^{n+1}$, or equivalently the space of equivalence classes of $n + 1$ complex numbers, not all zero, under

$$(Z^0, Z^1, \ldots, Z^n) \sim \lambda(Z^0, Z^1, \ldots, Z^n); \qquad \lambda \in \mathbb{C} \quad \lambda \neq 0. \tag{4.5}$$

We will use Greek indices to label the homogeneous coordinates. We can cover $\mathbb{CP}^n$ with affine coordinate patches like

$$Z^\alpha = (1, z^1, \ldots, z^n), \qquad Z^\alpha = (z^{n'}, 1, z^{1'}, \ldots, z^{(n-1)'}) \tag{4.6}$$

and so on. This leads to two important observations. The first is that one coordinate patch covers all of $\mathbb{CP}^n$ except that subset which has (say)

$$Z^\alpha = (0, Z^1, \ldots, Z^n). \tag{4.7}$$

But the set of all such rays is a $\mathbb{CP}^{n-1}$. Hence we conclude that topologically $\mathbb{CP}^n$ is like $\mathbb{C}^n$ with a $\mathbb{CP}^{n-1}$ attached 'at infinity'. Iterating this observation we obtain the *cell decomposition*

$$\mathbb{CP}^n = \mathbb{C}^n \cup \mathbb{C}^{n-1} \cup \cdots \cup \mathbb{C}^0. \tag{4.8}$$

Evidently $\mathbb{CP}^0$ is topologically a point and $\mathbb{CP}^1$ is a 2-sphere, while for $n > 1$ we get something new that we will have to get used to as we proceed. The second observation is that in a region where the coordinate systems overlap, they are related by (say)

$$z^{a'} = \frac{Z^{a+1}}{Z^1} = \frac{Z^0}{Z^1} \frac{Z^{a+1}}{Z^0} = \frac{z^{a+1}}{z^1}, \tag{4.9}$$

which is clearly an analytic function on the overlap region. Hence $\mathbb{CP}^n$ is a complex manifold.

The *linear subspaces* of $\mathbb{CP}^n$ are of major importance. They are defined as the images of the subspaces of $\mathbb{C}^{n+1}$ under the natural map from the vector space to the projective space. Equivalently they are given by a suitable number of linear equations in the homogeneous coordinates. Thus the *hyperplanes* are $n - 1$ dimensional submanifolds of $\mathbb{CP}^n$ defined by the equation

$$P_\alpha Z^\alpha = 0 \tag{4.10}$$

for some fixed set of $n + 1$ complex numbers P_α. This definition is unaffected by a change of scale for the homogeneous coordinates, and also by a change of scale in P_α. Hence the $n+1$ complex numbers P_α themselves are homogeneous coordinates for a $\mathbb{CP}^n$; in other words the hyperplanes in a projective n-space can be regarded as the points of another projective n-space which is dual to the original.

If we impose a set of $n - k$ independent linear equations on the Z^α we obtain a linear subspace of complex dimension k, also known as an k–*plane*, which is itself a $\mathbb{CP}^k$. Geometrically this is the intersection of m hyperplanes. The space whose points consist of all such k–planes is known as a *Grassmannian* – it is only in the case of hyperplanes (and the trivial case of points) that the Grassmannian is itself a $\mathbb{CP}^n$. A linear subspace of complex dimension one is known as a *complex projective line*, a linear subspace of dimension two is a complex projective plane and so on, while $\mathbb{CP}^0$ is just a point. The complex projective line is a $\mathbb{CP}^1$ – topologically this is a sphere and this may boggle some minds, but it is a line in the sense that one can draw a unique line between any pair of points (this is essentially the statement that two vectors in $\mathbb{C}^{n+1}$ determine a unique two-dimensional subspace). It also behaves like a line in the sense that two projective lines in a projective space intersect in a unique point if they intersect at all (this is the statement that a pair of two-dimensional subspaces are either disjoint or they share one common ray or they coincide). In general the intersection of two linear subspaces A and B is known as their *meet* $A \cap B$. We can also define their *join* $A \cup B$ by taking the linear span of the two subspaces of the underlying vector space to which A and B correspond, and then go back down to the projective space. These two operations – meet and join – turn the set of linear subspaces into a partially ordered structure known as a *lattice*, in which every pair of elements has a greatest lower bound (the meet) and a least upper bound (the join). This is the starting point of the subject known as quantum logic [497, 916]. It is also the second time we encounter a lattice – in Section 1.1 we came across the lattice of faces of a convex body. In Chapter 8 we will find a convex body whose lattice of faces agrees with the lattice of subspaces of a vector space. By then it will be a more interesting lattice, because we will have an inner product on $\mathbb{C}^N$, so that a given subspace can be associated with its orthogonal complement.

But we have not yet introduced any inner product in $\mathbb{C}^N$, or any metric in $\mathbb{CP}^n$. In fact we will continue to do without it for some time; projective geometry is precisely that part of geometry that can be done without a metric. There is an interesting group theoretical view of this, originated by Felix Klein. All statements about linear subspaces – such as when two linear subspaces intersect – are invariant under general linear transformations of $\mathbb{C}^N$. They form the group $GL(N, \mathbb{C})$, but only a subgroup acts *effectively* on $\mathbb{CP}^n$. (Recall that $N = n + 1$. A transformation is said to act effectively on some space if it moves at least one point.)

Changing the matrix with an overall complex factor does not change the transformation effected on $\mathbb{C}\mathbf{P}^n$, so we can set its determinant equal to one and multiply it with an extra complex Nth root of unity if we wish. The *projective group* is therefore $SL(N, \mathbb{C})/Z_N$.

According to Klein's conception, projective geometry is completely characterised by the projective group; its subgroups include the group of affine transformations that preserves the $\mathbb{C}\mathbf{P}^{n-1}$ at infinity and this subgroup characterises affine geometry. A helpful fact about the projective group is that any set of $n + 2$ points can be brought to the standard position $(1, 0, \ldots, 0), \ldots, (0, \ldots 0, 1)$, $(1, 1, \ldots, 1)$ by means of a projective transformation. For $\mathbb{C}\mathbf{P}^1$ this is the familiar statement that any triple of points on the complex plane can be transformed to $0, 1$ and ∞ by a Möbius transformation.

4.3 Complex curves, quadrics and the Segre embedding

The equation that defines a subspace of $\mathbb{C}\mathbf{P}^n$ does not have to be linear; any homogeneous equation in the homogeneous coordinates Z^α gives rise to a well defined submanifold. Hence in addition to the linear subspaces we have *quadrics, cubics, quartics* and so on, depending on the degree of the defining polynomial. The locus of a number of homogeneous equations

$$w_1(Z) = w_2(Z) = \ldots = w_m(Z) = 0 \tag{4.11}$$

is also a subspace known as an *algebraic* (or projective) *variety*, and we move from projective to *algebraic* geometry.[2] *Chow's theorem* states that every non-singular algebraic variety is a complex submanifold of $\mathbb{C}\mathbf{P}^n$, and conversely every compact complex submanifold is the locus of a set of homogeneous equations. On the other hand it is not true that every complex manifold can be embedded as a complex submanifold in $\mathbb{C}\mathbf{P}^n$.

There are two kinds of submanifolds that are of immediate interest in quantum mechanics. One of them is the *complex curve*; by definition this is a map of $\mathbb{C}\mathbf{P}^1$ into $\mathbb{C}\mathbf{P}^n$. In real terms it is a 2-surface in a $2n$-dimensional space. This does not sound much like a curve, but once it is accepted that $\mathbb{C}\mathbf{P}^1$ deserves the name of line it will be admitted that the name curve is reasonable here. Let us first choose $n = 2$, so that we are looking for a complex curve in the projective plane. Using (u, v) as homogeneous coordinates on $\mathbb{C}\mathbf{P}^1$ we clearly get a map into $\mathbb{C}\mathbf{P}^2$ if we set

$$(u, v) \quad \rightarrow \quad (u^2, uv, v^2). \tag{4.12}$$

---

[2] A standard reference, stressing the setting of complex manifolds, is the book by Griffiths and Harris [365]. For an alternative view see the book by Harris alone [407].

(In Section 6.4 we will adjust conventions a little.) This is a well defined map because the expression is homogeneous in the u and v. Evidently the resulting complex curve in $\mathbb{CP}^2$ obeys the equation

$$Z^0 Z^2 = Z^1 Z^1. \tag{4.13}$$

Hence it is a quadric hypersurface as well, and indeed any quadric can be brought to this form by projective transformations. In the projective plane a quadric is also known as a conic section.

In higher dimensions we encounter complex curves that are increasingly difficult to grasp since they are not quadrics. The next simplest case is the *twisted cubic* curve in $\mathbb{CP}^3$, defined by

$$(u, v) \quad \rightarrow \quad (u^3, u^2 v, uv^2, v^3). \tag{4.14}$$

We leave it aside for the moment though.

A class of submanifolds that is of particular interest in quantum mechanics arises in the following way. Suppose that the complex vector space is given as a *tensor product*

$$\mathbb{C}^{n+1} \otimes \mathbb{C}^{m+1} = \mathbb{C}^{(n+1)(m+1)}. \tag{4.15}$$

Then there should be an embedded submanifold

$$\Sigma_{n,m} = \mathbb{CP}^n \times \mathbb{CP}^m \subset \mathbb{CP}^{nm+n+m}. \tag{4.16}$$

Indeed this is true. In terms of homogeneous coordinates the submanifold can be parametrized as

$$Z^\alpha = Z^{\mu\mu'} = P^\mu Q^{\mu'}, \tag{4.17}$$

in a fairly obvious notation – the P^μ and $Q^{\mu'}$ are homogeneous coordinates on $\mathbb{CP}^n$ and $\mathbb{CP}^m$ respectively. The construction is known as the *Segre embedding*. The submanifold $\Sigma_{n,m}$ is known as the *Segre variety*. It is a Cartesian product of (complex) dimension $n + m$, and it follows from Chow's theorem that it can be defined as the locus of nm homogeneous equations in the large space. Indeed it is easy to see from the definition that the Segre variety is the intersection of the quadrics

$$Z^{\mu\mu'} Z^{\nu\nu'} - Z^{\mu\nu'} Z^{\nu\mu'} = 0. \tag{4.18}$$

The number of independent equations here is nm, as one can see by noting that the condition is that all the minors of a certain $(n + 1) \times (m + 1)$ matrix vanish. In quantum mechanics – and in Section 16.2 – the Segre variety reappears as the set of separable states of a composite system.

Let us consider the simplest case

$$\Sigma_{1,1} = \mathbb{CP}^1 \times \mathbb{CP}^1 \subset \mathbb{CP}^3. \tag{4.19}$$

Then only one equation is needed. Write

$$(Z^0, Z^1, Z^2, Z^3) = (Z^{00'}, Z^{01'}, Z^{10'}, Z^{11'}). \tag{4.20}$$

Then the submanifold $\mathbb{CP}^1 \times \mathbb{CP}^1$ is obtained as the quadric surface

$$Z^0 Z^3 - Z^1 Z^2 = 0. \tag{4.21}$$

The non-degenerate quadrics in $\mathbb{CP}^3$ are in one-to-one correspondence to all possible embeddings of $\mathbb{CP}^1 \times \mathbb{CP}^1$. It is interesting to consider the projection map from the product manifold to one of its factors. This means that we hold $Q^{\mu'}$ (say) fixed and vary P^μ. Then the fibre of the map – the set of points on the quadric that are projected to the point on $\mathbb{CP}^m$ that is characterised by that particular $Q^{\mu'}$ – will be given by the equations

$$\frac{Z^{00'}}{Z^{01'}} = \frac{P^0 Q^{0'}}{P^0 Q^{1'}} = \frac{Q^{0'}}{Q^{1'}} = \frac{P^1 Q^{0'}}{P^1 Q^{1'}} = \frac{Z^{10'}}{Z^{11'}}. \tag{4.22}$$

Since $Q^{\mu'}$ is fixed this implies that there is a complex number λ such that

$$Z^0 = \lambda Z^1 \qquad \text{and} \qquad Z^2 = \lambda Z^3. \tag{4.23}$$

This pair of linear equations defines a projective line in $\mathbb{CP}^3$. Projecting down to the other factor leads to a similar conclusion. In this way we see that the quadric is ruled by lines, or in other words that through any point on the quadric there goes a pair of straight lines lying entirely in the quadric.

For visualisation, let us consider the real Segre embedding

$$\mathbb{RP}^1 \times \mathbb{RP}^1 \subset \mathbb{RP}^3. \tag{4.24}$$

This time we choose to diagonalise the quadric; define

$$Z^0 = X + U, \quad Z^1 = X - U, \quad Z^2 = V + Y, \quad Z^3 = V - Y. \tag{4.25}$$

We then obtain

$$Z^0 Z^3 - Z^1 Z^2 = X^2 + Y^2 - U^2 - V^2 = 0. \tag{4.26}$$

Now let us choose affine coordinates by dividing through with V. The quadric becomes

$$x^2 + y^2 - u^2 = 1. \tag{4.27}$$

This is a hyperboloid of one sheet sitting in a three-dimensional real space. The fact that such a surface can be *ruled* by straight lines is a surprising fact of elementary geometry (first noted by Sir Christopher Wren).

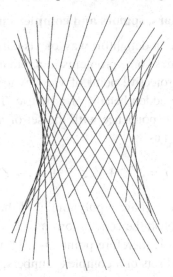

Figure 4.6 The real Segre embedding: a hyperboloid ruled by straight lines.

Algebraic geometers are very interested in how algebraic varieties intersect. Looking at Figure 4.6 we see that lines in $\mathbb{R}\mathbf{P}^3$ intersect the Segre variety either twice, once (if they are tangent to it), or not at all – and then there are exceptional cases where the entire line lies in the Segre variety. It turns out that the situation is easier to describe in $\mathbb{C}\mathbf{P}^3$. Then a complex projective line will always intersect $\Sigma_{1,1}$ in two points, except for the exceptional cases of lines that are tangent to the variety (which in a way is no exception: the two intersection points coincide), or when the line lies in the variety. This works in higher dimensions too. Having computed the 'degree of the Segre variety' one proves that a typical linear $(nm - n - m)$–plane in $\mathbb{C}\mathbf{P}^{nm+n+m}$ intersects $\Sigma_{n,m}$ in exactly $(n + m)!\,/n!\,m!$ points [407]. If the dimension of the plane is larger than this 'critical' dimension then its intersection with the Segre manifold will be an interesting submanifold in itself; thus a linear 3-plane in $\mathbb{C}\mathbf{P}^5$ intersects $\Sigma_{2,1}$ in a twisted cubic curve. If the dimension is smaller there is typically no intersection point at all.

In Chapter 16, where we will be seriously interested in this kind of thing, we will think in terms of linear subspaces of $\mathbb{C}^{NM}$. Then the statement is that a typical linear subspace $\mathbb{C}^{NM-N-M+2}$ of $\mathbb{C}^{NM}$ contains exactly

$$\binom{N + M - 2}{N - 1} = \frac{(N + M - 2)!}{(N - 1)!\,(M - 1)!} \tag{4.28}$$

product vectors of the form (4.17). A typical subspace of dimension $NM-N-M+1$ or less contains no product vectors at all.

4.4 Stars, spinors and complex curves

The *stellar representation* is a delightful way of visualizing $\mathbb{C}\mathbf{P}^n$ in real terms. It works for any n and simplifies some problems in a remarkable way.[3] Here we will develop it a little from a projective point of view, while Chapter 7 discusses the same construction with the added benefit of a metric. The idea is that vectors in $\mathbb{C}^{n+1}$ are in one-to-one correspondence with the set of nth degree polynomials in one complex variable z, such as

$$w(z) = Z^0 z^n + Z^1 z^{n-1} + \ldots + Z^n. \tag{4.29}$$

(The important point is that we have a polynomial of the nth degree; in Chapter 7 we will have occasion to polish the conventions a little.) If $Z^0 \neq 0$ we can rescale the vector Z^α so that $Z^0 = 1$; therefore points in $\mathbb{C}\mathbf{P}^n$ will be in one-to-one correspondence with unordered sets of n complex numbers, namely with the complex roots of the equation

$$Z^0 z^n + Z^1 z^{n-1} + \ldots + Z^{n+1} = 0 = Z^0 (z - z_1)(z - z_2) \cdots (z - z_n). \tag{4.30}$$

Multiple roots are allowed. If $Z^0 = 0$ then infinity counts as a root (of multiplicity m if $Z^1 = \ldots = Z^{m-1} = 0$). Finally, by means of a stereographic projection the roots can be represented as unordered sets of n points on an ordinary 2-sphere – the points are called 'stars' and thus we have arrived at the stellar representation, in which points in $\mathbb{C}\mathbf{P}^n$ are represented by n unordered stars on a 'celestial' sphere. As a mathematical aside it follows that $\mathbb{C}\mathbf{P}^1 = \mathbf{S}^2$, $\mathbb{C}\mathbf{P}^2 = \mathbf{S}^2 \times \mathbf{S}^2 / S_2$ and in general that $\mathbb{C}\mathbf{P}^n = \mathbf{S}^2 \times \mathbf{S}^2 \times \cdots \times \mathbf{S}^2 / S_n$, where S_n is the *symmetric group* of permutations of n objects. See Figure 4.7.

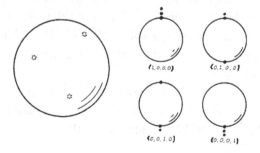

Figure 4.7 Any point in $\mathbb{C}\mathbf{P}^3$ may be represented by three stars on the sphere. Some special cases, where the stars coincide, are shown on the right.

---

[3] The first reference appears to be Majorana (1932) [639]. The result has been rediscovered many times within [80] and without [719] the context of quantum mechanics.

There is a piece of notation that is conveniently introduced at this juncture. Let us denote the homogeneous coordinates on $\mathbb{CP}^1$ by

$$\zeta^A = (\zeta^0, \zeta^1) = (u, v). \tag{4.31}$$

So we use a capital Latin letter for the index and we will refer to ζ^A as a *spinor*. The overall scale of the spinor is irrelevant to us, so we can introduce an affine coordinate z by

$$z = \frac{v}{u} : \quad \zeta^A \sim (1, z). \tag{4.32}$$

A spinor for which $u = 0$ then corresponds to the south pole on the Riemann sphere. We will use the totally anti-symmetric tensor $\epsilon_{AB} = -\epsilon_{BA}$ (the symplectic structure on $\mathbf{S}^2$, in fact) to raise and lower indices according to

$$\zeta_A \equiv \zeta^B \epsilon_{BA} \quad \Leftrightarrow \quad \zeta^A = \epsilon^{AB} \zeta_B. \tag{4.33}$$

Due to the fact that ϵ_{AB} is anti-symmetric there is a definite risk that sign errors will occur when one uses its inverse ϵ^{AB}. Just stick to the convention just made and all will be well. Note that $\zeta_A \zeta^A = 0$. So far, then, a spinor simply denotes a vector in $\mathbb{C}^2$. We do not think of this as a Hilbert space yet because there is no inner product. On the contrary the formalism is manifestly invariant under the full projective group $SL(2, \mathbb{C})/Z_2$. A special linear transformation on $\mathbb{C}^2$ gives rise to a *Möbius transformation* on the sphere;

$$\zeta^A \to \begin{bmatrix} u' \\ v' \end{bmatrix} = \begin{bmatrix} \alpha & \beta \\ \gamma & \delta \end{bmatrix} \begin{bmatrix} u \\ v \end{bmatrix} \quad \Rightarrow \quad z \to z' = \frac{\alpha z + \beta}{\gamma z + \delta}, \tag{4.34}$$

where $\alpha\delta - \beta\gamma = 1$ is the condition (on four complex numbers) that guarantees that the matrix belongs to $SL(2, \mathbb{C})$ and that the transformation preserves ϵ_{AB}. The group of Möbius transformations is exactly the group of projective transformations of $\mathbb{CP}^1$.

We can go on to consider totally symmetric *multispinors*

$$\Psi^{AB} = \Psi^{(AB)}, \qquad \Psi^{ABC} = \Psi^{(ABC)} \tag{4.35}$$

and so on. (The brackets around the indices mean that we are taking the totally symmetric part.) It is easy to see that when the index ranges from zero to one then the number of independent components of a totally symmetric multispinor carrying n indices is $n + 1$, just right so that the multispinor can serve as homogeneous coordinates for $\mathbb{CP}^n$. To see that this works, consider the equation

$$\Psi_{AB \dots M} \zeta^A \zeta^B \dots \zeta^M = 0. \tag{4.36}$$

If we now choose the scale so that $\zeta^A = (1, z)$ then the above equation turns into an nth degree polynomial in z, having n complex roots. We can use this fact together with Eq. (4.30) to translate between the $\Psi^{AB \cdots M}$ and the Z^α, should this be needed. We can also use the fundamental theorem of algebra to rewrite the polynomial as a product of n factors. In analogy with Eq. (4.30) we find that

$$\Psi_{AB \ldots M} \zeta^A \zeta^B \cdots \zeta^M = 0 = (\alpha_0 + \alpha_1 z)(\beta_0 + \beta_1 z) \cdots (\mu_0 + \mu_1 z). \quad (4.37)$$

The conclusion is that a multispinor with n indices can be written – uniquely except for an overall factor of no interest to us – as a symmetrised product of n spinors:

$$\Psi_{AB \ldots M} \zeta^A \zeta^B \cdots \zeta^M = \alpha_A \zeta^A \beta_B \zeta^B \cdots \mu_M \zeta^M$$

$$\Rightarrow \qquad\qquad\qquad\qquad\qquad\qquad\qquad (4.38)$$

$$\Psi_{AB \ldots M} = \alpha_{(A} \beta_B \cdots \mu_{M)}.$$

The factors are known as *principal spinors*, and they are of course the n unordered points (note the symmetrisation) on the sphere in slight disguise.

The stellar representation, or equivalently the spinor notation, deals with linear subspaces in an elegant way. Consider a complex projective line (a $\mathbb{CP}^1$) in $\mathbb{CP}^2$ for definiteness. A general point in $\mathbb{CP}^2$ is described by a pair of unordered points on the 2-sphere, or equivalently as a spinor

$$\Psi^{AB} = \alpha^{(A} \beta^{B)} \equiv \frac{1}{2} \left(\alpha^A \beta^B + \alpha^B \beta^A \right). \quad (4.39)$$

Evidently we get a complex projective line by holding one of the points fixed and letting the other vary, that is by holding one of the principal spinors (say β^A) fixed and letting the other vary.

The spinor notation also deals elegantly with complex curves in general. Thus we get a conic section in $\mathbb{CP}^2$ as the set of points for which the principal spinors coincide. That is to say that

$$\Psi^{AB} = \Psi^A \Psi^B \quad (4.40)$$

for some spinor Ψ^A. Through any point on the quadric (for which $\Psi^A = \alpha^A$ say) there goes a complex projective line

$$\Psi^{AB} = \zeta^{(A} \alpha^{B)} \quad (4.41)$$

(where ζ^A varies). This line is *tangent* to the quadric since it touches the quadric in a unique point $\alpha^A \alpha^B$. It is moreover rather easy to see that a pair of tangent lines always intersect in a unique point. See Figure 4.8.[4]

---

[4] For further explanations, see Brody and Hughston [177] and Penrose and Rindler [722].

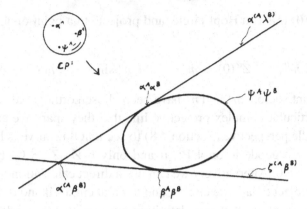

Figure 4.8 A conic section in $\mathbb{CP}^2$ and a pair of tangent lines.

4.5 The Fubini–Study metric

Now we want a notion of distance, and indeed a Riemannian metric, in $\mathbb{CP}^n$. How do we do it? Pick an arbitrary pair of points. The distance between them will be the length of the geodesic curve connecting them. On the other hand we know that there is a unique projective line connecting them; topologically this is $\mathbb{CP}^1 = \mathbf{S}^2$. Suppose that we insist that this 2-sphere is a round 2-sphere, with the ordinary round metric. Let us also insist that the metric on $\mathbb{CP}^n$ is such that the projective lines are *totally geodesic*. Technically a submanifold is said to be totally geodesic if a geodesic with respect to the induced metric on the submanifold is also a geodesic with respect to the metric on the embedding space, or equivalently if a geodesic that starts out parallel to the submanifold stays in the submanifold. But a geodesic on a complex projective line is simply a great circle on a 2-sphere, so we have found a way to define geodesics between two arbitrary points in $\mathbb{CP}^n$. The resulting notion of distance is called the *Fubini–Study distance*.[5]

It only remains to make this definition of the metric on $\mathbb{CP}^n$ explicit. Since the geodesic lives on some complex projective line, we can write down its equation using nothing but the results of Chapter 3. Let us recall Eq. (3.85) for a geodesic on an odd-dimensional sphere:

$$Z^\alpha(\sigma) = Z^\alpha(0) \cos \sigma + \dot{Z}^\alpha(0) \sin \sigma, \tag{4.42}$$

where

$$Z(0) \cdot \bar{Z}(0) = \dot{Z}(0) \cdot \dot{\bar{Z}}(0) = 1, \qquad Z(0) \cdot \dot{\bar{Z}}(0) + \dot{Z}(0) \cdot \bar{Z}(0) = 0. \tag{4.43}$$

---

[5] It was first studied by two leading geometers from a century ago: Fubini (1903) [318] and Study (1905) [865].

If $\dot{Z}^\alpha(0) = iZ^\alpha(0)$ this is a Hopf circle, and projects to a point on $\mathbb{CP}^1$. In general we can write

$$Z^\alpha(0) \equiv m^\alpha, \qquad \dot{Z}^\alpha(0) = m^\alpha \cos a + n^\alpha \sin a, \qquad m \cdot \bar{n} = 0. \qquad (4.44)$$

Since the constant vectors m^α and n^α have been chosen orthogonal, they lie antipodally on the particular complex projective line that they span. We can now appeal to the fibre bundle perspective (Section 4.8) to see that this curve is horizontal, and hence projects to a geodesic on $\mathbb{CP}^1$, if and only if $\dot{Z} \cdot \bar{Z} = 0$, that is to say if and only if $\cos a = 0$. Alternatively we can use a direct calculation to show that for general a they project to latitude circles, and to great circles if and only if $\cos a = 0$. Either way, we conclude that our definition implies that a geodesic on $\mathbb{CP}^n$ is given by

$$Z^\alpha(\sigma) = m^\alpha \cos \sigma + n^\alpha \sin \sigma, \qquad m \cdot \bar{m} = n \cdot \bar{n} = 1, \; m \cdot \bar{n} = 0. \qquad (4.45)$$

More precisely, this is the horizontal lift of a geodesic on $\mathbb{CP}^n$ to the odd-dimensional sphere $\mathbf{S}^{2n+1}$, in the fibre bundle $\mathbb{CP}^n = \mathbf{S}^{2n+1}/\mathbf{S}^1$.

We will now use this set-up to define the Fubini–Study distance D_{FS} between any two points on $\mathbb{CP}^n$. Since everything takes place within a complex projective line, which lifts to a 3-sphere, we can use the expression for the distance arrived at in our discussion of the Hopf fibration of $\mathbf{S}^3$, namely Eq. (3.95). The Fubini–Study distance D_{FS} must be given by

$$\cos^2 D_{\mathrm{FS}} = \kappa, \qquad (4.46)$$

where the *projective cross ratio* between two points P^α and Q^α is given by

$$\kappa = \frac{P \cdot \bar{Q} \, Q \cdot \bar{P}}{P \cdot \bar{P} \, Q \cdot \bar{Q}}. \qquad (4.47)$$

We give the same formula in standard quantum mechanical notation in Eq. (5.17). There is one new feature: in Section 3.5 we assumed that $P \cdot \bar{P} = Q \cdot \bar{Q} = 1$, but this assumption has now been dropped. As in Section 4.1 there is a polarity – that is to say a map from the space of points to the space of hyperplanes – hidden here: the polarity maps the point P^α to that special hyperplane consisting of all points Q^α with

$$Q \cdot \bar{P} \equiv Q^\alpha \bar{P}_\alpha = 0. \qquad (4.48)$$

Interestingly complex conjugation appears here for the first time. Another interesting fact is that our definitions imply that $\mathbb{CP}^n$ is a C_π manifold: all geodesics are closed and have the same circumference. This is a quite exceptional property, related to the fact that the isotropy group – the subgroup of isometries leaving a given point invariant – is so large that it acts transitively on the space of directions there. Spheres have this property too. The circumference of a geodesic on $\mathbb{C}\mathbf{P}^n$ equals π because $\mathbb{C}\mathbf{P}^1$ is a 2-sphere of radius $1/2$; the geodesic on $\mathbf{S}^{2n+1}$ has circumference 2π and doubly covers the geodesic on $\mathbb{C}\mathbf{P}^n$. We also note that the maximal distance between two points is equal to $\pi/2$ (when the two vectors are orthogonal), and that all geodesics meet again at a point of maximal distance from their point of origin.

To obtain the line element in a local form we assume that

$$Q^\alpha = P^\alpha + \mathrm{d}P^\alpha \tag{4.49}$$

and expand to second order in the vector $\mathrm{d}P^\alpha$ (and to second order in D_{FS}). The result is the *Fubini–Study metric*

$$\mathrm{d}s^2 = \frac{\mathrm{d}P \cdot \mathrm{d}\bar{P}\, P \cdot \bar{P} - \mathrm{d}P \cdot \bar{P}\, P \cdot \mathrm{d}\bar{P}}{P \cdot \bar{P}\, P \cdot \bar{P}}. \tag{4.50}$$

Before we are done we will become very familiar with this expression.

It is already clear that many formulae will be simplified if we normalise our vectors so that $P \cdot \bar{P} = 1$. In quantum mechanics – where the vectors are the usual Hilbert space state vectors – this is usually done. In this chapter we will use normalised vectors now and then in calculations, but on festive occasions – say when stating a definition – we do not.

An important definition follows immediately, namely that of the 2-form

$$\Omega = i\,\frac{P \cdot \bar{P}\, \mathrm{d}P \cdot \wedge \mathrm{d}\bar{P} - \mathrm{d}P \cdot \bar{P} \wedge P \cdot \mathrm{d}\bar{P}}{P \cdot \bar{P}\, P \cdot \bar{P}}. \tag{4.51}$$

This is clearly a relative of the metric and the suggestion is that it is a (closed) symplectic form and that $\mathbb{C}\mathbf{P}^n$ is a Kähler manifold for all n. And this is true. To prove the Kähler property it is helpful to use the affine coordinates from Section 4.2. When expressed in terms of them the Fubini–Study metric becomes

$$\mathrm{d}s^2 = \frac{1}{1+|z|^2}\left(\mathrm{d}z^a\mathrm{d}\bar{z}_a - \frac{\bar{z}_a\mathrm{d}z^a\mathrm{d}\bar{z}_b z^b}{1+|z|^2}\right) \equiv 2g_{a\bar{b}}\mathrm{d}z^a\mathrm{d}\bar{z}^{\bar{b}}, \qquad |z|^2 \equiv z^a\bar{z}_a. \tag{4.52}$$

(We are using $\delta_{a\bar{b}}$ to change $\bar{z}^{\bar{a}}$ into $\bar{z}_a$.) Now

$$2g_{a\bar{b}} = \frac{1}{1+|z|^2}\left(\delta_{a\bar{b}} - \frac{\bar{z}_a z_{\bar{b}}}{1+|z|^2}\right) = \partial_a\partial_{\bar{b}}\ln\left(1+|z|^2\right). \tag{4.53}$$

Or more elegantly

$$g_{a\bar{b}} = \frac{1}{2}\partial_a\partial_{\bar{b}}\ln Z \cdot \bar{Z}, \tag{4.54}$$

where it is understood that the homogeneous coordinates should be expressed in terms of the affine ones. As we know from Section 3.3 this proves that $\mathbb{CP}^n$ is a Kähler manifold. Every complex submanifold is Kähler too. For $n = 1$ we recognise the metric on $\mathbf{S}^2$ written in stereographic coordinates.

We will use a different choice of coordinate system to explore the space in detail, but the affine coordinates are useful for several purposes, for instance if one wants to perform an explicit check that the curve (4.45) is indeed a geodesic, or when one wants to compute the Riemann curvature tensor. Thus:

$$g^{a\bar{b}} = 2(1 + |z|^2)(\delta^{a\bar{b}} + z^a \bar{z}^{\bar{b}}) \tag{4.55}$$

$$\Gamma_{bc}{}^{a} = -\frac{1}{1 + |z|^2}(\delta_b^a \bar{z}_c + \delta_c^a \bar{z}_b) \tag{4.56}$$

$$R_{a\bar{b}c\bar{d}} = -2(g_{a\bar{b}}g_{c\bar{d}} + g_{a\bar{d}}g_{c\bar{b}}) \tag{4.57}$$

$$R_{a\bar{b}} = 2(n + 1)g_{a\bar{b}}. \tag{4.58}$$

All components not related to these through index symmetries or complex conjugation are zero. From the expression for the Ricci tensor $R_{a\bar{b}}$ we see that its traceless part vanishes; the Fubini–Study metric solves the Euclidean Einstein equations with a cosmological constant. The form of the full Riemann tensor $R_{a\bar{b}c\bar{d}}$ shows that the space has constant holomorphic sectional curvature (but it does not show that the curvature is constant, and in fact it is not – our result is weaker basically because $g_{a\bar{b}}$ is just one 'block' of the metric tensor). There is an important theorem that says that a simply connected complex manifold of constant holomorphic sectional curvature is necessarily isometric with $\mathbb{CP}^n$ or else it has the topology of an open unit ball in $\mathbb{C}^n$, depending on the sign of the curvature. (If it vanishes the space is $\mathbb{C}^n$.) The situation is clearly analogous to that of simply connected two-dimensional manifolds of constant sectional curvature, which are either spheres flat spaces or hyperbolic spaces with the topology of a ball.

We observe that both the Riemannian and the symplectic geometry relies on the Hermitian form $Z \cdot \bar{Z} = Z^\alpha \bar{Z}_\alpha$ in $\mathbb{C}^N$. Therefore distances and areas are invariant only under those projective transformations that preserve this form. These are precisely the unitary transformations, with anti-unitary transformations added if we allow the transformations to flip the sign of the symplectic form. In quantum mechanics this theorem goes under the name of Wigner's theorem [966]:

Wigner's theorem. *All isometries of $\mathbb{CP}^n$ arise from unitary or anti-unitary transformations of $\mathbb{C}^N$.*

Since only a part of the unitary group acts effectively, the connected component of the isometry group (that containing the identity map) is in fact $SU(N)/Z_N$. For $N = 2$ we are dealing with a sphere and $SU(2)/Z_2 = SO(3)$, that is the group of proper rotations. The full isometry group $O(3)$ includes reflections and arises when anti-unitary transformations are allowed.

We obtain an infinitesimal isometry by choosing an Hermitian matrix – a generator of $U(N)$ – and writing

$$i\dot{Z}^\alpha = H^\alpha{}_\beta Z^\beta. \tag{4.59}$$

This equation – to reappear as the *Schrödinger equation* in Section 5.1 – determines a Killing flow on $\mathbb{CP}^n$. A part of it represents 'gauge', that is changes in the overall phase of the homogeneous coordinates. Therefore we write the projective equation

$$iZ^{[\alpha}\dot{Z}^{\beta]} = Z^{[\alpha}H^{\beta]}{}_\gamma Z^\gamma, \tag{4.60}$$

where the brackets denote anti-symmetrisation. This equation contains all the information concerning the Killing flow on $\mathbb{CP}^n$ itself, and is called the *projective Schrödinger equation* [480]. The fixed points of the flow occur when the right-hand side vanishes. Because of the anti-symmetrised indices this happens when Z^α is an eigenvector of the Hamiltonian,

$$H^\alpha{}_\beta Z^\beta = EZ^\alpha. \tag{4.61}$$

A picture will be given later.

There are other ways of introducing the Fubini–Study metric. We usually understand $\mathbb{CP}^1$, that is the 2-sphere, as a subset embedded in $\mathbb{R}^3$. Through the Hopf fibration we can also understand it as the base space of a fibre bundle whose total space is either $\mathbf{S}^3$ or $\mathbb{C}^2$ with the origin excluded. Both of these pictures can be generalised to $\mathbb{CP}^n$ for arbitrary n. The fibre bundle picture will be discussed later. The embedding picture relies on an embedding into $\mathbb{R}^{(n+1)^2-1}$. The dimension is rather high but we will have to live with this. To see how it works, use homogeneous coordinates to describe $\mathbb{CP}^n$ and form the matrix

$$\rho^\alpha{}_\beta \equiv \frac{Z^\alpha \bar{Z}_\beta}{Z \cdot \bar{Z}}. \tag{4.62}$$

This is a nice way to represent a point in $\mathbb{CP}^n$ since the reduncancy of the homogeneous coordinates has disappeared. The resulting matrix is Hermitian and has trace unity. Hence we have an embedding of $\mathbb{CP}^n$ into the space of Hermitian $N \times N$ matrices with trace unity, which has the dimension advertised, and can be made into a flat space in a natural way. We simply set

$$D^2(A, B) = \frac{1}{2}\text{Tr}(A - B)^2. \tag{4.63}$$

This expression defines a distance D between the matrices A and B (equal to the *Hilbert–Schmidt distance* introduced in Chapter 8). It is analogous to the chordal distance between two points on the sphere when embedded in a flat space in the standard way.

To see that the embedding gives rise to the Fubini–Study metric we take two matrices representing different points of $\mathbb{CP}^n$, such as

$$\rho_{P\ \beta}^{\alpha} = \frac{P^{\alpha}\bar{P}_{\beta}}{P \cdot \bar{P}} \qquad \rho_{Q\ \beta}^{\alpha} = \frac{Q^{\alpha}\bar{Q}_{\beta}}{Q \cdot \bar{Q}}, \tag{4.64}$$

and compute

$$D^2(\rho_P, \rho_Q) = 1 - \kappa = 1 - \cos^2 D_{\mathrm{FS}}, \tag{4.65}$$

where κ is the projective cross ratio (4.47) and D_{FS} is the Fubini–Study distance between P and Q along a curve lying within the embedded $\mathbb{CP}^n$. If the points are very close we can expand the square of the cosine to second order and obtain

$$D = D_{\mathrm{FS}} + \text{higher order terms}. \tag{4.66}$$

This proves our point. The Riemannian metric

$$ds^2 = \frac{1}{2} \operatorname{Tr} d\rho \, d\rho, \tag{4.67}$$

is precisely the Fubini–Study metric, provided that it is evaluated at a point where the matrix ρ is an outer product of two vectors – and $d\rho$ is the 'infinitesimal' difference between two such matrices, or more precisely a tangent vector to $\mathbb{CP}^n$. This embedding will reappear in Chapter 8 as the embedding of pure states into the space of density matrices.

4.6 $\mathbb{CP}^n$ illustrated

At this point we still do not know what $\mathbb{CP}^n$ 'looks like', even in the modest sense that we feel that we know what an n-sphere 'looks like' for arbitrary n. There is a choice of coordinate system that turns out to be surprisingly helpful in this regard [97, 112]. Define

$$(Z^0, Z^1, \ldots, Z^n) = (n_0, n_1 e^{i\nu_1}, \ldots, n_n e^{i\nu_n}), \tag{4.68}$$

where $0 \leq \nu_i < 2\pi$ and the real numbers n_0, n_i are non-negative, $n_0 \geq 0, n_i \geq 0$, and obey the constraint

$$n_0^2 + n_1^2 + \cdots + n_n^2 = 1. \tag{4.69}$$

We call such coordinates *octant coordinates*, because $n_0, n_1, \ldots$ clearly form the positive hyperoctant of an n-sphere. The phases ν_i form an n-torus, so we already see a picture of the topology of $\mathbb{C}\mathbf{P}^n$ emerging: we have a set of tori parametrized by the points of a hyperoctant, with the proviso that the picture breaks down at the edges of the hyperoctant, where the phases are undefined.[6]

To go all the way to a local coordinate system we can set

$$
\begin{cases}
n_0 & = & \cos\vartheta_1 \sin\vartheta_2 \sin\vartheta_3 \ldots \sin\vartheta_n \\
n_1 & = & \sin\vartheta_1 \sin\vartheta_2 \sin\vartheta_3 \ldots \sin\vartheta_n \\
n_2 & = & \cos\vartheta_2 \sin\vartheta_3 \ldots \sin\vartheta_n \qquad 0 < \vartheta_i < \dfrac{\pi}{2}. \\
\ldots & & \ldots \\
n_n & = & \cos\vartheta_n
\end{cases}
\tag{4.70}
$$

This is just like Eq. (1.15), except that the range of the coordinates has changed. We can also use n_i as orthographic coordinates. Alternatively we can set

$$
p_i = n_i^2
\tag{4.71}
$$

and use the n coordinates p_i. This looks contrived at this stage but will suggest itself later. If we also define $p_0 = n_0^2$ we clearly have

$$
p_0 + p_1 + \cdots + p_n = 1.
\tag{4.72}
$$

A probability interpretation is not far behind.

To see how it all works, let us consider $\mathbb{C}\mathbf{P}^2$. The point is that octant coordinates are quite well adapted to the Fubini–Study metric, which becomes

$$
\mathrm{d}s^2 = \mathrm{d}n_0^2 + \mathrm{d}n_1^2 + \mathrm{d}n_2^2 +
$$
$$
+ n_1^2(1 - n_1^2)\,\mathrm{d}\nu_1^2 + n_2^2(1 - n_2^2)\,\mathrm{d}\nu_2^2 - 2n_1^2 n_2^2\,\mathrm{d}\nu_1\,\mathrm{d}\nu_2.
\tag{4.73}
$$

The first piece here, given Eq. (4.69), is recognizable as the ordinary round metric on the sphere. The second part is the metric on a flat torus, whose shape depends on where we are on the octant. Hence we are justified in thinking of $\mathbb{C}\mathbf{P}^2$ as a set of flat 2-tori parametrized by a round octant of a 2-sphere. See Figure 4.9. There is an evident generalisation to all n, and in fact we can draw accurate pictures of $\mathbb{C}\mathbf{P}^3$ if we want to.

For $n = 1$ we obtain a one parameter family of circles that degenerate to points at the end of the interval; a moment's thought will convince the reader that this is

---

[6] There is an entire branch of mathematics called 'toric geometry' whose subject matter, roughly speaking, consists of spaces that can be described in this way. See Ewald [305] for more on this.

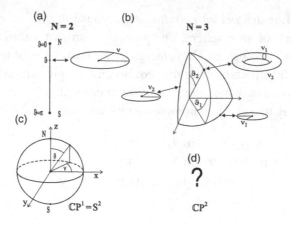

Figure 4.9 $\mathbb{C}P^2$ may be visualised as the positive octant of a 2-sphere. Each point inside the octant represents a torus T^2 spanned by the phases (v_1, v_2) (b). Each point at the edges of the octant denotes a circle, so each of three edges representes a 2-sphere. For comparison we plot $\mathbb{C}P^1$ in the same manner, in panel (a). A realistic view of $\mathbb{C}P^1$ is shown in panel (c); an analogous view of $\mathbb{C}P^2$ would be more difficult (d).

simply a way to describe a 2-sphere. Just to make sure, use the angular coordinates and find that (4.73) becomes

$$ds^2 = d\vartheta_1^2 + \frac{1}{4}\sin^2(2\vartheta_1)\,dv_1^2 = \frac{1}{4}(d\theta^2 + \sin^2\theta\,d\phi^2), \qquad (4.74)$$

where we used $\theta = 2\vartheta_1$ and $\phi = v_1$ in the second step.

To make the case $n = 2$ quite clear we make a map of the octant, using a stereographic or a gnomonic projection (Section 3.1). The latter is quite useful here and it does not matter that only half the sphere can be covered since we need to cover only one octant anyway. It is convenient to centre the projection at the centre of the octant and adjust the coordinate plane so that the coordinate distance between a pair of corners of the resulting triangle equals one. The result is shown in Figure 4.10.

This takes care of the octant. We obtain a picture of $\mathbb{C}P^2$ when we remember that each interior point really represents a flat torus, conveniently regarded as a parallelogram with opposite sides identified. The shape of the parallelogram – discussed in Section 3.3 – is relevant. According to Eq. (4.73) the lengths of the sides are

$$L_1 = \int_0^{2\pi} ds = 2\pi n_1\sqrt{1 - n_1^2} \quad \text{and} \quad L_2 = 2\pi n_2\sqrt{1 - n_2^2}. \qquad (4.75)$$

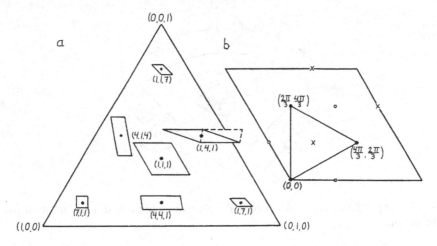

Figure 4.10 In (a) we indicate how the torus lying over each interior point changes with position in the octant. The position in the octant is given by an unnormalised vector. At the edges the tori degenerate to circles so edges are complex projective lines. The corners of the octant represent orthogonal points. It may be convenient to perform some cutting and gluing of the parallelogram before thinking about the shape of the torus it defines, as indicated with dashed lines for the torus lying over the point $(1, 4, 1)$. The size of the octant relative to that of the tori is exaggerated in the picture. To bring this home we show, in (b), the largest torus – the one sitting over $(1, 1, 1)$ – decorated with three times three points (marked with crosses and filled or unfilled dots). Each such triple corresponds to an orthogonal basis. The coordinates (ν_1, ν_2) are given for one of the triples.

The angle between them is given by

$$\cos \theta_{12} = -\frac{n_1 n_2}{\sqrt{1 - n_1^2}\sqrt{1 - n_2^2}}. \tag{4.76}$$

The point is that the shape depends on where we are on the octant. So does the total area of the torus,

$$A = L_1 L_2 \sin \theta_{12} = 4\pi^2 n_0 n_1 n_2. \tag{4.77}$$

The 'biggest' torus occurs at the centre of the octant. At the boundaries the area of the tori is zero. This is because there the tori degenerate to circles. In effect an edge of the octant is a one parameter family of circles, in other words it is a $\mathbb{C}P^1$.

It is crucial to realise that there is nothing special going on at the edges and corners of the octant, whatever the impression left by the map may be. Like the sphere, $\mathbb{C}P^n$ is a homogeneous space and looks the same from every point. To see this, note that any choice of an orthogonal basis in a three-dimensional Hilbert space gives rise to three points separated by the distance $\pi/2$ from each other in $\mathbb{C}P^2$.

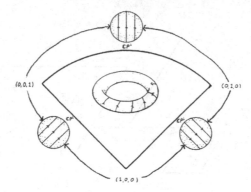

Figure 4.11 Killing vector flows on $\mathbb{C}\mathbf{P}^2$. Stereographic coordinates are used for the octant.

By an appropriate choice of coordinates we can make any such triplet of points sit at the corners of an octant in a picture identical to the one shown in Figure 4.11.

To get used to the picture we begin by looking at some one real dimensional curves in it. We choose to look at the flow lines of an isometry – a Killing field. Since the isometries are given by unitary transformations of $\mathbb{C}^N$, we obtain an infinitesimal isometry by choosing an Hermitian matrix. Any such matrix can be diagonalised, so we can choose a basis in which the given Hermitian 3×3 matrix takes the form

$$H^\alpha{}_\beta = \begin{bmatrix} E_0 & 0 & 0 \\ 0 & E_1 & 0 \\ 0 & 0 & E_2 \end{bmatrix}. \tag{4.78}$$

We can therefore arrange our octant picture so that the fixed points of the flow – determined by the eigenvectors – occur at the corners. If we exponentiate Eq. (4.59) we find that the isometry becomes

$$\begin{bmatrix} n^0 \\ n^1 e^{i\nu_1} \\ n^2 e^{i\nu_2} \end{bmatrix} \rightarrow \begin{bmatrix} e^{-iE_0 t} & 0 & 0 \\ 0 & e^{-iE_1 t} & 0 \\ 0 & 0 & e^{-iE_2 t} \end{bmatrix} \begin{bmatrix} n^0 \\ n^1 e^{i\nu_1} \\ n^2 e^{i\nu_2} \end{bmatrix} \tag{4.79}$$

where t is a parameter along the flow lines. Taking out an overall phase, we find that this implies that

$$n^I \rightarrow n^I, \qquad \nu_1 \rightarrow \nu_1 + (E_0 - E_1)t, \qquad \nu_2 \rightarrow \nu_2 + (E_0 - E_2)t. \tag{4.80}$$

Hence the position on the octant is preserved by the Killing vector flow; the movement occurs on the tori only. At the edges of the octant (which are spheres, determined by the location of the fixed points) the picture is the expected one. At a generic point the orbits wind around the tori and – unless the frequencies are rational multiples of each other – they will eventually cover the tori densely. Closed orbits are exceptional.

We go on to consider some submanifolds. Every pair of points in the complex projective plane defines a unique complex projective line, that is a $\mathbb{C}\mathbf{P}^1$, containing the pair of points. Conversely a pair of complex projective lines always intersect at a unique point. Through every point there passes a 2-sphere's worth of complex projective lines, conveniently parametrized by the way they intersect the line at infinity, that is the set of points at maximal distance from the given point. This is easily illustrated provided we arrange the picture so that the given point sits in a corner, with the line at infinity represented by the opposite edge. An interesting fact about $\mathbb{C}\mathbf{P}^2$ follows from this, namely that it contains *incontractible* 2-spheres – a complex projective line can be deformed so that its radius grows, but it cannot be deformed so that its radius shrinks because it has to intersect the line at infinity in a point. (The topological reasons for this can be seen in the picture.)

Another submanifold is the real projective plane $\mathbb{R}\mathbf{P}^2$. It is defined in a way analogous to the definition of $\mathbb{C}\mathbf{P}^2$ except that real rather than complex numbers are used. The points of $\mathbb{R}\mathbf{P}^2$ are therefore in one-to-one correspondence with the set of lines through the origin in a three-dimensional real vector space and also with the points of $\mathbf{S}^2/Z_2$, that is to say the sphere with antipodal points identified. In its turn this is a hemisphere with antipodal points on the equator identified. $\mathbb{R}\mathbf{P}^2$ is clearly a subset of $\mathbb{C}\mathbf{P}^2$. It is illuminating to see how the octant picture is obtained, starting from the stereographic projection of a hemisphere (a unit disc) and folding it twice. See Figure 4.12.

Next choose a point. Adjust the picture so that it sits at a corner of the octant. Surround the chosen point with a 3-sphere consisting of points at constant distance from that point. In the picture this will appear as a curve in the octant, with an entire torus sitting over each interior point on the curve. See Figure 4.13. This makes

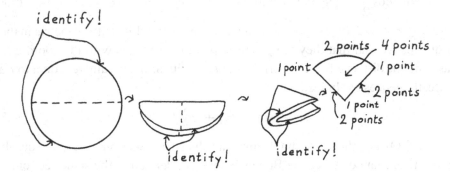

Figure 4.12 Using stereographic coordinates we show how the octant picture of the real submanifold $\mathbb{R}\mathbf{P}^2$ is related to the standard description as a hemisphere with antipodal points on the equator identified.

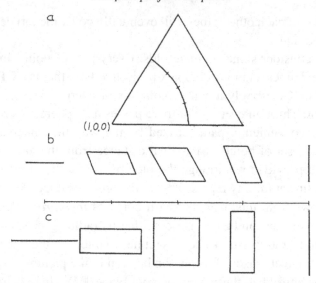

Figure 4.13 The set of points at constant distance from a corner form a squashed 3-sphere. In (a) we show how such a submanifold appears in the octant. All points in the torus lying over a point on the curve are included. In (b) we show how the size and shape of the torus change as we move along the curve in the octant; at the ends of the interval the tori collapse into circles. For comparison, in (c) we show the corresponding picture for a round 3-sphere.

sense: from the Hopf fibration we know that $\mathbf{S}^3$ can be thought of as a one-parameter family of tori with circles at the ends. The 3-sphere is round if the tori have a suitable rectangular shape. But as we let our 3-sphere grow, its tori get more and more 'squashed' by the curvature of $\mathbb{CP}^2$, and the roundness gradually disappears. When the radius reaches its maximum value of $\pi/2$ the 3-sphere has collapsed to a 2-sphere, namely to the projective line at infinity. In equations, set

$$n_1 = \sin r \cos \frac{\theta}{2}, \quad n_2 = \sin r \sin \frac{\theta}{2}, \quad \tau = v_1 + v_2, \quad \phi = v_1 - v_2, \quad (4.81)$$

where $0 \le r \le \pi/2, 0 \le \theta \le \pi$. When we express the Fubini–Study metric in these coordinates we find that they are geodesic polar coordinates with the coordinate r measuring the distance from the origin – curves with affine parameter equal to r are manifestly geodesics. Indeed

$$\mathrm{d}s^2 = \mathrm{d}r^2 + \sin^2 r \, (\Theta_1^2 + \Theta_2^2 + \cos^2 r \, \Theta_3^2), \quad (4.82)$$

where the Θ_I are the invariant forms introduced in Section 3.7; our squashed spheres at constant distance are Berger spheres, as defined in the same section.

Finally, a warning: the octant picture distorts distances in many ways. For instance, the true distance between two points in a given torus is shorter than it looks, because the shortest path between them is not the same as a straight line within the torus itself. Note also that we have chosen to exaggerate the size of the octant relative to that of the tori in the pictures. To realise how much room there is in the largest torus we note that one can find four sets of orthonormal bases in the Hilbert space $\mathbb{C}^3$ such that the absolute value of the scalar product between members of different bases is $1/\sqrt{3}$ – a kind of sphere packing problem in $\mathbb{CP}^2$. If one basis is represented by the corners of the octant, the remaining 3×3 basis vectors are situated in the torus over the centre of the octant,[7] as illustrated in Figure 4.10. The message is that the biggest torus is really big.

4.7 Symplectic geometry and the Fubini–Study measure

So far the symplectic form has not been illustrated, but the octant coordinates are very useful for this purpose too. If we use $n_1, n_2, \ldots, n_n$ as orthographic coordinates on the octant we find that the symplectic form (4.51) is

$$\Omega = 2(n_1 dn_1 \wedge d\nu_1 + \ldots + n_n dn_n \wedge d\nu_n). \tag{4.83}$$

Even better, we can use $p_i = n_i^2$ as coordinates. Then Ω takes the canonical form

$$\Omega = dp_1 \wedge d\nu_1 + \ldots + dp_n \wedge d\nu_n. \tag{4.84}$$

In effect (p_i, ν_i) are action-angle variables.

Given a symplectic form we can form a phase space volume by wedging it with itself enough times. This is actually simpler than to compute the determinant of the metric tensor, and the two notions of volume – symplectic and metric – agree because we are on a Kähler manifold. Thus, in form language, the *Fubini–Study volume element* on $\mathbb{CP}^n$ is

$$d\tilde{\Omega}_n = \frac{1}{n!} \left(\frac{1}{2}\Omega\right) \wedge \left(\frac{1}{2}\Omega\right) \wedge \ldots \wedge \left(\frac{1}{2}\Omega\right), \tag{4.85}$$

where we take n wedge products. Equivalently we just compute the square root of the determinant of Ω. We have decorated the volume element with a tilde because in Section 7.6 we will divide it with the total volume of $\mathbb{CP}^n$ to obtain the *Fubini–Study measure* $d\Omega_n$, which is a probability distribution whose integral over $\mathbb{CP}^n$ equals unity. Any measure related to the Fubini–Study volume element by a constant is distinguished by the fact that it gives the same volume to every ball of given radius, as measured by the Fubini–Study metric, and by the fact that it is unitarily invariant, in the sense that $\text{vol}(A) = \text{vol}(U(A))$ for any subset A of $\mathbb{CP}^n$.

---

[7] It is not known to what extent this property generalises to an arbitrary Hilbert space dimension. Ivanović (1981) [491] and Wootters and Fields (1989) [983] have shown that $N + 1$ orthonormal bases with the modulus of the scalar product between members of different bases always equal to $1/\sqrt{N}$ can be found if the dimension $N = p^k$, where p is a prime number. See Chapter 12.

It is a volume element worth studying in several coordinate systems. Using octant coordinates we obtain

$$d\tilde{\Omega}_n = n_0 n_1 \dots n_n \, dV_{S^n} \, dv_1 \dots dv_n, \quad 0 < v_i < 2\pi \quad (4.86)$$

where dV_{S^n} is the measure on the round unit n-sphere. If we use orthographic coordinates on the round octant, remember that $\sqrt{g} = 1/n_0$ in these coordinates, and use the coordinates $p_i = n_i^2$, we obtain

$$d\tilde{\Omega}_n = n_1 \dots n_n \, dn_1 \dots dn_n \, dv_1 \dots dv_n = \frac{1}{2^n} \, dp_1 \dots dp_n \, dv_1 \dots dv_n. \quad (4.87)$$

All factors in the measure have cancelled out! As far as calculations of volumes are concerned $\mathbb{CP}^n$ behaves like a Cartesian product of a flat simplex and a flat torus of a fixed size.

The angular coordinates ϑ can also be used on the octant. We just combine eqs. (4.86) with Eq. (1.16) for the measure on the round octant. Computing the total volume of $\mathbb{CP}^n$ with respect to its Fubini–Study volume element then leads to an easy integral:

$$\text{vol}(\mathbb{CP}^n) = \int_{\mathbb{CP}^n} d\tilde{\Omega}_n = \prod_{i=1}^{n} \int_0^{\frac{\pi}{2}} d\vartheta_i \int_0^{2\pi} dv_i \cos\vartheta_i \sin^{2i-1}\vartheta_i = \frac{\pi^n}{n!}. \quad (4.88)$$

We fixed the linear scale of the space by the requirement that a closed geodesic has circumference π. Then the volume goes to zero when n goes to infinity. Asymptotically the volume of $\mathbb{CP}^n$ even goes to zero somewhat faster than that of S^{2n}, but the comparison is not fair unless we rescale the sphere until the great circles have circumference π. Having done so we find that the volume of $\mathbb{CP}^n$ is always larger than that of the sphere (except when $n = 1$, when they coincide).

It is curious to observe that

$$\text{vol}(\mathbb{CP}^n) = \text{vol}(\Delta_n) \times \text{vol}(\mathbf{T}^n) = \frac{\text{vol}(S^{2n+1})}{\text{vol}(S^1)}. \quad (4.89)$$

The first equality expresses the volume as the product of the volumes of a flat simplex and a flat torus and is quite surprising, while the second is the volume of S^{2n+1}, given in Eq. (1.19), divided by the volume 2π of a Hopf circle as one would perhaps expect from the Hopf fibration. The lesson is that the volume does not feel any of the more subtle topological properties involved in the fibre bundle construction of $\mathbb{CP}^n$.

4.8 Fibre bundle aspects

Although we did not stress it so far, it is clear that $\mathbb{CP}^n$ is the base manifold of a bundle whose total space is $\mathbf{S}^{2n+1}$ or even $\mathbb{C}^N = \mathbb{C}^{n+1}$ with its origin deleted. The latter bundle is known as the *tautological bundle* for fairly obvious reasons. A lightning review of our discussion of fibre bundles in Chapter 3 consists, in effect, of the observation that our expression for the Fubini–Study metric is invariant under

$$\mathrm{d}P^\alpha \;\rightarrow\; \mathrm{d}P^\alpha + zP^\alpha, \tag{4.90}$$

where $z = x + iy$ is some complex number. But the vector xP^α is orthogonal to the sphere, while iyP^α points along its Hopf fibres. Because it is unaffected by changes in these directions the Fubini–Study metric really is a metric on the space of Hopf fibres. In some ways it is convenient to normalise our vectors to $Z \cdot \bar{Z} = 1$ here because then we are dealing with a principal bundle with gauge group $U(1)$. The tautological bundle is a principal bundle with gauge group $GL(1, \mathbb{C})$; technically it is called a *Hermitian line bundle*.

A fibre bundle can always be equipped with a connection that allows us to lift curves in the base manifold to the bundle space in a unique manner. Here the preferred choice is

$$\omega = \frac{i}{2}\,\frac{Z \cdot \mathrm{d}\bar{Z} - \mathrm{d}Z \cdot \bar{Z}}{Z \cdot \bar{Z}} = -i\,\mathrm{d}Z \cdot \bar{Z}. \tag{4.91}$$

In the last step we used normalised vectors. The equation $\omega = 0$ now expresses the requirement that $\mathrm{d}Z^\alpha$ be orthogonal to Z^α, so that the lifted curve will be perpendicular to the fibres of the bundle. A minor calculation confirms that

$$\mathrm{d}\omega = \Omega, \tag{4.92}$$

where Ω is the symplectic 2-form defined in Eq. (4.51).

Let us lift some curves on $\mathbb{CP}^n$ to Hilbert space. We have already done so for a geodesic on the base manifold – the result is given in Eq. (4.45). Since $\dot{Z} \cdot \bar{Z} = 0$ along this curve, it follows that $\omega = 0$ along it, so indeed it is a horizontal lift. Another example that we have encountered already: in Eq. (4.59) we wrote the Schrödinger equation, or equivalently the flowline of some isometry, as

$$i\mathrm{d}Z^\alpha = H^\alpha{}_\beta Z^\beta \mathrm{d}t. \tag{4.93}$$

Evidently

$$\omega = \frac{1}{2}(Z^\alpha \bar{H}_\alpha{}^\beta \bar{Z}_\beta - \bar{Z}_\alpha H^\alpha{}_\beta Z^\beta)\mathrm{d}t = 0 \tag{4.94}$$

along the resulting curve in the bundle, so that this is the horizontal lift of a curve defined in the base manifold by the projective Schrödinger equation (4.60).

Now consider an arbitrary curve $C(\sigma)$ in $\mathbb{C}\mathbf{P}^n$, and an arbitrary lift of it to $\mathbb{C}^N$. Let the curve run from σ_1 to σ_2. Define its *geometric phase* as

$$\phi_g = \arg\left(Z(\sigma_2) \cdot \bar{Z}(\sigma_1)\right) - \int_C \omega. \tag{4.95}$$

From now on we assume that the length of the curve is less than $\pi/2$ so that the argument that defines the *total phase* – the first term on the right-hand side – of the curve is well defined. Suppose that we change the lift of the curve by means of the transformation

$$Z^\alpha \to e^{i\lambda} Z^\alpha \quad \Rightarrow \quad \omega \to \omega + d\lambda, \tag{4.96}$$

where λ is some function of σ. Although both the connection and the total phase change under this transformation, the geometric phase does not. Indeed it is easy to see that

$$\phi_g \to \phi_g + \lambda(\sigma_2) - \lambda(\sigma_1) - \int_{\sigma_1}^{\sigma_2} d\lambda = \phi_g. \tag{4.97}$$

The conclusion is that the geometric phase does not depend on the particular lift that we use. Hence it is genuinely a function of the original curve in $\mathbb{C}\mathbf{P}^n$ itself. The total phase on the other hand is not – although it equals the geometric phase for a horizontal lift of the curve.

An interesting observation, with consequences, is that if we compute the total phase along the horizontal lift (4.45) of a geodesic in $\mathbb{C}\mathbf{P}^n$ we find that it vanishes. This means that the geometric phase is zero for any lift of a geodesic. A geodesic is therefore a *null phase curve*. The converse is not true – we will find further examples of null phase curves in Chapter 6. To see what the consequences are, consider a *geodesic triangle* in $\mathbb{C}\mathbf{P}^n$, that is three points P, Q and R connected with geodesic arcs not longer than $\pi/2$. In flat space the lengths of the sides determine the triangle uniquely up to isometries, but in curved spaces this need not be so. As it happens a geodesic triangle in $\mathbb{C}\mathbf{P}^n$ is determined up to isometries by its side lengths and its *symplectic area*, that is the integral of the symplectic form Ω over any area bounded by the three geodesic arcs [167]. To such a triangle we can associate the *Bargmann invariant* [8]

$$\Delta_3(P, Q, R) = \frac{P \cdot \bar{Q} Q \cdot \bar{R} R \cdot \bar{P}}{P \cdot \bar{P} Q \cdot \bar{Q} R \cdot \bar{R}} = \cos D_{PQ} \cos D_{QR} \cos D_{RP}\, e^{-i\Phi}. \tag{4.98}$$

---

[8] It is invariant under phase changes of the homogeneous coordinates and attracted Bargmann's (1964) [92] attention because, unlike the Fubini–Study distance, it is a complex valued function defined on $\mathbb{C}\mathbf{P}^n$. Blaschke and Terheggen (1939) [137] considered this early on. That its argument is the area of a triangle was proved by both physicists (Mukunda and Simon [676]) and mathematicians (Hangan and Masala [396]).

Here D_{PQ} is the Fubini–Study distance between the points whose homogeneous coordinates are P and Q, and so on. But what is the phase Φ? Since each side of the triangle is a geodesic arc the geometric phase of each side is zero, and by Eq. (4.95) the total phase of the side is equal to the integral along the arc of ω. Adding the contribution of all the sides (and normalising the vectors) we get

$$\arg P \cdot \bar{Q} Q \cdot \bar{R} R \cdot \bar{P} = -\arg Q \cdot \bar{P} - \arg R \cdot \bar{Q} - \arg P \cdot \bar{R} = -\oint_{\partial \Delta} \omega. \qquad (4.99)$$

It follows from Stokes' theorem that the phase Φ is the symplectic area of the triangle,

$$\Phi = \int_{\Delta} \Omega. \qquad (4.100)$$

We can go on to define Bargmann invariants for four, five and more points in the same cyclic manner and – by triangulation – we find that their phases are given by the symplectic area of the geodesic polygon spanned by the points.

The geometric phase also goes under the name of the *Berry phase*; the case that Berry studied was that of an eigenstate of a Hamiltonian H that is carried around a loop by adiabatic cycling of H.[9]

4.9 Grassmannians and flag manifolds

Projective space is only the first of its kind. Starting from an N-dimensional vector space V, real or complex, we can consider nested sequences of subspaces V_i of dimension d_i, such that $V_1 \subset V_2 \subset \cdots \subset V_r$. This is called a *flag* of subspaces (as explained in Figure 4.14). The space of all flags of a given kind is known as a *flag manifold* and denoted by $\mathbf{F}^{(N)}_{d_1 \ldots d_r}$; projective space is the easy case where a flag consists of a single one-dimensional subspace only. The next simplest case is that where the flags consist of a single M-dimensional subspace; such a flag manifold is known as the *Grassmannian* $\mathbf{F}^{(N)}_M = \mathbf{Gr}^{\mathbb{C}}_{M,N}$. The notation also tells us which number field ($\mathbb{R}$ or $\mathbb{C}$) we are using; if there is no label then the complex numbers are implied.

To see how this works let us consider the case of $\mathbf{Gr}_{2,4}$, which is the space of 2-planes in a four-dimensional vector space. We fix a 2-plane in $\mathbb{C}^4$ by fixing two

[9] The original paper by Berry (1984) [128] is reprinted in a book edited by Shapere and Wilczek [830]. See Anandan and Aharonov [33] and Mukunda and Simon [676] for the point of view that we take here, and Chruściński and Jamiołkowski [221] for much more.

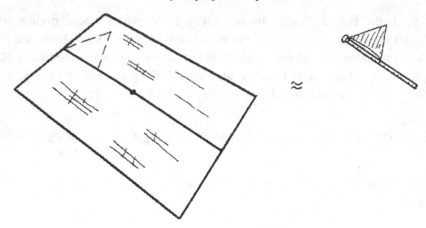

Figure 4.14 A flag may consist of a one-dimensional subspace of a two-dimensional subspace of a three-dimensional space – it is clear why it is called that.

linearly independent vectors spanning that plane, and collect them into a rank two $N \times 2$ matrix

$$
\begin{bmatrix}
Z_{0,0} & Z_{0,1} \\
Z_{1,0} & Z_{1,1} \\
Z_{2,0} & Z_{2,1} \\
Z_{3,0} & Z_{3,1}
\end{bmatrix}
\sim
\begin{bmatrix}
1 & 0 \\
0 & 1 \\
z_{2,0} & z_{2,1} \\
z_{3,0} & z_{3,1}
\end{bmatrix} .
\tag{4.101}
$$

The entries in the matrix on the left-hand side are homogeneous coordinates on the Grassmannian. Then we exercised our right to perform linear combinations and rescalings of the columns to get the matrix in a standard form. The remaining four complex numbers serve as affine coordinates on the four complex dimensional Grassmannian. Note that if the upper two rows were linearly dependent this form could not have been reached, but since the matrix as a whole has rank two we can introduce a similar coordinate system by singling out another pair of rows. (Indeed it is not hard to see that the Grassmannian is a complex manifold that can be completely covered by six coordinate patches of this kind.) From a geometrical point of view it is sometimes advantageous to think of the Grassmannian of 2-planes as the space of projective lines in a projective space of dimension $N - 1$; it is an interesting exercise to convince oneself directly that the space of lines in 3-space is four dimensional. On the other hand the use of an $N \times M$ matrix as homogeneous coordinates for $\mathbf{Gr}_{M,N}$ has advantages too and leads to an immediate proof that it is a complex manifold of $M(N - M)$ complex dimensions.[10]

---

[10] The affine coordinates are particularly useful if one wants to discuss the natural (Fubini–Study like) metric properties of the Grassmannians. See Wong [977] for this.

When we get to more involved examples of flag manifolds any kind of explicit description of the space will get involved, too. For this reason it is a good idea to fall back on a description as coset spaces. Let us begin with $\mathbf{F}_1^{(4)} = \mathbb{C}\mathbf{P}^3$. The group $GL(4, \mathbb{C})$ acts transitively on the underlying vector space. Pick any point, say $p = (1, 0, 0, 0)$ and find its isotropy group – the subgroup of the transitive group that leaves the given point invariant. Here the isotropy group consists of all matrices h of the form

$$h = \begin{bmatrix} \bullet & \bullet & \bullet & \bullet \\ 0 & \bullet & \bullet & \bullet \\ 0 & \bullet & \bullet & \bullet \\ 0 & \bullet & \bullet & \bullet \end{bmatrix}, \tag{4.102}$$

where $\bullet$ is any complex number. This defines a subgroup of $GL(4, \mathbb{C})$, and we can now define $\mathbb{C}\mathbf{P}^3$ as a coset space following the recipe in Section 3.8. We can use $SL(4, \mathbb{C})$ as our starting point instead; the resulting coset will be the same once the form of h is restricted so that it has unit determinant. Call the resulting isotropy group $P_1^{(4)}$. Indeed we can restrict ourselves to $U(N)$ since the unitary subgroup of $GL(4, \mathbb{C})$ also acts transitively; when h is restricted to $U(N)$ it becomes a block diagonal matrix representing $U(1) \times U(3)$. The argument clearly generalises to any dimension, so that we have proved that

$$\mathbf{F}_1^{(N)} \equiv \mathbb{C}\mathbf{P}^{N-1} = \frac{SL(N, \mathbb{C})}{P_1^{(4)}} = \frac{U(N)}{U(1) \times U(N-1)}. \tag{4.103}$$

For real projective space we just have to replace $SL(\mathbb{C})$ with $SL(\mathbb{R})$ and the unitary groups by orthogonal groups.

The same argument repeated for the Grassmannian $\mathbf{Gr}_{2,4}$ reveals that the isotropy group can be written as the set $P_2^{(4)}$ of special linear matrices of the form

$$h = \begin{bmatrix} \bullet & \bullet & \bullet & \bullet \\ \bullet & \bullet & \bullet & \bullet \\ 0 & 0 & \bullet & \bullet \\ 0 & 0 & \bullet & \bullet \end{bmatrix}, \tag{4.104}$$

or as block diagonal unitary matrices belonging to $U(2) \times U(2)$ if only unitary transformations are considered. For the flag manifold $\mathbf{F}_{1,2,3}^{(N)}$ the isotropy group is a *Borel subgroup* B of $SL(4, \mathbb{C})$; by definition the *standard Borel subgroup* is the group of upper triangular matrices and a general Borel subgroup is given by conjugation of the standard one. If the group is $U(N)$ the Borel subgroup is the subgroup $U(1) \times U(1) \times U(1) \times U(1)$, given by diagonal matrices in the standard case. The isotropy groups considered above are examples of *parabolic subgroups* P of the group G, represented by 'block upper triangular' matrices such that $B \subset P \subset G$. Thus we can write, in complete generality,

$$\mathbf{F}^{(N)}_{d_1 d_2 \ldots d_r} = \frac{SL(N, \mathbb{C})}{P^{(N)}_{d_1 d_2 \ldots d_r}} = \frac{U(N)}{[U(k_1) \times U(k_2) \times \cdots \times U(k_{r+1})]}, \tag{4.105}$$

where $k_1 = d_1$ and $k_{i+1} = d_{i+1} - d_i$ (and $d_{r+1} = N$). Given that the real dimension of $U(N)$ is N^2 the real dimension of an arbitrary flag manifold is

$$\dim(\mathbf{F}^{(N)}_{d_1 d_2 \ldots d_r}) = N^2 - \sum_{i=1}^{r+1} k_i^2. \tag{4.106}$$

Equivalently it is simply twice the number of zeros in the complex matrices representing the parabolic subgroup $P^{(N)}_{d_1 d_2 \ldots d_r}$. For real flag manifolds, replace the unitary groups with orthogonal groups.

The two descriptions of flag manifolds given in Eq. (4.105) are useful in different ways. To see why we quote two facts: the quotient of two complex groups is a complex manifold, and the quotient of two compact groups is a compact manifold. Then the conclusion is that all flag manifolds are compact complex manifolds. Indeed they are also Kähler manifolds. This has to do with another way of arriving at them, based on *coadjoint orbits* in a Lie algebra. This aspect of flag manifolds will be discussed at length in Chapter 8 but let us reveal the point already here. The Lie algebra of the unitary group is the set of Hermitian matrices, and such matrices can always be diagonalised by unitary transformations. The set of Hermitian matrices with a given spectrum $(\lambda_1, \lambda_2, \ldots, \lambda_N)$ can therefore be written as

$$H_{\lambda_1 \lambda_2 \ldots \lambda_N} = U H_{\text{diag}} U^{-1}. \tag{4.107}$$

This is a coadjoint orbit and a little bit of thinking reveals that it is also a flag manifold $\mathbf{F}^{(N)}$; which particular one depends on whether there are degeneracies in the spectrum. For the special case when $\lambda_1 = 1$ and all the others are zero we came across this fact when we embedded $\mathbb{C}\mathbf{P}^n$ in the flat vector space of Hermitian matrices; see Eq. (4.62). The theme will be further pursued in Section 8.5.[11]

A variation on the theme deserves mention too. A *Stiefel manifold* is by definition the space of sets of M orthonormal vectors in an N-dimensional vector space. It is not hard to see that these are the complex (real) homogeneous spaces

$$St^{(\mathbb{C})}_{M,N} = \frac{U(N)}{U(N-M)} \quad \text{and} \quad St^{(\mathbb{R})}_{M,N} = \frac{O(N)}{O(N-M)}. \tag{4.108}$$

[11] For further information on flag manifolds, in a form accessible to physicists, consult Picken [736].

This is not quite the same as a Grassmannian since the definition of the latter is insensitive to the choice of basis in the M dimensional subspaces that are the points of the Grassmannian. As special cases we get $\mathcal{S}t_{1,N}^{(\mathbb{C})} = \mathbf{S}^{2N-1}$ and $\mathcal{S}t_{1,N}^{(\mathbb{R})} = \mathbf{S}^{N-1}$; see Section 3.8.

Problems

4.1   Consider $n + 2$ ordered points on the plane, not all of which coincide. Consider two such sets equivalent if they can be transformed into each other by translations, rotations and scalings. Show that the topology of the resulting set is that of $\mathbb{C}\mathbf{P}^n$. What does $\mathbb{C}\mathbf{P}^n$ have to do with archeology?

4.2   If you manage to glue together a Möbius strip and a hemisphere you get $\mathbb{R}\mathbf{P}^2$. What will you obtain if you glue two Möbius strips together?

4.3   Carry through the argument needed to prove that a complex projective line cannot be shrunk to a point within $\mathbb{C}\mathbf{P}^n$, using formulae, and using the octant picture.

5

Outline of quantum mechanics

Quantum mechanics is like a pot: it is almost indestructible and extremely rigid, but also very flexible because you can use any ingredients for your soup.

—*Göran Lindblad*[1]

5.1 Quantum mechanics

Although our first four chapters have been very mathematical, quantum mechanics has never been very far away. Let us recall how the mathematical structure of quantum mechanics is usually summarised at the end of a first course in the subject. First of all the pure states are given by vectors in a Hilbert space. If that Hilbert space is finite dimensional it is simply the vector space $\mathbb{C}^N$ equipped with a scalar product of the particular kind that we called a Hermitian form in Eq. (3.82). Actually a pure state corresponds to an entire equivalence class of vectors; this is usually treated in such a way that the vectors are normalised to have length one and afterwards vectors differing by an overall phase $e^{i\phi}$ are regarded as physically equivalent. In effect then the space of pure states is the complex projective space $\mathbb{CP}^n$; as always in this book $n = N - 1$. The notation used in quantum mechanics differs from what we have used so far. We have denoted vectors in $\mathbb{C}^N$ by Z^α, while in quantum mechanics they are usually denoted by a *ket* vector $|\psi\rangle$. We can think of the index α as just a label telling us that Z^α is a vector, and then these two notations are in fact exactly equivalent. It is more common to regard the index as taking N different values, and then Z^α stands for the N components of the vector with respect to some chosen basis, that is

$$|\psi\rangle = \sum_{\alpha=0}^{n} Z^\alpha |e_\alpha\rangle. \tag{5.1}$$

---

[1] Göran Lindblad, remark made while lecturing to undergraduates. Reproduced with kind permission of Göran Lindblad.

To each vector Z^α there corresponds in a canonical way a dual vector $\bar{Z}_\alpha$; in Dirac's notation this is a *bra* vector $\langle\psi|$. Given two vectors there are then two kinds of objects that can be formed: a complex number $\langle\phi|\psi\rangle$, and an operator $|\psi\rangle\langle\phi|$ that can be represented by a square matrix of size N.

The full set of states includes both pure and mixed states. To form mixtures we first write all pure states as operators, that is we define

$$\rho = |\psi\rangle\langle\psi| \Leftrightarrow \rho^\alpha_{\ \beta} = Z^\alpha \bar{Z}_\beta, \tag{5.2}$$

where the state vector is assumed to be normalised to length one and our two equivalent notations have been used. This is a Hermitian matrix of trace one and rank one, in other words it is a projection operator onto the linear subspace spanned by the original state vector; note that the unphysical phase of the state vector has dropped out of this formula so that we have a one-to-one correspondence between pure states and projection operators onto rays. Next we take convex combinations of K pure states, and obtain expressions of the form

$$\rho = \sum_{i=1}^{K} p_i |\psi_i\rangle\langle\psi_i|. \tag{5.3}$$

This is a *density matrix*, written as a convex mixture of pure states.[2] The set $\mathcal{M}$ of all states, pure or mixed, coincides with the set of Hermitian matrices with non-negative eigenvalues and unit trace, and the pure states as defined above form the extreme elements of this set (in the sense of Section 1.1).

The time evolution of pure states is given by the *Schrödinger equation*. It is unitary in the sense that the scalar product of two vectors is preserved by it. In the two equivalent notations it is given by

$$i\hbar\,\partial_t|\psi\rangle = H|\psi\rangle \Leftrightarrow i\hbar\dot{Z}^\alpha = H^\alpha_{\ \beta}Z^\beta, \tag{5.4}$$

where H is a Hermitian matrix that has been chosen as the Hamiltonian of the system and Planck's constant $\hbar$ has been explicitly included. We will follow the usual custom of defining our units so that $\hbar = 1$. In the geometrical language of Section 4.6 the Hamiltonian is a generator of a Killing field on $\mathbb{CP}^n$. By linearity, the unitary time evolution of a density matrix is given by

$$i\dot{\rho} = [H, \rho] = H\rho - \rho H. \tag{5.5}$$

A system that interacts with the external world can change its state in other and non-unitary ways as well; how this happens is discussed in Chapter 10.

---

[2] The density matrix was introduced in 1927 by Landau [571] and by von Neumann [940]. In his book [941], the latter calls it the statistical operator. The word density matrix was first used by Dirac, in a slightly different sense. According to Coleman [232] quantum chemists and statistical physicists came to an agreement on nomenclature in the early 1960s.

It remains to extract physical statements out of this formalism. One way to do this is to associate the projection operators (5.2) to elementary 'yes/no' questions. In this view, the *transition probability* $|\langle\psi|\phi\rangle|^2$ is the probability that a system in the state $|\phi\rangle$ answers 'yes' to the question represented by $|\psi\rangle$ – or the other way around, the expression is symmetric in $|\phi\rangle$ and $|\psi\rangle$. Hermitian operators can be thought of as weighted sums of projection operators – namely those projection operators that project onto the eigenvectors of the operator. They can also be thought of as random variables. What one ends up with is an association between physical observables, or measurements, on the one hand, and Hermitian operators on the other hand, such that the possible outcomes of a measurement are in one-to-one correspondence with the eigenvalues of the operator, and such that the expectation value of the measurement is

$$\langle A\rangle = \mathrm{Tr}\rho A. \tag{5.6}$$

When the state is pure, as in Eq. (5.2), this reduces to

$$\langle A\rangle = \langle\psi|A|\psi\rangle. \tag{5.7}$$

There is more to be said about measurements – and we will say a little bit more about this tangled question in Section 10.1.

The role of the state is to assign probabilities to measurements. Note that, even if the state is pure, probabilities not equal to one or zero are present as soon as $[\rho, A] \neq 0$. There is a sample space for every non-degenerate Hermitian matrix, and the quantum state ρ has to assign a probability distribution for each and every one of these. A pure quantum state ρ is classically pure only with respect to the very special observables that commute with ρ. This is the reason (or at least the beginning of a reason) for why quantum mechanics is so much more subtle than classical statistics, where the set of pure states itself has the trivial structure of a discrete set of points.

Two disclaimers must be made. First, if infinite dimensional Hilbert spaces are allowed then one must be more careful with the mathematical formulation. Second, the interpretation of quantum mechanics is a difficult subject. For the moment we ignore both points – but we do observe that using optical devices it is possible to design an experimental realisation of arbitrary $N \times N$ unitary matrices [768]. Therefore the formalism must be taken seriously in its entirety.

5.2 Qubits and Bloch spheres

It is not enough to say that $\mathbb{CP}^n$ is the space of pure states. We have to know what physical states are being referred to. Like the sphere, $\mathbb{CP}^n$ is a homogeneous space in which a priori all points are equivalent. Therefore it cannot serve as a space of states without further embellishments, just as a sphere is a poor model of the surface of the Earth until we have decided which particular points are to represent Kraków, Stockholm, and so on. The claim is that every physical system can be modelled by $\mathbb{C}\mathbf{P}^n$ for some (possibly infinite) value of n, provided that a definite correspondence between the system and the points of $\mathbb{C}\mathbf{P}^n$ is set up.

But how do we set up such a correspondence, or in other words how do we tie the physics of some particular system to the mathematical framework? The answer depends very much on the system. One case which we understand very well is that of a particle of spin 1/2 that we can perform Stern–Gerlach experiments on. The idea is that to every pure state of this particle there corresponds a unique direction in space such that the spin is 'up' in that direction. Therefore its space of pure states is isomorphic to the set of all directions in space – ordinary physical space – and hence to $\mathbf{S}^2 = \mathbb{C}\mathbf{P}^1$. In this connection the sphere is known as the *Bloch sphere*; from the way it is defined it is evident that antipodal points on the Bloch sphere correspond to states that have spin up in opposite directions, that is to orthogonal states as usually defined. The *Bloch ball*, whose boundary is the Bloch sphere, is also of interest and in fact it corresponds to the space of density matrices in this case. Indeed an arbitrary Hermitian matrix of unit trace can be parametrized as

$$\rho = \begin{bmatrix} \frac{1}{2} + z & x - iy \\ x + iy & \frac{1}{2} - z \end{bmatrix}. \tag{5.8}$$

It is customary to regard this as an expansion in terms of the Pauli matrices $\vec{\sigma} = (\sigma_x, \sigma_y, \sigma_z)$, so that

$$\rho = \frac{1}{2}\mathbb{1} + \vec{\tau} \cdot \vec{\sigma}. \tag{5.9}$$

The vector $\vec{\tau}$ is known as the *Bloch vector*, and its components are coordinates in the space of matrices with unit trace. The matrix ρ is a density matrix if and only if its eigenvalues are non-negative, and this is so if and only if

$$x^2 + y^2 + z^2 \leq \frac{1}{4}. \tag{5.10}$$

This is indeed a ball, with the Bloch sphere as its surface. We will study density matrices in depth in Chapter 8; meanwhile let us observe that if the physics singles out some particular spatial direction as being of special importance – say because the particle is in a magnetic field – then the homogeneous Bloch sphere begins to acquire an interesting geography.

Note that we are clearly oversimplifying things here. A real spin 1/2 particle has degrees of freedom such as position and momentum that we ignore. Our description of the spin 1/2 particle is therefore in a way a caricature, but like all good caricatures it does tell us something essential about its subject.

Another well understood case is that of a photon of fixed momentum. Here we can measure the polarisation of the photon, and it is found that the states of the photon are in one-to-one correspondence to the set of all oriented ellipses (including the degenerate cases corresponding to circular and linear polarisation). This set at first sight does not seem to be a sphere. What we can do is regard every ellipse as the projection down to a plane of an oriented great circle on the sphere, and associate a point on the sphere to a vector through the origin that is orthogonal to the given great circle. Unfortunately to every oriented ellipse – except the two special ones representing circular polarisation – there now correspond two points on the sphere, so this cannot be quite right. The solution is simple. Start with the Riemann sphere and represent its points with a complex number

$$z = x + iy = \frac{\sin\theta e^{i\phi}}{1 + \cos\theta} = \tan\frac{\theta}{2} e^{i\phi}. \tag{5.11}$$

Here x and y are the stereographic coordinates defined in Section 3.1, while ϕ and θ are the latitude and longitude (counted from the north pole). Let 0 and ∞ – the north and south poles of the sphere – correspond to circular polarisation. Now take the square root $w = \sqrt{z}$ and introduce another Riemann sphere whose points are labelled with w. For every point except 0 and ∞ there are two distinct points w for every z. Finally associate an oriented ellipse with the second sphere as originally indicated. The pair of points w that correspond to the same z now give rise to the same ellipse. In this way we obtain a one-to-one correspondence between oriented ellipses, that is to the states of the photon, and the points z on the first of the Riemann spheres. Hence the state space is again $\mathbb{CP}^1$, in this connection known as the *Poincaré sphere*. Note that antipodal points on the equator of the Poincaré sphere correspond to states that are linearly polarised in perpendicular directions. For an arbitrary state the ratio of the minor to the major axis of the polarisation ellipse is (with the angle α defined in Figure 5.1)

$$\cos\alpha = \frac{1 - \tan^2\alpha}{1 + \tan^2\alpha} = \frac{1 - \tan\frac{\theta}{2}}{1 + \tan\frac{\theta}{2}} = \frac{1 - |z|}{1 + |z|}. \tag{5.12}$$

The reader may now recognise the *Stokes' parameters* from textbooks on optics [147].

The spin 1/2 particle and the photon are examples of *two level systems*, whose defining property is that a (special kind of) measurement that we can perform on the system yields one of two possible results. In this respect a two-level system behaves like the 'bit' of computer science. However, quantum mechanics dictates that the full space of states of such a system is a $\mathbb{CP}^1$, a much richer state space than that of a classical bit. It is a *qubit*. Two level systems often appear when a sufficiently drastic approximation of a physical system is made; typically we have

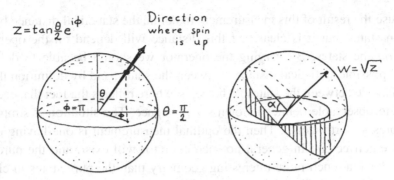

Figure 5.1 The Bloch ball, with a reminder about the coordinates that we are using. Also the auxiliary sphere used to establish the correspondence to photon polarisation.

a potential with two degenerate minima and a high barrier between them. There are then four points on the sphere that have a clear interpretation: the north and south poles correspond to states where the system sits in one of the minima of the potential, and the east and west poles correspond to eigenstates of the Hamiltonian. We may not have sufficient control over the system to interpret the remaining states of the qubit, but quantum mechanics dictates them to be there.

When the dimension of the Hilbert space goes up we encounter first *three-level systems* or *qutrits* and then *N-level systems* or *quNits*.[3] The quantum mechanical formalism treats all quNits in the same way, but its meaning depends on how the formalism is tied to the physics of the problem; it may be a spin system, it may be an atom with N relevant energy levels, or the number N may appear simply because we have binned our experimental data into N separate bins.

5.3 The statistical and the Fubini–Study distances

We now turn to the physical interpretation of the Fubini–Study metric. There is a very good reason to consider this particular geometry, namely that it gives the statistical distance between states, in the sense of Section 2.5. But the definition of statistical distance in quantum mechanics requires some thought. Suppose that we wish to distinguish between two quantum states by means of a finite set of experiments. It will then be necessary to choose some specific measurement to perform, or in mathematical terms to choose some Hermitian operator to describe it.

---

[3] The name 'qubit' was born in conversations between Wootters and Schumacher, and first used in print by the latter (1995) [808]. The name 'bit' is due to John Tukey, consulting for Bell Labs in the 1940s. Whether anyone deserves credit for the name 'quNit' is unclear. Authors who work with Hilbert spaces of dimension d talk of 'qudits', which is perhaps marginally better as a name.

We can use the result of this measurement to define the statistical distance between the given states, but it is clear that this distance will depend on the operator as well as on the states. By varying the operator we should be able to define the maximal possible statistical distance between the states, and by definition this will be the distance between the states. What we see here reflects the fact that each non-degenerate observable defines its own sample space. The situation is simple only if the states are orthogonal. Then the optimal measurement is one having the two states as eigenvectors. In general no such operator will exist, and the pure states therefore have a much more interesting geometry than the pure states in classical statistics, which are just a set of equidistant corners of a simplex.

Suppose that the two states are $|\psi\rangle$ and $|\phi\rangle$. Choose an operator A, and observe that it has $n + 1$ orthogonal eigenstates $|i\rangle$ in terms of which we can expand

$$|\psi\rangle = \sum_{i=0}^{n} \sqrt{p_i} e^{i\mu_i} |e_i\rangle, \qquad |\phi\rangle = \sum_{i=0}^{n} \sqrt{q_i} e^{i\nu_i} |e_i\rangle. \qquad (5.13)$$

Here p_i and q_i are non-negative real numbers such that

$$\sum_{i=0}^{n} p_i = \sum_{i=0}^{n} q_i = 1, \qquad (5.14)$$

that is to say our states are normalised. The phase factors μ_i, ν_i will be irrelevant to this argument. The probability to obtain a given outcome of the measurement is given by the standard interpretation of quantum mechanics – for the ith outcome to occur when the state is $|\psi\rangle$ it is p_i. According to Section 2.5 the statistical Bhattacharyya distance between the given states, given the operator, can be computed from the square roots of the probabilities:

$$\cos d_A = \sum_{i=0}^{n} \sqrt{p_i} \sqrt{q_i} = \sum_{i=0}^{n} |\langle\psi|i\rangle||\langle\phi|i\rangle|. \qquad (5.15)$$

According to the definition of distance between quantum mechanical states we should now choose the operator A in such a way that d_A becomes as large as possible, that is to say that the right-hand side should be made as small as it can be. But we have the inequality

$$\sum_{i=0}^{n} |\langle\psi|i\rangle||\langle\phi|i\rangle| \geq |\langle\psi|\phi\rangle| = \cos D_{\text{FS}}. \qquad (5.16)$$

The bound is determined by the Fubini–Study distance D_{FS} introduced in Eq. (4.46) and Eq. (4.47). Moreover the bound can be attained. As noted in Problem 5.1 there are several choices of the operator A that achieve this. One solution is to let A have either $|\psi\rangle$ or $|\phi\rangle$ as one of its eigenstates. So we conclude that the Fubini–Study distance between a pair of pure states can be interpreted as the largest possible Bhattacharyya distance between them.

The Fubini–Study distance was given in projectively invariant form in Eq. (4.46) and Eq. (4.47); in the present notation

$$\cos^2 D_{FS} = \kappa = \frac{|\langle \psi | \phi \rangle|^2}{\langle \psi | \psi \rangle \langle \phi | \phi \rangle}. \tag{5.17}$$

We have therefore established that the Fubini–Study metric measures the distinguishability of pure quantum states in the sense of statistical distance.[4] The projective cross ratio κ is more often referred to as the transition probability or, in quantum communication theory, as the *fidelity function* [509].

More precisely the Fubini–Study distance measures the experimental distinguishability of two quantum states under the assumption that there are no limitations on the kind of experiments we can do. In practice a laboratory may be equipped with measurement apparatuses corresponding to a small subset of Hermitian operators only, and this apparatuses may have various imperfections. An atomic physicist confronted with three orthogonal states of the hydrogen atom, say the ground state $n = 1$, and two states with $n = 100$ and $n = 101$, respectively, may justifiably feel that the latter two are in some sense closer even though all three are equidistant according to the Fubini–Study metric. (They are closer indeed with respect to the Monge metric discussed in Section 7.7.) The Fubini–Study geometry remains interesting because it concerns what we can know in general, without any knowledge of the specific physical system.

An instructive sidelight on the role of the Fubini–Study metric as a distinguishability measure is thrown by its appearance in the Aharonov–Anandan time–energy uncertainty relation [33]. In geometrical language this is a statement about the velocity of the Killing flow. Using homogeneous coordinates to express the metric as (4.50) together with the projective Schrödinger equation (4.60) we can write

$$\left(\frac{ds}{dt} \right)^2 = 2 \frac{Z^{[\alpha} d\dot{Z}^{\beta]} \bar{Z}_{[\alpha} d\dot{\bar{Z}}_{\beta]}}{\bar{Z} \cdot Z \bar{Z} \cdot Z} = \langle H^2 \rangle - \langle H \rangle^2, \tag{5.18}$$

where brackets denote anti-symmetrisation, a minor calculation precedes the last step and

$$\langle H \rangle \equiv \frac{\bar{Z}_\alpha H^\alpha{}_\beta Z^\beta}{\bar{Z} \cdot Z} \tag{5.19}$$

---

[4] This very precise interpretation of the Fubini–Study metric was first given by Wootters (1981) [978], although similar but less precise interpretations were given earlier.

and so on. The final result is nice:

$$\frac{ds}{dt} = \sqrt{\langle H^2 \rangle - \langle H \rangle^2}. \tag{5.20}$$

This is indeed a precise version of the time–energy uncertainty relation: the system is moving quickly through regions where the uncertainty in energy is large.

5.4 A real look at quantum dynamics

Why is the Hilbert space complex? This is a grand question that we will not answer. But an interesting point of view can be found if we 'take away' the imaginary number i and think of the complex vector space $\mathbb{C}^N$ as the real vector space $\mathbb{R}^{2N}$ with some extra structure on it.[5] The notation that is used for complex vector spaces – whether bras or kets or our index notation – actually hides some features that become transparent when we use real notation and keep careful track of whether the indices are upstairs or downstairs. The thing to watch is how the observables manage to play two different roles. They form an algebra, and in this respect they can cause transformations of the states, but they also provide a map from the states to the real numbers. Let us call the vectors X^I. A linear observable capable of transforming this vector to another vector must then be written as $A^I{}_J$ so that $X^I \rightarrow A^I{}_J X^J$. On the other hand the matrix should be able to transform the vector to a real number, too. This can be done if we regard the observable as a quadratic form, in which case it must have both its indices downstairs: $X^I \rightarrow X^I A_{IJ} X^J$. We therefore need a way of raising and lowering indices on the observable. A metric tensor g_{IJ} seems to be called for, but there is a problem with this. The metric tensor must be a very special object, and therefore it seems natural to insist that it is not changed by transformations caused by the observables. In effect then the observables should belong to the Lie algebra of the special orthogonal group $SO(2N)$. But using the metric to lower indices does not work because then we have

$$0 = \delta g_{IJ} = g_{IK} A^K{}_J + g_{KJ} A^K{}_I \Leftrightarrow A_{IJ} = -A_{JI}. \tag{5.21}$$

The problem is that in its other role as a quadratic form A_{IJ} needs to be symmetric rather than anti-symmetric in its indices. The metric will play a key role, but it cannot play this particular one!

But let us begin at the beginning. We split the complex numbers Z^α into real and imaginary parts and collect them into a $2N$-dimensional real vector X^I:

$$Z^\alpha = x^\alpha + i y^\alpha \rightarrow X^I = \begin{bmatrix} x^\alpha \\ y^\alpha \end{bmatrix}. \tag{5.22}$$

---

[5] This has been urged by, among others, Dyson [280], Gibbons and Pohle [331], and Hughston [480].

The imaginary number i then turns into a matrix $J^I{}_J$:

$$J^I{}_J = \begin{bmatrix} 0 & -1 \\ 1 & 0 \end{bmatrix} \quad \Rightarrow \quad J^2 = -1. \tag{5.23}$$

The existence of a matrix J that squares to minus one provides the $2N$-dimensional vector space with a *complex structure*. (This definition is equivalent to, and more straightforward than, the definition given in Section 3.3.)

If we write everything out in real terms further structure emerges. The complex valued scalar product of two vectors becomes

$$\langle X|Y \rangle = X^I g_{IJ} Y^J + i X^I \Omega_{IJ} Y^J, \tag{5.24}$$

where we had to introduce two new tensors

$$g_{IJ} = \begin{bmatrix} 1 & 0 \\ 0 & 1 \end{bmatrix} \qquad \Omega_{IJ} = \begin{bmatrix} 0 & 1 \\ -1 & 0 \end{bmatrix}. \tag{5.25}$$

The first is a metric tensor, the second is a symplectic form (in the sense of Section 3.4). Using the inverse of the symplectic form,

$$\Omega^{IK} \Omega_{KJ} = \delta^I_J, \tag{5.26}$$

we observe the key equation

$$J^I{}_J = \Omega^{IK} g_{KJ} \quad \Leftrightarrow \quad g_{IJ} = \Omega_{IK} J^K{}_J. \tag{5.27}$$

Thus, the complex, metric and symplectic structures depend on each other. In fact we are redoing (for the case of a flat vector space) the definition of Kähler manifolds that we gave in Section 3.3.

We can use the symplectic structure to define Poisson brackets according to Eq. (3.73). But we will be interested in Poisson brackets of a rather special set of functions on our phase space. Let $O_{IJ} = O_{JI}$ be a symmetric tensor and define

$$\langle O \rangle \equiv X^I O_{IJ} X^J. \tag{5.28}$$

Such tensors can then play one of the two key roles for observables, namely to provide a map from the state vectors to the real numbers. But observables play another role too. They transform states into new states. According to the rules for how tensor indices may be contracted this means that they must have one index upstairs. We therefore need a way of raising and lowering indices on the observables-to-be. We have seen that the metric is not useful for this purpose, so we try the symplectic form:

$$\tilde{O}^I{}_J \equiv \Omega^{IK} O_{KJ}. \tag{5.29}$$

The observables now form an algebra because they can be multiplied together, $\tilde{O}_1\tilde{O}_2 = \tilde{O}^I_{1K}\tilde{O}^K_{2J}$. Note that, according to Eq. (5.27), the complex structure is the transformation matrix corresponding to the metric tensor: $J = \tilde{g}$.

We can work out some Poisson brackets. Using Eq. (3.73) we find that

$$\{X^I, \langle O\rangle\} = 2\tilde{O}^I_{\,J}X^J = \Omega^{IJ}\partial_J\langle O\rangle \tag{5.30}$$

$$\{\langle O_1\rangle, \langle O_2\rangle\} = 2\langle\Omega[\tilde{O}_1, \tilde{O}_2]\rangle \quad \text{where} \quad [\tilde{O}_1, \tilde{O}_2] \equiv \tilde{O}_1\tilde{O}_2 - \tilde{O}_2\tilde{O}_1. \tag{5.31}$$

The Poisson bracket algebra is isomorphic to the commutator algebra of the matrices $\tilde{O}$ – even though there is no classical limit involved.

The observables as defined so far are too general. We know that we should confine ourselves to Hermitian operators, but why? What is missing is a requirement that is built into the complex formalism, namely that operators commute with the number i. In real terms this means that our observables should obey

$$[J, \tilde{O}] = 0 \quad \Leftrightarrow \quad O_{IJ} = \begin{bmatrix} m & -a \\ a & m \end{bmatrix}, \tag{5.32}$$

where m is a symmetric and a an anti-symmetric matrix. From now on only such observables are admitted. We call them *Hermitian*. There is a 'classical' way of looking at this condition. We impose the constraint that the states be normalised,

$$\langle g\rangle - 1 \equiv X^I g_{IJ}X^J - 1 = 0. \tag{5.33}$$

According to the rules of constrained Hamiltonian dynamics we must then restrict our observables to be those that Poisson commute with this constraint, that is to say that Eq. (5.31) forces us to impose Eq. (5.32).

Finally we select a specific observable, call it H, and write down Hamilton's equations:

$$\dot{X}^I = \{X^i, \langle H\rangle\} = 2\tilde{H}^I_{\,J}X^J = \Omega^{IJ}\partial_J\langle H\rangle. \tag{5.34}$$

Using Eqs (5.22) and (5.32) reveals that, in complex notation, this is precisely

$$\dot{Z}^\alpha = \dot{x}^\alpha + i\dot{y}^\alpha = 2i(m+ia)^\alpha_{\,\beta}\left(x^\beta + iy^\beta\right) \equiv -iH^\alpha_{\,\beta}Z^\beta. \tag{5.35}$$

This is Schrödinger's equation, with a complex matrix H that is Hermitian in the ordinary sense. In this way we learn that Schrödinger's equation is simply Hamilton's equation in disguise.[6] Note also that the constraint (5.33) generates a Hamiltonian flow

$$\dot{X}^I = \{X^I, \langle g\rangle - 1\} = 2J^I_{\,J}X^J \quad \Leftrightarrow \quad \dot{Z}^\alpha = 2iZ^\alpha. \tag{5.36}$$

---

[6] It appears that this interesting point was first made by Strocchi (1966) [864], and independently by Kibble (1979) [525]. Further studies were made by, among others, Gibbons [330], Ashtekar and Schilling [63], and Brody and Hughston [177].

According to the theory of constrained Hamiltonian systems this is an unobservable *gauge motion*, as indeed it is. It just changes the unphysical phase of the complex state vector.

In a sense then quantum mechanics is classical mechanics on $\mathbb{CP}^n$, formulated as a constrained Hamiltonian system on $\mathbb{C}^N$. But there is a key difference that we have so far glossed over. In classical mechanics arbitrary functions on phase space are used as observables, and each such function defines a Hamiltonian flow via Hamilton's equations. In our discussion we restricted ourselves at the outset to quadratic functions of the state vectors, leading to linear equations of motion. The reason for this restriction is that we require the Hamiltonian flow to leave not only the symplectic form but also the metric invariant. From Section 4.5 we know that the Killing vector fields of $\mathbb{CP}^n$ are generated by Hermitian matrices, and this is precisely the set of Hamiltonian flows that we ended up with. The unitary group $U(N)$, or the isometry group that acts on Hilbert space, is a subgroup of the orthogonal group $SO(2N)$; instead of the unfortunate Eq. (5.21) we now have

$$g_{IK}\, \tilde{O}^K_J + g_{KJ}\, \tilde{O}^K_I = 0 \quad \Leftrightarrow \quad [J, \tilde{O}] = 0. \tag{5.37}$$

Hence the transformations effected by our observables preserve the metric. From this point of view then quantum mechanics appears as a special case of classical mechanics, where the phase space is special ($\mathbb{CP}^n$) and the set of observables is restricted to be those that give rise to Hamiltonian vector fields that are also Killing vector fields. Unlike classical mechanics, quantum mechanics has a metric worth preserving.

The restriction that leads to quantum mechanics – that the Hamiltonian flow preserves the metric – can be imposed on any Kähler manifold, not just on $\mathbb{CP}^n$. It has often been asked whether this could lead to a viable generalisation of quantum mechanics. In fact several problems arise almost immediately. If the resulting formalism is to be of any interest one needs a reasonable large number of Hamiltonian Killing vector fields to work with. Some Kähler manifolds do not have any. If the dimension is fixed the maximal number of linearly independent Killing vector fields occurs if the Kähler space has constant holomorphic sectional curvature. But as noted in Section 4.5 if the space is also compact and simply connected this condition singles out $\mathbb{CP}^n$ uniquely.

There are other reasons why $\mathbb{CP}^n$ appears to be preferred over general Kähler spaces, notably the existence of the Segre embedding (Section 4.3). The fact that this embedding is always available makes it possible to treat composite systems in just the way that is peculiar to quantum mechanics, where the dimension of the state space of the composite system is surprisingly large. These dimensions are being used to keep track of *entanglement* – correlations between the subsystems that have no classical counterpart – and it is entanglement that gives quantum mechanics much of its special flavour (see Chapters 16 and 17). In some sense it is also the large dimension of the quantum mechanical state space that enables general symplectic manifolds to arise in the classical limit; although we have argued that quantum mechanics is (in a way) a special case of classical mechanics we must remember that quantum mechanics uses a much larger phase space than the classical theory – when the Hilbert space is finite dimensional the set of pure states of the corresponding classical theory is a finite set of points, while theories with finite dimensional classical phase spaces require an infinite dimensional quantum 'phase space'.

5.5 Time reversals

The operation of time reversal is represented by an anti-unitary rather than a unitary operator, that is to say by an isometry that is a reflection rather than a rotation.[7] A complex conjugation is then involved, and it is again helpful to think of the state space as a $2N$-dimensional real manifold. In some situations we can see that the choice of a complex structure on the real vector space is connected to the direction of time. The point is that a matrix J that squares to -1 can be chosen as a complex structure on a real $2N$-dimensional vector space. Regarded as an $SO(2N)$ transformation such a matrix can be thought of as producing quarter turns in N suitably chosen orthogonal two-dimensional planes. But the Hamiltonian is an $SO(2N)$ matrix as well, and it is chosen to commute with J, so that it generates rotations in the same N planes. If the Hamiltonian is also positive definite it defines a sense of direction for all these rotations, and it is natural to require that the quarter turns effected by J take place in the same directions. This then singles out a unique J. It follows that if we reverse the direction of time then the complex structure changes sign – which is why time reversal will be represented in quantum mechanics by an anti-unitary transformation.[8]

Wigner's theorem states that every isometry of $\mathbb{CP}^n$ can be effected by a transformation of $\mathbb{C}^N$ that is either unitary or anti-unitary. The latter possibility arises only where discrete isometries are concerned (since the square of an anti-unitary transformation is unitary), but as we have seen it includes the interesting case of time reversal. Let us see what this looks like in real terms. Since all we require is that the Fubini–Study metric is preserved it is enough to ensure that the projective

---

[7] The theory of anti-unitary operators was worked out by Wigner (1960) [967], and more recently reviewed by Uhlmann [913]. Dirac's notation was not designed to handle them. The usefulness of the real point of view was emphasised by Dyson (1962) [280].

[8] See Ashtekar and Magnon [62] and Gibbons and Pohle [331] for elaborations of this point.

cross ratio as derived from the Hermitian form is preserved. This will be so if the transformation is effected by a matrix which obeys either

$$UgU^T = g, \qquad U\Omega U^T = \Omega \qquad (5.38)$$

(the unitary case) or

$$\Theta g\Theta^T = g, \qquad \Theta\Omega\Theta^T = -\Omega \qquad (5.39)$$

(the anti-unitary case). Hence anti-unitary transformations are anti-canonical. Note by the way that the equation that defines an anti-unitary transformation, namely

$$\langle\Theta X|\Theta Y\rangle = \langle Y|X\rangle, \qquad (5.40)$$

indeed follows from the definition (5.24) of the Hermitian form.

Two options are still open for Θ:

$$\Theta^2 = \pm 1. \qquad (5.41)$$

The choice of sign depends on the system. For spin systems the choice of sign is made in an interesting way. The physical interpretation requires that the angular momentum operators $\mathbf{J}$ are odd under time reversals, so that

$$\Theta\mathbf{J} + \mathbf{J}\Theta = 0. \qquad (5.42)$$

An anti-unitary operator can always be written in the form

$$\Theta = UK, \qquad (5.43)$$

where K denotes the operation of complex conjugation and U is a unitary operator. In the standard representation of the angular momentum operators – Eqs (B.4)–(B.5), where J_y is imaginary and the others real – the unitary operator U must obey

$$UJ_x + J_xU = UJ_z + J_zU = UJ_y - J_yU = 0. \qquad (5.44)$$

These equations determine U uniquely up to an irrelevant phase factor; the answer is

$$U = \begin{bmatrix} 0 & 0 & 0 & 0 & (-1)^N \\ \cdots & \cdots & \cdots & \cdots & 0 \\ 0 & 0 & 1 & \cdots & 0 \\ 0 & -1 & 0 & \cdots & 0 \\ 1 & 0 & 0 & \cdots & 0 \end{bmatrix}. \qquad (5.45)$$

It then follows, with no ambiguity, that

$$N \text{ odd}, n \text{ even}; \quad \Rightarrow \quad \Theta^2 = 1 \qquad N \text{ even}, n \text{ odd} \quad \Rightarrow \quad \Theta^2 = -1. \quad (5.46)$$

Time reversals therefore work quite differently depending on whether the spin is integer (even n) or half-integer (odd n). For even n there will be a subspace of states that are left invariant by time reversal. In fact they form the real projective space $\mathbb{R}\mathbf{P}^n$.

Odd n is a different matter. There can be no Θ invariant states. This means that $|P\rangle$ and $\Theta|P\rangle$ must be distinct points in $\mathbb{C}\mathbf{P}^{2n+1}$ and therefore they define a unique projective line L_P; in fact they are placed on opposite poles of this projective line, regarded as a 2-sphere. Moreover it is not hard to show that the resulting projective line transforms onto itself under Θ. Now consider a point $|Q\rangle$ not lying on L_P. Together with $\Theta|Q\rangle$ it determines another projective line L_Q. But L_P and L_Q cannot have a point in common (because if they had then their intersection would also contain the time reversed point, and therefore they would coincide, contrary to assumption). It follows that $\mathbb{C}\mathbf{P}^{2n+1}$ can be foliated by complex projective lines that never touch each other. It is a little hard to see this directly, but a similar statement is true for real projective spaces of odd dimension. Then the real projective lines that never touch each other are precisely the Clifford parallels in the Hopf fibration of $\mathbf{S}^{2n+1}$, restricted to $\mathbb{R}\mathbf{P}^{2n+1}$; see Section 3.5.

Keeping n odd, suppose that we restrict ourselves throughout to time reversal invariant operators. This means that we decide to probe the system by means of time reversal invariant observables only. Such observables obey

$$[\tilde{O}, \Theta] = 0. \tag{5.47}$$

The theory when subject to a restriction of this kind is said to be obey a *superselection rule*. The first observation is that all such observables are degenerate, since all points on a time reversal invariant 2-sphere have the same eigenvalues of $\tilde{O}$. (This is known as *Kramer's degeneracy*.) The second observation is that this restriction of the observables effectively restricts the state space to be the base manifold of the fibre bundle whose bundle space is $\mathbb{C}\mathbf{P}^{2n+1}$ and whose fibres are $\mathbb{C}\mathbf{P}^1$. This is analogous to how $\mathbb{C}\mathbf{P}^n$ arises as the state space when we restrict ourselves to observables that commute with the complex structure.

What is this new state space? To see this, let us go back to the real vector space of states that we analysed in Section 5.4. We know that

$$\Theta J + J\Theta = 0, \tag{5.48}$$

where J is the complex structure. We define

$$\mathbf{i} = J \qquad \mathbf{j} = \Theta \qquad \mathbf{k} = J\Theta. \tag{5.49}$$

It follows that

$$\mathbf{i}^2 = \mathbf{j}^2 = \mathbf{k}^2 = -1, \qquad \mathbf{ij} = \mathbf{k} \qquad \mathbf{jk} = \mathbf{i} \qquad \mathbf{ki} = \mathbf{j}. \tag{5.50}$$

This is the algebra of *quaternions*.[9] Quaternions are best regarded as a natural generalisation of complex numbers. A general quaternion can be written in terms of four real numbers as

$$\mathbf{q} = a_0 + a_1\mathbf{i} + a_2\mathbf{j} + a_3\mathbf{k}. \tag{5.51}$$

For every quaternion there is a conjugate quaternion

$$\bar{\mathbf{q}} = a_0 - a_1\mathbf{i} - a_2\mathbf{j} - a_3\mathbf{k} \tag{5.52}$$

and the quantity $|\mathbf{q}|^2 \equiv \bar{\mathbf{q}}\mathbf{q} = \mathbf{q}\bar{\mathbf{q}}$ is a real number called the absolute value squared of the quaternion. The absolute value of a product is the product of the absolute values of the factors. In general two quaternions do not commute. Like the real numbers $\mathbb{R}$ and the complex numbers $\mathbb{C}$ the quaternions form a normed associative *division algebra*, that is to say that they share the algebraic structure of the real number field except that multiplication is not commutative. Moreover these three are the only division algebras over the real numbers that exist – a statement known as *Frobenius' theorem*. (Multiplication of *octonions* is not associative.) For the moment the main point is that we can form quaternionic vector spaces and quaternionic projective spaces too, provided we agree that multiplication with scalars – that are now quaternions – always takes place from the left (say); we must be careful to observe this rule since the scalars no longer commute with each other.

Since we are dealing with $\mathbb{CP}^{2n+2}$ we are working in a real vector space of dimension $4n+4$. We can regard this as a quaternionic vector space of quaternionic dimension $n + 1$. For $n = 1$ this means that an arbitrary vector is written as a two-component object

$$\mathbf{Z}^a = (Z^0 + Z^3\mathbf{j},\ Z^2 + Z^1\mathbf{j}), \tag{5.53}$$

where the Z^α are complex numbers and we must remember that the imaginary unit $\mathbf{i}$ anticommutes with $\mathbf{j}$. It is straightforward to check that time reversal as defined above is effected by a left multiplication of the vector by $\mathbf{j}$. Since we have imposed the superselection rule that multiplication of the vector with an overall quaternion leaves the state unchanged we can now form the quaternionic projective space $\mathbb{HP}^1$. Topologically this is $\mathbb{R}^4 + \infty = \mathbf{S}^4$.

Everything works out in such close analogy to the real and complex projective spaces that we need not give the details. Let us just quote the salient points: starting

---

[9] Denoted $\mathbb{H}$, for Hamilton who invented them. It was Dyson (1962) [280] who first noted that quaternions arise naturally in standard quantum mechanics when time reversals are considered. For further elaborations see Avron et al. [78] and Uhlmann [909]. For a review of just about every other aspect of quaternions, see Gürsey and Tze [378].

from a real vector space of dimension $4n+4$ we first normalise the vectors to unity, and then we impose the superselection rules

$$[\tilde{O}, \mathbf{i}] = 0 \quad \Rightarrow \quad \mathbb{C}\mathbf{P}^n \qquad [\tilde{O}, \mathbf{j}] = 0 \quad \Rightarrow \quad \mathbb{H}\mathbf{P}^n. \qquad (5.54)$$

This is known as the Hopf fibration of the $4n + 3$ sphere. In two steps $\mathbf{S}^{4n+3} \rightarrow \mathbb{C}\mathbf{P}^{2n+1}$ with fibre $\mathbf{S}^1$ and $\mathbb{C}\mathbf{P}^{2n+1} \rightarrow \mathbb{H}\mathbf{P}^n$ with fibre $\mathbf{S}^2$, or in one step $\mathbf{S}^{4n+3} \rightarrow \mathbb{H}\mathbf{P}^n$ with fibre $\mathbf{S}^3$. The base spaces inherit natural metrics from this construction. In the case of $\mathbb{H}\mathbf{P}^1$ this happens to be the round metric of the 4-sphere. When equipped with their natural metrics the projective spaces $\mathbb{R}\mathbf{P}^n$, $\mathbb{C}\mathbf{P}^n$ and $\mathbb{H}\mathbf{P}^n$ share some crucial features, notably that all their geodesics are closed.

Occasionally it is suggested that quaternionic quantum mechanics can offer an alternative to the standard formalism, but we see that the quaternionic projective space has a role to play as a state space also within the latter.

5.6 Classical and quantum states: a unified approach

Some remarks on how quantum mechanics can be axiomatised may not be out of place – even if they will be very incomplete. There is a choice whether observables or states should be regarded as primary. The same choice occurs when classical mechanics is defined, for the same reason: observables and states cooperate in producing the real numbers that constitute the predictions of the theory. These real numbers are the result of real valued linear maps applied to a vector space, and the space of such maps is itself a vector space. The end result is a duality between observables and states, so that either can be taken as primary. Just as one can take either the hen or the egg as primary. Let us try to summarise similarities and differences between states in classical and quantum mechanics, or perhaps more accurately in classical and quantum probability theory. A classical state is described by a probability vector or by an element from the set of all probability measures on the classical phase space. The space of quantum states consists of density matrices. What features are common for these different frameworks?

Let us start by considering a convex cone V^+ in a vector space V. Elements x belonging to V^+ will be called positive. Let $e : V \rightarrow \mathbb{R}$ be a linear functional on V. The space of all states is then defined as a cross-section of the positive cone, consisting of elements that obey the normalisation condition $e(x) = 1$. As shown in Table 5.1, both classical and quantum states fit well into this scheme. Let us be a little bit more explicit: in the classical case the vector space has arbitrary dimension and V^+ consists of positive vectors. The functional e is given by the l_1–norm of the vector $\vec{p}$, or by the integral over phase space in the continuous case; in the table Y denotes an arbitrary measurable subset of the classical phase space X. This is the case discussed at length in Chapter 2. In the quantum case the vector

Table 5.1 *Classical and quantum states: a comparison.*

Framework	States	Positivity	Normalisation
(i) Probability vectors	$\vec{p} \in \mathbb{R}^N$	$p_i \geq 0$	$\sum_i p_i = 1$
(ii) Probability measures	$\mu \in \Omega(X)$	$\mu(Y) \geq 0$	$\int_X d\mu(x) = 1$
(iii) Density operators	$\rho \in \mathcal{M}^{(N)}$	$\rho \geq 0$	$\text{Tr}\rho = 1$
(iv) States on C^* algebra	ω on $\mathcal{A}$	$\omega(x^*x) \geq 0$	$\omega(\mathbb{1}) = 1$

space is the space of Hermitian matrices, V^+ is the space of positive operators, and the functional e is the trace. Classical probability theory can be obtained from quantum mechanics through a restriction to diagonal matrices.

The similarities between the classical and quantum cases are made very transparent in Segal's axiomatic formulation.[10] Then the axioms are about a set $\mathcal{A}$ of objects called *observables*. The axioms determine what an observable is – reflection on what one observes in experiments comes in at a later stage. The set $\mathcal{A}$ has a distinguished subset $\mathcal{M}$ whose elements ρ are called *states*. We assume

(**I**) The observables form a real linear space.
(**II**) The observables form a commutative algebra.
(**III**) There exists a bilinear map from $\mathcal{M} \times \mathcal{A}$ to the real numbers.
(**IV**) The observables form a Lie algebra.

We see that a state can be thought of as a special kind of linear map from the observables to the real numbers – and this is characteristic of Segal's approach, even if at first sight it appears to be taking things in the wrong order!

These axioms do not characterise quantum mechanics completely. Indeed they are obeyed by both classical and quantum mechanics. (As far as we are aware no one knows if there is a third kind of mechanics also obeying Segal's axioms.) In classical mechanics states are probability distributions on phase space, observables are general functions on phase space, and their algebras are given by pointwise multiplication (Axiom II) and by Poisson brackets (Axiom IV). This means that the extreme points of the convex set form a symplectic manifold; unless rather peculiar measures are taken [979] this is a continuous manifold and the set of all states becomes infinite dimensional.

In quantum mechanics states are density matrices acting on some Hilbert space. The observables A are Hermitian matrices – since we assume that the Hilbert space is finite dimensional we need no further precision. The algebra in Axiom II is

---

[10] For further study we recommend the original paper by Segal (1947) [823]. For a simplified account see Currie et al. [240].

given by a commutative but non-associative *Jordan algebra*, to be discussed in Section 8.7. The Lie bracket of Axiom IV is given by the imaginary unit i times the commutator of the observables. The bilinear map required by Axiom III is the trace of the product of ρ and A; this is the quantum mechanical expectation value $\langle A \rangle = \text{Tr}\rho A$. When we think of states as real valued maps from the observables, we write $\rho(A) = \text{Tr}\rho A$. Note that $\text{Tr}\rho = 1$ then becomes the statement that $\rho(\mathbb{1}) = 1$. In both classical and quantum mechanics the states are defined so that they form a convex set, and it becomes important to identify the pure states. In quantum mechanics they are projection operators onto rays in a complex Hilbert space, and can therefore also be thought of as equivalence classes of vectors.

Segal's axioms are rather even handed in their choice between observables and states. In the C^*–algebra approach the algebra of observables occupy the centre of the stage. The states are defined as positive and normalised linear functionals ω on a suitable algebra $\mathcal{A}$. Both classical and quantum mechanics fit into this mould, since the states may also be viewed as functionals. In case (i) the state $\vec{p}$ maps the vector $\vec{y} \in \mathbb{R}^N$ to $\vec{p} \cdot \vec{y} \in \mathbb{R}$. In case (ii) the state μ acts on an arbitrary integrable function, $\mu : f(X) \rightarrow \int_X f(x)\, d\mu(x)$, while in case (iii) the state ρ acts on an Hermitian operator, $\rho : A \rightarrow \text{Tr}A\rho$. The most essential difference between the classical cases (i) or (ii) and the quantum case (iii) concerns commutativity: classical observables commute, but this is not so in quantum mechanics.[11]

Although we do not intend to pursue the algebraic approach, we observe that once states are defined as linear functionals on a suitable algebra, the set of all states will be a convex set. We are then inevitably drawn to the converse question: what properties must a convex set have, if it is to arise as the state space of an operator algebra? The answer to this question is known, but it is not a short answer and this is not the place to give it.[12]

To demonstrate further similarities between the classical and the quantum cases let us replace the sum in Eq. (5.3) by an integral over the space of pure states,

$$\hat{\rho} = \int_{\mathbb{CP}^{N-1}} \rho^s(\psi)\, |\psi\rangle\langle\psi|\, d\Omega(\psi). \tag{5.55}$$

Here $d\Omega(\psi)$ stands for the Fubini–Study measure (see Section 4.7), while $\rho^s(\psi)$ is a normalised probability distribution, and we decorated the density matrix itself with a hat – a notation that we use occasionally to distinguish operators. Thus a state may be considered as a probability measure on the space of pure states [658].

---

[11] For an account of the algebraic approach to quantum mechanics consult Emch [294]. See Section 8.7 for a glimpse.
[12] We refer to Alfsen and Shultz [20, 21]. Their answer takes 843 pages.

If $\rho^s(\psi)$ is a δ-function then $\hat{\rho}$ represents a pure state. The unitary time evolution of the density matrix,

$$\frac{d\hat{\rho}}{dt} = i[\hat{\rho}, H], \tag{5.56}$$

is equivalent to the Liouville equation for the distribution function ρ^s,

$$\frac{d\rho^s}{dt} = \{\langle H \rangle, \rho^s\}. \tag{5.57}$$

So far classical and quantum mechanics look similar.

But there are differences too. One key difference concerns the pure states. The classical set of pure states can be discrete. The set of pure quantum states is always continuous, and indeed it is always a symplectic manifold. In particular there always exists a continuous reversible transformation joining any two pure quantum states.[13] A related fact is that the representation (5.55) is not unique. There are many ρ^s for each $\hat{\rho}$; quantum theory identifies two probability distributions if their barycentres are equal. The result is a kind of projection of the infinite dimensional space of probability distributions ρ^s into the compact $N^2 - 1$ dimensional space $\mathcal{M}^{(N)}$ of density matrices. This is completely foreign to the classical theory, and it happens because of the severe restriction on allowed observables that is imposed in quantum mechanics. It is similar to the situation encountered in the theory of colours: the detector system of the eye causes a projection of the infinite dimensional space of all possible spectral distributions down to the three-dimensional cone of colours.

Note that we just studied a *hidden variable theory*. By definition, this is a classical model from which quantum mechanics arises through some projection reminiscent of the projection in colour theory. Indeed we studied this model in detail when we developed quantum mechanics as a form of classical mechanics in Section 5.4. Quantum mechanics arose once we declared that most classical variables were hidden. But for various reasons this model is not the model that people dream about.[14]

5.7 Gleason and Kochen–Specker

To end our slightly unsystematic tour of fundamentals, we point out that the amount of choice that one has in choosing how probability distributions emerge from the formalism is severely limited. Once it has been decided that we are in the Hilbert

---

[13] An axiomatic approach to quantum mechanics developed by Hardy [403] takes this observation as its starting point.
[14] For an interesting discussion of hidden variables from this point of view, see Holevo [447].

space framework a theorem comes into play when we try to assign a probability distribution over subspaces. We asssume that the probabilities are assigned to the rays (one-dimensional subspaces) of Hilbert space, or equivalently to the points in (real or complex) projective space. We let each ray be represented by some unit vector $|\psi\rangle$, or by a projector $|\psi\rangle\langle\psi|$. To prove the theorem two further assumptions are needed:

- **Normalisation**. For every orthonormal basis, its elements $|e_i\rangle$ are assigned probabilities $p_{|e_i\rangle} \geq 0$ such that

$$\sum_{i=1}^{N} p_{|e_i\rangle} = 1. \tag{5.58}$$

- **Non-contextuality**. When $N > 2$ every vector $|\psi\rangle$ is an element of many orthonormal bases, but the probability $p_{|\psi\rangle}$ assigned to its ray is independent of how the remaining vectors of the basis are chosen.

That is all. Nevertheless we now have:

Gleason's theorem. *Under the conditions stated, and provided the dimension N of the Hilbert space obeys $N > 2$, there exists a density matrix ρ such that $p_{|\psi\rangle} = \mathrm{Tr}\rho\,|\psi\rangle\langle\psi|$ for all (unit) vectors $|\psi\rangle$.*

So the density matrix and the trace rule (5.6) have been forced upon us – probability can enter the picture only in precisely the way that it does enters in conventional quantum mechanics. The proof proceeds by first reducing the problem to real three-dimensional subspaces, and then reaches its aim by means of a long string of elementary arguments which are famously difficult to follow. The remarkable thing is that no further assumptions are needed to prove the theorem – it is proved, not assumed, that the probability distribution is a continuous function on the projective space.[15]

Each ray can be represented by a projector, or in physical terms as an observable having 'yes' or 'no' as its answer. Orthogonal rays then correspond to commuting projectors and to compatible measurements that – at least for the theorist – can be made in such a way that they do not affect each other's outcomes. Every orthonormal basis gives rise to a probability distribution over N mutually exclusive events, and defines a *context*. The delicate way in which different contexts are interlocked force Gleason's conclusion.

The non-contextuality assumption is quite strong, since the experimental arrangement needed to measure a projector in one context differs from that used to measure

---

[15] The theorem is due to Gleason (1957) [345]. A proof that is comparatively easy to follow was given by Pitowsky [739].

it in another context. But it looks natural if we think that the properties of the system are independent of any decision taken by the observer. In hidden variable theories it is assumed that the system has a definite property corresponding to the 'yes' or 'no' answer it returns if the question corresponding to a projector is asked, and the system has this property regardless of whether the question is asked or not. The quantum state of the system may be such that the best prediction we can make from it is a probability assignment, but this may be (the hidden variable theorist thinks) a sign that the quantum mechanical description is incomplete.

We are thus led to ask if it makes sense to refer to the outcomes of all measurements that could have been performed, even though they were not. The answer is 'no' [551]:

The Kochen–Specker theorem. *It is impossible to assign to all rays of an $N > 2$ dimensional Hilbert space a truth value (true or false), in such a way that exactly one vector in each orthonormal basis is true.*

This is in fact a corollary of Gleason's theorem, because if we replace true and false by one and zero we have an example of a probability assignment obeying Gleason's assumptions. But probability distributions that come from density matrices are not of this form, so his theorem rules them out. The corollary is very interesting, since the outcome of a projective measurement does provide such truth value assignments to the vectors in the given orthonormal basis. Apparently then these assignments did not exist prior to the decision to perform that particular measurement. To the extent that one can talk about them all – which is not obvious because many of them correspond to measurements that could have happened, but in fact did not – they are necessarily contextual even though the probabilities are not.

The conclusion is often stated in the form of a colouring problem. By convention red corresponds to true, and green corresponds to false. The theorem then says that it is impossible to colour the rays of Hilbert space using two colours only, consistently with the Kochen–Specker rules, namely that exactly one vector in each orthonormal basis must be coloured red. It follows that two orthogonal rays cannot both be red. It is clear why $N = 2$ is an exception from the theorem: we can divide the Bloch sphere into two sets of antipodal points, and colour one set red and the other green. In two dimensions each ray has a unique negation.

The proof of the Kochen–Specker theorem is pleasantly easy, and proceeds by exhibiting a finite number of uncolourable vectors, including some orthonormal bases. Such sets are described by their *orthogonality graphs*, in which each node represents a vector and two nodes are joined by a link if and only if the corresponding vectors are orthogonal. The only connected graph that can be realised with two-dimensional vectors is the graph ○──○. In three dimensions there are many possibilities, including

The first example shows one ray belonging to three different contexts (and in fact the number of rays orthogonal to any given ray is infinite). The second example is a *complete graph*, that is every pair of vertices are connected by a link. A complete graph with N vertices represents a complete orthogonal basis in dimension N. (Some further terminology – not needed right now – is introduced in Figure 16.7.) Every graph can be realised as an orthogonality graph in $\mathbb{R}^N$, provided N is chosen large enough.[16]

An interesting example of a graph is [551]

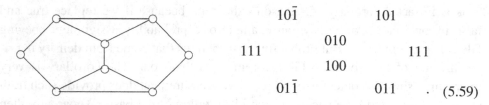

$$
\begin{array}{ccc}
& 10\bar{1} & 101 \\
& 010 & \\
111 & 100 & 111 \\
& & \\
01\bar{1} & 011 &
\end{array}
\qquad (5.59)
$$

It can be realised by vectors in $\mathbb{R}^3$, and a suggestion for how to do this is given on the right, in a condensed notation that should be clear once the reader is told that the rays are not normalised, and $\bar{1}$ stands for -1. Let us denote a unit vector corresponding to 111 by $|\psi\rangle$ and one corresponding to $11\bar{1}$ by $|\phi\rangle$. Then we obtain

$$
|\langle\psi|\phi\rangle|^2 = \frac{1}{9}. \qquad (5.60)
$$

The standard interpretation of this equation is that if we prepare the system in the state $|\psi\rangle$, and then ask if it is in state $|\phi\rangle$, the answer is 'yes' with probability $1/9$. But this is a little alarming. We have assigned the truth value 'true', or equivalently the colour red, to the ray at the far left of the graph. It is then a simple exercise to show that the Kochen–Specker colouring rules force the ray on the far right to be coloured green, and to be assigned the truth value 'false'. So the probability must equal zero if the assumptions behind the Kochen–Specker colouring rules are true. In effect we have derived a probabilistic contradiction between quantum mechanics and a certain class of hidden variable theories.

---

[16] Unlike many problems in graph theory, the problem of how to obtain an orthogonal representation of a given graph can be solved in a computationally efficient way. See Lovász [622].

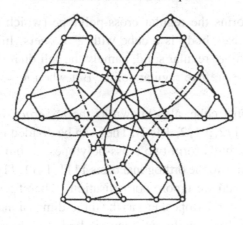

Figure 5.2 An orthogonality graph that cannot be coloured consistently with the Kochen–Specker rules, thus leading to metaphysical conclusions. Problem 5.2 reveals that it describes three interlocking cubes. Intercube orthogonalities are shown dashed.

To prove the Kochen–Specker theorem we need an absolute contradiction. Figure 5.2 gives an example of a graph that does just that, involving 33 vectors in three dimensions [725]. The 33 nodes in the graph represent a set of 33 measurements that can be performed on a qutrit, although only nodes connected by a link can be jointly measured. This time the predictions of quantum mechanics are logically inconsistent with the idea that the 33 possible outcomes were determined before the choice of what measurements to perform has been made.

The theorems of this section exclude all hidden variables of a certain kind, and make it absolutely transparent that quantum mechanics forces us to think very carefully about the relation between observations and what is being observed. However, the non-contextuality assumption can certainly be questioned.[17] It becomes much better motivated if the choice of the projector to be measured, and the choice of the compatible measurements that define its context, are made at separate locations. This observation led Bell to his inequalities [105, 107], which are violated by quantum mechanics and experiments [64, 343, 429, 827] alike.

Let us deviate from these deep questions into a simple geometrical one. There is a proof of the Kochen–Specker theorem that leads into the subject of Platonic solids in four dimensions, and we cannot resist the elegant details. In $\mathbb{R}^n$ each ray defines two unit vectors and two antipodal points on the unit sphere. The n rays defined by an orthonormal basis meets the unit sphere in $2n$ points

---

[17] It was, by Bell (1966) [106] and by Mermin (1993) [654], in non-commissioned reviews that restated the position with such clarity and simplicity that all previous discussions were eclipsed.

whose convex hull forms the regular cross-polytope (which is an octahedron if $n = 3$). The dual convex body is a cube with 2^n corners. In dimensions $n > 4$ there exists only one more regular solid, namely the self dual simplex. Plato found that when $n = 3$ there are five regular solids. But when $n = 4$ there are actually six of them.

The four-dimensional cube, also known as the *tesseract*, is formed by the 16 unit vectors $(\pm 1/2, \pm 1/2, \pm 1/2, \pm 1/2)$. They can be divided into two sets of eight vectors whose convex hulls form two cross-polytopes. In our condensed notation one such set corresponds to the orthogonal basis $1111, 1\bar{1}1\bar{1}, 11\bar{1}\bar{1}, 1\bar{1}\bar{1}1$ (remember that $\bar{1}$ stands for -1, and we ignore normalisation). Therefore the tesseract is the convex hull of two cross-polytopes, just as the three-dimensional cube is the convex hull of two simplices. Incidentally, all Platonic bodies in four dimensions can be visualised if we observe that all their vertices sit on a 3-sphere, and if we recall that Chapter 3 offered various ways in which the 3-sphere can be literally visualised. Indeed constructing a Platonic body is a question of tesselating a sphere in a suitable way [238].

In quantum mechanics an orthonormal basis can be thought of as the joint eigen-basis of a suitable set of commuting operators. Now consider a *Mermin square*, a set of nine operators that can be divided into sets of three mutually commuting Hermitian operators in six different ways [654]. The operators will be tensor products of the Pauli matrices (B.3). We arrange them as a square, where each row and each column contains such a commuting set, and for each set we give the corresponding eigenbasis in condensed notation:

$$
\begin{array}{ccc}
\mathbb{1} \otimes \sigma_z & \sigma_z \otimes \mathbb{1} & \sigma_z \otimes \sigma_z \\
\sigma_x \otimes \mathbb{1} & \mathbb{1} \otimes \sigma_x & \sigma_x \otimes \sigma_x \\
\sigma_x \otimes \sigma_z & \sigma_z \otimes \sigma_x & \sigma_y \otimes \sigma_y \\
\downarrow & \downarrow & \downarrow \\
1010 & 1100 & 1001 \\
10\bar{1}0 & 1\bar{1}00 & 100\bar{1} \\
0101 & 0011 & 0110 \\
010\bar{1} & 001\bar{1} & 01\bar{1}0
\end{array}
\qquad
\begin{array}{l}
\rightarrow \quad 1000, 0100, 0010, 0001 \\
\rightarrow \quad 1111, 1\bar{1}1\bar{1}, 11\bar{1}\bar{1}, 1\bar{1}\bar{1}1 \\
\rightarrow \quad 111\bar{1}, 1\bar{1}11, 11\bar{1}1, 1\bar{1}\bar{1}\bar{1}
\end{array}
\tag{5.61}
$$

Consider the three bases defined by the three rows. If the vectors are normalised to length one it will be seen that the scalar product of any vector in one of the bases with any vector in one of the other two bases equals $\pm 1/2$. Using the language to be adopted in Chapter 12, we say that these are three *mutually unbiased bases*. The convex hull of the 24 points on the unit sphere defined by these 12 rays is a polytope known as a *24-cell* (or as an icositetrachoron, if you prefer a name in Greek). It is formed by three symmetrically placed cross-polytopes, or equivalently by three interlocking tesseracts. Examining the scalar products among the 12 vectors defined by the columns of the square we find the same pattern of scalar products there, so in fact the Mermin square defines two 24-cells.

Now pick any unit vector coming from the columns, say $\mathbf{a}_1 = 1/\sqrt{2}(1, 1, 0, 0)$, and calculate its scalar product with all the 24 unit vectors $\mathbf{x}$ coming from the rows. By inspection

$$\mathbf{a} \cdot \mathbf{x} = \begin{cases} 1/\sqrt{2} & \text{6 times} \\ 0 & \text{8 times} \\ -1/\sqrt{2} & \text{6 times.} \end{cases} \tag{5.62}$$

The entire polytope is confined between two hyperplanes, $-1/\sqrt{2} \leq \mathbf{a} \cdot \mathbf{x} \leq 1/\sqrt{2}$. Thus we have found two of its facets containing six vertices each. Examining the Euclidean distances between the vertices of such a facet we find that the facet is, in fact, a regular three-dimensional octahedron. By the symmetry of the configuration we conclude that the 24-cell is a self dual Platonic body with 24 corners and 24 facets forming as many regular octahedra.

We seem to have lost sight of the Kochen–Specker theorem, but now we can come back to it. The two 24-cells defined by the Mermin square define a set of 24 rays that can be used to prove the Kochen–Specker theorem. The reason is that each pair of orthogonal rays from one of the 24-cells forms an orthonormal basis together with an orthonormal pair from the other. To prove uncolourability we begin by colouring one ray from each of two bases in the first 24-cell red. The symmetry of the configuration ensures that it does not matter which pair we choose. After some inspection we see that this forces the colouring of all the rays in the other 24-cell, and finally that this forces all the vectors in the remaining cross-polytope in the first 24-cell to be green – which is against the rules. So we have a set of 24 vectors that is indeed uncolourable in the Kochen–Specker sense.

Evidently many Kochen–Specker sets form seducingly beautiful configurations.[18] It remains to add that there is much ongoing research concerning orthogonality graphs and what they mean for quantum mechanics.[19]

---

[18] Kochen and Specker (1967) [551] used a set of 117 vectors for their first proof. The smallest number forcing the conclusion appears to be 18 vectors in four dimensions, and is obtained [186] by deleting six vectors from our four-dimensional example – which is itself due to Mermin [654] and Peres [725]. Readers who are curious about the two four-dimensional regular solids that have gone unmentioned here may consult Waegell and Aravind [943].

[19] See, in particular, Cabello, Severini and Winter [187], and take note of the preassigned harmony between quantum mechanics and mathematics [622].

Problems

5.1　Two pure states sit on the Bloch sphere separated by an angle θ. Choose an operator A whose eigenstates sit on the same great circle as the two pure states; the diameter defined by A makes an angle θ_A with the nearest of the pure states. Compute the Bhattacharyya distance between the two states for the measurement associated with A.

5.2　Show that the orthogonality graph in Figure 5.2 is uncolourable in the Kochen–Specker sense, and clarify its relation to the Escher print reproduced in Mermin's review [654].

6

Coherent states and group actions

Coherent states are the natural language of quantum theory.

—*John R. Klauder*[1]

In this chapter we study how groups act on the space of pure states, and especially the coherent states that can be associated to the group.

6.1 Canonical coherent states

The term 'coherent state' means different things to different people, but all agree that the *canonical coherent states*[2] form the canonical example, and with this we begin – even though it is somewhat against our rules in this book because now the Hilbert space is infinite dimensional. A key feature is that there is a group known as the *Heisenberg–Weyl group* that acts irreducibly on our Hilbert space.[3] The coherent states form a subset of states that can be reached from a special *reference* state by means of transformations belonging to the group. The group theoretical way of looking at things has the advantage that generalised coherent states can be defined in an analogous way as soon as one has a group acting irreducibly on a Hilbert space.

But let us begin at the beginning. The *Heisenberg algebra* of the operators $\hat{q}$ and $\hat{p}$ together with the unit operator $\mathbb{1}$ is defined by

$$[\hat{q}, \hat{p}] = i\hbar \mathbb{1}. \tag{6.1}$$

---

[1] Adapted from J. R. Klauder. Are coherent states the natural language of quantum theory? In V. Gorini and A. Frigerio, editors, *Fundamental Aspects of Quantum Theory*. Plenum, 1985.

[2] Also known as Glauber states. They were first described by Schrödinger (1926) [801], and then – after an interval – by Glauber (1963) [344]. For their use in quantum optics see Klauder and Sudarshan [541] and Mandel and Wolf [642].

[3] This is the point of view of Perelomov and Gilmore, the inventors of generalised coherent states. Useful reviews include Perelomov [724], Zhang et al. [1001], and Ali et al. [22]; a valuable reprint collection is Klauder and Skagerstam [540].

It acts on the infinite dimensional Hilbert space $\mathcal{H}_\infty$ of square integrable functions on the real line. In this chapter we will have hats on all the operators (because we will see much of the c-numbers q and p as well). Planck's constant $\hbar$ is written explicitly because in this chapter we will be interested in the limit in which $\hbar$ cannot be distinguished from zero. It is a dimensionless number because it is assumed that we have fixed units of length and momentum and use these to rescale the operators $\hat{q}$ and $\hat{p}$ so that they are dimensionless as well. If the units are chosen so that the measurement precision relative to this scale is of order unity, and if $\hbar$ is very small, then $\hbar$ can be safely set to zero. In SI units $\hbar = 1.054 \cdot 10^{-34}$ joule seconds. Here our policy is to set $\hbar = 1$, which means that classical behaviour may set in when measurements can distinguish only points with a very large separation in phase space. Classical phase space itself has no scale.

Equation (6.1) gives the Lie algebra of a group. First we recall the Baker–Hausdorff formula

$$e^{\hat{A}}e^{\hat{B}} = e^{\frac{1}{2}[\hat{A},\hat{B}]}e^{\hat{A}+\hat{B}} = e^{[\hat{A},\hat{B}]}e^{\hat{B}}e^{\hat{A}}, \tag{6.2}$$

which is valid whenever $[\hat{A},\hat{B}]$ commutes with $\hat{A}$ and $\hat{B}$. Thus equipped we form the unitary group elements

$$\hat{U}(q,p) \equiv e^{i(p\hat{q}-q\hat{p})}. \tag{6.3}$$

To find out what group they belong to we use the Baker–Hausdorff formula to find that

$$\hat{U}(q_1,p_1)\hat{U}(q_2,p_2) = e^{-i(q_1p_2-p_1q_2)}\,\hat{U}(q_2,p_2)\,\hat{U}(q_1,p_1). \tag{6.4}$$

This equation effectively defines a faithful representation of the *Heisenberg group*.[4] This group acts irreducibly on the Hilbert space $\mathcal{H}_\infty$ (and according to the Stone–von Neumann theorem it is the only unitary and irreducible representation that exists, but this is incidental to our purposes). Since the phase factor is irrelevant in the underlying projective space of states it is also a *projective representation* of the abelian group of translations in two dimensions.

Now we can form creation and annihilation operators in the standard way,

$$\hat{a} = \frac{1}{\sqrt{2}}(\hat{q}+i\hat{p}), \qquad \hat{a}^\dagger = \frac{1}{\sqrt{2}}(\hat{q}-i\hat{p}), \qquad [a,a^\dagger] = 1, \tag{6.5}$$

we can define the vacuum state $|0\rangle$ as the state that is annihilated by $\hat{a}$, and finally we can consider the two-dimensional manifold of states of the form

$$|q,p\rangle = \hat{U}(q,p)|0\rangle. \tag{6.6}$$

---

[4] This is really a three-dimensional group, including a phase factor. See Problem 12.2. The name *Weyl group* is more appropriate, but too many things are named after Weyl already. The Heisenberg algebra was discovered by Born and is engraved on his tombstone.

These are the canonical coherent states with the vacuum state serving as the reference state, and q and p serve as coordinates on the space of coherent states. Our question is: why are coherent states interesting? To answer it we must get to know them better.

Two important facts follow immediately from the irreducibility of the representation. First, the coherent states are complete in the sense that any state can be obtained by superposing coherent states. Indeed they form an *overcomplete* set because they are much more numerous than the elements of an orthonormal set would be – hence they are not orthogonal and do overlap. Second, we have the resolution of the identity

$$\frac{1}{2\pi} \int dq\, dp\, |q,p\rangle\langle q,p| = 1. \tag{6.7}$$

The overall numerical factor must be calculated, but otherwise this equation follows immediately from the easily ascertained fact that the operator on the left-hand side commutes with $\hat{U}(q,p)$; because the representation of the group is irreducible Schur's lemma implies that the operator must be proportional to the identity. Resolutions of identity will take on added importance in Section 10.1, where they will be referred to as 'POVMs'.

The coherent states form a Kähler manifold (see Section 3.3). To see this we first bring in a connection to complex analyticity that is very helpful in calculations. We trade $\hat{q}$ and $\hat{p}$ for the creation and annihilation operators and define the complex coordinate

$$z = \frac{1}{\sqrt{2}}(q + ip). \tag{6.8}$$

With some help from the Baker–Hausdorff formula, the submanifold of coherent states becomes

$$|q,p\rangle = |z\rangle = e^{z\hat{a}^\dagger - z^*\hat{a}}|0\rangle = e^{-|z|^2/2} \sum_{n=0}^{\infty} \frac{z^n}{\sqrt{n!}}|n\rangle. \tag{6.9}$$

We assume that the reader is familiar with the orthonormal basis spanned by the *number* or *Fock states* $|n\rangle$ [587].

We have reached a convenient platform from which to prove a number of elementary facts about coherent states. We can check that the states $|z\rangle$ are eigenstates of the annihilation operator:

$$\hat{a}|z\rangle = z|z\rangle. \tag{6.10}$$

Their usefulness in quantum optics has to do with this fact since light is usually measured by absorption of photons. In fact a high quality laser produces coherent states. A low quality laser produces a statistical mixture of coherent states – producing anything else is rather more difficult.

In x-space a coherent state wave function labelled by q and p is

$$\psi(x; q, p) = \langle x | q, p \rangle = \pi^{-1/4} e^{-ipq/2 + ipx - (x-q)^2/2}. \tag{6.11}$$

The shape is a Gaussian centred at $x = q$. The overlap between two coherent states is

$$\langle q_2, p_2 | q_1, p_1 \rangle = e^{-i(q_1 p_2 - p_1 q_2)/2} e^{-[(q_2 - q_1)^2 + (p_2 - p_1)^2]/4}. \tag{6.12}$$

It shrinks rapidly as the coordinate distance between the two points increases.

Let us now think about the space of coherent states itself. The choice of labels (q and p) is not accidental because we intend to regard the space of coherent states as being in some sense an embedding of the phase space of the classical system – whose quantum version we are studying – into the space of quantum states. Certainly the coherent states form a two-dimensional space embedded in the infinite dimensional space of quantum states and it will therefore inherit both a metric and a symplectic form from the latter. We know that the absolute value of the overlap is the cosine of the Fubini–Study distance D_{FS} between the two states (see Section 5.3), and for infinitesimally nearby coherent states we can read off the intrinsic metric ds^2 on the embedded submanifold. From Eq. (6.12) we see that the metric on the space of coherent states is

$$ds^2 = dz d\bar{z} = \frac{1}{2}(dq^2 + dp^2). \tag{6.13}$$

It is a flat space – indeed a flat vector space since the vacuum state forms a natural point of origin. From the phase of the overlap we can read off the symplectic form induced by the embedding on the submanifold of coherent states. It is non-degenerate:

$$\Omega = i dz \wedge d\bar{z} = dq \wedge dp. \tag{6.14}$$

It is the non-degenerate symplectic form that enables us to write down Poisson brackets and think of the space of coherent states as a phase space, isomorphic to the ordinary classical phase space spanned by q and p. The metric and the symplectic form are related to each other in precisely the way that is required for a Kähler manifold – although in a classical phase space the metric plays no particular role in the formalism. It is clearly tempting to try to argue that in some sense the space of coherent states is the classical phase space, embedded in the state space of its quantum version. A point in the classical phase space corresponds to a coherent state. The metric on phase space has a role to play here because Eq. (6.12) allows us to say that if the distance between the two points is large as measured by the metric,

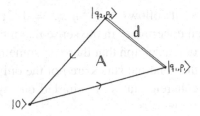

Figure 6.1 The overlap of two coherent states is determined by geometry: its modulus by the Euclidean distance d between the states and its phase by the (oriented) Euclidean area A of the triangle formed by the two states together with the reference state.

then the overlap between the two coherent states is small so that they interfere very little. Classical behaviour is clearly setting in; we will elaborate this point later on. Meanwhile we observe that the overlap can be written

$$\langle q_2, p_2 | q_1, p_1 \rangle = e^{-i \, 2(\text{area of triangle})} \, e^{-\frac{1}{2}(\text{distance})^2}, \qquad (6.15)$$

where the triangle is defined by the two states together with the reference state. See Figure 6.1. This is clearly reminiscent of the properties of geodesic triangles in $\mathbb{CP}^n$ that we studied in Section 4.8, but the present triangle lies within the space of coherent states itself. The reason why the phase equals an area is the same in both cases, namely that geodesics in $\mathbb{CP}^n$ as well as geodesics within the embedded subspace of canonical coherent states share the property of being null phase curves (in the sense of Section 4.8) [756].

There is a large class of observables – self-adjoint operators on Hilbert space – that can be associated with functions on phase space in a natural way. In general we define the *covariant symbol*[5] of the operator $\hat{A}$ as

$$A(q, p) = \langle q, p | \hat{A} | q, p \rangle. \qquad (6.16)$$

This is a function on the space of coherent states, that is on the would-be classical phase space. It is easy to compute the symbol of any operator that can be expressed as a polynomial in the creation and annihilation operators. In particular

$$\langle q, p | \hat{q} | q, p \rangle = q \qquad \langle q, p | \hat{q}^2 | q, p \rangle = q^2 + \frac{1}{2} \qquad (6.17)$$

(and similarly for $\hat{p}$). This implies that the variance, when the state is coherent, is

$$(\Delta q)^2 = \langle \hat{q}^2 \rangle - \langle \hat{q} \rangle^2 = \frac{1}{2} \qquad (6.18)$$

---

[5] The contravariant symbol will appear presently.

and similarly for $(\Delta p)^2$, so it follows that $\Delta q \Delta p = 1/2$; in words, the coherent states are states of minimal uncertainty in the sense that they saturate Heisenberg's inequality. This confirms our suspicion that there is 'something classical' about the coherent states. Actually the coherent states are not the only states that saturate the uncertainty relation; the coherent states are singled out by the extra requirement that $\Delta q = \Delta p$.

We have not yet given a complete answer to the question why coherent states are interesting – to get such an answer it is necessary to see how they can be used in some interesting application – but we do have enough hints. Let us try to gather together some key features:

- The coherent states form a complete set and there is a resolution of unity.
- There is a one-to-one mapping of a classical phase space onto the space of coherent states.
- There is an interesting set of observables whose expectation values in a coherent state match the values of the corresponding classical observables.
- The coherent states saturate an uncertainty relation and are in this sense as classical as they can be.

These are properties that we want any set of states to have if they are to be called coherent states. The generalised coherent states defined by Perelomov [724] and Gilmore [1001] do share these properties. The basic idea is to identify a group G that acts irreducibly on the Hilbert space and to define the coherent states as a particular *orbit* of the group. If the orbit is chosen suitably the resulting space of coherent states is a Kähler manifold. We will see how later; meanwhile let us observe that there are many alternative ways to define generalised coherent states. Sometimes any Kähler submanifold of $\mathbb{C}\mathbf{P}^n$ is referred to as coherent, regardless of whether there is a group in the game or not. Other times the group is there but the coherent states are not required to form a Kähler space. Here we require both, simply because it is interesting to do so. Certainly a *coherence group* formed by all the observables of interest for some particular problem arises in many applications, and the irreducible representation is just the minimal Hilbert space that allows us to describe the action of that group. Note that the coherence group is basically of kinematical origin; it is not supposed to be a symmetry group.

6.2 Quasi-probability distributions on the plane

The covariant symbol of an operator, as defined in Eq. (6.16), gives us the means to associate a function on phase space to any 'observable' in quantum mechanics. In classical physics an observable is precisely a function on phase space, and moreover the classical state of the system is represented by a particular such

Figure 6.2 Music scores resemble Wigner functions.

function – namely by a probability distribution on phase space. Curiously similar schemes can work in quantum mechanics too. It is interesting to think of music in this connection. See Figure 6.2. Music, as produced by orchestras and sold by record companies, is a certain function of time. But this is not how it is described by composers, who think of music[6] as a function of both time and frequency. Like the classical physicist, the composer can afford to ignore the limitations imposed by the uncertainty relation that holds in Fourier theory. The various quasi-probability distributions that we will discuss in this section are musical scores for quantum mechanics, and – remarkably – nothing is lost in this transcription. For the classical phase space we use the coordinates q and p. They may denote the position and momentum of a particle, but they may also define an electromagnetic field through its 'quadratures'.[7]

Quantum mechanics does provide a function on phase space that gives the probability distribution for obtaining the value q in a measurement associated with the special operator $\hat{q}$. This is just the familiar probability density for a pure state, or $\langle q|\hat{\rho}|q\rangle$ for a general mixed state. We ask for a function $W(q,p)$ such that this probability distribution can be recovered as a marginal distribution, in the sense that

$$\langle q|\hat{\rho}|q\rangle = \frac{1}{2\pi} \int_{-\infty}^{\infty} dp\, W(q,p). \tag{6.19}$$

This can be rewritten as an equation for the probability to find that the value q lies in an infinite strip bounded by the parallel lines $q = q_1$ and $q = q_2$, namely

$$P(q_1 \leq q \leq q_2) = \frac{1}{2\pi} \int_{strip} dq\, dp\, W(q,p). \tag{6.20}$$

In classical physics (as sketched in Section 5.6) we would go on to demand that the probability to find that the values of q and p are confined to an arbitrary phase space region Ω is given by the integral of $W(q,p)$ over Ω, and we would end up with a function $W(q,p)$ that serves as a joint probability distribution for both variables. This cannot be done in quantum mechanics. But it turns out that a requirement

---

[6] The sample attached is due to Józef Życzkowski, 1895–1967.
[7] A good general reference for this section is the survey of phase space methods given in the beautifully illustrated book by Leonhardt [587]. Note that factors of 2π are distributed differently throughout the literature.

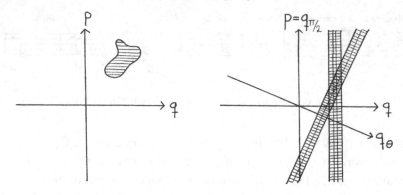

Figure 6.3 Left: in classical mechanics there is a phase space density such that we obtain the probability that p and q are confined to any region in phase space by integrating the density over that region. Right: in quantum mechanics we obtain the probability that p and q are confined to any infinite strip by integrating the Wigner function over that strip.

somewhat in-between Eq. (6.20) and the classical requirement can be met, and indeed uniquely determines the function $W(q, p)$, although the function will not qualify as a joint probability distribution because it may fail to be positive.

For this purpose consider the operators

$$\hat{q}_\theta = \hat{q} \cos\theta + \hat{p} \sin\theta \qquad \hat{p}_\theta = -\hat{q} \sin\theta + \hat{p} \cos\theta. \tag{6.21}$$

Note that $\hat{q}_\theta$ may be set equal to either $\hat{q}$ or $\hat{p}$ through a choice of the phase θ, and also that the commutator is independent of θ. The eigenvalues of $\hat{q}_\theta$ are denoted by q_θ. These operators gain in interest when one learns that the phase can actually be controlled in quantum optics experiments. We now have [129]

The theorem of Bertrand and Bertrand. *The function $W(q, p)$ is uniquely determined by the requirement that*

$$\langle q_\theta | \hat{\rho} | q_\theta \rangle = \frac{1}{2\pi} \int_{-\infty}^{\infty} dp_\theta \, W_\theta(q_\theta, p_\theta) \tag{6.22}$$

for all values of θ. Here $W_\theta(q_\theta, p_\theta) = W(q(q_\theta, p_\theta), p(q_\theta, p_\theta))$.

That is to say, as explained in Figure 6.3, we now require that all infinite strips are treated on the same footing. We will not prove uniqueness here, but we will see that the *Wigner function $W(q, p)$* has the stated property.

A convenient definition of the Wigner function is

$$W(q, p) = \frac{1}{2\pi} \int_{-\infty}^{\infty} \int_{-\infty}^{\infty} du \, dv \, \tilde{W}(u, v) \, e^{iuq + ivp}, \tag{6.23}$$

where the *characteristic function* is a 'quantum Fourier transformation' of the density matrix,

$$\tilde{W}(u,v) = \mathrm{Tr}\hat{\rho}\, e^{-iu\hat{q}-iv\hat{p}}. \tag{6.24}$$

To express this in the q-representation we use the Baker–Hausdorff formula and insert a resolution of unity to deduce that

$$e^{-iu\hat{q}-iv\hat{p}} = e^{\frac{i}{2}uv}e^{-iu\hat{q}}e^{-iv\hat{p}} = \int_{-\infty}^{\infty} \mathrm{d}q\, e^{-iuq}|q+\frac{v}{2}\rangle\langle q-\frac{v}{2}|. \tag{6.25}$$

We can just as well work with the operators in Eq. (6.21) and express everything in the q_θ-representation. We assume that this has been done – effectively it just means that we add a subscript θ to the eigenvalues. We then arrive, in a few steps, to Wigner's formula[8]

$$W_\theta(q_\theta, p_\theta) = \int_{-\infty}^{\infty} \mathrm{d}x\, \langle q_\theta - \frac{x}{2}|\hat{\rho}|q_\theta + \frac{x}{2}\rangle\, e^{ixp_\theta}. \tag{6.26}$$

Integration over p_θ immediately yields Eq. (6.22).

It is interesting to play with the definition a little. Let us look for a *phase space point operator* $\hat{A}_{qp}$ such that

$$W(q,p) = \mathrm{Tr}\hat{\rho}\hat{A}_{qp} = \int_{-\infty}^{\infty} \mathrm{d}q' \int_{-\infty}^{\infty} \mathrm{d}q''\, \langle q''|\hat{\rho}|q'\rangle\, \langle q'|\hat{A}_{qp}|q''\rangle. \tag{6.27}$$

That is to say that we will define the phase point operator through its matrix elements in the q-representation. The solution is

$$\langle q'|\hat{A}_{qp}|q''\rangle = \delta\left(q - \frac{q'+q''}{2}\right) e^{i(q'-q'')p}. \tag{6.28}$$

This permits us to write the density matrix in terms of the Wigner function as

$$\hat{\rho} = \frac{1}{2\pi} \int \mathrm{d}q\, \mathrm{d}p\, W(q,p)\, \hat{A}_{qp} \tag{6.29}$$

(as one can check by looking at the matrix elements). Hence the fact that the density matrix and the Wigner function determine each other has been made manifest. This is interesting because, given an ensemble of identically prepared systems, the various marginal probability distributions tied to the rotated position (or quadrature) operators in Eq. (6.21) can be measured – or at least a sufficient number of them

---

[8] Wigner (1932) [965] originally introduced this formula, with $\theta = 0$, as 'the simplest expression' that he (and Szilard) could think of.

can, for selected values of the phase θ – and then the Wigner function can be reconstructed using a tool known as the inverse Radon transformation. This is known as *quantum state tomography* and is actually being performed in laboratories.[9]

The Wigner function has some remarkable properties. First of all it is clear that we can associate a function W_A to an arbitrary operator $\hat{A}$ if we replace the operator $\hat{\rho}$ by the operator $\hat{A}$ in Eq. (6.24). This is known as the *Weyl symbol* of the operator and is very important in mathematics. If no subscript is used it is understood that the operator $\hat{\rho}$ is meant, that is $W \equiv W_\rho$. Now it is straightforward to show that the expectation value for measurements associated to the operator $\hat{A}$ is

$$\langle \hat{A} \rangle_{\hat{\rho}} \equiv \mathrm{Tr}\hat{\rho}\hat{A} = \frac{1}{2\pi} \int \mathrm{d}q\,\mathrm{d}p\, W_A(q,p)W(q,p). \tag{6.30}$$

This is the *overlap formula*. Thus the formula for computing expectation values is the same as in classical physics: integrate the function corresponding to the observable against the state distribution function over the classical phase space. Classical and quantum mechanics are nevertheless very different. To see this, choose two pure states $|\psi_1\rangle\langle\psi_1|$ and $|\psi_2\rangle\langle\psi_2|$ and form the corresponding Wigner functions. It follows (as a special case of the preceding formula, in fact) that

$$|\langle\psi_1|\psi_2\rangle|^2 = \frac{1}{2\pi} \int \mathrm{d}q\,\mathrm{d}p\, W_1(q,p)W_2(q,p). \tag{6.31}$$

If the two states are orthogonal the integral has to vanish. From this we conclude that the Wigner function cannot be a positive function in general. Therefore, even though it is normalised, it is not a probability distribution. But somehow it is 'used' by the theory as if it were.

On the other hand the Wigner function is subject to restrictions in ways that classical probability distributions are not – this must be so since we are using a function of two variables to express the content of the wave function, which depends on only one variable. For instance, using the Cauchy–Schwarz inequality one can show that

$$|W(q,p)| \leq 2. \tag{6.32}$$

It seems at first sight as if the only economical way to state all the restrictions is to say that the Wigner function arises from a Hermitian operator with trace unity and positive eigenvalues via Eq. (6.24). To formulate quantum mechanics in terms of the Wigner function is one thing. Making this formulation stand on its own legs is another, and for this we have to refer to the literature [241].

---

[9] The Wigner function was first measured experimentally by Smithey et al. [848] in 1993, and negative values of W were reported soon after. See Nogues et al. [697] and references therein.

To clarify how Wigner's formulation associates operators with functions we look at the moments of the characteristic function. Specifically, we observe that

$$\mathrm{Tr}\hat{\rho}(u\hat{q}+v\hat{p})^k = i^k \left(\frac{d}{d\sigma}\right)^k \mathrm{Tr}\hat{\rho}\, e^{-i\sigma(u\hat{q}+v\hat{p})}|_{\sigma=0} = i^k \left(\frac{d}{d\sigma}\right)^k \widetilde{W}(\sigma u, \sigma v)|_{\sigma=0}. \quad (6.33)$$

But if we undo the Fourier transformation we can conclude that

$$\mathrm{Tr}\hat{\rho}(u\hat{q} + v\hat{p})^k = \frac{1}{2\pi} \int dq\, dp\, (uq + vp)^k\, W(q,p). \quad (6.34)$$

By comparing the coefficients we see that the moments of the Wigner function give the expectation values of symmetrised products of operators, that is to say that

$$\mathrm{Tr}\hat{\rho}(\hat{q}^m\hat{p}^n)_{\mathrm{sym}} = \frac{1}{2\pi} \int dq\, dp\, W(q,p)\, q^m p^n, \quad (6.35)$$

where $(\hat{q}\hat{p})_{\mathrm{sym}} = (\hat{q}\hat{p}+\hat{p}\hat{q})/2$ and so on. Symmetric ordering is also known as *Weyl ordering*, and the precise statement is that Weyl ordered polynomials in $\hat{q}$ and $\hat{p}$ are associated to polynomials in q and p.

Finally, let us take a look at some examples. For the special case of a coherent state $|q_0, p_0\rangle$, with the wavefunction given in Eq. (6.11), the Wigner function is a Gaussian,

$$W_{|q_0,p_0\rangle}(q,p) = 2\, e^{-(q-q_0)^2-(p-p_0)^2}. \quad (6.36)$$

That the Wigner function of a coherent state is positive again confirms that there is 'something classical' about coherent states. Actually the coherent states are not the only states for which the Wigner function is positive – this property is characteristic of a class of states known as *squeezed states*.[10] If we superpose two coherent states the Wigner function will show two roughly Gaussian peaks with a 'wavy' structure in-between, where both positive and negative values occur; in the quantum optics literature such states are known (perhaps somewhat optimistically) as *Schrödinger cat states*. For the number (or Fock) states $|n\rangle$, the Wigner function is

$$W_{|n\rangle}(q,p) = 2\,(-1)^n\, e^{-q^2-p^2} L_n(2q^2 + 2p^2), \quad (6.37)$$

where the L_n are Laguerre polynomials. They have n zeros, so we obtain $n + 1$ circular bands of alternating signs surrounding the origin, concentrated within a radius of about $q^2 + p^2 = 2n + 1$. Note that both examples saturate the bound (6.32) somewhere.

---

[10] This is *Hudson's theorem* (1974) [479]. As the name suggests squeezed states are Gaussian, but 'squeezed'. A precise definition is $|\eta, z\rangle \equiv \exp[\eta(\hat{a}^\dagger)^2 - \eta^*\hat{a}^2]|z\rangle$, with $\eta \in \mathbb{C}$ [587].

To see how the Wigner function relates to other quasi-probability distributions that are in use we again look at its characteristic function, and introduce a one-parameter family of characteristic functions by changing the high frequency behaviour [188]:

$$\tilde{W}^{(s)}(u, v) = \tilde{W}(u, v) \, e^{s(u^2+v^2)/4}.\tag{6.38}$$

This leads to a family of phase space distributions

$$W^{(s)}(q,p) = \frac{1}{2\pi} \int du \, dv \, \tilde{W}(u, v) \, e^{iuq+ivp}.\tag{6.39}$$

For $s = 0$ we recover the Wigner function, $W = W^{(0)}$. We are also interested in the two 'dual' cases $s = -1$, leading to the *Husimi* or Q-function [485], and $s = 1$, leading to the *Glauber–Sudarshan* or P-function [344, 866]. Note that, when $s > 0$, the Fourier transformation of the characteristic function may not converge to a function, so the P-function will have a distributional meaning. Using the Baker–Hausdorff formula, and the definition (6.24) of $\tilde{W}(u, v)$, it is easily seen that the characteristic functions of the Q and P-functions are, respectively,

$$\tilde{Q}(u, v) \equiv \tilde{W}^{(-1)}(u, v) = \mathrm{Tr}\hat{\rho} \, e^{-i\eta^*\hat{a}} e^{-i\eta\hat{a}^\dagger}\tag{6.40}$$

$$\tilde{P}(u, v) \equiv \tilde{W}^{(1)}(u, v) = \mathrm{Tr}\hat{\rho} \, e^{-i\eta\hat{a}^\dagger} e^{-i\eta^*\hat{a}},\tag{6.41}$$

where $\eta \equiv (u + iv)/\sqrt{2}$. Equation (6.35) is now to be replaced by

$$\mathrm{Tr}\hat{\rho} \, \hat{a}^n \hat{a}^{\dagger m} = \frac{1}{2\pi} \int dq \, dp \, Q(z, \bar{z}) \, z^n \bar{z}^m\tag{6.42}$$

$$\mathrm{Tr}\hat{\rho} \, \hat{a}^{\dagger m} \hat{a}^n = \frac{1}{2\pi} \int dq \, dp \, P(z, \bar{z}) \, z^n \bar{z}^m,\tag{6.43}$$

where $z \equiv (q+ip)/\sqrt{2}$ as usual. Thus the Wigner, Q and P-functions correspond to different ordering prescriptions (symmetric, anti-normal and normal, respectively). The Q-function is a smoothed Wigner function,

$$Q(q,p) = \frac{1}{2\pi} \int dq' \, dp' \, W(q',p') \, 2 \, e^{-(q-q')^2-(p-p')^2},\tag{6.44}$$

as was to be expected because its high frequency behaviour was suppressed. It is also a familiar object. Using Eq. (6.36) for the Wigner function of a coherent state we see that

$$Q(q,p) = \frac{1}{2\pi} \int dq' \, dp' \, W_\rho(q',p') \, W_{|q,p\rangle}(q',p').\tag{6.45}$$

Using the overlap formula (6.30) this is

$$Q(q,p) = \text{Tr}\hat{\rho}\,|q,p\rangle\langle q,p| = \langle q,p|\hat{\rho}|q,p\rangle. \qquad (6.46)$$

This is the symbol of the density operator as defined in Eq. (6.16). The Q-function has some desirable properties that the Wigner function does not have, in particular it is everywhere positive. Actually, as should be clear from the overlap formula together with the fact that density matrices are positive operators, we can use the Wigner function of an arbitrary state to smooth a given Wigner function and we will always obtain a positive distribution. We concentrate on the Q-function because in that case the smoothing has been done with a particularly interesting reference state.

Since the integral of Q over phase space equals one the Husimi function is a genuine probability distribution. But it is a probability distribution of a somewhat peculiar kind, since it is not a probability density for mutually exclusive events. Instead $Q(q,p)$ is the probability that the system, if measured, would be found in a coherent state whose probability density has its mean at (q,p). Such 'events' are not mutually exclusive because the coherent states overlap. This has in its train that the overlap formula is not as simple as (6.30). If Q_A is the Q-function corresponding to an operator $\hat{A}$, and P_B the P-function corresponding to an operator $\hat{B}$, then

$$\text{Tr}\hat{A}\hat{B} = \frac{1}{2\pi} \int dq\,dp\, Q_A(q,p)\, P_B(q,p). \qquad (6.47)$$

This explains why the Q-function is known as a *covariant symbol* – it is dual to the P-function which is then the *contravariant symbol* of the operator. The relation of the P-function to the density matrix is now not hard to see (although unfortunately not in a constructive way). It must be true that

$$\hat{\rho} = \frac{1}{2\pi} \int dq\,dp\, |q,p\rangle P(q,p)\langle q,p|. \qquad (6.48)$$

This does not mean that the density matrix is a convex mixture of coherent states since the P-function may fail to be positive. Indeed in general it is not a function, and may fail to exist even as a tempered distribution. Apart from this difficulty we can think of the P-function as analogous to the barycentric coordinates introduced in Section 1.1.

Compared to the Wigner function the Q-function has the disadvantage that one does not recover the right marginals, say $|\psi(q)|^2$ by integrating over p. Moreover the definition of the Q-function (and the P-function) depends on the definition of the coherent states, and hence on some special reference state in the Hilbert space. This is clearly seen in Eq. (6.45), where the Wigner function of the reference state appears as a filter that is smoothing the Wigner function. But this peculiarity can be turned to an advantage. The Q-function may be the relevant probability distribution to use in a situation where the measurement device introduces a 'noise' that can be modelled by the reference state used to define the Q-function.[11] And the Q-function does have the advantage that it is a probability distribution. Unlike classical probability distributions, which obey no further constraints, it is also bounded from above by $1/2\pi$. This is a nice property that can be exploited to define an entropy associated to any given density matrix, namely

$$S_W = -\frac{1}{2\pi} \int dq \, dp \, Q(q,p) \ln Q(q,p). \tag{6.49}$$

This is the *Wehrl entropy* [949]. It is a concave function of the density matrix ρ, as it should be, and it has a number of other desirable properties as well. Unlike the classical Boltzmann entropy, which may assume the value $-\infty$, the Wehrl entropy obeys $S_W \geq 1$, and attains its lower bound if and only if the density matrix is a coherent pure state.[12] If we take the view that coherent states are classical states then this means that the Wehrl entropy somehow quantifies the departure from classicality of a given state. It will be compared to the quantum mechanical von Neumann entropy in Section 13.4.

6.3 Bloch coherent states

We will now study the *Bloch coherent states*.[13] In fact we have already done so – they are the complex curves, with topology $\mathbf{S}^2$, that were mentioned in Section 4.3. But this time we will develop them along the same lines as the canonical coherent states were developed. Our coherence group will be $SU(2)$, and our Hilbert space will be any finite dimensional Hilbert space in which $SU(2)$ acts irreducibly. The physical system may be a spin system of total spin j, but it can also be a collection of n two-level atoms. The mathematics is the same, provided that $n = 2j$; a good example of the flexibility of quantum mechanics. In the latter application the angular momentum eigenstates $|j, m\rangle$ are referred to as *Dicke states* [261], and the quantum number m is interpreted as half the difference between the number of excited and unexcited atoms. The dimension of Hilbert space is N, and throughout $N = n + 1 = 2j + 1$.

---

[11] There is a discussion of this, with many references, in Leonhardt's book [587] (and a very brief glimpse in our Section 10.1).

[12] This was conjectured by Wehrl (1978) [949] and proved by Lieb (1978) [601]. The original proof is quite difficult and depends on some hard theorems in Fourier analysis. The simplest proof so far is due to Luo [627], who relied on properties of the Heisenberg group. For some explicit expressions for selected states, see Orłowski [702].

[13] Also known as spin, atomic or $SU(2)$ coherent states. They were first studied by Klauder (1960) [538] and Radcliffe (1971) [759]. Bloch had, as far as we know, nothing to do with them, but we call them 'Bloch' so as to not prejudge which physical application we have in mind.

We need a little group theory to get started. We will choose our reference state to be $|j,j\rangle$, that is it has spin up along the z-axis. Then the coherent states are all states of the form $\mathcal{D}|j,j\rangle$, where $\mathcal{D}$ is a Wigner rotation matrix. Using our standard representation of the angular momentum operators (in Appendix B) the reference state is described by the vector $(1, 0, \ldots, 0)$, so the coherent states are described by the first column of $\mathcal{D}$. The rotation matrix can still be coordinatised in various ways. The Euler angle parametrization is a common choice, but we will use a different one that brings out the complex analyticity that is waiting for us. We set

$$\mathcal{D} = e^{zJ_-}e^{-\ln(1+|z|^2)J_3}e^{-\bar{z}J_+}e^{i\tau J_3}. \tag{6.50}$$

Because our group is defined using two by two matrices, we can prove this statement using two by two matrices; it will be true for all representations. Using the Pauli matrices from Appendix B we can see that

$$e^{zJ_-}e^{-\ln(1+|z|^2)J_3}e^{-\bar{z}J_+} = \frac{1}{\sqrt{1+|z|^2}}\begin{bmatrix} 1 & -\bar{z} \\ z & 1 \end{bmatrix} \tag{6.51}$$

and we just need to multiply this from the right with $e^{i\tau J_3}$ to see that we have a general $SU(2)$ matrix. Of course the complex number z is going to be a stereographic coordinate on the 2-sphere.

The final factor in Eq. (6.50) is actually irrelevant: we observe that when $e^{i\tau J_3}$ is acting on the reference state $|j,j\rangle$ it just contributes an overall constant phase to the coherent states. In $\mathbb{CP}^n$ the reference state is a *fixed point* of the $U(1)$ subgroup represented by $e^{i\tau J_3}$. In the terminology of Section 3.8 the isotropy group of the reference state is $U(1)$, and the $SU(2)$ orbit that we obtain by acting on it is the coset space $SU(2)/U(1) = \mathbf{S}^2$. This coset space will be coordinatised by the complex stereographic coordinate z. We choose the overall phase of the reference state to be zero. Since the reference state is annihilated by J_+ the complex conjugate $\bar{z}$ does not enter, and the coherent states are

$$|z\rangle = e^{zJ_-}e^{-\ln(1+|z|^2)J_3}e^{-\bar{z}J_+}|j,j\rangle = \frac{1}{(1+|z|^2)^j}e^{zJ_-}|j,j\rangle. \tag{6.52}$$

Using $z = \tan\frac{\theta}{2}e^{i\phi}$ we are always ready to express the coherent states as functions of the polar angles; $|z\rangle = |\theta, \phi\rangle$.

Since J_- is a lower triangular matrix, that obeys $(J_-)^{2j+1} = (J_-)^{n+1} = 0$, it is straightforward to express the unnormalised state in components. Using Eqs (B.4)–(B.6) from Appendix B we get

$$e^{zJ_-}|j,j\rangle = \sum_{m=-j}^{j} z^{j+m}\sqrt{\binom{2j}{j+m}}|j,m\rangle. \tag{6.53}$$

That is, the homogeneous coordinates – that do not involve the normalisation factor – for coherent states are

$$Z^\alpha = (1, \sqrt{2j}z, \ldots, \sqrt{\binom{2j}{j+m}}z^{j+m}, \ldots, z^{2j}).\qquad(6.54)$$

We can use this expression to prove that the coherent states form a sphere of radius $\sqrt{j/2}$, embedded in $\mathbb{C}\mathbf{P}^{2j}$. There is an intelligent way to do so, using the Kähler property of the metrics (see Section 3.3). First we compare with Eq. (4.6) and read off the affine coordinates $z^a(z)$ of the coherent states regarded as a complex curve embedded in $\mathbb{C}\mathbf{P}^{2j}$. For the coherent states we obtain

$$Z(z) \cdot \bar{Z}(\bar{z}) = (1 + |z|^2)^{2j}\qquad(6.55)$$

(so that, with the normalisation factor included, $\langle z|z\rangle = 1$). The Kähler potential for the metric is the logarithm of this expression, and the Kähler potential determines the metric as explained in Section 4.5. With no effort therefore, we see that on the coherent states the Fubini–Study metric induces the metric

$$ds^2 = \partial_a\bar{\partial}_b \ln\left(Z(z)\cdot\bar{Z}(\bar{z})\right) dz^a d\bar{z}^b = \partial\bar{\partial}\ln\left(1 + |z|^2\right)^{2j}dzd\bar{z},\qquad(6.56)$$

where we used the chain rule to see that $dz^a\partial_a = dz\partial_z$. This is $j/2$ times the metric on the unit 2-sphere written in stereographic coordinates, so we have proved that the embedding into $\mathbb{C}\mathbf{P}^{2j}$ turns the space of coherent states into a sphere of radius $\sqrt{j/2}$, as was to be shown. It is indeed a complex curve as defined in Section 4.3. The symplectic form on the space of coherent states is obtained from the same Kähler potential.

This is an important observation. At the end of Section 6.1 we listed four requirements that we want a set of coherent states to meet. One of them is that we should be able to think of the coherent states as forming a classical phase space embedded in the space of quantum mechanical pure states. In the case of canonical coherent states that phase space is the phase space spanned by q and p. The 2-sphere can also serve as a classical phase space because, as a Kähler manifold, it has a symplectic form that can be used to define Poisson brackets between any pair of functions on $\mathbf{S}^2$ (see Section 3.4). So all is well on this score. We also note that as j increases the sphere grows, and will in some sense approximate the flat plane with better and better accuracy.

Another requirement, listed in Section 6.1, is that there should exist a resolution of unity, so that an arbitrary state can be expressed as a linear combination of coherent states. This also works here. Using Eq. (6.53), this time with the normalisation factor from Eq. (6.52) included, we can prove that

$$\frac{2j+1}{4\pi}\int d\Omega|z\rangle\langle z| = \frac{2j+1}{4\pi}\int_0^\infty dr\int_0^{2\pi}d\phi\,\frac{4r}{(1+r^2)^2}|z\rangle\langle z| = \mathbb{1},\qquad(6.57)$$

where $d\Omega$ is the round measure on the unit 2-sphere, that we wrote out using the fact that z is a stereographic coordinate and the definition $z = re^{i\phi}$. It follows that the coherent states form a complete, and indeed an overcomplete, set. Next we require a correspondence between selected quantum mechanical observables on the one hand and classical observables on the other. Here we can use the symbol of an operator, defined in analogy with the definition for canonical coherent states. In particular the symbols of the generators of the coherence group are the classical phase space functions

$$J_i(\theta, \phi) = \langle z|\hat{J}_i|z \rangle = jn_i(\theta, \phi), \tag{6.58}$$

where $n_i(\theta, \phi)$ is a unit vector pointing in the direction labelled by the angles θ and ϕ. This is the symbol of the operator $\hat{J}_i$.

Our final requirement is that coherent states saturate an uncertainty relation. But there are several uncertainty relations in use for spin systems, a popular choice being

$$(\Delta J_x)^2 (\Delta J_y)^2 \geq -\frac{1}{4} \langle [\hat{J}_x, \hat{J}_y] \rangle^2 = \frac{\langle \hat{J}_z \rangle^2}{4}. \tag{6.59}$$

(Here $(\Delta J_x)^2 \equiv \langle \hat{J}_x^2 \rangle - \langle \hat{J}_x \rangle^2$ as usual.) States that saturate this relation are known [47] as *intelligent* states – but since the right-hand side involves an operator this does not mean that the left-hand side is minimised. The relation itself may be interesting if, say, a magnetic field singles out the z-direction for attention. We observe that a coherent state that has spin up in the z-direction satisfies this relation, but for a general coherent state the uncertainty relation itself has to be rotated before it is satisfied. Another measure of uncertainty is

$$\Delta^2 \equiv (\Delta J_x)^2 + (\Delta J_y)^2 + (\Delta J_z)^2 = \langle \hat{J}^2 \rangle - \langle \hat{J}_i \rangle \langle \hat{J}_i \rangle. \tag{6.60}$$

This has the advantage that Δ^2 is invariant under $SU(2)$, and takes the same value on all states in a given $SU(2)$ orbit in Hilbert space. This follows because $\langle \hat{J}_i \rangle$ transforms like an $SO(3)$ vector when the state is subject to an $SU(2)$ transformation.

One can now prove[14] that

$$j \leq \Delta^2 \leq j(j+1). \tag{6.61}$$

It is quite simple. We know that $\langle \hat{J}^2 \rangle = j(j+1)$. Moreover, in any given orbit we can use $SU(2)$ rotations to bring the vector $\langle \hat{J}_i \rangle$ to the form

$$\langle \hat{J}_i \rangle = \langle \hat{J}_z \rangle \delta_{i3}. \tag{6.62}$$

---

[14] As was done by Delbourgo [254] and further considered by Barros e Sá [98].

Expanding the state we see that

$$|\psi\rangle = \sum_{m=-j}^{j} c_m |m\rangle \;\Rightarrow\; \langle \hat{J}_z \rangle = \sum_{m=-j}^{j} m|c_m|^2 \;\Rightarrow\; 0 \le \langle \hat{J}_z \rangle \le j \qquad (6.63)$$

and the result follows in easy steps. It also follows that the lower bound in Eq. (6.61) is saturated if and only if the state is a Bloch coherent state, for which $\langle \hat{L}_z \rangle = j$. The upper bound will be saturated by states in the $\mathbf{RP}^2$ orbit when it exists, and also by some other states. It can be shown that Δ^2 when averaged over $\mathbf{CP}^n$, using the Fubini–Study measure from Section 4.7, is $j(j + \frac{1}{2})$. Hence for large values of j the average state is close to the upper bound in uncertainty.

In conclusion, Bloch coherent states obey all four requirements that we want coherent states to obey. There are further analogies to canonical coherent states to be drawn. Remembering the normalisation we obtain the overlap of two coherent states as

$$\langle z'|z\rangle = \frac{\sum_{k=0}^{n} \binom{n}{k}(z\bar{z}')^k}{(\sqrt{1+|z|^2}\sqrt{1+|z'|^2})^n} = \left(\frac{1+z\bar{z}'}{\sqrt{1+|z|^2}\sqrt{1+|z'|^2}} \right)^{2j}. \qquad (6.64)$$

The factorisation is interesting. On the one hand we can write

$$\langle z'|z\rangle = e^{-2iA} \cos D_{\mathrm{FS}}, \qquad (6.65)$$

where D_{FS} is the Fubini–Study distance between the two states and A is a phase factor that, for the moment, is not determined. On the other hand the quantity within brackets has a natural interpretation for $j = 1/2$, that is, on $\mathbf{CP}^1$. Indeed

$$\langle z'|z\rangle = \left(\langle z'|z\rangle_{|j=\frac{1}{2}} \right)^{2j}. \qquad (6.66)$$

But for the phase factor inside the brackets it is true that

$$\arg\langle z'|z\rangle_{|j=\frac{1}{2}} = \arg\langle z'|z\rangle_{|j=\frac{1}{2}} \langle z|+\rangle_{|j=\frac{1}{2}} \langle +|z'\rangle_{|j=\frac{1}{2}} = -2A_1, \qquad (6.67)$$

where $|+\rangle$ is the reference state for spin $1/2$, Eq. (4.98) was used, and A_1 is the area of a triangle on $\mathbf{CP}^1$ with vertices at the three points indicated. Comparing the two expressions we see that $A = 2jA_1$ and it follows that A is the area of a triangle on a sphere of radius $\sqrt{2j}/2 = \sqrt{j/2}$, that is the area of a triangle on the space of coherent states itself. The analogy with Eq. (6.15) for canonical coherent states is evident. This is a non-trivial result and has to do with our quite special choice of reference state; technically it happens because geodesics within the embedded 2-sphere of coherent states are null phase curves in the sense of Section 4.8, as a pleasant calculation confirms.[15]

---

[15] This statement remains true for the $SU(3)$ coherent states discussed in Section 6.4; Berceanu [125] has investigated things in more generality.

Table 6.1 *Comparison of the canonical coherent states on the plane and the Bloch coherent states on the sphere, defined by the Wigner rotation matrix $\mathcal{D}^{(j)}_{\theta,\phi}$. The overlap between two canonical (Bloch) coherent states is a function of the distance between two points on the plane (sphere), while the phase is determined by the area A of a flat (spherical) triangle.*

Hilbert space $\mathcal{H}$	Infinite	Finite, $N = 2j + 1$
phase space	plane $\mathbb{R}^2$	sphere S^2
commutation relations	$[\hat{q}, \hat{p}] = i$	$[J_i, J_j] = i\,\epsilon_{ijk}\,J_k$
basis	Fock $\{|0\rangle, |1\rangle, \ldots\}$	J_z eigenstates $|j, m\rangle$, $m = (-j, \ldots, j)$
reference state	vacuum $|0\rangle$	north pole $|\kappa\rangle = |j, j\rangle$
coherent states	$|q, p\rangle = \exp[i(p\hat{q} - q\hat{p})]\,|0\rangle$	$|\theta, \phi\rangle = \mathcal{D}^{(j)}_{\theta,\phi}\,|j, j\rangle$
POVM	$\frac{1}{2\pi}\int_{\mathbb{R}^2}|q, p\rangle\langle q, p|\,dq\,dp = \mathbb{1}$	$\frac{2j+1}{4\pi}\int_{\Omega}|\theta, \phi\rangle\langle\theta, \phi|\,d\Omega = \mathbb{1}$
overlap	$e^{-2iA}\exp\left[-\frac{1}{2}(D_E)^2\right]$	$e^{-2iA}\left[\cos\left(\frac{1}{2}D_R\right)\right]^{2j}$
Husimi representation	$Q_\rho(q, p) = \langle q, p|\rho|q, p\rangle$	$Q_\rho(\theta, \phi) = \langle\theta, \phi|\rho|\theta, \phi\rangle$
Wehrl entropy S_W	$-\frac{1}{2\pi}\int_{\mathbb{R}^2}dq\,dp\,Q_\rho\ln[Q_\rho]$	$-\frac{2j+1}{4\pi}\int_{\Omega}d\Omega\,Q_\rho\ln[Q_\rho]$
Lieb-Solovej theorem	$S_W(|\psi\rangle\langle\psi|) \geq 1$	$S_W(|\psi\rangle\langle\psi|) \geq \frac{2j}{2j+1}$

Quasi-probability distributions on the sphere can be defined in analogy to those on the plane. In particular, the Wigner function can be defined, and it is found to have similar properties as that on the plane. For instance, a Bloch coherent state $|\theta, \phi\rangle$ has a positive Wigner function centred around the point (θ, ϕ). We refer to the literature for details [10, 275]. The Husimi Q-function on the sphere will be given a detailed treatment in the next chapter. The Glauber–Sudarshan P-function exists for the Bloch coherent states whatever the dimension of the Hilbert space [675]; again a non-trivial statement because it requires the complex curve to 'wiggle around' enough inside $\mathbb{CP}^{2j}$ so that, once the latter is embedded in the flat vector space of Hermitian matrices, it fills out all the dimensions of the latter. It is like the circle that one can see embedded in the surface of a tennis ball.[16] The P-function will be positive for all mixed states in the convex cover of the Bloch coherent states.

To wind up the story so far we compare the canonical and the $SU(2)$ Bloch coherent states in Table 6.1.

---

[16] And when does the convex hull of an orbit of quantum states have full dimension in this sense? See Grabowski et al. [358].

6.4 From complex curves to $SU(K)$ coherent states

In the previous section we played down the viewpoint that regards the Bloch coherent states as a complex curve, but now we come back to it. Physically, what we have to do (for a spin system) is to assign a state to each direction in space. These states then serve as 'spin up' states for given directions. Mathematically this is a map from $\mathbf{S}^2 = \mathbb{CP}^1$ into $\mathbb{CP}^n$, with $n = 2j = N - 1$, that is a complex curve in the sense of Section 4.3. To describe the sphere of directions in space we use the homogeneous coordinates

$$(u, v) \sim \left(\cos \frac{\theta}{2}, \sin \frac{\theta}{2} e^{i\phi} \right) \sim \left(1, \tan \frac{\theta}{2} e^{i\phi} \right). \tag{6.68}$$

As we know already, the solution to our problem is

$$(u, v) \quad \rightarrow \quad \left(u^n, \sqrt{n} u^{n-1} v, \ldots, \sqrt{\binom{n}{k}} u^{n-k} v^k, \ldots, v^n \right). \tag{6.69}$$

As we also know, the Fubini–Study metric on $\mathbb{CP}^n$ induces the metric

$$ds^2 = \frac{n}{4} (d\theta^2 + \sin^2 \theta d\phi^2) \tag{6.70}$$

on this curve, so it is a round sphere with a radius of curvature equal to $\sqrt{n}/2$. The fact that already for modest values of n the radius of curvature becomes larger than the longest geodesic distance between two points in $\mathbb{CP}^n$ is not a problem since this sphere cannot be deformed to a point, and therefore it has no centre.

We have now specified the $m = j$ eigenstate for each possible spatial direction by means of a specific complex curve. It is remarkable that the location of all the other eigenstates is determined by the projective geometry of the curve. The location of the $m = -j$ state is evidently the antipodal point on the curve, as defined by the metric just defined. The other states lie off the curve and their location requires more work to describe. In the simple case of $n = 2$ (that is $j = 1$) the complex curve is a conic section, and the $m = 0$ state lies at the intersection of the unique pair of lines that are tangent to the curve at $m = \pm 1$, as described in Section 4.3. Note that the space of singlet states is an $\mathbb{RP}^2$, since any two antipodal points on the $m = 1$ complex curve defines the same $m = 0$ state. The physical interpretation of the points of $\mathbb{CP}^2$ is now fixed. Unfortunately it becomes increasingly difficult to work out the details for higher n.[17]

So far we have stuck to the simplest compact Lie algebra $SU(2)$. But, since the full isometry group of $\mathbb{CP}^n$ is $SU(n + 1)/Z_{n+1}$, it is clear that all the special unitary groups are of interest to us. For any K, a physical application of $SU(K)$ coherent

---

[17] See Brody and Hughston [177] for the $n = 3$ case.

states may be a collection of K-level atoms.[18] For a single K-level atom we would use a K-dimensional Hilbert space, and for the collection the dimension can be much larger. But how do we find the orbits under, say, $SU(3)$ in some $\mathbb{CP}^n$, and more especially what is the $SU(3)$ analogue of the Bloch coherent states? The simplest answer [342] is obtained by a straightforward generalisation of the approach just taken for $SU(2)$: since $SU(3)$ acts naturally on $\mathbb{CP}^2$ this means that we should ask for an embedding of $\mathbb{CP}^2$ into $\mathbb{CP}^n$. Let the homogeneous coordinates of a point in $\mathbb{CP}^2$ be $P^\alpha = (u, v, w)$. We embed this into $\mathbb{CP}^n$ through the equation

$$(u, v, w) \rightarrow \left(u^m, \ldots, \sqrt{\frac{m!}{k_1! k_2! k_3!}} u^{k_1} v^{k_2} w^{k_3}, \ldots \right); \qquad k_1 + k_2 + k_3 = m. \quad (6.71)$$

Actually this puts a restriction on the allowed values of n, namely

$$N = n + 1 = \frac{1}{2}(m + 1)(m + 2). \quad (6.72)$$

For these values of n we can choose the components of a symmetric tensor of rank m as homogeneous coordinates for $\mathbb{CP}^n$. The map from $P^\alpha \in \mathbb{CP}^2$ to a point in $\mathbb{CP}^5$ is then defined by

$$P^\alpha \rightarrow T^{\alpha\beta} = P^{(\alpha} P^{\beta)}. \quad (6.73)$$

In effect we are dealing with the symmetric tensor representation of $SU(3)$. (The brackets mean that we are taking the totally symmetric part; compare this to the symmetric multispinors used in Section 4.4.)

Anyway we now have an orbit of $SU(3)$ in $\mathbb{CP}^n$ for special values of n. To compute the intrinsic metric on this orbit (as defined by the Fubini–Study metric in the embedding space) we again take the short cut via the Kähler potential. We first observe that

$$Z \cdot \bar{Z} = \sum_{k_1 + k_2 + k_3 = m} \frac{m!}{k_1! k_2! k_3!} |u|^{2k_1} |v|^{2k_2} |w|^{2k_3} = (|u|^2 + |v|^2 + |w|^2)^m = (P \cdot \bar{P})^m.$$

$$(6.74)$$

Since the logarithm of this expression is the Kähler potential expressed in affine coordinates, we find that the induced metric on the orbits becomes

$$ds^2 = \partial_a \partial_{\bar{b}} \ln \left(P(z) \cdot \bar{P}(\bar{z}) \right)^m dz^a d\bar{z}^{\bar{b}} = m d\hat{s}^2, \quad (6.75)$$

where $d\hat{s}^2$ is the Fubini–Study metric on $\mathbb{CP}^2$ written in affine coordinates. Hence, just as for the Bloch coherent states, we find that the intrinsic metric on the orbit is just a rescaled Fubini–Study metric. Since the space of coherent states is Kähler the

---

[18] Here we refer to $SU(K)$ rather than $SU(N)$ because the letter $N = n + 1$ is otherwise engaged.

symplectic form can be obtained from the same Kähler potential, using the recipe in Section 3.3.

The generalisation to $SU(K)$ with an arbitrary K should be obvious. The Hilbert spaces in which we can represent $SU(K)$ using symmetric tensors of rank m have dimension

$$N_{K,m} \equiv \dim(\mathcal{H}_{K,m}) = \binom{K+m-1}{m} = \frac{(K+m-1)!}{m!\,(K-1)!}, \qquad (6.76)$$

which is the number of ways of distributing m identical objects in K boxes. For $K = 3$ it reduces to Eq. (6.72). The coherent states manifold itself is now $\mathbb{C}\mathbf{P}^{K-1}$, and the construction embeds it into $\mathbb{C}\mathbf{P}^{N_{K,m}-1}$.

But the story of coherent states for $SU(K)$ is much richer than this for every $K > 2$.

6.5 $SU(3)$ coherent states

Let us recall some group theory. In this book we deal mostly with the *classical groups* $SU(K)$, $SO(K)$, $Sp(K)$, and – in particular – with the special unitary groups $SU(K)$. There are several reasons for this. For one thing the isometry group of $\mathbb{C}\mathbf{P}^{K-1}$ is $SU(K)/Z_K$, for $\mathbb{R}\mathbf{P}^{K-1}$ it is the special orthogonal group $SO(K)$, and for the quaternionic projective space $\mathbb{H}\mathbf{P}^{K-1}$ it is the symplectic group $Sp(K)/Z_2$, so these groups are always there. Also they are all, in the technical sense, simple and compact groups and have in many respects analogous properties. In particular, most of their properties can be read off from their Lie algebras, and their complexified Lie algebras can be brought to the standard form

$$[H_i, H_j] = 0, \qquad [H_i, E_\alpha] = \alpha_i E_\alpha, \qquad (6.77)$$

$$[E_\alpha, E_\beta] = N_{\alpha\beta} E_{\alpha+\beta}, \qquad [E_\alpha, E_{-\alpha}] = \alpha^i H_i, \qquad (6.78)$$

where α_i is a member of the set of *root vectors* and $N_{\alpha\beta} = 0$ if $\alpha_i + \beta_i$ is not a root vector. The H_i form a maximal commuting set of operators and span what is known as the *Cartan subalgebra*. Of course α_i and $N_{\alpha\beta}$ depend on the group; readers not familiar with group theory will at least be able to see that $SU(2)$ fits into this scheme (if necessary, consult Appendix B, or a book on group theory [334]). A catalogue of the irreducible unitary representations can now be made by specifying a *highest weight vector* $|\mu\rangle$ in the Hilbert space, with the property that it is annihilated by the 'raising operators' E_α (for all positive roots), and it is labelled by the eigenvalues of the commuting operators H_i. Thus

$$E_\alpha|\mu\rangle = 0, \quad \alpha > 0; \qquad H_i|\mu\rangle = \frac{1}{2} m_i|\mu\rangle, \qquad (6.79)$$

where m_i are the components of a *weight vector*. For $SU(2)$ we expect every reader to be familiar with this result. For $SU(3)$ we obtain representations labelled by two integers m_1 and m_2, with the dimension of the representation being

$$\dim \mathcal{H}_{[m_1, m_2]} = \frac{1}{2}(m_1 + 1)(m_2 + 1)(m_1 + m_2 + 2). \tag{6.80}$$

We will concentrate on $SU(3)$ because the main conceptual novelty – compared to $SU(2)$ – can be seen already in this case.

In accordance with our general scheme we obtain $SU(3)$ coherent states by acting with $SU(3)$ on some reference state. It turns out that the resulting orbit is a Kähler manifold if and only if the reference state is a highest weight vector of the representation [724, 1001] – indeed the reason why the $\mathbf{S}^2$ orbit of $SU(2)$ is distinguished can now be explained as a consequence of its reference state being a highest weight vector. What is new compared to $SU(2)$ is that there are several qualitatively different choices of highest weight vectors. There is more news on a practical level: whereas the calculations in the $SU(2)$ case are straightforward, they become quite lengthy already for $SU(3)$. For this reason we confine ourselves to a sketch.[19] We begin in the defining representation of $SU(3)$ by introducing the 3×3 matrices

$$S_{ij} = |i\rangle\langle j|. \tag{6.81}$$

If $i < j$ we have a 'raising operator' E_α with positive root, if $i > j$ we have a 'lowering operator' $E_{-\alpha}$ with negative root. We exponentiate the latter and define

$$b_-(z) = e^{z_3 S_{31}} e^{z_1 S_{21}} e^{z_2 S_{32}}, \tag{6.82}$$

where the γ_i are complex numbers and no particular representation is assumed. In the defining representation this is the lower triangular 3×3 matrix,

$$b_-(z) = \begin{bmatrix} 1 & 0 & 0 \\ z_1 & 1 & 0 \\ z_3 & z_2 & 1 \end{bmatrix}. \tag{6.83}$$

Upper triangular matrices b_+ are defined analogously (or by Hermitian conjugation) and will annihilate the reference state that we are about to choose. Then we use that the fact that almost all (in the sense of the Haar measure) $SU(3)$ matrices can be written in the Gauss form

$$A = b_- D b_+, \tag{6.84}$$

---

[19] For full details consult Gnutzmann and Kuś [347], from whom everything that we say here has been taken.

where D is a diagonal matrix (obtained by exponentiating the elements of the Cartan subalgebra). Finally we define the coherent states by

$$|z\rangle = \mathcal{N}(z)\, b_-(z)\, |\mu_{[m_1,m_2]}\rangle, \tag{6.85}$$

where the reference state is a highest weight vector for the irreducible representation that we have chosen. This formula is clearly analogous to Eq. (6.52). The calculation of the normalising factor $\mathcal{N}$ is again straightforward but is somewhat cumbersome compared to the calculation in the $K = 2$ case. Let us define

$$\gamma_1 \equiv 1 + |z_1|^2 + |z_3|^2 \tag{6.86}$$
$$\gamma_2 \equiv 1 + |z_2|^2 + |z_3 - z_1 z_2|^2. \tag{6.87}$$

Then the result is

$$\mathcal{N}(z) = \sqrt{\gamma_1^{-m_1} \gamma_2^{-m_2}}. \tag{6.88}$$

With a view to employing affine coordinates in $\mathbb{C}\mathbf{P}^n$ we may prefer to write the state vector in the form $Z^\alpha = (1, \dots)$ instead. Then we find

$$Z(z) \cdot \bar{Z}(\bar{z}) = \gamma_1^{m_1} \gamma_2^{m_2}. \tag{6.89}$$

Equipped with this Kähler potential we can easily write down the metric and the symplectic form that is induced on the submanifold of coherent states.

There is, however, a subtlety. For degenerate representations, namely when either m_1 or m_2 equals zero, the reference state is annihilated not only by S_{ij} for $i < j$ but also by an additional operator. Thus

$$j_2 = 0 \quad \Rightarrow \quad S_{31}|\mu\rangle = 0 \quad \text{and} \quad m_1 = 0 \quad \Rightarrow \quad S_{32}|\mu\rangle = 0. \tag{6.90}$$

Readers who are familiar with the representation theory of $SU(3)$ see this immediately from Figure 6.4. This means that the isotropy subgroup is larger for the degenerate representations than in the generic case. Generically the state vector is left invariant up to a phase by group elements belonging to the Cartan subgroup $U(1) \times U(1)$, but in the degenerate case the isotropy subgroup grows to $SU(2) \times U(1)$.

The conclusion is that for a general representation the space of coherent states is the six-dimensional space $SU(3)/U(1) \times U(1)$, with metric and symplectic form derived from the Kähler potential given in Eq. (6.89). However, if $m_2 = 0$ the space of coherent states is the four-dimensional space $SU(3)/SU(2) \times U(1) = \mathbb{C}\mathbf{P}^2$, with metric and symplectic form again given by Eq. (6.89). This agrees with what we found in the previous section, where the Kähler potential was given by Eq. (6.74). For $m_1 = 0$ the space of coherent states is again $\mathbb{C}\mathbf{P}^2$; the Kähler potential is obtained from Eq. (6.89) by setting $z_1 = 0$. Interestingly, the 'classicality' of

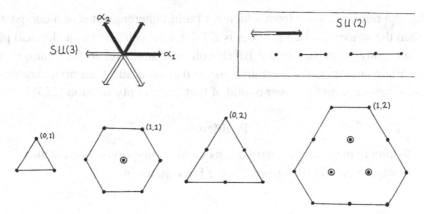

Figure 6.4 A reminder about representation theory: we show the root vectors of $SU(3)$ and four representations – the idea is that one can get to every point (state) by subtracting one of the simple root vectors α_1 or α_2 from the highest weight vector. For the degenerate representations $(0, m)$ there is no way to go when subtracting α_1; this is the reason why Eq. (6.90) holds. The corresponding picture for $SU(2)$ is shown inserted.

coherent states now gives rise to classical dynamics on manifolds of either four or six dimensions [346].

The partition of unity – or, the POVM – becomes

$$\mathbb{1} = \frac{(m_1 + 1)(m_2 + 1)(m_1 + m_2 + 2)}{\pi^3} \int d^2z_1 \, d^2z_2 \, d^2z_3 \, \frac{1}{\gamma_1^2 \gamma_2^2} |z\rangle\langle z| \qquad (6.91)$$

in the generic case and

$$\mathbb{1} = \frac{(m_1 + 1)(m_1 + 2)}{\pi^2} \int d^2z_1 \, d^2z_3 \, \frac{1}{\gamma_1^3} |z\rangle\langle z| \qquad \text{for } m_2 = 0, \qquad (6.92)$$

$$\mathbb{1} = \frac{(m_2 + 1)(m_2 + 2)}{\pi^2} \int d^2z_2 \, d^2z_3 \, \frac{1}{\gamma_2^3} |z\rangle\langle z| \qquad \text{for } m_1 = 0, \qquad (6.93)$$

where the last integral is evaluated at $z_1 = 0$.

In conclusion the $SU(3)$ coherent states differ from the $SU(2)$ coherent states primarily in that there is more variety in the choice of representation, and hence more variety in the possible coherent state spaces that occur. As is easy to guess, the same situation occurs in the general case of $SU(K)$ coherent states. Let us also emphasise that it may be useful to define generalised coherent states for some more complicated groups. For instance, in Chapter 16 we analyse pure product states of a composite $N \times M$ system, which may be regarded as coherent with respect to the group $SU(N) \times SU(M)$. The key point to observe is that if we use a maximal weight vector as the reference state from which we build coherent states of a compact Lie group then the space of coherent states is Kähler, so it can serve as a classical phase space, and many properties of the Bloch coherent states recur, for example there is an invariant measure of uncertainty using the quadratic Casimir operator, and coherent states saturate the lower bound of that uncertainty relation [255].

Problems

6.1   Compute the Q and P distributions for the one-photon Fock state.
6.2   Compute the Wehrl entropy for the Fock states $|n\rangle$.

7

The stellar representation

We are all in the gutter, but some of us are looking at the stars.

—*Oscar Wilde*[1]

We have already, in Section 4.4, touched on the possibility to regard points in complex projective space $\mathbb{CP}^{N-1}$ as unordered sets of $n = N - 1$ stars on a 'celestial sphere'. There is an equivalent description in terms of the n zeros of the Husimi function. Formulated either way, the stellar representation illuminates the orbits of $SU(2)$, the properties of the Husimi function, and the nature of 'typical' and 'random' quantum states.

7.1 The stellar representation in quantum mechanics

Our previous discussion of the stellar representation was based on projective geometry only. When the points are thought of as quantum states we will wish to take the Fubini–Study metric into account as well. This means that we will want to restrict the transformations acting on the celestial sphere, from the Möbius group $SL(2, \mathbb{C})/Z_2$ to the distance preserving subgroup $SO(3) = SU(2)/Z_2$. So the transformations that we consider are

$$z \to z' = \frac{\alpha z - \beta}{\beta^* z + \alpha^*}, \tag{7.1}$$

where it is understood that $z = \tan \frac{\theta}{2} e^{i\phi}$ is a stereographic coordinate on the 2-sphere. Recall that the idea in Section 4.4 was to associate a polynomial to each vector in $\mathbb{C}^N$, and the roots of that polynomial to the corresponding point in $\mathbb{CP}^{N-1}$. The roots are the stars. We continue to use this idea, but this time we want to make sure that an $SU(2)$ transformation really corresponds to an ordinary rotation of the

---

[1] Reproduced from Oscar Wilde, *The Importance of Being Earnest*. Methuen & Co, 1917.

sphere on which we have placed our stars. For this purpose we need to polish our conventions a little: we want a state of spin 'up' in the direction given by the unit vector $\mathbf{n}$ to be represented by $n = 2j$ points sitting at the point where $\mathbf{n}$ meets the sphere, and more generally a state that is an eigenstate of $\mathbf{n} \cdot \mathbf{J}$ with eigenvalue m to be represented by $j + m$ points at this point and $j - m$ points at the antipode. Now consider a spin $j = 1$ particle and place two points at the east pole of the sphere. In stereographic coordinates the east pole is at $z = \tan \frac{\pi}{4} = 1$, so the east pole polynomial is

$$w(z) = (z - 1)^2 = z^2 - 2z + 1. \tag{7.2}$$

The eigenvector of J_x with eigenvalue $+1$ is $Z^\alpha = (1, \sqrt{2}, 1)$, so if we are to read off this vector from (7.2) we must set

$$w(z) \equiv Z^0 z^2 - \sqrt{2} Z^1 z + Z^2. \tag{7.3}$$

After a little experimentation like this it becomes clear how to choose the conventions.

The correspondence between a vector in Hilbert space and the n roots of the polynomial is given by

$$|\psi\rangle = \sum_{\alpha=0}^{n} Z^\alpha |e_\alpha\rangle$$

$$\leftrightarrow \tag{7.4}$$

$$w(\zeta) = \sum_{\alpha=0}^{n} (-1)^\alpha Z^\alpha \sqrt{\binom{n}{\alpha}} \zeta^{n-\alpha} = N \sum_{\alpha=0}^{n} (-1)^\alpha s_\alpha \zeta^{n-k}.$$

We use ζ for the variable to avoid notational confusion later, $s_\alpha = s_\alpha(z_1, \ldots, z_n)$ is the αth symmetric function of the n roots $z_1, \ldots, z_n$, and $N = N(z_1, \ldots, z_n)$ is a normalising factor that also depends on the roots. The convention for when ∞ counts as a root is as described in Section 4.4.

The action of $SU(2)$ is now completely transparent. An $SU(2)$ transformation in Hilbert space corresponds to a rotation of the sphere where the stars sit. An eigenstate of the operator $\mathbf{n} \cdot \mathbf{J}$ with eigenvalue m, where $\mathbf{n} = (\sin \theta \cos \phi, \sin \theta \sin \phi, \cos \theta)$, is represented by $j + m$ stars at $z_i = \tan \frac{\theta_i}{2} e^{i\phi_i}$ and $j - m$ stars at the antipodal point (see Figure 4.7).[2]

The anti-unitary operation of time reversal Θ, as defined in Section 5.5, is nicely described too. See Figure 7.1. We insist that

$$\Theta i + i\Theta = \Theta \mathbf{J} + \mathbf{J}\Theta = 0 \tag{7.5}$$

---

[2] The resemblance to Schwinger's (1952) [816] harmonic oscillator representation of $SU(2)$ is not accidental – he was led to his representation by Majorana's (1932) [639] description of the stellar representation. See Appendix B.2.

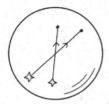 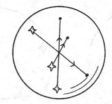 

Figure 7.1 In the stellar representation a time reversal moves the stars to their antipodal positions; time reversal invariant states can therefore occur only when the number of stars is even (as it is in the rightmost case, representing a point in $\mathbb{RP}^4$).

and find that Θ is represented by moving all stars to their antipodal points,

$$z_i \rightarrow \omega_i = -\frac{1}{\bar{z}_i}. \tag{7.6}$$

Since no configuration of an odd number of points can be invariant under such a transformation it immediately follows that there are no states invariant under time reversal when n is odd. For even n there will be a subspace of states left invariant under time reversal. For $n = 2$ it is evident that this subspace is the real projective space $\mathbb{RP}^2$, because the stellar representation of a time reversal invariant state is a pair of points in antipodal position on $\mathbf{S}^2$. This is not the $\mathbb{RP}^2$ that we would obtain by choosing all coordinates real, rather it is the $\mathbb{RP}^2$ consisting of all possible $m = 0$ states. For higher n the story is less transparent, but it is still true that the invariant subspace is $\mathbb{RP}^n$, and we obtain a stellar representation of real projective space into the bargain – a point in the latter corresponds to a configuration of stars on the sphere that is symmetric under inversion through the origin.

The action of a general unitary transformation, not belonging to the $SU(2)$ subgroup that we have singled out for attention, is of course not transparent in the stellar representation. Nor is it easy to see if two constellations of stars represent orthogonal states, except perhaps for small n. Some special cases are easy though. For instance, two constellations of stars represent orthogonal states if one state has p coinciding stars at some points, the other has q stars at the antipodal point, and $p + q > n$. This endows the location of the stars with an operational significance. A spin system, say, cannot be observed to have spin up along a direction that points away from a star on our celestial sphere. This is so because its state is necessarily orthogonal to a state with n coinciding stars at the antipode.

One special orthogonal basis is easily recognised: the one consisting of the angular momentum eigenstates, conveniently denoted $|j + m, j - m\rangle = |n - k, k\rangle$ and now referred to as Dicke states. They appear as $n - k$ stars sitting at the north

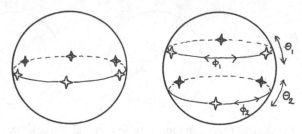

Figure 7.2 To the left we see a noon state, a superposition of two coherent states. Changing the latitude and the orientation of the n-gon we sweep out a two-dimensional subspace. To the right we see a superposition of three Dicke states (guess which!). Changing the two latitudes and the orientations of the two n-gons we sweep out a three-dimensional subspace.

pole and k stars sitting at the south pole. Other states resulting in easily recognised constellations of stars include the *noon state*

$$|\text{noon}\rangle = |n, 0\rangle + |0, n\rangle. \tag{7.7}$$

It appears as n stars making up the vertices of a regular n-gon on a circle at constant latitude. An interesting feature of noon states is that a rotation through an angle of only π/n radians suffice to turn it into an orthogonal state, suggesting that it may be useful in quantum metrology [143]. The reader is invited to think up further examples of interesting-looking states. See Figure 7.2.

7.2 Orbits and coherent states

The stellar representation shows its strength when we decide to classify all possible orbits of $SU(2)/Z_2 = SO(3)$ in $\mathbb{C}\mathbf{P}^n$.[3] The general problem is that of a group G acting on a manifold M; the set of points that can be reached from a given point is called a G-orbit. In itself the orbit is the coset space G/H, where H is the subgroup of transformations leaving the given point invariant. H is called the *little group*. We want to know what kind of orbits there are, and how M is partitioned into G-orbits. The properties of the orbits will depend on the isotropy group H. A part of the manifold M where all orbits are similar is called a *stratum*, and in general M becomes a *stratified manifold* foliated by orbits of different kinds. We can also define the *orbit space* as the space whose points are the orbits of G in M; a function of M is called *G-invariant* if it takes the same value at all points of a given orbit, which means that it is automatically a well defined function on the orbit space.

In Section 6.3 we had $M = \mathbb{C}\mathbf{P}^n$, and – by selecting a particular reference state – we selected a particular orbit as the space of Bloch coherent states. This orbit had

---

[3] This was done by Bacry (1974) [80]. By the way his paper contains no references whatsoever to prior work.

the intrinsic structure of $SO(3)/SO(2) = \mathbf{S}^2$. It was a successful choice, but it is interesting to investigate if other choices would have worked as well. For $\mathbb{C}\mathbf{P}^1$ the problem is trivial: we have one star and can rotate it to any position, so there is only one orbit, namely $\mathbb{C}\mathbf{P}^1$ itself. For $\mathbb{C}\mathbf{P}^2$ it gets more interesting. We now have two stars on the sphere. Suppose first that they coincide. The little group – the subgroup of rotations that leaves the configuration of stars invariant – consists of rotations around the axis where the pair is situated. Therefore the orbit becomes $SO(3)/SO(2) = O(3)/O(2) = \mathbf{S}^2$. Every state represented by a pair of coinciding points lies on this orbit. Referring back to Section 6.4 we note that the states on this orbit can be regarded as states that have spin up in some direction, and we already know that these form a sphere inside $\mathbb{C}\mathbf{P}^2$. But now we know it in a new way. The next case to consider is when the pair of stars are placed antipodally on the sphere. This configuration is invariant under rotations around the axis defined by the stars, but also under an extra turn that interchanges the two points. Hence the little group is $SO(2) \times Z_2 = O(2)$ and the orbit is $\mathbf{S}^2/Z_2 = O(3)/[O(2) \times O(1)] = \mathbb{R}\mathbf{P}^2$. For any other pair of stars the little group has a single element, namely a discrete rotation that interchanges the two. Hence the generic orbit is $SO(3)/Z_2 = O(3)/[O(1) \times O(1)]$. Since $SO(3) = \mathbb{R}\mathbf{P}^3 = \mathbf{S}^3/Z_2$ we can also think of this as a space of the form $\mathbf{S}^3/\Gamma$, where Γ is a discrete subgroup of the isometry group of the 3-sphere. Spaces of this particular kind are called *lens spaces* by mathematicians.

To solve the classification problem for arbitrary n we first recall that the complete list of subgroups of $SO(3)$ consists of e (the trivial subgroup consisting of just the identity), the discrete groups C_p (the cyclic groups, with p some integer), D_p (the dihedral groups), T (the symmetry group of the tetrahedron), O (the symmetry group of the octahedron and the cube), Y (the symmetry group of the icosahedron and the dodecahedron) and also the continuous groups $SO(2)$ and $O(2)$. This is well known to crystallographers and to mathematicians who have studied the regular polyhedra. Recall moreover that the tetrahedron has four vertices, six edges and four faces so that we may denote it by $\{4,6,4\}$. Similarly the octahedron is $\{6,12,8\}$, the cube $\{8,12,6\}$, the dodecahedron $\{12,30,20\}$ and the icosahedron is $\{20,30,12\}$. The question is: given the number n, does there exist a configuration of n stars on the sphere invariant under the subgroup Γ? The most interesting case is for $\Gamma = SO(2)$ which occurs for all n, for instance when all the stars coincide. For $O(2)$ the stars must divide themselves equally between two antipodal positions, which can happen for all even n. The cyclic group C_p occurs for all $n \geq p$, the groups D_2 and D_4 for all even $n \geq 4$, and the remaining dihedral groups D_p when $n = p + pa + 2b$ with a and b non-negative integers. For the tetrahedral group T, we must have $n = 4a + 6b$ with a non-negative (this may be a configuration of a stars at each corner of the tetrahedron and b stars sitting 'above' each edge midpoint – if only the latter stars are present the symmetry group becomes O). Similarly the octahedral group O occurs when $n = 6a + 8b$ and the icosahedral group Y when $n = 12a + 20b + 30c$, a, b and c being integers. Finally configurations with no symmetry at all appear for all $n \geq 3$. The possible orbits are of the form $SO(3)/\Gamma$; if Γ is one of the discrete groups this is a three-dimensional manifold. The only orbit which is a Kähler manifold is the orbit arising from coinciding stars, which is one of the $SO(3)/SO(2) = S^2$ orbits. The Kähler orbit can serve as a classical phase space, and we will use it to form coherent states.

Since the orbits have rather small dimensions the story of how $\mathbb{CP}^n$ can be partitioned into $SU(2)$ orbits is rather involved when n is large, but for $n = 2$ it can be told quite elegantly. There will be a one-parameter family of three-dimensional orbits, and correspondingly a one-parameter choice of reference vectors. A possible choice is

$$Z_0^\alpha(\sigma) = \begin{pmatrix} \cos \sigma \\ 0 \\ \sin \sigma \end{pmatrix}, \qquad 0 \leq \sigma \leq \frac{\pi}{4}. \tag{7.8}$$

The corresponding polynomial is $w(z) = z^2 + \tan \sigma$ and its roots will go from coinciding to antipodally placed as σ grows. If we act on this vector with a general 3×3 rotation matrix $\mathcal{D}$ – parametrized say by Euler angles as in Eq. (3.143), but this time for the three-dimensional representation – we will obtain the state vector

$$Z^\alpha(\sigma, \tau, \theta, \phi) = \mathcal{D}^\alpha{}_\beta(\tau, \theta, \phi) Z_0^\beta(\sigma), \tag{7.9}$$

where σ labels the orbit and the Euler angles serve as coordinates within the orbit. The range of τ turns out to be $[0, \pi[$. Together these four parameters serve as a coordinate system for $\mathbb{CP}^2$.

By means of lengthy calculations we can express the Fubini–Study metric in these coordinates; in the notation of Section 3.7

$$ds^2 = d\sigma^2 + 2(1 + \sin 2\sigma)\Theta_1^2 + 2(1 - \sin 2\sigma)\Theta_2^2 + 4\sin^2 2\sigma\,\Theta_3^2. \tag{7.10}$$

On a given orbit σ is constant and the metric becomes the metric of a 3-sphere that has been squashed in a particular way. It is in fact a lens space rather than a 3-sphere because of the restricted range of the periodic coordinate τ. When $\sigma = 0$ the orbit degenerates to a (round) 2-sphere, and when $\sigma = \pi/4$ to real projective 2-space. The parameter σ measures the distance to the $\mathbf{S}^2$ orbit.

Another way to look at this is to see what the orbits look like in the octant picture (Section 4.6). The answer turns out to be quite striking [97] and is given in Figure 7.3.

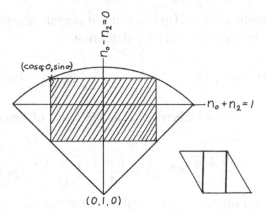

Figure 7.3 An orbit of $SU(2)$ acting on $\mathbb{C}\mathbf{P}^2$. We use orthographic coordinates to show the octant in which the orbit fills out a two-dimensional rectangle. Our reference state is at its upper left-hand corner. In the tori there is a circle and we show how it winds around its torus; its precise position varies. When $\sigma \to 0$ the orbit collapses to $\mathbf{S}^2$ and when $\sigma \to \pi/4$ it collapses to $\mathbb{R}\mathbf{P}^2$.

7.3 The Husimi function

Scanning our list of orbits we see that the only even-dimensional orbits – on which a non-degenerate symplectic form can be defined – are the exceptional $SO(3)/SO(2)$ $= \mathbf{S}^2$ orbits. Of those, the most interesting one, and the one that will serve us as a classical phase space, is the orbit consisting of states with all stars coinciding. These are the Bloch coherent states. We will now proceed to the Husimi or Q function for such coherent states. In Section 6.2 it was explained that the Husimi function is a genuine probability distribution on the classical phase space and that, at least in theory, it allows us to reconstruct the quantum state. Our notation will differ somewhat from that of Section 6.3, so before we introduce it we recapitulate what we know so far. The dimension of the Hilbert space $\mathcal{H}_N$ is $N = n+1 = 2j+1$. Using the basis states $|e_\alpha\rangle = |j, m\rangle$, a general pure state can be written in the form

$$|\psi\rangle = \sum_{\alpha=0}^{n} Z^\alpha |e_\alpha\rangle. \tag{7.11}$$

Using the correspondence in Eq. (7.4) we find that the (normalised) Bloch coherent states, with n coinciding stars at $z = \tan \frac{\theta}{2} e^{i\phi}$, take the form

$$|z\rangle = \frac{1}{(1 + |z|^2)^{\frac{n}{2}}} \sum_{\alpha=0}^{n} \sqrt{\binom{n}{\alpha}} z^\alpha |e_\alpha\rangle. \tag{7.12}$$

(The notation here is inconsistent – Z^α is a component of a vector, while z^α is the complex number z raised to a power – but quite convenient.) At this point we introduce the *Bargmann function*. Up to a factor it is again an nth order polynomial uniquely associated to any given state. By definition

$$\psi(z) = \langle \psi | z \rangle = \frac{1}{(1+|z|^2)^{\frac{n}{2}}} \sum_{\alpha=0}^{n} \bar{Z}_\alpha \sqrt{\binom{n}{\alpha}} z^\alpha. \qquad (7.13)$$

It is convenient to regard our Hilbert space as the space of functions of this form, with the scalar product

$$\langle \psi | \phi \rangle = \frac{n+1}{4\pi} \int d\Omega \, \bar{\psi} \phi \equiv \frac{n+1}{4\pi} \int \frac{4 \, d^2 z}{(1+|z|^2)^2} \, \bar{\psi} \, \bar{\phi}. \qquad (7.14)$$

So $d\Omega$ is the usual measure on the unit 2-sphere. That this is equivalent to the usual scalar product follows from our formula (6.57) for the resolution of unity.[4]

Being a polynomial the Bargmann function can be factorised. Then we obtain

$$\psi(z) = \frac{\bar{Z}_n}{(1+|z|^2)^{\frac{n}{2}}} (z-\omega_1)(z-\omega_2) \cdots \cdots (z-\omega_n). \qquad (7.15)$$

The state vector is uniquely characterised by the zeros of the Bargmann function, so again we have stars on the celestial sphere to describe our states. But the association is not quite the same that we have used so far. For instance, we used to describe a coherent state by the polynomial

$$w(z) = (z-z_0)^n. \qquad (7.16)$$

(And we know how to read off the components Z^α from this expression.) But the Bargmann function of the same coherent state $|z_0\rangle$ is

$$\psi_{z_0}(z) = \langle z_0 | z \rangle = \frac{\bar{z}_0^n}{(1+|z|^2)^{\frac{n}{2}}(1+|z_0|^2)^{\frac{n}{2}}} \left(z + \frac{1}{\bar{z}_0} \right)^n. \qquad (7.17)$$

Hence $\omega_0 = -1/\bar{z}_0$. In general the zeros of the Bargmann function are antipodally placed with respect to our stars. As long as there is no confusion, no harm is done.

With the Husimi function for the canonical coherent states in mind, we rely on the Bloch coherent states to define the Husimi function as

$$Q_\psi(z) = |\langle \psi | z \rangle|^2 = \frac{|Z^n|^2}{(1+|z|^2)^n} |z-\omega_1|^2 |z-\omega_2|^2 \cdots \cdots |z-\omega_n|^2. \qquad (7.18)$$

It is by now obvious that the state $|\psi\rangle$ is uniquely defined by the zeros of its Husimi function. It is also obvious that Q is positive and, from Eqs. (7.14) and (6.57), that it is normalised to one:

$$\frac{n+1}{4\pi} \int d\Omega \, Q_\psi(z) = 1. \qquad (7.19)$$

---

[4] This Hilbert space was presented by V. Bargmann (1961) [91] in an influential paper.

Hence it provides a genuine probability distribution on the 2-sphere. It is bounded from above. Its maximum value has an interesting interpretation: the Fubini–Study distance between $|\psi\rangle$ and $|z\rangle$ is given by $D_{FS} = \arccos\sqrt{\kappa}$, where $\kappa = |\langle\psi|z\rangle|^2 = Q_\psi(z)$, so the maximum of $Q_\psi(z)$ determines the minimum distance between $|\psi\rangle$ and the orbit of coherent states.

A convenient way to rewrite the Husimi function is

$$Q(z) = k_n\, \sigma(z, \omega_1)\sigma(z, \omega_2) \cdots \cdots \sigma(z, \omega_n), \tag{7.20}$$

where

$$\sigma(z, \omega) \equiv \frac{|z - \omega|^2}{(1 + |z|^2)(1 + |\omega|^2)} = \frac{1 - \cos d}{2} = \sin^2 \frac{d}{2} = \frac{d_{ch}^2}{4}, \tag{7.21}$$

and d is the geodesic, d_{ch} the chordal distance between the two points z and ω. (To show this, note that the function is invariant under rotations. Set $z = \tan\frac{\theta}{2}e^{i\phi}$ and $\omega = 0$. Simple geometry now tells us that $\sigma(z, \omega)$ is one-quarter of the square of the chordal distance d_{ch} between the two points, assuming the sphere to be of unit radius. (See Figure 7.4.) The factor k_n in Eq. (7.20) is a z-independent normalising factor. Unfortunately it is a somewhat involved business to actually calculate k_n when n is large. For low values of n one finds – using the notation $\sigma_{kl} \equiv \sigma(\omega_k, \omega_l)$ – that

$$k_2^{-1} = 1 - \frac{1}{2}\sigma_{12} \tag{7.22}$$

$$k_3^{-1} = 1 - \frac{1}{3}(\sigma_{12} + \sigma_{23} + \sigma_{31}) \tag{7.23}$$

$$k_4^{-1} = 1 - \frac{1}{4}(\sigma_{12} + \sigma_{23} + \sigma_{31} + \sigma_{14} + \sigma_{24} + \sigma_{34}) +$$

$$+ \frac{1}{12}(\sigma_{12}\sigma_{34} + \sigma_{13}\sigma_{24} + \sigma_{14}\sigma_{23}). \tag{7.24}$$

For general n the answer can be given as a sum of a set of symmetric functions of the squared chordal distances σ_{kl} [581].

Figure 7.4 We know that $\sigma(z, \omega) = \sin^2\frac{d}{2}$; here we see that $\sin\frac{d}{2}$ equals one-half of the chordal distance d_{ch} between the two points.

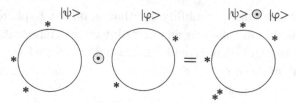

Figure 7.5 Composing states by adding stars.

To get some preliminary feeling for Q we compute it for the Dicke states $|\psi_k\rangle$, that is, for states that have the single component $Z^k = 1$ and all others zero. We find

$$Q_{|\psi_k\rangle}(z) = \binom{n}{k} \frac{|z|^{2k}}{(1+|z|^2)^n} = \binom{n}{k} \left(\cos\left(\frac{\theta}{2}\right)\right)^{2(n-k)} \left(\sin\left(\frac{\theta}{2}\right)\right)^{2k}, \quad (7.25)$$

where we switched to polar coordinates on the sphere in the last step. The zeros sit at $z = 0$ and at $z = \infty$, that is, at the north and south poles of the sphere – as we knew they would. When $k = 0$ we have an $m = j$ state, all the zeros coincide, and the function is concentrated around the north pole. This is a coherent state. If n is even and $k = n/2$ we have an $m = 0$ state, and the function is concentrated in a band along the equator. The indication, then, is that the Husimi function tends to be more spread out the more the state differs from being a coherent one. As a first step towards confirming this conjecture we will compute the moments of the Husimi function. But before doing so, let us discuss how it can be used to *compose* two states.

The tensor product of an $N = n + 1$ dimensional Hilbert space with itself is

$$\mathcal{H}_N \otimes \mathcal{H}_N = \mathcal{H}_{2N-1} \oplus \mathcal{H}_{2N-3} \oplus \cdots \oplus \mathcal{H}_1. \quad (7.26)$$

Given two states $|\psi_1\rangle$ and $|\psi_2\rangle$ in $\mathcal{H}_N$ we can define a state $|\psi_1\rangle \odot |\psi_2\rangle$ in the tensor product space by the equation

$$Q_{|\psi_1\rangle \odot |\psi_2\rangle} \propto Q_{|\psi_1\rangle} Q_{|\psi_2\rangle}. \quad (7.27)$$

We simply add the stars of the original states to obtain a state described by $2(N-1)$ stars. This state clearly sits in the subspace $\mathcal{H}_{2N-1}$ of the tensor product space. This operation becomes particularly interesting when we compose a coherent state with itself. The result is again a state with all stars coinciding, and moreover all states with $2N-1$ coinciding stars arise in this way. Reasoning along this line one quickly sees that

$$\frac{2n+1}{4\pi} \int d\Omega \, |z\rangle|z\rangle\langle z|\langle z| = \mathbb{1}_{2N-1}, \quad (7.28)$$

where $\mathbb{1}_{2N-1}$ is the projector, in $\mathcal{H}_N \otimes \mathcal{H}_N$, onto $\mathcal{H}_{2N-1}$. Similarly

$$\frac{3n+1}{4\pi} \int d\Omega \, |z\rangle |z\rangle |z\rangle \langle z| \langle z| \langle z| = \mathbb{1}_{3N-2}. \tag{7.29}$$

And so on. These are useful facts.

Thus equipped we turn to the second moment of the Husimi function:

$$\frac{n+1}{4\pi} \int d\Omega \, Q^2 = \frac{n+1}{2n+1} \langle \psi | \langle \psi | \frac{2n+1}{4\pi} \int d\Omega \, |z\rangle |z\rangle \langle z| \langle z| \psi \rangle | \psi \rangle \le \frac{n+1}{2n+1} \tag{7.30}$$

with equality if and only if $|\psi\rangle |\psi\rangle \in \mathcal{H}_{2N-1}$, that is by the preceding argument if and only if $|\psi\rangle$ is a coherent state. For the higher moments one shows in the same way that

$$\frac{n+1}{4\pi} \int d\Omega \, Q^p \le \frac{n+1}{pn+1} \quad \text{i.e.} \quad \frac{pn+1}{4\pi} \int d\Omega \, Q^p \le 1. \tag{7.31}$$

We will use these results later.[5] For the moment we simply observe that if we define the *Wehrl participation number* as

$$R = \left(\frac{n+1}{4\pi} \int d\Omega \, Q^2 \right)^{-1}, \tag{7.32}$$

and if we take this as a first measure of delocalisation, then the coherent states have the least delocalised Husimi functions [348, 813].

The Husimi function can be defined for mixed states as well, by

$$Q_\rho(z) = \langle z | \rho | z \rangle. \tag{7.33}$$

The density matrix ρ can be written as a convex sum of projectors $|\psi_i\rangle \langle \psi_i|$, so the Husimi function of a mixed state is a sum of polynomials up to a common factor. It has no zeros, unless there is a zero that is shared by all the polynomials in the sum. Let us order the eigenvalues of ρ in descending order, $\lambda_1 \ge \lambda_2 \ge \cdots \ge \lambda_N$. The largest (smallest) eigenvalue gives a bound for the largest (smallest) projection onto a pure state. Therefore it will be true that

$$\max_{z \in S^2} Q_\rho(z) \le \lambda_1 \quad \text{and} \quad \min_{z \in S^2} Q_\rho(z) \ge \lambda_N. \tag{7.34}$$

These inequalites are saturated if the eigenstate of ρ corresponding to the largest (smallest) eigenvalue happens to be a coherent state. The main conclusion is that the Husimi function of a mixed state tends to be flatter than that of a pure state, and generically it is nowhere zero.

---

[5] Note that a somewhat stronger result is available; see Bodmann [140].

7.4 Wehrl entropy and the Lieb conjecture

We can go on to define the Wehrl entropy [949, 950] of the state $|\psi\rangle$ by

$$S_W(|\psi\rangle\langle\psi|) \equiv -\frac{n+1}{4\pi} \int_\Omega d\Omega\, Q_\psi(z) \ln Q_\psi(z). \tag{7.35}$$

One of the key properties of the Q-function on the plane was that – as proved by Lieb – the Wehrl entropy attains its minimum for the coherent states. Clearly we would like to know whether the same is true for the Q-function on the sphere. Consider the coherent state $|z_0\rangle = |j,j\rangle$ with all stars at the south pole. Its Husimi function is given in Eq. (7.25), with $k = 0$. The integration (7.35) is easily performed (try the substitution $x = \cos^2(\frac{\theta}{2})$!) and one finds that the Wehrl entropy of a coherent state is $n/(n+1)$. It is also easy to see that the Wehrl entropy of the maximally mixed state $\rho_* = \frac{1}{n+1}\mathbb{1}$ is $S_W(\rho_*) = \ln(n+1)$; given that that the Wehrl entropy is concave in ρ this provides us with a rough upper bound. The **Lieb conjecture** [601] stated that

$$S_W(|\psi\rangle\langle\psi|) \geq \frac{n}{n+1} = \frac{2j}{2j+1} \tag{7.36}$$

with equality if and only if $|\psi\rangle$ is a coherent state. It took a long time to prove this conjecture, but it was done eventually, by Lieb and Solovej [604]. See Section 13.6. In this chapter we do not have all the ingredients for the proof available, but we can make it clear that the result is non-trivial.

The integral that defines S_W can be calculated because the logarithm factorises the integral.[6] In effect

$$S_W = -\frac{n+1}{4\pi} \int_\Omega d\Omega\, Q(z) \left(\ln k_n + \sum_{i=1}^{n} \ln\big(\sigma(z,\omega_i)\big) \right). \tag{7.37}$$

The answer is again given in terms of various symmetric functions of the squares of the chordal distances. We make the definitions:

$$① = \sum_{i<j} \sigma_{ij} \tag{7.38}$$

$$② = \sum_{k=1}^{n} \sum_{i<j} \sigma_{ik}\sigma_{jk} \tag{7.39}$$

$$③ = \sum_{l=1}^{n} \sum_{i<j<k} \sigma_{il}\sigma_{jl}\sigma_{kl}. \tag{7.40}$$

---

[6] This feat was performed by Lee in 1988 [581]. Unfortunately the answer looks so complicated that we do not quote it in full here. We will however sketch his proof that S_W assumes a local minimum at the coherent states.

The notation (due to Schupp [813]) is intended to make it easier to remember the structure of the various functions. As n grows we will need more of them. For $n = 4$:

$$\text{①} = \sigma_{12}\sigma_{34} + \sigma_{13}\sigma_{24} + \sigma_{14}\sigma_{23}. \tag{7.41}$$

For arbitrary n, sum over all quadratic terms such that all indices are different; but it is becoming evident that it will be labourious even to write down all the symmetric functions that occur for high values of n [581]. Anyway, with this notation

$$n = 2: \quad S_W = k_2 \left(\frac{2}{3} + \frac{1}{6}\sigma_{12} \right) - \ln k_2 = k_2 \left(\frac{2}{3} + \frac{1}{6}\text{①} \right) - \ln k_2, \tag{7.42}$$

$$n = 3: \quad S_W = k_3 \left(\frac{3}{4} + \frac{1}{12}\text{①} - \frac{1}{6}\oslash \right) - \ln k_3, \tag{7.43}$$

$$n = 4: \quad S_W = k_4 \left(\frac{4}{5} + \frac{1}{20}\text{①} - \frac{13}{180}\text{①} - \frac{1}{12}\oslash - \frac{1}{24}\oplus \right) - \ln k_4. \tag{7.44}$$

For $n = 2$ it is easy to see that S_W assumes its minimum when $\sigma_{12} = 0$, that is when the zeros coincide and the state is coherent [822]. Also $n = 3$ is doable [813]. The first non-trivial case is $n = 4$: we are facing a very difficult optimisation problem because the σ_{kl} are constrained by the requirement that they can be given in terms of the chordal distances between n points on a sphere.

From a different direction Bodmann [140] showed that

$$S_W \geq n \ln \left(1 + \frac{1}{n+1} \right). \tag{7.45}$$

The limit of large n is therefore taken care of, not surprisingly since in some sense it then goes over to the known result for canonical coherent states.

A complementary problem is to ask for states that maximise the Wehrl entropy. For $n = 1$ all pure states are coherent, so the question does not arise; $S_{\min} = S_{\max} = 1/2$. For $n = 2$ the maximal Wehrl entropy $S_{\max} = 5/3 - \ln 2$ is achieved for states whose stars are placed antipodally on the sphere. For $n = 3$ states whose stars are located on an equilateral triangle inscribed in a great circle give $S_{\max} = 21/8 - 2\ln 2$. Problem 7.2 provides some further information, but the general problem of finding such maximally delocalised states for arbitrary n is still open.

There are quite a few related problems, for instance finding states that saturate the upper limit of the uncertainty relation (6.61) [250], and a more difficult one concerning quantum polarisation of light [134]. One typically finds that three stars placed on an equilateral triangle on a great circle, or four stars placed at the corners at a regular tetrahedron, is optimal – but also that the story becomes more complicated when $n \geq 5$.

Considerable effort has been spent on a classical problem with a somewhat similar flavour, namely that of optimising the potential energy

$$E_s = \sum_{k<l} \sigma_{kl}^s \tag{7.46}$$

for n points on the sphere. The case $s = -\frac{1}{2}$ corresponds to electrostatic interaction and is of interest both to physicists concerned with Thomson's plum pudding model of the atom (assuming anyone is left) and to chemists concerned with Buckminster Fullerenes (molecules like C_{60}).[7] Also this problem has many open ends. The minima do tend to be regular configurations, but as a matter of fact neither the cube (for $n = 8$) nor the dodecahedron (for $n = 12$) are minima. It is also known that when the number of point charges is large there tends to be many local minima of nearly degenerate energy. If the experience gained from this problem, and others like it, is to be trusted then we expect states that maximise the Wehrl entropy to form interesting and rather regular patterns when looked at in the stellar representation.

With this background information in mind we can sketch Lee's proof that the coherent states provide local minima of the Wehrl entropy. For this it is enough to expand S_W to second order in the σ_{ij}, and this can be done for all n. The answer is

$$S_W = \frac{n}{n+1} + \frac{1}{2n^2} \oslash \oslash - \frac{n-2}{n(n-1)^2} \oslash - \frac{1}{n(n-1)} \oslash + o(\sigma_{il}^3). \tag{7.47}$$

Next we expand the position (θ_i, ϕ_i) of the n zeros around their average position (θ_0, ϕ_0), using polar angles as coordinates. Thus

$$\Delta\theta_i = \theta_i - \theta_0 \Rightarrow \sum_i \Delta\theta_i = \sum_i \theta_i - n\theta_0 = 0, \tag{7.48}$$

and similarly for $\Delta\phi_i$. To lowest non-trivial order a short calculation gives

$$\sigma_{il} = \sin^2 \frac{d_{il}}{2} \approx \frac{1}{4}(\Delta\theta_i - \Delta\theta_l)^2 + \frac{1}{4}\sin^2\theta_0(\Delta\phi_i - \Delta\phi_l)^2. \tag{7.49}$$

Finally a long calculation gives [581]

$$S_W \approx \frac{n}{n+1} + \frac{1}{32(n-1)^2}(F_1 - F_3)^2 + \frac{1}{8(n-1)^2}F_2^2, \tag{7.50}$$

where

$$F_1 = \sum_i (\Delta\theta_i)^2, \quad F_2 = \sin\theta_0 \sum_i (\Delta\theta_i)(\Delta\phi_i), \quad F_3 = \sin^2\theta_0 \sum_i (\Delta\phi_i)^2. \tag{7.51}$$

Evidently Eq. (7.50) implies that the coherent states form a – quite shallow – local minimum of S_W.

---

[7] For a review of this problem, with entries to the literature, see Saff and Kuijlaars [786]. For much more detail, see Erber and Hockney [298].

7.5 Generalised Wehrl entropies

One can also formulate the Lieb conjecture for the generalised entropies discussed in Section 2.7, which provide alternative measures of localisation of a quantum state in the phase space. All generalised entropies depend on the shape of the Husimi function only, and not on where it may be localised.

For instance, one may consider the Rényi–Wehrl entropies, defined according to (2.79),

$$S_q^{RW}(\psi) = \frac{1}{1-q} \ln \left[\frac{n+1}{4\pi} \int_\Omega d\Omega \, (Q_\psi(z))^q \right]. \tag{7.52}$$

Two different proofs that their minima are attained for coherent states are available when $q = 2, 3, \ldots$ [348, 813]. The easy way is to rely on Eq. (7.31). For q positive the Rényi–Wehrl entropy is smallest when the qth moment of the Q-function is maximal, and for $q = 2, 3, \ldots$, we already know that this happens if and only if the state is coherent. In this way we get

$$S_q^{RW}(|z\rangle) \geq \frac{1}{1-q} \ln \left(\frac{n+1}{qn+1} \right), \qquad q = 2, 3, \ldots, \tag{7.53}$$

with equality if and only if the state is coherent.

Before we go on, let us define the *digamma function*

$$\Psi(x) = \frac{\Gamma'(x)}{\Gamma(x)}. \tag{7.54}$$

For any integer $m > n$ it enjoys the property that

$$\Psi(m) - \Psi(n) = \sum_{k=n}^{m-1} \frac{1}{k}. \tag{7.55}$$

When x is large it is true that $\Psi(x+1) \sim \ln x + 1/2x$ [852] .

To get some feeling for how the Wehrl entropies behave, let us look at the eigenstates $|j, m\rangle$ of the angular momentum operator J_z. We computed their Husimi functions in Eq. (7.25). They do not depend on the azimuthal angle ϕ, which simplifies things. Direct integration [348] gives

$$S_W(|j, m\rangle) = \frac{2j}{2j+1} - \ln \binom{2j}{j-m} + 2j\,\Psi(2j+1) \tag{7.56}$$
$$- (j+m)\Psi(j+m+1) - (j-m)\Psi(j-m+1)$$

and

$$S_q^{RW}(|j, m\rangle) = \frac{1}{1-q} \ln \left[\frac{2j+1}{2qj+1} \binom{2j}{j-m}^q \frac{\Gamma[q(j+m)+1]\Gamma[q(j-m)+1]}{\Gamma(2qj+1)} \right]. \tag{7.57}$$

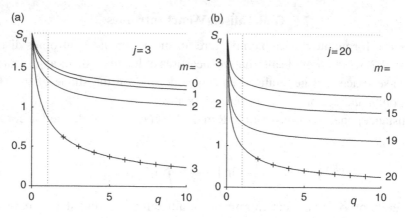

Figure 7.6 Rényi–Wehrl entropy S_q of the J_z eigenstates $|j, m\rangle$ for (a) $j = 3$ and (b) $j = 20$. For $m = j$ the states are coherent, and the crosses show integer values of q.

For $m = 0$ the state is localised on the equator of the Bloch sphere. A Stirling-like expansion [852], $\ln(k!) \approx (k + 1/2) \ln k - k + \ln \sqrt{2\pi}$, allows us to find the asymptotics, $S_{|m=0\rangle} \sim \frac{1}{2} \ln N + \frac{1}{2}(1 + \ln \pi/2)$. Interestingly, the mean Wehrl entropy of the eigenstates of J_z behaves similarly:

$$\frac{1}{2j+1} \sum_{m=-j}^{m=j} S_W(|j, m\rangle) = j - \frac{1}{2j+1} \sum_{m=-j}^{m=j} \ln \binom{2j}{j-m} \approx \frac{1}{2} \ln N + \ln(2\pi) - \frac{1}{2}.$$

(7.58)

A plot of some Rényi–Wehrl entropies obtained for the angular momentum eigenstates is shown in Figure 7.6. Any Rényi entropy is a continuous, non-increasing function of the Rényi parameter q. The fact that it attains minimal values for coherent states may suggest that for these very states the absolute value of the derivative $\partial S_q / \partial q |_{q=0}$ is maximal.

So far we had $SU(2)$ coherent states in mind, but the treatment is easily generalised to the (rather special) $SU(K)$ coherent states $|z^{(K)}\rangle$ from Section 6.4. Then our 'skies' are $\mathbb{CP}^{K-1}$, and they contain m stars if the dimension of the representation is $N_{K,m}$, as defined in Eq. (6.76). The Husimi function is normalised by

$$N_{K,m} \int_{\Omega_{K-1}} d\Omega_{K-1}(z) \, |\langle \psi | z^{(K)} \rangle|^2 = 1,$$

(7.59)

where $d\Omega_{K-1}$ is the FS measure on $\mathbb{CP}^{K-1}$, normalised so that $\int_{\Omega_n} d\Omega_n = 1$.

The moments of the Husimi function can be bounded from above, using the same method that was used for $SU(2)$. In particular

$$M_2(\psi) \equiv N_{K,n} \int_{\Omega_{K-1}} d\Omega_{K-1}(z) |\langle z|\psi\rangle|^4 \tag{7.60}$$

$$= \frac{N_{K,n}}{N_{K,2n}} N_{K,2n} \int_{\Omega_{K-1}} d\Omega_{K-1}(z) |\langle z \otimes z|\psi \otimes \psi\rangle|^2 = \frac{N_{K,n}}{N_{K,2n}} \|P_{2n}|\psi \otimes \psi\rangle\|^2,$$

where P_{2n} projects the space $N_{K,n} \otimes N_{K,n}$ into $N_{K,2n}$. The norm of the projection $\|P_{2n}|\psi \otimes \psi\rangle\|^2$ is smaller than $\||\psi \otimes \psi\rangle\|^2 = 1$ unless $|\psi\rangle$ is coherent, in which case $|\psi \otimes \psi\rangle = |\psi \odot \psi\rangle \in \mathcal{H}_{N_{K,2n}}$. Therefore $M_2(|\psi\rangle) \leq N_{K,n}/N_{K,2n}$. The same trick [813, 871, 872] works for any integer $q \geq 2$, and we obtain

$$M_q(\psi) = N_{K,n} \int_{\Omega_{K-1}} d\Omega_{K-1}(z) |\langle z|\psi\rangle|^{2q} \leq \frac{N_{K,n}}{N_{K,qn}} = \frac{\binom{K+n-1}{n}}{\binom{K+qn-1}{qn}}, \tag{7.61}$$

with equality if and only if the state $|\psi\rangle$ is coherent. By analytical continuation one finds that the Rényi–Wehrl entropy of an $SU(K)$ coherent state is

$$S_q^{RW}\left(|z^{(K)}\rangle\right) = \frac{1}{1-q} \ln\left[\frac{\Gamma(K+n)\Gamma(qn+1)}{\Gamma(K+qn)\Gamma(n+1)}\right]. \tag{7.62}$$

In the limit $q \to 1$ one obtains the Wehrl entropy of a coherent state [506, 847],

$$S_W\left(|z^{(K)}\rangle\right) = n\left[\Psi(n+K) - \Psi(n+1)\right]. \tag{7.63}$$

Using Eq. (7.55) for the digamma function, the right-hand side equals $n/(n+1)$ for $K = 2$, as we knew already.

Lieb and Solovej have shown that the coherent states provide a lower bound for the Wehrl entropy also for $K > 2$. See Section 13.6.

7.6 Random pure states

Up until now we were concerned with properties of individual pure states. But one may also define an ensemble of pure states and ask about the properties of a typical (random) state. In effect we are asking for the analogue of Jeffrey's prior (Section 2.6) for pure quantum states. There is such an analogue, determined uniquely by unitary invariance, namely the Fubini–Study measure that we studied in Section 4.7, this time normalised so that its integral over $\mathbb{CP}^n$ equals one. Interestingly the measure splits into two factors, one describing a flat simplex spanned by the moduli squared $p_i = |\langle e_i|\psi\rangle|^2$ of the state vector, and another describing a flat torus of a fixed size, parametrized by the phases. In fact

$$d\Omega_n = \frac{n!}{(2\pi)^n} dp_1 \ldots dp_n \, dv_1 \ldots dv_n, \tag{7.64}$$

or using our octant coordinates (4.70),

$$d\Omega_n = \frac{n!}{\pi^n} \prod_{k=1}^{n} \cos\vartheta_k \, (\sin\vartheta_k)^{2k-1} \, d\vartheta_k \, d\nu_k. \tag{7.65}$$

It is convenient to use the independent random variables ϑ_k, ν_k, and to introduce

$$P(\vartheta_k) = 2k\sin(\vartheta_k)(\sin\vartheta_k)^{2k-1}, \qquad P(\nu_k) = \frac{1}{2\pi}. \tag{7.66}$$

Then we set

$$\xi_k = (\sin\vartheta_k)^{2k} \Leftrightarrow \vartheta_k = \arcsin\left(\xi_k^{1/2k}\right) \tag{7.67}$$

and obtain the probability measure

$$d\Omega_n = d\xi_1 \ldots d\xi_n \frac{d\nu_1}{2\pi} \ldots \frac{d\nu_n}{2\pi}. \tag{7.68}$$

The advantage, compared to Eq. (7.64), is that the range of integration for the variables ξ_k is always $[0, 1]$. To generate a random quantum state all we have to do is to choose independent random variables ξ_k distributed uniformly in this range (your computer knows how to do this!), do the same for ν_k, and then use Eqs. (7.67) and (4.70).

Integrating out all but one variable we obtain the probability distribution $P(p) = n(1 - y)^{n-1}$. We can use this to compute the Fubini–Study average of $|\langle\Psi|\Phi\rangle|^{2t}$, where $|\Phi\rangle$ is some fixed unit vector. The calculation is simple:

$$\int_{\mathbb{CP}^n} d\Omega_n |\langle\Psi|\Phi\rangle|^{2t} = \int_0^1 dy \, n(1 - y)^{n-1} y^t = \frac{n! \, t!}{(n+t)!}. \tag{7.69}$$

We made use of the unitary invariance of the measure to arrange the coordinates so that the vector $|\Phi\rangle$ has only one non-vanishing component (equal to 1). Note that when the dimension n is large the average is close to zero; two random pure states are likely to be almost orthogonal. The distribution $P(y)$ is a special, $\beta = 2$ case of the general formula

$$P_\beta(y) = \frac{\Gamma((n+1)\beta/2)}{\Gamma(\beta/2)\Gamma(n\beta/2)} y^{\beta/2-1}(1 - y)^{n\beta/2-1}. \tag{7.70}$$

For $\beta = 1$ this gives the distribution of components of real random vectors distributed according to the round measure on $\mathbb{RP}^n$, while for $\beta = 4$ and $n + 1$ even it describes the quaternionic case [564]. For large $N = n + 1$ these distributions approach the χ^2 distributions with β degrees of freedom,

$$P_\beta(y) \approx N\chi_\beta^2(Ny) = \left(\frac{\beta N}{2}\right)^{\beta/2} \frac{1}{\Gamma(\beta/2)} y^{\beta/2-1} \exp\left(-\frac{\beta N}{2}y\right). \tag{7.71}$$

Another way to generate pure states at random aims at a uniform distribution on the unit sphere in Hilbert space, and ignores the resulting phase factor. A uniform distribution on the unit sphere is easily obtained by a variant of the trick we used to compute its volume in Section 1.2. We use a normal distribution for vectors in $\mathbb{R}^N$, depending on the length of the vectors only (but not on their direction). This distribution splits into independent normal distributions for the N components of the vectors, so we can choose each component of the vector at random using the normal distribution (again your computer knows how to do this). At the end we normalise the vector [677]. In Hilbert space the real and imaginary parts of the components are generated independently.

Frequently one wants to generate unitary matrices, and not only vectors, at random. The Haar measure on $U(N)$ is defined such that $\mu(A) = \mu(AU)$, that is to say it is invariant under the action of the group, and this measure defines what we mean by 'at random' here. We can proceed by means of a coordinate description of the group manifold [1011], but in practice the second procedure described above is more convenient: choose two real numbers at random using the normal distribution, and let them be the real and imaginary parts of a complex number. Repeat N^2 times, and use the resulting complex numbers as the entries of an $N \times N$ matrix. Finally, apply Gram–Schmidt orthogonalisation to this matrix. The result is a unitary matrix chosen at random according the Haar measure. Random orthogonal or symplectic matrices can be obtained similarly.[8]

To summarise, random pure states may be generated by:

- selecting $2n$ random variables according to (7.66) and using them as octant coordinates;
- picking a set of N independent complex random numbers z_i, drawn according to the normal distribution, and normalising the resulting vector.

They appear as

- a row (column) of a random unitary matrix U distributed according to the Haar measure on $U(N)$;
- eigenvectors of a random Hermitian (unitary) matrix pertaining to GUE[9] or CUE[10] multiplied with a random phase;
- a row (column) of a random Hermitian matrix pertaining to GUE, normalising it as above, and multiplying by a random phase.

---

[8] This is not the place to delve into the subject of random matrices. The articles by Diaconis [260] and by Mezzadri [657] provide entry points. See the classic book of Mehta [651] or the more recent opera by Forrester [317] or Anderson, Guionnet and Zeitouni [34] for more. There are many applications in quantum information theory [233].

[9] Gaussian unitary ensemble of Hermitian matrices consisting of independent Gaussian entries; such a probability measure is invariant with respect to unitary rotations.

[10] Circular unitary ensemble of unitary matrices distributed according to Haar measure on $U(N)$.

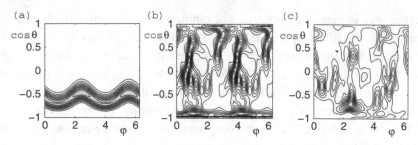

Figure 7.7 Husimi functions on the sphere (in Mercator's projection) for a typical eigenstate of a unitary operator describing (a) regular dynamics, (b) chaotic dynamics and (c) a random pure state for $N = 60$. The Wehrl entropies read 2.05, 3.67 and 3.68, respectively. The maximal entropy is $\ln N \approx 4.094$.

Random pure states may be analysed in the Husimi representation. If the Fubini–Study measure on $\mathbb{CP}^n$ is used, the states should not distinguish any part of the sphere; therefore the averaged density of zeros of the Husimi function for a random state will be uniform on the sphere [141, 142, 577]. However, this does not imply that each zero may be obtained as an independent random variable distributed uniformly on the sphere. On the contrary, the zeros of a random state are correlated as shown by Hannay [397]. In particular, the probability to find two zeros at small distance s behaves as $P(s) \sim s^2$ [578], while for independent random variables the probability grows linearly, $P(s) \sim s$.

The statistical properties of the Husimi zeros for random states are shared by the zeros representing eigenstates of unitary operators that give the one-step time evolution of quantised versions of classically chaotic maps [577]. Such states are delocalised, and their statistical properties [390] coincide with the properties of random pure states. Eigenstates of unitary operators giving the one-step time evolution of a regular dynamical system behave very differently, and their Husimi functions tend to be concentrated along curves in phase space [579, 1008]. Figure 7.7 shows a selected eigenstate of the unitary operator $U = \exp(ipJ_z) \exp(ikJ_x^2/2j)$, representing a periodically kicked top [390, 564, 1008] for $j = 29,5$, $p = 1.7$ and kicking strengths $k = 0.7$ (a) and $k = 10.0$ (b). When $k = 0$ we have an angular momentum eigenstate, in which case the Husimi function assumes its maximum on a latitude circle; as the kicking strength grows the curve where the maximum occurs begins to wiggle and the zeros start to diffuse away from the poles – and eventually the picture becomes similar to that for a random state.[11]

If the system in question enjoys a time reversal symmetry, the evolution operator pertains to the circular orthogonal ensemble (COE), and the distribution of Husimi

---

[11] Similar observations were reported for quantisation of dynamical systems on the torus [579, 698].

zeros of the eigenstates is no more uniform. The time reversal symmetry induces a symmetry in the coefficients of the Bargmann polynomial (7.15), causing some zeros to cluster along a symmetry line on the sphere [141, 160, 753].[12]

The time evolution of a pure state may be translated into a set of equations of motion for each star [577]. Parametric statistics of stars was initiated by Prosen [754], who found the distribution of velocities of stars of random pure states, while Hannay has shown [398] how the Berry phase (discussed in Section 4.8) can be related to the loops formed by the stars during a cyclic variation of the state. Let us emphasise that the number of the stars of any quantum state in a finite dimensional Hilbert space is constant during the time evolution (although they may coalesce), while for the canonical harmonic oscillator coherent states some zeros of the Husimi function may 'jump to infinity', so that the number of zeros may vary in time (see e.g. [553]).

Let us now compute the generalised Wehrl entropy of a typical random pure state $|\psi\rangle \in \mathcal{H}_N$. In other words we are going to average the entropy of a pure state with respect to the Fubini–Study measure on $\mathbb{C}\mathbf{P}^n$. The normalisation constant $N_{n+1,1}$ is defined in (6.76) and the Husimi function is computed with respect to the standard $SU(2)$ coherent states, so the average Wehrl moments are

$$\bar{M}_q = N_{n+1,1} \int_{\mathbb{C}\mathbf{P}^n} d\Omega_n(\psi) M_q(\psi) = N_{n+1,1} \int_{\mathbb{C}\mathbf{P}^n} d\Omega_n(\psi) \left[N_{2,n} \int_{\mathbb{C}\mathbf{P}^1} d\Omega_1(z) |\langle z|\psi\rangle|^{2q} \right].$$
$$(7.72)$$

It is now sufficient to change the order of integrations

$$\bar{M}_q = N_{2,n} \int_{\mathbb{C}\mathbf{P}^1} d\Omega_1(z) \left[N_{n+1,1} \int_{\mathbb{C}\mathbf{P}^n} d\Omega_n(\psi) |\langle z|\psi\rangle|^{2q} \right] = M_q \left(|z^{(K)}\rangle \right), \qquad (7.73)$$

and to notice that the integrals factorise: the latter gives the value of the Rényi moment (7.61) of a $SU(K)$ coherent state with $K = n+1$, while the former is equal to unity due to the normalisation condition. This result has a simple explanation: any state in an $N = n+1$ dimensional Hilbert space can be obtained by the action of $SU(N)$ on the maximal weight state. Indeed all states of size N are $SU(N)$ coherent. Yet another clarification: the state $|\psi\rangle$ is specified by $n = N - 1$ points on the sphere, or by a single point on $\mathbb{C}\mathbf{P}^n$. Equation (7.73)) holds for any integer Rényi parameter q and due to concavity of the logarithmical function we obtain an upper bound for the average entropies of random pure states in $\mathbb{C}\mathbf{P}^n$

$$\bar{S}_q \equiv \langle S_q(|\psi\rangle)\rangle_{\mathbb{C}\mathbf{P}^{N-1}} \leq S_q(|z^{(N)}\rangle) = \frac{1}{1-q} \ln \left[\frac{\Gamma(N+1)\Gamma(q+1)}{\Gamma(N+q)} \right], \qquad (7.74)$$

---

[12] Mark Kac (1943) [512] has considered a closely related problem: what is the expected number of real roots of a random polynomial with real coefficients? For a more recent discussion of this issue consult the very nice paper by Edelman and Kostlan [282].

where the explicit formula (7.62) was used. Thus the Wehrl participation number of a typical random state is $1/\bar{M}_2 = (N+1)/2$. In the limit $q \to 1$ we can actually perform the averaging analytically by differentiating the averaged moment with respect to the Rényi parameter q,

$$\bar{S}_W = \left\langle -\lim_{q \to 1} \frac{\partial}{\partial q} M_q(\psi) \right\rangle = \frac{\partial}{\partial q}|_{q=1} \langle M_q(\psi) \rangle = \Psi(N+1) - \Psi(2) = \sum_{k=2}^{N} \frac{1}{k}.$$

(7.75)

This result gives the mean Wehrl entropy of a random pure state [847]. In the asymptotic limit $N \to \infty$ the mean entropy $\bar{S}_W$ behaves as $\ln N + \gamma - 1$, where $\gamma \approx 0.5772\ldots$ is the Euler constant. Hence the average Wehrl entropy comes close to its maximal value $\ln N$, attained for the maximally mixed state.

The Wehrl entropy allows us to quantify the *global* properties of a state in the classical phase space, say of an eigenstate $|\phi_k\rangle$ of an evolution operator U. To get information concerning local properties one may expand a given coherent state $|z\rangle$ in the eigenbasis of U. The Shannon entropy $S(\vec{y}_z)$ of the vector $y_k = |\langle z|\phi_k\rangle|^2$ provides some information on the character of the dynamics in the vicinity of the classical point z [390, 980, 1006]. As shown by Jones [506] the mean entropy, $\langle S \rangle_{\mathbb{CP}^{N-1}}$, calculated using the FS measure, is just equal to (7.75), which happens to be the continuous Boltzmann entropy (2.40) of the distribution (7.70) with $\beta = 2$. Since the variance $\langle (\Delta S)^2 \rangle_{\mathbb{CP}^{N-1}}$ behaves like $(\pi^2/3 - 3)/N$ [667, 980], the relative fluctuations of the entropy decrease with the size N of the Hilbert space. Based on these results one may conjecture that if the corresponding classical dynamics

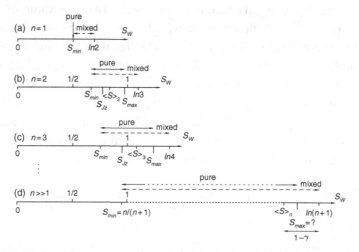

Figure 7.8 Range of the Wehrl entropy S_W for different dimensions $N = n+1$ of the Hilbert space; γ is Euler's constant ≈ 0.58.

in the vicinity of the point z is chaotic, then the Shannon entropy of the expansion coefficients of the coherent state $|z\rangle$ is of order $\langle S \rangle_{\mathbb{CP}^{N-1}}$. In the opposite case, a value $S(\bar{y}_z)$ much smaller than the mean value minus $\Delta S \sim 1/\sqrt{N}$ may be considered as a signature of a classically regular dynamics in the fragment of the phase space distinguished by the point $z \in \Omega$.

A generic evolution operator U has a non-degenerate spectrum, and the entropy $S(\bar{y}_z)$ acquires another useful interpretation [667, 895]: it is equal to the von Neumann entropy of the mixed state $\bar{\rho}$ obtained by the time average over the trajectory initiated from the coherent state,

$$\bar{\rho} = \lim_{T \to \infty} \frac{1}{T} \sum_{t=1}^{T} U^t |z\rangle \langle z| (U^\dagger)^t. \tag{7.76}$$

To show this, it is sufficient to expand the coherent state $|z\rangle$ in the eigenbasis of the evolution operator U and observe that the diagonal terms only do not dephase.

Let us return to the analysis of global properties of pure states in the classical phase space Ω. We have shown that for $N \gg 1$ the random pure states are delocalised in Ω. One can ask, if the Husimi distribution of a random pure state tends to the uniform distribution on the sphere in the semiclassical limit $N \to \infty$. Since strong convergence is excluded by the presence of exactly $N - 1$ zeros of the Husimi distribution, we will consider weak convergence only. To characterise this convergence quantitatively we introduce the L_2 distance between the Husimi distribution of the state $|\psi\rangle$ and the uniform distribution $Q_* = 1/N$, normalised as $N \int Q_* d\Omega_1 = 1$ and representing the maximally mixed state $\rho_* = \mathbb{1}/N$,

$$L_2(\psi) \equiv L_2(Q_\psi, Q_*) = \left(N \int_\Omega \left[Q_\psi(z) - \frac{1}{N} \right]^2 d\Omega_1(z) \right)^{1/2} = M_2(\psi) - \frac{1}{2j+1}. \tag{7.77}$$

Applying the previous result (7.73) we see that the mean L_2 distance to the uniform distribution tends to zero in the semiclassical limit,

$$\langle L_2(\psi) \rangle_{\mathbb{CP}^{2j}} = \frac{2j}{(2j+1)(2j+2)} \to 0 \qquad \text{as} \qquad j \to \infty. \tag{7.78}$$

Thus the Husimi distribution of a typical random state tends, in the weak sense, to the uniform distribution on the sphere.

The L_2 distance to Q_* may be readily computed for any coherent state $|z\rangle$,

$$L_2(|z\rangle) \equiv L_2(Q_{|z\rangle}, Q_*) = \frac{j}{2j+1} \to \frac{1}{2}, \qquad \text{as} \qquad j \to \infty. \tag{7.79}$$

This outcome differs qualitatively from (7.78), which emphasises the fact that coherent states are exceedingly non-typical states in a large dimensional Hilbert space. For random states the n stars cover the sphere uniformly, but the stars must coalesce in a single point to give rise to a coherent state. From the generalised Lieb conjecture proved for $q = 2$ it follows that the L_2 distance achieves its maximum for coherent states.

7.7 From the transport problem to the Monge distance

The stellar representation allows us to introduce a simple metric in the manifold of pure states with an interesting 'classical' property: the distance between two spin coherent states is equal to the Riemannian distance between the two points on the sphere where the stars of the states are situated. We begin the story be formulating a version of the famous **transport problem**:

Let n stars be situated at n not necessarily distinct points x_i. The stars are moved so that they occupy n not necessarily distinct points y_i. The cost of moving a star from x_i to y_j is c_{ij}. How should the stars be moved if we want to minimise the total cost

$$T = \sum_{i=1}^{n} c_{i\pi(i)} \tag{7.80}$$

where $\pi(i)$ is some permutation telling us which star goes where?[13]

To solve this problem it is convenient to relax it, and consider the linear programming problem of minimizing

$$\tilde{T} = \sum_{i,j=1}^{N} c_{ij} B_{ij}, \tag{7.81}$$

where B_{ij} is any bistochastic matrix (Section 2.1). The minimum always occurs at a corner of the convex polytope of bistochastic matrices, that is for some permutation matrix, so a solution of the relaxed problem automatically solves the original problem. Since there are $n!$ permutation matrices altogether one may worry about the growth of computer time with n. However, there exist efficient algorithms of finding it where the amount of computer time grows as n^3 only [986].

We can now try to define the distance between two pure states in a finite dimensional Hilbert space as the minimal total cost $T_{\min}$ of transporting their stars into each other, where the cost c_{ij} is given by the geodesic distance between the stars on a sphere of an appropriate radius (chosen to be $1/2j$ if the dimension of the Hilbert space is $2j+1$). If we wish we can formulate this as a distance between two discrete probability distributions,

$$D_{\text{dM}}(|\psi\rangle, |\phi\rangle) \equiv T_{\min}\big(P_1(|\psi_1\rangle), P_2(|\psi_2\rangle)\big), \tag{7.82}$$

[13] The problem is often referred to as the assignment problem. Clearly 'cost' suggests some application to economics; during World War II the setting was transport of munitions to fighting brigades.

where any state is associated with the distribution $P = \sum_{i=1}^{n} \frac{1}{n} \delta(z - z_i)$, where z_i are the zeros of the Husimi function Q_ψ. This is a legitimate notion of distance because – by construction – it obeys the triangle inequality [758]. We refer to this as the *discrete Monge distance* D_{dM} between the states [1014], since it is a discrete analogue of the continuous Monge distance that we will soon introduce. We observe that the discrete Monge distance between two coherent states (each represented by $2j$ coinciding stars) then becomes equal to the usual distance between two points on the unit sphere. In fact, locally the set of pure states will now have the same geometry as a product of 2-spheres almost everywhere, although the global properties are quite different.

Some further properties are worth noting. The discrete Monge distance between the states $|j, m\rangle$ and $|j, m'\rangle$ becomes

$$D_{dM}(|j, m\rangle, |j, m'\rangle) = \frac{\pi}{2j} |m - m'|. \tag{7.83}$$

The set of eigenstates of J_z therefore form a metric line with respect to D_{dM}, while they form the corners of a simplex with respect to the Fubini–Study distance. It is also easy to see that all the eigenstates of J_x are situated at the distance $\pi/2$ from any eigenstate of J_z. Finally, consider two uncorrelated random states $|\psi_{rand}\rangle$ and $|\phi_{rand}\rangle$. Since the states are random their respective stars, carrying the weight $1/2j$ each, will be evenly distributed on the sphere. Its surface can be divided into $2j$ cells of diameter $\sim (2j)^{-1/2}$ and given a star from one of the states the distance to the nearest star of the other will be of the same order. Hence the discrete Monge distance between two random states will behave as

$$D_{dM}(|\psi_{rand}\rangle, |\phi_{rand}\rangle) \sim \frac{2j}{(2j)^{3/2}} = \frac{1}{(2j)^{1/2}} \to 0 \quad \text{as} \quad j \to \infty. \tag{7.84}$$

The discrete Monge distance between two random states goes to zero in the semiclassical limit.

Using similar ideas, a more sophisticated notion of distance can be introduced between continuous probability distributions. The original **Monge problem**, formulated in 1781, emerged from studying the most effective way of transporting soil [757]:

Split two equally large volumes of soil into infinitely small particles and then associate them with each other so that the sum of the paths of the particles over the volume is least. Along which paths must the particles be transported and what is the smallest transportation cost?

Let $P_1(x, y)$ and $P_2(x, y)$ denote two probability densities, defined on the Euclidean plane, that describe the initial and the final location of the soil. Let the sets Ω_1 and Ω_2 denote the supports of both probability distributions. Consider

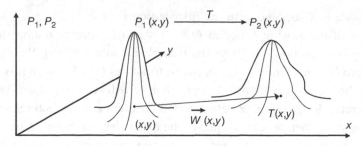

Figure 7.9 Monge transport problem: how to shovel a pile of soil $P_1(x, y)$ into a new location $P_2(x, y)$, minimizing the work done?.

smooth one-to-one maps $T : \Omega \to \Omega$ which generate volume preserving transformations Ω_1 into Ω_2, that is,

$$P_1(x, y) = P_2(T(x, y)) |T'(x, y)| \tag{7.85}$$

for all points in Ω, where $T'(x, y)$ denotes the Jacobian of the map T at the point (x, y). We shall look for a transformation giving the minimal displacement integral (see Figure 7.9), and then we define the *Monge distance* between two probability distributions as [757]

$$D_M(P_1, P_2) \equiv \inf \int_\Omega |\vec{W}(x, y)| \, P_1(x, y) \, dx \, dy, \tag{7.86}$$

where $|\vec{W}(x, y)| = |(x, y) - T(x, y)|$ denotes the length of the path travelled – the Euclidean distance between a point in Ω_1 and its image. The infimum is taken over all volume preserving transformations T. The optimal transformation need not be unique; its existence can be guaranteed under reasonably general conditions. The generalisation to curved spaces is straightforward; we just replace the Euclidean distance with the appropriate Riemannian one.

Note that in this formulation of the problem the vertical component of the soil movement is neglected, so perhaps the definition does not capture the essence of moving soil around. But we can use it to define the Monge distance between two probability distributions, as well as the Monge distance between pairs of quantum states as the Monge distance between their Husimi functions. A decided advantage is that – since the Husimi function exists for any density matrix – the definition applies to pure and mixed states alike:

$$D_{Mon}(\rho, \sigma) \equiv D_M(Q_\rho(z), Q_\sigma(z)). \tag{7.87}$$

The definition depends on the choice of reference state for a set of coherent states, and it is not unitarily invariant. But in some circumstances this may be a desirable feature.

A price we have to pay is that the Monge distance is, in general, difficult to compute [757, 758]. Some things can be said right away however. If two Husimi functions have the same shape then the optimal transformation T is just a rigid translation, and the Monge distance becomes equal to the Riemannian distance with which our classical phase space is equipped. In particular this will be the case for pairs of coherent states. In some cases one may use symmetry to reduce the two-dimensional Monge problem to the one-dimensional case, for which an analytical solution due to Salvemini is known (see Problem 7.5). In this way it is possible to compute the Monge distance between two arbitrary Fock states [1013]. An asymptotic result, $D_{\text{Mon}}(|n\rangle, |m\rangle) \approx \frac{1}{2} \sum_{k=n+1}^{m} 1/\sqrt{k}$, has an intuitive interpretation: although all Fock states are orthogonal, and hence equidistant according to the Fubini–Study metric, the Monge distance shows that the state $|100\rangle$ is much closer to $|101\rangle$ than to the vacuum state $|0\rangle$. Using Bloch coherent states to define the Monge distance in finite dimensional state spaces, one finds that the eigenstates of J_z form a metric line in the sense that

$$
D_{\text{Mon}}(|j, -j\rangle, |j, j\rangle) = \sum_{m=-j+1}^{j} D_{\text{Mon}}(|j, m-1\rangle, |j, m\rangle) = \pi \left[1 - \binom{2N}{N} 2^{1-2N} \right]. \quad (7.88)
$$

This is similar to the discrete Monge distance defined above (and very different from the Fubini–Study distance).

As in the discrete case the task of computing the Monge distance is facilitated by a relaxation of the problem (in this case due to Kantorovitch [514]), which is explicitly symmetric with respect to the two distributions. Using discretisation one can make use of the efficient numerical methods that have been developed for the discrete Monge problem.

So what is the point? One reason why the Monge distance may be of interest is precisely the fact that it is not invariant under unitary transformations. This resembles the classical situation, in which two points in the phase space may drift away under the action of a given Hamiltonian system. Hence the discrete Monge distance (for pure states) and the Monge distance (for pure and mixed states) may be useful to construct a quantum analogue of the classical Lyapunov exponent and to elucidate various aspects of the quantum–classical correspondence.[14]

Problems

7.1   For the angular momentum eigenstates $|j, m\rangle$, find the minimal Fubini–Study distance to the set of coherent states.

7.2   For $N = 2, 3, 4, 5$, compute the Wehrl entropy S_W and the Wehrl participation number R for the J_z eigenstates, and for the states $|\psi_\triangle\rangle$, three stars in an equilateral triangle, and $|\psi_{\text{tetr.}}\rangle$, four stars forming a regular tetrahedron.

[14] For further details consult Życzkowski and Słomczyński [1014].

7.3 Show that the mean Fubini–Study distance of a random state $|\psi_{rand}\rangle$ to any selected pure state $|1\rangle$ reads

$$\left\langle D_{FS}(|1\rangle, |\psi_{rand}\rangle)\right\rangle_{\mathbb{CP}^n} = \frac{\pi}{2} - \frac{\sqrt{\pi}\,\Gamma(n+1/2)}{2\,\Gamma(n+1)}, \qquad (7.89)$$

which equals $\pi/4$ for $n = 1$ and tends to $\pi/2$ when $n \to \infty$.

7.4 Show that the average discrete Monge distance of a random state $|\psi_{rand}\rangle$ to the eigenstates $|j, m\rangle$ of J_z reads [1014]

$$\left\langle D_{dM}(|j,m\rangle, |\psi_{rand}\rangle)\right\rangle_{\mathbb{CP}^{2j}} = \chi \sin \chi + \cos \chi, \quad \text{where} \quad \chi = \frac{m\pi}{2j}. \ (7.90)$$

7.5 Prove that the Monge distance, in the one-dimensional case, $\Omega = \mathbb{R}$, is given by the *Salvemini solution* [757, 787, 1013].

$$D_M(P_1, P_2) = \int_{-\infty}^{+\infty} |F_1(x) - F_2(x)| \, dx, \qquad (7.91)$$

where the distribution functions are $F_i(x) \equiv \int_{-\infty}^{x} P_i(t) \, dt$ for $i = 1, 2$.

7.6 Making use of the Salvemini formula (7.91) analyse the space of $N = 2$ mixed states from the point of view of the Monge distance: (a) prove that $D_{Mon}(\rho_+, \rho_-) = \pi/4$, $D_{Mon}(\rho_+, \rho_*) = \pi/8$, and in general (b) prove that the Monge distance between any two mixed states of a qubit is proportional to their Hilbert–Schmidt distance and generates the geometry of the Bloch ball [1014].

8

The space of density matrices

Over the years, the mathematics of quantum mechanics has become more abstract and, consequently, simpler.

—*V. S. Varadarajan*[1]

We have already introduced density matrices and made use of some of their properties. In general a complex $N \times N$ matrix is a density matrix if it is

$$
\begin{array}{lll}
\text{(i)} & \text{Hermitian,} & \rho = \rho^\dagger, \\
\text{(ii)} & \text{positive,} & \rho \geq 0, \\
\text{(iii)} & \text{normalised,} & \text{Tr}\rho = 1.
\end{array}
\tag{8.1}
$$

The middle equation is a shorthand for the statement that all the eigenvalues of ρ are non-negative. The set of density matrices will be denoted $\mathcal{M}^{(N)}$. It is a convex set sitting in the vector space of Hermitian matrices, and its pure states are density matrices obeying $\rho^2 = \rho$. As explained in Section 4.5 the pure states form a complex projective space.

Why should we consider this particular convex set for our state space? One possible answer is that there is a point in choosing vectors that are also matrices, or – a different way of saying the same thing – a vector space that is also an algebra. In effect our vectors now have an intrinsic structure, namely their spectra when regarded as matrices, and this enables us to define a positive cone and a set of pure states in a more interesting way than that used in classical probability theory. We will cast a glance at the algebraic framework towards the end of this chapter, but first we explore the structure as it is given to us.

[1] Reproduced from V. S. Varadarajan. *Geometry of Quantum Theory*. Springer, 1985.

8.1 Hilbert–Schmidt space and positive operators

This section will be devoted to some basic facts about complex matrices, leading up to the definition of the space $\mathcal{M}^{(N)}$. We begin with an N complex dimensional Hilbert space $\mathcal{H}$. There is then always a dual Hilbert space $\mathcal{H}^*$ defined as the space of linear maps from $\mathcal{H}$ to the complex numbers; in the finite dimensional case these two spaces are isomorphic. Another space that is always available is the space of operators on $\mathcal{H}$. This is actually a Hilbert space in its own right when it is equipped with the Hermitian form

$$\langle A, B \rangle = c \operatorname{Tr} A^\dagger B, \tag{8.2}$$

where c is a real number that sets the scale. This is *Hilbert–Schmidt space $\mathcal{HS}$*; an alternative notation is $\mathcal{B}(\mathcal{H})$ where $\mathcal{B}$ stands for 'bounded operators'. All our operators are bounded, and all traces exist, but this is so only because all our Hilbert spaces are finite dimensional. The scalar product gives rise to a Euclidean distance, the *Hilbert–Schmidt distance*

$$D_2^2(A, B) \equiv \frac{1}{2} \operatorname{Tr}[(A - B)(A^\dagger - B^\dagger)] \equiv \frac{1}{2} D_{\text{HS}}^2(A, B). \tag{8.3}$$

In this chapter we set the scale with $c = 1/2$, and work with the distance D_2, while in Chapter 9 we use D_{HS}. As explained in Section 4.5 this ensures that we agree with standard conventions in quantum mechanics.

Chapter 9 will hinge on the fact that

$$\mathcal{HS} = \mathcal{H} \otimes \mathcal{H}^*. \tag{8.4}$$

This is a way of saying that any operator can be written in the form

$$A = a|P\rangle\langle Q| + b|R\rangle\langle S| + \ldots, \tag{8.5}$$

provided that enough terms are included in the sum. Given $\mathcal{H}$, the N^2 complex dimensional space $\mathcal{HS}$ – also known as the algebra of complex matrices – is always with us.

Some brief reminders about (linear) operators may prove useful. First we recall that an operator A can be diagonalised by a unitary change of bases, if and only if it is *normal*, that is if and only if $[A, A^\dagger] = 0$. Examples include *Hermitian* operators for which $A^\dagger = A$ and *unitary* operators for which $A^\dagger = A^{-1}$. A normal operator is Hermitian if and only if it has real eigenvalues. Any normal operator can be written in the form

$$A = \sum_{i=1}^{r} z_i |e_i\rangle\langle e_i|, \tag{8.6}$$

where the sum includes orthogonal projectors corresponding to the r non-vanishing eigenvalues z_i. The eigenvectors $|e_i\rangle$ span a linear subspace supp(A) known as the *support* or *range* of the operator (and these notions coincide for normal operators). The set of all vectors $|\psi\rangle$ such that $A|\psi\rangle = 0$ forms a linear subspace kern(A) called the *kernel* of A. For normal operators the kernel is orthogonal to the range, and forms the subspace spanned by all eigenvectors with zero eigenvalues. It has dimension $N - r$. The full Hilbert space may be expressed as the direct sum $\mathcal{H} = $ kern(A) $\oplus$ supp(A).

The vector space $\mathcal{HM}$ of Hermitian operators is an N^2 real dimensional subspace of Hilbert–Schmidt space. It can also – and this will turn out to be important – be thought of as the Lie algebra of $U(N)$. The $N^2 - 1$ dimensional subspace of Hermitian operators with zero trace is the Lie algebra of the group $SU(N)$. In Appendix B we give an explicit basis for the latter vector space, orthonormal with respect to the scalar product (8.2). If we add the unit matrix we see that a general Hermitian matrix can be written in the form

$$A = \tau_0 \sqrt{\frac{2}{N}} \mathbb{1} + \sum_{i=1}^{N^2-1} \tau_i \sigma_i \quad \Leftrightarrow \quad \tau_0 = \frac{\mathrm{Tr}A}{\sqrt{2N}}, \quad \tau_i = \frac{1}{2}\mathrm{Tr}\sigma_i A, \qquad (8.7)$$

where σ_i are generators of $SU(N)$ obeying

$$\sigma_i \sigma_j = \frac{2}{N}\delta_{ij} + d_{ijk}\sigma_k + if_{ijk}\sigma_k. \qquad (8.8)$$

Here f_{ijk} is totally anti-symmetric in its indices and d_{ijk} is totally symmetric (and vanishing if and only if $N = 2$). The use of the notation (τ_0, τ_i) for the Cartesian coordinates in this basis is standard. Of course there is nothing sacred about the basis that we have chosen; an orthonormal basis consisting of N^2 elements of equal trace may be a better choice.

We will now use the 'internal structure' of our vectors to define a positive cone. By definition, a *positive* operator P is an operator such that $\langle\psi|P|\psi\rangle$ is real and non-negative for all vectors $|\psi\rangle$ in the Hilbert space. This condition actually implies that the operator is Hermitian, so we can equivalently define a positive operator as a Hermitian matrix with non-negative eigenvalues. (To see this, observe that any matrix P can be written as $P = X + iY$, where X and Y are Hermitian – the argument fails if the vector space is real, in which case a positive operator is usually defined as a symmetric matrix with non-negative eigenvalues.) The condition that an operator be positive can be rewritten in a useful way:

$$P \geq 0 \quad \Leftrightarrow \quad \langle\psi|P|\psi\rangle \geq 0 \quad \Leftrightarrow \quad P = AA^\dagger, \qquad (8.9)$$

for all vectors $|\psi\rangle$ and for some matrix A. The set $\mathcal{P}$ of positive operators obeys

$$\mathcal{P} \subset \mathcal{HM} \qquad \dim[\mathcal{P}] = \dim[\mathcal{HM}] = N^2. \qquad (8.10)$$

Since it is easy to see that a convex combination of positive operators is again a positive operator, the set $\mathcal{P}$ is a convex cone (in the sense of Section 1.1).

A positive operator admits a unique positive square root $\sqrt{P}$,

$$(\sqrt{P})^2 = P. \tag{8.11}$$

For an arbitrary (not necessarily Hermitian) operator A we can define the positive operator $AA^\dagger$, and then the *absolute value* $|A|$:

$$|A| \equiv \sqrt{AA^\dagger}. \tag{8.12}$$

This is in itself a positive operator. Furthermore, any linear operator A can be decomposed into polar form, which means that

$$A = |A|U = \sqrt{AA^\dagger}\,U, \tag{8.13}$$

where U is a unitary operator which is unique if A is invertible. This *polar decomposition* is analogous to the representation $z = re^{i\phi}$ of a complex number. The modulus r is non-negative; so are the eigenvalues of the positive operator $|A|$, which are known as the *singular values* of A. (There exists an alternative left polar decomposition, $A = U\sqrt{A^\dagger A}$, which we will not use. It leads to the same singular values.) The polar decomposition can be used to prove that any operator, normal or not, admits a *singular value decomposition* (SVD)

$$A = UDV, \tag{8.14}$$

where U and V are unitary, and D is diagonal with non-negative entries which are precisely the singular values of A.[2]

Evidently the question whether a given Hermitian matrix is positive is raising its head – an ugly head since generally speaking it is not easy to diagonalise a matrix and investigate its spectrum. A helpful fact is that the spectrum can be determined by the traces of the first N powers of the matrix. To see how, look at the *characteristic equation*

$$\det(\lambda\mathbb{1} - A) = \lambda^N - s_1\lambda^{N-1} + s_2\lambda^{N-2} - \cdots + (-1)^N s_N = 0. \tag{8.15}$$

If the matrix is in diagonal form the coefficients s_k – that clearly determine the spectrum – are the elementary symmetric functions of the eigenvalues,

$$s_1 = \sum_i \lambda_i, \quad s_2 = \sum_{i<j} \lambda_i\lambda_j, \quad s_3 = \sum_{i<j<k} \lambda_i\lambda_j\lambda_k, \quad \cdots \tag{8.16}$$

---

[2] In popular numerical routines for computing the SVD of an arbitrary matrix one obtains the vector of singular values as well as the matrices U and V. An interesting review of properties of SVD may be found in both books by Horn and Johnson [450, 451].

and so on. If the matrix is not in diagonal form we can still write

$$s_1 = \mathrm{Tr}A \qquad s_2 = \frac{1}{2}\left(s_1\mathrm{Tr}A - \mathrm{Tr}A^2\right) \qquad (8.17)$$

and in general, iteratively,

$$s_k = \frac{1}{k}\left(s_{k-1}\mathrm{Tr}A - s_{k-2}\mathrm{Tr}A^2 + \cdots + (-1)^{k-1}\mathrm{Tr}A^k\right). \qquad (8.18)$$

(Proof: Diagonalise the matrix. Since the traces are not affected by diagonalisation it is only a matter of comparing our two expressions for the coefficients.) Returning to the question whether A is positive, it can be proved that A is positive if and only if all the coefficients s_k in the characteristic equation are positive. This is encouraging since the criterion only requires the calculation of traces, but it remains a lengthy business to apply it to a given A. There is no easy way.

8.2 The set of mixed states

Finally we come to the density matrices. The set of density matrices consists of all positive operators ρ with unit trace, $\mathrm{Tr}\rho = 1$. We denote it by $\mathcal{M}$, or by $\mathcal{M}^{(N)}$ if we want to emphasise that it consists of $N \times N$ matrices. $\mathcal{M}$ is the intersection, in the space of Hermitian matrices, of the positive cone $\mathcal{P}$ with a hyperplane parallel to the linear subspace of traceless operators. It is a convex set in its own right, whose pure states are projectors onto one-dimensional subspaces in $\mathcal{H}$, that is density matrices of the form

$$\rho = |\psi\rangle\langle\psi| \quad \Leftrightarrow \quad \rho^\alpha_\beta = Z^\alpha \bar{Z}_\beta \Leftrightarrow \rho^2 = \rho. \qquad (8.19)$$

In this chapter we assume that the vectors are normalised. As we observed in Section 4.5 this is an isometric embedding of $\mathbb{CP}^{N-1}$ in $\mathcal{HM}$.

Two different ways of coordinatizing $\mathcal{M}^{(N)}$ spring to mind. We can set

$$\rho = \frac{1}{N}\mathbb{1} + \sum_{i=1}^{N^2-1} \tau_i\sigma_i. \qquad (8.20)$$

What we have done is to identify $\mathcal{M}^{(N)}$ with a subset of the Lie algebra of $SU(N)$, by shifting the origin of $\mathcal{HM}$ from the zero matrix to the matrix

$$\rho_\star \equiv \frac{1}{N}\mathbb{1}. \qquad (8.21)$$

This is a special density matrix known as the *maximally mixed state*; it is also known as the *tracial state* or, familiarly, as the 'matrix of ignorance'. The components τ_i of the *Bloch vector* serve as Cartesian coordinates in *Bloch space*, and are also known as *mixture coordinates*. A convex combination of two density matrices lies on a manifestly straight line when these coordinates are used. Moreover the definition (8.3), with (B.13), implies that

$$D_2\left(\sum_i \tau_i \sigma_i, \sum_j \tau_j' \sigma_j\right) = \sqrt{\sum_i (\tau_i - \tau_i')^2}, \qquad (8.22)$$

which is the familiar formula in a Euclidean space.

There is another way of identifying $\mathcal{M}^{(N)}$ with a subset of the Lie algebra of $SU(N)$, namely to set

$$\rho = \frac{e^{-\beta H}}{\mathrm{Tr} e^{-\beta H}}; \qquad H = \sum_{i=1}^{N^2-1} x_i \sigma_i. \qquad (8.23)$$

The coordinates x_i are known as *exponential coordinates*; the real number β is known as the inverse temperature and serves as a reminder of the role these coordinates play in statistical mechanics. They are consistent with their own notion of straight line – the analogue of the exponential families of classical statistics that we studied in Section 3.2.

We would like to draw a more detailed picture of the set of density matrices than that offered in Figure 8.1. If we choose to work over the real rather than the complex numbers, then we can investigate the space of real symmetric 2×2 matrices. This has only three dimensions. The condition that the trace be unity defines a two-dimensional plane in this space, and we can literally see how it intersects the convex cone of positive matrices in a circular disc, which is the two-dimensional space of *rebits*. Physics requires complex qubits, but rebits are simpler to look at. But already for $N = 3$ the set of real density matrices has dimension five, which is too high for easy visualisation.

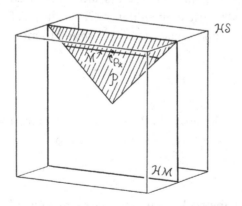

Figure 8.1 The spaces discussed in this chapter. Hilbert–Schmidt space $\mathcal{HS}$ only serves as a background for the linear subspace $\mathcal{HM}$ of Hermitian matrices that contains the positive cone $\mathcal{P}$ and the set of density matrices $\mathcal{M}$. But compare to Figure 9.1 in the next chapter.

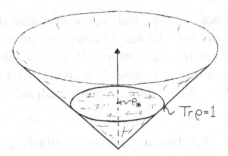

Figure 8.2 The cone of positive real symmetric 2×2 matrices and its intersection with the plane of Hermitian matrices of unit trace.

Moving on to qubits, we find that the space $\mathcal{M}^{(2)}$ has three real dimensions. This is the Bloch ball discussed in Section 5.2. There is no particular difficulty in understanding a ball as a convex set. Physically however there is much to think about, because we now have two different ways of adding two pure states together. We can form a complex superposition – another pure state – and we can form a statistical mixture – a mixed state. In the other direction, any given point in the interior can be obtained as a mixture of pure states in many different ways. We are confronted with an issue that does not arise in classical statistics at all: any mixed state can be expressed as a mixture of pure states in many different ways, indeed in as many ways as a point in a ball can be thought of as the 'centre of mass' of a mass distribution on the surface of the ball. Physically it is a basic tenet of quantum mechanics that there does not exist an operational procedure to distinguish different ensembles of pure states if they yield the same density matrix – otherwise quantum correlations between separated systems could be used for instantaneous signalling [430].

Quantum mechanics is a significant generalisation of classical probability theory. When $N = 2$ there are two possible outcomes of any measurement described by a Hermitian operator, or put in another way the sample space belonging to a given observable consists of two points. They correspond to two orthogonal pure states, placed antipodally on the surface of the Bloch ball. Each pair of antipodal points on the surface corresponds to a new sample space coexisting with the original.

The Bloch ball is a convenient example to keep in mind since it is easy to visualise, but in some respects it is quite misleadingly simple. We begin to see this if we ask what conditions one has to put on the Bloch vector in order for a density matrix to describe a pure state. Using Eq. (8.8) it is straightforward to deduce that

$$
\rho^2 = \rho \quad \Leftrightarrow \quad
\begin{cases}
\tau^2 = \frac{N-1}{2N} \\[2mm]
(\vec{\tau} \star \vec{\tau})_i \equiv d_{ijk} \tau_j \tau_k = \frac{N-2}{N} \tau_i
\end{cases}.
\tag{8.24}
$$

The first condition implies that the Bloch vector of a pure state is confined to an $N^2 - 2$ dimensional outsphere. The second condition arises only for $N > 2$, and it says that the pure states form a certain well defined subset of the surface of the outsphere.[3] We know that this subset is a complex projective space, of real dimension $2(N - 1)$. When $N > 2$ this is a rather small subset of the outsphere.

8.3 Unitary transformations

It is convenient to think of the set of density matrices as a rigid body in $\mathbb{R}^{N^2-1}$. We think of $\mathbb{R}^{N^2-1}$ as a Euclidean space. When $N > 2$ our rigid body is not spherical, and we must try to understand its shape. The first question one asks about a rigid body is: what is its symmetry? What subgroup of rotations leaves it invariant? Unless the body is spherical, the symmetry group is a proper subgroup of the group of rotations. For the Platonic solids, the symmetry groups are discrete subgroups of $SO(3)$. Our case is subtler; although pure quantum states form only a small subset of the outsphere, they do form a continuous manifold, and the symmetry group is not discrete. As it happens there is an analogue of Wigner's theorem (Section 4.5) that applies to density matrices, and answers our question. Its assumptions concern only the convex structure, not the Euclidean distance:

Kadison's theorem. *Let there be a map Φ from $\mathcal{M}$ to $\mathcal{M}$ which is one-to-one and onto, and which preserves the convex structure in the sense that*

$$\Phi(p\rho_1 + (1 - p)\rho_2) = p\Phi(\rho_1) + (1 - p)\Phi(\rho_2). \tag{8.25}$$

Then the map must take the form

$$\Phi(\rho) = U\rho U^{-1}, \tag{8.26}$$

where the operator U is either unitary or anti-unitary.

In infinitesimal form a unitary transformation takes the form

$$\dot{\rho} = i[\rho, H], \tag{8.27}$$

where H is a Hermitian operator (say, a Hamiltonian) that determines the one-parameter family of unitary operators $U(t) = e^{-iHt}$.

It is easy to see why something like this must be true.[4] Because the map preserves the convex structure it must be an affine map, and it must map pure states to pure states. For $N = 2$ the pure states form a sphere, and the only affine maps that

---

[3] The elegant 'star product' used here occurs now and then in the literature, for instance in Arvind et al. [60]. As a general reference for this section we suggest Mahler and Weberruß [636].
[4] We will not give a strict proof of Kadison's theorem; a complete and elementary proof was given by Hunziker [484].

preserve the sphere are rotations; the map must belong to the orthogonal group $O(3)$, that is, it must take the form (8.26). For $N > 3$ the pure states form just a subset of a sphere, and the rotation must be a rather special one – in particular, it must give rise to an isometry of the space of its pure states. The last condition points right at the unitary group.

In a sense Kadison's theorem answers all our questions, but we must make sure that we understand the answer. Because the body of density matrices sits in a vector space that we have identified with the Lie algebra of $SU(N)$, and because the only unitary transformations that have any effect on density matrices are those that belong to $SU(N)$, we have here an example of an *adjoint action* of a group on its own Lie algebra. What sets the scene is that $SU(N)/Z_N$ is a subgroup of $SO(N^2-1)$. An explicit way to see this goes as follows: let ρ be defined as in Eq. (8.20) and let

$$\rho' = U\rho U^\dagger = \frac{1}{N}\mathbb{1} + \sum_i \tau_i U\sigma_i U^\dagger \equiv \frac{1}{N}\mathbb{1} + \sum_i \tau_i' \sigma_i. \tag{8.28}$$

Here τ_i' is the rotated Bloch vector. We compute that

$$\tau_i' = \frac{1}{2}\mathrm{Tr}\rho'\sigma_i = \frac{1}{2}\sum_j \mathrm{Tr}\left(\sigma_i U\sigma_j U^\dagger\right)\tau_j \equiv \sum_j O_{ij}\tau_j, \tag{8.29}$$

where the matrix O by definition has the matrix elements

$$O_{ij} \equiv \mathrm{Tr}\left(\sigma_i U\sigma_j U^\dagger\right). \tag{8.30}$$

This must be an orthogonal matrix, and indeed it is. It is easy to check that the elements are real. Using the completeness relation (B.14) for the generators it is fairly easy to check also that

$$(OO^T)_{ij} = \sum_k O_{ik} O_{jk} = \delta_{ij}. \tag{8.31}$$

In this way we have exhibited $SU(N)/Z_N$ as a subgroup of the rotation group $SO(N^2 - 1)$.

Another way to see what goes on is to observe that rotations preserve the distance to the origin – in this case, to ρ_*. This means that they preserve

$$D_2^2\left(\rho, \rho_*\right) = \frac{1}{2}\mathrm{Tr}\left(\rho - \rho_*\right)^2 = \frac{1}{2}\mathrm{Tr}\rho^2 - \frac{1}{2N}. \tag{8.32}$$

Therefore all rotations preserve $\mathrm{Tr}\rho^2$. But unitary transformations preserve the additional traces $\mathrm{Tr}\rho$, $\mathrm{Tr}\rho^3$, $\mathrm{Tr}\rho^4$ and so on, up to the last independent trace $\mathrm{Tr}\rho^N$. We are dealing with very special rotations that preserve the spectrum of every density matrix.

Since rotations in arbitrary dimensions may be unfamiliar, let us first discuss the action of a generic rotation matrix. With a suitable choice of basis it can always be block diagonalised, that is to say it can be written with 2×2 rotation matrices occurring along the diagonal and zeros elsewhere. If the dimension is odd, we add an extra 1 on the diagonal. What this means is that for any rotation of $\mathbb{R}^n$ we can choose a set of totally orthogonal 2-planes such that the rotation can be described as independent rotations in these 2-planes; if n is odd there will always be – as Euler pointed out – an axis that is not affected by the rotation at all. A typical flow line is not a circle. If the dimension is either $2n$ or $2n + 1$ it will wind around a flat torus of dimension n, since there are n totally orthogonal 2-planes involved. Unless the rotations in these 2-planes are carefully adjusted, the resulting curve will not even be closed. (We came across this kind of thing in Section 4.6, and tried to draw it in Figure 4.11. Now we offer Figure 8.3.) A generic rotation has only one fixed point if the dimension of the space is even, and only one fixed axis if it is odd.

The $SU(N)/Z_N$ subgroup of $SO(N^2 - 1)$ that we are dealing with does not describe generic rotations, and the picture changes as follows: after choosing a basis in which the $SU(N)$ matrix is diagonal, the latter belongs to a Cartan subgroup of dimension $N-1$. Generically therefore its flow lines will lie on a torus of dimension $N-1$; quite a bit smaller than the torus that occurs for a generic $SO(N^2-1)$ rotation. The set of fixed points consists of all density matrices that commute with the given $SU(N)$ matrix. When all the eigenvalues of the latter are different, the set of fixed points consists of all density matrices that are diagonal in the chosen basis. This set is the $(N-1)$-dimensional *eigenvalue simplex*, to be discussed in the next section. The eigenvalue simplex contains only N pure states: the eigenstates of the $SU(N)$ matrix that describes the rotation. The set of fixed points is of larger dimension if degeneracies occur in the spectrum of the $SU(N)$ matrix.

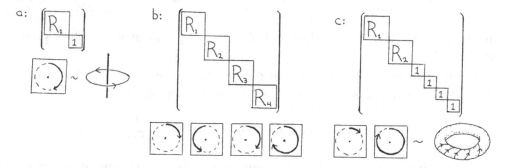

Figure 8.3 Rotations can be represented with 2×2 rotation matrices along the diagonal, or pictorially as rotations in totally orthogonal 2-planes. We show (a) an $SO(3)$ rotation, (b) a generic $SO(8)$ rotation and (c) a generic $SU(3) \in SO(8)$ rotation.

The action of $SU(N)$ on $\mathcal{M}^{(N)}$ contains a number of intricacies that will occupy us later in this chapter, but at least now we have made a start.

8.4 The space of density matrices as a convex set

Let us state some general facts and definitions.[5] The dimension of $\mathcal{M}^{(N)}$ is $N^2 - 1$. The pure states are the projectors

$$\rho^2 = \rho \quad \Leftrightarrow \quad \rho = |\psi\rangle\langle\psi|. \tag{8.33}$$

The space of pure states is $\mathbb{C}\mathbf{P}^{N-1}$. As we observed in Section 4.5 this space is isometrically embedded in the set of Hermitian matrices provided that we define distance as the Hilbert–Schmidt distance (8.3). The pure states form a $2(N-1)$ dimensional subset of the $N^2 - 2$ dimensional boundary of $\mathcal{M}^{(N)}$. To see whether a given density matrix belongs to the boundary or not, we diagonalise it and check its eigenvalues. If one of them equals zero we are on the boundary.

Any density matrix can be diagonalised. The set of density matrices that are diagonal in a given basis $\{|e_i\rangle\}$ can be written as

$$\rho = \sum_{i=1}^{N} \lambda_i |e_i\rangle\langle e_i|, \quad \rho|e_i\rangle = \lambda_i|e_i\rangle, \quad \sum_{i=1}^{N} \lambda_i = 1. \tag{8.34}$$

This set is known as the *eigenensemble* or as the *eigenvalue simplex*. It forms a special $N-1$ dimensional cut through the set of density matrices, and every density matrix sits in some eigenvalue simplex. It is a simplex since the eigenvalues are positive and add up to one – indeed it is a copy of the $N-1$ dimensional simplex with N corners that we studied in classical probability theory. It is centred at the maximally mixed state $\rho_\star$.

In Section 1.1 we defined the rank of a point in a convex set as the minimum number r of pure points that are needed to express it as a convex combination of pure states. It is comforting to observe that this definition coincides with the usual definition of the *rank* of a Hermitian matrix: a density matrix of matrix rank r can be written as a convex sum of no less than r projectors (as is obvious when the matrix is diagonalised). Hence the maximal rank of a mixed state is equal to N, much less than the upper bound N^2 provided by Carathéodory's theorem (Section 1.1).

The distance between an arbitrary density matrix ρ and the centre $\rho_\star$ was given in Eq. (8.32), where we saw that $\mathrm{Tr}\rho^2$ determines the distance from $\rho_\star$. It happens that $\mathrm{Tr}\rho^2 \leq 1$ with equality if and only if the state is pure, so as expected the pure

---

[5] Some elementary properties of the convex set of density matrices were discussed by Bloore (1976) [139] and by Harriman (1978) [406].

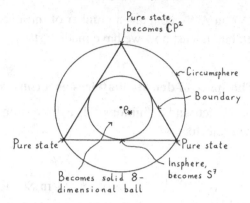

Figure 8.4 This picture shows the eigenvalue simplex for $N = 3$ with its outsphere and its insphere, and indicates what happens if we apply $SU(3)/Z_3$ transformations to obtain the eight-dimensional body of density matrices.

states lie at maximal distance from $\rho_\star$. This observation determines the radius R_{out} of the outsphere (the smallest ball containing $\mathcal{M}$). To compute the radius r_{in} of the insphere (the largest ball inscribed in $\mathcal{M}$), we observe that every density matrix belongs to some eigenvalue simplex. It follows that the radius of the insphere of $\mathcal{M}^{(N)}$ will equal that of a simplex Δ_{N-1}, and this we computed in Section 1.2. So we find

$$R_{out} = \sqrt{\frac{N-1}{2N}} \quad \text{and} \quad r_{in} = \sqrt{\frac{1}{2N(N-1)}}. \tag{8.35}$$

The maximally mixed state $\rho_\star$ is surrounded by a ball of radius r_{in} consisting entirely of density matrices. These deliberations are illustrated in Figure 8.4, which shows the eigenvalue simplex for a qutrit.

In Eq. (1.28) we computed the angle subtended by two corners of a simplex and found that it approaches 90° when N becomes large. The corners of the eigenvalue simplex represent pure states at maximal distance from each other, so if χ denotes the angle subtended by two pure states at $\rho_\star$ then

$$\cos \chi \leq -\frac{1}{N-1}, \tag{8.36}$$

with equality if and only if the states are at maximal distance from each other, that is to say if and only if they are orthogonal. When N is large this means that all pure states in the hemisphere opposite to a given pure state (with respect to $\rho_\star$) lie very close to the equator. This is not surprising since almost all the area of the circumsphere is concentrated around the equator for large N (see Section 1.2). But it is very different from the $N = 2$ case, where to every pure state there corresponds an antipodal pure state.

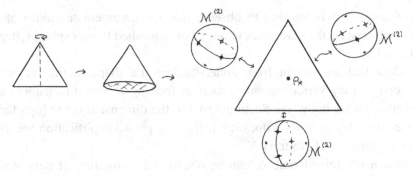

Figure 8.5 An attempt to visualise $\mathcal{M}^{(3)}$. We rotate the eigenvalue simplex to obtain a cone, then we rotate it in another dimension to turn the base of the cone into a Bloch ball rather than a disc; that is a maximal face of $\mathcal{M}^{(3)}$. On the right, we imagine that we have done this to all the three edges of the simplex. In each maximal face we have placed three equidistant points – it happens that when these points are placed correctly on all the three spheres, they form a regular simplex inscribed in $\mathcal{M}^{(3)}$.

For the $N = 3$ case we have to think in eight dimensions. This is not easy, but we can try. By looking at Figure 8.4 we can see the in-spheres and outspheres, but because $\mathcal{M}^{(3)}$ is left invariant only under quite special rotations – those that move a given corner of the simplex through a two complex dimensional projective space embedded in $\mathbb{R}^8$ – it is not so easy to imagine what the full structure looks like. An example of an allowed rotation is a rotation of the eigenvalue simplex around an axis joining a corner to ρ_*. This turns the simplex into a cone (see Figure 8.5). In fact, if we could imagine just one more dimension, we could see a four-dimensional slice of $\mathcal{M}^{(3)}$, which would be a cone whose base is a three-dimensional ball, having one of the edges of the simplex as its diameter. This ball is one of the *faces* of $\mathcal{M}^{(3)}$.

Perhaps we should recall, at this point, that a face of a convex body is defined as a convex subset stable under purification and mixing. Contemplation of Figure 8.5 shows that there is a face opposite to each pure state, consisting of all density matrices that are mixtures of pure states that are orthogonal to the given pure state. This is to say that $\mathcal{M}^{(3)}$ has faces that in themselves are Bloch balls. There is a useful way to look at this, which goes as follows. Pick a pure state $|Z\rangle$ and study the equation

$$\mathrm{Tr}\rho\,|Z\rangle\langle Z| = \bar{Z}\rho Z = c, \qquad c \in [0, 1]. \tag{8.37}$$

For fixed $|Z\rangle$ this is an affine functional representing a family of parallel hyperplanes in the space of Hermitian matrices of unit trace; when $c = 1$ the hyperplane intersects $\mathcal{M}^{(3)}$ in a single pure state and when $c = 0$ in the face opposite to that pure state. It is interesting to observe that for intermediate values of c the hyperplane intersects the pure states in one of the squashed Berger spheres, depicted in Figure 4.13.

It is clear that we are far from understanding the shape of the set of qutrit states – one can try various devices, such as looking at low dimensional cross-sections through the body (see Section 8.6), but the dimension is too high for easy comprehension. Since $N = 3$ does not bring any great simplification we may as well discuss the case of arbitrary N.

The maximally mixed state $\rho_\star$ can be obtained as a mixture of pure states by putting equal weight (in the sense of the Fubini–Study measure) on all pure states, or by putting equal weight on each corner in an eigenvalue simplex, and also in many other ways. A similar non-uniqueness afflicts all mixed states. Interestingly this non-uniqueness can be expressed in a precise way:[6]

Schrödinger's mixture theorem. *A density matrix ρ having the diagonal form*

$$\rho = \sum_{i=1}^{N} \lambda_i \, |e_i\rangle \langle e_i| \tag{8.38}$$

can be written in the form

$$\rho = \sum_{i=1}^{M} p_i |\psi_i\rangle \langle \psi_i|, \qquad \sum_{i=1}^{M} p_i = 1, \qquad p_i \geq 0 \tag{8.39}$$

if and only if there exists a unitary $M \times M$ matrix U such that

$$|\psi_i\rangle = \frac{1}{\sqrt{p_i}} \sum_{j=1}^{N} U_{ij} \sqrt{\lambda_j} \, |e_j\rangle. \tag{8.40}$$

Here all states are normalised to unit length but they need not be orthogonal to each other.

Given ρ, this theorem supplies all the ways in which ρ can be expressed as an ensemble of pure states. Observe that the matrix U does not act on the Hilbert space but on vectors whose components are state vectors, and also that we may well have $M > N$. But only the first N columns of U appear in the equation – the remaining $M - N$ columns are just added in order to allow us to refer to the matrix U as a unitary matrix. What the theorem basically tells us is that the pure states that make up an ensemble are linearly dependent on the N vectors $|e_i\rangle$ that make up the eigenensemble. Moreover an arbitrary state in that linear span can be included.

---

[6] This theorem has an interesting history. Schrödinger (1936) [805] published it with no claim to priority. When the time was ripe it was rediscovered by (among others) Gisin (1989) [338] and Hughston, Jozsa and Wootters (1993) [481]; it is now often known as the GHJW lemma.

For definiteness we assume that all density matrices have rank N so that we consider ensembles of M pure states in an N dimensional Hilbert space.

It is straightforward to prove that Eq. (8.39) and Eq. (8.40) imply Eq. (8.38). To prove the converse, define the first N columns of the unitary matrix U by

$$U_{ij} \equiv \frac{\sqrt{p_i}}{\sqrt{\lambda_j}} \langle e_j | \psi_i \rangle \quad \Rightarrow \quad \sum_{j=1}^{N} U_{ij} \sqrt{\lambda_j} | e_j \rangle = \sqrt{p_i} | \psi_i \rangle. \qquad (8.41)$$

The remaining $N - M$ columns can be chosen at will, consistent with unitarity of the matrix. The matrix will be unitary because

$$\sum_{i=1}^{M} U_{ki}^{\dagger} U_{ij} = \sum_{i=1}^{M} \frac{p_i}{\sqrt{\lambda_j \lambda_k}} \langle e_j | \psi_i \rangle \langle \psi_i | e_k \rangle = \frac{1}{\sqrt{\lambda_j \lambda_k}} \langle e_j | \rho | e_k \rangle = \delta_{jk}. \qquad (8.42)$$

This concludes the proof of the mixture theorem. Since only a rectangular submatrix of U is actually used in the theorem we could leave it like that and refer to U as a 'right unitary' or *isometry* matrix if we wanted to, but the extra columns do no harm. In fact they are helpful. We can deduce that

$$p_i = \sum_{j=1}^{N} |U_{ij}|^2 \lambda_j. \qquad (8.43)$$

Thus $\vec{p} = B\vec{\lambda}$, where B is a unistochastic, and hence bistochastic, matrix. In the language of Section 2.1, the vector $\vec{p}$ is majorised by the eigenvalue vector $\vec{\lambda}$.

To see how useful the mixture theorem is, consider the face structure of $\mathcal{M}^{(N)}$. Recall (from Section 1.1) that a face is a convex subset of a convex set that is stable under mixing and purification. But the mixture theorem tells us that a density matrix is always a mixture of pure states belonging to that subspace of Hilbert space that is spanned by its eigenvectors. What this means is that every face is a copy of $\mathcal{M}^{(K)}$, the body of density matrices for a system whose Hilbert space has dimension K. If $K < N$ the face is a proper face, and belongs to the boundary of $\mathcal{M}^{(N)}$.

On closer inspection, we see that the face structure of $\mathcal{M}^{(N)}$ reveals 'crystalline' regularities. A given face corresponds to some subspace $\mathcal{H}_K$ of Hilbert space. Introduce a projector E that projects down to that subspace. In full analogy to Eq. (8.37) we can consider the affine functional

$$\text{Tr}\rho E = c, \qquad c \in [0, 1]. \qquad (8.44)$$

Again, this defines a family of parallel hyperplanes in the space of Hermitian matrices of unit trace. For $c = 1$ it defines a face of density matrices with support in $\mathcal{H}_K$, and for $c = 0$ it defines an opposing face with support in the orthogonal complement of that subspace. There is a straight line between the 'centres of mass' of these two faces, passing through the 'centre of mass' of $\mathcal{M}^{(N)}$ (i.e. through $\rho_\star$).

Next, we recall from Section 1.1 that the faces of a convex body always form a partially ordered structure known as a lattice, and from Section 4.2 that the set of subspaces of a Hilbert space also forms a lattice. The following is true [21]:

Theorem. *The lattice of faces of $\mathcal{M}^{(N)}$ is identical to the orthocomplemented lattice of subspaces in $\mathcal{H}_N$.*

The identity of the two lattices is by now obvious, but interesting nevertheless. A lattice is said to be *orthocomplemented* if and only if there is a map $a \to a'$ of the lattice L onto itself, such that

$$(a')' = a \qquad a \leq b \Rightarrow b' \leq a' \qquad a \cap a' = 0 \qquad a \cup a' = L \qquad (8.45)$$

for all $a, b \in L$, where $\cap$ and $\cup$ denote respectively the greatest lower and the smallest upper bound of a pair of elements (for the lattice of subspaces of $\mathcal{H}_N$, they are respectively the intersection and the linear span of the union of a pair of subspaces). In our lattice, two opposing faces of $\mathcal{M}^{(N)}$ are indeed related by a map $a \to a'$. It is possible to single out further properties of this lattice, and to use them for an axiomatic formulation of quantum mechanics – this is the viewpoint of quantum logic.[7]

Let us mention in passing that there is another angle from which we can try to view the structure: we choose a convex polytope that we feel comfortable with, let it have the same outradius as the body of density matrices itself, and ask if it can be inscribed in $\mathcal{M}^{(N)}$. The obvious choice is a regular simplex. The simplex Δ_8 can be inscribed in $\mathcal{M}^{(3)}$, and Figure 8.5 indicates how. (To avoid misunderstanding: this Δ_8 does not have edge lengths of unit length. The largest simplex with edge lengths equal to one that can be placed inside $\mathcal{M}^{(N)}$ is Δ_{N-1}.) Oddly enough it is not as easy to do this for $N > 3$, but at least for moderately small N it can be done, and so our intuition has a little more material to work with. See Section 12.8 for the details, as far as they are known.

8.5 Stratification

Yet another way to organise our impressions of $\mathcal{M}^{(N)}$ is to study how it is partitioned into orbits of the unitary group (recall Section 7.2). We will see that each individual orbit is a flag manifold (Section 4.9) and that the space of orbits has a transparent structure.[8] We begin from the fact that any Hermitian matrix can be diagonalised by a unitary rotation,

$$\rho = V \Lambda V^\dagger. \qquad (8.46)$$

---

[7] Further details can be found in the books by Jauch [497] and Varadarajan [916]; for a version of the story that emphasises the geometry of the convex set one can profitably consult Mielnik [659].
[8] A general reference for this section is Adelman, Corbett and Hurst [7].

where Λ is a diagonal density matrix that fixes a point in the eigenvalue simplex. We obtain a $U(N)$ orbit from Eq. (8.46) by letting the matrix V range over $U(N)$. Before we can tell what the result is, we must see to what extent different choices of V can lead to the same ρ. Let B be a diagonal unitary matrix. It commutes with Λ, so

$$\rho = V\Lambda V^\dagger = VB\Lambda B^\dagger V^\dagger. \tag{8.47}$$

In the case of a non-degenerate spectrum this is all there is; the matrix V is determined up to the N arbitrary phases entering B, and the orbit will be the coset space $U(N)/U(1) \times U(1) \times \cdots \times U(1)$. From Section 4.9 we recognise this as the flag manifold $\mathbf{F}^{(N)}_{1,2,\dots,N-1}$. If degeneracies occur in the spectrum of ρ, the matrix B need not be diagonal in order to commute with Λ, and various special cases ensue. In the language of Section 7.2, the isotropy group changes, and so does the nature of the orbit.

Let us discuss the case of $N = 3$ in detail to see what happens; our deliberations are illustrated in Figure 8.6b. The space of diagonal density matrices is the simplex Δ_2. However, using unitary permutation matrices we can change the order in which the eigenvalues occur, so that without loss of generality we may assume that $\lambda_1 \geq \lambda_2 \geq \lambda_3 \geq 0$. This corresponds to dividing the simplex Δ_2 into 3! parts, and to picking one of them. Denote it by $\widetilde{\Delta}_2$. We call it a *Weyl chamber*, with a terminology borrowed from group theory. It is the Weyl chamber that forms the space of orbits of $U(3)$ in $\mathcal{M}^{(3)}$.

The nature of the orbit will depend on its location in the Weyl chamber. Depending on the degeneracy of the spectrum, we decompose $\widetilde{\Delta}_2$ into four parts [7, 1014] (see also Figure 8.6b):

(a) A point K_3 representing the state ρ_* with triple degeneracy $\{1/3, 1/3, 1/3\}$; the isotropy group is $U(3)$;
(b) A One-dimensional line K_{12} representing the states with double degeneracy, $\lambda_2 = \lambda_3$; the isotropy group is $U(1) \times U(2)$;
(c) A One-dimensional line K_{21} representing the states with double degeneracy, $\lambda_1 = \lambda_2$; the isotropy group is $U(2) \times U(1)$;
(d) The two-dimensional part K_{111} of the generic points of the simplex, for which no degeneracy occurs; the isotropy group is $U(1) \times U(1) \times U(1)$.

Since the degeneracy of the spectrum determines the isotropy group it also determines the topology of the $U(3)$ orbit. In case (a) the orbit is $U(3)/U(3)$, that is, a single point, namely the maximally mixed state ρ_*. In cases (b) and (c) the orbit is $U(3)/[U(1) \times U(2)] = \mathbf{F}^{(3)}_1 = \mathbb{C}\mathbf{P}^2$. In the generic case (d) we obtain the generic flag manifold $\mathbf{F}^{(3)}_{1,2}$.

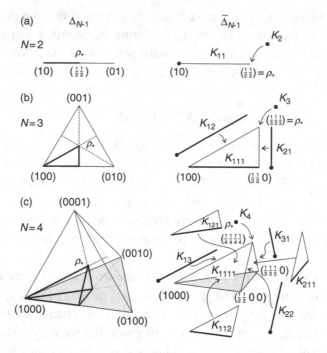

Figure 8.6 The eigenvalue simplex and the Weyl chamber for $N = 2, 3$ and 4. The Weyl chamber $\tilde{\Delta}_{N-1}$, enlarged on the right-hand side, can be decomposed according to the degeneracy into 2^{N-1} parts.

Now we are ready to tackle the general problem of $N \times N$ density matrices. There are two things to watch: the boundary of $\mathcal{M}^{(N)}$, and the stratification of $\mathcal{M}^{(N)}$ by qualitatively different orbits under $U(N)$. It will be easier to follow the discussion if Figure 8.6 is kept in mind.

The diagonal density matrices form a simplex Δ_{N-1}. It can be divided into $N!$ parts depending on the ordering of the eigenvalues, and we can select one of these parts to be the Weyl chamber $\tilde{\Delta}_{N-1}$. The Weyl chamber is the $N-1$ dimensional space of orbits under $U(N)$. The nature of the orbits is determined by the degeneracy of the spectrum, so we decompose the Weyl chamber into parts $K_{k_1,\dots,k_m}$ where the largest eigenvalue has degeneracy k_1, the second largest degeneracy k_2, and so on. Clearly $k_1 + \cdots + k_m = N$. Each of these parts parametrize a stratum (see Section 7.2) of $\mathcal{M}^{(N)}$, where each orbit is a flag manifold

$$\mathbf{F}^{(N)}_{k_1, k_2, \dots, k_{m-1}} = \frac{U(N)}{U(k_1) \times \cdots \times U(k_m)}. \tag{8.48}$$

See Section 4.9 for the notation. The generic case is $K_{1,1,\dots,1}$ consisting of the interior of $\tilde{\Delta}_{N-1}$ together with one of its (open) facets corresponding to the case of one vanishing eigenvalue. This means that, except for a set of measure zero, the space $\mathcal{M}^{(N)}$ is equal to

$$\mathcal{M}_{1,...,1} \sim \left[\frac{U(N)}{T^N}\right] \times K_{1,1,...,1} = \mathbf{F}^{(N)}_{1,2,...,N-1} \times G_N. \tag{8.49}$$

Here we used T^N to denote the product of N $U(1)$ factors, topologically a torus, and we also used $G_N \equiv K_{1,1,...,1}$. The equality holds in a topological sense only, but – as we will see in Chapter 15 – it is also accurate when computing volumes in $\mathcal{M}$.

There are exceptional places in the Weyl chamber where the spectrum is degenerate. In fact there are 2^{N-1} different possibilities for $K_{k_1,...,k_m}$, because there are $N-1$ choices between 'larger than' or 'equal' when the eigenvalues are ordered. Each $K_{k_1,...,k_m}$ forms an $m-1$ dimensional (irregular) simplex that we denote by G_m. Each G_m can be realised in $\binom{N-1}{m-1}$ different ways (e.g., for $N=4$ the set G_2 can be realised as $K_{3,1}$, $K_{2,2}$, $K_{1,3}$). In this way we get a decomposition of the Weyl chamber as

$$\tilde{\Delta}_{N-1} = \bigcup_{k_1+\cdots+k_m=N} K_{k_1,...,k_m}, \tag{8.50}$$

and a topological decomposition of the space of density matrices as

$$\mathcal{M}^{(N)} \sim \bigcup_{k_1+\cdots+k_m=N} \left[\frac{U(N)}{(U(k_1) \times \cdots \times U(k_m))}\right] \times K_{k_1,...,k_m}, \tag{8.51}$$

where the sum ranges over all partitions of N into sums of positive integers. The total number of such partitions is 2^{N-1}. However, the orbits sitting over, say, $K_{1,2}$ and $K_{2,1}$ will be flag manifolds of the same topology. To count the number of qualitatively different flag manifolds that appear, we must count the number of partitions of N with no regard to ordering, that is we must compute the number $p(N)$ of different representations of the number N as the sum of positive integers. For $N = 1, 2, \ldots, 10$ the number $p(N)$ is equal to $1, 2, 3, 5, 7, 11, 15, 22, 30$ and 42, while for large N the asymptotic Hardy–Ramanujan formula [402] gives $p(N) \simeq \exp\left(\pi\sqrt{2N/3}\right)/4\sqrt{3}N$. Figure 8.6 and Table 8.1 summarise these deliberations for $N \leq 4$.

Let us now take a look at the boundary of $\mathcal{M}^{(N)}$. It consists of all density matrices of less than maximal rank. The boundary as a whole consists of a continuous family of maximal faces, and each maximal face is a copy of $\mathcal{M}^{(N-1)}$. To every pure state there corresponds an opposing maximal face, so the family of maximal faces can be parametrized by $\mathbb{CP}^{N-1}$. It is simpler to describe the orbit space of the boundary, because it is the base of the Weyl chamber and has dimension $N-2$. It is clear from Figure 8.6 that the orbit space of the boundary $\partial\mathcal{M}^{(N)}$ is the same as the orbit space of $\mathcal{M}^{(N-1)}$, but the orbits are different because the group that acts is larger.

Table 8.1 *Stratification of $\mathcal{M}^{(N)}$. $U(N)$ is the unitary group, T^k is a k-dimensional torus and G_m stands for a part of the eigenvalue simplex defined in the text. The dimension d of the strata equals $d_F + d_S$, where d_F is the even dimension of the complex flag manifold $\mathbf{F}$, while $d_S = m - 1$ is the dimension of G_m.*

N	Label	Subspace	Part of the Weyl chamber	Topological structure	Flag manifold	Dimension $d = d_F + d_S$
1	$\mathcal{M}_1$	λ_1	point	$[U(1)/U(1)] \times G_1 = \{\rho_*\}$	$\mathbf{F}_0^{(1)}$	$0 = 0 + 0$
2	$\mathcal{M}_{11}$	$\lambda_1 > \lambda_2$	line with left edge	$[U(2)/T^2] \times G_2$	$\mathbf{F}_1^{(2)}$	$3 = 2 + 1$
	$\mathcal{M}_2$	$\lambda_1 = \lambda_2$	right edge	$[U(2)/U(2)] \times G_1 = \{\rho_*\}$	$\mathbf{F}_0^{(2)}$	$0 = 0 + 0$
3	$\mathcal{M}_{111}$	$\lambda_1 > \lambda_2 > \lambda_3$	triangle with base without corners	$[U(3)/T^3] \times G_3$	$\mathbf{F}_{12}^{(3)}$	$8 = 6 + 2$
	$\mathcal{M}_{12}$	$\lambda_1 > \lambda_2 = \lambda_3$	edges with	$[U(3)/(U(2) \times T)] \times G_2$	$\mathbf{F}_1^{(3)}$	$5 = 4 + 1$
	$\mathcal{M}_{21}$	$\lambda_1 = \lambda_2 > \lambda_3$	lower corners		$\mathbf{F}_2^{(3)}$	
	$\mathcal{M}_3$	$\lambda_1 = \lambda_2 = \lambda_3$	upper corner	$[U(3)/U(3)] \times G_1 = \{\rho_*\}$	$\mathbf{F}_0^{(3)}$	$0 = 0 + 0$
4	$\mathcal{M}_{1111}$	$\lambda_1 > \lambda_2 > \lambda_3 > \lambda_4$	interior of tetrahedron with bottom face	$[U(4)/T^4] \times G_4$	$\mathbf{F}_{123}^{(4)}$	$15 = 12 + 3$
	$\mathcal{M}_{112}$	$\lambda_1 > \lambda_2 > \lambda_3 = \lambda_4$			$\mathbf{F}_{12}^{(4)}$	
	$\mathcal{M}_{121}$	$\lambda_1 > \lambda_2 = \lambda_3 > \lambda_4$	faces without side edges	$[U(4)/(U(2) \times T^2)] \times G_3$	$\mathbf{F}_{13}^{(4)}$	$12 = 10 + 2$
	$\mathcal{M}_{211}$	$\lambda_1 = \lambda_2 > \lambda_3 > \lambda_4$			$\mathbf{F}_{23}^{(4)}$	
	$\mathcal{M}_{13}$	$\lambda_1 > \lambda_2 = \lambda_3 = \lambda_4$			$\mathbf{F}_1^{(4)}$	$7 = 6 + 1$
	$\mathcal{M}_{31}$	$\lambda_1 = \lambda_2 = \lambda_3 > \lambda_4$	edges with lower corners	$[U(4)/(U(3) \times T)] \times G_2$	$\mathbf{F}_3^{(4)}$	
	$\mathcal{M}_{22}$	$\lambda_1 = \lambda_2 > \lambda_3 = \lambda_4$		$[U(4)/(U(2) \times U(2))] \times G_2$	$\mathbf{F}_2^{(4)}$	$9 = 8 + 1$
	$\mathcal{M}_4$	$\lambda_1 = \lambda_2 = \lambda_3 = \lambda_4$	upper corner	$[U(4)/U(4)] \times G_1 = \{\rho_*\}$	$\mathbf{F}_0^{(4)}$	$0 = 0 + 0$

Generically there will be no degeneracies in the spectrum, so except for a set of measure zero the boundary has the structure $(U(N)/T^N) \times G_{N-1}$.

Alternatively, the boundary can be decomposed into sets of matrices with different rank. It is not hard to show that the dimension of the set of states of rank $r = N - k$ is equal to $N^2 - k^2 - 1$.

The main message of this section has been that the Weyl chamber gives a good picture of the set of density matrices, because it represents the space of orbits under the unitary group. It is a very good picture, because the Euclidean distance between two points within a Weyl chamber is equal to the minimal Hilbert–Schmidt distance between the pair of orbits that they represent. In equations, let U and V denote arbitrary unitary matrices of size N. Then

$$D_{HS}(Ud_1 U^\dagger, Vd_2 V^\dagger) \geq D_{HS}(d_1, d_2), \tag{8.52}$$

where d_1 and d_2 are two diagonal matrices with their eigenvalues in decreasing order. The proof of this attractive observation is simple, once we know something about the majorisation order for matrices. For this reason its proof is deferred to Problem 13.5.

8.6 Projections and cross-sections

One way to try to 'see' the high-dimensional body that we have been discussing is to study two- and three-dimensional cross-sections through it, or the shadows it is casting on low-dimensional screens (although our position will be less favourable than that of the inhabitants of Plato's cave, who were looking at the shadows cast by three-dimensional objects only).[9]

In Chapter 1 we encountered the notion of dual convex sets. The set of quantum states $\mathcal{M}^{(N)}$ – the main hero of the entire book – is self dual for any dimension N. To see this note that ρ is a valid density matrix if the inequality

$$\mathrm{Tr}\, \sigma\rho \geq 0 \tag{8.53}$$

holds for every density matrix σ. This condition uses the scalar product in the Hilbert–Schmidt space of operators. Thus the set of positive operators forms a self dual cone with its apex at the origin, and its base, defined by $\mathrm{Tr}\rho = 1$, is self dual too. Shifting the entire set by $\rho_* = \mathbb{1}/N$ we obtain a set satisfying the self duality condition[10] analogous to (1.10),

$$(\mathcal{M}^{(N)} - \rho_*) = \{\rho \mid 1/N + \mathrm{Tr}(\rho\sigma) \geq 0 \; \forall \sigma \in (\mathcal{M}^{(N)} - \rho_*)\}. \tag{8.54}$$

---

[9] Cross-sections through the body of qutrit states were first studied by Bloore (1976) [139]. Later on they were systematically studied by many authors [529, 355, 792].

[10] Self duality is a key property of the state space which distinguishes quantum mechanics from other generalised theories [678, 968].

Taking the dual exchange faces of rank r (copies of $\mathcal{M}^{(r)}$) and faces of rank $N - r$ (copies of $\mathcal{M}^{(N-r)}$).

In the qubit case the set $\mathcal{M}^{(2)}$ forms the self dual Bloch ball. For higher dimensions the convex set $\mathcal{M}^{(N)} \subset \mathbb{R}^{N^2-1}$ is difficult to imagine, as some properties may conflict if we try to realise them in dimension three. To reduce the dimension one can investigate cross-sections of the eight-dimensional set $\mathcal{M}^{(3)}$ with a plane of dimension two or three, as well as its orthogonal projections on these planes – the shadows cast by the body on the planes, when illuminated by a very distant light source. Clearly the cross-sections will always be contained in the projections, but in exceptional cases they may coincide.

What kind of cross-sections appear? In the classical case it is known that every convex polytope arises as a cross-section of a simplex of sufficiently high dimension [368]. It is also true that every convex polytope arises as the projection of a simplex. But what are the cross-sections and the projections of $\mathcal{M}^{(3)}$?

Consider a linear subspace V of Bloch space, that is the space of traceless Hermitian matrices. Selecting a shifted subspace, $V + \rho_*$, which contains the maximally mixed state $\rho_* = \mathbb{1}/N$, we denote by S_V its cross-section with the set of the density matrices. Let P_V denote the orthogonal projection of $\mathcal{M}^{(N)}$ onto the subspace $V + \rho_*$. There exists a beautiful relation between projections and cross-sections, holding for all self dual convex bodies [764, 213, 955]. For them, cross-sections and projections are dual to each other, in the sense that

$$(S_V - \rho_*) = \{u \mid 1/N + \mathrm{Tr}(uv) \geq 0 \; \forall v \in (P_V - \rho_*)\} \tag{8.55}$$

and

$$(P_V - \rho_*) = \{u \mid 1/N + \mathrm{Tr}(uv) \geq 0 \; \forall v \in (S_V - \rho_*)\}. \tag{8.56}$$

These relations are illustrated in Figure 8.7.

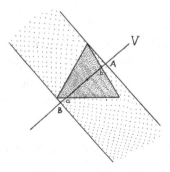

Figure 8.7 We intersect the self dual triangle with a one-dimensional subspace V, and obtain a cross-section extending from a to b. The dual of this segment is a two-dimensional strip, and when we project this onto V we obtain a projection extending from A to B, which is dual to the cross-section within V.

To appreciate what we see in cross-sections and projections we will first discuss two-dimensional screens. One can compute two-dimensional projections using the fact that they are dual to a cross-section. But we can also use the notion of the *numerical range* (or *field of values*) W of a given operator A [451, 382]. This is a non-empty subset of the complex plane consisting of expectation values of A among any normalised pure states of size N,

$$W(A) = \left\{ z \in \mathbb{C} : z = \langle \psi | A | \psi \rangle, \ |\psi\rangle\langle\psi| \in \mathcal{M}^{(N)} \right\}. \tag{8.57}$$

In the above definition one can replace $\langle \psi | A | \psi \rangle$ by $\mathrm{Tr}\rho A$ for $\rho \in \mathcal{M}^{(N)}$. If the matrix A is Hermitean its numerical range reduces to a segment of the real axis. A crucial property of $W(A)$ was established in a theorem with a long history [896, 411] and important consequences [249, 383]:

Toeplitz–Hausdorff theorem. *For an arbitrary linear operator A acting on a finite Hilbert space its numerical range $W(A)$ forms a convex set.*

To see the connection to projections, observe that changing the trace of A gives rise to a translation of the whole set, so we may as well fix the trace to equal unity. Then we can write

$$A = \lambda \mathbb{1} + H_1 + iH_2, \tag{8.58}$$

where H_1 and H_2 are traceless Hermitian matrices and $\lambda \in \mathbb{C}$. It follows that the set of all possible numerical ranges $W(A)$ of arbitrary matrices A of order N is affinely equivalent to the set of orthogonal projections of $\mathcal{M}^{(N)}$ on a 2-plane [427, 276]. This observation provides a simple way to show the Toeplitz–Hausdorff theorem, as any projection of a convex set is convex.

Thus to understand the structure of projections of $\mathcal{M}$ onto a plane it is enough to analyse the geometry of numerical ranges. The numerical range of a matrix A of order $N = 2$ forms an elliptical disc [681, 596], with focal points at the eigenvalues λ_1, λ_2 and the minor axis $d = \sqrt{\mathrm{Tr}AA^\dagger - |\lambda_1|^2 - |\lambda_2|^2}$. See Figure 8.9a. The ellipse

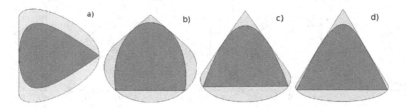

Figure 8.8 The drawings are dual pairs of planar cross-sections $S_V - \mathbb{1}/3$ (dark) and projections $P_V - \mathbb{1}/3$ (bright) of the convex body of qutrit quantum states $\mathcal{M}^{(3)}$. The cross-sections in (b)–(d) have an elliptic, parabolic and hyperbolic boundary piece, respectively.

degenerates to a circle if both eigenvalues of A coincide and to an interval if A is normal. These are the possible projections of the Bloch ball onto a plane.

In the case of a 3×3 matrix A, its numerical range is compact and can be characterised algebraically [532, 518]. The possible shapes can be divided into four cases according to the number of flat boundary parts: the boundary ∂W

(1) has *no* flat parts. Then W is strictly convex, it is bounded by an ellipse or equals the convex hull of a planar algebraic curve of order six.
(2) has *one* flat part. Then W is the convex hull of a planar quartic curve – e.g. W is the convex hull of a trigonometric curve known as the cardioid.
(3) has *two* flat parts. Then W is the convex hull of an ellipse and a point outside it.
(4) has *three* flat parts. Then W is a triangle with corners at eigenvalues of A.

Looking at the planar projections of $\mathcal{M}^{(3)}$ shown in Figure 8.8 we recognise cases 2 and 3. Case 4 occurs when the matrix A is normal, $AA^{\dagger} = A^{\dagger}A$. To find more examples we use the third root of unity, $\omega = e^{2\pi i/3}$, to construct four matrices of size three,

$$\begin{bmatrix} 1 & 1 & 1 \\ 0 & \omega & 1 \\ 0 & 0 & \omega^2 \end{bmatrix}, \quad \begin{bmatrix} 1 & 1 & 1 \\ -1 & \omega & 1 \\ -1 & -1 & \omega^2 \end{bmatrix}, \quad \begin{bmatrix} 1 & 1 & 0 \\ 0 & \omega & 0 \\ 0 & 0 & \omega^2 \end{bmatrix}, \quad \begin{bmatrix} 1 & 0 & 0 \\ 0 & \omega & 0 \\ 0 & 0 & \omega^2 \end{bmatrix}.$$

$$(8.59)$$

Their numerical ranges belong to each of four classes defined above. See Figure 8.9(b)–(e). The diagonal matrix is normal, so its numerical range forms a triangle. In general, the numerical range of a normal matrix A of order N forms the convex hull of its spectrum and is equal to a projection of a simplex Δ_{N-1} on a plane.

Any matrix A can be decomposed into its Hermitian and anti-Hermitian parts. Thus its numerical range $W(A)$ can be considered as the joint numerical range of

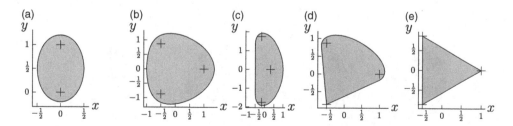

Figure 8.9 Numerical range of: (a) non-normal matrix $A_2 = [0, 1; 0, i]$ – a projection of the Bloch ball onto a plane; (b)–(e) of four matrices of order three defined in (8.59) which exemplify projections of the set $\mathcal{M}^{(3)}$ onto a plane with $0, 1, 2$ and 3 segments in the boundary. Crosses denote eigenvalues of the matrix.

two Hermitian operators H_1 and H_2. The general definition of *joint numerical range* of a collection of L Hermitian operators reads

$$W(H_1, \ldots, H_L) \equiv \left\{ (x_1, \ldots, x_L) \in \mathbb{R}^L : x_j = \langle \psi | H_j | \psi \rangle, \; |\psi\rangle\langle\psi| \in \mathcal{M}^{(N)} \right\}, \quad (8.60)$$

for $j = 1, \ldots, L$. Its convex hull is a projection of the set $\mathcal{M}^{(N)}$ of mixed states onto an L dimensional subspace V. In the case $N = 2$ and $L = 3$ linearly independent traceless Hermitian matrices their joint numerical range forms a hollow ellipsoid – an affine transformation of the Bloch sphere. In general, for $L = 3$ and $N \geq 3$ this set is convex [597], while for $L \geq 4$ matrices of size $N \geq 3$ this is not the case [384]. In particular for any $N \geq 2$ and $L = N^2 - 1$ generic matrices their joint numerical range is affinely isomorphic to the set of pure states, embedded in $\mathcal{M}^{(N)}$ [385].

Expression (8.60) implies a direct physical interpretation of $W(H_1, \ldots, H_L)$. This is the set of all possible results $(x_1, \ldots, x_L)$ of measurements of L observables H_j performed on several copies of various pure quantum states. The case $L = 2$ corresponds to the standard numerical range of the operator $A = H_1 + iH_2$. For instance, the set of possible results of measuring expectation values of two angular momentum operators J_x and J_z for a one-qubit state ρ forms a circular disc – a projection of the Bloch ball.

To actually calculate a two-dimensional projection $P \equiv \{(\text{Tr} H_1 \rho, \text{Tr} H_2 \rho)^{\mathrm{T}} \in \mathbb{R}^2 \mid \rho \in \mathcal{M}^{(3)}\}$ of the set $\mathcal{M}^{(3)}$ determined by two traceless Hermitian matrices H_1 and H_2 one may proceed as follows [451]. For every non-zero matrix X in the real span of H_1 and H_2 we calculate the maximal eigenvalue λ and the corresponding normalised eigenvector $|\psi\rangle$ with $X|\psi\rangle = \lambda|\psi\rangle$. Then $(\langle\psi|H_1|\psi\rangle, \langle\psi|H_2|\psi\rangle)^{\mathrm{T}}$ belongs to the projection $P = W(H_1, H_2)$, and these points cover all exposed points of P.

Let us now choose the vector space V to be three-dimensional and consider the joint numerical range of three Hermitian operators. In other words we analyse triples of expectation values of three measurements performed on several copies of the same state. In Figure 8.10 we show three-dimensional printouts of joint numerical ranges – projections of the eight-dimensional set $\mathcal{M}^{(3)}$ into a 3-space, which differ by the number of flat faces in their boundary. Taking the number of edges in the boundary into account, one can divide the totality of non-degenerate joint numerical ranges of three Hermitian matrices of size three into ten classes [882]. For $N = 4$ things get more involved as the list of possible cases becomes longer. For large N the complicated structure of $\mathcal{M}^{(N)}$ is difficult to observe, as typical low-dimensional cross-sections become close to spherical, in agreement with the Dvoretzky theorem [279] to be discussed in Section 15.8.

A closed convex set arising when an affine subspace intersects the cone of positive matrices is known as a *spectrahedron*. To see how rich the possibilities are we

Figure 8.10 Examples of joint numerical range $W(H_1, H_2, H_3)$ for three Hermitian matrices of size three – projections of $\mathcal{M}^{(3)}$ into three dimensions – the boundary of which contains (a) no flat parts (generic case), or (b) one, (c) two, (d) three, or (e) four elliptical discs. Any further projection of these convex bodies into a 2-plane has zero, one, two or three flat parts of the boundary, as they belong to one of four classes of standard numerical range for $N = 3$.

mention the curious fact that the convex hull of any space curve whose coordinate functions are trigonometric polynomials (of finite order) can be obtained in this way [428]. There is a branch of mathematics called convex algebraic geometry, where matters such as those discussed here are treated in depth.

We have seen some shadows on the walls of our cave, but after several years of research it is fair to admit that we do not know what the set of mixed quantum states really looks like, even for $N = 3$. It is neither a ball nor a polytope. The best three-dimensional model we can suggest right now is the convex hull of the stitching of a tennis ball (see Figure C.3) in which the stitching represents the connected manifold of pure quantum states. But this model is far from perfect. Every pure qutrit state belongs to a cross-section of the set of mixed states forming an equilateral triangle – a property not shared by the points on the stitching of the tennis ball. It is easier to follow an apophatic approach, and to emphasise what the set $\mathcal{M}^{(3)}$ *does not* look like [114].

8.7 An algebraic afterthought

Quantum mechanics is built around the fact that the set of density matrices is a convex set of a very special shape. From the perspective of Chapter 1 it seems a strange set to consider. There are many convex sets to choose from. The simplex is evidently in some sense preferred, and leads to classical probability theory and – ultimately, once the pure states are numerous enough to form a symplectic manifold – to classical mechanics. But why the convex set of density matrices? A standard answer is that we want the vector space that contains the convex set to have further structure, turning it into an *algebra*.[11] (By definition, an algebra is a vector

---

[11] The algebraic viewpoint was invented by Jordan; key mathematical results were derived by Jordan, von Neumann and Wigner in 1934 [508]. To see how far it has developed, see Emch [294] and Alfsen and Shultz [21].

space where vectors can be multiplied as well as added.) At first sight this looks like an odd requirement: the observables may form an algebra, but why should the states sit in one? We get an answer of sorts if we think of the states as linear maps from the algebra to the real numbers, because then we will obtain a vector space that is dual to the algebra and can be identified with the algebra. Some further hints will emerge in Chapter 11. For now, let us simply accept it. The point is that if the algebra has suitable properties then this will give rise to new ways of defining positive cones – more interesting ways than the simple requirement that the components of the vectors be positive.

To obtain an algebra we must define a product $A \circ B$. We may want the algebra to be *real* in the sense that

$$A \circ A + B \circ B = 0 \quad \Rightarrow \quad A = B = 0. \tag{8.61}$$

We do not insist on associativity, but we insist that polynomials of operators should be well defined. In effect we require that $A^n \circ A^m = A^{n+m}$ where $A^n \equiv A \circ A^{n-1}$. Call this *power associativity*. With this structure in hand we do have a natural definition of *positive* vectors, as vectors A that can be written as $A = B^2$ for some vector B, and we can define *idempotents* as vectors obeying $A^2 = A$. If we can also define a trace, we can define pure states as idempotents of unit trace. But by now there are not that many algebras to choose from. To make the algebra real in the above sense we would like the algebra to consist of Hermitian matrices. Ordinary matrix multiplication will not preserve Hermiticity, and therefore matrix algebras will not do as they stand. However, because we can square our operators we can define the *Jordan product*

$$A \circ B \equiv \frac{1}{4}(A + B)^2 - \frac{1}{4}(A - B)^2. \tag{8.62}$$

There is no obvious physical interpretation of this composition law, but it does turn the space of Hermitian matrices into a (commutative) *Jordan algebra*. If A and B are elements of a matrix algebra this product is equal to one-half of their anti-commutator, but we need not assume this. Jordan algebras have all the properties we want, including power associativity. Moreover all simple Jordan algebras have been classified, and there are only four kinds (and one exceptional case). A complete list is given in Table 8.2.[12]

The case that really concerns us is $J_N^{\mathbb{C}}$. Here the Jordan algebra is the space of complex valued Hermitian N by N matrices, and the Jordan product is given by one-half of the anti-commutator. This is the very algebra that we have – implicitly – been using, and with whose positive cone we are by now reasonably familiar. We can easily define the trace of any element in the algebra, and the pure states

---

[12] For a survey of Jordan algebras, see McCrimmon [650].

Table 8.2 *Jordan algebras.*

Jordan algebra	Dimension	Norm	Positive cone	Pure states
$J_N^{\mathbb{R}}$	$N(N+1)/2$	$\det M$	Self dual	$\mathbb{R}\mathbf{P}^{N-1}$
$J_N^{\mathbb{C}}$	N^2	$\det M$	Self dual	$\mathbb{C}\mathbf{P}^{N-1}$
$J_N^{\mathbb{H}}$	$N(2N-1)$	$\det M$	Self dual	$\mathbb{H}\mathbf{P}^{N-1}$
$J_2(V_n)$	$n+1$	$\eta_{IJ}X^I X^J$	Self dual	$\mathbf{S}^{n-1}$
$J_3^{\mathbb{O}}$	27	$\det M$	Self dual	$\mathbb{O}\mathbf{P}^2$

in the table are assumed to be of unit trace. One can replace the complex numbers with real or quaternionic numbers, giving two more families of Jordan algebras. The state spaces that result from them occur in special quantum mechanical situations, as we saw in Section 5.5. The fourth family of Jordan algebras are the *spin factors* $J_2(V_n)$. They can also be embedded in matrix algebras, their norm uses a Minkowski space metric η_{IJ}, and their positive cones are the familiar forward light cones in Minkowski spaces of dimension $n+1$. Their state spaces occur in special quantum mechanical situations too [909], but this is not the place to go into that. (Finally there is an exceptional case based on octonions, that needs not concern us.)

So what is the point? One point is that very little in addition to the quantum mechanical formalism turned up in this way. This is to say: once we have committed ourselves to finding a self dual positive cone in a finite dimensional real algebra, then we are almost (but not quite) restricted to the standard quantum mechanical formalism already. Another point is that the positive cone now acquires an interesting geometry. Not only is it self dual (see Section 1.1), it is also foliated in a natural way by hypersurfaces for which the determinant of the matrix is constant. These hypersurfaces turn out to be symmetric coset spaces $SL(N, \mathbb{C})/SU(N)$, or relatives of that if we consider a Jordan algebra other than $J_N^{\mathbb{C}}$. Given the norm $\mathcal{N}$ on the algebra, there is a natural looking metric

$$g_{ij} = -\frac{1}{d}\partial_i\partial_j \ln \mathcal{N}^{d/N}. \tag{8.63}$$

where d is the dimension of the algebra. (Since the norm is homogeneous of order N, the exponent ensures that the argument of the logarithm is homogeneous of order d.) This metric is positive definite for all the Jordan algebras, and it makes the boundary of the cone sit at an infinite distance from any point in the interior. If we specialise to diagonal matrices – which means that the Jordan algebra is no longer simple – we recover the positive orthant used in classical probability theory, and the natural metric turns out to be flat, although it differs from the 'obvious' flat metric on $\mathbb{R}^N$.

We doubt that the reader feels compelled to accept the quantum mechanical formalism only because it looms large in a Jordan algebra framework. Another way of arguing for quantum mechanics from (hopefully) simple assumptions is provided by quantum logic. This language can be translated into convex set theory [659], and turns out to be equivalent to setting conditions on the lattice of faces that one wants the underlying convex set to have. From a physical point of view it concerns the kind of 'yes/no' experiments that one expects to be able to perform; choosing one's expectations suitably [48] one can argue along such lines that the state spaces that will emerge are necessarily state spaces of Jordan algebras, and we are back where we started.

But there is a further algebraic structure waiting in the wings, and it is largely this additional structure that Chapters 9 and *passim* are about.

8.8 Summary

Let us try to summarise basic properties of the set of mixed quantum states $\mathcal{M}^{(N)}$. As before Δ_{N-1} is an $N-1$ dimensional simplex, $\widetilde{\Delta}_{N-1}$ is a Weyl chamber in Δ_{N-1}, and $\mathbf{F}^{(N)}$ is the complex flag manifold $\mathbf{F}^{(N)}_{12\ldots N-1}$.

1. $\mathcal{M}^{(N)}$ is a convex set of $N^2 - 1$ dimensions. It is topologically equivalent to a ball and does not have pieces of lower dimensions ('no hairs').
2. The set $\mathcal{M}^{(N)}$ is inscribed in a ball of radius $R_{\mathrm{out}} = \sqrt{(N-1)/2N}$ and contains a maximal ball of radius $r_{\mathrm{in}} = 1/\sqrt{2N(N-1)}$.
3. The set $\mathcal{M}^{(N)}$ is self dual.
4. It is neither a polytope nor a smooth body. Its faces are copies of $\mathcal{M}^{(K)}$ with $K < N$.
5. It is partitioned into orbits of the unitary group, and the space of orbits is a Weyl chamber $\widetilde{\Delta}_{N-1}$.
6. The full measure of $\mathcal{M}^{(N)}$ has locally the structure of $\mathbf{F}^{(N)} \times \Delta_{N-1}$.
7. The boundary $\partial\mathcal{M}^{(N)}$ contains all states of less than maximal rank. Almost everywhere it has the local structure of $\mathbf{F}^{(N)} \times \Delta_{N-2}$.
8. The set of pure states is a connected $2N - 2$ dimensional set, which has zero measure with respect to the $N^2 - 2$ dimensional boundary $\partial\mathcal{M}^{(N)}$.
9. Explicit formulae for the volume V and the area A of $\mathcal{M}^{(N)}$ are known; see Chapter 15.

In this summary we have not mentioned the remarkable way in which composite systems are handled by quantum theory. The discussion of this topic starts in the next chapter and culminates in Chapters 16 and 17.

Problems

8.1   Prove that the polar decomposition of an invertible operator is unique.

8.2   Consider a square matrix A. Perform an arbitrary permutation of its rows and/or columns. Will its eigenvalues or singular values change?

8.3   What are the singular values of (a) a Hermitian matrix, (b) a unitary matrix, (c) any normal matrix A (such that $[A, A^\dagger] = 0$)?

8.4   A unitary similarity transformation does not change the eigenvalues of any matrix. Show that this is true for the singular values as well.

8.5   Show that $\mathrm{Tr}(AA^\dagger)\,\mathrm{Tr}(BB^\dagger) \geq |\mathrm{Tr}(AB^\dagger)|^2$, always.

8.6   Show that the diagonal elements of a positive operator are positive.

8.7   Take a generic vector in $\mathbb{R}^{N^2-1}$. How many of its components can you set to zero, if you are allowed to act only with an $SU(N)$ subgroup of the rotation group?

8.8   Transform a density matrix ρ of size 2 into $\rho' = U\rho\,U^\dagger$ by a general unitary matrix $U = \begin{bmatrix} \cos\vartheta\,e^{i\phi} & \sin\vartheta\,e^{i\psi} \\ -\sin\vartheta\,e^{-i\psi} & \cos\vartheta\,e^{-i\phi} \end{bmatrix}$. What is the orthogonal matrix $O \in SO(3)$ which transforms the Bloch vector $\vec{\tau}$? Find the rotation angle t and the vector $\vec{\Omega}$ determining the orientation of the rotation's axis.

8.9   Let ρ be a Hermitian matrix obeying $\mathrm{Tr}\rho^3 = \mathrm{Tr}\rho^2 = 1$. Prove that ρ describes a pure quantum state.

9

Purification of mixed quantum states

In this significant sense quantum theory subscribes to the view that '*the whole is greater than the sum of its parts*'.

—*Hermann Weyl*[1]

In quantum mechanics the whole, built from parts, is described using the *tensor product* that defines the composition of an N dimensional vector space V and an M dimensional one V' as the NM dimensional vector space $V \otimes V'$. One can go on, using the tensor product to define an infinite dimensional tensor algebra. The interplay between the tensor algebra and the other algebraic structures is subtle indeed. In this chapter we study the case of two subsystems only. The arena is Hilbert–Schmidt space (real dimension $2N^2$), but now regarded as the Hilbert space of a composite system. We will use a partial trace to take ourselves from Hilbert–Schmidt space to the space of density matrices acting on an N dimensional Hilbert space. The result is the quantum analogue of a marginal probability distribution. It is also like a projection in a fibre bundle, with Hilbert–Schmidt space as the bundle space and the group $U(N)$ acting on the fibres, while the positive cone serves as the base space (real dimension $2N^2 - N^2 = N^2$). Physically, the important idea is that of *purification*; a density matrix acting on $\mathcal{H}$ is regarded as a pure state in $\mathcal{H} \otimes \mathcal{H}^*$, with some of its details forgotten. We could now start an argument whether all mixed quantum states are really pure states in some larger Hilbert space, but we prefer to focus on the interesting geometry that is created on the space of mixed states by this construction.[2]

---

[1] Reproduced from H. Weyl. *Group Theory and Quantum Mechanics*. E. P. Dutton, 1932.

[2] For an eloquent defence of the point of view that regards density matrices as primary, see Mermin [655]. With equal eloquence, Penrose [721] takes the opposite view.

9.1 Tensor products and state reduction

The *tensor product* of two vector spaces is not all that easy to define. The easiest way is to rely on a choice of basis in each of the factors.[3] We are interested in the tensor product of two Hilbert spaces $\mathcal{H}_1$ and $\mathcal{H}_2$, with dimensions N_1 and N_2, respectively. The tensor product space will be denoted $\mathcal{H}_{12}$, and it will have dimension $N_1 N_2$. The statement that the whole is greater than its parts is related to the fact that $N_1 N_2 > N_1 + N_2$ (unless $N_1 = N_2 = 2$).

We expect the reader to be familiar with the basic features of the tensor product, but to fix our notation let us choose the bases $\{|m\rangle\}_{m=1}^{N_1}$ in $\mathcal{H}_1$, and $\{|\mu\rangle\}_{\mu=1}^{N_2}$ in $\mathcal{H}_2$. Then the Hilbert space $\mathcal{H}_{12} \equiv \mathcal{H}_1 \otimes \mathcal{H}_2$ is spanned by the basis formed by the $N_1 N_2$ elements $|m\rangle \otimes |\mu\rangle = |m\rangle|\mu\rangle$, where the sign $\otimes$ will be written explicitly only on special occasions. The basis vectors are direct products of vectors in the factor Hilbert spaces, but by taking linear combinations we will obtain vectors that cannot be written in such a form – which explains why the composite Hilbert space $\mathcal{H}_{12}$ is so large. Evidently we can go on to define the Hilbert space $\mathcal{H}_{123}$, starting from three-factor Hilbert spaces, and indeed the procedure never stops. By taking tensor products of a vector space with itself, we will end up with an infinite dimensional *tensor algebra*. Here our concern is with *bipartite* systems that use only the Hilbert space $\mathcal{H}_{12}$. In many applications of quantum mechanics, a further elaboration of the idea is necessary: it may be that the subsystems are indistinguishable from each other, in which case one must take symmetric or anti-symmetric combinations of $\mathcal{H}_{12}$ and $\mathcal{H}_{21}$, leading to bosonic or fermionic subsystems, or perhaps utilise some less trivial representation of the symmetric group that interchanges the subsystems. But we will avoid this elaboration.

The matrix algebra of operators acting on a given Hilbert space is itself a vector space – the Hilbert–Schmidt vector space $\mathcal{HS}$ studied in Section 8.1. We can take tensor products also of algebras. If A_1 acts on $\mathcal{H}_1$, and A_2 acts on $\mathcal{H}_2$, then their *tensor* or *Kronecker product* $A_1 \otimes A_2$ is defined by its action on the basis elements:

$$(A_1 \otimes A_2)|m\rangle \otimes |\mu\rangle \equiv A_1|m\rangle \otimes A_2|\mu\rangle. \tag{9.1}$$

Again, this is not the most general operator in the tensor product algebra since we can form linear combinations of operators of this type. For a general operator we can form matrix elements according to

$$A_{n\nu}^{m\mu} = \langle m| \otimes \langle \mu|A|n\rangle \otimes |\nu\rangle. \tag{9.2}$$

On less special occasions we may write this as $A_{m\mu,n\nu}$.

---

[3] This is the kind of procedure that mathematicians despise; the basis independent definition can be found in Kobayashi and Nomizu [550]. Since we tend to think of operators as explicit matrices, the simple minded definition is good enough for us.

Everything works best if the underlying field is that of the complex numbers [48]: let the space of observables, that is Hermitian operators, on a Hilbert space $\mathcal{H}$ be denoted $\mathcal{HM}(\mathcal{H})$. The dimensions of the spaces of observables on a pair of complex Hilbert spaces $\mathcal{H}_1$ and $\mathcal{H}_2$ obey

$$\dim[\mathcal{HM}(\mathcal{H}_1 \otimes \mathcal{H}_2)] = \dim[\mathcal{HM}(\mathcal{H}_1)] \dim[\mathcal{HM}(\mathcal{H}_2)]. \tag{9.3}$$

That is, $(N_1 N_2)^2 = N_1^2 N_2^2$. If we work over the real numbers the left-hand side of (9.3) is larger than the right-hand side, and if we work over quaternions (using a suitable definition of the tensor product) the right-hand side is the largest. As an argument for why we should choose to work over the complex numbers, this observation may not be completely compelling. But the tensor algebra over the complex numbers has many wonderful properties.

Most of the time we think of vectors as columns of numbers, and of operators as explicit matrices – in finite dimensions nothing is lost, and we gain concreteness. We organise vectors and matrices into arrays, using the obvious lexicographical order. At least, the order should be obvious if we write it out for a simple example: if

$$A = \begin{bmatrix} A_{11} & A_{12} \\ A_{21} & A_{22} \end{bmatrix} \quad \text{and} \quad B = \begin{bmatrix} B_{11} & B_{12} \\ B_{21} & B_{22} \end{bmatrix}$$

then

$$A \otimes B = \begin{bmatrix} A_{11}B & A_{12}B \\ A_{21}B & A_{22}B \end{bmatrix} = \begin{bmatrix} A_{11}B_{11} & A_{11}B_{12} & A_{12}B_{11} & A_{12}B_{12} \\ A_{11}B_{21} & A_{11}B_{22} & A_{12}B_{21} & A_{12}B_{22} \\ A_{21}B_{11} & A_{21}B_{12} & A_{22}B_{11} & A_{22}B_{12} \\ A_{21}B_{21} & A_{21}B_{22} & A_{22}B_{21} & A_{22}B_{22} \end{bmatrix}. \tag{9.4}$$

Contemplation of this expression should make it clear what lexicographical ordering we are using. At first sight one may worry that A and B are treated quite asymmetrically here, but on reflection one sees that this is only a matter of basis changes, and does not affect the spectrum of $A \otimes B$. See Problem 9.4, and Problems 9.2–9.5 for further information about tensor products.

The tensor product is a main theme in quantum mechanics. We will use it to split the world into two parts; a part 1 that we study and another part 2 that we may refer to as the environment. This may be a physical environment that must be taken into account when doing experiments, but not necessarily so. It may also be a mathematical device that enables us to prove interesting theorems about the system under study, with no pretense of realism as far as the environment is concerned. Either way, the split is more subtle than it used to be in classical physics, precisely because the composite Hilbert space $\mathcal{H}_{12} = \mathcal{H}_1 \otimes \mathcal{H}_2$ is so large. Most of its vectors are not direct products of vectors in the factor spaces. If not, the subsystems are said to be *entangled*.

Let us view the situation from the Hilbert space $\mathcal{H}_{12}$. To compute the expectation value of an arbitrary observable we need the density matrix ρ_{12}. It is assumed that we know exactly how $\mathcal{H}_{12}$ is defined as a tensor product $\mathcal{H}_1 \otimes \mathcal{H}_2$, so the representation (9.2) is available for all its operators. Then we can define *reduced density matrices* ρ_1 and ρ_2, acting on $\mathcal{H}_1$ and $\mathcal{H}_2$ respectively, by taking *partial traces*. Thus

$$\rho_1 \equiv \mathrm{Tr}_2\, \rho_{12} \quad \text{where} \quad (\rho_1)^m_{\ n} = \sum_{\mu=1}^{N_2} (\rho_{12})^{m\mu}_{\ n\mu}, \tag{9.5}$$

and similarly for ρ_2. (Readers who need some practice may consult Problem 9.1.) This construction is interesting, because it could be that experiments are performed exclusively on the first subsystem, in which case we are only interested in observables of the form

$$A = A_1 \otimes \mathbb{1}_2 \quad \Leftrightarrow \quad A^{m\mu}_{\ n\nu} = (A_1)^m_{\ n}\, \delta^\mu_\nu. \tag{9.6}$$

Then the state ρ_{12} is more than we need; the reduced density matrix ρ_1 acting on $\mathcal{H}_1$ is enough, because

$$\langle A \rangle = \mathrm{Tr}\rho_{12}A = \mathrm{Tr}_1\rho_1 A_1. \tag{9.7}$$

Here Tr_1 denotes the trace taken over the first subsystem only. Moreover $\rho_1 = \mathrm{Tr}_2\rho_{12}$ is the only operator that has this property for every operator of the form $A = A_1 \otimes \mathbb{1}_2$; this observation can be used to give a basis-independent definition of the partial trace.

Even if ρ_{12} is a pure state, the state ρ_1 will in general be a mixed state. Interestingly, it is possible to obtain any mixed state as a partial trace over a pure state in a suitably enlarged Hilbert space. To make this property transparent, we need some further preparation.

9.2 The Schmidt decomposition

An exceptionally useful fact is the following:[4]

Schmidt's theorem. *Every pure state in the Hilbert space* $\mathcal{H}_{12} = \mathcal{H}_1 \otimes \mathcal{H}_2$ *can be expressed in the form*

$$|\Psi\rangle = \sum_{i=1}^{N} \sqrt{\lambda_i}|e_i\rangle \otimes |f_i\rangle, \tag{9.8}$$

where $\{|e_i\rangle\}_{i=1}^{N_1}$ *is an orthonormal basis for* $\mathcal{H}_1$, $\{|f_i\rangle\}_{i=1}^{N_2}$ *is an orthonormal basis for* $\mathcal{H}_2$, *and* $N \leq \min\{N_1, N_2\}$.

---

[4] The original Schmidt's theorem, that appeared in 1907 [799], concerns infinite dimensional spaces. The present formulation was used by Schrödinger (1936) in his analysis of entanglement [805], by Everett (1957) in his relative state (or many worlds) formulation of quantum mechanics [304], and in the 1960s by Carlson and Keller [193] and Coleman [232]. Simple expositions of the Schmidt decomposition are provided by Ekert and Knight [292], and by Aravind [51].

This is known as the *Schmidt decomposition* or *Schmidt's polar form* of a *bipartite pure state*. It should come as a surprise, because there is only a single sum; what is obvious is that only a pure state can be expressed in the form

$$|\Psi\rangle = \sum_{i=1}^{N_1} \sum_{j=1}^{N_2} C_{ij} |\hat{e}_i\rangle \otimes |\hat{f}_j\rangle, \tag{9.9}$$

where C is some complex valued matrix and the bases are arbitrary. The Schmidt decomposition becomes more reasonable when it is observed that the theorem concerns a special state $|\Psi\rangle$; changing the state may force us to change the bases used in Eq. (9.8).

To deduce the Schmidt decomposition we assume, without loss of generality, that $N_1 \leq N_2$. Then we observe that we can rewrite Eq. (9.9) by introducing the states $|\hat{\phi}_i\rangle = \sum_j C_{ij} |\hat{f}_j\rangle$; these will not be orthonormal states but they certainly exist, and permit us to write the state in $\mathcal{H}_{12}$ as

$$|\Psi\rangle = \sum_{i=1}^{N_1} |\hat{e}_i\rangle |\hat{\phi}_i\rangle. \tag{9.10}$$

Taking a partial trace of $\rho_\Psi = |\Psi\rangle\langle\Psi|$ with respect to the second subsystem, we find

$$\rho_1 = \mathrm{Tr}_2 (|\Psi\rangle\langle\Psi|) = \sum_{i=1}^{N_1} \sum_{j=1}^{N_1} \langle\hat{\phi}_j|\hat{\phi}_i\rangle \, |\hat{e}_i\rangle\langle\hat{e}_j|. \tag{9.11}$$

Now comes the trick. We can always perform a unitary transformation to a new basis $|e_i\rangle$ in $\mathcal{H}_1$, so that ρ_1 takes the diagonal form

$$\rho_1 = \sum_{i=1}^{N_1} \lambda_i |e_i\rangle\langle e_i|, \tag{9.12}$$

where the coefficients λ_i are real and non-negative. Finally we go back and repeat the argument, using this basis from the start. Taking the hats away, we find

$$\langle\phi_j|\phi_i\rangle = \lambda_i \delta_{ij}. \tag{9.13}$$

That is to say, we can set $|\phi_i\rangle = \sqrt{\lambda_i}|f_i\rangle$. The result is precisely the Schmidt decomposition.

An alternative way to obtain the Schmidt decomposition is to rely on the singular value decomposition (8.14) of the matrix C in Eq. (9.9). In Section 8.1 we considered square matrices, but since the singular values are really the square roots of the eigenvalues of the matrix $CC^\dagger$ – which is square in any case – we can lift that restriction here. Let the singular values of C be $\sqrt{\lambda_i}$. There exist two unitary matrices U and V such that

$$C_{ij} = \sum_{k,l} U_{ik} \sqrt{\lambda_k}\, \delta_{kl} V_{lj}. \tag{9.14}$$

Using U and V to effect changes of the bases in $\mathcal{H}_1$ and $\mathcal{H}_2$ we recover the Schmidt decomposition (9.8). Indeed

$$\rho_1 \equiv \mathrm{Tr}_2 \rho = CC^\dagger \quad \text{and} \quad \rho_2 \equiv \mathrm{Tr}_1 \rho = C^T C^*. \tag{9.15}$$

In the generic case all the singular values λ_i are different, and the Schmidt decomposition is unique up to phases which are free parameters determined by any specific choice of the eigenvectors of U and V. The bases used in the Schmidt decomposition are distinguished because they are precisely the eigenbases of the reduced density matrices, one of them given in Eq. (9.12) and the other being

$$\rho_2 = \mathrm{Tr}_1 \left(|\Psi\rangle\langle\Psi| \right) = \sum_i \lambda_i |f_i\rangle\langle f_i|. \tag{9.16}$$

When the spectra of the reduced density matrices are degenerate the bases may be rotated in the corresponding subspace.

At this point we introduce some useful terminology. The real numbers λ_i that occur in the Schmidt decomposition (9.8) are called *Schmidt coefficients*,[5] and they obey

$$\sum_i \lambda_i = 1, \qquad \lambda_i \geq 0. \tag{9.17}$$

The set of all possible vectors $\vec{\lambda}$ forms an $(N-1)$-dimensional simplex, known as the *Schmidt simplex*. The number r of non-vanishing λ_i is called the *Schmidt rank* of the state $|\Psi\rangle$. It is equal to the rank of the reduced density matrix. The latter describes a pure state if and only if $r = 1$. If $r > 1$ the state $|\Psi\rangle$ is an entangled state of its two subsystems (see Chapter 16).

A warning concerning whether the Schmidt decomposition is appropriate: there is no similar strong result available for Hilbert spaces that are direct products of more than two factor spaces. This is evident because if there are M factor spaces, all of dimension N, then the number of parameters describing a general state grows like N^M, while the number of unitary transformations one can use to choose basis vectors within the factors grows only like $M \times N^2$. But we can look at the Schmidt decomposition through different glasses. The Schmidt coefficients are not changed by *local unitary transformations*, that is to say – in the Hilbert space $\mathcal{H} \otimes \mathcal{H}$,

---

[5] We find this definition convenient. Others [692] use this name for $\sqrt{\lambda_i}$.

where both factors have dimension N – transformations belonging to the subgroup $U(N) \otimes U(N)$, acting on each factor separately. In Chapter 17 we will have many such factors, and we will ask for invariants under the action of $U(N) \otimes U(N) \otimes \cdots \otimes U(N)$, characterising the orbits of that group.

9.3 State purification and the Hilbert–Schmidt bundle

With the Schmidt decomposition in hand we can discuss the opposite of state reduction: given any density matrix ρ on a Hilbert space $\mathcal{H}$, we can use Eq. (9.8) to write down a pure state on a larger Hilbert space whose reduction down to $\mathcal{H}$ is ρ. The key statements are the following:

Reduction lemma. *Let ρ_{12} be a pure state on $\mathcal{H}_{12}$. Then the spectra of the reduced density matrices ρ_1 and ρ_2 are identical, except possibly for the degeneracy of any zero eigenvalue.*

Purification lemma. *Given a density matrix ρ_1 on a Hilbert space $\mathcal{H}_1$, there exists a Hilbert space $\mathcal{H}_2$ and a pure state ρ_{12} on $\mathcal{H}_1 \otimes \mathcal{H}_2$ such that $\rho_1 = \text{Tr}_2 \rho_{12}$.*

These statements follow trivially from Schmidt's theorem, but they have far-reaching consequences. It is notable that any density matrix ρ acting on a Hilbert space $\mathcal{H}$ can be purified in the Hilbert–Schmidt space $\mathcal{HS} = \mathcal{H} \otimes \mathcal{H}^*$, introduced in Section 8.1. Any attempt to use a smaller Hilbert space will fail in general, and, mathematically, there is no point in choosing a larger space since the purified density matrices will always belong to a subspace that is isomorphic to the Hilbert–Schmidt space. Hence Hilbert–Schmidt space provides a canonical arena for the purification of density matrices. We will try to regard it as a fibre bundle, along the lines of Chapter 3. Let us see if we can.

The vectors of $\mathcal{HS}$ can be represented as operators A acting on $\mathcal{H}$, and there is a projection down to the cone $\mathcal{P}$ of positive operators defined by

$$\Pi : \quad A \longrightarrow \rho = AA^{\dagger}. \tag{9.18}$$

The fibres will consist of operators projecting to the same positive operator, and the unitary group acts on the fibres as

$$A \longrightarrow A' = AU. \tag{9.19}$$

We could have used the projection $A \longrightarrow \rho' = A^{\dagger}A$ instead. More interestingly, we could have used the projection $A \longrightarrow \rho' = AA^{\dagger}/\text{Tr}AA^{\dagger}$. This would take us all the way down to the density matrices (of unit trace), but the projection (9.18) turns out to be more convenient to work with.

Do we have a fibre bundle? Not quite, because the fibres are not isomorphic. We do have a fibre bundle if we restrict the bundle space to be the open set of

Hilbert–Schmidt operators with a trivial kernel. The boundary of the base manifold is not really lost, since it can be recovered by continuity arguments. And the fibre bundle perspective is really useful, so we will adopt it here.[6] The structure group of the bundle is $U(N)$ and the base manifold is the interior of the positive cone. The bundle projection is given by Eq. (9.18). From a topological point of view this is a trivial bundle, admitting a global section

$$\tau : \quad \rho \longrightarrow \sqrt{\rho}. \tag{9.20}$$

The map τ is well defined because a positive operator admits a unique positive square root, it is a section because $\Pi(\tau(\rho)) = (\sqrt{\rho})^2 = \rho$, and it is global because it works everywhere.

What is interesting about our bundle is its geometry. We want to think of Hilbert–Schmidt space as a real vector space, so we adopt the metric

$$X \cdot Y = \frac{1}{2}(\langle X, Y \rangle + \langle Y, X \rangle) = \frac{1}{2}\text{Tr}(X^\dagger Y + Y^\dagger X), \tag{9.21}$$

where X and Y are tangent vectors. (Because we are in a vector space, the tangent spaces can be identified with the space itself.) This is the *Hilbert–Schmidt bundle*. A matrix in the bundle space will project to a properly normalised density matrix if and only if it sits on the unit sphere in $\mathcal{HS}$. The whole setting is quite similar to that encountered for the 3-sphere in Chapter 3. Like the 3-sphere, the Hilbert–Schmidt bundle space has a preferred metric, and therefore there is a preferred connection and a preferred metric on the base manifold.

According to Section 3.6, a connection is equivalent to a decomposition of the bundle tangent space into vertical and horizontal vectors. The vertical tangent vectors pose no problem. By definition they point along the fibres; since any unitary matrix U can be obtained by exponentiating a Hermitian matrix H, a curve along a fibre is given by

$$A\,U(t) = A\,e^{iHt}. \tag{9.22}$$

Therefore every vertical vector takes the form iAH for some Hermitian matrix H. The horizontal vectors must be defined somehow, and we do so by requiring that they are orthogonal to the vertical vectors under our metric. Thus, for a horizontal vector X, we require

$$\text{Tr}X(iAH)^\dagger + \text{Tr}(iAH)X^\dagger = i\,\text{Tr}(X^\dagger A - A^\dagger X)H = 0 \tag{9.23}$$

for all Hermitian matrices H. Hence X is a horizontal tangent vector at the point A if and only if

$$X^\dagger A - A^\dagger X = 0. \tag{9.24}$$

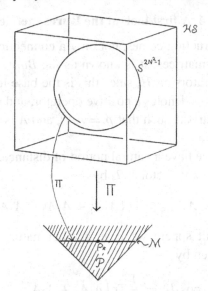

Figure 9.1 The Hilbert–Schmidt bundle. It is the unit sphere in $\mathcal{HS}$ that projects down to density matrices $\mathcal{M}$. Compare with Figure 8.1.

Thus equipped, we can lift curves in the base manifold to horizontal curves in the bundle.

In particular, suppose that we have a curve $\rho(s)$ in $\mathcal{M}^{(N)}$. We are looking for a curve $A(s)$ such that $AA^{\dagger}(s) = \rho(s)$, and such that its tangent vector $\dot{A}$ is horizontal, that is to say that

$$\dot{A}^{\dagger}A = A^{\dagger}\dot{A}. \tag{9.25}$$

It is easy to see that the latter condition is fulfilled if

$$\dot{A} = GA, \tag{9.26}$$

where G is a Hermitian matrix. To find the matrix G, we observe that

$$AA^{\dagger}(\sigma) = \rho(\sigma) \quad \Rightarrow \quad \dot{\rho} = G\rho + \rho G. \tag{9.27}$$

As long as ρ is a strictly positive operator this equation determines G uniquely [875, 131], and it follows that the horizontal lift of a curve in the base space is uniquely determined. We could go on to define a mixed state generalisation of the geometric phase discussed in Section 4.8, but in fact we will turn to somewhat different matters.[7]

---

[7] Geometric phases were among Uhlmann's motivations for developing the material in this chapter [906, 908]. Other approaches to geometric phases for mixed states exist [301]; for a good review see Chruściński and Jamiołkowski [221].

9.4 A first look at the Bures metric

Out of our bundle construction comes, not only a connection, but a natural metric on the space of density matrices. It is known as the *Bures metric*, and it lives on the cone of positive operators on $\mathcal{H}$, since this is the base manifold of our bundle. Until further notice then, ρ denotes a positive operator, and we allow $\mathrm{Tr}\rho \neq 1$. The purification of ρ is a matrix A such that $\rho = AA^{\dagger}$, and A is regarded as a vector in the Hilbert–Schmidt space.

In the bundle space, we have a natural notion of distance, namely the Euclidean distance defined (without any factor $1/2$) by

$$d_B^2(A_1, A_2) = ||A_1 - A_2||_{\mathrm{HS}}^2 = \mathrm{Tr}\left(A_1 A_1^{\dagger} + A_2 A_2^{\dagger} - A_1 A_2^{\dagger} - A_2 A_1^{\dagger}\right). \tag{9.28}$$

If A_1, A_2 lie on the unit sphere we have another natural distance, namely the geodesic distance d_A given by

$$\cos d_A = \frac{1}{2}\mathrm{Tr}\left(A_1 A_2^{\dagger} + A_2 A_1^{\dagger}\right). \tag{9.29}$$

Unlike the Euclidean distance, which measures the length of a straight chord, the second distance measures the length of a curve that projects down, in its entirety, to density matrices of unit trace. In accordance with the philosophy of Chapter 3, we define the distance between two density matrices ρ_1 and ρ_2 as the length of the shortest path, in the bundle, that connects the two fibres lying over these density matrices. Whether we choose to work with d_A or d_B, the technical task we face is to calculate the *root fidelity*[8]

$$\sqrt{F}(\rho_1, \rho_2) \equiv \frac{1}{2}\max \mathrm{Tr}\left(A_1 A_2^{\dagger} + A_2 A_1^{\dagger}\right) = \max|\mathrm{Tr}A_1 A_2^{\dagger}|. \tag{9.30}$$

The optimisation is with respect to all possible purifications of ρ_1 and ρ_2. Once we have done this, we can define the *Bures distance* D_B,

$$D_B^2(\rho_1, \rho_2) = \mathrm{Tr}\rho_1 + \mathrm{Tr}\rho_2 - 2\sqrt{F}(\rho_1, \rho_2), \tag{9.31}$$

and the *Bures angle* D_A,

$$\cos D_A(\rho_1, \rho_2) = \sqrt{F}(\rho_1, \rho_2). \tag{9.32}$$

The Bures angle is a measure of the length of a curve within $\mathcal{M}^{(N)}$, while the Bures distance measures the length of a curve within the positive cone. By construction, they are Riemannian distances – and indeed they are consistent with the same

---

[8] Its square F was called *fidelity* by Jozsa (1994) [509]. Later several authors, including Nielsen and Chuang [692], began to refer to our root fidelity as fidelity. We have chosen to stick with the original names, partly to avoid confusion, partly because experimentalists prefer a fidelity to be some kind of a probability – and fidelity is a kind of transition probability, as we will see.

Riemannian metric. Moreover they are both monotoneously decreasing functions of the root fidelity.[9]

Root fidelity is a useful concept in its own right, and will be discussed in some detail in Section 14.3. It is so useful that we state its evaluation as a theorem:

Uhlmann's fidelity theorem. *The root fidelity, defined as the maximum of $|\text{Tr}A_1A_2^\dagger|$ over all possible purifications of two density matrices ρ_1 and ρ_2, is*

$$\sqrt{F}(\rho_1, \rho_2) = \text{Tr}|\sqrt{\rho_2}\sqrt{\rho_1}| = \text{Tr}\sqrt{\sqrt{\rho_2}\,\rho_1\sqrt{\rho_2}}. \tag{9.33}$$

To prove this, we first use the polar decomposition to write

$$A_1 = \sqrt{\rho_1}\,U_1 \qquad \text{and} \qquad A_2 = \sqrt{\rho_2}\,U_2. \tag{9.34}$$

Here U_1 and U_2 are unitary operators that move us around the fibres. Then

$$\text{Tr}A_1A_2^\dagger = \text{Tr}(\sqrt{\rho_1}\,U_1U_2^\dagger\sqrt{\rho_2}) = \text{Tr}(\sqrt{\rho_2}\sqrt{\rho_1}\,U_1U_2^\dagger). \tag{9.35}$$

We perform yet another polar decomposition

$$\sqrt{\rho_2}\sqrt{\rho_1} = |\sqrt{\rho_2}\sqrt{\rho_1}|V, \qquad VV^\dagger = 1. \tag{9.36}$$

We define a new unitary operator $U \equiv VU_1U_2^\dagger$. The final task is to maximise

$$\text{Tr}(|\sqrt{\rho_2}\sqrt{\rho_1}|U) + \text{complex conjugate} \tag{9.37}$$

over all possible unitary operators U. In the eigenbasis of the positive operator $|\sqrt{\rho_2}\sqrt{\rho_1}|$ it is easy to see that the maximum occurs when $U = 1$. This proves the theorem; the definition of the Bures distance, and of the Bures angle, is thereby complete.

The catch is that root fidelity is difficult to compute. Because of the square roots, we must go through the labourious process of diagonalizing a matrix twice. Indeed, although our construction makes it obvious that $\sqrt{F}(\rho_1, \rho_2)$ is a symmetric function of ρ_1 and ρ_2, not even this property is obvious just by inspection of the formula – although in Section 14.3 we will give an elegant direct proof of this property. To come to grips with root fidelity, we work it out in two simple cases, beginning with the case when $\rho_1 = \text{diag}(p_1, p_2, \ldots, p_N)$ and $\rho_2 = \text{diag}(q_1, q_2, \ldots, q_N)$, that is when both matrices are diagonal. We also assume that they have trace one. This is an easy case: we get

$$\sqrt{F}(\rho_1, \rho_2) = \sum_{i=1}^{N} \sqrt{p_iq_i}. \tag{9.38}$$

---

[9] The Bures distance was introduced, in an infinite dimensional setting, by Bures (1969) [183], and then shown to be a Riemannian distance by Uhlmann [906]. Our Bures angle was called *Bures length* by Uhlmann [908], and *angle* by Nielsen and Chuang [692].

It follows that the Bures angle D_A equals the classical Bhattacharyya distance (2.56), while the Bures distance is given by

$$D_B^2(\rho_1, \rho_2) = 2 - 2 \sum_{i=1}^{N} \sqrt{p_i q_i} = \sum_{i=1}^{N} (\sqrt{p_i} - \sqrt{q_i})^2 = D_H^2(P, Q), \qquad (9.39)$$

where D_H is the Hellinger distance between two classical probability distributions. These distances are familiar from Section 2.5. Both of them are consistent with the Fisher–Rao metric on the space of classical probability distributions, so this is our first hint that what we are doing will have some statistical significance.

The second easy case is that of two pure states. The nice thing about a pure density matrix is that it squares to itself and therefore equals its own square root. For a pair of pure states a very short calculation shows that

$$\sqrt{F}\big(|\psi_1\rangle\langle\psi_1|, |\psi_2\rangle\langle\psi_2|\big) = |\langle\psi_1|\psi_2\rangle| = \sqrt{\kappa}, \qquad (9.40)$$

where κ is the projective cross ratio, also known as the transition probability. It is therefore customary to refer to *fidelity*, that is the square of root fidelity, also as the *Uhlmann transition probability*, regardless of whether the states are pure or not. Anyway, we can conclude that the Bures angle between two pure states is equal to their Fubini–Study distance.

With some confidence that we are studying an interesting definition, we turn to the Riemannian metric defined by the Bures distance. This admits a compact description that we will derive right away, although we will not use it until Section 15.1. It will be convenient to use an old fashioned notation for tangent vectors, so that dA is a tangent vector on the bundle, projecting to $d\rho$, which is a tangent vector on $\mathcal{M}^{(N)}$. The length squared of $d\rho$ is then defined by

$$ds^2 = \min\big[\mathrm{Tr}\, dA\, dA^\dagger\big], \qquad (9.41)$$

where the minimum is sought among all vectors dA that project to $d\rho$, and achieved if dA is a horizontal vector (orthogonal to the fibres). According to Eq. (9.26) this happens if and only if $dA = GA$, where G is a Hermitian matrix. As we know from Eq. (9.27), as long as ρ is strictly positive, G will be determined uniquely by

$$d\rho = G\rho + \rho G. \qquad (9.42)$$

Pulling the strings together, we find that

$$ds^2 = \mathrm{Tr}\, GAA^\dagger G = \mathrm{Tr} G\rho G = \frac{1}{2}\, \mathrm{Tr}\, Gd\rho. \qquad (9.43)$$

This is the *Bures metric*. Its definition is somewhat implicit. It is difficult to do better though: explicit expressions in terms of matrix elements tend to become so complicated that they seem useless – except when ρ and $d\rho$ commute, in which case $G = d\rho/(2\rho)$, and except for the special case $N = 2$ to which we now turn.[10] A head on attack on Eq. (9.43) will be made in Section 15.1.

9.5 Bures geometry for $N = 2$

It happens that for the qubit case, $N = 2$, we can get fully explicit results with elementary means. The reason is that every two by two matrix M obeys

$$M^2 - M\,\mathrm{Tr}M + \det M = 0. \tag{9.44}$$

Hence

$$(\mathrm{Tr}M)^2 = \mathrm{Tr}M^2 + 2\det M. \tag{9.45}$$

If we set

$$M \equiv \sqrt{\sqrt{\rho_1}\rho_2\sqrt{\rho_1}}, \tag{9.46}$$

we find – as a result of an elementary calculation – that

$$F = (\mathrm{Tr}M)^2 = \mathrm{Tr}\rho_1\rho_2 + 2\sqrt{\det \rho_1 \det \rho_2}, \tag{9.47}$$

where the fidelity F is used for the first time! The $N = 2$ Bures distance is now given by

$$D_{\mathrm{B}}^2(\rho_1, \rho_2) = \mathrm{Tr}\rho_1 + \mathrm{Tr}\rho_2 - 2\sqrt{\mathrm{Tr}\rho_1\rho_2 + 2\sqrt{\det \rho_1 \det \rho_2}}. \tag{9.48}$$

What is nice is that no square roots of operators appear in this expression.

It is now a question of straightforward calculation to obtain an explicit expression for the Riemannian metric on the positive cone, for $N = 2$. To do so, we set

$$\rho_1 = \frac{1}{2}\begin{bmatrix} t+z & x-iy \\ x+iy & t-z \end{bmatrix}, \qquad \rho_2 = \rho_1 + \frac{1}{2}\begin{bmatrix} dt+dz & dx-idy \\ dx+idy & dt-dz \end{bmatrix}. \tag{9.49}$$

It is elementary – although admittedly a little laborious – to insert this in Eq. (9.48) and expand to second order. The final result, for the Bures line element squared, is

$$ds^2 = \frac{1}{4t}\left(dx^2 + dy^2 + dz^2 + \frac{(xdx + ydy + zdz - tdt)^2}{t^2 - x^2 - y^2 - z^2}\right). \tag{9.50}$$

In the particular case that t is constant, so that we are dealing with matrices of constant trace, this is recognizable as the metric on the upper hemisphere of the

---

[10] In the $N = 2$ case we follow Hübner (1992) [477]. Actually Dittmann [263] has provided an expression valid for all N, which is explicit in the sense that it depends only on matrix invariants, and does not require diagonalisation of any matrix.

3-sphere, of radius $1/2\sqrt{t}$, in the orthographic coordinates introduced in Eq. (3.2). Indeed we can introduce the coordinates

$$X^0 = \sqrt{t^2 - x^2 - y^2 - z^2}, \quad X^1 = x, \quad X^2 = y, \quad X^3 = z. \tag{9.51}$$

Then the Bures metric on the positive cone is

$$ds^2 = \frac{1}{4t} \left(dX^0 dX^0 + dX^1 dX^1 + dX^2 dX^2 + dX^3 dX^3 \right), \tag{9.52}$$

where

$$\text{Tr}\rho = t = \sqrt{(X^0)^2 + (X^1)^2 + (X^2)^2 + (X^3)^2}. \tag{9.53}$$

Only the region for which $X^0 \geq 0$ is relevant.

Let us set $t = 1$ for the remainder of this section, so that we deal with matrices of unit trace. We see that, according to the Bures metric, they form a hemisphere of a 3-sphere of radius $1/2$; the pure states sit at its equator, which is a 2-sphere isometric with $\mathbb{CP}^1$. Unlike a 2-sphere in Euclidean space, the equator of the 3-sphere is a totally geodesic surface – by definition, a surface such that a geodesic within the surface itself is also a geodesic in the embedding space. We can draw a picture (Figure 9.2) that summarises the Bures geometry of the qubit. Note that the set of diagonal density matrices appears as a semi-circle in this picture, not as the quarter circle that we had in Figure 2.13. Actually, because this set is

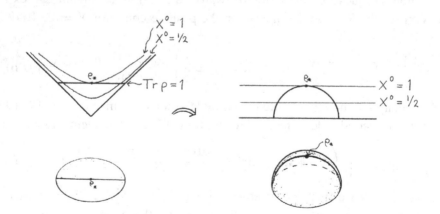

Figure 9.2 Left: a faithful illustration of the Hilbert–Schmidt geometry of a rebit (a flat disc, compare Figure 8.2). Right: the same for its Bures geometry (a round hemisphere). Above the rebit we show exactly how it sits in the positive cone. On the right the latter appears very distorted, because we have adapted its coordinates to the Bures geometry.

one-dimensional, the intrinsic geometries on the two circle segments are the same, the length is $\pi/2$ in both cases, and there is no contradiction.

Finally, the qubit case is instructive, but it is also quite misleading in some respects – in particular the case $N = 2$ is especially simple to deal with.

9.6 Further properties of the Bures metric

When $N > 2$ it does not really pay to proceed as directly as we did for the qubit, but the fibre bundle origins of the Bures metric mean that much can be learned about it with indirect means. First, what is a geodesic with respect to the Bures metric? The answer is that it is a projection of a geodesic in the unit sphere embedded in the bundle space $\mathcal{HS}$, with the added condition that the latter geodesic must be chosen to be orthogonal to the fibres of the bundle. We know what a geodesic on a sphere looks like, namely (Section 3.1)

$$A(s) = A(0) \cos s + \dot{A}(0) \sin s, \tag{9.54}$$

where

$$\mathrm{Tr} A(0) A^\dagger(0) = \mathrm{Tr} \dot{A}(0) \dot{A}^\dagger(0) = 1, \quad \mathrm{Tr}\left(A(0)\dot{A}^\dagger(0) + \dot{A}(0)A^\dagger(0)\right) = 0. \tag{9.55}$$

The second equation just says that the tangent vector of the curve is orthogonal to the vector defining the starting point on the sphere. In addition the tangent vector must be horizontal; according to Eq. (9.24) this means that we must have

$$\dot{A}^\dagger(0)A(0) = A^\dagger(0)\dot{A}(0). \tag{9.56}$$

That is all. An interesting observation – we will see why in a moment – is that if we start the geodesic at a point where A, and hence $\rho = AA^\dagger$, is block diagonal, and if the tangent vector $\dot{A}$ at that point is block diagonal too, then the entire geodesic will consist of block diagonal matrices. The conclusion is that block diagonal density matrices form totally geodesic submanifolds in the space of density matrices.

Now let us consider a geodesic that joins the density matrices ρ_1 and ρ_2, and let them be projections of A_1 and A_2, respectively. The horizontality condition says that $A_1^\dagger A_2$ is a Hermitian operator, and in fact a positive operator if the geodesic does not hit the boundary in-between. From this one may deduce that

$$A_2 = \frac{1}{\sqrt{\rho_1}} \sqrt{\sqrt{\rho_1} \rho_2 \sqrt{\rho_1}} \frac{1}{\sqrt{\rho_1}} A_1. \tag{9.57}$$

The operator in front of A_1 is known as the *geometric mean* of ρ_1^{-1} and ρ_2; see Section 13.1. It can also be proved that the geodesic will bounce N times from the boundary of $\mathcal{M}^{(N)}$, before closing on itself [908]. The overall conclusion is that we do have control over geodesics and geodesic distances with respect to the Bures metric.

Concerning symmetries, it is known that any bijective transformation of the set of density matrices into itself which conserves the Bures distance (or angle) is implemented by a unitary or an anti-unitary operation [670]. This result is a generalisation of Wigner's theorem concerning the transformations of pure states that preserve the transition probabilities (see Section 4.5).

For further insight we turn to a cone of density matrices in $\mathcal{M}^{(3)}$, having a pure state for its apex and a Bloch ball of density matrices with orthogonal support for its base. This can be coordinatised as

$$
\rho = \begin{bmatrix} t\rho^{(2)} & 0 \\ 0 & 1-t \end{bmatrix} = \begin{bmatrix} t(1+z)/2 & t(x-iy)/2 & 0 \\ t(x+iy)/2 & t(1-z)/2 & 0 \\ 0 & 0 & 1-t \end{bmatrix}. \tag{9.58}
$$

This is a submanifold of block diagonal matrices. It is also simple enough so that we can proceed directly, as in Section 9.5. Doing so, we find that the metric is

$$
ds^2 = \frac{dt^2}{4t(1-t)} + \frac{t}{4}d^2\Omega, \tag{9.59}
$$

where $d^2\Omega$ is the metric on the unit 3-sphere (in orthographic coordinates, and only one-half of the 3-sphere is relevant). As $t \to 0$, that is as we approach the tip of our cone, the radii of the 3-spheres shrink and their intrinsic curvature diverges. This does not sound very dramatic, but in fact it is, because by our previous argument about block diagonal matrices these 3-hemispheres are totally geodesic submanifolds of the space of density matrices. Now it is a fact from differential geometry that if the intrinsic curvature of a totally geodesic submanifold diverges, then the curvature of the entire manifold also diverges. (More precisely, the sectional curvatures, evaluated for 2-planes that are tangent to the totally geodesic submanifold, will agree. We hope that this statement sounds plausible, even though we will not explain it further.) The conclusion is that $\mathcal{M}^{(3)}$, equipped with the Bures metric, has conical curvature singularities at the pure states. The general picture is as follows [262]:

Dittmann's theorem. *For $N \geq 2$, the Bures metric is everywhere well defined on submanifolds of density matrices with constant rank. However, the sectional curvature of the entire space diverges in the neighbourhood of any submanifold of rank less than $N - 1$.*

For $N > 2$, this means that it is impossible to embed $\mathcal{M}^{(N)}$ into a Riemannian manifold of the same dimension, such that the restriction of the embedding to submanifolds of density matrices of constant rank is isometric. The problem does not arise for the special case $N = 2$, and indeed we have seen that $\mathcal{M}^{(2)}$ can be embedded into the 3-sphere.

Some further facts are known. Thus, the curvature scalar R assumes its global minimum at the maximally mixed state $\rho_* = \mathbb{1}/N$. It is then natural to conjecture that the scalar curvature is monotone, in the sense that if $\rho_1 \prec \rho_2$, that is if ρ_1 is majorised by ρ_2, then $R(\rho_1) \leq R(\rho_2)$. But this is false.[11]

This is perhaps a little disappointing. To recover our spirits, let us look at the Bures distance in two cases where it is very easy to compute. The Bures distance to the maximally mixed state is

$$D_{\mathrm{B}}^2(\rho, \rho_*) = 1 + \mathrm{Tr}\rho - \frac{2}{\sqrt{N}} \mathrm{Tr} \sqrt{\rho}. \tag{9.60}$$

To compute this it is enough to diagonalise ρ. The distance from an arbitrary density matrix ρ to a pure state is even easier to compute, and is given by a single matrix element of ρ. Figure 9.3 shows where density matrices equidistant to a pure state lie on the probability simplex, for some of the metrics that we have considered. In particular, the distance between a face and its complementary, opposite face – that

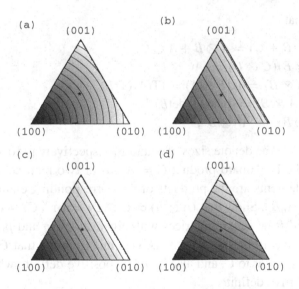

Figure 9.3 The eigenvalue simplex for $N = 3$. The curves consist of points equidistant from the pure state $(1, 0, 0)$ with respect to (a) Hilbert–Schmidt distance, (b) Bures distance, (c) trace distance (Section 14.2) and (d) Monge distance (Section 7.7).

---

[11] The result here is due to Dittmann [264], who also found a counter-example to the conjecture (but did not publish it, as far as we know).

is, between density matrices of orthogonal support – is constant and maximal, when the Bures metric is used.

We are not done with fidelity and the Bures metric. We will come back to these things in Section 14.3, and place them in a wider context in Section 15.1. This context is – as we have hinted already – that of statistical distinguishability and monotonicity under appropriate stochastic maps. Precisely what is appropriate here will be made clear in the next two chapters.

Problems

9.1   Consider a matrix ρ of size 4 written in the standard lexicographical basis in terms of four 2×2 blocks, and then calculate its partial traces. That is, show that

$$\rho = \begin{bmatrix} A & B \\ C & D \end{bmatrix} \quad \Rightarrow \quad \mathrm{Tr}_2\rho = \begin{bmatrix} \mathrm{Tr}A & \mathrm{Tr}B \\ \mathrm{Tr}C & \mathrm{Tr}D \end{bmatrix}, \quad \mathrm{Tr}_1\rho = A + D.$$

(9.61)

Also consider two local unitary operations, $V_1 = U \otimes \mathbb{1}$ and $V_2 = \mathbb{1} \otimes U$, where U is an arbitrary unitary matrix of size 2. Calculate $V_1\rho V_1^\dagger$ and $V_2\rho V_2^\dagger$.

9.2   Check that

(a) $A \otimes (B + C) = A \otimes B + A \otimes C$,
(b) $(A \otimes B)(C \otimes D) = (AC \otimes BD)$,
(c) $\mathrm{Tr}(A \otimes B) = \mathrm{Tr}(B \otimes A) = (\mathrm{Tr}A)(\mathrm{Tr}B)$
(d) $\det(A \otimes B) = (\det A)^N (\det B)^M$
(e) $(A \otimes B)^T = A^T \otimes B^T$,

where N and M denote sizes of A and B, respectively [450, 451].

9.3   Define the Hadamard product $C = A \circ B$ of two matrices as the matrix whose elements are the products of the corresponding elements of A and B, $C_{ij} = A_{ij}B_{ij}$. Show that $(A \otimes B) \circ (C \otimes D) = (A \circ C) \otimes (B \circ D)$.

9.4   Let A and B be square matrices with eigenvalues α_i and β_i, respectively. Find the spectrum of $C = A \otimes B$. Use this to prove that $C' = B \otimes A$ is unitarily similar to C, and also that C is positive definite whenever A and B are positive definite.

9.5   Show that the singular values of a tensor product satisfy the relation $\mathrm{sv}(A \otimes B) = \{\mathrm{sv}(A)\} \times \{\mathrm{sv}(B)\}$.

9.6   Let ρ be a density matrix and A and B denote any matrices of the same size. Show that $|\mathrm{Tr}(\rho AB)|^2 \leq \mathrm{Tr}(\rho AA^\dagger) \times \mathrm{Tr}(\rho BB^\dagger)$.

10

Quantum operations

There is no measurement problem. Bohr cleared that up.

—Stig Stenholm[1]

So far we have described the space of quantum states. Now we will allow some action in this space: we shall be concerned with quantum dynamics. At first sight this seems to be two entirely different issues – it is one thing to describe a given space, and another to characterise the way you can travel in it – but we will gradually reveal an intricate link between them.

In this chapter we draw on results from the research area known as *open quantum systems*. Our aim is to understand the quantum analogues of the classical stochastic maps, because with their help we reach a better understanding of the structure of the space of states. Stochastic maps can also be used to provide a kind of stroboscopic time evolution; much of the research on open quantum systems is devoted to understanding how continuous time evolution takes place, but for this we have to refer to the literature.[2]

10.1 Measurements and POVMs

Throughout, the system of interest is described by a Hilbert space $\mathcal{H}_N$ of dimension N. All quantum operations can be constructed by composing four kinds of transformations.

The dynamics of an isolated quantum system is given by **(i) unitary transformations**. But quantum theory for open systems admits non-unitary processes as

[1] Stig Stenholm. Remark made in a discussion with Anton Zeilinger. Reproduced with kind permission of Stig Stenholm.
[2] Pioneering results in this direction were obtained by Gorini, Kossakowski and Sudarshan (1976) [351] and by Lindblad (1976) [611]. Good books on the subject include Alicki and Lendi [26], Streater [863], Ingarden, Kossakowski and Ohya [487], Breuer and Petruccione [170], and Alicki and Fannes [24].

well. We can **(ii) extend the system** and define a new state in an extended Hilbert space $\mathcal{H} = \mathcal{H}_N \otimes \mathcal{H}_K$,

$$\rho \to \rho' = \rho \otimes \sigma. \tag{10.1}$$

The auxiliary system is described by a Hilbert space $\mathcal{H}_K$ of dimension K (as yet unrelated to N). This represents an environment, and is often referred to as the *ancilla*.[3] The reverse of this operation is given by the **(iii) partial trace** and leads to a reduction of the size of the Hilbert space,

$$\rho \to \rho' = \text{Tr}_K \rho \quad \text{so that} \quad \text{Tr}_K(\rho \otimes \sigma) = \rho. \tag{10.2}$$

This corresponds to discarding the redundant information concerning the fate of the ancilla. Transformations that can be achieved by a combination of these three kinds of transformation are known as *deterministic* or *proper quantum operations*.

Finally we have the **(iv) selective measurement**, in which a concrete result of a measurement is specified. This is called a *probabilistic quantum operation*.

Let us see where we get using tranformations of the first three kinds. Let us assume that the ancilla starts out in a pure state $|v\rangle$, while the system we are analysing starts out in the state ρ. The entire system including the ancilla remains isolated and evolves in a unitary fashion. Adding the ancilla to the system (ii), evolving the combined system unitarily (i), and tracing out the ancilla at the end (iii), we find that the state ρ is changed to

$$\rho' = \text{Tr}_K \left[U \left(\rho \otimes |v\rangle\langle v| \right) U^\dagger \right] = \sum_{\mu=1}^{K} \langle \mu | U | v \rangle \rho \langle v | U^\dagger | \mu \rangle, \tag{10.3}$$

where $\{|\mu\rangle\}_{\mu=1}^{K}$ is a basis in the Hilbert space of the ancilla – and we use Greek letters to denote its states. We can then define a set of operators in the Hilbert space of the original system through

$$A_\mu \equiv \langle \mu | U | v \rangle. \tag{10.4}$$

We observe that

$$\sum_{\mu=1}^{K} A_\mu^\dagger A_\mu = \sum_\mu \langle v | U^\dagger | \mu \rangle \langle \mu | U | v \rangle = \langle v | U^\dagger U | v \rangle = \mathbb{1}_N, \tag{10.5}$$

where $\mathbb{1}_N$ denotes the unit operator in the Hilbert space of the system of interest.

In conclusion: first we assumed that an isolated quantum system evolves through a unitary transformation,

$$\rho \to \rho' = U \rho U^\dagger, \quad U^\dagger U = \mathbb{1}. \tag{10.6}$$

---

[3] In Latin an *ancilla* is a maidservant. This not 100 per cent politically correct expression was imported to quantum mechanics by Helstrom [422], and has become widely accepted.

By allowing ourselves to add an ancilla, later removed by a partial trace, we were led to admit operations of the form

$$\rho \to \rho' = \sum_{i=1}^{K} A_i \rho A_i^\dagger, \qquad \sum_{i=1}^{K} A_i^\dagger A_i = \mathbb{1}, \tag{10.7}$$

where we dropped the subscript on the unit operator and switched to Latin indices, since we are not interested in the environment per se. Formally, this is the *operator sum representation* of a *completely positive map*. Although a rather special assumption was slipped in – a kind of *Stoßzahlansatz* whereby the combined system started out in a product state – we will eventually adopt this expression as the most general quantum operation that we are willing to consider.

The process of *quantum measurement* remains somewhat enigmatic. Here we simply accept without proof a postulate concerning the *collapse of the wave function*. It has the virtue of generality, not of preciseness:

Measurement postulate. Let the space of possible measurement outcomes consist of k elements, related to k *measurement operators* A_i, which satisfy the *completeness relation*

$$\sum_{i=1}^{k} A_i^\dagger A_i = \mathbb{1}. \tag{10.8}$$

The quantum measurement performed on the initial state ρ produces the ith outcome with probability p_i and transforms ρ into ρ_i according to

$$\rho \to \rho_i = \frac{A_i \rho A_i^\dagger}{\mathrm{Tr}(A_i \rho A_i^\dagger)} \quad \text{with} \quad p_i = \mathrm{Tr}\left(A_i \rho A_i^\dagger\right). \tag{10.9}$$

The probabilities are positive and sum to unity due to the completeness relation. Such measurements, called *selective* since concrete results labelled by i are recorded, cannot be obtained by the transformations (i)–(iii) and form another class of transformations (iv) on their own. If no selection is made based on the outcome of the measurement, the initial state is transformed into a convex combination of all possible outcomes – namely that given by Eq. (10.7).

Note that the 'collapse' happens in the statistical description that we are using. Similar 'collapses' occur also in classical probability theory. Suppose that we know that either Alice or Bob is in jail, but not both. Let the probability that Bob is in jail be p. If this statement is accepted as meaningful, we find that there is a collapse of the probability distribution associated with Bob as soon as we run into Alice in the cafeteria – even though nothing happened to Bob. This is not a real philosophical difficulty, but the quantum case is subtler. Classically the pure states are safe from collapse, but in quantum mechanics there are no safe havens. Also, a classical probability distribution $P(X)$ can collapse to a conditional probability distribution $P(X|Y_i)$, but if no selection according to the outcomes Y_i is made classical probability theory informs us that

$$\sum_i P(X|Y_i)\,P(Y_i) \;=\; P(X). \tag{10.10}$$

Thus nothing happens to the probability distribution in a *non-selective* measurement, while the quantum state is severely affected also in this case. A non-selective quantum measurement is described by Eq. (10.7), and this is a mixed state even if the initial state ρ is pure. In general one cannot receive any information about the fate of a quantum system without performing a measurement that perturbs its unitary time evolution.

In a *projective measurement* the measurement operators are orthogonal projectors, so $A_i \;=\; P_i \;=\; A_i^\dagger$, and $P_i P_j \;=\; \delta_{ij} P_i$ for $i,j \;=\; 1,\dots,N$. A projective measurement is described by an observable – a Hermitian operator O. Possible outcomes of the measurement are labelled by the eigenvalues of O, which for now we assume to be non-degenerate. Using the spectral decomposition $O \;=\; \sum_i \lambda_i P_i$ we obtain a set of orthogonal measurement operators $P_i \;=\; |e_i\rangle\langle e_i|$, satisfying the completeness relation (10.8). In a non-selective projective measurement, the initial state is transformed into the mixture

$$\rho \;\to\; \rho' \;=\; \sum_{i=1}^{N} P_i \rho P_i. \tag{10.11}$$

The state has been forced to commute with O.

In a selective projective measurement the outcome labelled by λ_i occurs with probability p_i; the initial state is transformed as

$$\rho \;\to\; \rho_i \;=\; \frac{P_i \rho P_i}{\mathrm{Tr}(P_i \rho P_i)}, \quad \text{where} \quad p_i = \mathrm{Tr}\,(P_i \rho P_i) = \mathrm{Tr}\,(P_i \rho). \tag{10.12}$$

The expectation value of the observable reads

$$\langle O \rangle \;=\; \sum_{i=1}^{N} p_i \lambda_i \;=\; \sum_{i=1}^{N} \lambda_i \mathrm{Tr} P_i \rho \;=\; \mathrm{Tr}(O\rho). \tag{10.13}$$

A key feature of projective measurements is that they are repeatable, in the sense that the state in (10.12) remains the same – and gives the same outcome – if the measurement is repeated.[4]

----

[4] Projective measurements are also called *Lüders–von Neumann* measurements (of the first kind), because of the contributions by von Neumann (1932) [941] and Lüders (1951) [626].

Most measurements are not repeatable. The formalism deals with this by relaxing the orthogonality constraint on the measurement operators. This leads to *Positive Operator Valued Measures* (POVM), which are defined by any partition of the identity operator into a set of k positive operators E_i acting on an N dimensional Hilbert space $\mathcal{H}_N$. These operators, called *effects*, satisfy

$$\sum_{i=1}^{k} E_i = \mathbb{1} \quad \text{and} \quad E_i = E_i^{\dagger}, \quad E_i \geq 0, \quad i = 1, \ldots .k. \tag{10.14}$$

A POVM measurement applied to the state ρ produces the i-th outcome with probability $p_i = \text{Tr} E_i \rho$. Note that the elements of the POVM – the operators E_i – need not commute. The name POVM refers to any set of operators satisfying (10.14), and suggests correctly that the discrete sum may be replaced by an integral over a continuous index set, thus defining a measure in the space of positive operators. Indeed the coherent state resolution of unity (6.7) is the paradigmatic example, yielding the Husimi Q-function as the resulting probability distribution. POVMs fit into the general framework of the measurement postulate, since one may choose $E_i = A_i^{\dagger} A_i$. Note however that the POVM does not determine the measurement operators A_i uniquely (except in the special case of a projective measurement). Exactly what happens to the state when a measurement is made depends on how the POVM is implemented in the laboratory.[5]

The definition of the POVM ensures that the probabilities $p_i = \text{Tr} E_i \rho$ sum to unity, but the probability distribution that one obtains is a constrained one. (We came across this phenomenon in Section 6.2, when we observed that the Q-function is a very special probability distribution.) This is so because the POVM defines an affine map from the set of density matrices $\mathcal{M}^{(N)}$ to the probability simplex Δ_{k-1}. To see this, use the Bloch vector parametrization

$$\rho = \frac{1}{N}\mathbb{1} + \sum_a \tau_a \sigma_a \quad \text{and} \quad E_i = e_{i0}\mathbb{1} + \sum_a e_{ia}\sigma_a. \tag{10.15}$$

Then an easy calculation yields

$$p_i = \text{Tr } E_i \rho = 2\sum_a e_{ia}\tau_a + e_{i0}. \tag{10.16}$$

This is an affine map. Conversely, any affine map from $\mathcal{M}^{(N)}$ to Δ_{k-1} defines a POVM. We know from Section 1.1 that an affine map preserves convexity. Therefore the resulting probability vector $\vec{p}$ must belong to a convex subset of the

---

[5] POVMs were introduced by Jauch and Piron (1967) [498] and they were explored in depth in the books by Davies (1976) [248] and Holevo (1982) [445]. Holevo's book is the best source of knowledge that one can imagine. For a more recent discussion, see Peres and Terno [727].

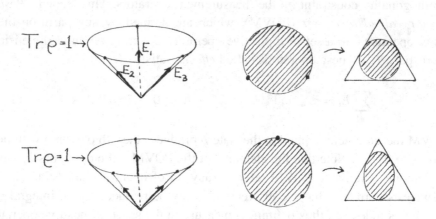

Figure 10.1 Two informationally complete POVMs for a rebit; we show first a realistic picture in the space of real Hermitian matrices, and then the affine map from the rebit to the probability simplex.

probability simplex. For qubits $\mathcal{M}^{(2)}$ is a ball. Therefore its image is an ellipsoid, degenerating to a line segment if the measurement is projective. Figure 10.1 illustrates the case of real density matrices, for which we can draw the positive cone. For $N > 2$ illustration is no longer an easy matter.

A POVM is called *informationally complete* if the statistics of the POVM uniquely determines the density matrix. This requires that the POVM has N^2 elements – a projective measurement will not do. A POVM is called *pure* if each operator E_i is of rank one, so there exists a pure state $|\phi_i\rangle$ such that E_i is proportional to $|\phi_i\rangle\langle\phi_i|$. An impure POVM can always be turned into a pure POVM by replacing each operator E_i by its spectral decomposition. Observe that a set of k pure states $|\phi_i\rangle$ defines a pure POVM if and only if the maximally mixed state $\rho_* = \mathbb{1}/N$ can be decomposed as $\rho_* = \sum_{i=1}^{k} p_i |\phi_i\rangle\langle\phi_i|$, where $\{p_i\}$ form a suitable set of positive coefficients. Indeed any ensemble of pure or mixed states representing ρ_* defines a POVM [481]. For any set of operators E_i defining a POVM we take quantum states $\rho_i = E_i/\mathrm{Tr}E_i$ and mix them with probabilities $p_i = \mathrm{Tr}E_i/N$ to obtain the maximally mixed state:

$$\sum_{i=1}^{k} p_i \rho_i = \sum_{i=1}^{k} \frac{1}{N} E_i = \frac{1}{N}\mathbb{1} = \rho_*. \tag{10.17}$$

Conversely, any such ensemble of density matrices defines a POVM (see Figure 10.2).

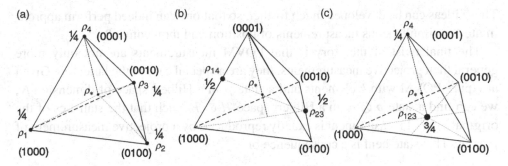

Figure 10.2 Any POVM is equivalent to an ensemble representing the maximally mixed state. For $N = 4$ ρ_* is situated at the centre of the tetrahedron of diagonal density matrices; (a) a pure POVM – in the picture a combination of four projectors with weights $1/4$, (b) and (c) unpure POVMs.

Arguably the most famous of all POVMs is the one based on coherent states. Assume that a classical phase space Ω has been used to construct a family of coherent states, $x \in \Omega \rightarrow |x\rangle \in \mathcal{H}$. The POVM is given by the resolution of the identity

$$\int_\Omega |x\rangle \langle x| \, dx = \mathbb{1}, \tag{10.18}$$

where dx is a natural measure on Ω. Examples of this construction were given in Chapter 6, and include the 'canonical' phase space where $x = (q, p)$. Any POVM can be regarded as an affine map from the set of quantum states to a set of classical probability distributions; in this case the resulting probability distributions are precisely those given by the Q-function. A discrete POVM can be obtained by introducing a partition of phase space into cells,

$$\Omega = \Omega_1 \cup \ldots \cup \Omega_k. \tag{10.19}$$

This partition splits the integral into k terms and defines k positive operators E_i that sum to unity. They are not projection operators since the coherent states overlap, and thus

$$E_i \equiv \int_{\Omega_i} dx \, |x\rangle \langle x| \neq \int_{\Omega_i} dx \int_{\Omega_i} dy \, |x\rangle \langle x|y\rangle \langle y| = E_i^2. \tag{10.20}$$

Nevertheless they do provide a notion of localisation in phase space; if the state is ρ the particle will be registered in cell Ω_i with probability

$$p_i = \text{Tr}(E_i \rho) = \int_{\Omega_i} \langle x|\rho|x\rangle \, dx = \int_{\Omega_i} Q_\rho(x) \, dx. \tag{10.21}$$

These ideas can be developed much further, so that one can indeed perform approximate but simultaneous measurements of position and momentum.[6]

The final twist of the story is that POVM measurements are not only more general than projective measurements, they are a special case of the latter too. Given any pure POVM with k elements and a state ρ in a Hilbert space of dimension N, we can find a state $\rho \otimes \rho_0$ in a Hilbert space $\mathcal{H} \otimes \mathcal{H}'$ such that the statistics of the original POVM measurement is exactly reproduced by a projective measurement of $\rho \otimes \rho_0$. This statement is a consequence of:

Naimark's theorem. *Any POVM $\{E_i\}$ in the Hilbert space $\mathcal{H}$ can be dilated to an orthogonal resolution of identity $\{P_i\}$ in a larger Hilbert space in such a way that $E_i = \Pi P_i \Pi$, where Π projects down to $\mathcal{H}$.*

For a proof see Problem 10.1. The next idea is to choose a pure state ρ_0 such that, in a basis in which Π is diagonal,

$$\rho \otimes \rho_0 = \left[\begin{array}{c|c} \rho & 0 \\ \hline 0 & 0 \end{array} \right] = \Pi \rho \otimes \rho_0 \Pi. \tag{10.22}$$

It follows that $\mathrm{Tr} P_i \rho \otimes \rho_0 = \mathrm{Tr} P_i \Pi \rho \otimes \rho_0 \Pi = \mathrm{Tr} E_i \rho$.

We are left somewhat at a loss to say which notion of measurement is the fundamental one. Let us just observe that classical statistics contains the notion of *randomised experiments*: equip the experimenter with a random number generator and surround his or her laboratory with a black box. He or she has a choice between different experiments and will perform them with different probabilites p_i. It may not sound like a useful notion, but it is. We can view a POVM measurement as a randomised experiment in which the source of randomness is a quantum mechanical ancilla. Again the quantum case is more subtle than its classical counterpart; the set of all possible POVMs forms a convex set whose extreme points include the projective measurements, but there are other extreme points as well. The POVMs shown in Figure 10.1, perhaps reinterpreted as POVMs for qubits, may serve as examples.

10.2 Algebraic detour: matrix reshaping and reshuffling

Before proceeding with our analysis of quantum operations, we will discuss some simple algebraic transformations that one can perform on matrices. We also introduce a notation that we sometimes find convenient for work in the composite Hilbert space $\mathcal{H}_N \otimes \mathcal{H}_M$, or in the Hilbert–Schmidt (HS) space of linear operators $\mathcal{H}_{\mathrm{HS}}$.

---

[6] A pioneering result is due to Arthurs and Kelly (1965) [57]; for more, see the books by Holevo [445], Busch, Lahti and Mittelstaedt [184], and Leonhardt [587].

Consider a rectangular matrix A_{ij}, $i = 1, \ldots, M$ and $j = 1, \ldots, N$. The matrix may be *reshaped* by putting its elements in lexicographical order (row after row)[7] into a vector $\vec{a}_k$ of size MN,

$$\vec{a}_k = A_{ij} \quad \text{where} \quad k = (i-1)N + j, \quad i = 1, \ldots, M, \quad j = 1, \ldots N. \quad (10.23)$$

Conversely, any vector of length MN may be reshaped into a rectangular matrix. The simplest example of such a vectorial notation for matrices reads

$$A = \begin{bmatrix} A_{11} & A_{12} \\ A_{21} & A_{22} \end{bmatrix} \quad \longleftrightarrow \quad \vec{a} = (A_{11}, A_{12}, A_{21}, A_{22}). \quad (10.24)$$

The scalar product in HS space (matrices of size N) now looks like an ordinary scalar product between two vectors of size N^2,

$$\langle A|B \rangle \equiv \text{Tr}\, A^\dagger B = \vec{a}^* \cdot \vec{b} = \langle a|b \rangle. \quad (10.25)$$

Thus the HS norm of a matrix is equal to the norm of the associated vector, $||A||_{\text{HS}}^2 = \text{Tr}\, A^\dagger A = |\vec{a}|^2$.

Sometimes we will label a component of $\vec{a}$ by a_{ij}. This vector of length MN may be linearly transformed into $a' = Ca$ by a matrix C of size $MN \times MN$. Its elements may be denoted by $C_{kk'}$ with $k, k' = 1, \ldots, MN$, but it is also convenient to use a four-index notation, $C_{m\mu \atop n\nu}$ where $m, n = 1, \ldots, N$ while $\mu, \nu = 1, \ldots, M$. In this notation the elements of the transposed matrix are $C_{m\mu \atop n\nu}^T = C_{n\nu \atop m\mu}$, since the upper pair of indices determines the row of the matrix, while the lower pair determines its column. The matrix C may represent an operator acting in a composite space $\mathcal{H} = \mathcal{H}_N \otimes \mathcal{H}_M$. The tensor product of any two bases in both factors provides a basis in $\mathcal{H}$, so that

$$C_{m\mu \atop n\nu} = \langle e_m \otimes f_\mu | C | e_n \otimes f_\nu \rangle, \quad (10.26)$$

where Latin indices refer to the first subsystem, $\mathcal{H}_N$, and Greek indices to the second, $\mathcal{H}_M$. For instance the elements of the identity operator $\mathbb{1}_{NM} \equiv \mathbb{1}_N \otimes \mathbb{1}_M$ are $\mathbb{1}_{m\mu \atop n\nu} = \delta_{mn}\delta_{\mu\nu}$. The trace of a matrix reads $\text{Tr}C = C_{m\mu \atop m\mu}$, where summation over repeating indices is understood. The operation of partial trace over the second subsystem produces the matrix $C^A \equiv \text{Tr}_B C$ of size N, while tracing over the first subsystem leads to an $M \times M$ matrix $C^B \equiv \text{Tr}_A C$,

$$C_{mn}^A = C_{m\mu \atop n\mu}, \quad \text{and} \quad C_{\mu\nu}^B = C_{m\mu \atop m\nu}. \quad (10.27)$$

---

[7] Some programs like MATLAB offer a built-in matrix command *reshape*, which performs such a task. Storing matrix elements column after column leads to the anti-lexicographical order.

If $C = A \otimes B$ then $C_{\substack{m\mu \\ n\nu}} = A_{mn}B_{\mu\nu}$. This form should not be confused with a product of two matrices $C = AB$, the elements of which are given by a double sum over repeating indices, $C_{\substack{m\mu \\ n\nu}} = A_{\substack{m\mu \\ l\lambda}}B_{\substack{l\lambda \\ n\nu}}$. Observe that the standard product of three matrices may be rewritten by means of an object Φ,

$$ABC = \Phi B \qquad \text{where} \qquad \Phi = A \otimes C^T. \tag{10.28}$$

This is a telegraphic notation; since Φ is a linear map acting on B we might write $\Phi(B)$ on the right-hand side, and the left-hand side could be written as $A \cdot B \cdot C$ to emphasise that matrix multiplication is being used there. Equation (10.28) is a concise way of saying all this. It is unambiguous once we know the nature of the objects that appear in it.

Consider a unitary matrix $\tilde{U}$ of size N^2. Its N^2 columns reshaped into square matrices $\tilde{U}_k$ of size N form an orthogonal basis in $\mathcal{H}_{HS}$. Using the HS norm we normalise them according to $A_k = \tilde{U}_k/(\mathrm{Tr}\tilde{U}_k\tilde{U}_k^\dagger)^{1/2} = \tilde{U}_k/\|\tilde{U}_k\|_{HS}$, and obtain an orthonormal basis $\langle A_k|A_j\rangle \equiv \mathrm{Tr}A_k^\dagger A_j = \delta_{kj}$. Alternatively, in a double index notation with $k = (m-1)N + \mu$ and $j = (n-1)N + \nu$ this orthogonality relation reads $\langle A^{m\mu}|A^{n\nu}\rangle = \delta_{mn}\delta_{\mu\nu}$. Note that in general the matrices A_k are not unitary.

Let X denote an arbitrary matrix of size N^2. This may be represented as a double (quadruple) sum,

$$|X\rangle = \sum_{k=1}^{N^2}\sum_{j=1}^{N^2} C_{kj}|A_k\rangle \otimes |A_j\rangle = C_{\substack{m\mu \\ n\nu}}|A^{m\mu}\rangle \otimes |A^{n\nu}\rangle, \tag{10.29}$$

where $C_{kj} = \mathrm{Tr}((A_k \otimes A_j)^\dagger X)$. The matrix X may be considered as a vector in the composite HS space $\mathcal{H}_{HS} \otimes \mathcal{H}_{HS}$, so applying its Schmidt decomposition (9.8) we arrive at

$$|X\rangle = \sum_{k=1}^{N^2} \sqrt{\lambda_k}\, |A_k'\rangle \otimes |A_k''\rangle, \tag{10.30}$$

where $\sqrt{\lambda_k}$ are the singular values of C, that is the square roots of the non-negative eigenvalues of $CC^\dagger$. The sum of their squares is determined by the norm of the operator, $\sum_{k=1}^{N^2} \lambda_k = \mathrm{Tr}(XX^\dagger) = \|X\|_{HS}^2$.

Since the Schmidt coefficients do not depend on the initial basis, let us choose the basis in $\mathcal{H}_{HS}$ obtained from the identity matrix, $U = \mathbb{1}$ of size N^2, by reshaping its columns. Then each of the N^2 basis matrices of size N consists of only one non-zero element which equals unity, $A_k = A^{m\mu} = |m\rangle\langle\mu|$, where $k = N(m-1) + \mu$. Their tensor products form an orthonormal basis in $\mathcal{H}_{HS} \otimes \mathcal{H}_{HS}$ and allow us to represent an arbitrary matrix X in the form (10.29). In this case the matrix of the coefficients C has a particularly simple form, $C_{\substack{m\mu \\ n\nu}} = \mathrm{Tr}[(A^{m\mu} \otimes A^{n\nu})X] = X_{\substack{mn \\ \mu\nu}}$.

This particular reordering of a matrix deserves a name, so we shall write $X^R \equiv C(X)$ and call it *reshuffling*.[8] Using this notion our findings may be summarised in the following lemma:

Operator Schmidt decomposition. *The Schmidt coefficients of an operator X acting on a bipartite Hilbert space are equal to the squared singular values of the reshuffled matrix, X^R.*

More precisely, the Schmidt decomposition (10.30) of an operator X of size MN may be supplemented by a set of three equations

$$\begin{cases} \{\lambda_k\}_{k=1}^{N^2} = \left\{ sv(X^R) \right\}^2 & : & \text{eigenvalues of } (X^R)^\dagger X^R \\ |A'_k\rangle & : & \text{reshaped eigenvectors of } (X^R)^\dagger X^R , \\ |A''_k\rangle & : & \text{reshaped eigenvectors of } X^R (X^R)^\dagger \end{cases} \tag{10.31}$$

where sv denotes singular values and we have assumed that $N \leq M$. The initial basis is transformed by a local unitary transformation $W_a \otimes W_b$, where W_a and W_b are matrices of eigenvectors of matrices $(X^R)^\dagger X^R$ and $X^R (X^R)^\dagger$, respectively. If and only if the rank of $X^R (X^R)^\dagger$ equals one, the operator can be factorised into a product form, $X = X_1 \otimes X_2$, where $X_1 = \text{Tr}_2 X$ and $X_2 = \text{Tr}_1 X$.

To get a better feeling for the reshuffling transformation, observe that reshaping each row of an initially square matrix X of size MN according to Eq. (10.23) into a rectangular $M \times N$ submatrix, and placing it in lexicographical order block after block, produces the reshuffled matrix X^R. Let us illustrate this procedure for the simplest case $N = M = 2$, in which any row of the matrix X is reshaped into a 2×2 matrix

$$C_{kj} = X^R_{kj} \equiv \begin{bmatrix} \mathbf{X_{11}} & \mathbf{X_{12}} & X_{21} & X_{22} \\ X_{13} & X_{14} & \mathbf{X_{23}} & \mathbf{X_{24}} \\ \hline \mathbf{X_{31}} & \mathbf{X_{32}} & X_{41} & X_{42} \\ X_{33} & X_{34} & \mathbf{X_{43}} & \mathbf{X_{44}} \end{bmatrix}. \tag{10.32}$$

In the symmetric case with $M = N$, N^3 elements of X (typeset **boldface**) do not change position during reshuffling, while the remaining $N^4 - N^3$ elements do. Thus the space of complex matrices with the reshuffling symmetry $X = X^R$ is $2N^4 - 2(N^4 - N^3) = 2N^3$ dimensional.

The operation of reshuffling can be defined in an alternative way; for example the reshaping of the matrix A from (10.23) could be performed column after column

---

[8] In general one may reshuffle square matrices, if its size K is not prime. The symbol X^R has a unique meaning if a concrete decomposition of the size $K = MN$ is specified. If $M \neq N$ the matrix X^R is an $N^2 \times M^2$ rectangular matrix. Since $(X^R)^R = X$ we see that one may also reshuffle rectangular matrices, provided both dimensions are squares of natural numbers. Similar reorderings of matrices were considered by Oxenrider and Hill [711] and Yopp and Hill [989].

into a vector $\vec{a}'$. In the four-indices notation introduced above the two reshuffling operations take the form

$$X^R_{\substack{m\mu \\ n\nu}} \equiv X_{\substack{mn \\ \mu\nu}} \qquad \text{and} \qquad X^{R'}_{\substack{m\mu \\ n\nu}} \equiv X_{\substack{\nu\mu \\ nm}}. \tag{10.33}$$

Two reshuffled matrices are equivalent up to permutation of rows and columns and transposition, so the singular values of $X^{R'}$ and X^R are equal.

For comparison we provide analogous formulae showing the action of *partial transposition*: with respect to the first subsystem, $T_A \equiv T \otimes \mathbb{1}$ and with respect to the second, $T_B \equiv \mathbb{1} \otimes T$,

$$X^{T_A}_{\substack{m\mu \\ n\nu}} = X_{\substack{n\mu \\ m\nu}} \qquad \text{and} \qquad X^{T_B}_{\substack{m\mu \\ n\nu}} = X_{\substack{m\nu \\ n\mu}}. \tag{10.34}$$

Note that all these operations consist of exchanging a given pair of indices. However, while partial transposition (10.34) preserves Hermiticity, the reshuffling (10.33) does not. There is a related *swap* transformation among the two subsystems, $X^S_{\substack{m\mu \\ n\nu}} \equiv X_{\substack{\mu m \\ \nu n}}$, the action of which consists in relabelling certain rows (and columns) of the matrix, so its spectrum remains preserved. Note that for a tensor product $X = Y \otimes Z$ one has $X^S = Z \otimes Y$. Alternatively, define a SWAP operator

$$S \equiv \sum_{i,j=1}^{N} |i,j\rangle \langle j, i| \quad \text{so that} \quad S_{\substack{m\mu \\ n\nu}} = \delta_{m\nu}\delta_{n\mu}. \tag{10.35}$$

Observe that S is symmetric, Hermitian and unitary, and the identity $X^S = SXS$ holds. In full analogy to partial transposition we use also two operations of *partial swap*, $X^{S_1} = SX$ and $X^{S_2} = XS$.

All transformations listed in Table 10.1 are involutions, since performed twice they are equal to identity. It is not difficult to find relations between them, for example $X^{S_1} = [(X^{R'})^{T_A}]^{R'} = [(X^R)^{T_B}]^R$. Since $X^{R'} = [(X^R)^S]^T = [(X^R)^T]^S$, while $X^{T_B} = (X^{T_A})^T$ and $X^{S_1} = (X^{S_2})^S$, the spectra and singular values of the reshuffled (partially transposed, partially swapped) matrices do not depend on the way each operation has been performed, that is $\text{eig}(X^R) = \text{eig}(X^{R'})$ and $\text{sv}(X^R) = \text{sv}(X^{R'})$, while $\text{eig}(X^{S_1}) = \text{eig}(X^{S_2})$ and $\text{sv}(X^{S_1}) = \text{sv}(X^{S_2})$.

10.3 Positive and completely positive maps

Thus equipped, we return to physics. We will use the notation of Section 10.2 freely, so an alternative title for this section is 'complete positivity as an exercise in index juggling'. Let $\rho \in \mathcal{M}^{(N)}$ be a density matrix acting on an N-dimensional Hilbert space. What conditions need to be fulfilled by a map $\Phi : \mathcal{M}^{(N)} \to \mathcal{M}^{(N)}$ if it is to represent a physical operation? One class of maps that we will admit are those given in Eq. (10.7). We will now argue that nothing else is needed.

Our first requirement is that the map should be a linear one. It is always hard to argue for linearity, but at this level linearity is also hard to avoid, since we do not

Table 10.1 *Reorderings of a matrix X representing an operator which acts on a composite Hilbert space. The arrows denote the indices exchanged.*

Transformation	Definition	Symbol	Preserves Hermiticity	Preserves spectrum
Transposition	$X^T_{m\mu \atop n\nu} = X_{n\nu \atop m\mu}$	$\updownarrow\updownarrow$	yes	yes
Partial	$X^{T_A}_{m\mu \atop n\nu} = X_{n\mu \atop m\nu}$	$\updownarrow\,\cdot$	yes	no
Transpositions	$X^{T_B}_{m\mu \atop n\nu} = X_{m\nu \atop n\mu}$	$\cdot\,\updownarrow$	yes	no
Reshuffling	$X^{R}_{m\mu \atop n\nu} = X_{mn \atop \mu\nu}$	$\nearrow$	no	no
Reshuffling $'$	$X^{R'}_{m\mu \atop n\nu} = X_{\nu\mu \atop nm}$	$\searrow$	no	no
Swap	$X^{S}_{m\mu \atop n\nu} = X_{\mu m \atop \nu n}$	$\overset{\leftrightarrow}{\leftrightarrow}$	yes	yes
Partial	$X^{S_1}_{m\mu \atop n\nu} = X_{\mu m \atop n\nu}$	$\overset{\leftrightarrow}{\cdot}$	no	no
Swaps	$X^{S_2}_{m\mu \atop n\nu} = X_{m\mu \atop \nu n}$	$\dot{\leftrightarrow}$	no	no

want the image of ρ to depend on the way in which ρ is presented as a mixture of pure states – the entire probabilistic structure of the theory is at stake here.[9] We are thus led to postulate the existence of a *linear superoperator* Φ,

$$\rho' = \Phi\rho \qquad \text{or} \qquad \rho'_{m\mu} = \Phi_{m\mu \atop n\nu}\,\rho_{n\nu}. \tag{10.36}$$

Summation over repeated indices is understood to apply throughout this section. Inhomogeneous maps $\rho' = \Phi\rho + \sigma$ are automatically included, since

$$\Phi_{m\mu \atop n\nu}\,\rho_{n\nu} + \sigma_{m\mu} = \left(\Phi_{m\mu \atop n\nu} + \sigma_{m\mu}\delta_{n\nu}\right)\rho_{n\nu} = \Phi'_{m\mu \atop n\nu}\,\rho_{n\nu} \tag{10.37}$$

due to $\text{Tr}\rho = 1$. We deal with affine maps of density matrices.

The map should take density matrices to density matrices. This means that whenever ρ is (i) Hermitian, (ii) of unit trace and (iii) positive, its image ρ' must share these properties.[10] These three conditions impose three constraints on the matrix Φ:

$$\text{(i)} \qquad \rho' = (\rho')^\dagger \quad \Leftrightarrow \quad \Phi_{m\mu \atop n\nu} = \Phi^*_{\mu m \atop \nu n} \quad \text{so} \quad \Phi^* = \Phi^S, \tag{10.38}$$

$$\text{(ii)} \qquad \text{Tr}\rho' = 1 \quad \Leftrightarrow \quad \Phi_{mm \atop n\nu} = \delta_{n\nu} \tag{10.39}$$

$$\text{(iii)} \qquad \rho' \geq 0 \quad \Leftrightarrow \quad \Phi_{m\mu \atop n\nu}\,\rho_{n\nu} \geq 0 \quad \text{when } \rho > 0. \tag{10.40}$$

As they stand, these conditions are not very illuminating.

---

[9] Non-linear quantum mechanics is actually a lively field of research; see Mielnik [660] and references therein.
[10] Any map $\Phi\rho$ can be normalised according to $\rho \rightarrow \Phi\rho/\text{Tr}[\Phi\rho]$. It is sometimes convenient to work with unnormalised maps, but the a posteriori normalisation procedure may spoil linearity.

The meaning of our three conditions becomes much clearer if we reshuffle Φ according to (10.33) and define the *dynamical matrix* [11]

$$D_\Phi \equiv \Phi^R \quad \text{so that} \quad D_{\substack{mn\\\mu\nu}} = \Phi_{\substack{m\mu\\n\nu}}. \tag{10.41}$$

The dynamical matrix D_Φ uniquely determines the map Φ. It obeys

$$D_{a\Phi+b\Psi} = aD_\Phi + bD_\Psi, \tag{10.42}$$

that is to say it is a linear function of the map.

In terms of the dynamical matrix our three conditions become

(i) $\quad \rho' = (\rho')^\dagger \quad \Leftrightarrow \quad D_{\substack{mn\\\mu\nu}} = D^\dagger_{\substack{mn\\\mu\nu}} \quad \text{so} \quad D_\Phi = D^\dagger_\Phi, \tag{10.43}$

(ii) $\quad \text{Tr}\rho' = 1 \quad \Leftrightarrow \quad D_{\substack{mn\\m\nu}} = \delta_{n\nu} \tag{10.44}$

(iii) $\quad \rho' \geq 0 \quad \Leftrightarrow \quad D_{\substack{mn\\\mu\nu}}\rho_{n\nu} \geq 0 \quad \text{when } \rho > 0. \tag{10.45}$

Condition (i) holds if and only if D_Φ is Hermitian. Condition (ii) also takes a familiar form – its partial trace with respect to the first subsystem is the unit operator for the second subsystem:

$$D_{\substack{mn\\m\nu}} = \delta_{n\nu} \quad \Leftrightarrow \quad \text{Tr}_A D = \mathbb{1} \tag{10.46}$$

Only condition (iii), for positivity, requires further unravelling.

The map is said to be a *positive map* if it takes positive matrices to positive matrices. To see if a map is positive, we must test whether condition (iii) holds. Let us first assume that the original density matrix is a pure state, so that $\rho_{n\nu} = z_n z_\nu^*$. Then its image will be positive if and only if, for all vectors x_m,

$$x_m \rho'_{m\mu} x_\mu^* = x_m z_n D_{\substack{mn\\\mu\nu}} x_\mu^* z_\nu^* \geq 0. \tag{10.47}$$

This means that the dynamical matrix itself must be positive when it acts on product states in $\mathcal{H}_{N^2}$. This property is called *block-positivity*. We have arrived at [494]:

Jamioɫkowski's theorem. *A linear map Φ is positive if and only if the corresponding dynamical matrix D_Φ is block-positive.*

The converse holds since condition (10.47) is strong enough to ensure that condition (10.45) holds for all mixed states ρ as well.

Interestingly, the condition for positivity has not only one but two drawbacks. First, it is difficult to work with. Second, it is not enough from a physical point of view. Any quantum state ρ may be extended by an ancilla to a state $\rho \otimes \sigma$ of a larger composite system. The mere possibility that an ancilla may be added

---

[11] This concept was introduced by Sudarshan, Mathews and Rau (1961) [867], and even earlier – in the mathematics literature – by Schatten (1950) [797].

requires us to check that the map $\Phi \otimes \mathbb{1}$ is positive as well. Since the map leaves the ancilla unaffected this may seem like a foregone conclusion. Classically it is so, but quantum mechanically it is not. Let us state this condition precisely: a map Φ is said to be *completely positive* if and only if for an arbitrary K dimensional extension

$$\mathcal{H}_N \rightarrow \mathcal{H}_N \otimes \mathcal{H}_K \quad \text{the map} \quad \Phi \otimes \mathbb{1}_K \quad \text{is positive.} \quad (10.48)$$

This is our final condition on a physical map.[12]

In order to see what the condition of complete positivity says about the dynamical matrix we will backtrack a little and introduce a canonical form for the latter. Since the dynamical matrix is an Hermitian matrix acting on $\mathcal{H}_{N^2}$, it admits a spectral decomposition

$$D_\Phi = \sum_{i=1}^{r} d_i |\chi_i\rangle \langle\chi_i| \quad \text{so that} \quad D_{mn \atop \mu\nu} = \sum_{i=1}^{N^2} d_i \chi^i_{mn} \bar{\chi}^i_{\mu\nu}. \quad (10.49)$$

The eigenvalues d_i are real, and the notation emphasises that the matrices χ^i_{mn} are (reshaped) vectors in $\mathcal{H}_{N^2}$.

Now we are in a position to investigate the conditions that ensure that the map $\Phi \otimes \mathbb{1}$ preserves positivity when it acts on matrices in $\mathcal{HS} = \mathcal{H} \otimes \mathcal{H}'$. We pick an arbitrary vector $z_{nn'}$ in $\mathcal{HS}$, and act with our map on the corresponding pure state:

$$\rho'_{mm'\mu\mu'} = \Phi_{mu \atop n\nu} \delta_{m'\mu'} z_{nn'} z^*_{\nu\nu'} = D_{mn \atop \mu\nu} z_{nm'} z^*_{\nu\mu'} = \sum_i d_i \chi^i_{mn} z_{nm'} (\chi^i_{\mu\nu} z_{\nu\mu'})^*. \quad (10.50)$$

Then we pick another arbitrary vector $x_{mm'}$ and test whether ρ' is a positive operator:

$$x_{mm'} \rho'_{mm'\mu\mu'} x^*_{\mu\mu'} = \sum_i d_i |\chi^i_{mn} x_{mn'} z_{nm'}|^2 \geq 0. \quad (10.51)$$

This must hold for arbitrary $x_{mm'}$ and $z_{mm'}$, and therefore all the eigenvalues d_i must be positive (or zero). In this way we have arrived at:

Choi's theorem. *A linear map Φ is completely positive if and only if the corresponding dynamical matrix D_Φ is positive.*

There is some fine print. If condition (10.48) holds for a fixed K only, the map is said to be *K-positive*. The map will be completely positive if and only if it is N–positive, which is the condition that we actually investigated.[13] The dynamical matrix D_Φ is sometimes called the *Choi matrix*.

---

[12] The mathematical importance of complete positivity was first noted by Stinespring (1955) [859]; its importance in quantum theory was emphasised by Kraus (1971) [559], Accardi (1976) [4], and Lindblad (1976) [611].

[13] 'Choi's theorem' is theorem 2 in Choi (1975) [217]. Theorem 1 (the existence of the operator sum representation) and theorem 5 are given below. The paper contains no theorems 3 or 4.

It is striking that we obtain such a simple result when we strengthen the condition on the map from positivity to complete positivity. The set of completely positive maps is isomorphic to the set of positive matrices D_Φ of size N^2. When the map is also trace preserving we add the extra condition (10.46), which implies that $\mathrm{Tr} D_\Phi = N$. We can therefore think of the set of trace-preserving completely positive maps as a subset of the set of density matrices in $\mathcal{H}_{N^2}$, albeit with an unusual normalisation. This analogy will be further pursued in Chapter 11, where (for reasons that will become clear later) we will also occupy ourselves with understanding the way in which the set of completely positive maps forms a proper subset of the set of all positive maps.

The dynamical matrix is positive if and only if it can be written in the form

$$D_\Phi = \sum_i |A_i\rangle \langle A_i| \qquad \text{so that} \qquad D_{mn}_{\mu\nu} = \sum_i A^i_{mn} \bar{A}^i_{\mu\nu}, \qquad (10.52)$$

where the vectors A_i are arbitrary to an extent given by Schrödinger's mixture theorem (see Section 8.4). In this way we obtain an alternative characterisation of completely positive maps. They are the maps that can be written in the

Operator sum representation. *A linear map Φ is completely positive if and only if it is of the form*

$$\rho \to \rho' = \sum_i A_i \rho A_i^\dagger. \qquad (10.53)$$

This is also known as the *Kraus* or *Stinespring form*, since its existence follows from the *Stinespring dilation theorem*.[14] The operators A_i are known as *Kraus operators*. The map will be *trace preserving* if and only if condition (10.44) holds, which translates itself to

$$\sum_i A_i^\dagger A_i = \mathbb{1}_N. \qquad (10.54)$$

We have recovered the class of operations that we introduced in Eq. (10.7), but our new point of view has led us to the conclusion that this is the most general class that we need to consider. Trace-preserving completely positive maps go under various names: *deterministic* or *proper quantum operations*, *quantum channels*, or *stochastic maps*. They are the sought for analogue of classical stochastic maps.

The convex set of proper quantum operations is denoted $\mathcal{CP}_N$. To find its dimension we note that the dynamical matrices belong to the positive cone in the space of Hermitian matrices of size N^2, which has dimension N^4; the dynamical matrix

---

[14] In physics the operator sum representation was introduced by Kraus (1971) [559], based on an earlier (somewhat more abstract) theorem by Stinespring (1955) [859], and independently by Sudarshan et al. (1961) [867]. See also Kraus (1983) [560] and Evans (1984) [303].

corresponds to a trace-preserving map only if its partial trace (10.46) is the unit operator, so it is subject to N^2 conditions. Hence the dimension of $\mathcal{CP}_N$ equals $N^4 - N^2$.

Since the operator sum representation does not determine the Kraus operators uniquely we would like to bring it to a canonical form. The problem is quite similar to that of introducing a canonical form for a density matrix – in both cases, the solution is to present a Hermitian matrix as a mixture of its eigenstates. Such a decomposition of the dynamical matrix was given in Eq. (10.49). A set of canonical Kraus operators can be obtained by setting $A_i = \sqrt{d_i}\chi_i$. The result is the

Canonical Kraus form. *A completely positive map* $\Phi : \mathcal{M}^{(N)} \to \mathcal{M}^{(N)}$ *can be represented as*

$$\rho \;\to\; \rho' \;=\; \sum_{i=1}^{r \leq N^2} d_i\, \chi_i \rho\, \chi_i^\dagger \;=\; \sum_{i=1}^{r} A_i \rho A_i^\dagger, \tag{10.55}$$

where

$$\mathrm{Tr} A_i^\dagger A_j \;=\; \sqrt{d_i d_j}\, \langle \chi_i | \chi_j \rangle \;=\; d_i\, \delta_{ij}. \tag{10.56}$$

If the map is also trace preserving then

$$\sum_i A_i^\dagger A_i \;=\; \mathbb{1}_N \qquad \Rightarrow \qquad \sum_{i=1}^{r} d_i \;=\; N. \tag{10.57}$$

If D_Φ is non-degenerate the canonical form is unique up to phase choices for the Kraus operators. The *Kraus rank* of the map is the number of Kraus operators that appear in the canonical form, and equals the rank r of the dynamical matrix.

The operator sum representation can be written

$$\Phi \;=\; \sum_{i=1}^{N^2} A_i \otimes \bar{A}_i \;=\; \sum_{i=1}^{N^2} d_i\, \chi^i \otimes \bar{\chi}^i. \tag{10.58}$$

The CP map can be described in the notation of Eq. (10.28), and the operator sum representation may be considered as a Schmidt decomposition (10.30) of Φ, with Schmidt coefficients $\lambda_i = d_i^2$.

10.4 Environmental representations

We began this chapter by adding an ancilla in the state σ to the principal system $\mathcal{S}$, evolving the composite system unitarily, and then removing the ancilla through a partial trace at the end. This led us to the *environmental representation* of the map Φ, that is to

$$\rho \;\to\; \rho' \;=\; \Phi(\rho) \;=\; \mathrm{Tr}_{\mathrm{env}}\left[U\left(\rho \otimes \sigma\right) U^\dagger \right]. \tag{10.59}$$

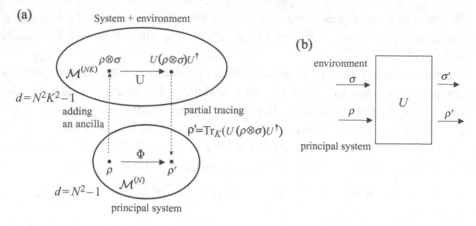

Figure 10.3 Quantum operations represented by (a) a unitary operator U of size NK in an enlarged system including the environment, (b) a black box picture.

See Figure 10.3. We have shown that the resulting map can be written in the Kraus form, and now we know that this means that it is a completely positive map. What was missing from the argument was a proof that any CP map admits an environmental representation, and indeed one in which the ancilla starts out in a pure state $\sigma = |v\rangle\langle v|$. This we will now supply.[15] We are given a set of K Kraus operators A_μ (equipped with Greek indices because we use such letters to denote states of the ancilla). Due to the completeness relation (10.54) we can regard them as defining N orthogonal columns in a matrix U with NK rows,

$$A^\mu_{mn} = \langle m, \mu|U|n, v\rangle = U_{\substack{m\mu \\ nv}} \quad \Leftrightarrow \quad A_\mu = \langle\mu|U|v\rangle. \tag{10.60}$$

Here v is fixed, but we can always find an additional set of columns that turns U into a unitary matrix of size NK. By construction then, for an ancilla of dimension K,

$$\rho' = \mathrm{Tr}_{\mathrm{env}}\left[U\left(\rho \otimes |v\rangle\langle v|\right)U^\dagger\right] = \sum_{\mu=1}^{K}\langle\mu|U|v\rangle\rho\langle v|U^\dagger|\mu\rangle = \sum_{\mu=1}^{K}A_\mu\rho A^\dagger_\mu. \tag{10.61}$$

This is the Kraus form, since the operators A_μ satisfy the completeness relation

$$\sum_{\mu=1}^{K}A^\dagger_\mu A_\mu = \sum_{\mu=1}^{K}\langle v|U^\dagger|\mu\rangle\langle\mu|U|v\rangle = \langle v|U^\dagger U|v\rangle = \mathbb{1}_N. \tag{10.62}$$

Note that the 'extra' columns that we added to the matrix U do not influence the quantum operation in any way.

---

[15] Originally this was noted by Arveson (1969) [59] and Lindblad (1975) [610].

Although we may choose an ancilla that starts out in a pure state, we do not have to do it. If the initial state of the environment in the representation (10.59) is a mixture $\sigma = \sum_{\nu=1}^{r} q_\nu |\nu\rangle\langle\nu|$, we obtain an operator sum representation with rK terms,

$$\rho \to \rho' = \text{Tr}_{\text{env}} \left[U \left(\rho \otimes \sum_{\nu=1}^{r} q_\nu |\nu\rangle\langle\nu| \right) U^\dagger \right] = \sum_{l=1}^{rK} A_l \rho A_l^\dagger \qquad (10.63)$$

where $A_l = \sqrt{q_\nu} \langle \mu | U | \nu \rangle$ and $l = \mu + \nu(K - 1)$.

If the initial state of the ancilla is pure the dimension of its Hilbert space need not exceed N^2, the maximal number of Kraus operators required. More precisely its dimension may be set equal to the Kraus rank of the map. If the environment is initially in a mixed state, its weights q_ν are needed to specify the operation. Counting the number of parameters, one could thus speculate that the action of any quantum operation may be simulated by a coupling with a mixed state of an environment of size N. However, this is not the case: already for $N = 2$ there exist operations which have to be simulated with a three-dimensional environment [889, 992]. The general question of the minimal size of $\mathcal{H}_{\text{env}}$ remains open.

It is illuminating to discuss the special case in which the initial state of the N-dimensional environment is maximally mixed, $\sigma = \mathbb{1}_N/N$. The unitary matrix U of size N^2, defining the map, may be treated as a vector in the composite HS space $\mathcal{H}_{\text{HS}}^A \otimes \mathcal{H}_{\text{HS}}^B$ and represented in its Schmidt form $U = \sum_{i=1}^{N^2} \sqrt{\lambda_i} |\tilde{A}_i\rangle \otimes |\tilde{A}'_i\rangle$, where λ_i are eigenvalues of $(U^R)^\dagger U^R$. Since the operators $\tilde{A}'_i$ (reshaped eigenvectors of $(U^R)^\dagger U^R$) form an orthonormal basis in $\mathcal{H}_{\text{HS}}$, the procedure of partial tracing leads to a Kraus form with N^2 terms:

$$\rho' = \Phi_U \rho = \text{Tr}_{\text{env}} \left[U(\rho \otimes \frac{1}{N} \mathbb{1}_N) U^\dagger \right] \qquad (10.64)$$

$$= \text{Tr}_{\text{env}} \left[\sum_{i=1}^{N^2} \sum_{j=1}^{N^2} \sqrt{\lambda_i \lambda_j} \left(\tilde{A}_i \rho \tilde{A}_j^\dagger \right) \otimes \left(\frac{1}{N} \tilde{A}'_i \tilde{A}_j'^\dagger \right) \right] = \frac{1}{N} \sum_{i=1}^{N^2} \lambda_i \tilde{A}_i \rho \tilde{A}_i^\dagger.$$

The standard Kraus form is obtained by rescaling the operators, $A_i = \sqrt{\lambda_i/N} \tilde{A}_i$. Operations for which there exist a unitary matrix U providing a representation in the above form, we shall call *unistochastic channels*.[16] Note that the matrix U is determined up to a local unitary matrix V of size N, in the sense that U and $U' = U(\mathbb{1}_N \otimes V)$ generate the same unistochastic map, $\Phi_U = \Phi_{U'}$.

The physical motivation is simple: not knowing anything about the environment (whose dimension is negotiable, and is sometimes taken to be a power of N), one assumes that it is initially in the maximally mixed state. Unistochastic operations

---

[16] In analogy to classical transformations given by unistochastic matrices, $\vec{p}' = T\vec{p}$, where $T_{ij} = |U_{ij}|^2$.

appear in the context of quantum information processing [548, 749], in studies of thermodynamics inspired by quantum information theory [796], and, under the name 'noisy maps', when studying reversible transformations from pure to mixed states [461].

A debatable point remains, namely that the combined system started out in a product state. This may look like a very special initial condition. However, in general it not so easy to present a well defined procedure for how to assign a state of the composite system, given only a state of the system of interest to start with. Suppose $\rho \rightarrow \omega$ is such an assignment, where ω acts on the composite Hilbert space. Ideally one wants the assignment map to obey three conditions: (i) it preserves mixtures, (ii) $\text{Tr}_{\text{env}}\omega = \rho$, and (iii) ω is positive for all positive ρ. But it is known[17] that these conditions are so stringent that the only solution is of the form $\omega = \rho \otimes \sigma$.

To describe what happens to the environment after the discrete interaction described in (10.59) one defines a map $\tilde{\Phi}$ complementary to Φ

$$\rho \rightarrow \sigma' = \tilde{\Phi}(\rho) = \text{Tr}_S\left[U(\rho \otimes \sigma)U^\dagger\right]. \tag{10.65}$$

This equation differs from (10.59) only by the choice of the subsystem S traced away. A complementary map transforms an initial state ρ of size N into a state σ' of size K, the dimension of the environment. This output state can be expressed by a family of Kraus operators (Problem 10.3) and its entropy is used in the Lindblad inequality (13.104) to characterise how the map Φ changes the entropy of a state ρ.

10.5 Some spectral properties

A quantum operation Φ is uniquely characterised by its dynamical matrix, but the spectra of these matrices are quite different. The dynamical matrix is Hermitian, but Φ is not and its eigenvalues z_i are complex. Let us order them according to their moduli, $|z_1| \geq |z_2| \geq \cdots |z_{N^2}| \geq 0$. The operation Φ sends the convex compact set $\mathcal{M}^{(N)}$ into itself. Therefore, due to a fixed point theorem, the transformation has a fixed point – an invariant state σ_1 such that $\Phi\sigma_1 = \sigma_1$. Thus $z_1 = 1$ and all eigenvalues fulfil $|z_i| \leq 1$, since otherwise the assumption that Φ is positive would be violated. These spectral properties – see Table 10.2 – are similar to those enjoyed by classical stochastic matrices due to the Frobenius–Perron theorem (Section 2.1).

The trace preserving condition, applied to the equation $\Phi\sigma_i = z_i\sigma_i$, implies that if $z_i \neq 1$ then $\text{Tr}(\sigma_i) = 0$. If $R = |z_2| < 1$ then the matrix Φ is *primitive* [645]; under repeated applications of the map all states converge to the invariant state σ_1. If Φ is diagonalizable (its Jordan decomposition has no non-trivial blocks, so that

---

[17] See the exchange between Pechukas [730] and Alicki [23].

Table 10.2 *Quantum operations* $\Phi : \mathcal{M}^{(N)} \to \mathcal{M}^{(N)}$: *properties of the superoperator* Φ *and the dynamical matrix* $D_\Phi = \Phi^R$. *For coarse graining map consult Eq. (13.77) while for the entropy S see Section 13.7.*

Matrices	Superoperator $\Phi = (D_\Phi)^R$	Dynamical matrix D_Φ
Hermiticity	No	Yes
Trace	spectrum is symmetric $\Rightarrow$ $\mathrm{Tr}\ \Phi \in \mathbb{R}$	$\mathrm{Tr}\ D_\Phi = N$
Eigenvectors (right)	invariant states or transient corrections	Kraus operators
Eigenvalues	$|z_i| \leq 1$, $-\ln|z_i| =$ decay rates	Weights of Kraus operators, $d_i \geq 0$
Unitary evolution $D_U = (U \otimes U^*)^R$	$\|\Phi_U\|_{\mathrm{HS}}^2 = N^2$	$S(D_U) = 0$
Coarse graining	$\|\Phi_{\mathrm{CG}}\|_{\mathrm{HS}}^2 = N$	$S(D_{\mathrm{CG}}) = \ln N$
Complete depolarisation, $D_{\Phi_*} = \mathbb{1}_{N^2}/N$	$\|\Phi_*\|_{\mathrm{HS}}^2 = 1$	$S(D_*) = 2\ln N$

the number of right eigenvectors σ_i is equal to the size of the matrix), then any initial state ρ_0 may be expanded in the eigenbasis of Φ,

$$\rho_0 = \sum_{i=1}^{N^2} c_i \sigma_i \qquad \text{while} \qquad \rho_t = \Phi^t \rho_0 = \sum_{i=1}^{N^2} c_i z_i^t \sigma_i. \tag{10.66}$$

Therefore ρ_0 converges exponentially fast to the invariant state σ_1 with a decay rate not smaller than $-\ln R$ and the right eigenstates σ_i for $i \geq 2$ play the role of the transient traceless corrections to ρ_0. The superoperator Φ sends Hermitian operators to Hermitian operators, $\rho_1^\dagger = \rho_1 = \Phi\rho_0 = \Phi\rho_0^\dagger$, so

$$\text{if} \qquad \Phi\chi = z\chi \qquad \text{then} \qquad \Phi\chi^\dagger = z^*\chi^\dagger, \tag{10.67}$$

and the spectrum of Φ (contained in the unit circle) is symmetric with respect to the real axis. Thus the trace of Φ is real, as follows also from the Hermiticity of $D_\Phi = \Phi^R$. Using (10.58) we obtain[18]

$$\mathrm{Tr}\ \Phi = \sum_{i=1}^{r} (\mathrm{Tr}A_i)(\mathrm{Tr}\bar{A}_i) = \sum_{i=1}^{r} |\mathrm{Tr}A_i|^2 \geq 0. \tag{10.68}$$

The trace is equal to zero for a unitary map corresponding to any cyclic permutation of all states, to N^2 for the identity map and to unity for a map given by the rescaled

---

[18] This trace determines the *mean operation fidelity* $\langle F(\rho_\psi, \Phi\rho_\psi)\rangle_\psi$ averaged over random pure states ρ_ψ [691, 994].

identity matrix, $D_* = \mathbb{1}_{N^2}/N$. The latter map describes the *completely depolarizing channel* Φ_*, which transforms any initial state ρ into the maximally mixed state, $\Phi_*\rho = \rho_* = \mathbb{1}_N/N$.

Given a set of Kraus operators A_i for a quantum operation Φ, and any two unitary matrices V and W of size N, the operators $A'_i = VA_iW$ will satisfy the relation (10.54) and define the operation

$$\rho \to \rho' = \Phi_{VW}\rho = \sum_{i=1}^k A'_i \rho A'^{\dagger}_i = V\left(\sum_{i=1}^k A_i(W\rho W^{\dagger})A^{\dagger}_i\right)V^{\dagger}. \qquad (10.69)$$

The operations Φ and Φ_{VW} are in general different, but *unitarily similar*, in the sense that their dynamical matrices have the same spectra. The equality $||\Phi||_{HS} = ||\Phi_{VW}||_{HS}$ follows from the transformation law

$$\Phi_{VW} = (V \otimes V^*)\,\Phi\,(W \otimes W^*), \qquad (10.70)$$

which is a consequence of (10.58). This implies that the dynamical matrix transforms by a local unitary, $D_{VW} = (V \otimes W^T)D(V \otimes W^T)^{\dagger}$.

10.6 Unital and bistochastic maps

A trace-preserving completely positive map is called a *bistochastic map* if it is also *unital*, that is to say if it leaves the maximally mixed state invariant.[19] Evidently this is the quantum analogue of a bistochastic matrix – a stochastic matrix that leaves the uniform probability vector invariant. The composition of two bistochastic maps is bistochastic. In the operator sum representation the condition that the map be bistochastic reads

$$\rho \to \rho' = \sum_i A_i\rho A^{\dagger}_i, \quad \sum_i A^{\dagger}_i A_i = \mathbb{1}, \quad \sum_i A_i A^{\dagger}_i = \mathbb{1}. \qquad (10.71)$$

For the dynamical matrix this means that $\mathrm{Tr}_A D = \mathrm{Tr}_B D = \mathbb{1}$.

The channel is bistochastic if all the Kraus operators obey $[A_i, A^{\dagger}_i] = 0$. Indeed the simplest example is a unitary transformation. A more general class of bistochastic channels is given by convex combinations of unitary operations, also called [26] *random external fields* (REF),

$$\rho' = \Phi_{REF}\rho = \sum_{i=1}^k p_i V_i\rho V^{\dagger}_i, \quad \text{with} \quad p_i > 0 \quad \text{and} \quad \sum_{i=1}^k p_i = 1, \qquad (10.72)$$

where each operator V_i is unitary. The Kraus form (10.53) can be reproduced by setting $A_i = \sqrt{p_i}\,V_i$.

---

[19] See the book by Alberti and Uhlmann [16].

The set of all bistochastic CP maps, denoted $\mathcal{B}_N$, is a convex set in itself. The set of all bistochastic matrices is, as we learned in Section 2.1, a convex polytope with permutation matrices as its extreme points. Reasoning by analogy one would guess that the extreme points of $\mathcal{B}_N$ are unitary transformations, in which case $\mathcal{B}_N$ would coincide with the set of random external fields. This happens to be true for qubits, as we will see in detail in Section 10.7, but it fails for all $N > 2$. Computable criteria, which allow one to identify bistochastic maps which are not mixtures of unitaries, were given by Mendl and Wolf [653].

There is a theorem that characterises the extreme points of the set of stochastic maps [217]:

Choi's lemma. *A stochastic map* Φ *is extreme in* $\mathcal{CP}_N$ *if and only if it admits a canonical Kraus form for which the matrices* $A_i^\dagger A_j$ *are linearly independent.*

We prove that the condition is sufficient: assume that $\Phi = p\Psi_1 + (1 - p)\Psi_2$. If so it will be true that

$$\Phi\rho = \sum_i A_i \rho A_i^\dagger = p\Psi_1\rho + (1 - p)\Psi_2\rho$$

$$= p \sum_i B_i \rho B_i^\dagger + (1 - p) \sum_i C_i \rho C_i^\dagger. \tag{10.73}$$

The right-hand side is not in the canonical form, but we assume that the left-hand side is. Therefore there is a unique way of writing

$$B_i = \sum_j m_{ij} A_j. \tag{10.74}$$

In fact this is Schrödinger's mixture theorem in slight disguise. Next we observe that

$$\sum_i B_i^\dagger B_i = \mathbb{1} = \sum_i A_i^\dagger A_i \quad \Rightarrow \quad \sum_{ij} \left((m^\dagger m)_{ij} - \delta_{ij}\right) A_i^\dagger A_j = 0. \tag{10.75}$$

Because of the linear independence condition this means that $(m^\dagger m)_{ij} = \delta_{ij}$. This is what we need in order to show that $\Phi = \Psi_1$, and it follows the map Φ is indeed pure.

The matrices $A_i A_j^\dagger$ are of size N, so there can be at most N^2 linearly independent ones. This means that there can be at most N Kraus operators occurring in the canonical form of an extreme stochastic map.

It remains to find an example of an extreme bistochastic map which is not unitary. Using the $N = 2j + 1$ dimensional representation of $SU(2)$, we take the three Hermitian angular momentum operators J_i and define the map

$$\rho \to \rho' = \frac{1}{j(j+1)} \sum_{i=1}^3 J_i \rho J_i, \quad J_1^2 + J_2^2 + J_3^2 = j(j+1). \tag{10.76}$$

Table 10.3 *Quantum maps acting on density matrices and given by a positive definite dynamical matrix D versus classical Markov dynamics on probability vectors defined by transition matrix T with non-negative elements.*

Quantum	Completely positive maps	Classical	Markov chains given by
S_1^Q	Trace preserving, $\mathrm{Tr}_A D = \mathbb{1}$	S_1^{Cl}	Stochastic matrices T
S_2^Q	Unital, $\mathrm{Tr}_B D = \mathbb{1}$	S_2^{Cl}	T^T is stochastic
S_3^Q	Unital and trace preserving maps	S_3^{Cl}	Bistochastic matrices B
S_4^Q	Maps with $A_i = A_i^\dagger \Rightarrow D = D^T$	S_4^{Cl}	Symmetric stochastic matrices, $B = B^T$
S_5^Q	Unistochastic operations, $D = \left(U^R\right)^\dagger U^R$	S_5^{Cl}	Unistochastic matrices, $B_{ij} = |U_{ij}|^2$
S_6^Q	Unitary transformations	S_6^{Cl}	Permutations

Choi's condition for an extreme map is that the set of matrices $J_i J_j^\dagger = J_i J_j$ must be linearly independent. By angular momentum addition our set spans a $9 = 5 + 3 + 1$ dimensional representation of $SO(3)$, and all the matrices will be linearly independent provided that they are non-zero. The example fails for $N = 2$ only – in that case $J_i J_j + J_j J_i = 0$, the J_i are both Hermitian and unitary, and we do not get an extreme point.[20]

For any quantum channel Φ one defines its *dual channel* Φ°, such that the HS scalar product satisfies $\langle \Phi \sigma | \rho \rangle = \langle \sigma | \Phi^\circ \rho \rangle$ for any states σ and ρ. If a CP map is given by the Kraus form $\Phi \rho = \sum_i A_i \rho A_i^\dagger$, the dual channel reads $\Phi^\circ \rho = \sum_i A_i^\dagger \rho A_i$. This provides a link between the dynamical matrices representing dual channels,

$$\Phi^\circ = (\Phi^T)^S = (\Phi^S)^T \qquad \text{and} \qquad D_{\Phi^\circ} = (D_\Phi^T)^S = (D_\Phi^S)^T = \overline{D_\Phi^S}. \qquad (10.77)$$

Since neither the transposition nor the swap modify the spectrum of a matrix, the spectra of the dynamical matrices for dual channels are the same.

If channel Φ is trace preserving, its dual Φ° is unital, and conversely if Φ is unital then Φ° is trace preserving. Thus the channel dual to a bistochastic one is bistochastic.

Let us now analyse in some detail the set $\mathcal{BU}_N$ of all *unistochastic operations* [684], for which the representation (10.64) exists. The initial state of the environment is maximally mixed, $\sigma = \mathbb{1}/N$, so the map Ψ_U is determined by a unitary matrix U of size N^2. The Kraus operators A_i are eigenvectors of the dynamical matrix D_{Ψ_U}. On the other hand, they also enter the Schmidt decomposition (10.30)

---

[20] This example is due to Landau and Streater [572].

of U as shown in (10.64), and are proportional to the eigenvectors of $(U^R)^\dagger U^R$. Therefore

$$D_{\Psi_U} = \frac{1}{N}(U^R)^\dagger U^R \quad \text{so that} \quad \Psi_U = \frac{1}{N}\left[(U^R)^\dagger U^R\right]^R. \tag{10.78}$$

We have thus arrived at an important result: for any unistochastic map the spectrum of the dynamical matrix is given by the Schmidt coefficients, $d_i = \lambda_i/N$, of the unitary matrix U treated as an element of the composite HS space. For any local operation, $U = U_1 \otimes U_2$ the superoperator is unitary, $\Psi_U = U_1 \otimes U_1^*$ so $||\Psi_U||^2_{HS} = \text{Tr}\Psi_U \Psi_U^\dagger = N^2$. The resulting unitary operation is an isometry and can be compared with a permutation S_6^{Cl} acting on classical probability vectors. The spaces listed in Table 10.3 satisfy the relations $S_1 \cap S_2 = S_3$ and $S_3 \supset S_5 \supset S_6$ in both the classical and the quantum setup. However, the analogy is not exact since the inclusion $S_3^{Cl} \supset S_4^{Cl}$ does not have a quantum counterpart.

10.7 One qubit maps

When $N = 2$ the quantum operations are called *binary channels*. In general, the space $\mathcal{CP}_N$ is $N^4 - N^2$ dimensional. Hence the set of binary channels has 12 dimensions. To parametrize it we begin with the observation that a binary channel is an affine map of the Bloch ball into itself – subject to restrictions that we will deal with later.[21] If we describe density matrices through their Bloch vectors, as in Eq. (5.9), this means that the map $\rho' = \Phi\rho$ can be written in the form

$$\vec{\tau}' = t\vec{\tau} + \vec{\kappa} = O_1 \eta O_2^T \vec{\tau} + \vec{\kappa}, \tag{10.79}$$

where t denotes a real matrix of size 3 which we bring to diagonal form by two orthogonal transformations O_1 and O_2. Actually we permit only rotations belonging to the $SO(3)$ group, which means that some of the elements of the diagonal matrix η may be negative – the restriction is natural because it corresponds to using unitarily similar quantum operations, see Eq. (10.69). The elements of the diagonal matrix η are collected into a vector $\vec{\eta} = (\eta_x, \eta_y, \eta_z)$, called the *distortion vector* because the transformation $\vec{\tau}' = \eta\vec{\tau}$ takes the Bloch ball to an ellipsoid given by

$$\frac{1}{4} = \tau_x^2 + \tau_y^2 + \tau_z^2 = \left(\frac{\tau_x'}{\eta_x}\right)^2 + \left(\frac{\tau_y'}{\eta_y}\right)^2 + \left(\frac{\tau_z'}{\eta_z}\right)^2. \tag{10.80}$$

Finally the vector $\vec{\kappa} = (\kappa_x, \kappa_y, \kappa_z)$ is called the *translation vector*, because it moves the centre of mass of the ellipsoid. We can now see where the 12 dimensions come

---

[21] The explicit description given below is due to Fujiwara and Algoet [324] and to King and Ruskai [531]. Geometric properties of the set of positive one-qubit maps were also studied in [700, 974]. There is a beautiful relation with Lorentz transformations (the group $SL(2, \mathbb{C})$), explained in [55, 583].

from: there are three parameters $\vec{\eta}$ that determine the shape of the ellipsoid, three parameters $\vec{\kappa}$ that determine its centre of mass, three parameters to determine its orientation, and three parameters needed to rotate the Bloch ball to a standard position relative to the ellipsoid (before it is subject to the map described by $\vec{\eta}$).

The map is positive whenever the ellipsoid lies within the Bloch ball. The map is unital if the centre of the Bloch ball is a fixed point of the map, which means that $\vec{\kappa} = 0$. But the map is completely positive only if the dynamical matrix is positive definite, which means that not every ellipsoid inside the Bloch ball can be the image of a completely positive map. We are not so interested in the orientation of the ellipsoid, so as our canonical form of the affine map we choose

$$\vec{\tau}' = \eta\vec{\tau} + \vec{\kappa}. \tag{10.81}$$

It is straightforward to work out the superoperator Φ of the map. Reshuffling this according to (10.41) we obtain the dynamical matrix

$$D = \frac{1}{2} \begin{bmatrix} 1 + \eta_z + \kappa_z & 0 & \kappa_x + i\kappa_y & \eta_x + \eta_y \\ 0 & 1 - \eta_z + \kappa_z & \eta_x - \eta_y & \kappa_x + i\kappa_y \\ \kappa_x - i\kappa_y & \eta_x - \eta_y & 1 - \eta_z - \kappa_z & 0 \\ \eta_x + \eta_y & \kappa_x - i\kappa_y & 0 & 1 + \eta_z - \kappa_z \end{bmatrix}. \tag{10.82}$$

Adding two diagonal blocks of D we see that $\text{Tr}_A D = \mathbb{1}$, as required. But the parameters $\vec{\eta}$ and $\vec{\kappa}$ must now be chosen so that D is positive definite, otherwise the transformation is not completely positive.

We will study the simple case when $\vec{\kappa} = 0$, in which case the totally mixed state is invariant and the map is unital (bistochastic). Then the matrix D splits into two blocks and its eigenvalues are

$$d_{0,3} = \frac{1}{2}[1 + \eta_z \pm (\eta_x + \eta_y)] \quad \text{and} \quad d_{1,2} = \frac{1}{2}[1 - \eta_z \pm (\eta_x - \eta_y)]. \tag{10.83}$$

Hence, if the *Fujiwara–Algoet conditions* [324],

$$(1 \pm \eta_z)^2 \geq (\eta_x \pm \eta_y)^2, \tag{10.84}$$

hold, the dynamical matrix is positive definite and the corresponding positive map $\Phi_{\vec{\eta}}$ is CP. There are four inequalities: they define a convex polytope and indeed a regular tetrahedron whose extreme points are $\vec{\eta} = (1, 1, 1)$, $(1, -1, -1)$, $(-1, 1, -1)$, $(-1, -1, 1)$. All maps within the cube defined by $|\eta_i| \leq 1$ are positive, so the tetrahedron of completely positive unital maps is a proper subset $\mathcal{B}_2$ of the set of all positive unital maps.

Note that dynamical matrices of unital maps of the form (10.82) commute. In effect, if we think of dynamical matrices as rescaled density matrices, our tetrahedron can be regarded as an eigenvalue simplex in $\mathcal{M}^{(4)}$. The eigenvectors consist of the identity $\sigma_0 = \mathbb{1}_2$ and the three Pauli matrices. Our conclusion is that any map

Table 10.4 *Some one-qubit channels: distortion vector $\vec{\eta}$, translation vector $\vec{\kappa}$ equal to zero for unital channels, Kraus spectrum $\vec{d}$, and Kraus rank r.*

Channels	$\vec{\eta}$	$\vec{\kappa}$	Unital	$\vec{d}$	r
Rotation	$(1,1,1)$	$(0,0,0)$	yes	$(2,0,0,0)$	1
Phase flip	$(1-p,1-p,1)$	$(0,0,0)$	yes	$(2-p,p,0,0)$	2
Decaying	$(\sqrt{1-p},\sqrt{1-p},1-p)$	$(0,0,p)$	no	$(2-p,p,0,0)$	2
Depolarizing	$[1-x](1,1,1)$	$(0,0,0)$	yes	$\frac{1}{2}(4-3x,x,x,x)$	4
Linear	$(0,0,q)$	$(0,0,0)$	yes	$\frac{1}{2}(1+q,1-q,$ $1-q,1+q)$	4
Planar	$(0,s,q)$	$(0,0,0)$	yes	$\frac{1}{2}(1+q+s,1-q-s,$ $1-q+s,1+q-s)$	4

$\Phi \in \mathcal{B}_2$ can be brought by means of unitary rotations (10.69) into the **canonical form of one-qubit bistochastic maps**:

$$\rho \rightarrow \rho' = \frac{1}{2}\sum_{i=0}^{3} d_i\,\sigma_i\rho\,\sigma_i \quad \text{with} \quad \sum_{i=0}^{3} d_i = 2. \quad (10.85)$$

This explains the name *Pauli channels*. The factor of $1/2$ compensates the normalisation of the Pauli matrices, $\mathrm{Tr}\,\sigma_i^2 = 2$. The Kraus operators are $A_i = \sqrt{d_i/2}\,\sigma_i$. For the Pauli matrices $\sigma_j = -i\exp(i\pi\sigma_j/2)$ and the overall phase is not relevant, so the extreme points that they represent are rotations of the Bloch ball around the corresponding axis by the angle π. This confirms that the set of binary bistochastic channels is the convex hull of the unitary operations, which is no longer true when $N > 2$.

For concreteness let us distinguish some one-qubit channels; the following list should be read in conjunction with Table 10.4, which gives the distortion vector $\vec{\eta}$ and the Kraus spectrum $\vec{d}$ for each map. Figure 10.4 illustrates the action of the maps.

Unital channels (with $\vec{\kappa} = 0$):

- *Identity* which is our canonical form of a unitary rotation.
- *Phase flip* (or *phase-damping channel*), $\vec{\eta} = (1 - 2p, 1 - 2p, 1)$. This channel turns the Bloch ball into an ellipsoid touching the Bloch sphere at the north and south poles. When $p = 1/2$ the image degenerates to a line. The analogous channel with $\vec{\eta} = (1, 1 - 2p, 1 - 2p)$ is called a *bit flip*, while the channel with

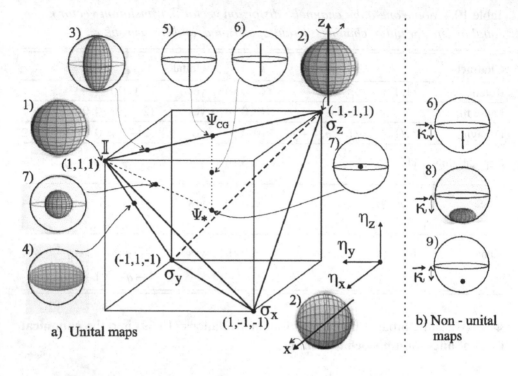

Figure 10.4 One-qubit maps: (a) unital Pauli channels: (1) identity, (2) rotation, (3) phase flip, (4) bit-phase flip, (5) coarse graining, (6) linear channel, (7) depolarizing channel; (b) non-unital maps ($\kappa \neq 0$): (6) linear channel, (8) decaying channel, (9) complete contraction.

$\vec{\eta} = (1 - 2p, 1, 1 - 2p)$ is called a *bit-phase flip*. To understand these names we observe that, with probability p, a bit flip exchanges the states $|0\rangle$ and $|1\rangle$.

- *Linear channel*, $\vec{\eta} = (0, 0, q)$. This sends the entire Bloch ball into a line segment of length $2q$. For $q = 0$ and $q = 1$ we arrive at the *completely depolarizing channel* Ψ_* and the *coarse graining* operation, $\Psi_{CG}(\rho) = \text{diag}(\rho)$, respectively.
- *Planar channel*, $\vec{\eta} = (0, s, q)$, sends the Bloch ball into an ellipse with semi-axis s and q. Complete positivity requires $s \leq 1 - q$.
- *Depolarizing channel*, $\vec{\eta} = [1 - x](1, 1, 1)$. This simply shrinks the Bloch ball. When $x = 1$ we again arrive at the centre of the tetrahedron, that is at the completely depolarizing channel. Note that the Kraus spectrum has a triple degeneracy, so there is an extra freedom in choosing the canonical Kraus operators.

If we drop the condition that the map be unital our canonical form gives a six-dimensional set; it is again a convex set but considerably more difficult to analyse.[22] A map of $\mathcal{CP}_2$, the canonical form of which consists of two Kraus

---

[22] A complete treatment of the qubit case was given by Ruskai et al. [785]. For a compact generalization of the Fujiwara–Algoet conditions see Braun et al. [159].

operators, is either extremal or bistochastic (if $A_1^\dagger A_1 \sim A_2^\dagger A_2 \sim \mathbb{1}$). Table 10.4 gives one example of a non-unital channel, namely the decaying channel.

- *Decaying channel* (also called *amplitude-damping channel*), defined by $\vec{\eta} = (\sqrt{1-p}, \sqrt{1-p}, 1-p)$ and $\vec{\kappa} = (0, 0, p)$. The Kraus operators are

$$A_1 = \begin{bmatrix} 1 & 0 \\ 0 & \sqrt{1-p} \end{bmatrix} \quad \text{and} \quad A_2 = \begin{bmatrix} 0 & \sqrt{p} \\ 0 & 0 \end{bmatrix}. \tag{10.86}$$

Physically this is an important channel – and it exemplifies that a quantum operation can take a mixed state to a pure state.

Problems

10.1 Prove Naimark's theorem.

10.2 A map Φ is called *diagonal* if all Kraus operators A_i mutually commute, so in a certain basis they are diagonal, $\vec{d}^i = \text{diag}(U^\dagger A_i U)$. Show that such a dynamics is given by the Hadamard product, from Problem 9.3, $\Phi(\tilde{\rho}) = H \circ \tilde{\rho}$, where $H_{mn} = \sum_{i=1}^k d_m^i \bar{d}_n^i$ with $m, n = 1, \ldots, N$ while $\tilde{\rho} = U^\dagger \rho\, U$ [572, 413].

10.3 Let ρ be an arbitrary density operator acting on $\mathcal{H}_N$ and $\{A_i\}_{i=1}^K$ be a set of Kraus operators representing a given CP map Φ in its canonical Kraus form. Prove that the matrix

$$\sigma'_{ij} \equiv \langle A_i | \rho | A_j \rangle = \text{Tr} \rho A_j A_i^\dagger \tag{10.87}$$

forms a state acting on an extended Hilbert space, $\mathcal{H}_{N^2}$. Show that, if $\rho = \mathbb{1}/N$, then this state is proportional to the dynamical matrix represented in its eigenbasis.

10.4 Show that a binary, unital map $\Phi_{\eta_x, \eta_y, \eta_z}$ defined by (10.82) transforms any density matrix ρ in the following way [531]:

$$\Phi_{\eta_x, \eta_y, \eta_z} \begin{bmatrix} a & z \\ \bar{z} & 1-a \end{bmatrix} = \frac{1}{2} \begin{bmatrix} 1 + (2a-1)\eta_z & (z+\bar{z})\eta_x + (z-\bar{z})\eta_y \\ (z+\bar{z})\eta_x - (z-\bar{z})\eta_y & 1 - (2a-1)\eta_z \end{bmatrix}.$$

10.5 What qubit channel is described by the Kraus operators

$$A_1 = \begin{bmatrix} 1 & 0 \\ 0 & \sqrt{1-p} \end{bmatrix} \quad \text{and} \quad A_2 = \begin{bmatrix} 0 & 0 \\ 0 & \sqrt{p} \end{bmatrix}? \tag{10.88}$$

10.6 Let $\rho \in \mathcal{M}^{(N)}$. Show that the operation Φ_ρ defined by $D_\Phi \equiv \rho \otimes \mathbb{1}_N$ acts as a complete one-step contraction, $\Phi_\rho(\sigma) = \rho$ for any $\sigma \in \mathcal{M}^{(N)}$.

11

Duality: maps versus states

Good mathematicians see analogies. Great mathematicians see analogies between analogies.

—*Stefan Banach*[1]

We have already discussed the static structure of our QuantumTown – the set of density matrices – on the one hand, and the set of all physically realizable processes which may occur in it on the other hand. Now we are going to reveal a quite remarkable property: the set of all possible ways to travel in the QuantumTown is equivalent to a QuantumCountry – an appropriately magnified copy of the initial QuantumTown! More precisely, the set of all transformations which map the set of density matrices of size N into itself (dynamics) is identical with a subset of the set of density matrices of size N^2 (kinematics). From a mathematical point of view this relation is based on the *Jamiołkowski isomorphism*, analysed later in this chapter. Before discussing this intriguing duality, let us leave the friendly set of quantum operations and pay a short visit to an as yet unexplored, neighbouring land of maps, which are positive but not completely positive.

11.1 Positive and decomposable maps

Quantum transformations which describe physical processes are represented by completely positive maps. Why should we care about maps which are not completely positive? On the one hand it is instructive to realise that seemingly innocent transformations are not CP, and thus do not correspond to any physical process. On the other hand, as discussed in Chapter 16, positive but not completely positive maps provide a crucial tool in the investigation of quantum entanglement.

Consider the transposition of a density matrix in a fixed basis, $T : \rho \to \rho^T$. (Since ρ is Hermitian this is equivalent to complex conjugation.) The superoperator

---

[1] Reproduced from S. M. Ulam. *Adventures of a Mathematician*. University of California Press, 1991.

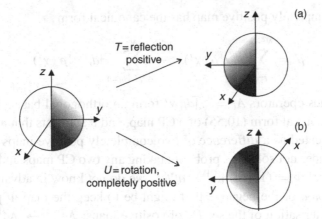

Figure 11.1 Non-contracting transformations of the Bloch ball: (a) transposition (reflection with respect to the $x - z$ plane), not completely positive; (b) rotation by π around z axis, completely positive.

entering (10.36) is the SWAP operator, $(T)_{m\mu \atop n\nu} = \delta_{m\nu}\delta_{n\mu} = S_{m\mu \atop n\nu}$. Hence it is symmetric with respect to reshuffling, $T = T^R = D_T$. This permutation matrix contains N diagonal entries equal to unity and $N(N-1)/2$ blocks of size two. Thus its spectrum consists of $N(N+1)/2$ eigenvalues equal to unity and $N(N-1)/2$ eigenvalues equal to -1, consistent with the constraint $\text{Tr}D = N$. The matrix D_T is not positive, so the transposition T is not completely positive. Another way to reach this conclusion is to act with the extended map of partial transposition on the maximally entangled state (11.21) and to check that $[T \otimes \mathbb{1}](|\phi^+\rangle\langle\phi^+|)$ has negative eigenvalues.

The transposition of an N-dimensional Hermitian matrix changes the signs of the imaginary part of the elements D_{ij}. This is a reflection in an $N(N+1)/2 - 1$ dimensional hyperplane. As shown in Figure 11.1 this is simple to visualise for $N = 2$: when we use the representation (5.8) the transposition reflects the Bloch ball in the (x, z) plane. Note that a unitary rotation of the Bloch ball around the z-axis by the angle π also exchanges the 'western' and the 'eastern' hemispheres, but is completely positive.

As discussed in Section 10.3, a map fails to be CP if its dynamical matrix D contains at least one negative eigenvalue. Let $m \geq 1$ denote the number of the negative eigenvalues (in short, the *neg rank*[2] of D). Ordering the spectrum of D decreasingly allows us to rewrite its spectral decomposition

$$D = \sum_{i=1}^{N^2-m} d_i |\chi_i\rangle\langle\chi_i| - \sum_{i=N^2-m+1}^{N^2} |d_i| \, |\chi_i\rangle\langle\chi_i|. \tag{11.1}$$

[2] For transposition, the neg rank of D_T is $m = N(N-1)/2$.

Thus a not completely positive map has the canonical form

$$\rho' = \sum_{i=1}^{N^2-m} d_i \chi^i \rho (\chi^i)^\dagger - \sum_{i=N^2-m+1}^{N^2} |d_i| \chi^i \rho (\chi^i)^\dagger, \tag{11.2}$$

where the Kraus operators $A_i = \sqrt{|d_i|}\, \chi^i$ form an orthogonal basis. This is analogous to the canonical form (10.55) of a CP map, and it suggests that a positive map may be represented as a *difference* of two completely positive maps [868]. While this is true, it does not solve the problem: taking any two CP maps and constructing a quasi-mixture[3] $\Phi = (1+a)\Psi_1^{CP} - a\Psi_2^{CP}$, we do not know in advance how large the contribution a of the negative part might be to keep the map Φ positive...[4] In fact the characterisation of the set $\mathcal{P}_N$ of positive maps: $\mathcal{M}^{(N)} \to \mathcal{M}^{(N)}$ for $N > 2$ is by far not simple.[5] By definition, $\mathcal{P}_N$ contains the set CP_N of all CP maps as a proper subset. To learn more about the set of positive maps we will need some other features of the operation of transposition T. For any operation Φ the modifications of the dynamical matrix induced by a composition with T may be described by the transformation of partial transpose (see Table 10.1),

$$T\Phi = \Phi^{S_1}, \quad D_{T\Phi} = D_\Phi^{T_A}, \quad \text{and} \quad \Phi T = \Phi^{S_2}, \quad D_{\Phi T} = D_\Phi^{T_B}. \tag{11.3}$$

To demonstrate this it is enough to use the explicit form of Φ_T and the observation that

$$D_{\Psi\Phi} = [D_\Psi^R D_\Phi^R]^R. \tag{11.4}$$

Positivity of $D_{\Psi\Phi}$ follows also from the fact that the composition of two CP maps is completely positive. This follows directly from the identity $(\Psi\Phi) \otimes \mathbb{1} = (\Psi \otimes \mathbb{1}) \cdot (\Phi \otimes \mathbb{1})$ and implies the

Reshuffling lemma. *Consider two Hermitian matrices A and B of the same size KN.*

$$\textit{If} \quad A \geq 0 \quad \textit{and} \quad B \geq 0 \quad \textit{then} \quad (A^R B^R)^R \geq 0. \tag{11.5}$$

For a proof [412] see Problem 11.1.

Sandwiching Φ between two transpositions does not influence the spectrum of the dynamical matrix, $T\Phi T = \Phi^S = \Phi^*$ and $D_{T\Phi T} = D_\Phi^T = D_\Phi^*$. Thus if Φ is a CP map, so is $T\Phi T$ (if D_Φ is positive so is D_Φ^T). See Figure 11.2.

---

[3] The word *quasi* is used here to emphasise that some weights are negative.

[4] Although some criteria for positivity are known [861, 495, 637], they do not lead to a practical test of positivity. A technique of extending the system (and the map) a certain number of times gives a constructive test for positivity for a large class of maps [270].

[5] This issue was a subject of mathematical interest many years ago [861, 216, 985] and more recently [566, 638].

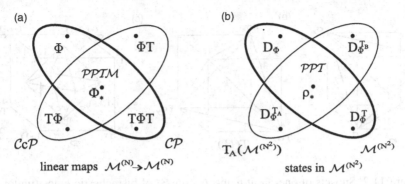

(a) linear maps $\mathcal{M}^{(N)} \to \mathcal{M}^{(N)}$

(b) states in $\mathcal{M}^{(N^2)}$

Figure 11.2 (a) The set of CP maps, its image with respect to transposition $CcP = T(CP)$, and their intersection $PPTM \equiv CP \cap CcP$; (b) the isomorphic sets $\mathcal{M}^{(N)}$ of quantum states (dynamical matrices), its image $T_A(\mathcal{M}^{(N)})$ under the action of partial transposition, and the set of states with positive partial transpose (PPT).

The not completely positive transposition map T allows one to introduce the following definition [861, 217, 219]:

A map Φ is called **completely co-positive** *(CcP), if the map $T\Phi$ is CP.*

Properties (11.3) of the dynamical matrix imply that the map ΦT could be used instead to define the same set of CcP maps. Thus any CcP map Φ may be written in a Kraus-like form

$$\rho' = \Phi(\rho) = \sum_{i=1}^{k} A_i \rho^T A_i^\dagger. \tag{11.6}$$

Moreover, as shown in Figure 11.2, the set CcP may be understood as the image of CP with respect to the transposition. Since we have already identified the transposition with a reflection, it is rather intuitive to observe that the set CcP is a twin copy of CP with the same shape and volume. This property is easiest to analyse for the set $\mathcal{B}_2$ of one-qubit bistochastic maps [700], written in the canonical form (10.85). Then the dual set of CcP unital one qubit maps, $T(\mathcal{B}_2)$, forms a tetrahedron spanned by four maps $T\sigma_i$ for $i = 0, 1, 2, 3$. This is the reflection of the set of bistochastic maps with respect to its centre – the completely depolarizing channel Φ_*. See Figure 11.3b. Observe that the corners of $\mathcal{B}_2$ are formed by proper rotations while the extreme points of the set of CcP maps represent reflections. The intersection of the tetrahedra forms an octahedron of positive partial transpose (PPT) preserving maps ($PPTM$).

A positive map Φ is called *decomposable* if it can be expressed as a convex combination of a CP map and a CcP map,

$$\Phi = a\Phi_{CP} + (1 - a)\Phi_{CcP} \quad \text{with} \quad a \in [0, 1]. \tag{11.7}$$

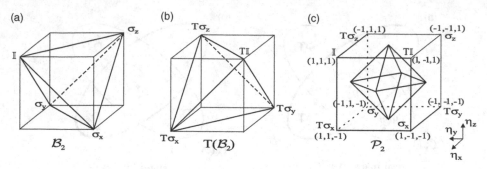

Figure 11.3 Subsets of one-qubit maps: (a) set B_2 of bistochastic maps (unital & CP), (b) set $T(B_2)$ of unital & CcP maps, (c) set of positive (decomposable) unital maps. The intersection of the two tetrahedra forms an octahedron of super-positive maps.

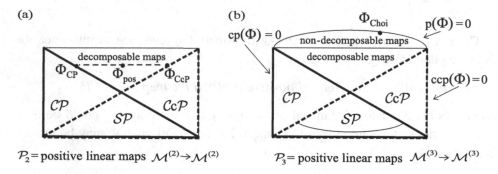

Figure 11.4 Sketch of the set positive maps: (a) for $N = 2$ all maps are decomposable so $SP = CP \cap CcP = PPTM$, (b) for $N > 2$ there exist non-decomposable maps and $SP \subset CP \cap CcP$ – see section 11.2.

A relation between CP maps acting on quaternion matrices and the decomposable maps defined on complex matrices was found by Kossakowski [554]. An important characterisation of the set P_2 of positive maps acting on (complex) states of one qubit follows from [861, 984]

Størmer–Woronowicz theorem. *Every one-qubit positive map* $\Psi \in P_2$ *is decomposable.*

In other words, the set of $N = 2$ positive maps can be represented by the convex hull of the set of CP and CcP maps. This property is illustrated for unital maps (in canonical form) in Figure 11.3, where the cube of positive maps forms the convex hull of the two tetrahedra, and schematically in Figure 11.4a. It holds also for the maps $\mathcal{M}^{(2)} \to \mathcal{M}^{(3)}$ and $\mathcal{M}^{(3)} \to \mathcal{M}^{(2)}$ [985], but is not true in higher dimensions, see Figure 11.4b.

Consider a map defined on $\mathcal{M}^{(3)}$, depending on three non-negative parameters,

$$\Psi_{a,b,c}(\rho) = \begin{bmatrix} a\rho_{11} + b\rho_{22} + c\rho_{33} & 0 & 0 \\ 0 & c\rho_{11} + a\rho_{22} + b\rho_{33} & 0 \\ 0 & 0 & b\rho_{11} + c\rho_{22} + a\rho_{33} \end{bmatrix} - \rho. \tag{11.8}$$

The map $\Psi_{2,0,2} \in \mathcal{P}_3$ was a first example of an indecomposable map, found by Choi in 1975 [218]. As denoted schematically in Figure 11.4b this map is extremal and belongs to the boundary of the convex set $\mathcal{P}_3$. The *Choi map* was generalised in [220] and in [215], where it was shown that the map (11.8) is positive if and only if

$$a \geq 1, \quad a + b + c \geq 3, \quad 1 \leq a \leq 2 \Longrightarrow bc \geq (2 - a)^2, \tag{11.9}$$

while it is decomposable if and only if

$$a \geq 1, \quad 1 \leq a \leq 3 \Longrightarrow bc \geq (3 - a)^2/4. \tag{11.10}$$

In particular, $\Psi_{2,0,c}$ is positive but not decomposable for $c \geq 1$. All generalised indecomposable Choi maps are known to be *atomic* [386], that is they cannot be written as a convex sum of 2-positive and 2-co-positive maps [887]. Examples of indecomposable maps belonging to $\mathcal{P}_4$ were given in [985] and [775]. A family of indecomposable maps for an arbitrary finite dimension $N \geq 3$ was found by Kossakowski [555]. They consist of an affine contraction of the set $\mathcal{M}^{(N)}$ of density matrices into a ball inscribed in it, followed by a generic rotation from $O(N^2 - 1)$. Although several other methods of construction of indecomposable maps were proposed [888, 887, 704, 526], some of them in the context of quantum entanglement [892, 389], the general problem of describing them remains open. For large dimensions a typical positive map becomes indecomposable, while a typical decomposable map is not completely positive [879].

Analysing the set $\mathcal{P}_N$ of positive maps acting on N dimensional states one may ask whether there exists a finite collection of K positive maps $\{\Psi_j\}$, such that $\mathcal{P}_N = $ conv hull $\left(\cup_{j=1}^{K} \Psi_j(\mathcal{CP}_N)\right)$. Due to the theorem of Størmer and Woronowicz the answer is known for $N = 2$, for which $K = 2$, $\Psi_1 = \mathbb{1}$ and $\Psi_2 = T$. As we shall see in Chapter 16 these properties of the set $\mathcal{P}_2$ are decisive for the separability problem: the separability criterion based on positivity of $(\mathbb{1} \otimes T)\rho$ is conclusive for the system of two qubits [455]. It happens that the case of a qutrit, $N = 3$, is entirely different: as shown by Skowronek [841] the cone of positive maps $\mathcal{P}_3$ cannot be described with the help of a finite number of positive maps $\{\Psi_j\}$. This implies that it is not possible to characterise the set of separable states of the two-qutrit system, by means of any finite number of separability criteria, $(\mathbb{1} \otimes \Psi_j)\rho \geq 0$, with $j = 1, \ldots, K$.

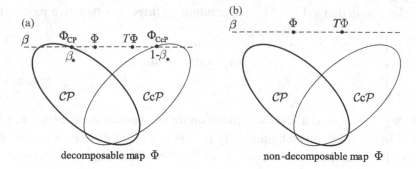

Figure 11.5 Geometric criterion to verify decomposability of a map Φ: (a) if the line passing through Φ and ΦT crosses the set of completely positive maps, a decomposition of Φ is explicitly constructed.

Indecomposable maps are worth investigating, since each such map provides a criterion for separability (Section 16.6). Conditions for a positive map Φ to be decomposable were found some time ago by Størmer [862]. Since this criterion is not a constructive one, we describe here a simple test. Assume first that the map is not symmetric with respect to the transposition,[6] $\Phi \neq T\Phi$. These two points determine a line in the space of maps, parametrized by β, along which we check if the dynamical matrix

$$D_{\beta\Phi + (1-\beta)T\Phi} = \beta D_\Phi + (1 - \beta)D_\Phi^{T_A} \qquad (11.11)$$

is positive. If it is found to be positive for some $\beta_* < 0$ (or $\beta_* > 1$) then the line (11.11) crosses the set of completely positive maps (see Figure 11.5a). Since $D(\beta_*)$ represents a CP map Ψ_{CP}, then $D(1 - \beta_*)$ defines a completely co-positive map Ψ_{CcP}, and we find an explicit decomposition, $\Phi = [-\beta_* \Psi_{CP} + (1 - \beta_*)\Psi_{CcP}]/(1 - 2\beta_*)$. In this way decomposability of Φ may be established, but one cannot confirm that a given map is indecomposable.

To study the geometry of the set of positive maps one may work with the Hilbert–Schmidt (HS) distance $d(\Psi, \Phi) = ||\Psi - \Phi||_{HS}$. Since reshuffling of a matrix does not influence its HS norm, the distance can be measured directly in the space of dynamical matrices, $d(\Psi, \Phi) = D_{HS}(D_\Psi, D_\Phi)$. Note that for unital one-qubit maps, (10.82) with $\vec{\kappa} = 0$, one has $d(\Phi_1, \Phi_2) = |\vec{\eta}_1 - \vec{\eta}_2|$, so Figure 11.3 represents correctly the HS geometry of the space of $N = 2$ unital maps.

In order to characterise to what extent a given map is close to the boundary of the set of positive (CP or CcP) maps, let us define the quantities

---

[6] If this is the case, one may perform this procedure with a perturbed map, $\Phi' = (1 - \epsilon)\Phi + \epsilon\Psi$, for which $\Phi' \neq T\Phi'$, and study the limit $\epsilon \to 0$.

    (a)   **complete positivity,** $cp(\Phi) \equiv \min_{\mathcal{M}^{(N)}} \langle \rho | D_\Phi | \rho \rangle,$ (11.12)

    (b)   **complete co-positivity,** $ccp(\Phi) \equiv \min_{\mathcal{M}^{(N)}} \langle \rho | D_\Phi^{T_A} | \rho \rangle,$ (11.13)

    (c)   **positivity,** $p(\Phi) \equiv \min_{|x\rangle, |y\rangle \in \mathbb{C}P^{N-1}} [\langle x \otimes y | D_\Phi | x \otimes y \rangle].$ (11.14)

The first two quantities are easily found by diagonalisation, since $cp(\Phi) = \min \{\text{eig}(D_\Phi)\}$ and $ccp(\Phi) = \min\{\text{eig}(D_\Phi^{T_A})\}$. Although $p(\Phi) \geq cp(\Phi)$ by construction,[7] the evaluation of positivity is more involved, since one needs to perform the minimisation over the space of all product states, that is the Cartesian product $\mathbb{C}P^{N-1} \times \mathbb{C}P^{N-1}$. No straightforward method of computing this minimum is known, so one has to rely on numerical minimisation.[8]

A given map Φ is completely positive (CcP, positive) if and only if the complete positivity (ccp, positivity) is non-negative. As marked in Figure 11.4b, the relation $cp(\Phi) = 0$ defines the boundary of the set $\mathcal{CP}_N$, while $ccp(\Phi) = 0$ and $p(\Phi) = 0$ define the boundaries of $\mathcal{CcP}_N$ and $\mathcal{P}_N$. By direct diagonalisation of the dynamical matrix we find that $cp(\mathbb{1}) = ccp(T) = 0$ and $ccp(\mathbb{1}) = cp(T) = -1$.

For any not completely positive map Φ_{nCP} one may look for its best approximation with a physically realizable[9] CP map Φ_{CP}, that is by minimizing their HS distance $d(\Phi_{\text{nCP}}, \Phi_{\text{CP}})$ – see Figure 11.8a. To see a simple application of complete positivity, consider a non-physical positive map with $cp(\Phi_{\text{nCP}}) = -x < 0$. One – in general not optimal – CP approximation may be constructed out of its convex combination with the completely depolarizing channel Ψ_*. Diagonalizing the dynamical matrix representing the map $\Psi_x = a\Phi_{\text{nCP}} + (1-a)\Psi_*$ with $a = 1/(Nx+1)$ we see that its smallest eigenvalue is equal to zero, so Ψ_x belongs to the boundary of $\mathcal{CP}_N$. Hence the distance $d(\Phi_{\text{nCP}}, \Phi_x)$, which is a function of the complete positivity $cp(\Phi_{\text{nCP}})$, gives an upper bound for the distance of Φ_{nCP} from the set $\mathcal{CP}$. In a similar way one may use $ccp(\Phi)$ to obtain an upper bound for the distance of an analysed non-CcP map Φ_{nCcP} from the set $\mathcal{CcP}$ – compare Problem 11.6. As discussed in further sections and, in more detail, in Chapter 16, the solution of the analogous problem in the space of density matrices allows one to characterise the entanglement of a two-qubit mixed state ρ by its minimal distance to the set of separable states.

---

[7] Both quantities are equal if D is a product matrix, which occurs if only one singular value ξ of the superoperator $L = D^R$ is positive.

[8] In certain cases this quantity was estimated analytically by Terhal [892] and numerically by Gühne et al. [371] when characterising entanglement witnesses.

[9] Such *structural physical approximations* were introduced in [465] to propose an experimentally feasible scheme of entanglement detection and later studied in [314].

11.2 Dual cones and super-positive maps

Since a CP map Φ is represented by a positive dynamical matrix D_Φ, the trace $\mathrm{Tr} P D_\Phi$ is non-negative for any projection operator P. Furthermore, for any two CP maps, the HS scalar product of their dynamical matrices satisfies $\mathrm{Tr} D_\Phi D_\Psi \geq 0$. If such a relation is fulfilled for any CP map Ψ, it implies complete positivity of Φ. More formally, we define a pairing between maps,

$$(\Phi, \Psi) \equiv \langle D_\Phi, D_\Psi \rangle = \mathrm{Tr}\, D_\Phi^\dagger D_\Psi = \mathrm{Tr}\, D_\Phi D_\Psi, \tag{11.15}$$

and obtain the following characterisation of the set $\mathcal{CP}$ of CP maps,

$$\{\Phi \in \mathcal{CP}\} \Leftrightarrow (\Phi, \Psi) \geq 0 \quad \text{for all} \quad \Psi \in \mathcal{CP}. \tag{11.16}$$

This property is illustrated in Figure 11.6a – the angle formed by any two positive dynamical matrices at the zero map 0 will not be greater than $\pi/2$. Thus the set of CP maps is self dual and is represented as a right-angled cone. All trace preserving maps belong to the horizontal line given by the condition $\mathrm{Tr} D_\Phi = N$.

In a similar way one may define the cone dual to the set $\mathcal{P}$ of positive maps. *A linear map* $\Phi : \mathcal{M}^{(N)} \to \mathcal{M}^{(N)}$ *is called* **super-positive**[10] (SP) [39], *if*

$$\{\Phi \in \mathcal{SP}\} \Leftrightarrow (\Phi, \Psi) \geq 0 \quad \text{for all} \quad \Psi \in \mathcal{P}. \tag{11.17}$$

Once $\mathcal{SP}$ is defined as a set containing all SP maps, one may write a dual condition to characterise the set of positive maps,

$$\{\Phi \in \mathcal{P}\} \Leftrightarrow (\Phi, \Psi) \geq 0 \quad \text{for all} \quad \Psi \in \mathcal{SP}. \tag{11.18}$$

The cones $\mathcal{SP}$ and $\mathcal{P}$ are dual by construction and any boundary line of $\mathcal{P}$ determines the perpendicular (dashed) boundary line of $\mathcal{SP}$. The self dual set of CP maps is included in the set of positive maps $\mathcal{P}$, and includes the dual set of super-positive maps. All the three sets are convex. See Figure 11.6. For large dimensions the relative volume of each of these subsets tends to zero [879].

The dynamical matrix of a positive map Ψ is block positive, so it is clear that condition (11.17) implies that a map Φ is super-positive if its dynamical matrix admits a tensor product representation

$$D_\Phi = \sum_i^k A_i \otimes B_i, \quad \text{with} \quad A_i \geq 0, \quad B_i \geq 0; \quad i = 1, \dots, k. \tag{11.19}$$

As we shall see in Chapter 16 this very condition is related to the separability of the state ρ associated with $D_\Phi \in \mathcal{CP}$. In particular, if the angle α between the

---

[10] SP maps are also called *entanglement breaking channels* (see Section 16.6).

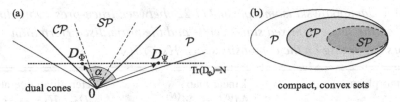

Figure 11.6 (a) Dual cones $\mathcal{P} \leftrightarrow \mathcal{SP}$ and self dual $\mathcal{CP} \leftrightarrow \mathcal{CP}$; (b) the corresponding compact sets of trace preserving positive, completely positive and super-positive maps, $\mathcal{P} \supset \mathcal{CP} \supset \mathcal{SP}$.

vectors pointing to D_Φ and a block positive D_Ψ is obtuse, the state $\rho = D_\Phi/N$ is entangled–compare the notion of entanglement witness in Section 16.6.

In general it is not easy to describe the set $\mathcal{SP}$ explicitly. The situation simplifies for one-qubit maps: due to the theorem by Størmer and Woronowicz any positive map may be represented as a convex combination of a CP map and a CcP map. The sets $\mathcal{CP}$ and $\mathcal{CcP}$ share similar properties and are both self dual. Hence a map Φ is super-positive if it is simultaneously CP and CcP. But the neat relation $\mathcal{SP} = \mathcal{CP} \cap \mathcal{CcP}$ holds for $N = 2$ only. For $N > 2$ the set $\mathcal{P}$ of positive maps is larger (there exist non-decomposable maps), so the dual set becomes smaller, $\mathcal{SP} \subset \mathcal{CP} \cap \mathcal{CcP} \equiv \mathcal{PPTM}$. See Figure 11.4.

11.3 Jamiołkowski isomorphism

Let $\mathcal{CP}_N$ denote the convex set of all trace preserving, completely positive maps $\Phi : \mathcal{M}^{(N)} \to \mathcal{M}^{(N)}$. Any such map may be uniquely represented by its dynamical matrix D_Φ of size N^2. It is a positive, Hermitian matrix and its trace is equal to N. Hence the rescaled matrix $\rho_\Phi \equiv D_\Phi/N$ represents a mixed state in $\mathcal{M}^{(N^2)}$. In fact rescaled dynamical matrices form only a subspace of this set, determined by the trace preserving conditions (10.46), which impose N^2 constraints. Let us denote this $N^4 - N^2$ dimensional set by $\mathcal{M}_1^{(N^2)}$. Since any trace preserving CP map has a dynamical matrix, and vice versa, the correspondence between maps in $\mathcal{CP}_N$ and states in $\mathcal{M}_1^{(N^2)}$ is one to one. In Table 11.1 this isomorphism is labelled J_{II}.

Let us find the dynamical matrix for the identity operator:

$$\mathbb{1}_{\substack{m\mu \\ n\nu}} = \delta_{mn}\delta_{\mu\nu} \quad \text{so that} \quad D^{\mathbb{1}}_{\substack{m\mu \\ n\nu}} = (\mathbb{1}_{\substack{m\mu \\ n\nu}})^R = \delta_{m\mu}\delta_{n\nu} = N\rho^{\phi}_{\substack{m\mu \\ n\nu}}, \quad (11.20)$$

where $\rho^\phi = |\phi^+\rangle\langle\phi^+|$ represents the operator of projection on a maximally entangled state of the composite system, viz.

$$|\phi^+\rangle = \frac{1}{\sqrt{N}} \sum_{i=1}^{N} |i\rangle \otimes |i\rangle. \quad (11.21)$$

Table 11.1 *Jamiołkowski isomorphism (11.22) between trace-preserving, linear maps Φ on the space of mixed states $\mathcal{M}^{(N)}$ and the normalised Hermitian operators D_Φ acting on the composite space $\mathcal{H}_N \otimes \mathcal{H}_N$.*

Isomorphism	Linear map $\Phi : \mathcal{M}^{(N)} \to \mathcal{M}^{(N)}$	Hermitian operators $D_\Phi : \mathcal{H}_{N^2} \to \mathcal{H}_{N^2}$
J_I	set $\mathcal{P}$ of positive maps Φ	block positive operators D
J_{II}	set $\mathcal{CP}$ of completely positive Φ	positive operators D: subset $\mathcal{M}_1$ of quantum states
J_{III}	set $\mathcal{SP}$ of super-positive Φ	subset of separable quantum states
E_{III}	unitary rotations $\Phi(\rho) = U\rho U^\dagger,$ $D_\Phi = (U \otimes U^*)^R$	maximally entangled pure states $(U \otimes \mathbb{1})|\phi^+\rangle$
$N = 2$ example of E_{III}	$\mathbb{1} \leftrightarrow (1, 0, 0, 1)$	$|\phi^+\rangle \equiv \frac{1}{\sqrt{2}}(|00\rangle + |11\rangle)$
Pauli matrices	$\sigma_x \leftrightarrow (0, 1, 1, 0)$	$|\psi^+\rangle \equiv \frac{1}{\sqrt{2}}(|01\rangle + |10\rangle)$
versus	$i\sigma_y \leftrightarrow (0, 1, -1, 0)$	$|\psi^-\rangle \equiv \frac{1}{\sqrt{2}}(|01\rangle - |10\rangle)$
Bell states $\rho_\phi = |\phi\rangle\langle\phi|$	$\sigma_z \leftrightarrow (1, 0, 0, -1)$	$|\phi^-\rangle \equiv \frac{1}{\sqrt{2}}(|00\rangle - |11\rangle)$
J_{IV}	completely depolarizing channel Φ_*	maximally mixed state $\rho_* = \mathbb{1}/N$

This state is written in its Schmidt form (9.8), and we see that all its Schmidt coefficients are equal, $\lambda_1 = \lambda_i = \lambda_N = 1/N$. Thus we have found that the identity operator corresponds to the maximally entangled pure state $|\phi^+\rangle\langle\phi^+|$ of the composite system. Interestingly, this correspondence may be extended for other operations, or in general, for arbitrary linear maps. The **Jamiołkowski isomorphism** [11]

$$\Phi : \mathcal{M}^{(N)} \to \mathcal{M}^{(N)} \qquad \longleftrightarrow \qquad \rho_\Phi \equiv D_\Phi/N = [\Phi \otimes \mathbb{1}](|\phi^+\rangle\langle\phi^+|) \qquad (11.22)$$

allows us to associate a linear map Φ acting on the space of mixed states $\mathcal{M}^{(N)}$ with an operator acting in the enlarged Hilbert state $\mathcal{H}_N \otimes \mathcal{H}_N$ – see Figure 11.7. To show this relation, write the operator $\Phi \otimes \mathbb{1}$ as an eight-indices matrix [12] and study its action on the state ρ^ϕ expressed by two Kronecker's δ as in (11.20),

$$\Phi \begin{smallmatrix} mn \\ m'n' \end{smallmatrix} \mathbb{1} \begin{smallmatrix} \mu\nu \\ \mu'\nu' \end{smallmatrix} \rho^\phi_{\begin{smallmatrix} m'\mu' \\ n'\nu' \end{smallmatrix}} = \frac{1}{N}\Phi \begin{smallmatrix} mn \\ \mu\nu \end{smallmatrix} = \frac{1}{N}D \begin{smallmatrix} m\mu \\ n\nu \end{smallmatrix}. \qquad (11.23)$$

---

[11] This refers to the contribution of Jamiołkowski (1972) [494]. Various aspects of the duality between maps and states were investigated in [412, 56, 234].

[12] An analogous operation $\mathbb{1} \otimes \Phi$ acting on ρ^ϕ leads to the matrix D^S with the same spectrum.

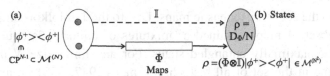

Figure 11.7 Duality (11.22) between a quantum map Φ acting on a part of the maximally entangled state $|\phi^+\rangle$ and the resulting density matrix $\rho = \frac{1}{N} D_\Phi$.

Conversely, for any positive matrix D we find the corresponding map Φ by diagonalisation. The reshaped eigenvectors of D, rescaled by the roots of the eigenvalues, give the canonical Kraus form (10.55) of the operation Φ. If $\mathrm{Tr}_A \rho_\Phi = \mathbb{1}/N$ so that $\rho_\Phi \in \mathcal{M}_1^{(N^2)}$, the map Φ is trace preserving.

Consider now a more general case in which ρ denotes a state acting on a composite Hilbert space $\mathcal{H}_N \otimes \mathcal{H}_N$. Let Φ be an arbitrary map which sends $\mathcal{M}^{(N)}$ into itself and let $D_\Phi = \Phi^R$ denote its dynamical matrix (of size N^2). Acting with the extended map on ρ we find its image $\rho' = [\Phi \otimes \mathbb{1}](\rho)$. Writing down the explicit form of the corresponding linear map in analogy to (11.23) and contracting over four indices which represent $\mathbb{1}$ we obtain

$$(\rho')^R = \Phi \rho^R \quad \text{so that} \quad \rho' = (D_\Phi^R \rho^R)^R. \tag{11.24}$$

In the above formula the standard multiplication of square matrices takes place, in contrast to Eq. (10.36) in which the state ρ acts on a simple Hilbert space and is treated as a vector.

Note that Eq. (11.22) may be obtained as a special case of (11.24) if one takes for ρ the maximally entangled state (11.21), for which $(\rho^\phi)^R = \mathbb{1}/N$. Formula (11.24) provides a useful application of the dynamical matrix corresponding to a map Φ, which acts on a subsystem. Since the normalisation of matrices does not influence positivity, this result implies the reshuffling lemma (11.5).

Formula (11.22) may also be used to find operators D associated with positive maps Φ which are neither trace preserving nor complete positive. The Jamiołkowski isomorphism relates thus the set of positive linear maps with dynamical matrices acting in the composite space and positive on product states. Let us mention explicitly some special cases of this isomorphism, labelled J_I in Table 11.1. The set of completely positive maps Φ is isomorphic to the set of all positive matrices D.

The case J_{II} concerns quantum operations which correspond to quantum states $\rho = D/N$ fulfilling an additional constraint,[13] $\mathrm{Tr}_A D = \mathbb{1}$. States satisfying $\mathrm{Tr}_B D = \mathbb{1}$ correspond to unital maps, while the set of states satisfying both conditions

---

[13] An apparent asymmetry between the role of both subsystems is due to the particular choice of the relation (11.22); if the operator $\mathbb{1} \otimes \Phi$ is used instead, the subsystems A and B need to be interchanged.

corresponds to the set of bistochastic maps. Due to the Jamiołkowski isomorphism (11.22) its subset of maps obtained as mixtures of unitaries is equivalent to the convex hull of maximally entangled states. For any finite size N this set has a positive volume in the set of all bistochastic maps [947, 358], in analogy to the positive volume of the set of separable states – see Section 16.7.

An important case J_{III} of the isomorphism concerning the super-positive maps which for $N = 2$ are isomorphic with the PPT states ($\rho^{T_A} \geq 0$) will be further analysed in Section 16.6, but we are now in a position to comment on item E_{III}. If the map Φ is a unitary rotation, $\rho' = \Phi(\rho) = U\rho U^\dagger$ then (11.22) results in the pure state $(U\otimes\mathbb{1})|\phi^+\rangle$. The local unitary operation $(U\otimes\mathbb{1})$ preserves the purity of a state and its Schmidt coefficients. As shown in Section 16.2 the set of unitary matrices U of size N – or more precisely $SU(N)/Z_N$ – is isomorphic to the set of maximally entangled pure states of the composite $N \times N$ system. In particular, vectors obtained by reshaping the Pauli matrices σ_i represent the Bell states in the computational basis, as listed in Table 11.1. Eventually, case J_{IV} consists of a single, distinguished point in both spaces: the completely depolarizing channel $\Phi_* : \mathcal{M}^{(N)} \to \rho_*^{(N)}$ and the corresponding maximally mixed state $\rho_*^{(N^2)}$.

Table 11.1 deserves one more comment: the key word *duality* may be used here in two different meanings. The 'vertical' duality between its two columns describes the isomorphism between maps and states, while the 'horizontal' duality between the rows J_I and J_{III} follows from the dual cones construction. Note the inclusion relations $J_I \supset J_{II} \supset J_{III} \supset J_{IV}$ and $J_{II} \supset E_{III}$, valid for both columns of Table 11.1 and visualised in Figure 11.8.

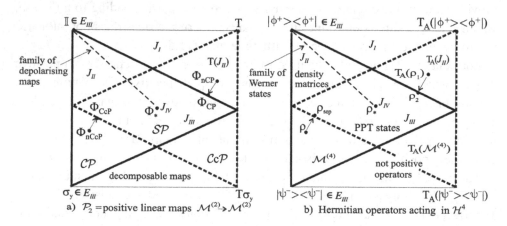

Figure 11.8 Isomorphism between objects, sets, and problems: (a) linear one–qubit maps, (b) linear operators acting in two–qubit Hilbert space $\mathcal{H}_4$. Labels J_i refer to the sets defined in Table 11.1.

11.4 Quantum maps and quantum states

The relation (11.22) links an arbitrary linear map Φ with the corresponding linear operator given by the dynamical matrix D_Φ. Expressing the maximally entangled state $|\phi^+\rangle$ in (11.22) in its Schmidt form (11.21) we may compute the matrix elements of D_Φ in the product basis consisting of the states $|i \otimes j\rangle$. Due to the factorisation of the right-hand side we see that the double sum describing $\rho_\Phi = D_\Phi/N$ drops out and the result reads

$$\langle k \otimes i|D_\Phi|l \otimes j\rangle \;=\; \langle k|\Phi(|i\rangle\langle j|)|l\rangle. \tag{11.25}$$

This equation may also be understood as a definition of a map Φ related to the linear operator D_Φ. Its special case, $k = l$ and $i = j$, proves the isomorphism J_I from Table 11.1: if D_Φ is block positive, then the corresponding map Φ sends positive projection operators $|i\rangle\langle i|$ into positive operators [494].

As listed in Table 11.1 and shown in Figure 11.8 the Jamiołkowski isomorphism (11.25) may be applied in various setups. Relating linear maps from $\mathcal{P}_N$ with operators acting on an extended space $\mathcal{H}_N \otimes \mathcal{H}_N$ we may compare:

1. Individual objects, e.g. completely depolarizing channel Φ_* and the maximally mixed state ρ_*;
2. Families of objects, e.g. the depolarizing channels and generalised Werner states,
3. entire sets, e.g. the set of $\mathcal{CP} \cap \mathcal{C}c\mathcal{P}$ maps and the set of PPT states,
4. entire problems, e.g. finding the SP map closest to a given CP map versus finding the separable state closest to a given state, and
5. their solutions . . .

For a more comprehensive discussion of the issues related to quantum entanglement see Chapter 16. Some general impression may be had by comparing both sides of Figure 11.8, in which a drawing of both spaces is presented. Note that this illustration may also be considered as a strict representation of a fragment of the space of one–qubit unital maps (a) or the space of two-qubits density matrices in the HS geometry (b). It is nothing but the cross-section of the cube representing the positive maps in Figure 11.3 along the plane determined by $\mathbb{1}, T$ and Φ_*.

The maps–states duality is particularly fruitful in investigating *quantum gates*: unitary operations U performed on an N-level quantum system. Since the overall phase is not measurable, we may fix the determinant of U to unity, restricting our attention to the group $SU(N)$. For instance, the set of $SU(4)$ matrices may be considered as

Table 11.2 *Quantum states, gates and effects ordered by the*
normalisation conditions involving L_p norms with $p = 1, 2$ and ∞.

Objects studied	condition	normalisation	norm
a) Quantum states:	$\rho^\dagger = \rho \geq 0,$	$\|\rho\|_1 = \mathrm{Tr}\rho = 1$	L_1
b) Quantum gates:	$UU^\dagger = \mathbb{1},$	$\|U\|_2 = \sqrt{\mathrm{Tr}\, UU^\dagger} = \sqrt{N}$	L_2
c) Quantum effects:	$E^\dagger = E \geq 0,$	$\|E\|_\infty = \max \mathrm{eig}(E) \leq 1$	L_∞

- the space of maximally entangled states of a composite, 4×4 system, $|\psi\rangle \in \mathbb{CP}^{15} \subset \mathcal{M}^{(16)}$,
- the set of two-qubit unitary gates,[14] acting on $\mathcal{M}^{(4)}$,
- the set $\mathcal{BU}_2$ of one qubit unistochastic operations (10.64), $\Psi_U \in \mathcal{BU}_2 \subset \mathcal{B}_2$.

The first item on this list is connected to quantum states and kinematics, the second to unitary dynamics, and the third to quantum operations. In general, in our story we are concerned about the sets of quantum states ρ, quantum gates U and quantum effects E, as they form basic tools daily used by a quantum mechanician. The first and the third set contain positive operators and are dual, as $p_i = \mathrm{Tr}\rho E_i \geq 0$. The set of gates is rather different, as it consists of unitary operators. However, if one considers the normalisation conditions, these sets can be ordered according to the index of the norm used. See Table 11.2. Interestingly, the norm L_1, related to quantum states is dual to the norm L_∞ corresponding to effects, while the norm L_2 is self dual.

We have seen some analogies and dualities between different notions used in quantum theory. There exists also a classical analogue of the Jamiołkowski isomorphism. The space of all classical states forms the $N-1$ dimensional simplex Δ_{N-1}. A discrete dynamics in this space is given by a stochastic transition matrix $T_N : \Delta_{N-1} \to \Delta_{N-1}$. Its entries are non–negative, and due to stochasticity (2.4.ii), the sum of all its elements is equal to N. Hence the reshaped transition matrix is a vector $\vec{t}$ of length N^2. The rescaled vector $\vec{t}/N$ may be considered as a probability vector. The classical states defined in this way form an $N(N-1)$ dimensional, convex subset of Δ_{N^2-1}. Consider, for instance, the set of $N=2$ stochastic matrices, which

can be parametrized as $T_2 = \begin{bmatrix} a & b \\ 1-a & 1-b \end{bmatrix}$ with $a, b \in [0, 1]$. The set of the corresponding probability vectors $\vec{t}/2 = (a, b, 1-a, 1-b)/2$ forms a square – the maximal square one can inscribe into the unit tetrahedron of all $N=4$ probability vectors.

---

[14] As shown by DiVincenzo [267] and Lloyd [616] such gates are *universal* for quantum computing, which means that their suitable composition can produce an arbitrary unitary transformation. Such gates may be realised experimentally [671, 257].

Classical dynamics may be considered as a (very) special subclass of quantum dynamics, defined on the set of diagonal density matrices. Hence the classical and quantum duality between maps and states may be succinctly summarised in a commutative diagram:

$$\text{quantum}: \quad \left[\Phi : \mathcal{M}^{(N)} \to \mathcal{M}^{(N)}\right] \quad \longrightarrow \quad \tfrac{1}{N}D_\Phi \in \mathcal{M}^{(N^2)}$$

$$\downarrow \Psi_{\text{CG}} \qquad\qquad \downarrow \; \text{maps} \qquad\qquad \downarrow \; \text{states} \qquad (11.26)$$

$$\text{classical}: \quad \left[T : \Delta_{N-1} \to \Delta_{N-1}\right] \quad \longrightarrow \quad \tfrac{1}{N}\vec{t} \in \Delta_{N^2-1}.$$

Alternatively, vertical arrows may be interpreted as the action of the coarse graining operation Ψ_{CG} to be defined in Eq. (13.77). For instance, for the trivial (do nothing) one–qubit quantum map $\Phi_{\mathbb{1}}$, the superoperator $\mathbb{1}_4$ restricted to diagonal matrices gives the identity matrix, $T = \mathbb{1}_2$, and the classical state $\vec{t}/2 = (1,0,0,1)/2 \in \Delta_3$. But this very vector represents the diagonal of the maximally entangled state $\tfrac{1}{2}D_\Phi = |\phi^+\rangle\langle\phi^+|$. To prove commutativity of the diagram (11.26) in the general case define the stochastic matrix T as a submatrix of the superoperator (10.36), $T_{mn} = \Phi_{\substack{mm \\ nn}}$ (left vertical arrow). Note that the vector $\vec{t}$ obtained by its reshaping satisfies $\vec{t} = \text{diag}(\Phi^R) = \text{diag}(D_\Phi)$. Hence, as denoted by the right vertical arrow, it represents the diagonal of the dynamical matrix, which completes the reasoning.

Problems

11.1 Prove the reshuffling lemma (11.5).

11.2 a) Find the Kraus spectrum of the (non-positive) dynamical matrix representing transposition acting in $\mathcal{M}^{(N)}$.

b) Show that the canonical Kraus representation of the transposition of one qubit is given by a *difference* between two CP maps,

$$\rho^T = (1+a)\Phi_{\text{CP}}(\rho) - a\Phi'_{\text{CP}}(\rho) = \frac{1}{2}\left(\sigma_0\rho\sigma_0 + \sigma_x\rho\sigma_x + \sigma_z\rho\sigma_z - \sigma_y\rho\sigma_y\right) \tag{11.27}$$

11.3 Show that the map $\Psi_r(\rho) \equiv (N\rho_* - \rho)/(N-1)$, acting on $\mathcal{M}^{(N)}$, is not completely positive. Is it positive or completely co–positive?

11.4 Show that $\Phi_T \equiv \tfrac{2}{3}\Phi_* + \tfrac{1}{3}T$ is the best structural physical approximation (SPA) of the non–CP map T of the transposition of a qubit [465]. How does such a SPA look like for the transposition of a quNit?

11.5 Compute complete positivity (complete co-positivity) of the generalised Choi map (11.8). Find the conditions for $\Psi_{a,b,c}$ to be CP (CcP).

11.6 Show that the minimal distance of a positive (but not CcP) map Φ_{nCcP} from the set $\mathcal{C}c\mathcal{P}_N$ is smaller than $Nx\sqrt{\mathrm{Tr}(D^{T_A})^2 - 1}/(Nx+1)$, where D_Φ represents the dynamical matrix, and the positive number x is opposite to the negative, minimal eigenvalue of $D_{\Phi T} = D_\Phi^{T_A}$.

11.7 Consider an arbitrary map in its Kraus form $\Phi(\rho) = \sum_i A_i \rho A_i^\dagger$. What is the meaning of a matrix T of order N defined by the following sum of the Hadamard products (see Problem 9.3),

$$T \equiv \sum_i A_i \circ \bar{A}_i. \tag{11.28}$$

12
Discrete structures in Hilbert space

There is a much greater kinship between the complex and the combinatorial than between the real and the combinatorial.

—*Roger Penrose*[1]

Quantum state spaces are continuous, but they have some intriguing realisations of discrete structures hidden inside. We will discuss some of them, starting with unitary operator bases, a notion of strategic im portance in the theory of entanglement, signal processing, quantum computation, and more. The structures we are aiming at are known under strange acronyms such as 'MUB' and 'SIC'. They will be spelled out in due course, but in most of the chapter we let the Heisenberg groups occupy the centre stage. It seems that the Heisenberg groups understand what is going on.

12.1 Unitary operator bases and the Heisenberg groups

Starting from a Hilbert space $\mathcal{H}$ of dimension N we have another Hilbert space of dimension N^2 for free, namely the Hilbert–Schmidt space of all complex operators acting on $\mathcal{H}$, canonically isomorphic to the Hilbert space $\mathcal{H} \otimes \mathcal{H}^*$. It was introduced in Section 8.1 and further explored in Chapter 9. Is it possible to find an orthonormal basis in $\mathcal{H} \otimes \mathcal{H}^*$ consisting solely of unitary operators? A priori this looks doubtful, since the set of unitary matrices has real dimension N^2, which is only one-half of the real dimension of $\mathcal{H} \otimes \mathcal{H}^*$. But physical observables are naturally associated with unitary operators, so if such bases exist they are likely to be important. They are called *unitary operator bases*.

Unitary operator bases do in fact exist, in great abundance.[2] And we can ask for more [547]. We can insist that the elements of the basis form a group. More

---

[1] Reproduced from Roger Penrose. Twistors and Particles: an Outline. In L. Castell, M. Drieschner, and Carl Friedrich Weizsäcker, editors, *Quantum Theory and the Structures of Time and Space*. Carl Hanser, 1975.
[2] Unitary operator bases were introduced by Schwinger (1960) [815], and heavily used by him [817].

precisely, let $\bar{G}$ be a finite group of order N^2, with identity element e. Let U_g be unitary operators giving a projective representation of $\bar{G}$, such that

1. U_e is the identity matrix.
2. $g \neq e \quad \Rightarrow \quad \mathrm{Tr} U_g = 0$.
3. $U_g U_h = \lambda(g, h) U_{gh}$, where $|\lambda(g, h)| = 1$.

(So λ is a phase factor.) Then this collection of unitary matrices is a unitary operator basis. To see this, observe that

$$U_g^\dagger = \lambda(g^{-1}, g)^{-1} U_{g^{-1}}. \tag{12.1}$$

It follows that

$$g^{-1} h \neq e \quad \Rightarrow \quad \mathrm{Tr} U_g^\dagger U_h = 0, \tag{12.2}$$

and moreover that $\mathrm{Tr} U_g^\dagger U_g = \mathrm{Tr} U_e = N$. Hence these matrices are orthogonal with respect to the Hilbert–Schmidt inner product from Section 8.1. Unitary operator bases arising from a group in this way are known as unitary operator bases *of group type*, or as *nice error bases* – a name that comes from the theory of quantum computation (where they are used to discretise errors, thus making the latter correctable – as we will see in Section 17.7).

   The question of the existence of nice error bases is a question in group theory. First of all we note that there are two groups involved in the construction, the group G which is faithfully represented by the above formulas, and the *collineation group* $\bar{G}$ which is the group G with all phase factors ignored. The group $\bar{G}$ is also known, in this context, as the *index group*. Unless N is a prime number (in which case the nice error bases are essentially unique), there is a long list of possible index groups. An abelian index group is necessarily of the form $H \times H$, where H is an abelian group. Non-abelian index groups are more difficult to classify, but it is known that every index group must be soluble.[3] *Soluble groups* will reappear in Section 12.8; for the moment let us just mention that all abelian groups are soluble.

   The paradigmatic example of a group G giving rise to a unitary operator basis is the *Weyl–Heisenberg group*[4] $H(N)$. This can be presented as follows. Introduce three group elements X, Z, and ω. Declare them to be of order N:

$$X^N = Z^N = \omega^N = \mathbb{1}. \tag{12.3}$$

Insist that ω belongs to the centre of the group (it commutes with everything):

$$\omega X = X\omega, \qquad \omega Z = Z\omega. \tag{12.4}$$

---

[3] The classification problem has been studied by Klappenecker and Rötteler [534], making use of the classification of finite groups. They also maintain an on line catalogue.

[4] This group appeared in many different contexts, starting in nineteenth-century algebraic geometry, and in the beginnings of matrix theory [876]. In the twentieth century it took on a major role in the theory of elliptic curves [679]. Weyl (1932) [961] studied its unitary representations in his book on quantum mechanics.

Then we impose one further key relation:

$$ZXZ^{-1}X^{-1} = \omega \quad \Leftrightarrow \quad ZX = \omega XZ. \tag{12.5}$$

The Weyl–Heisenberg group consists of all 'words' that can be written down using the three 'letters' ω, X, Z, subject to the above relations. It requires no great effort to see that it suffices to consider N^3 words of the form $\omega^t X^r Z^s$, where t, r, s are integers modulo N.

The Weyl–Heisenberg group admits an essentially unique unitary representation in dimension N. First we represent ω as multiplication with a phase factor which is a primitive root of unity, conveniently chosen to be

$$\omega = e^{2\pi i/N}. \tag{12.6}$$

If we further insist that Z be represented by a diagonal operator we are led to the *clock-and-shift* representation

$$Z|i\rangle = \omega^i |i\rangle, \quad X|i\rangle = |i+1\rangle. \tag{12.7}$$

The basis kets are labelled by integers modulo N. A very important area of application for the Weyl–Heisenberg group is that of time-frequency analysis of signals; then the operators X and Z may represent time delays and Doppler shifts of a returning radar wave form. But here we stick to the language of quantum mechanics.[5]

To orient ourselves we first write down the matrix form of the generators for $N = 3$, which is a good choice for illustrative purposes:

$$Z = \begin{pmatrix} 1 & 0 & 0 \\ 0 & \omega & 0 \\ 0 & 0 & \omega^2 \end{pmatrix}, \quad X = \begin{pmatrix} 0 & 0 & 1 \\ 1 & 0 & 0 \\ 0 & 1 & 0 \end{pmatrix}. \tag{12.8}$$

In two dimensions Z and X become the Pauli matrices σ_z and σ_x respectively, see Eqs. (B.3), which explains the notation. We note the resemblance between Eq. (12.5) and a special case of the equation that defines the original Heisenberg group, Eq. (6.4). This explains why Weyl took this finite group as a toy model of the latter. We also note that although the Weyl–Heisenberg group has order N^3 its collineation group – the group modulo phase factors, which is the group acting on projective space – has order N^2. In fact the collineation group is an abelian product of two cyclic groups, $Z_N \times Z_N$. The slight departure from commutativity ensures an interesting representation theory.

There is a complication to notice at this point: because $\det Z = \det X = (-1)^{N+1}$, it matters if N is odd or even. If N is odd the Weyl–Heisenberg group is a subgroup of $SU(N)$, but if N is even it is a subgroup of $U(N)$ only. Moreover, if N is odd

---

[5] We refer to Howard et al. [472] for an introduction to signal processing and radar applications.

the Nth power of every group element is the identity, but if N is even we must go to the $2N$th power to say as much. (For $N = 2$ we find $X^2 = Z^2 = \mathbb{1}$ but $(ZX)^2 = (i\sigma_y)^2 = -\mathbb{1}$.) These annoying facts make even dimensions significantly more difficult to handle, and leads to the definition

$$\bar{N} = \begin{cases} N & \text{if } N \text{ is odd} \\ 2N & \text{if } N \text{ is even.} \end{cases} \tag{12.9}$$

Keeping the complication in mind we turn to the problem of choosing suitable phase factors for the N^2 words that will make up our nice error basis. The peculiarities of even dimensions suggest an odd-looking move. We introduce the phase factor

$$\tau \equiv -e^{\pi i/N} = -\sqrt{\omega}, \qquad \tau^{\bar{N}} = 1. \tag{12.10}$$

Note that τ is an Nth root of unity only if N is odd. Then we define the *displacement operators*

$$D_{\mathbf{p}} \equiv D_{r,s} = \tau^{rs} X^r Z^s, \qquad 0 \leq r, s \leq \bar{N} - 1. \tag{12.11}$$

These are the words to use in the error basis. The double notation – using either $D_{\mathbf{p}}$ or $D_{r,s}$ – emphasises that it is convenient to view r, s as the two components of a 'vector' $\mathbf{p}$. Because of the phase factor τ the displacement operators are not actually in the Weyl–Heisenberg group, as originally defined, if N is even. So there is a price to pay for this, but in return we get the group law in the form

$$D_{r,s} D_{u,v} = \omega^{us-vr} D_{u,v} D_{r,s} = \tau^{us-vr} D_{r+u,s+v}$$

$$\Leftrightarrow \tag{12.12}$$

$$D_{\mathbf{p}} D_{\mathbf{q}} = \omega^{\Omega(\mathbf{p},\mathbf{q})} D_{\mathbf{q}} D_{\mathbf{p}} = \tau^{\Omega(\mathbf{p},\mathbf{q})} D_{\mathbf{p}+\mathbf{q}}.$$

The expression in the exponent of the phase factors is anti-symmetric in the 'vectors' that label the displacement operators. In fact $\Omega(\mathbf{p}, \mathbf{q})$ is a symplectic form (see Section 3.4), and a very nice object to encounter.

A desirable by-product of our conventions is

$$D_{r,s}^{-1} = D_{r,s}^{\dagger} = D_{-r,-s}. \tag{12.13}$$

Another nice feature is that the phase factor ensures that all displacement operators are of order N. On the other hand we have $D_{r+N,s} = \tau^{Ns} D_{r,s}$. This means that we have to live with a treacherous sign if N is even, since the displacement operators are indexed by integers modulo $2N$ in that case. Even dimensions are unavoidably difficult to deal with.[6]

---

[6] We do not know who first commented that 'even dimensions are odd', but he or she had a point.

Finally, and importantly, we observe that

$$\mathrm{Tr}D_{r,s} = \tau^{rs} \sum_{i=0}^{N-1} \omega^{si} \langle i | i + r \rangle = N\delta_{r,0}\delta_{s,0}. \tag{12.14}$$

Thus all the displacement operators except the identity are represented by traceless matrices, which means that the Weyl–Heisenberg group does indeed provide a unitary operator basis of group type, a nice error basis. Any complex operator A on $\mathcal{H}_N$ can be written, uniquely, in the form

$$A = \sum_{r=0}^{N-1}\sum_{s=0}^{N-1} a_{rs}D_{r,s}, \tag{12.15}$$

where the expansion coefficients a_{rs} are complex numbers given by

$$a_{rs} = \frac{1}{N}\mathrm{Tr}D_{r,s}^{\dagger}A = \frac{1}{N}\mathrm{Tr}D_{-r,-s}A. \tag{12.16}$$

Again we are using the Hilbert–Schmidt scalar product from Section 8.1. Such expansions were called 'quantum Fourier transformations' in Section 6.2.

12.2 Prime, composite and prime power dimensions

The Weyl–Heisenberg group cares deeply whether the dimension N is given by a prime number (denoted p), or by a composite number (say $N = p_1 p_2$ or $N = p^K$). If $N = p$ then every element in the group has order N (or $\bar{N}$ if $N = 2$), except of course for the identity element. If on the other hand $N = p_1 p_2$ then $(Z^{p_1})^{p_2} = \mathbb{1}$, meaning that the element Z^{p_1} has order p_2 only. This is a striking difference between prime and composite dimensions.

The point is that we are performing arithmetic modulo N, which means that we regard all integers that differ by multiples of N as identical. With this understanding, we can add, subtract and multiply the integers freely. We say that integers modulo N form a *ring*, just as the ordinary integers do. If N is composite it can happen that $xy = 0$ modulo N, even though the integers x and y are non-zero. For instance, $2 \cdot 2 = 0$ modulo 4. As Problem 12.2 should make clear, a ring is all we need to define a Heisenberg group, so we can proceed anyway. However, things work more smoothly if N equals a prime number p, because of the striking fact that every non-zero integer has a multiplicative inverse modulo p. Integers modulo p form a *field*, which by definition is a ring whose non-zero members form an abelian group under multiplication, so that we can perform division as well. In a field – the set of rational numbers is a standard example – we can perform addition, subtraction, multiplication, and division. In a ring – such as the set of all the integers – division sometimes fails. The distinction becomes important in Hilbert space once the latter is being organised by the Weyl–Heisenberg group.

The field of integers modulo a prime p is denoted $\mathbb{Z}_p$. When the dimension $N = p$ the operators $D_{\mathbf{p}}$ can be regarded as indexed by elements $\mathbf{p}$ of a two-dimensional vector space. (We use the same notation for arbitrary N, but in general we need quotation marks around the word 'vector'. In a true vector space the scalar numbers must belong to a field.) Note that this vector space contains p^2 vectors only. Now, whenever we have encountered a vector space in this book, we have tended to focus on the set of lines through its origin. This is a fruitful thing to do here as well. Each such line consists of all vectors $\mathbf{p}$ obeying the equation $\mathbf{a} \cdot \mathbf{p} = 0$ for some fixed vector $\mathbf{a}$. Since $\mathbf{a}$ is determined only up to an overall factor we obtain $p + 1$ lines in all, given by

$$\mathbf{a} = \begin{pmatrix} a_1 \\ a_2 \end{pmatrix} \in \left\{ \begin{pmatrix} 0 \\ 1 \end{pmatrix}, \begin{pmatrix} 1 \\ 0 \end{pmatrix}, \begin{pmatrix} 1 \\ 1 \end{pmatrix}, \cdots, \begin{pmatrix} 1 \\ p-1 \end{pmatrix} \right\}. \tag{12.17}$$

This set of lines through the origin is a projective space with only $p + 1$ points. Of more immediate interest is the fact that these lines through the origin correspond to cyclic subgroups of the Weyl–Heisenberg group, and indeed to its maximal abelian subgroups. (Choosing $\mathbf{a} = (1, x)$ gives the cyclic subgroup generated by $D_{-x,1}$.) The joint eigenbases of such subgroups are related in an interesting way, which will be the subject of Section 12.4.

Readers who want a simple story are advised to ignore everything we say about non-prime dimensions. With this warning, we ask: what happens when the dimension is not prime? On the physical side this is often the case: we may build a Hilbert space of high dimension by taking the tensor product of a large number of Hilbert spaces which individually have a small dimension (perhaps to have a Hilbert space suitable for describing many atoms). This immediately suggests that it might be interesting to study the direct product $H(N_1) \times H(N_2)$ of two Weyl–Heisenberg groups, acting on the tensor product space $\mathcal{H}_{N_1} \otimes \mathcal{H}_{N_2}$. (Irreducible representations of a direct product of groups always act on the tensor product of their representation spaces.) Does this give something new?

On the group theoretical side it is known that the cyclic groups $Z_{N_1 N_2}$ and $Z_{N_1} \times Z_{N_2}$ are isomorphic if and only if the integers N_1 and N_2 are relatively prime, that is to say if they do not contain any common factor. To see why, look at the examples $N = 2 \cdot 2$ and $N = 2 \cdot 3$. Clearly $Z_2 \times Z_2$ contains only elements of order 2, hence it cannot be isomorphic to Z_4. On the other hand it is easy to verify that $Z_2 \times Z_3 = Z_6$. This observation carries over to the Weyl–Heisenberg group: the groups $H(N_1 N_2)$ and $H(N_1) \times H(N_2)$ are isomorphic if and only if N_1 and N_2 are relatively prime. Thus, in many composite dimensions including the physically interesting case $N = p^K$ we have a choice of more than one Heisenberg group to play with. They all form nice error bases. In applications to signal processing one sticks to $H(N)$ also when N is large and composite, but in many-body physics and in the theory of quantum computation – carried out on tensor products of K qubits, say – it is the *multipartite Heisenberg group* $H(p)^{\otimes K} = H(p) \times H(p) \times \cdots \times H(p)$ that comes to the fore.

There is a way of looking at the group $H(p)^{\otimes K}$ which is quite analogous to our way of looking at $H(p)$. This employs finite fields with p^K elements, and de-emphasises the tensor product structure of the representation space – which in some situations is a disadvantage, but it is worth explaining anyway, especially since we will be touching on the theory of fields in Section 12.8. We begin by recalling how the field of complex numbers is constructed. One starts from the real field $\mathbb{R}$, now called the *ground field*, and observes that the polynomial equation $P(x) \equiv x^2 + 1 = 0$ does not have a real solution. To remedy this the number i is introduced as a root of the equation $P(x) = 0$. With this new number in hand the complex number field $\mathbb{C}$ is constructed as a two-dimensional vector space over $\mathbb{R}$, with 1 and i as basis vectors. To multiply two complex numbers together we calculate modulo the polynomial $P(x)$, which simply amounts to setting i^2 equal to -1 whenever it occurs. The finite fields $GF(p^K)$ are defined in a similar way using the finite field $\mathbb{Z}_p$ as a ground field.[7]

For an example, we may choose $p = 2$. The polynomial $P(x) = x^2 + x + 1$ has no zeros in the ground field $\mathbb{Z}_2$, so we introduce an 'imaginary' number α which is declared to obey $P(\alpha) = 0$. Adding and multiplying in all possible ways, and noting that $\alpha^2 = \alpha + 1$ (using binary arithmetic), we obtain a larger field having 2^2 elements of the form $x_1 + x_2\alpha$, where x_1 and x_2 are integers modulo 2. This is the finite field $GF(p^2)$. Interestingly its three non-zero elements can also be described as α, $\alpha^2 = \alpha + 1$ and $\alpha^3 = 1$. The first representation is convenient for addition, the second for multiplication. We have a third description as the set of binary sequences $\{00, 01, 10, 11\}$, and consequently a way of adding and multiplying binary sequences together. This is very useful in the theory of error-correcting codes [745]. But this is by the way.

By pursuing this idea it has been proven that there exists a finite field $GF(N)$ with N elements if and only if $N = p^K$ is a power of prime number p, and moreover that this finite field is unique for a given N (which is by no means obvious because there will typically exist several polynomials of the same degree having no solution modulo p). These fields can be regarded as K dimensional vector spaces over the ground field $\mathbb{Z}_p$, which is useful when we do addition. When we do multiplication it is helpful to observe the (non-obvious) fact that finite fields always contain a *primitive element* in terms of which every non-zero element of the field can be

---

[7] 'G' stands for Galois. For a lively introduction to finite fields we refer to Arnold [54]. For quantum mechanical applications we recommend the review by Vourdas [942].

Table 12.1 *The field with eight elements has
irreducible polynomials $P_1(x) = x^3 + x + 1$ and
$P_2(x) = x^3 + x^2 + 1$; α is a root of $P_1(x)$.*

Element	Polynomial	tr$x$	trx^2	Order
$0 = 0$	x	0	0	–
$\alpha^7 = 1$	$x + 1$	1	1	1
$\alpha = \alpha$	$P_1(x)$	0	0	7
$\alpha^2 = \alpha^2$	$P_1(x)$	0	0	7
$\alpha^3 = \alpha + 1$	$P_2(x)$	1	1	7
$\alpha^4 = \alpha^2 + \alpha$	$P_1(x)$	0	0	7
$\alpha^5 = \alpha^2 + \alpha + 1$	$P_2(x)$	1	1	7
$\alpha^6 = \alpha^2 + 1$	$P_2(x)$	1	1	7

written as the primitive element raised to some integer power, so the non-zero elements form a cyclic group. Some further salient facts are:

- Every element obeys $x^{p^K} = x$.
- Every non-zero element obeys $x^{p^K - 1} = 1$.
- $GF(p^{K_1})$ is a subfield of $GF(p^{K_2})$ if and only if K_1 divides K_2.

The field with 2^3 elements is presented in Table 12.1, and the field with 3^2 elements is given as Problem 12.4.

By definition the field theoretic *trace* is

$$\text{tr } x = x + x^p + x^{p^2} + \cdots + x^{p^{K-1}}. \tag{12.18}$$

If x belongs to the finite field $\mathbb{F}_{p^K}$ its trace belongs to the ground field $\mathbb{Z}_p$. Like the trace of a matrix, the field theoretic trace enjoys the properties that $\text{tr}(x + y) = \text{tr}x + \text{tr}y$ and $\text{tr}ax = a\text{tr}x$ for any integer a modulo p. It is used to define the concept of *dual bases* for the field. A basis is simply a set of elements such that any element in the field can be expressed as a linear combination of this set, using coefficients in $\mathbb{Z}_p$. Given a basis e_i the dual basis $\tilde{e}_j$ is defined through the equation

$$\text{tr}(e_i \tilde{e}_j) = \delta_{ij}. \tag{12.19}$$

For any field element x we can then write, uniquely,

$$x = \sum_{i=1}^{K} x_i e_i \quad \text{where} \quad x_i = \text{tr}(x \tilde{e}_i). \tag{12.20}$$

From Table 12.1 we can deduce that the basis $(1, \alpha, \alpha^2)$ is dual to $(1, \alpha^2, \alpha)$, while the basis $(\alpha^3, \alpha^5, \alpha^6)$ is dual to itself.

Let us now apply what we have learned to the Heisenberg groups.[8] Let x, u, v be elements of the finite field $GF(p^K)$. Introduce an orthonormal basis and label its vectors by the field elements,

$$|x\rangle : \quad |0\rangle, |1\rangle, |\alpha\rangle, \ldots, |\alpha^{p^K-2}\rangle. \quad (12.21)$$

Here α is a primitive element of the field, so there are p^K basis vectors altogether. Using this basis we define the operators X_u, Z_u by

$$X_u|x\rangle = |x + u\rangle, \quad Z_u|x\rangle = \omega^{\mathrm{tr}(xu)}|x\rangle, \quad \omega = e^{2\pi i/p}. \quad (12.22)$$

Note that X_u is not equal to X raised to the power u – this would make no sense, while the present definition does. In particular the phase factor ω is raised to an exponent that is just an ordinary integer modulo p. Due to the linearity of the field trace it is easily checked that

$$Z_u X_v = \omega^{\mathrm{tr}(vu)} X_v Z_u. \quad (12.23)$$

Note that it can happen that X and Z commute – it does happen for $GF(2^2)$, for which $\mathrm{tr}(1) = 0$ – so the definition takes some getting used to.

We can go on to define displacement operators

$$D_{\mathbf{u}} = \tau^{\mathrm{tr}\, u_1 u_2} X_{u_1} Z_{u_2}, \quad \tau = -e^{i\pi/p}, \quad \mathbf{u} = \begin{pmatrix} u_1 \\ u_2 \end{pmatrix}. \quad (12.24)$$

The phase factor has been chosen so that we obtain the desirable properties

$$D_{\mathbf{u}} D_{\mathbf{v}} = \tau^{\langle \mathbf{u}, \mathbf{v} \rangle} D_{\mathbf{u}+\mathbf{v}}, \quad D_{\mathbf{u}} D_{\mathbf{v}} = \omega^{\langle \mathbf{u}, \mathbf{v} \rangle} D_{\mathbf{v}} D_{\mathbf{u}}, \quad D_{\mathbf{u}}^\dagger = D_{-\mathbf{u}}. \quad (12.25)$$

Here we have introduced the symplectic form

$$\langle \mathbf{u}, \mathbf{v} \rangle = \mathrm{tr}(u_2 v_1 - u_1 v_2). \quad (12.26)$$

So the formulae are arranged in parallel with those used to describe $H(N)$. It remains to be shown that the resulting group is isomorphic to the one obtained by taking K-fold products of the group $H(p)$.

There do exist isomorphisms between the two groups, but there does not exist a canonical isomorphism. Instead we begin by choosing a pair of dual bases for the field, obeying $\mathrm{tr}(e_i \tilde{e}_j) = \delta_{ij}$. We can then expand a given element of the field in two different ways,

$$x = \sum_{i=1}^K x_i e_i = \sum_{r=1}^K \tilde{x}_i \tilde{e}_i \quad \Leftrightarrow \quad \begin{cases} x_i = \mathrm{tr}(x\tilde{e}_i) \\ \tilde{x}_i = \mathrm{tr}(xe_i). \end{cases} \quad (12.27)$$

[8] For more details see Vourdas [942], Gross [366], and Appleby [41].

We then introduce an isomorphism between $\mathcal{H}_{p^K}$ and $\mathcal{H}_p \otimes \cdots \otimes \mathcal{H}_p$,

$$S|x\rangle = |x_1\rangle \otimes |x_2\rangle \otimes \cdots \otimes |x_K\rangle. \tag{12.28}$$

In each p dimensional factor space we have the group $H(p)$ and the displacement operators

$$D_{ij}^{(p)} = \tau^{rs} X^r Z^s, \qquad r, s \in \mathbb{Z}_p. \tag{12.29}$$

To set up an isomorphism between the two groups we expand

$$\mathbf{u} = \begin{pmatrix} u_1 \\ u_2 \end{pmatrix} = \begin{pmatrix} \sum_i u_{1i} e_i \\ \sum_i \tilde{u}_{2i} \tilde{e}_i \end{pmatrix}. \tag{12.30}$$

Then the isomorphism is given by

$$D_{\mathbf{u}} = S^{-1} \left(D_{u_{11}, \tilde{u}_{21}}^{(p)} \otimes D_{u_{12}, \tilde{u}_{22}}^{(p)} \otimes \cdots \otimes D_{u_{1K}, \tilde{u}_{2K}}^{(p)} \right) S. \tag{12.31}$$

The verification consists of a straightforward calculation showing that

$$D_{\mathbf{u}} D_{\mathbf{v}} = \tau^{\sum_i (\tilde{u}_{2i} v_{1i} - u_{1i} \tilde{v}_{2i})} D_{\mathbf{u}+\mathbf{v}} = \tau^{\langle \mathbf{u}, \mathbf{v}\rangle} D_{\mathbf{u}+\mathbf{v}}. \tag{12.32}$$

It must of course be kept in mind that the isomorphism inherits the arbitrariness involved in choosing a field basis. Nevertheless this reformulation has its uses, notably because we can again regard the set of displacement operators as a vector space over a field, and we can obtain $N + 1 = p^K + 1$ maximal abelian subgroups from the set of lines through its origin. However, unlike in the prime dimensional case, we do not obtain every maximal abelian subgroup from this construction [543].

12.3 More unitary operator bases

Do all interesting things come from groups? For unitary operator bases the answer is a resounding 'no'. We begin with a slight reformulation of the problem. Instead of looking for special bases in the ket/bra Hilbert space $\mathcal{H} \otimes \mathcal{H}^*$ we look for them in the ket/ket Hilbert space $\mathcal{H} \otimes \mathcal{H}$. We relate the two spaces with a map that interchanges their computational bases, $|i\rangle\langle j| \leftrightarrow |i\rangle|j\rangle$, while leaving the components of the vectors unchanged. A unitary operator U with matrix elements U_{ij} then corresponds to the state

$$|U\rangle = \frac{1}{\sqrt{N}} \sum_{i,j=0}^{N-1} U_{ij} |i\rangle|j\rangle \in \mathcal{H} \otimes \mathcal{H}. \tag{12.33}$$

States of this form are said to be maximally entangled, and we will return to discuss them in detail in Section 16.3. A special example, obtained by setting $U_{ij} = \delta_{ij}$, has appeared already in Eq. (11.21). For now we just observe that the task of finding a unitary operator basis for $\mathcal{H} \otimes \mathcal{H}^*$ is equivalent to that of finding a maximally entangled basis for $\mathcal{H} \otimes \mathcal{H}$.

A rich supply of maximally entangled bases can be obtained using two concepts imported from discrete mathematics: Latin squares and (complex) Hadamard matrices. We explain the construction for $N = 3$, starting with the special state

$$|\Omega\rangle = \frac{1}{\sqrt{3}}(|0\rangle|0\rangle + |1\rangle|1\rangle + |2\rangle|2\rangle). \tag{12.34}$$

Now we bring in a *Latin square*. By definition this is an array of N columns and N rows containing a symbol from an alphabet of N letters in each position, subject to the restriction that no symbol occurs twice in any row or in any column.[9] We use a Latin square to expand our maximally entangled state into N orthonormal maximally entangled states. An example with $N = 3$ makes the idea transparent:

$$|\Omega_0\rangle = \tfrac{1}{\sqrt{3}}(|0\rangle|0\rangle + |1\rangle|1\rangle + |2\rangle|2\rangle)$$

0	1	2
1	2	0
2	0	1

$\rightarrow \quad |\Omega_1\rangle = \tfrac{1}{\sqrt{3}}(|0\rangle|1\rangle + |1\rangle|2\rangle + |2\rangle|0\rangle) \tag{12.35}$

$$|\Omega_2\rangle = \tfrac{1}{\sqrt{3}}(|0\rangle|2\rangle + |1\rangle|0\rangle + |2\rangle|1\rangle).$$

The fact that the three states (in $\mathcal{H}_9$) are mutually orthogonal is an automatic consequence of the properties of the Latin square. But we want 3^2 orthonormal states. To achieve this we bring in a *complex Hadamard matrix*, that is to say a unitary matrix, each of whose elements have the same modulus. The *Fourier matrix F*, whose matrix elements are

$$F_{jk} = \frac{1}{\sqrt{N}}\omega^{jk} = \frac{1}{\sqrt{N}}(e^{\frac{2\pi i}{N}})^{jk}, \qquad 0 \leq j, k \leq N - 1. \tag{12.36}$$

provides an example that works for every N. For $N = 3$ it is an essentially unique example.[10] We use it to expand the vector $|\Omega_0\rangle$ according to the pattern

$$|\Omega_{00}\rangle = \tfrac{1}{\sqrt{3}}(|0\rangle|0\rangle + |1\rangle|1\rangle + |2\rangle|2\rangle)$$

$$\frac{1}{\sqrt{3}}\begin{bmatrix} 1 & 1 & 1 \\ 1 & \omega & \omega^2 \\ 1 & \omega^2 & \omega \end{bmatrix} \rightarrow \quad |\Omega_{01}\rangle = \tfrac{1}{\sqrt{3}}(|0\rangle|0\rangle + \omega|1\rangle|1\rangle + \omega^2|2\rangle|2\rangle) \tag{12.37}$$

$$|\Omega_{02}\rangle = \tfrac{1}{\sqrt{3}}(|0\rangle|0\rangle + \omega^2|1\rangle|1\rangle + \omega|2\rangle|2\rangle).$$

---

[9] The study of Latin squares goes back to Euler; Stinson [860] provides a good account. If the reader has spent time on sudokus her or she has worked within this tradition already. Serious applications of Latin squares to randomisation of agricultural experiments were promoted by Fisher [313].

[10] For complex Hadamard matrices in general, see Tadej and Życzkowski [883], and Szöllősi [880].

The orthonormality of these states is guaranteed by the properties of the Hadamard matrix, and they are obviously maximally entangled. Repeating the construction for the remaining states in (12.35) yields a full orthonormal basis of maximally entangled states. In fact for $N = 3$ we have obtained nothing new; we have simply reconstructed the unitary operator basis provided by the Weyl–Heisenberg group. The same is true for $N = 2$, where the analogous basis is known as the Bell basis, and will reappear in Eq. (16.1).

The generalisation to any N should be clear, especially if we formalise the notion of Latin squares a little. This will also provide some clues as to how the set of all Latin squares can be classified. First of all Latin squares exist for any N, because the multiplication table of a finite group is a Latin square. But most Latin squares do not arise in this way. So how many Latin squares are there? To count them one may first agree to present them in reduced form, which means that the symbols appear in lexicographical order in the first row and the first column. This can always be arranged by permutations of rows and columns. But there are further natural equivalences in the problem. A Latin square can be presented as N^2 triples (r, c, s), for 'row, column, and symbol'. The rule is that in this collection all pairs (r, c) are different, and so are all pairs (r, s) and (s, c). So we have N non-attacking rooks on a cubic chess board of size N^3. In this view the symbols are on the same footing as the rows and columns, and can be permuted. A formal way of saying this is to introduce a map $\lambda : \mathbb{Z}_N \times \mathbb{Z}_N \to \mathbb{Z}_N$, where $\mathbb{Z}_N$ denotes the integers modulo N, such that the maps

$$i \to \lambda(i, j), \qquad i \to \lambda(j, i), \tag{12.38}$$

are injective for all values of j. Two Latin squares are said to be *isotopic* if they can be related by permutations within the three copies of $\mathbb{Z}_N$ involved in the map. The classification of Latin squares under these equivalences was completed up to $N = 6$ by Fisher and his collaborators [313], but for higher N the numbers grow astronomical. See Table 12.2.

The second ingredient in the construction, complex Hadamard matrices, also raises a difficult classification problem. The appropriate equivalence relation for this classification includes permutation of rows and columns, as well as acting with diagonal unitaries from the left and from the right. Thus we adopt the equivalence relation

$$H' \sim PDHD'P', \tag{12.39}$$

where D, D' are diagonal unitaries and P, P' are permutation matrices. For $N = 2, 3$ and 5, all complex Hadamard matrices are equivalent to the Fourier matrix. For $N = 4$ there exists a one-parameter family of inequivalent examples, including

Table 12.2 *Counting Latin squares and equivalence classes of complex Hadamard matrices of order N. The result for Hadamard matrices of size N = 6 is quite recent [146, 881], for N = 7 we give a lower bound only, and little is known beyond that. For N = 8 there are 1676257 isotopy classes of Latin squares, so in both cases we have good reasons to break off the table.*

N	Latin squares	Reduced squares	Isotopic squares	Hadamards
2	2	1	1	1
3	12	1	1	1
4	576	4	2	∞
5	161280	56	2	1
6	812851200	9408	22	$\infty^4 + 1$
7	61479419904000	16942080	564	$\geq \infty + 1$

a purely real Hadamard matrix.[11] Again see Table 12.2, and also Problem 12.3. Real Hadamard matrices (having entries ± 1) can exist only if $N = 2$ or $N = 4k$. Paley conjectured [714] that in these cases they always exist, but his conjecture remains open.[12]

With these ingredients in hand we can write down the vectors in a maximally entangled basis as

$$|\Omega_{ij}\rangle = \frac{1}{\sqrt{N}} \sum_{k=0}^{N-1} H_{jk} |k\rangle |\lambda(i,k)\rangle, \tag{12.40}$$

where H_{ik} are the entries in a complex Hadamard matrix and the function λ defines a Latin square. This construction is due to Werner [960]. Since it relies on arbitrary Latin squares and arbitrary complex Hadamard matrices we get an enormous supply of unitary operator bases out of it.[13] This many groups do not exist, so most of these bases cannot be obtained from group theory. Still some nice error bases – in particular, the ones coming from the Weyl–Heisenberg group – do turn up (as, in fact, it did in our $N = 3$ example). The converse question, whether all nice error bases come from Werner's construction, has an answer, namely 'no'. The examples constructed in this section are all *monomial*, meaning that the unitary operators can be represented by matrices having only one non-zero entry in each row and in each

---

[11] The freely adjustable phase factor was discovered by Hadamard [391]. It was adjusted in experiments performed many years later [568]. Karlsson wrote down a three-parameter family of $N = 6$ complex Hadamard matrices in fully explicit and remarkably elegant form [516]. Karlsson's family is qualitatively more interesting than the $N = 4$ example.

[12] The smallest unsolved case is $N = 668$. Real Hadamard matrices have many uses, and discrete mathematicians have spent much effort constructing them [448].

[13] A quantum variation on the theme of Latin squares, giving an even richer supply, is known [683].

column. Nice error bases not of this form are known. Still it is interesting to observe that the operators in a nice error basis can be represented by quite sparse matrices – they always admit a representation in which at least one-half of the matrix elements equal zero [537].

12.4 Mutually unbiased bases

Two orthonormal bases $\{|e_i\rangle\}_{i=0}^{N-1}$ and $\{|f_j\rangle\}_{j=0}^{N-1}$ are said to be *complementary* or *unbiased* if

$$|\langle e_i|f_j\rangle|^2 = \frac{1}{N} \qquad (12.41)$$

for all possible pairs of vectors consisting of one vector from each basis. If a system is prepared in a state belonging to one of the bases, then no information whatsoever is available about the outcome of a von Neumann measurement using the complementary basis. The corresponding observables are complementary in the sense of Niels Bohr, whose main concern was with the complementarity between position and momentum as embodied in the formula

$$|\langle x|p\rangle|^2 = \frac{1}{2\pi}. \qquad (12.42)$$

The point is that the right-hand side is a constant. Its actual value is determined by the probabilistic interpretation only when the dimension of Hilbert space is finite.

To see why a set of mutually unbiased bases may be a desirable thing to have, suppose we are given an unlimited supply of identically prepared N level systems, and that we are asked to determine the $N^2 - 1$ parameters in the density matrix ρ. That is, we are asked to perform quantum state tomography on ρ. Performing the same von Neumann measurement on every copy will determine a probability vector with $N - 1$ parameters. But

$$N^2 - 1 = (N + 1)(N - 1). \qquad (12.43)$$

Hence we need to perform $N+1$ different measurements to fix ρ. If – as is likely to happen in practice – each measurement can be performed a finite number of times only, then there will be a statistical spread, and a corresponding uncertainty in the determination of ρ. Figure 12.1 is intended to suggest (correctly as it turns out) that the best result will be achieved if the $N + 1$ measurements are performed using *mutually unbiased bases* (abbreviated *MUB* from now on).

MUB have numerous other applications, notably to quantum cryptography. The original BB84 protocol for quantum key distribution [116] used a pair of qubit MUB. Going to larger sets of MUB in higher dimensions yields further advantages

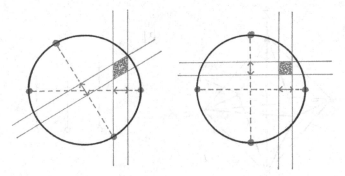

Figure 12.1 Quantum state tomography for a real qubit (a 'rebit'), using pairs of bases. The statistics of each measurement determine the position of the density matrix only up to a statistical spread. Choosing the two bases representing the measurement to be unbiased – this is done for the rebit on the right – will minimise the resulting uncertainty (represented by the shaded area). Readers who recall Figure 2.11 will realise that we suppress some complications here.

[202]. The bottom line is that MUB are of interest when one is trying to find or hide information.[14]

Our concern will be to find out how many MUB exist in a given dimension N. The answer will tell us something about the shape of the convex body of mixed quantum states. To see this, note that when $N = 2$ the pure states making up the three MUB form the corners of a regular octahedron inscribed in the Bloch sphere. See Figure 12.2. Now let the dimension of Hilbert space be N. The set of mixed states has real dimension $N^2 - 1$, and it has the maximally mixed state as its natural origin. An orthonormal basis corresponds to a regular simplex Δ_{N-1} centred at the origin. This spans an $(N-1)$-dimensional plane through the origin. Using the trace inner product we find that

$$|\langle e_i | f_j \rangle|^2 = \frac{1}{N} \quad \Rightarrow \quad \mathrm{Tr}\left(|e_i\rangle\langle e_i| - \frac{1}{N}\mathbb{1}\right)\left(|f_j\rangle\langle f_j| - \frac{1}{N}\mathbb{1}\right) = 0. \qquad (12.44)$$

Hence the condition that two such simplices represent a pair of MUB translates into the condition that the two $(N-1)$-planes be totally orthogonal, in the sense that every Bloch vector in one of them is orthogonal to every Bloch vector in the other. But the central equation of this section, namely (12.43), implies that there is room for at most $N + 1$ totally orthogonal $(N-1)$-planes. It follows that there can exist at most $N + 1$ MUB. But it does not at all follow that this many MUB actually exist. Our collection of $N + 1$ simplices form an interesting convex polytope with $N(N+1)$ vertices, and what we are asking is whether this *complementarity polytope* can be inscribed into the convex body $\mathcal{M}^{(N)}$ of density matrices. In fact, given our

---

[14] Further applications include entanglement detection in the laboratory [853], a famous retrodiction problem [296, 771], and more.

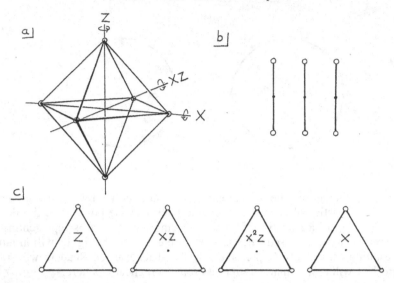

Figure 12.2 (a) When $N = 2$ a complete set of MUB forms a regular octahedron inscribed in the Bloch sphere. Each of the bases is an eigenbasis for some element in the Weyl–Heisenberg group, represented in the figure by a π-rotation through an axis. (b) A poor man's picture of the octahedron, shown as three 1-simplices, assumed to be mutually orthogonal and to intersect only at the maximally mixed state. (c) A poor man's picture of four MUB in three dimensions – in the eight-dimensional Bloch space it consists of four 2-simplices spanning totally orthogonal subspaces, intersecting only at the maximally mixed state. Each basis is the eigenbasis of a cyclic subgroup generated by some element of the Weyl–Heisenberg group.

caricature of this body, as the stitching found on a tennis ball (Section 8.6), this does seem a little unlikely (unless $N = 2$).

Anyway a set of $N + 1$ MUB in N dimensions is referred to as a *complete set*. Do such sets exist? If we think of a basis as given by the column vectors of a unitary matrix, and if the basis is to be unbiased relative to the computational basis, then that unitary matrix must be a complex Hadamard matrix. Classifying pairs of MUB is equivalent to classifying such matrices. In Section 12.3 we saw that they exist for all N, often in abundance. To be specific, let the identity matrix and the Fourier matrix (12.36) represent a pair of MUB. Can we find a third basis, unbiased with respect to both? Using $N = 3$ as an illustrative example we find that Figure 4.10 gives the story away. A column vector in a complex Hadamard matrix corresponds to a point on the maximal torus in the octant picture. The twelve column vectors

$$
\begin{bmatrix}
1 & 0 & 0 & 1 & \omega^2 & \omega^2 & 1 & \omega & \omega & 1 & 1 & 1 \\
0 & 1 & 0 & \omega^2 & 1 & \omega^2 & \omega & 1 & \omega & 1 & \omega & \omega^2 \\
0 & 0 & 1 & \omega^2 & \omega^2 & 1 & \omega & \omega & 1 & 1 & \omega^2 & \omega
\end{bmatrix}
\tag{12.45}
$$

form four MUB, and it is clear from the picture that this is the largest number that can be found. (For convenience we did not normalise the vectors. The columns of the Fourier matrix F were placed on the right.)

We multiplied the vectors in the two bases in the middle with phase factors in a way that helps to make the pattern memorable. Actually there is a bit more to it. They form *circulant matrices* of size 3, meaning that they can be obtained by cyclic permutations of their first row. Circulant matrices have the nice property that $F^\dagger C F$ is a diagonal matrix for every circulant matrix C.

The picture so far is summarised in Figure 12.2. The key observation is that each of the bases is an eigenbasis for a cyclic subgroup of the Weyl–Heisenberg group. As it turns out this generalises straightforwardly to all dimensions such that $N = p$, where p is an odd prime. We gave a list of $N + 1$ cyclic subgroups in Eq. (12.17). Each cyclic subgroup consists of a complete set of commuting observables, and they determine an eigenbasis. We denote the a-th vector in the x-th eigenbasis as $|x, a\rangle$, and we have to solve

$$D_{0,1}|0, a\rangle = \omega^a|0, a\rangle, \quad D_{x,1}|x, a\rangle = \omega^a|x, a\rangle, \quad D_{-1,0}|\infty, a\rangle = \omega^a|\infty, a\rangle. \quad (12.46)$$

Here $x = 1, \ldots, p - 1$, but in the spirit of projective geometry we can extend the range to include 0 and ∞ as well. The solution, with $\{|e_r\rangle\}_{r=0}^{p-1}$ denoting the computational basis, is

$$|0, a\rangle = |e_a\rangle, \quad |x, a\rangle = \frac{1}{\sqrt{p}} \sum_{r=0}^{p-1} \omega^{\frac{(r-a)^2}{2x}} |e_r\rangle, \quad |\infty, a\rangle = \frac{1}{\sqrt{p}} \sum_{r=0}^{p-1} \omega^{ar} |e_r\rangle. \quad (12.47)$$

It is understood that if '1/2' occurs in an exponent it denotes the inverse of 2 in arithmetic modulo p (and similarly for '1/x'). There are $p - 1$ bases presented as columns of circulant matrices, and we use ∞ to label the Fourier basis for a reason (Problem 12.5). One can show directly[15] that these bases form a complete set of MUB (Problem 12.6), but a simple and remarkable theorem will save us from this effort.

Interestingly there is an alternative way to construct the complete set. When $N = 2$ it is clear that we can start with the eigenbasis of the group element Z (say), choose a point on the equator, and then apply all the transformations effected by the Weyl–Heisenberg group. The resulting orbit will consist of N^2 points, and if the starting point is judiciously chosen they will form N MUB, all of them unbiased to the eigenbasis of Z. Again the construction works in all prime dimensions $N = p$, and indeed the resulting complete set is equivalent to the previous one in the sense that there exists a unitary transformation taking the one to the other.[16]

---

[15] This construction was found by Ivanović (1981), whose interest was in state tomography [491].
[16] This construction is due to Alltop (1980), whose interest was in radar applications [28]. Later references have made the construction more transparent and more general [136, 535].

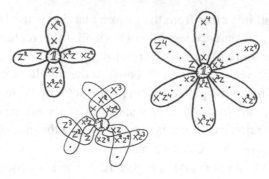

Figure 12.3 For $N = 3$ and 5 the cyclic subgroups of the Weyl–Heisenberg group form a flower. For $N = 4$ the petals get in each other's way and the construction fails.

The central theorem about MUB existence is:[17]

The BBRV theorem. *A complete set of MUB exists in $\mathbb{C}^N$ if and only if there exists a unitary operator basis which can be divided into $N + 1$ sets of N commuting operators such that the sets have only the unit element in common.*

Let us refer to unitary operator bases of this type as *flowers*, and the sets into which they are divided as *petals*. Note that it is Eq. (12.43) that makes flowers possible. There can be at most N mutually commuting orthogonal unitaries since, once they are diagonalised, the vectors defined by their diagonals must form orthogonal vectors in N dimensions. The Weyl–Heisenberg groups are flowers if and only if N is a prime number – as exemplified in Figure 12.3. So the fact that (12.47) gives a complete set of $N + 1$ MUB whenever N is prime follows from the theorem.

We prove the BBRV theorem one way. Suppose a complete set of MUB exists. We obtain a maximal set of commuting Hilbert–Schmidt orthogonal unitaries by carefully choosing N unitary matrices U_r, with r between 0 and $N - 1$:

$$U_r = \sum_r \omega^{ri} |e_i\rangle \langle e_i| \quad \Rightarrow \quad \mathrm{Tr} U_r^\dagger U_s = \sum_i \omega^{i(r-s)} = N\delta_{rs}. \tag{12.48}$$

If the bases $\{|e_i\rangle\}_{i=0}^{N-1}$ and $\{|f_i\rangle\}_{i=0}^{N-1}$ are unbiased we can form two such sets, and it is easy to check that

$$V_r = \sum_r \omega^{ri} |f_i\rangle \langle f_i| \quad \Rightarrow \quad \mathrm{Tr} V_r^\dagger U_s = \frac{1}{N} \sum_{i,j} \omega^{-ir} \omega^{js} = N\delta_{r,0}\delta_{s,0}. \tag{12.49}$$

Hence $U_r \neq V_s$ unless $r = s = 0$. It may seem as if we have constructed two cyclic subgroups of the Weyl–Heisenberg group, but in fact we have not since we have

---

[17] This is due to Bandyopadhyay, Boykin, Roychowdhury and Vatan [88].

said nothing about the phase factors that enter into the scalar products $\langle e_i|f_j\rangle$. But this is as may be. It is still clear that if we go in this way, we will obtain a flower from a collection of $N + 1$ MUB. Turning this into an 'if and only if' statement is not very hard [88].

We have found flowers in all prime dimensions. What about $N = 4$? At this point we recall the Mermin square (5.61). This defines (as it must, if one looks at the proof of the BBRV theorem) two distinct triplets of MUB. It turns out, however, that these are examples of *unextendible* sets of MUB, that cannot be completed to complete sets (a fact that can be ascertained without calculations, if the reader has the prerequisites needed to solve Problem 12.7). A more constructive observation is that the operators occurring in the square belong to the 2-partite Heisenberg group $H(2) \times H(2)$. This group is also a unitary operator basis, and it contains no less than 15 maximal abelian subgroups, or petals in our language. Denoting the elements of the collineation group of $H(2)$ as $X = \sigma_x$, $Y = \sigma_y$, $Z = \sigma_z$, and the elements of the 2-partite group as (say) $XY = X \otimes Y$, we label the petals as

$$
\begin{aligned}
1 &= \{1Z, Z1, ZZ\} \quad 2 = \{X1, 1X, XX\} \quad 3 = \{XZ, ZX, YY\} \\
4 &= \{1Z, X1, XZ\} \quad 5 = \{Z1, 1X, ZX\} \quad 6 = \{ZZ, XX, YY\}
\end{aligned}
\tag{12.50}
$$

(these are the petals occurring in the Mermin square), and

$$
\begin{aligned}
7 &= \{1Z, Y1, YZ\} \quad\;\; 8 = \{Z1, 1Y, ZY\} \quad\;\; 9 = \{X1, 1Y, XY\} \\
10 &= \{1X, Y1, YX\} \quad 11 = \{1Y, Y1, YY\} \quad 12 = \{XY, YX, ZZ\} \\
13 &= \{XZ, YX, ZY\} \quad 14 = \{XY, YZ, ZX\} \quad 15 = \{XX, YZ, ZY\}.
\end{aligned}
\tag{12.51}
$$

After careful inspection one finds that the unitary operator basis can be divided into disjoint petals in six distinct ways, namely

$$
\begin{aligned}
&\{1, 2, 11, 13, 14\} \quad \{4, 6, 8, 10, 15\} \quad \{2, 3, 7, 8, 12\} \\
&\{1, 3, 9, 10, 15\} \quad \{4, 5, 11, 12, 14\} \quad \{5, 6, 7, 9, 13\}.
\end{aligned}
\tag{12.52}
$$

So we have six flowers, each of which contains exactly two Mermin petals (not by accident, as Problem 12.7 reveals). The pattern is summarised in Figure 12.4.

This construction can be generalised to any prime power dimension $N = p^K$. The multipartite Heisenberg group $H(p)^{\otimes K}$ gives many interlocking flowers, and hence many complete sets of MUB. The finite Galois fields are useful here: the formulae can be made to look quite similar to (12.47), except that the field theoretic trace is used in the exponents of ω. We do not go into details here.[18] The question whether all complete sets of MUB can be transformed into each other by means of some unitary transformation arises at this point. It is a difficult one. For $N \leq 5$ one can show that all complete sets are unitarily equivalent to each other, but for $N = 2^5$

---

[18] Complete sets of MUB were constructed for prime power dimensions by Wootters and Fields (1989) [983]. Calderbank et al. [190] gave a more complete list, still relying on Heisenberg groups. All known constructions are unitarily equivalent to one on their list [349].

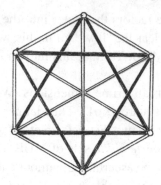

Figure 12.4 The 15 edges of the complete 6-graph represent the bases defined by the altogether 15 maximal abelian subgroups of $H(2) \times H(2)$. The vertices are complete sets of MUB. The five edges meeting at a vertex represent the bases in that set. The bases occurring in the Mermin square are painted black.

this is no longer true. In this case the 5-partite Heisenberg group can be partitioned into flowers in several unitarily inequivalent ways.[19]

What about dimensions that are not prime powers? Unitary operator bases exist in abundance and defy classification, so it is not easy to judge whether there may be a flower among them. Nice error bases are easier to deal with (because groups can be classified). There are Heisenberg groups in every dimension, and if the dimension is $N = p_1^{K_1} p_2^{K_2} \ldots p_n^{K_n}$ one can use them to construct $\min_i(p_i^{K_i}) + 1$ MUB in the composite dimension N [535, 996]. What is more, it is known that this is the largest number one can obtain from the partitioning of any nice error basis into petals [61]. However, the story does not end there, because – making use of the combinatorial concepts introduced in the next section – one can show that in certain square dimensions larger, but still incomplete, sets do exist. In particular, in dimension $N = 26^2 = 2^2 13^2$ group theory would suggest at most five MUB, but Wocjan and Beth found a set of six MUB. Moreover, in square dimensions $N = m^2$ the number of MUB grows at least as fast as $m^{1/14.8}$ (with finitely many exceptions) [973].

This leaves us somewhat at sea. We do not know whether a complete set of MUB exists if the dimension is $N = 6, 10, 12, \ldots$. How is the question to be settled? Numerical computer searches with carefully defined error bounds could settle the matter in low dimensions, but the only case that has been investigated in any depth is that of $N = 6$. In this case there is convincing evidence that at most three MUB can be found, but there is still no proof.[20] And $N = 6$ may be a very special case.

---

[19] This was noted by Kantor, who has also reviewed the full story [513].
[20] Especially convincing evidence was found by Brierley and Weigert [174], Jaming et al. [493], and Raynal et al. [767]. See also Grassl [360], who studies the set of vectors unbiased relative to both the computational and Fourier basis. There are 48 such vectors.

A close relative of the MUB existence problem arises in Lie algebra theory, and is unsolved there as well.[21] One can imagine that it has do with harmonic (Fourier) analysis [649]. Or perhaps with symplectic topology: there exists an elegant geometrical theorem which says that, given any two bases in $\mathbb{C}^N$, not necessarily unbiased, there always exist at least 2^{N-1} vectors that are unbiased relative to both bases. But there is no information about whether these vectors form orthogonal bases (and in non-generic cases the vectors may coincide).[22] We offer these suggestions as hints and return to a brief summary of known facts at the end of Section 12.5. Then we will have introduced some combinatorial ideas which – whether they have any connection to the MUB existence problem or not – have a number of applications to physics.

12.5 Finite geometries and discrete Wigner functions

A combinatorial structure underlying the usefulness of mutually unbiased bases is that of *finite affine planes*. A finite plane is just like an ordinary plane, but the number of its points is finite. A finite affine plane contains lines too, which by definition are subsets of the set of points. It is required that for any pair of points there is a unique line containing them. It is also required that for every point not contained in a given line there exists a unique line having no point in common with the given line. (Devoted students of Euclid will recognise this as the parallel axiom, and will also recognise that disjoint lines deserve to be called parallel.) Finally, to avoid degenerate cases, it is required that there are at least two points in each line, and that there are at least two distinct lines. With these three axioms one can prove that two lines intersect either exactly once or not at all, and also that for every finite affine plane there exists an integer N such that

1. There are N^2 points;
2. Each line contains N points;
3. Each point is contained in $N + 1$ lines;
4. There are altogether $N + 1$ sets of N disjoint lines.

The proofs of these theorems are exercises in pure combinatorics [124], and appear at first glance quite unconnected to the geometry of quantum states.

It is much harder to decide whether a finite affine plane of order N actually exists. If N is a power of a prime number, finite planes can be constructed using coordinates, just like the ordinary plane (where lines are defined by linear equations

---

[21] But at least it received a nice name: the Winnie-the-Pooh problem. The reason cannot be fully rendered into English [556].

[22] See [36] for a description of this theorem, and [53] for a description of this area of mathematics.

in the coordinates that label the points), with the difference that the coordinates are elements of a finite field of order N. Thus a point is given by a pair (x, y), where x, y belong to the finite field, and a line consists of all points obeying either $y = ax + b$ or $x = c$, where a, b, c belong to the field. This is not quite the end of the story of the finite affine planes because examples have been constructed that do not rely on finite fields, though the order N of all these examples is a power of some prime number. Whether there exist finite affine planes for any other N is not known.

Let us go a little deeper into the combinatorics before we explain what it has to do with us. A finite plane can clearly be thought of as a grid of N^2 points, and its rows and columns provide us with two sets of N disjoint or parallel lines, such that each line in one of the sets intersect each line in the other set exactly once. But what of the next set of N parallel lines? We can label its lines with letters from an alphabet containing N symbols, and the requirement that any two lines meet at most once translates into the observation that finding the third set is equivalent to finding a Latin square. As we saw in Section 12.3, there are many Latin squares to choose from. The difficulty comes in the next step, when we ask for two Latin squares describing two different sets of parallel lines. Use Latin letters as the alphabet for the first, and Greek letters for the second. Then each point in the array will be supplied with a pair of letters, one Latin and one Greek. Since two lines are forbidden to meet more than once, a given pair of letters, such as (A, α) or (B, γ), is allowed to occur only once in the array. In other words the letters from the two alphabets serve as alternative coordinates for the array. Pairs of Latin squares enjoying this property are known as *Graeco-Latin* or *orthogonal Latin squares*. For $N = 3$ it is easy to find Graeco-Latin pairs, such as

$$\left(\begin{array}{|c|c|c|} \hline A & B & C \\ \hline B & C & A \\ \hline C & A & B \\ \hline \end{array} , \begin{array}{|c|c|c|} \hline \alpha & \beta & \gamma \\ \hline \gamma & \alpha & \beta \\ \hline \beta & \gamma & \alpha \\ \hline \end{array} \right) = \begin{array}{|c|c|c|} \hline A\alpha & B\beta & C\gamma \\ \hline B\gamma & C\alpha & A\beta \\ \hline C\beta & A\gamma & B\alpha \\ \hline \end{array} . \tag{12.53}$$

An example for $N = 4$, using alphabets that may appeal to bridge players, is

$$\begin{array}{|c|c|c|c|} \hline A\spadesuit & K\clubsuit & Q\diamondsuit & J\heartsuit \\ \hline K\heartsuit & A\diamondsuit & J\clubsuit & Q\spadesuit \\ \hline Q\clubsuit & J\spadesuit & A\heartsuit & K\diamondsuit \\ \hline J\diamondsuit & Q\heartsuit & K\spadesuit & A\clubsuit \\ \hline \end{array} . \tag{12.54}$$

Graeco-Latin pairs can be found for all choices of $N > 2$ except (famously) $N = 6$. This is so even though Table 12.2 shows that there is a very large supply of Latin squares for $N = 6$.

To define a complete affine plane, with $N + 1$ sets of parallel lines, requires us to find a set of $N - 1$ *mutually orthogonal Latin squares*, or *MOLS*. For $N = 6$ and $N = 10$ this is impossible, and in fact an infinite number of possibilities (beginning with $N = 6$) are ruled out by the *Bruck–Ryser theorem*, which says that if an affine plane of order N exists, and if $N = 1$ or 2 modulo 4, then N must be the sum of two squares. Note that 10 is a sum of two squares, but this case has been ruled out by different means.[23] If $N = p^k$ for some prime number p a solution can easily be found using analytic geometry over finite fields. There remain an infinite number of instances, beginning with $N = 12$, for which the existence of a finite affine plane is an open question.[24]

At this point we recall that complete sets of MUB exist when $N = p^k$, but quite possibly not otherwise. Moreover such a complete set is naturally described by $N+1$ sets of N vectors. The total number of vectors is the same as the number of lines in a finite affine plane, so the question is whether we can somehow associate N^2 'points' to a complete set of MUB, in such a way that the incidence structure of the finite affine plane becomes useful. One way to do this is to start with the picture of a complete set of MUB as a polytope in Bloch space. We do not have to assume that a complete set of MUB exists. We simply introduce $N(N + 1)$ Hermitian matrices P_v of unit trace, denoted P_v, and obeying

$$\text{Tr}P_vP_{v'} = \begin{cases} 1 & \text{if } v = v' \\ 1/N & \text{if } v \text{ and } v' \text{ sit in different planes} \\ 0 & \text{if } v \neq v' \text{ sit in the same plane} \end{cases} \quad (12.55)$$

The condition that $\text{Tr}P_v^2 = 1$ ensures that P_v lies on the outsphere of the set of quantum states. If their eigenvalues are non-negative these are projectors, and then they actually are quantum states, but this is not needed for the definition of the polytope. To understand the face structure of the polytope we begin by noting that the convex hull of one vertex from each of the $N + 1$ individual simplices forms a face. (This is fairly obvious, and anyway we are just about to prove it.) Using the matrix representation of the vertices we can then form the Hermitian unit trace matrix

$$A_f = \sum_{\text{face}} P_v - \mathbb{1}, \quad (12.56)$$

where the sum runs over the $N + 1$ vertices in the face. This is called a *face point operator* (later to be subtly renamed as a *phase point operator*). If $N = 3$ we can think of it pictorially as $\triangle \, \triangle \, \triangle \, \triangle$, say – each triangle represents a basis. See Figure 12.2. It is easy to see that $0 \leq \text{Tr}\rho A_f \leq 1$ for any matrix ρ that lies in the complementarity polytope, which means that the latter is confined between two

[23] The story behind these non-existence results goes back to Euler, who was concerned with arranging 36 officers belonging to six regiments, six ranks and six arms in a square. Lam describes the computer based non-existence proof for $N = 10$ in a thought-provoking way [569].

[24] See the books by Bennett [124] and Stinson [860], for more information, and for proofs of the statements we have made so far.

Figure 12.5 All the points on a line determine a P_v, and all the lines through a point determine an A_{ij}. In fact we are illustrating Eq. (12.60).

parallel hyperplanes. There is a facet defined by $\text{Tr}\rho A_f = 0$ (pictorially, this would be △ △ △ △) and an opposing face containing one vertex from each simplex. Every vertex is included in one of these two faces. There are N^{N+1} operators A_f altogether and equally many facets.

The idea is to select N^2 phase point operators and use them to represent the points of an affine plane. The $N + 1$ vertices P_v that appear in the sum (12.56) are to be regarded as the $N + 1$ lines passing through the point A_f. A set of N parallel lines in the affine plane will represent a complete set of orthonormal projectors.

To do so, recall that each A_f is defined by picking one P_v from each basis. Let us begin by making all possible choices from the first two and arrange them in an array:

$$\triangle \; \triangle \; \triangle\triangle \qquad \triangle \; \triangle \; \triangle\triangle \qquad \triangle \; \triangle \; \triangle\triangle$$

$$\triangle \; \triangle \; \triangle\triangle \qquad \triangle \; \triangle \; \triangle\triangle \qquad \triangle \; \triangle \; \triangle\triangle \qquad (12.57)$$

$$\triangle \; \triangle \; \triangle\triangle \qquad \triangle \; \triangle \; \triangle\triangle \qquad \triangle \; \triangle \; \triangle\triangle$$

We set $N = 3$ – enough to make the idea come through – in this illustration. Thus the $N+1$ simplices in the totally orthogonal $(N-1)$-planes appear as four triangles, two of which have been used up to make the array. We use the vertices of the remaining $N - 1$ simplices to label the lines in the remaining $N - 1$ pencils of parallel lines. To ensure that non-parallel lines intersect exactly once, a pair, such as △ △ (picked from any two out of the four triangles), must occur exactly once in the array. This problem can be solved because an affine plane of order N is presumed to exist. One solution is

$$\triangle \; \triangle \; \triangle \; \triangle \qquad \triangle \; \triangle \; \triangle \; \triangle \qquad \triangle \; \triangle \; \triangle \; \triangle$$

$$\triangle \; \triangle \; \triangle \; \triangle \qquad \triangle \; \triangle \; \triangle\triangle \qquad \triangle \; \triangle \; \triangle \; \triangle \qquad (12.58)$$

$$\triangle \; \triangle \; \triangle \; \triangle \qquad \triangle \; \triangle \; \triangle\triangle \qquad \triangle \; \triangle \; \triangle \; \triangle$$

We have now singled out N^2 face point operators for attention, and the combinatorics of the affine plane guarantees that any pair of them have exactly one P_v in common. Equation (12.55) then enables us to compute that

$$\text{Tr} A_f A_{f'} = N\delta_{ff'}. \tag{12.59}$$

This is a regular simplex in dimension $N^2 - 1$.

We used only N^2 out of the N^{N+1} face point operators for this construction, but a little reflection shows that the set of all of them can be divided into N^{N-1} disjoint face point operator simplices. In effect we have inscribed the complementarity polytope into this many regular simplices, which is an interesting datum about the former. Each such simplex forms an orthogonal operator basis, although not necessarily a unitary operator basis. Let us focus on one of them, and label the operators occurring in it as $A_{i,j}$. It is easy to see that Eq. (12.56) can be rewritten (in the language of the affine plane), and supplemented, so that we have the two equations

$$P_v = \frac{1}{N} \sum_{\text{line } v} A_{i,j}, \qquad A_{i,j} = \sum_{\text{point } (i,j)} P_v - \mathbb{1}. \tag{12.60}$$

The summations extend through all points on the line, respectively all lines passing through the point. Using the fact that the phase point operators form an operator basis we can then define a discrete Wigner function by

$$W_{i,j} = \frac{1}{N} \text{Tr} A_{i,j} \rho. \tag{12.61}$$

Knowledge of the N^2 real numbers $W_{i,j}$ is equivalent to knowledge of the density matrix ρ. Each line in the affine plane is now associated with the number

$$p_v = \sum_{\text{line } v} W_{i,j} = \text{Tr} P_v \rho. \tag{12.62}$$

Clearly the sum of these numbers over a pencil of parallel lines equals unity. However, so far we have used only the combinatorics of the complementarity polytope, and we have no right to expect that the operators P_v have positive spectra. They will be projectors onto pure states if and only if the complementarity polytope has been inscribed into the set of density matrices, which is a difficult thing to achieve. If it is achieved we conclude that $p_v \geq 0$, and then we have an elegant discrete Wigner function – giving all the correct marginals – on our hands [979]. This will receive further polish in the next section.

Meanwhile, now that we have the concepts of mutually orthogonal Latin squares and finite planes on the table, we can discuss some interesting but rather abstract analogies to the MUB existence problem. Fix $N = p_1^{K_1} p_2^{K_2} \ldots p_n^{K_n}$. Let $\#_{\text{MUB}}$ be the number of MUB, and let #$_{\text{MOLS}}$ be the number of MOLS. (We need only $N - 1$ MOLS to construct a finite affine plane.) Then

$$\min_i(p_i^{K_i}) - 1 \leq \#_{\text{MOLS}} \leq N - 1$$

$$\min_i(p_i^{K_i}) + 1 \leq \#_{\text{MUB}} \leq N + 1. \tag{12.63}$$

The lower bound for MOLS is known as the MacNeish bound [634]. Moreover, if $N = 6$, we know that Latin squares cannot occur in orthogonal pairs, and we believe that there exist only three MUB. Finally, it is known that if there exist $N - 2$ MOLS there necessarily exist $N - 1$ MOLS, and if there exist N MUB there necessarily exist $N + 1$ MUB [954]. This certainly encourages the speculation that the existence problem for finite affine planes is related to the existence problem for complete sets of MUB in some unknown way. However, the idea fades a little if one looks carefully into the details [718, 953].

12.6 Clifford groups and stabilizer states

To go deeper into the subject we need to introduce the *Clifford group*. We use the displacement operators $D_{\mathbf{p}}$ from Section 12.1 to describe the Weyl–Heisenberg group. By definition the Clifford group consists of all unitary operators U such that

$$U D_{\mathbf{p}} U^{-1} \sim D_{f(\mathbf{p})}, \tag{12.64}$$

where $\sim$ means 'equal up to phase factors'. (Phase factors will appear if U itself is a displacement operator, which is allowed.) Thus we ask for unitaries that permute the displacement operators, so that the conjugate of an element of the Weyl–Heisenberg group is again a member of the Weyl–Heisenberg group. The technical term for this is that the Clifford group is the *normaliser* of the Weyl–Heisenberg group within the unitary group, or that the Weyl–Heisenberg group is an *invariant subgroup* of the Clifford group. If we change $H(N)$ into a non-isomorphic multipartite Heisenberg group we obtain another Clifford group, but at first we stick with $H(N)$.[25]

The hard part of the argument is to show that

$$U_G D_{\mathbf{p}} U_G^{-1} \sim D_{G\mathbf{p}}, \tag{12.65}$$

where G is a two-by-two matrix with entries in $\mathbb{Z}_{\tilde{N}}$, the point being that the map $\mathbf{p} \rightarrow f(\mathbf{p})$ has to be linear [42]. We take this on trust here. To see the consequences

---

[25] The origin of the name 'Clifford group' is a little unclear to us. It seems to be connected to Clifford's gamma matrices rather than to Clifford himself [144].

we return to the group law (12.12). In the exponent of the phase factor we encounter the symplectic form

$$\Omega(\mathbf{p}, \mathbf{q}) = p_2 q_1 - p_1 q_2. \tag{12.66}$$

When strict equality holds in Eq. (12.65) it follows from the group law that

$$U_G D_{\mathbf{p}} D_{\mathbf{q}} U_G^{-1} = \tau^{\Omega(\mathbf{p},\mathbf{q})} U D_{\mathbf{p}+\mathbf{q}} U_G^{-1} \;\Rightarrow\; D_{G\mathbf{p}} D_{G\mathbf{q}} = \tau^{\Omega(\mathbf{p},\mathbf{q})} D_{G(\mathbf{p}+\mathbf{q})}. \tag{12.67}$$

On the other hand we know that

$$D_{G\mathbf{p}} D_{G\mathbf{q}} = \tau^{\Omega(G\mathbf{p},G\mathbf{q})} D_{G(\mathbf{p}+\mathbf{q})}. \tag{12.68}$$

Consistency requires that

$$\Omega(\mathbf{p}, \mathbf{q}) = \Omega(G\mathbf{p}, G\mathbf{q}) \quad \mod \bar{N}. \tag{12.69}$$

The two by two matrix G must leave the symplectic form invariant. The arithmetic in the exponent is modulo $\bar{N}$, where $\bar{N} = N$ if N is odd and $\bar{N} = 2N$ if N is even. We have deplored this unfortunate complication in even dimensions already in Section 12.1.

Let us work with explicit matrices

$$\Omega = \begin{pmatrix} 0 & -1 \\ 1 & 0 \end{pmatrix}, \qquad G = \begin{pmatrix} \alpha & \beta \\ \gamma & \delta \end{pmatrix}, \qquad \alpha, \beta, \gamma, \delta \in \mathbb{Z}_{\bar{N}}. \tag{12.70}$$

Then Eq. (12.69) states that

$$\begin{pmatrix} 0 & -1 \\ 1 & 0 \end{pmatrix} = \begin{pmatrix} 0 & -\alpha\delta + \beta\gamma \\ \alpha\delta - \beta\gamma & 0 \end{pmatrix}. \tag{12.71}$$

Hence the matrix G must have a determinant equal to 1 modulo $\bar{N}$. Such matrices form the group $SL(2, \mathbb{Z}_{\bar{N}})$, where $\mathbb{Z}_{\bar{N}}$ stands for the ring of integers modulo $\bar{N}$. It is also known as a *symplectic group*, because it leaves the symplectic form invariant.

The full structure of the Clifford group $C(N)$ is complicated by the phase factors, and rather difficult to describe in general. Things are much simpler when $\bar{N} = N$ is odd, so let us restrict our description to this case.[26] Then the symplectic group is a subgroup of the Clifford group. Another subgroup is evidently the Weyl–Heisenberg group itself. Moreover, if we consider the Clifford group modulo its centre, $C(N)/I(N)$, that is to say if we identify group elements differing only by phase factors – which we would naturally do if we are interested only in how it transforms the quantum states – then we find that $C(N)/I(N)$ is a semi-direct product of the symplectic rotations given by $SL(2, \mathbb{Z}_N)$ and the translations given by $H(N)$ modulo its centre.

[26] A complete, and clear, account of the general case is given by Appleby [42].

In every case – also when N is even – the unitary representation of the Clifford group is uniquely determined by the unitary representation of the Weyl–Heisenberg group. The easiest case to describe is that when N is an odd prime number p. Then the symplectic group is defined over the finite field $\mathbb{Z}_p$ consisting of integers modulo p, and it contains $p(p^2 - 1)$ elements altogether. Insisting that there exists a unitary matrix U_G such that $U_G D_{\mathbf{p}} U_G^{-1} = D_{G\mathbf{p}}$ we are led to the representation

$$G = \begin{pmatrix} \alpha & \beta \\ \gamma & \delta \end{pmatrix} \rightarrow \begin{cases} U_G = \frac{e^{i\theta}}{\sqrt{p}} \sum_{i,j} \omega^{\frac{1}{2\beta}(\delta i^2 - 2ij + \alpha j^2)} |i\rangle\langle j| & \beta \neq 0 \\[2mm] U_G = \pm \sum_j \omega^{\frac{\alpha\gamma}{2}j^2} |\alpha j\rangle\langle j| & \beta = 0. \end{cases} \tag{12.72}$$

In these formulas '$1/\beta$' stands for the multiplicative inverse of the integer β in arithmetic modulo p (and since $1/2$ occurs it is obvious that special measures must be taken if $p = 2$). An overall phase factor is left undetermined: it can be pinned down by insisting on a faithful representation of the group [41, 686], but in many situations it is not needed. It is noteworthy that the representation matrices are either complex Hadamard matrices, or monomial matrices. It is interesting to see how they act on the set of mutually unbiased bases. In the affine plane a symplectic transformation takes lines to lines, and indeed parallel lines to parallel lines. If one works this out one finds that the symplectic group acts like Möbius transformations on a projective line whose points are the individual bases. See Problem 12.5.

Even though $N = 2$ is even, it is the easiest case to understand. The collineation group $C(2)/I(2)$ is just the group of rotations that transforms the polytope formed by the MUB states into itself, or in other words it is the symmetry group S_4 of the octahedron. In higher prime dimensions the Clifford group yields only a small subgroup of the symmetry group of the complementarity polytope. When N is a composite number, and especially if N is an even composite number, there are some significant complications for which we refer elsewhere [42]. These complications do sit in the background in Section 12.8, where the relevant group is the *extended Clifford group* obtained by allowing also two-by-two matrices of determinant ± 1. In Hilbert space this doubling of the group is achieved by representing the extra group elements by anti-unitary transformations [42].

To cover MUB in prime power dimensions we need to generalise in a different direction. The relevant Heisenberg group is the multipartite Heisenberg group. We can still define the Clifford group as the normaliser of this Heisenberg group. We recall that the latter contains many maximal abelian subgroups, and we refer to the joint eigenvectors of these subgroups as *stabilizer states*. The Clifford group acts by permuting the stabilizer states, and every such permutation can be built as a sequence of operations on no more than two qubits (or quNits as the case may be) at a time. In one standard blueprint for universal quantum computing [352], the quantum computer is able to perform such permutations in a fault-tolerant way, and the stabilizer states play a role reminiscent of that played by the separable states (to be defined in Chapter 16) in quantum communication.

The total number M of stabilizer states in $N = p^K$ dimensions is

$$M = p^K \prod_{i=1}^{K} (p^i + 1). \tag{12.73}$$

Dividing out a factor p^K we obtain the number of maximal abelian subgroups of the Heisenberg group. In dimension $N = 2^2$ there are altogether 60 stabilizer states forming 15 bases and 6 interlocking complete sets of MUB, because there are 6 different ways in which the group $H(2) \times H(2)$ can be displayed as a flower. See Figure 12.4. The story in higher dimensions is complicated by the appearance of complete sets that fail to be unitarily equivalent to each other. We must refer elsewhere for the details [513], but it is worth remarking that, for the 'canonical' choice of a complete set written down by Ivanović [491] and by Wootters and Fields [983], there exists a very interesting subgroup of the Clifford group leaving this set invariant as a set. It is known as the *restricted Clifford group* [41], and has an elegant description in terms of finite fields. Moreover (with an exception in dimension 3) the set of vectors that make up this set of MUB is distinguished by the property that it provides the smallest orbit under this group [1004]. For both Clifford groups, the quotient of their collineation groups with the discrete translation group provided by their Heisenberg groups is a symplectic group. If we start with the full Clifford group the symplectic group acts on a $2K$-dimensional vector space over $\mathbb{Z}_p$, while in the case of the restricted Clifford group it can be identified with the group $SL(2, \mathbb{F}_{p^K})$ acting on a two-dimensional vector space over the finite field $\mathbb{F}_{p^K} = GF(p^K)$ [366].

Armed with these group theoretical facts we can return to the subject of discrete Wigner functions. If we are in a prime power dimension it is evident that we can produce a phase point operator simplex by choosing any phase point operator A_f, and act on it with the appropriate Heisenberg group. But we can ask for more. We can ask for a phase point operator simplex that transforms into itself when acted on by the Clifford group. If we succeed, we will have an affine plane that behaves like a true phase space, since it will be equipped with a symplectic structure. This turns out to be possible in odd prime power dimensions, but not quite possible when the dimension is even. We confine ourselves to odd prime dimensions here. Then the Clifford group contains a unique element of order two, whose unitary representative we call $A_{0,0}$ [979]. Using Eq. (12.72) it is

$$G = \begin{pmatrix} -1 & 0 \\ 0 & -1 \end{pmatrix} \quad \Rightarrow \quad A_{0,0} = U_G = \sum_{i=0}^{N-1} |N - i\rangle\langle i|. \tag{12.74}$$

Incidentally we observe that $A_{0,0}^2 = F$, where F is the Fourier matrix. Making use of Eqs. (12.15)–(12.16) we find

$$A_{0,0} = \frac{1}{N} \sum_{r,s} D_{r,s}. \tag{12.75}$$

To perform this sum, split it into a sum over the $N + 1$ maximal abelian subgroups and subtract $\mathbb{1}$ to avoid overcounting. Diagonalise each individual generator of these subgroups, say

$$D_{0,1} = Z = \omega^a \sum_a |0, a\rangle\langle 0, a| \quad \Rightarrow \quad \sum_{i=0}^{N-1} Z^i = N|0, 0\rangle\langle 0, 0|. \tag{12.76}$$

All subgroups work the same way, so this is sufficient. We conclude that

$$A_{0,0} = \sum_z |z, 0\rangle\langle z, 0| - \mathbb{1}. \tag{12.77}$$

(The range of the label z is extended to cover also the bases that we have labelled by 0 and ∞.) Since we are picking one projector from each of the $N + 1$ bases this is in fact a phase point operator.

Starting from $A_{0,0}$ we can build a set of N^2 order two phase point operators

$$A_{r,s} = D_{r,s} A D_{r,s}^{-1}. \tag{12.78}$$

Their eigenvalues are ± 1, so these operators are both Hermitian and unitary. The dimension N is odd, so we can write $N = 2m - 1$. Each phase point operator splits Hilbert space into a direct sum of eigenspaces,

$$\mathcal{H}_N = \mathcal{H}_m^{(+)} \oplus \mathcal{H}_{m-1}^{(-)}. \tag{12.79}$$

Altogether we have N^2 subspaces of dimension m, each of which contain $N + 1$ MUB vectors. Conversely, one can show that each of the $N(N + 1)$ MUB vectors belongs to N such subspaces. This intersection pattern was said to be '*une configuration très-remarquable*' when it was first discovered.[27]

The operators $A_{r,s}$ form a phase point operator simplex which enjoys the twin advantages of being both a unitary operator basis and an orbit under the Clifford group. A very satisfactory discrete Wigner function can be obtained from it [366, 979]. The situation in even prime power dimensions is somewhat less satisfactory since covariance under the full Clifford group cannot be achieved in this case.

The set of phase point operators forms a particularly interesting unitary operator basis, existing in odd prime power dimensions only. Its symmetry group acts on it

---

[27] By Segre (1886) [824], who was studying elliptic normal curves. From the present point of view it was first discovered by Wootters (1987) [979].

in such a way that any pair of elements can be transformed to any other pair. This is at the root of its usefulness: from it we obtain a discrete Wigner function on a phase space lacking any kind of scale, just as the ordinary symplectic vector spaces used in classical mechanics lack any kind of scale. Moreover (with two exceptions, one in dimension two and and one in dimension eight) it is uniquely singled out by this property [1005].

12.7 Some designs

To introduce our next topic let us say something obvious. We know that

$$\mathbb{1}_N = \frac{1}{N}\sum_{i=1}^{N}|e_i\rangle\langle e_i| = \int_{\mathbb{CP}^n} d\Omega_\Psi |\Psi\rangle\langle\Psi|, \tag{12.80}$$

where $d\Omega_\Psi$ is the unitarily invariant Fubini–Study measure. Let A be any operator acting on $\mathbb{C}^N$. It follows that

$$\frac{1}{N}\sum_{i=0}^{N-1}\langle e_i|A|e_i\rangle = \int_{\mathbb{CP}^n} d\Omega_\Psi \langle\Psi|A|\Psi\rangle = \langle\langle\Psi|A|\Psi\rangle\rangle_{\text{FS}}. \tag{12.81}$$

On the right-hand side we are averaging an (admittedly special) function over all of $\mathbb{CP}^{N-1}$. On the left-hand side we take the average of its values at N special points. In statistical mechanics this equation allows us to evaluate the average expectation value of the energy by means of the convenient fiction that the system is in an energy eigenstate – which at first sight is not obvious at all.

To see how this can be generalised we recall the mean value theorem, which says that for every continuous function defined on the closed interval [0, 1] there exists a point x in the interval such that

$$f(x) = \int_0^1 ds f(s). \tag{12.82}$$

Although it is not obvious, this can be generalised to the case of sets of functions f_i [826]. Given such a set of functions one can always find an *averaging set* consisting of K different points x_I such that, for all the f_i,

$$\frac{1}{K}\sum_{I=1}^{K}f_i(x_I) = \int_0^1 ds f_i(s). \tag{12.83}$$

Of course the averaging set (and the integer K) will depend on the set of functions $\{f_i\}$ one wants to average. We can generalise even more by replacing the interval with a connected space, such as $\mathbb{CP}^{N-1}$, and by replacing the real valued functions with, say, the set $\text{Hom}(t, t)$ of all complex valued functions homogeneous of order

t in the homogeneous coordinates and their complex conjugates alike. (The restriction on the functions is needed in order to ensure that we get functions on $\mathbb{CP}^{N-1}$. Note that the expression $\langle\psi|A|\psi\rangle$ belongs to $\mathrm{Hom}(1,1)$.) This too can always be achieved, with an averaging set being a collection of points represented by the unit vectors $|\Psi_I\rangle$, $1 \leq I \leq K$, for some sufficiently large integer K [826]. We define a *complex projective t–design*, or *t–design* for short, as a collection of unit vectors $\{|\Psi_I\rangle\}_{I=1}^K$ such that

$$\frac{1}{K}\sum_{I=1}^K f(|\Psi_I\rangle) = \int_{\mathbb{CP}^n} d\Omega_\psi f(|\Psi\rangle) \tag{12.84}$$

for all polynomials $f \in \mathrm{Hom}(t,t)$ with the components of the vector, and their complex conjugates, as arguments. Formulae like this are called *cubature formulas*, since – like quadratures – they give explicit solutions of an integral, and they are of practical interest – for many signal processing and quantum information tasks – provided that K can be chosen to be reasonably small.

Eq. (12.81) shows that orthonormal bases are 1-designs. More generally, every POVM is a 1-design. Let us also note that functions $f \in \mathrm{Hom}(t-1,t-1)$ can be regarded as special cases of functions in $\mathrm{Hom}(t,t)$, since they can be rewritten as $f = \langle\Psi|\Psi\rangle f \in \mathrm{Hom}(t,t)$. Hence a t-design is automatically a $(t-1)$-design. But how do we recognise a t-design when we see one?

The answer is quite simple. In Eq. (7.69) we calculated the Fubini–Study average of $|\langle\Phi|\Psi\rangle|^{2t}$ for a fixed unit vector $|\Phi\rangle$. Now let $\{|\Psi_I\rangle\}_{I=1}^K$ be a t-design. It follows that

$$\frac{1}{K}\sum_J |\langle\Psi_I|\Psi_J\rangle|^{2t} = \langle|\langle\Psi_I|\Psi\rangle|^{2t}\rangle_{\mathrm{FS}} = \frac{t!\,(N-1)!}{(N-1+t)!}. \tag{12.85}$$

If we multiply by $1/K$ and then sum over I we obtain

$$\frac{1}{K^2}\sum_{I,J} |\langle\Psi_I|\Psi_J\rangle|^{2t} = \frac{t!\,(N-1)!}{(N-1+t)!}. \tag{12.86}$$

We have proved one direction of the following theorem.

Design theorem. *The set of unit vectors* $\{|\Psi\rangle\}_{I=1}^K$ *forms a t-design if and only if Eq. (12.86) holds.*

In the other direction a little more thought is needed [536]. Take any vector $|\Psi\rangle$ in $\mathbb{C}^N$ and construct a vector in $(\mathbb{C}^N)^{\otimes t}$ by taking the tensor product of the vector with itself t times. Do the same with $|\bar\Psi\rangle$, a vector whose components are the complex conjugates of the components, in a fixed basis, of the given vector. A final tensor product leads to the vector

$$|\Psi\rangle^{\otimes t} \otimes |\bar\Psi\rangle^{\otimes t} \in (\mathbb{C}^N)^{\otimes 2t}. \tag{12.87}$$

In the given basis the components of this vector are

$$(z_0 \ldots z_0 \bar{z}_0 \ldots \bar{z}_0, z_0 \ldots z_0 \bar{z}_1 \ldots \bar{z}_1, \ldots \ldots, z_n \ldots z_n \bar{z}_n \ldots \bar{z}_n). \qquad (12.88)$$

In fact the components consists of all possible monomials in $\mathrm{Hom}(t, t)$. Thus, to show that a set of unit vectors forms a t-design it is enough to show that the vector

$$|\Phi\rangle = \frac{1}{K} \sum_I |\Psi_I\rangle^{\otimes t} \otimes |\bar{\Psi}_I\rangle^{\otimes t} - \int d\Omega_\Psi |\Psi\rangle^{\otimes t} \otimes |\bar{\Psi}\rangle^{\otimes t} \qquad (12.89)$$

is the zero vector. This will be so if its norm vanishes. We observe preliminarily that

$$\langle \Psi_I^{\otimes t} | \Psi_J^{\otimes t} \rangle = \langle \Psi_I | \Psi_J \rangle^t. \qquad (12.90)$$

If we make use of the ubiquitous Eq. (7.69) we find precisely that

$$||\Phi||^2 = \frac{1}{K^2} \sum_{I,J} |\langle \Psi_I | \Psi_J \rangle|^{2t} - \frac{t! \, (N-1)!}{(N-1+t)!}. \qquad (12.91)$$

This vanishes if and only if Eq. (12.86) holds. But $|\Phi\rangle = 0$ is a sufficient condition for a t-design, and the theorem is proven.

This result is closely related to the *Welch bound* [956], which holds for every collection of K vectors in $\mathbb{C}^N$. For any positive integer t

$$\binom{N+t-1}{t} \sum_{I,J} |\langle x_I | x_J \rangle|^{2t} \geq \left(\sum_I \langle x_I | x_I \rangle^t \right)^2. \qquad (12.92)$$

Evidently a collection of unit vectors forms a t-design if and only if the Welch bound is saturated. The binomial coefficient occurring here is the number of ways in which t identical objects can be distributed over N boxes, or equivalently it is the dimension of the symmetric subspace $\mathcal{H}_{\mathrm{sym}}^{\otimes t}$ of the t-partite Hilbert space $\mathcal{H}_N^{\otimes t}$. This is not by accident. Introduce the operator

$$F = \sum_I |\Psi_I^{\otimes t}\rangle \langle \Psi_I^{\otimes t}|. \qquad (12.93)$$

It is then easy to see, keeping Eq. (12.90) in mind, that

$$\mathrm{Tr} F = \sum_I \langle \Psi_I^{\otimes t} | \Psi_I^{\otimes t} \rangle = K \qquad (12.94)$$

$$\mathrm{Tr} F^2 = \sum_{I,J} \langle \Psi_I^{\otimes t} | \Psi_J^{\otimes t} \rangle \langle \Psi_J^{\otimes t} | \Psi_I^{\otimes t} \rangle = \sum_{I,J} |\langle \Psi_I | \Psi_J \rangle|^{2t}. \qquad (12.95)$$

Now we can minimise $\mathrm{Tr}F^2$ under the constraint that $\mathrm{Tr}F = K$. In fact this means that all the eigenvalues λ_i of F have to be equal, namely equal to

$$\lambda_i = \frac{K}{\dim(\mathcal{H}_{\mathrm{sym}}^{\otimes t})}. \tag{12.96}$$

So we have rederived the inequality

$$\sum_{I,J} |\langle \Psi_I | \Psi_J \rangle|^{2t} = \mathrm{Tr}F^2 \geq \frac{K^2}{\dim(\mathcal{H}_{\mathrm{sym}}^{\otimes t})}. \tag{12.97}$$

Moreover we see that the operator F projects onto the symmetric subspace.

Although t-designs exist in all dimensions, for all t, it is not so easy to find examples with small number of vectors. A lower bound on the number of vectors needed is [441]

$$\text{(number of vectors)} \geq \binom{N + (t/2)_+ - 1}{(t/2)_+} \binom{N + (t/2)_- - 1}{(t/2)_-}, \tag{12.98}$$

where $(t/2)_+$ is the smallest integer not smaller than $t/2$ and $(t/2)_-$ is the largest integer not larger than $t/2$. The design is said to be *tight* if the number of its vectors saturates this bound. Can the bound be achieved? For dimension $N = 2$ much is known [400]. A tight 2-design is obtained by inscribing a regular tetrahedron in the Bloch sphere. A tight 3-design is obtained by inscribing a regular octahedron, and a tight 5-design by inscribing a regular icosahedron. The icosahedron is also the smallest 4-design, so tight 4-designs do not exist in this dimension. A cube gives a 3-design and a dodecahedron gives a 5-design. For dimensions $N > 2$ it is known that tight t-designs can exist at most for $t = 1, 2, 3$. Every orthonormal basis is a tight 1-design. A tight 2-design needs N^2 vectors, and the question whether they exist is the subject of Section 12.8. Meanwhile we observe that tthe $N(N + 1)$ vectors in a complete set of MUB saturate the Welch bound for $t = 1, 2$. Hence complete sets of MUB are 2-designs, and much of their usefulness stems from this fact. Tight 3–designs exist in dimensions 2, 4, and 6. In general it is not known how many vectors that are needed for minimal t–designs in arbitrary dimensions, which is why the terminology 'tight' is likely to be with us for some time.[28]

The name 'design' is used for more than one concept. One example, closely related to the one we have been discussing, is that of a *unitary t–design*.[29] By definition this is a set of unitary operators $\{U_I\}_{I=1}^K$ with the property that

$$\frac{1}{K} \sum_I U_I^{\otimes t} A (U_I^{\otimes t})^\dagger = \int_{U(N)} dU \, U^{\otimes t} A (U^{\otimes t})^\dagger, \tag{12.99}$$

---

[28] A particularly nice account of all these matters is in the University of Waterloo Master's thesis by Belovs (2008) [108]. For more results, and references that we have omitted, see Scott [820].

[29] Although there was a prehistory, the name seems to stem from a University of Waterloo Master's thesis by Dankert (2005) [246]. The idea was further developed in papers to which we refer for proofs, applications, and details [367, 778].

where A is any operator acting on the t-partite Hilbert space and dU is the normalised Haar measure on the group manifold. In the particularly interesting case $t = 2$ the averaging operation performed on the right hand side is known as *twirling*. Condition (12.86) for when a collection of vectors forms a projective t–design has a direct analogue: the necessary and sufficient condition for a collection of K unitary matrices to form a unitary t–design is that

$$\frac{1}{K^2} \sum_{I,J} |\mathrm{Tr} U_I^\dagger U_J|^2 = \int_{U(N)} dU |\mathrm{Tr} U|^{2t} = \begin{cases} \frac{(2t)!}{t!(t+1)!}, & \text{if } N = 2 \\ t!, & \text{if } N \geq t. \end{cases} \qquad (12.100)$$

When $t > N$ the right-hand side looks more complicated.

It is natural to ask for the operators U_I to form a finite group. The criterion for (a projective unitary representation of) a finite group to serve as a unitary t–design is that it should have the same number of irreducible components in the t-partite Hilbert space as the group $U^{\otimes t}$ itself. Thus a nice error basis, such as the Weyl–Heisenberg group, is always a 1-design because any operator commuting with all the elements of a nice error basis is proportional to the identity matrix. When $t = 2$ the group $U \otimes U$ splits the bipartite Hilbert space into its symmetric and its anti-symmetric subspace. In prime power dimensions both the Clifford group and the restricted Clifford group are unitary 2-designs [266]. In fact, it is enough to use a particular subgroup of the Clifford group [203]. For qubits, the minimal unitary 2-design is the tetrahedral group, which has only 12 elements. In even prime power dimensions 2^k, the Clifford group, but not the restricted Clifford group, is a unitary 3–design as well [562, 948, 1002]. Interestingly, every orbit of a group yielding a unitary t–design is a projective t–design. This gives an alternative proof that a complete set of MUB is a 2–design (in those cases where it is an orbit under the restricted Clifford group). In even prime power dimensions the set of all stabilizer states is a 3–design. In dimension 4 it consists of 60 vectors, while a tight 3–design (which actually exists in this case) has 40 vectors only.

12.8 SICs

At the end of Section 8.4 we asked the seemingly innocent question: is it possible to inscribe a regular simplex of full dimension into the convex body of density matrices? A tight 2-design in dimension N, if it exists, has N^2 vectors only, and our question can be restated as: do tight 2-designs exist in all dimensions? In Hilbert space language the question is: can we find an informationally complete POVM made up of *equiangular vectors*? Since absolute values of the scalar products are taken the word 'vector' really refers to a ray (a point in $\mathbb{CP}^n$). That is, we ask for N^2 vectors $|\psi_I\rangle$ such that

$$\frac{1}{N} \sum_{I=1}^{N^2} |\psi_I\rangle\langle\psi_I| = \mathbb{1} \qquad (12.101)$$

$$|\langle\psi_I|\psi_J\rangle|^2 = \begin{cases} 1 & \text{if } I = J \\ \frac{1}{N+1} & \text{if } I \neq J. \end{cases} \qquad (12.102)$$

We need N^2 unit vectors to have informational completeness (in the sense of Section 10.1), and we are assuming that the mutual fidelities are equal. The precise number $1/(N+1)$ follows by squaring the expression on the left-hand side of Eq. (12.101), and then taking the trace. Such a collection of vectors is called a SIC, so the final form of the question is: do SICs exist?[30]

If they exist, SICs have some desirable properties. First of all they saturate the Welch bound, and hence they are 2-designs with the minimal number of vectors. Moreover, also for other reasons, they are theoretically advantageous in quantum state tomography [820], and they provide a preferred form for informationally complete POVMs. Indeed an entire philosophy can be built around them [322].

But the SIC existence problem is unsolved. Perhaps we should begin by noting that there is a crisp non-existence result for the real Hilbert spaces $\mathbb{R}^N$. Then Bloch space has dimension $N(N + 1)/2 - 1$, so the number of equiangular vectors in a real SIC is $N(N + 1)/2$. For $N = 2$ the rays of a real SIC pass through the vertices of a regular triangle, and for $N = 3$ the six diagonals of an icosahedron will serve. However, for $N > 3$ it can be shown that a real SIC cannot exist unless $N + 2$ is a square of an odd integer. In particular $N = 4$ is ruled out. In fact SICs do not exist in real dimension $47 = 7^2 - 2$ either, so there are further obstructions.[31]

In $\mathbb{C}^N$ exact solutions are available in all dimensions $2 \leq N \leq 21$, and in a handful of dimensions higher than that. Numerical solutions to high precision are available in all dimensions given by two-digit numbers and a bit beyond that.[32] The existing solutions support a three-pronged conjecture:

---

[30] The acronym is short for Symmetric Informationally Complete Positive Operator Valued Measure [772], and is rarely spelled out. We prefer to use 'SIC' as a noun. When pronounced as 'seek' it serves to remind us that the existence problem may well be hiding their most important message.

[31] The non-existence result is due to Neumann, and reported by Lemmens and Seidel (1973) [586]. Since then more has been learned [873]. Incidentally the SIC in $\mathbb{R}^3$ has been proposed as an ideal configuration for an interferometric gravitational wave detector [154].

[32] For the state of the art in 2009, see Scott and Grassl [821]. For recent numerical solutions, see [321, 818]. For recent exact solutions, see [43]. The first two parts of the following conjecture are due to Zauner (1999) [996], the third to Appleby et al. (2013) [46]. We are restating it a little for convenience.

1. In every dimension there exists a SIC which is an orbit of the Weyl–Heisenberg group;
2. Moreover, there exists a SIC of this kind, where the individual vectors are invariant under a Clifford group element of order 3;
3. When $N > 3$ the overlaps of the SIC vectors are algebraic units in an abelian extension of the real quadratic field $\mathbb{Q}(\sqrt{(N-3)(N+1)})$.

Let us sort out what this means, beginning with the easily understood part 1.

In two dimensions a SIC forms a tetrahedron inscribed in the Bloch sphere. If we orient it so that its corners lie right on top of the faces of the octahedron whose corners are the stabilizer states it is easy to see (look at Figure 12.2a) that the Weyl–Heisenberg group can be used to reach any corner of the tetrahedron starting from any fixed corner. In other words, when $N = 2$ we can always write the N^2 SIC vectors in the form

$$|\psi_{r,s}\rangle = D_{r,s}|\psi_0\rangle, \qquad 0 \le r,s < N-1, \qquad (12.103)$$

where $|\psi_0\rangle$ is known as the *fiducial vector* for the SIC (and has to be chosen carefully, in a fixed representation of the Weyl–Heisenberg group). Conjecture 1 says that it is possible to find such a fiducial vector in every dimension. Numerical searches are based on this, and basically proceed by minimizing the function

$$f_{\text{SIC}} = \sum_{r,s} \left(|\langle \psi_0 | \psi_{r,s}\rangle|^2 - \frac{1}{N+1} \right)^2, \qquad (12.104)$$

where the sum runs over all pairs $(r,s) \ne (0,0)$. The arguments of the function are the components of the fiducial vector $|\psi_0\rangle$. This is a fiducial vector for a SIC if and only if $f_{\text{SIC}} = 0$. Solutions have been found in all dimensions that have been looked at – even though the presence of many local minima of the function makes the task difficult. SICs arising in this way are said to be *covariant* under the Weyl–Heisenberg group. It is believed that the numerical searches for such WH-SICs are exhaustive up to dimension $N = 50$ [821]. They necessarily fall into orbits of the Clifford group, extended to include anti-unitary symmmetries. For $N \le 50$ there are six cases where there is only one such orbit (namely $N = 2, 4, 5, 10$, or 22), while as many as ten orbits occur in two cases ($N = 35$ or 39).

Can SICs not covariant under a group exist? The only publicly available answer to this question is that if $N \le 3$ then all SICs are orbits under the Weyl–Heisenberg group [482]. Can any other group serve the purpose? If $N = 8$ there exists an elegant SIC covariant under $H(2)^{\otimes 3}$ [442], as well as two Clifford orbits of SICs covariant under the Weyl–Heisenberg group $H(8)$. No other examples of a SIC not covariant under $H(N)$ are known, and indeed it is known that for prime N the Weyl–Heisenberg group is the only group that can yield SICs [1003]. Since the mutually unbiased bases rely on the multipartite Heisenberg group this means that there can be no obvious connection between MUB and SICs, except in prime dimensions. In prime dimensions it is known that the Bloch vector of a SIC projector, when projected onto any one of the MUB eigenvalue simplices, has the same length for all the $N + 1$ simplices defined by a complete set of MUB [44, 523]. If $N = 2, 3$, every state having this property is a SIC fiducial [35], but when $N \geq 5$ this is far from being the case. The two lowest dimensions have a very special status.

The second part of the conjecture is due to Zauner. It clearly holds if $N = 2$. Then the Clifford group, the group that permutes the stabilizer states, is the symmetry group of the octahedron. This group contains elements of order 3, and by inspection we see that such elements leave some corner of the SIC-tetrahedron invariant. The conjecture says that such a symmetry is shared by all SIC vectors in all dimensions, and this has been found to hold true for every solution found so far. The sizes of the Clifford orbits shrink accordingly. In many dimensions – very much so in 19 and 48 dimensions – there are SICs left invariant by larger subgroups of the Clifford group, but the order 3 symmetry is always present and appears to be universal. There is no understanding of why this should be so.

To understand the third part of the conjecture, and why it is interesting, it is necessary to go into the methods used to produce solutions in the first place. Given conjecture 1, the straightforward way to find a SIC is to solve the equations

$$|\langle \psi_0 | D_{r,s} | \psi_0 \rangle|^2 = \frac{1}{N+1} \tag{12.105}$$

for $(r, s) \neq (0, 0)$. Together with the normalisation this is a set of N^2 multivariate quartic polynomial equations in the $2N$ real variables needed to specify the fiducial vector. To solve them one uses the method of Gröbner bases to reduce the set of equations to a single polynomial equation in one variable [821]. This is a task for computer programs such as MAGMA, and a clever programmer. The number of equations greatly exceeds the number of variables, so it would not be surprising if they did not have a solution. But solutions do exist.

As an example, here is a fiducial vector for a SIC in 4 dimensions [772, 996]:

$$\psi_0 = \begin{pmatrix} \frac{1-\tau}{2} \sqrt{\frac{5+\sqrt{5}}{10}} \\ \frac{1}{20} \left(i\sqrt{50 - 10\sqrt{5}} + (1+i)\sqrt{5(5 + 3\sqrt{5})} \right) \\ -\frac{1+\tau}{2} \sqrt{\frac{5+\sqrt{5}}{10}} \\ \frac{1}{20} \left(i\sqrt{50 - 10\sqrt{5}} - (1-i)\sqrt{5(5 + 3\sqrt{5})} \right) \end{pmatrix} \tag{12.106}$$

(with $\tau = -e^{i\pi/4}$). This does not look memorable at first sight. Note though that all components can be expressed in terms of nested square roots. This means that they are numbers that can be constructed by means of rulers and compasses, just as the ancient Greeks would have wished. This is not at all what one would expect to come out from a Gröbner basis calculation, which in the end requires one to solve a polynomial equation in one variable but of high degree. Galois showed long ago that generic polynomial equations cannot be solved by means of nested root extractions if their degree exceeds four.

And we can simplify the expression. Using the fact that the Weyl–Heisenberg group is a unitary operator basis, Eqs. (12.15)–(12.16), we can write the fiducial projector as

$$|\psi_0\rangle\langle\psi_0| = \frac{1}{N}\sum_{r,s=0}^{N-1} D_{r,s}\langle\psi_0|D^\dagger_{r,s}|\psi_0\rangle. \tag{12.107}$$

But the modulus of the overlaps is fixed by the SIC conditions, so we can define the phase factors

$$e^{i\theta_{r,s}} = \sqrt{N+1}\langle\psi_0|D_{r,s}|\psi_0\rangle, \qquad (r,s) \neq (0,0). \tag{12.108}$$

These phase factors are independent of the choice of basis in $\mathbb{C}^N$, and if we know them we can reconstruct the SIC. The number of independent phase factors is limited by the Zauner symmetry (and by any further symmetry that the SIC may have). For the $N = 4$ example we find

$$
\begin{bmatrix}
\times & e^{i\theta_{0,1}} & e^{i\theta_{0,2}} & e^{i\theta_{0,3}} \\
e^{i\theta_{1,0}} & e^{i\theta_{1,1}} & e^{i\theta_{1,2}} & e^{i\theta_{1,3}} \\
e^{i\theta_{2,1}} & e^{i\theta_{2,1}} & e^{i\theta_{2,2}} & e^{i\theta_{2,3}} \\
e^{i\theta_{3,0}} & e^{i\theta_{3,1}} & e^{i\theta_{3,2}} & e^{i\theta_{3,3}}
\end{bmatrix}
=
\begin{bmatrix}
\times & u & -1 & 1/u \\
u & 1/u & -1/u & 1/u \\
-1 & -u & -1 & 1/u \\
1/u & u & u & u
\end{bmatrix},
\tag{12.109}
$$

where

$$u = \frac{\sqrt{5}-1}{2\sqrt{2}} + \frac{i\sqrt{\sqrt{5}+1}}{2}. \tag{12.110}$$

The pattern in Eq. (12.109) is forced upon us by the Zauner symmetry, so once we know the number u it is straightforward to reconstruct the entire SIC.

What is this number? To answer this question one computes (usually by means of a computer) the *minimal polynomial*, the lowest degree polynomial with coefficients among the integers satisfied by the number u. In this case it is

$$p(t) = t^8 - 2t^6 - 2t^4 - 2t^2 + 1. \tag{12.111}$$

Because the minimal polynomial exists, we say that u is an *algebraic number*. Because its leading coefficient equals 1, we say that u is an *algebraic integer*.

Algebraic integers form a ring, just as the ordinary integers do. (Algebraic number theorists refer to ordinary integers as 'rational integers'. This is the special case when the polynomial is of first order.) Finally we observe that $1/u$ is an algebraic integer too – in fact it is another root of the same equation – so we say that u is an *algebraic unit*.[33]

Hence u is a very special number, and the question arises which number field it belongs to. This is obtained by adding the roots of (12.111) to the field of rational numbers, $\mathbb{Q}$. We did provide a thumbnail sketch of field extensions in Section 12.2, but now we are interested in fields with an infinite number of elements. When we adjoined a root of the equation $x^2 + 1 = 0$ to the real field $\mathbb{R}$ we obtained the complex field $\mathbb{R}(i) = \mathbb{C}$. In fact there are two roots of the equation, and there is a group – in this case the abelian group Z_2 – that permutes them. This group is called the *Galois group* of the extension. It can also be regarded as the group of automorphisms of the extended field $\mathbb{C}$ that leaves the ground field $\mathbb{R}$ invariant. A Galois group arises whenever a root of an irreducible polynomial is added to a field. Galois proved that a polynomial with rational coefficients can be solved with root extractions if and only if the Galois group is soluble (as is the case for a generic quartic, but not for a generic quintic). This is the origin of the name 'soluble'. For a group to be soluble it must have a particular pattern of invariant subgroups.

What one finds in the case at hand is that the Galois group is non-abelian, but only barely so. The field for the $N = 4$ SIC can be obtained by first extending $\mathbb{Q}$ to $\mathbb{Q}(\sqrt{5})$, that is to say to consider the *real quadratic field* consisting of all numbers of the form $x + \sqrt{5}y$, where x, y are rational. A further extension then leads to the field $\mathbb{Q}(u, r)$, where r is an additional (real) root of (12.111). The Galois group arising in the second step, considered by itself, is abelian, and the second extension is therefore said to be an *abelian extension*.

Thus the mysterious number u does not only belong to the field whose construction we sketched, it has a very special status in it, and moreover the field is of an especially important kind, technically known as a ray class field.

Things are not quite so simple in higher dimensions, but almost so – in terms of principles that is, not in terms of the calculations that need to be done.[34] The SIC overlaps, and the SIC vectors themselves if expressed in the natural basis, are still given by nested radicals, although no longer by square roots only so they are not constructible with ruler and compass. Indeed the third part of the SIC conjecture holds in all cases that have been looked at. What is more, the SIC overlaps continue

---

[33] Neither algebraic number theory nor Galois theory (which we are coming to) lend themselves to thumbnail sketches. For first encounters we recommend the books by Alaca and Williams [13] and by Howie [473].

[34] The number theory of SICs, so far as it is known, was developed by Appleby, Flammia, McConnell, and Yard (2016) [45].

to yield algebraic units. But it is not understood why the polynomial equations that define the SICs have this property.

A field extension is said to be abelian whenever its Galois group is abelian. In the nineteenth century Kronecker and Weber studied abelian extensions of the rational field $\mathbb{Q}$, and they proved that all such extensions are subfields of a *cyclotomic field*, which is what one obtains by adjoining a root of unity to the rational numbers. (It will be observed that MUB vectors have all their overlaps in a cyclotomic field.) Kronecker's *Jugendtraum* was to extend this result to much more general ground fields. For instance, there are quadratic extensions of the rationals, such as $\mathbb{Q}(\sqrt{5})$, consisting of all numbers of the form $x_1 + x_2\sqrt{5}$, where x_1, x_2 are rational numbers. More generally one can replace the square root of five with the square root of any integer. If that integer is negative the extension is an imaginary quadratic extension, otherwise it is a real quadratic extension. For the case of abelian extensions of an imaginary quadratic extension of $\mathbb{Q}$, Kronecker's dream led to brilliant successes, with deep connections to – among other things – the theory of elliptic curves. Such numbers turn up for special choices of the arguments of elliptic and modular functions, much as the numbers in a cyclotomic field turn up when the function $e^{i\pi x}$ is evaluated at rational values of x. As the 12th on his famous list of problems for the twentieth century, Hilbert asked for the restriction to imaginary quadratic fields to be removed.[35] The natural first step in solving the 12th problem would seem to be to find a framework for the abelian extensions of the real quadratic fields. This remains unsolved, but it seems to be in these deep waters that the SICs are swimming.

This chapter may have left the reader a little bewildered, and appalled by equations like (12.106). In defence of the chapter, we observe that practical applications (to signal processing, adaptive radar, and more) lie close to it. But the last word goes to Hilbert [770]:

There still lies an abundance of priceless treasures hidden in this domain, belonging as a rich reward to the explorer who knows the value of such treasures and, with love, pursues the art to win them.

Problems

12.1   Prove that, for every unitary operator basis $\{U_I\}_{I=0}^{N^2-1}$ and every operator A, there holds

$$\sum_{I=0}^{N^2-1} U_I A U_I^\dagger = N\operatorname{Tr}A \, \mathbb{1}. \tag{12.112}$$

---

[35] For an engaging account of this piece of history see the book by Gray [362].

12.2 Show that matrices of the form

$$\begin{pmatrix} 1 & p & s \\ 0 & 1 & x \\ 0 & 0 & 1 \end{pmatrix} \qquad (12.113)$$

form a group whenever x, p, s belong to a ring. Show that all the various Heisenberg groups in our book can be represented in this way.

12.3 Write down the most general 4×4 complex Hadamard matrix up to the natural equivalences (rows and columns can be permuted arbitrarily, and afterwards we can multiply with diagonal unitaries until the first row and the first column have all entries equal to 1).

12.4 Construct finite fields with eight and nine elements, respectively. Show that they contain a primitive element, namely an element α such that all the non-zero numbers in the field can be written as α raised to some integer power. Also compute the trace of each element.

12.5 In prime dimensions, show that symplectic transformations act like Möbius transformations on the set of MUB.

12.6 Sums of roots of unity are well known. Gauss went on to prove that

$$\sum_{a=0}^{N-1} \omega^{a^2} = \sqrt{N+1} \frac{1 + i^{-N}}{1 + i^{-1}}. \qquad (12.114)$$

Using this example of a *Gauss sum*, prove directly that the collection of bases in (12.47) are MUB.

12.7 (For readers who have studied Section 16.9.) Use the 2–design property to prove that the triplets of MUB coming from the Mermin square (5.61) cannot be extended to complete sets.

12.8 Making use of three Pauli matrices σ_j, Hadamard matrix H and the phase gate $V = \sqrt{\sigma_z} = \mathrm{diag}(1, i)$ write down all 24 matrices from $U(2)$ which form the octahedral group. By applying Eq. (12.100) show that they form a unitary 3-design. Identify the subgroup of 12 elements which forms a 2-design.

12.9 Find SICs in dimension 3, and place them in Figure 4.10. How are they related to MUB in this dimension?

13

Density matrices and entropies

A given object of study cannot always be assigned a unique value, its 'entropy'. It may have many different entropies, each one worthwhile.

– Harold Grad[1]

In quantum mechanics, the von Neumann entropy

$$S(\rho) = -\mathrm{Tr}\rho \ln \rho \tag{13.1}$$

plays a role analogous to that played by the Shannon entropy in classical probability theory. They are both functionals of the state, they are both monotone under a relevant kind of mapping, and they can be singled out uniquely by natural requirements. In Section 2.2 we recounted the well known anecdote according to which von Neumann helped to christen Shannon's entropy. Indeed von Neumann's entropy is older than Shannon's, and it reduces to the Shannon entropy for diagonal density matrices. But in general the von Neumann entropy is a subtler object than its classical counterpart. So is the quantum relative entropy that depends on two density matrices that perhaps cannot be diagonalised at the same time. Quantum theory is a non-commutative probability theory. Nevertheless, as a rule of thumb we can pass between the classical discrete, classical continuous and quantum cases by choosing between sums, integrals and traces. While this rule of thumb has to be used cautiously, it will give us quantum counterparts of most of the concepts introduced in Chapter 2, and conversely we can recover Chapter 2 by restricting the matrices of this chapter to be diagonal.

---

[1] Reproduced from Harold Grad. The many faces of entropy. *Commun. Pure Appl. Math.* 14:3, 1961.

13.1 Ordering operators

The study of quantum entropy is to a large extent a study in inequalities, and this is where we begin. We will be interested in extending inequalities that are valid for functions defined on $\mathbb{R}$ to functions of operators. This is a large morsel. But it is at least straightforward to define *operator functions*, that is functions of matrices, as long as our matrices can be diagonalised by unitary transformations: then, if $A = U\text{diag}(\lambda_i)U^\dagger$, we set $f(A) \equiv U\text{diag}(f(\lambda_i))U^\dagger$, where f is any function on $\mathbb{R}$. Our matrices will be Hermitian and therefore they admit a partial order; $B \geq A$ if and only if $B - A$ is a positive operator. It is a difficult ordering relation to work with though, ultimately because it does not define a lattice – the set $\{X : X \geq A$ and $X \geq B\}$ has no minimum point in general.

With these observations in hand we can define an *operator monotone function* as a function such that

$$A \leq B \quad \Rightarrow \quad f(A) \leq f(B). \tag{13.2}$$

Also, an *operator convex function* is a function such that

$$f(pA + (1 - p)B) \leq pf(A) + (1 - p)f(B), \qquad p \in [0, 1]. \tag{13.3}$$

Finally, an *operator concave function* f is a function such that $-f$ is operator convex. In all three cases it is assumed that the inequalities hold for all matrix sizes (so that an operator monotone function is always monotone in the ordinary sense, but the converse may fail).[2]

The definitions are simple, but we have entered deep waters, and we will be submerged by difficulties as soon as we evaluate a function at two operators that do not commute with each other. Quite innocent looking monotone functions fail to be operator monotone. An example is $f(t) = t^2$. Moreover the function $f(t) = e^t$ is neither operator monotone nor operator convex. To get serious results in this subject some advanced mathematics, including frequent excursions into the complex domain, are needed. We will confine ourselves to stating a few facts. Operator monotone functions are characterised by:

Löwner's theorem. *A function $f(t)$ on an open interval is operator monotone if and only if it can be extended analytically to the upper half plane and transforms the upper half plane into itself.*

---

[2] The theory of operator monotone functions was founded by Löwner (1934) [623]. An interesting early paper is by Bendat and Sherman (1955) [111]. For a survey see Bhatia's book [130], and (for matrix means) see Ando [38].

Therefore the following functions are operator monotone:

$$f(t) = t^\gamma, \quad t \geq 0 \quad \text{if and only if} \quad \gamma \in [0, 1]$$

$$f(t) = \frac{at+b}{ct+d}, \quad t \neq -d/c, \quad ad - bc > 0 \tag{13.4}$$

$$f(t) = \ln t, \quad t > 0.$$

This small supply can be enlarged by the observation that the composition of two operator monotone functions is again operator monotone; so is $f(t) = -1/g(t)$ if $g(t)$ is operator monotone. The set of all operator monotone functions is convex, as a consequence of the fact that the set of positive operators is a convex cone.

A continuous function f mapping $[0, \infty)$ into itself is operator concave if and only if f is operator monotone. Operator convex functions include $f(t) = -\ln t$, and $f(t) = t \ln t$ when $t > 0$; we will use the latter function to construct entropies. More generally $f(t) = tg(t)$ is operator convex if $g(t)$ is operator monotone.

Finally we define the *mean A#B* of two operators. We require that $A\#A = A$, as well as homogeneity, $\alpha(A\#B) = (\alpha A)\#(\alpha B)$, and monotonicity, $A\#B \geq C\#D$ if $A \geq C$ and $B \geq D$. Moreover we require that $(TAT^\dagger)\#(TBT^\dagger) \geq T(A\#B)T^\dagger$ for every matrix T, as well as a suitable continuity property. It turns out [38] that every mean obeying these demands takes the form

$$A\#B = \sqrt{A} f\left(\frac{1}{\sqrt{A}} B \frac{1}{\sqrt{A}}\right) \sqrt{A}, \tag{13.5}$$

where $A > 0$ and f is an operator monotone function on $[0, \infty)$ with $f(1) = 1$. The mean will be symmetric in A and B if and only if f is *self inversive*, that is if and only if

$$f(1/t) = f(t)/t. \tag{13.6}$$

Special cases of symmetric means include the *arithmetic mean* for $f(t) = (1+t)/2$, the *geometric mean* for $f(t) = \sqrt{t}$ and the *harmonic mean* for $f(t) = 2t/(1 + t)$. It can be shown that the arithmetic mean is maximal among symmetric means, while the harmonic mean is minimal [561]. And means for more than two operators can be defined [735].

We will find use for these results throughout the next three chapters. But to begin with we will get by with inequalities that apply, not to functions of operators directly, but to their traces. The subject of convex trace functions is somewhat more manageable than that of operator convex functions. A key result is that the inequality (1.13) for convex functions can be carried over in this way.[3]

---

[3] The original statement here is due to Oskar Klein (1931) [542].

Klein's inequality. *If f is a convex function and A and B are Hermitian operators, then*

$$\text{Tr}[f(A) - f(B)] \geq \text{Tr}[(A - B)f'(B)]. \tag{13.7}$$

As a special case

$$\text{Tr}(A \ln A - A \ln B) \geq \text{Tr}(A - B) \tag{13.8}$$

with equality if and only if A = B.

To prove this, use the eigenbases:

$$A|e_i\rangle = a_i|e_i\rangle \qquad B|f_i\rangle = b_i|f_i\rangle \qquad \langle e_i|f_j\rangle = c_{ij}. \tag{13.9}$$

A calculation then shows that

$$\langle e_i|f(A) - f(B) - (A - B)f'(B)|e_i\rangle = f(a_i) - \sum_j |c_{ij}|^2[f(b_j) - (a_i - b_j)f'(b_j)]$$

$$= \sum_j |c_{ij}|^2[f(a_i) - f(b_j) - (a_i - b_j)f'(b_j)]. \tag{13.10}$$

This is positive by Eq. (1.13). The special case follows if we specialise to $f(t) = t \ln t$. It is true that the condition for equality requires some extra attention.

Another useful result is:

Peierl's inequality. *If f is a strictly convex function and A is a Hermitian operator, then*

$$\text{Tr} f(A) \geq \sum_i f(\langle f_i|A|f_i\rangle), \tag{13.11}$$

where $\{|f_i\rangle\}$ is any complete set of orthonormal vectors, or more generally a resolution of the identity. Equality holds if and only if $|f_i\rangle = |e_i\rangle$ for all i, where $A|e_i\rangle = a_i|e_i\rangle$.

To prove this, observe that for any vector $|f_i\rangle$ we have

$$\langle f_i|A|f_i\rangle = \sum_j |\langle f_i|e_j\rangle|^2 f(a_j) \geq f\left(\sum_j |\langle f_i|e_j\rangle|^2 a_j\right) = f(\langle f_i|A|f_i\rangle). \tag{13.12}$$

Summing over all i gives the result.

We quote without proofs two further trace inequalities, the *Golden Thompson inequality*

$$\text{Tr} e^A e^B \geq \text{Tr} e^{A+B}, \tag{13.13}$$

with equality if and only if the Hermitian matrices A and B commute, and its more advanced cousin, the *Lieb inequality*

$$\text{Tr } e^{\ln A - \ln C + \ln B} \geq \text{Tr } \int_0^\infty A \frac{1}{C + u\mathbb{1}} B \frac{1}{C + u\mathbb{1}} \, du, \qquad (13.14)$$

where A, B, C are all positive.[4]

13.2 Von Neumann entropy

Now we can begin. First we establish some notation. In Chapter 2 we used S to denote the Shannon entropy $S(\vec{p})$ of a probability distribution. Now we use S to denote the von Neumann entropy $S(\rho)$ of a density matrix, but we may want to mention the Shannon entropy too. When there is any risk of confusing these entropies, they are distinguished by their arguments. We will also use $S_i \equiv S(\rho_i)$ to denote the von Neumann entropy of a density matrix ρ_i acting on the Hilbert space $\mathcal{H}_i$.

In classical probability theory a state is a probability distribution and the Shannon entropy is a distinguished function of a probability distribution. In quantum theory a state is a density matrix, and a given density matrix ρ can be associated with many probability distributions because there are many possible POVMs. Also any density matrix can arise as a mixture of pure states in many different ways. From Section 8.4 we recall that if we write our density matrix as a mixture of normalised states,

$$\rho = \sum_{i=1}^{M} p_i |\psi_i\rangle \langle \psi_i|, \qquad (13.15)$$

then a large amount of arbitrariness is present, even in the choice of the number M. So if we define the *mixing entropy* of ρ as the Shannon entropy of the probability distribution $\vec{p}$ then this definition inherits a large amount of arbitrariness. But on reflection it is clear that there is one such definition that is more natural than the other. The point is that the density matrix itself singles out one preferred mixture, namely

$$\rho = \sum_{i=1}^{N} \lambda_i |e_i\rangle \langle e_i|, \qquad (13.16)$$

where $|e_i\rangle$ are the eigenvectors of ρ and N is the rank of ρ. The *von Neumann entropy* is[5]

---

[4] Golden was a person [350]. The Lieb inequality was proved in [599].
[5] The original reference is to von Neumann (1927) [940], whose main concern at the time was with statistical mechanics. His book [941], Wehrl's [949] review and the (more advanced) book by Ohya and Petz [699] serve as useful general references for this chapter.

$$S(\rho) \equiv -\mathrm{Tr}\rho \ln \rho = -\sum_{i=1}^{N} \lambda_i \ln \lambda_i. \tag{13.17}$$

Hence the von Neumann entropy is the Shannon entropy of the spectrum of ρ, and varies from zero for pure states to $\ln N$ for the maximally mixed state $\rho_* = \mathbb{1}/N$.

Further reflection shows that the von Neumann entropy has a very distinguished status among the various mixing entropies. While we were proving Schrödinger's mixture theorem in Section 8.4 we observed that any vector $\vec{p}$ occurring in Eq. (13.15) is related to the spectral vector $\vec{\lambda}$ by $\vec{p} = B\vec{\lambda}$, where B is a bistochastic (and indeed a unistochastic) matrix. Since the Shannon entropy is a Schur concave function, we deduce from the discussion in Section 2.1 that

$$S_{\mathrm{mix}} \equiv -\sum_{i=1}^{M} p_i \ln p_i \geq -\sum_{i=1}^{N} \lambda_i \ln \lambda_i = S(\rho). \tag{13.18}$$

Hence the von Neumann entropy is the smallest possible among all the mixing entropies S_{mix}.

The von Neumann entropy is a continuous function of the eigenvalues of ρ, and it can be defined in an axiomatic way as the only such function that satisfies a suitable set of axioms. In the classical case, the key axiom that singles out the Shannon entropy is the recursion property. In the quantum case this becomes a property that concerns *disjoint* states – two density matrices are said to be disjoint if they have orthogonal support, that is if their respective eigenvectors span orthogonal subspaces of the total Hilbert space.

- **Recursion property.** If the density matrices ρ_i have support in orthogonal subspaces $\mathcal{H}_i$ of a Hilbert space $\mathcal{H} = \oplus_{i=1}^{M} \mathcal{H}_i$, then the density matrix $\rho = \sum_i p_i \rho_i$ has the von Neumann entropy

$$S(\rho) = S(\vec{p}) + \sum_{i=1}^{M} p_i S(\rho_i). \tag{13.19}$$

Here $S(\vec{p})$ is a classical Shannon entropy. It is not hard to see that the von Neumann entropy has this property; if the matrix ρ_i has eigenvalues λ_{ij} then the eigenvalues of ρ are $p_i \lambda_{ij}$, and the result follows from the recursion property of the Shannon entropy (Section 2.2).

As with the Shannon entropy, the von Neumann entropy is interesting because of the list of the properties that it has and the theorems that can be proved using this list. So, instead of presenting a list of axioms we present a selection of these properties in Table 13.1, where we also compare the von Neumann entropy to the Shannon and Boltzmann entropies. Note that most of the entries concern a situation

Table 13.1 *Properties of entropies.*

Property	Equation	Von Neumann	Shannon	Boltzmann
Positivity	$S \geq 0$	Yes	Yes	No
Concavity	(13.22)	Yes	Yes	Yes
Monotonicity	$S_{12} \geq S_1$	No	Yes	No
Subadditivity	$S_{12} \leq S_1 + S_2$	Yes	Yes	Yes
Araki–Lieb inequality	$|S_1 - S_2| \leq S_{12}$	Yes	Yes	No
Strong subadditivity	$S_{123} + S_2 \leq S_{12} + S_{23}$	Yes	Yes	Yes

where ρ_1 is defined on a Hilbert space $\mathcal{H}_1$, ρ_2 on another Hilbert space $\mathcal{H}_2$, and ρ_{12} on their tensor product $\mathcal{H}_{12} = \mathcal{H}_1 \otimes \mathcal{H}_2$, or even more involved situations involving the tensor product of three Hilbert spaces. Moreover ρ_1 is always the reduced density matrix obtained by taking a partial trace of ρ_{12}, thus

$$S_1 \equiv S(\rho_1), \qquad S_{12} \equiv S(\rho_{12}), \qquad \rho_1 \equiv \mathrm{Tr}_2\rho_{12}, \qquad (13.20)$$

and so on (with the obvious modifications in the classical cases). As we will see, even relations that involve only one Hilbert space are conveniently proved through a detour into a larger Hilbert space. We can say more. In Section 9.3 we proved a purification lemma that states that the ancilla can always be chosen so that, for any ρ_1, it is true that $\rho_1 = \mathrm{Tr}_2\rho_{12}$ where ρ_{12} is a pure state. Moreover we proved that in this situation the reduced density matrices ρ_1 and ρ_2 have the same spectra (up to vanishing eigenvalues). This means that

$$\rho_{12}\rho_{12} = \rho_{12} \quad \Rightarrow \quad S_1 = S_2. \qquad (13.21)$$

If the ancilla purifies the system, the entropy of the ancilla is equal to the entropy of the original system.

Let us begin by taking note of the property (monotonicity) that the von Neumann entropy does not have. As we know very well from Chapter 9 a composite system can be in a pure state, so that $S_{12} = 0$, while its subsystems are mixed, so that $S_1 > 0$. In principle – although it might be a very contrived universe – your own entropy can increase without limit while the entropy of the world remains zero.

It is clear that the von Neumann entropy is positive. To convince ourselves of the truth of the rest of the entries in the table[6] we must rely on Section 13.1.

---

[6] These entries have a long history. Concavity and subadditivity were first proved by Delbrück and Molière (1936) [256]. Lanford and Robinson (1968) [573] observed that strong subadditivity is used in classical statistical mechanics and conjectured that it holds in the quantum case as well. Araki and Lieb (1970) [49] were unable to prove this but found other inequalities that were enough to complete the work of Lanford and Robinson. Eventually strong subadditivity was proved by Lieb and Ruskai (1973) [603].

Concavity and subadditivity are direct consequences of Klein's inequality, for the special case when A and B are density matrices, so that the right-hand side of Eq. (13.8) vanishes. First out is concavity, where all the density matrices live in the same Hilbert space.

- **Concavity.** If $\rho = p\sigma + (1 - p)\omega$, $0 \le p \le 1$, then

$$S(\rho) \ge pS(\sigma) + (1 - p)S(\omega). \tag{13.22}$$

In the proof we use Klein's inequality, with $A = \sigma$ or ω and $B = \rho$:

$$\mathrm{Tr}\rho \ln \rho = p\mathrm{Tr}\sigma \ln \rho + (1 - p)\mathrm{Tr}\omega \ln \rho \le p\mathrm{Tr}\sigma \ln \sigma + (1 - p)\mathrm{Tr}\omega \ln \omega. \tag{13.23}$$

Sign reversion gives the result, which is a lower bound on $S(\rho)$.

Using Peierl's inequality we can prove a much stronger result. Let f be any convex function. With $0 \le p \le 1$ and A and B any Hermitian operators, it will be true that

$$\mathrm{Tr}f(pA + (1 - p)B) \le p\mathrm{Tr}f(A) + (1 - p)\mathrm{Tr}B. \tag{13.24}$$

Namely, let $|e_i\rangle$ be the eigenvectors of $pA + (1 - p)B$. Then

$$\mathrm{Tr}f(pA + (1 - p)B) = \sum_i \langle e_i|f(pA + (1 - p)B)|e_i\rangle = \sum_i f(\langle e_i|pA + (1 - p)B|e_i\rangle)$$

$$\le p\sum_i f(\langle e_i|A|e_i\rangle) + (1 - p)\sum_i f(\langle e_i|B|e_i\rangle)$$

$$\le p\mathrm{Tr}f(A) + (1 - p)\mathrm{Tr}f(B) \tag{13.25}$$

where Peierl's inequality was used in the last step.

The recursion property (13.19) for disjoint states can be turned into an inequality that, together with concavity, nicely brackets the von Neumann entropy:

$$\rho = \sum_{a=1}^{K} p_a \rho_a \quad \Rightarrow \quad \sum_{a=1}^{K} p_a S(\rho_a) \le S(\rho) \le S(\vec{p}) + \sum_{a=1}^{K} p_a S(\rho_a). \tag{13.26}$$

(The index a is used because, in this chapter, i labels different Hilbert spaces.) To prove this, one first observes that for a pair of positive operators A, B one has the trace inequality

$$\mathrm{Tr}\,A[\ln(A + B) - \ln A] \ge 0. \tag{13.27}$$

This is true because $\ln t$ is operator monotone, and the trace of the product of two positive operators is positive. When $K = 2$ the upper bound in (13.26) follows if we first set $A = p_1\rho_1$ and $B = p_2\rho_2$, then $A = p_2\rho_2$, $B = p_1\rho_1$, add the resulting inequalities, and reverse the sign at the end. The result for arbitrary K follows if we use the recursion property (2.19) for the Shannon entropy $S(\vec{p})$.

The remaining entries in Table 13.1 concern density matrices defined on different Hilbert spaces, and the label on the density matrix tells us which Hilbert space.

- **Subadditivity.**

$$S(\rho_{12}) \leq S(\rho_1) + S(\rho_2), \qquad (13.28)$$

with equality if and only if $\rho_{12} = \rho_1 \otimes \rho_2$.

To prove this, use Klein's inequality with $B = \rho_1 \otimes \rho_2 = (\rho_1 \otimes \mathbb{1})(\mathbb{1} \otimes \rho_2)$ and $A = \rho_{12}$. Then

$$\mathrm{Tr}_{12}\rho_{12} \ln \rho_{12} \geq \mathrm{Tr}_{12}\rho_{12} \ln \rho_1 \otimes \rho_2 = \mathrm{Tr}_{12}\rho_{12}(\ln \rho_1 \otimes \mathbb{1} + \ln \mathbb{1} \otimes \rho_2)$$
$$= \mathrm{Tr}_1\rho_1 \ln \rho_1 + \mathrm{Tr}_2\rho_2 \ln \rho_2, \qquad (13.29)$$

which becomes subadditivity when we reverse the sign. It is not hard to see that equality holds if and only if $\rho_{12} = \rho_1 \otimes \rho_2$.

We can now give a third proof of concavity, since it is in fact a consequence of subadditivity. The trick is to use a two-level system, with orthogonal basis vectors $|a\rangle$ and $|b\rangle$, as an ancilla. The original density matrix will be written as the mixture $\rho_1 = p\rho_a + (1 - p)\rho_b$. Then we define

$$\rho_{12} = p\rho_a \otimes |a\rangle\langle a| + (1 - p)\rho_b \otimes |b\rangle\langle b|. \qquad (13.30)$$

By the recursion property

$$S_{12}(\rho_{12}) = S(p, 1 - p) + pS_1(\rho_a) + (1 - p)S_1(\rho_b). \qquad (13.31)$$

But $S_2 = S(p, 1 - p)$, so that subadditivity implies that S_1 is concave, as advertised.

- **The Araki–Lieb triangle inequality.**

$$|S(\rho_1) - S(\rho_2)| \leq S(\rho_{12}). \qquad (13.32)$$

This becomes a triangle inequality when combined with subadditivity.

The proof is a clever application of the fact that if a bipartite system is in a pure state (with zero von Neumann entropy) then the von Neumann entropies of the factors are equal. Of course the inequality itself quantifies how much the entropies of the factors can differ if this is not the case. But we can consider a purification of the state ρ_{12} using a Hilbert space $\mathcal{H}_{123}$. From subadditivity we know that $S_3 + S_1 \geq S_{13}$. By construction $S_{123} = 0$, so that $S_{13} = S_2$ and $S_3 = S_{12}$. A little rearrangement gives the result.

• **Strong subadditivity.**

$$S(\rho_{123}) + S(\rho_2) \leq S(\rho_{12}) + S(\rho_{23}).$$ (13.33)

This is a deep result, and we will not prove it – although it follows fairly easily from Lieb's inequality[7] (13.14). Let us investigate what it says, however. First, it is equivalent to the inequality

$$S(\rho_1) + S(\rho_2) \leq S(\rho_{13}) + S(\rho_{23}).$$ (13.34)

To see this, purify the state ρ_{123} by factoring with a fourth Hilbert space. Then we have

$$S_{1234} = 0 \quad \Rightarrow \quad S_{123} = S_4 \ \& \ S_{12} = S_{34}.$$ (13.35)

Inserting this in (13.33) yields (13.34), and conversely. This shows that strong subadditivity of the Shannon entropy is a rather trivial thing, since in that case monotonicity implies that $S_1 \leq S_{13}$ and $S_2 \leq S_{23}$. In the quantum case these inequalities do not hold separately, but their sum does!

The second observation is that strong subadditivity implies subadditivity – to see this, let the Hilbert space $\mathcal{H}_2$ be one dimensional, so that $S_2 = 0$. It implies much more, though. It is tempting to say that every deep result follows from it; we will meet with an example in the next section. Meanwhile we can ask whether this is the end of the story. Suppose we have a state acting on the Hilbert space $\mathcal{H}_1 \otimes \mathcal{H}_2 \otimes \cdots \otimes \mathcal{H}_n$. Taking partial traces in all possible ways we get a set of $2^n - 1$ non-trivial density matrices, and hence $2^n - 1$ possible entropies constrained by the inequalities in Table 13.1. These inequalities define a convex cone in an $2^n - 1$ dimensional space, and we can then ask whether the possible entropies fill out this cone. The answer is no. There are points on the boundary of the cone that cannot be reached in this way, and there may be further inequalities waiting to be discovered [615].

To end on a somewhat different note, we recall that the operational significance of the Shannon entropy was made crystal clear by Shannon's noiseless coding theorem. There is a corresponding quantum noiseless coding theorem. To state it, we imagine that Alice has a string of pure states $|\psi_i\rangle$, generated by the probabilities p_i. She codes her states in qubit states, using a *channel system C*. The qubits are sent to Bob, who decodes them and produces a string of output states ρ_i. The question is: how many qubits must be sent over the channel if the message is to go through undistorted? More precisely, we want to know the *average fidelity*

---

[7] For a proof – in fact several proofs – and more information on entropy inequalities generally, we recommend two reviews written by experts, Lieb (1975) [600] and Ruskai (2002) [784]. The structure of states which saturate strong subadditivity was studied in [416].

$$\bar{F} = \sum_i p_i \langle \psi_i | \rho_i | \psi_i \rangle \tag{13.36}$$

that can be achieved. The quantum problem is made more subtle by the fact that generic pure states cannot be distinguished with certainty, though the answer is given by

Schumacher's noiseless coding theorem. *Let*

$$\rho = \sum_i p_i |\psi_i\rangle\langle\psi_i| \quad \text{and} \quad S(\rho) = -\text{Tr}\rho \log_2 \rho. \tag{13.37}$$

Also let $\epsilon, \delta > 0$ and let $S(\rho) + \delta$ qubits be available in the channel per input state. Then, for large $\mathcal{N}$, it is possible to transmit blocks of $\mathcal{N}$ states with average fidelity $\bar{F} > 1 - \epsilon$.

This theorem marks the beginning of quantum information theory.[8]

13.3 Quantum relative entropy

In the classical case the relative entropy of two probability distributions played a key role, notably as a measure of how different two such distributions are from each other. There is a quantum relative entropy too, and for roughly similar reasons it plays a key role in the description of the quantum state space. In some ways it is a deeper concept than the von Neumann entropy itself, and we will see several uses of it as we proceed. The definition looks deceptively simple: for any pair of quantum states ρ and σ their *relative entropy* is[9]

$$S(\rho||\sigma) \equiv \text{Tr}[\rho(\ln\rho - \ln\sigma)]. \tag{13.38}$$

If σ has zero eigenvalues this may diverge, otherwise it is a finite and continuous function. The quantum relative entropy reduces to the classical Kullback–Leibler relative entropy for diagonal matrices, but is not as easy to handle in general. Using the result of Problem 13.1, it can be rewritten as

$$S(\rho||\sigma) = \int_0^\infty \text{Tr}\rho \frac{1}{\sigma + u\mathbb{1}}(\rho - \sigma)\frac{1}{\rho + u\mathbb{1}} du. \tag{13.39}$$

This is convenient for some manipulations that one may want to perform.

---

[8] The quantum noiseless coding theorem is due to Schumacher (1995) [808] and Jozsa and Schumacher (1994) [511]; a forerunner is due to Holevo (1973) [443]. For Shannon's theorem formulated in the same language, see section 3.2 of Cover and Thomas [237].

[9] The relative entropy was introduced into quantum mechanics by Umegaki (1962) [914] and resurfaced in a paper by Lindblad (1973) [608]. A general reference for this section is Ohya and Petz [699]; for valuable reviews see Schumacher and Westmoreland [811], and Vedral [917].

Two of the properties of relative entropy are immediate:

- **Unitary invariance.** $S(\rho_1||\rho_2) = S(U\rho_1 U^\dagger||U\rho_2 U^\dagger)$.
- **Positivity.** $S(\rho||\sigma) \geq 0$ with equality if and only if $\rho = \sigma$.

The second property is immediate only because it is precisely the content of Klein's inequality – and we proved that in Section 13.1. More is true:

$$S(\rho||\sigma) \geq \frac{1}{2}\mathrm{Tr}(\rho - \sigma)^2 = D_2^2(\rho,\sigma). \tag{13.40}$$

This is as in the classical case, Eq. (2.30); in both cases a stronger statement can be made, and we will come to it in Chapter 14. In general $S(\rho||\sigma) \neq S(\sigma||\rho)$ which is also as in the classical case.

Three deep properties of relative entropy are:

- **Joint convexity.** For any $p \in [0, 1]$ and any four states

$$S(p\rho_a + (1-p)\rho_b||p\sigma_c + (1-p)\sigma_d) \leq pS(\rho_a||\sigma_c) + (1-p)S(\rho_b||\sigma_d). \tag{13.41}$$

- **Monotonicity under partial trace.**

$$S(\mathrm{Tr}_2\rho_{12}||\mathrm{Tr}_2\sigma_{12}) \leq S(\rho_{12}||\sigma_{12}). \tag{13.42}$$

- **Monotonicity under CP maps.** For any completely positive map Φ

$$S(\Phi\rho||\Phi\sigma) \leq S(\rho||\sigma). \tag{13.43}$$

Any of these properties imply the other two, and each is equivalent to strong sub-additivity of the von Neumann entropy.[10] The importance of monotonicity is obvious – it implies everything that monotonicity under stochastic maps implies for the classical Kullback–Leibler relative entropy.

It is clear that monotonicity under CP maps implies monotonicity under partial trace – taking a partial trace is a special case of a CP map. To see the converse, use the environmental representation of a CP map given in Eq. (10.61); we can always find a larger Hilbert space in which the CP map is represented as

$$\rho' = \Phi(\rho) = \mathrm{Tr}_2(U\rho \otimes P_\nu U^\dagger), \tag{13.44}$$

where P_ν is a projector onto a pure state of the ancilla. A simple calculation ensures:

$$S\left(\mathrm{Tr}_2(U\rho \otimes P_\nu U^\dagger)||\mathrm{Tr}_2(U\sigma \otimes P_\nu U^\dagger)\right)$$
$$\leq S\left(U\rho \otimes P_\nu U^\dagger||U\sigma \otimes P_\nu U^\dagger\right) = S(\rho \otimes P_\nu||\sigma \otimes P_\nu) = S(\rho||\sigma), \tag{13.45}$$

---

[10] Again the history is intricate. Monotonicity of relative entropy was proved from strong subadditivity by Lindblad [610]. A proof from first principles is due to Uhlmann [905].

where we used monotonicity under partial trace, unitary invariance and the easily proved additivity property that

$$S(\rho_1 \otimes \rho_2||\sigma_1 \otimes \sigma_2) = S(\rho_1||\rho_2) + S(\rho_2||\sigma_2). \tag{13.46}$$

To see that monotonicity under partial trace implies strong subadditivity, we introduce a third Hilbert space and consider

$$S(\rho_{23}||\rho_2 \otimes \mathbb{1}) \leq S(\rho_{123}||\rho_{12} \otimes \mathbb{1}). \tag{13.47}$$

Now we just apply the definition of relative entropy and rearrange terms to arrive at Eq. (13.33). The converse statement, that strong subadditivity implies monotonicity under partial trace, is true as well. One proof proceeds via the Lieb inequality (13.14).

The close link between the relative entropy and the von Neumann entropy can be unearthed as follows: the relative entropy between ρ and the maximally mixed state ρ_* is

$$S(\rho||\rho_*) = \ln N - S(\rho). \tag{13.48}$$

This is the quantum analogue of (2.37) and shows that the von Neumann entropy $S(\rho)$ is implicit in the definition of the relative entropy. In a sense the link goes the other way too. Form the one-parameter family of states

$$\rho_p = p\rho + (1 - p)\sigma, \qquad p \in [0, 1]. \tag{13.49}$$

Then define the function

$$f(p) \equiv S(\rho_p) - pS(\rho) - (1 - p)S(\sigma). \tag{13.50}$$

With elementary manipulation this can be rewritten as

$$f(p) = pS(\rho||\rho_p) + (1 - p)S(\sigma||\rho_p) = pS(\rho||\sigma) - S(\rho_p||\sigma). \tag{13.51}$$

From the strict concavity of the von Neumann entropy we conclude that $f(p) \geq 0$, with equality if and only if $p = 0, 1$. This further implies that the derivative of f is positive at $p = 0$. We are now in position to prove that

$$\lim_{p \to 0} \frac{1}{p} f(p) = S(\rho||\sigma). \tag{13.52}$$

This is so because the limit exists [608] and because Eq. (13.51) implies, first, that the limit is greater than or equal to $S(\rho||\sigma)$, and, second, that it is smaller than or equal to $S(\rho||\sigma)$. In this sense the definition of relative entropy is implicit in the definition of the von Neumann entropy. If we recall Eq. (1.13) for convex

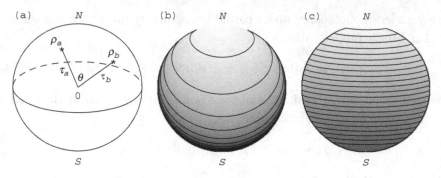

Figure 13.1 Relative entropy between $N = 2$ mixed states depends on the lengths of their Bloch vectors τ_i and the angle θ between them (a). Relative entropies with respect to the north pole ρ_N: (b) $S(\rho||\rho_N)$ and (c) $S(\rho_N||\rho)$.

functions – and reverse the sign because $f(p)$ is concave – we can also express the conclusion as

$$S(\rho||\sigma) = \sup_p \frac{1}{p}\Big(S(\rho_p) - pS(\rho) - (1-p)S(\sigma)\Big). \tag{13.53}$$

The same argument applies in the classical case, and in Section 14.1 we will see that the symmetric special case $f(1/2)$ deserves attention for its own sake.

For $N = 2$ Cortese [235] found an explicit formula for the relative entropy between any two mixed states,

$$S(\rho_a||\rho_b) = \frac{1}{2}\ln\left(\frac{1-\tau_a^2}{1-\tau_b^2}\right) + \frac{\tau_a}{2}\ln\left(\frac{1+\tau_a}{1-\tau_a}\right) - \frac{\tau_a\cos\theta}{2}\ln\left(\frac{1+\tau_b}{1-\tau_b}\right), \tag{13.54}$$

where the states are represented by their Bloch vectors, for example $\rho_a = \frac{1}{2}(\mathbb{1}+\vec{\tau}_a \cdot \vec{\sigma})$, τ_a is the length of a Bloch vector, and θ is the angle between the two. See Figure 13.1, in which the relative entropy with respect to a pure state is shown. Note that the data along the polar axis, representing diagonal matrices, coincide with those plotted at the vertical axis in Figure 2.8c and d for the classical case.

Now we would like to argue that the relative entropy, as defined above, is indeed a quantum analogue of the classical Kullback–Leibler relative entropy. We could have tried a different way of defining quantum relative entropy, starting from the observation that there are many probability distributions associated with each density matrix, in fact one for every POVM $\{E_i\}$. Since we expect relative entropy to serve as an information divergence, that is to say that it should express 'how far' from each other two states are in the sense of statistical hypothesis testing, this suggests that we should define a relative entropy by taking the supremum over all possible POVMs:

$$S_1(\rho||\sigma) = \sup_E \sum_i p_i \ln \frac{p_i}{q_i}, \quad \text{where} \quad p_i = \text{Tr}E_i\rho \quad \text{and} \quad q_i = \text{Tr}E_i\sigma. \quad (13.55)$$

Now it can be shown [609] (using monotonicity of relative entropy) that

$$S_1(\rho||\sigma) \leq S(\rho||\sigma). \quad (13.56)$$

We can go on to assume that we have several independent and identically distributed systems that we can make observations on, that is to say that we can make measurements on states of the form $\rho^{\mathcal{N}} \equiv \rho \otimes \rho \otimes \cdots \otimes \rho$ (with $\mathcal{N}$ identical factors altogether, and with a similar definition for $\sigma^{\mathcal{N}}$). We optimise over all POVMs $\{\tilde{E}_i\}$ on the tensor product Hilbert space, and define

$$S_{\mathcal{N}}(\rho||\sigma) = \sup_{\tilde{E}} \frac{1}{\mathcal{N}} \sum_i p_i \ln \frac{p_i}{q_i}, \quad p_i = \text{Tr}\tilde{E}_i\rho^{\mathcal{N}}, \quad q_i = \text{Tr}\tilde{E}_i\sigma^{\mathcal{N}}. \quad (13.57)$$

In the large Hilbert space we have many more options for making collective measurements, so this ought to be larger than $S_1(\rho||\sigma)$. Nevertheless we have the bound [272]

$$S_{\mathcal{N}}(\rho||\sigma) \leq S(\rho||\sigma). \quad (13.58)$$

Even more is true. In the limit when the number of copies of our states go to infinity, it turns out that

$$\lim_{\mathcal{N}\to\infty} S_{\mathcal{N}} = S(\rho||\sigma). \quad (13.59)$$

This limit can be achieved by projective measurements. We do not intend to prove these results here, we only quote them in order to inspire some confidence in Umegaki's definition of quantum relative entropy.[11]

13.4 Other entropies

In the classical case we presented a wide selection of alternative definitions of entropy, some of which are quite useful. When we apply our rule of thumb – turn sums into traces – to Section 2.7, we obtain (among others) the *quantum Rényi entropy*, labelled by a non-negative parameter q,

$$S_q(\rho) \equiv \frac{1}{1-q} \ln[\text{Tr}\rho^q] = \frac{1}{1-q} \ln \left[\sum_{i=1}^{N} \lambda_i^q \right]. \quad (13.60)$$

---

[11] The final result here is due to Hiai and Petz (1991) [433]. And the reader should be warned that our treatment is somewhat simplified.

This is a function of the L_q-norm of the density matrix, $||\rho||_q = \left(\frac{1}{2}\text{Tr}\rho^q\right)^{1/q}$. It is non-negative and in the limit $q \to 1$ it tends to the von Neumann entropy $S(\rho)$. The logarithm is used in the definition to ensure additivity for product states:

$$S_q(\rho_1 \otimes \rho_2) = S_q(\rho_1) + S_q(\rho_2) \tag{13.61}$$

for any real q. This is immediate, given the spectrum of a product state (see Problem 9.4). The quantum Rényi entropies fulfil properties already discussed for their classical versions. In particular, for any value of the coefficient q the generalised entropy S_q equals zero for pure states and achieves its maximum $\ln N$ for the maximally mixed state ρ_*. In analogy to (2.80), the Rényi entropy is a continuous, non-increasing function of its parameter q.

Some special cases of S_q are often encountered. The quantity $\text{Tr}\rho^2$, called the *purity* of the quantum state, is frequently used since it is easy to compute. The larger the purity, the more pure the state (or more precisely, the larger is its Hilbert–Schmidt distance from the maximally mixed state). Obviously one has $S_2(\rho) = -\ln[\text{Tr}\rho^2]$. The *Hartley entropy* S_0 is a function of the rank r of ρ; $S_0(\rho) = \ln r$. In the other limiting case the entropy depends on the largest eigenvalue of ρ; $S_\infty = -\ln \lambda_{\max}$. For any positive, finite value of q the generalised entropy is a continuous function of the state ρ. The Hartley entropy is not continuous at all. The concavity relation (13.22) holds at least for $q \in (0, 1]$, and the quantum Rényi entropies for different values of q are correlated in the same way as their classical counterparts (see Section 2.7). They are additive for product states, but not subadditive. A weak version of subadditivity holds [915]:

$$S_q(\rho_1) - S_0(\rho_2) \leq S_q(\rho_{12}) \leq S_q(\rho_1) + S_0(\rho_2), \tag{13.62}$$

where S_0 denotes the Hartley entropy – the largest of the Rényi entropies.

The entropies considered so far have been unitarily invariant, and they take the value zero for any pure state. This is not always an advantage. An interesting alternative is the Wehrl entropy, that is the classical Boltzmann entropy of the Husimi function $Q(z) = \langle z|\rho|z\rangle$. This is not unitarily invariant because it depends on the choice of a special set of coherent states $|z\rangle$ (see Sections 6.2 and 7.4). The Wehrl entropy is important in situations where this set is physically singled out, say as 'classical states'. A key property is [950]

Wehrl's inequality. *For any state ρ the Wehrl entropy is bounded from below by the von Neumann entropy,*

$$S_W(\rho) \geq S(\rho). \tag{13.63}$$

To prove this it is sufficient to use the continuous version of Peierls' inequality (13.11): for any convex function f convexity implies

$$\text{Tr}f(\rho) = \int_\Omega \langle z|f(\rho)|z\rangle \, \text{d}^2 z \geq \int_\Omega f(\langle z|\rho|z\rangle) \, \text{d}^2 z = \int_\Omega f(Q(z)) \, \text{d}^2 z. \tag{13.64}$$

Setting $f(t) = t \ln t$ and reverting the sign of the inequality we get Wehrl's result. Rényi–Wehrl entropies can be defined similarly, and the argument applies to them as well, so that for any $q \geq 0$ and any state ρ the inequality $S_q^{RW}(\rho) \geq S_q(\rho)$ holds.

For composite systems we can define a Husimi function by

$$Q(z_1, z_2) = \langle z_1 | \langle z_2 | \rho_{12} | z_1 \rangle | z_2 \rangle \tag{13.65}$$

and analyse its Wehrl entropy (see Problem 13.4). For a pure product state the Husimi function factorises and its Wehrl entropy is equal to the sum of the Wehrl entropies of both subsystems. There are two possible definitions of the marginal Husimi distribution, and happily they agree, in the sense that

$$Q(z_1) \equiv \int_{\Omega_2} Q(z_1, z_2) \, \text{d}^2 z_2 = \langle z_1 | \text{Tr}_2 \rho_{12} | z_1 \rangle \equiv Q(z_1). \tag{13.66}$$

The Wehrl entropy can then be shown to be very well behaved, in particular it is monotone in the sense that $S_{12} \geq S_1$. Like the Shannon entropy, but unlike the Boltzmann entropy when the latter is defined over arbitrary distributions, the Wehrl entropy obeys all the inequalities in Table 13.1.

Turning to relative entropy we find many alternatives to Umegaki's definition. Many of them reproduce the classical relative entropy (2.25) when their two arguments commute. An example is the *Belavkin–Staszewski relative entropy* [103]

$$S_{BS}(\rho||\sigma) = \text{Tr}\left(\rho \ln \rho^{1/2} \sigma^{-1} \rho^{1/2}\right). \tag{13.67}$$

This is monotone, and it can be shown that $S_{BS}(\rho||\sigma) \geq S(\rho||\sigma)$ [433].

The classical relative entropy itself is highly non-unique. We gave a very general class of monotone classical relative entropies in Eq. (2.74). In the quantum case we insist on monotonicity under completely positive stochastic maps, but not really on much else besides. A straightforward attempt to generalise the classical definition to the quantum case encounters the difficulty that the operator $\frac{\rho}{\sigma}$ is ambiguous in the non-commutative case. There are various ways of circumventing this difficulty, and then one can define a large class of monotone relative entropies. Just to be specific, let us mention a one-parameter family of monotone and jointly convex relative entropies:

$$S_\alpha(\rho, \sigma) = \frac{4}{1 - \alpha^2} \text{Tr}\left(\mathbb{1} - \sigma^{(\alpha+1)/2} \rho^{(\alpha-1)/2}\right)\rho, \quad -1 < \alpha < 1. \tag{13.68}$$

Umegaki's definition (13.38) is recovered a limiting case. In fact in the limit $\alpha \to -1$ we obtain $S(\rho||\sigma)$, while we get $S(\sigma||\rho)$ when $\alpha \to 1$. Many more monotone relative entropies exist.[12]

13.5 Majorisation of density matrices

The von Neumann entropy (like the Rényi entropies, but unlike the Wehrl entropy) provides a measure of the 'degree of mixing' of a given quantum state. A more sophisticated ordering of quantum states, with respect to the degree of mixing, is provided by the theory of majorisation (Section 2.1). In the classical case the majorisation order is really between orbits of probability vectors under permutations of their components – a fact that is easily missed since in discussing majorisation one tends to represent these orbits with a representative $\vec{p}$, whose components appear in non-increasing order. When we go from probability vectors to density matrices, majorisation will provide an ordering of the orbits under the unitary group. By definition the state σ is majorised by the state ρ if and only if the eigenvalue vector of σ is majorised by the eigenvalue vector of ρ,

$$\sigma \prec \rho \quad \Leftrightarrow \quad \vec{\lambda}(\sigma) \prec \vec{\lambda}(\rho). \tag{13.69}$$

This ordering relation between matrices has many advantages; in particular it forms a lattice.[13]

The first key fact to record is that if $\sigma \prec \rho$ then σ lies in the convex hull of the unitary orbit to which ρ belongs. We state this as a theorem:

Uhlmann's majorisation theorem. *If two density matrices of size N are related by $\sigma \prec \rho$, then there exists a probability vector $\vec{p}$ and unitary matrices U_I such that*

$$\sigma = \sum_I p_I U_I \rho U_I^\dagger. \tag{13.70}$$

Despite its importance, this theorem is easily proved. Suppose σ is given in diagonal form. We can find a unitary matrix U_I such that $U_I \rho U_I^\dagger$ is diagonal too; in fact we can find altogether $N!$ such unitary matrices since we can permute the eigenvalues. But now all matrices are diagonal and we are back to the classical case. From the classical theory we know that the eigenvalue vector of σ lies in the convex hull of the $N!$ different eigenvalue vectors of ρ. This provides one way of realizing Eq. (13.70).

---

[12] See Petz [732] and Lesniewski and Ruskai [588] for the full story here.
[13] In physics, the subject of this section was begun by Uhlmann (1971) [904]; the work by him and his school is summarised by Alberti and Uhlmann (1982) [16]. A more recent survey is by Ando [38].

a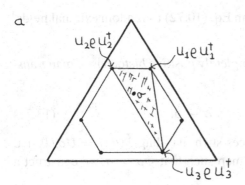

b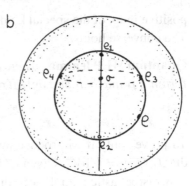

Figure 13.2 Left: $N = 3$, and we show majorisation in the eigenvalue simplex. Right: $N = 2$, and we show two different ways of expressing a given σ as a convex sum of points on the orbit (itself a sphere!) of a majorizing ρ.

There are many ways of realizing σ as a convex sum of states on the orbit of ρ, as we can see from Figure 13.2. In fact it is known that we can arrange things so that all components of p_I become equal. The related but weaker statement that any density matrix can be realised as a uniform mixture of pure states is very easy to prove [113]. Let $\sigma = \text{diag}(\lambda_i)$. For $N = 3$, say, form a closed curve of pure state vectors by

$$z^\alpha(\tau) = \begin{bmatrix} e^{in_1\tau} & 0 & 0 \\ 0 & e^{in_2\tau} & 0 \\ 0 & 0 & e^{in_3\tau} \end{bmatrix} \begin{bmatrix} \sqrt{\lambda_1} \\ \sqrt{\lambda_2} \\ \sqrt{\lambda_3} \end{bmatrix}, \tag{13.71}$$

where the n_i are integers. Provided that the n_i are chosen so that $n_i - n_j$ is non-zero when $i \neq j$, it is easy to show that

$$\sigma = \frac{1}{2\pi} \int_0^{2\pi} d\tau \, Z^\alpha \bar{Z}_\beta(\tau). \tag{13.72}$$

The off-diagonal terms are killed by the integration, so that σ is realised by a mixture of pure states distributed uniformly on the circle. The argument works for any N. Moreover a finite set of points on the curve will do as well, but we need at least N points since then we must ensure that $n_i - n_j \neq 0$ modulo N. When $N > 2$ these results are somewhat surprising – it was not obvious that one could find such a curve consisting only of pure states, since the set of pure states is a small subset of the outsphere.

Return to Uhlmann's theorem: in the classical case bistochastic matrices made their appearance at this point. This is true in the quantum case also; the theorem explicitly tells us that σ can be obtained from ρ by a bistochastic completely positive map, of the special kind known from Eq. (10.72) as random external fields. The converse holds:

Quantum HLP lemma. *There exists a completely positive bistochastic map transforming ρ into σ if and only if $\sigma \prec \rho$,*

$$\rho \xrightarrow{\text{bistochastic}} \sigma \quad \Leftrightarrow \quad \sigma \prec \rho. \tag{13.73}$$

To prove 'only if', introduce unitary matrices such that $\text{diag}(\vec{\lambda}(\sigma)) = U\sigma U^{\dagger}$ and $\text{diag}(\vec{\lambda}(\rho)) = V\rho V^{\dagger}$. Given a bistochastic map such that $\Phi\rho = \sigma$ we construct a new bistochastic map Ψ according to

$$\Psi X \equiv U[\Phi(V^{\dagger}XV)]U^{\dagger}. \tag{13.74}$$

By construction $\Psi\left(\text{diag}(\vec{\lambda}(\rho))\right) = \text{diag}(\vec{\lambda}(\sigma))$. Next we introduce a complete set of projectors P_i onto the basis vectors, in the basis we are in. We use them to construct a matrix whose elements are given by

$$B_{ij} \equiv \text{Tr}P_i\Psi P_j. \tag{13.75}$$

We recognise that B is a bistochastic matrix, and finally we observe that

$$\lambda_i(\sigma) = \text{Tr}P_i \, \text{diag}(\vec{\lambda}(\sigma)) = \text{Tr}P_i\Psi\left(\sum_j P_j\lambda_j(\rho)\right) = \sum_j B_{ij}\lambda_j(\rho), \tag{13.76}$$

where we used linearity in the last step. An appeal to the classical HLP lemma concludes the proof.

An interesting example of a completely positive and bistochastic map is the operation of coarse graining with respect to a given Hermitian operator H (e.g. a Hamiltonian). We denote it by Φ_{CG}^H, and define it by

$$\rho \rightarrow \Phi_{CG}^H(\rho) = \sum_{i=1}^{N} P_i\rho P_i = \sum_{i=1}^{N} p_i \, |h_i\rangle\langle h_i|, \tag{13.77}$$

where the P_i project onto the eigenvectors $|h_i\rangle$ of H (assumed to be non-degenerate for simplicity) – see Figure 13.3. In more mundane terms, this is the map that deletes all off-diagonal elements from a density matrix. It obeys

The Schur–Horn theorem. *Let ρ be a Hermitian matrix, $\vec{\lambda}$ its spectrum and $\vec{p}$ its diagonal elements in a given basis. Then*

$$\vec{p} \prec \vec{\lambda}. \tag{13.78}$$

Conversely, if this equation holds then there exists a Hermitian matrix with spectrum $\vec{\lambda}$ whose diagonal elements are given by $\vec{p}$.

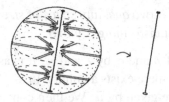

Figure 13.3 Coarse graining a density matrix with a map Φ_{CG}.

We prove this one way. There exists a unitary matrix that diagonalises the matrix, so we can write

$$p_i = \rho_{ii} = \sum_{j,k} U_{ij} \lambda_j \delta_{jk} U_{ki}^\dagger = \sum_j |U_{ij}|^2 \lambda_j. \tag{13.79}$$

The vector $\vec{p}$ is obtained by acting on $\vec{\lambda}$ with a unistochastic, hence bistochastic, matrix, and the result follows from Horn's lemma (Section 2.1).[14]

The Schur–Horn theorem has weighty consequences. It is clearly of interest when one tries to quantify decoherence, since the entropy of the coarse grained density matrix $\Phi_H \rho$ will be greater than that of ρ. It also leads to an interesting definition of the von Neumann entropy that again brings out the distinguished status of the latter. Although we did not bring it up in Section 13.2, we could have defined the entropy of a density matrix relative to any pure POVM $\{E_i\}$, as the Shannon entropy of the probability distribution defined cooperatively by the POVM and the density matrix. That is, $S(\rho) \equiv S(\vec{p})$ where $p_i = \mathrm{Tr} E_i \rho$. To make the definition independent of the POVM, we could then minimise the resulting entropy over all possible pure POVMs, so a possible definition that depends only on ρ itself would be

$$S(\rho) \equiv \min_{\mathrm{POVM}} S(\vec{p}), \qquad p_i = \mathrm{Tr} E_i \rho. \tag{13.80}$$

But the entropy defined in this way is equal to the von Neumann entropy. The Schur–Horn theorem shows this for the special case where we minimise only over projective measurements, and the argument can be extended to cover the general case. Note that Wehrl's inequality (13.63) is really a special case of this observation, since the Wehrl entropy is the entropy that we get from the POVM defined by the coherent states.

Now recall the definition of unistochastic maps, Eq. (10.64). Inspired by the original Horn's lemma (Section 2.1) one may ask if the word 'bistochastic' in the quantum HLP lemma can be replaced by 'unistochastic'. Indeed it can.[15]

---

[14] This part of the theorem is due to Schur (1923) [814]. Horn (1954) [449] proved the converse.
[15] An early version of this result [461] had to assume an ancilla of a large size. This restriction was removed by Scharlau and Müller [796], whose proof we follow.

Quantum Horn's lemma. *If two quantum states of size N satisfy $\sigma \prec \rho$ then there exists a unistochastic map transforming ρ into σ.*

Let us denote the spectra of ρ and σ by $\vec{\lambda}$ and $\vec{\mu}$. It is assumed that $\vec{\mu} \prec \vec{\lambda}$, so due to the Schur–Horn theorem there exists a unitary $V_1 \in U(N)$ such that the diagonal elements of $\rho' = V_1 \rho V_1^\dagger$ are given by $\vec{\mu}$. We then construct a coarse graining map using a cyclic permutation matrix X defined by $X|i\rangle = |i+1\rangle$, where the basis vectors are labelled by integers modulo N. This allows us to define a unitary matrix of size N^2,

$$W = \sum_{m=1}^{N} |m\rangle\langle m| \otimes X^m. \tag{13.81}$$

We transform the state ρ' into $\sigma' = \Phi_W(\rho') = \mathrm{Tr}_B[W(\rho' \otimes \frac{\mathbb{1}}{N})W^\dagger]$. For the matrix elements of σ' we find $\sigma'_{ij} = \frac{1}{N}\langle i|\rho'|j\rangle\mathrm{Tr}(X^{i-j})$. But $\mathrm{Tr}(X^{i-j}) = N\delta_{ij}$, so all non-diagonal elements have been removed. A final unitary transformation $V_2 \in U(N)$ turns the state σ' into the target state, $\sigma = V_2\sigma'V_2^\dagger$. Thus the unistochastic map defined by $U = (V_2 \otimes \mathbb{1})W(V_1 \otimes \mathbb{1})$ performs the required task, $\sigma = \Phi_U(\rho)$.

The Schur–Horn theorem is indeed useful and, from a mathematical point of view, much more interesting than it appears to be at first sight. To begin to see why, we can restate it: consider the map that takes a Hermitian matrix to its diagonal entries. Then the theorem says that the image of the space of isospectral Hermitian matrices, under this map, is a convex polytope whose corners are the $N!$ fixed points of the map. Already it sounds more interesting! Starting from this example, mathematicians have developed a theory that deals with maps from connected symplectic manifolds and conditions under which the image will be the convex hull of the image of the set of fixed points of a group acting on the manifold.[16]

13.6 Proof of the Lieb conjecture

The time has come to go back to the inequality that formed the subject of Section 7.4. There we called it the Lieb conjecture, but 35 years after its formulation it became a theorem [604]. The original conjecture stated that the Wehrl entropy, that is Boltzmann's entropy for the Husimi function $Q_\psi(z, \bar{z}) = \langle\psi|z\rangle\langle z|\psi\rangle$, assumes its minimum for the $SU(2)$ coherent states, that is for the maximal weight states of $SU(2)$. An example is the special Dicke state $|n, 0\rangle$. In the notation used here, a general Dicke state is written as $|n_+, n_-\rangle = |j+m, j-m\rangle$. Should it be necessary, the reader can refer to Appendix B.2 to refresh his or her recollections of Schwinger's oscillator representation of $SU(2)$. It will be used freely in this section.

---

[16] For an overview of this theory, and its connections to symplectic geometry and to interesting problems of linear algebra, see Knutson [549]. We return to these matters in Section 17.6.

The theorem that eventually emerged covers more than the original Lieb conjecture since it concerns general concave functions, and not the function $f(x) = -x \ln x$ only. To be precise:

The Lieb–Solovej theorem. *Let $f : [0, 1] \to \mathbb{R}$ be any concave function. Then it holds that*

$$\frac{n+1}{4\pi} \int d\Omega\, f(\langle z|\rho|z\rangle) \geq \frac{n+1}{4\pi} \int d\Omega\, f(\langle z|n, 0\rangle\langle n, 0|z\rangle). \qquad (13.82)$$

The measure $d\Omega$ is the usual measure on the round unit sphere. The theorem will hold for all ρ if it holds for all pure states, so the statement to be proved can be expressed in terms of the Husimi function as the statement that

$$\frac{n+1}{4\pi} \int d\Omega\, f(Q_\psi) \geq \frac{n+1}{4\pi} \int d\Omega\, f(Q_{|n,0\rangle}). \qquad (13.83)$$

The proof is somewhat lengthy, but pleasantly so, and traverses ground that we have covered after closing Chapter 7.

Lieb and Solovej introduced a CP map that takes density matrices acting on $\mathcal{H}^{n+1}$ to density matrices acting on a Hilbert space of one more dimension. The map is defined in terms of Schwinger's oscillators by

$$\Phi^1(\rho) = \frac{1}{n+2} \left(a_+^\dagger \rho a_+ + a_-^\dagger \rho a_- \right). \qquad (13.84)$$

This is a CP map written in Kraus form, which is obviously trace preserving and has the key property that

$$\Phi^1([J_i, \rho]) = [J_i, \Phi^1(\rho)], \qquad (13.85)$$

where J_i are the angular momentum operators defined in Eq. (B.9). What this equation says is that states on the same $SU(2)$ orbit in $\mathcal{H}^{n+1}$ will map to states on the same $SU(2)$ orbit in $\mathcal{H}^{n+2}$. Thus the map Φ^1 talks nicely to $SU(2)$.

The map Φ^1 can be iterated as many times as one wants, and it will remain a trace preserving CP map taking $SU(2)$ orbits to $SU(2)$ orbits. The result is

$$\Phi^{m-n}(\rho) = \frac{(n+1)!}{(m+1)!} \sum a^\dagger \ldots a^\dagger \rho a \ldots a$$

$$= \frac{(n+1)!}{(m+1)!} \sum_{k=0}^{m-n} \binom{m-n}{k} (a_-^\dagger)^k (a_+^\dagger)^{m-n-k} \rho a_-^k a_+^{m-n-k}. \qquad (13.86)$$

The map Φ^{m-n} takes density matrices acting on $\mathcal{H}^{n+1}$ to density matrices acting on $\mathcal{H}^{m+1}$. Throughout it is understood that $m > n$. In a sense this map allows us to approach the classical limit in stages.[17] There is a reverse Lieb–Solovej map taking us from a larger Hilbert space to a smaller:

$$\Phi^{n-m}(\rho) = \frac{n!}{m!} \sum a \dots a\rho a^\dagger \dots a^\dagger. \tag{13.87}$$

It is easy to convince oneself that the image, under this reverse map, of a pure state will be pure if and only if that pure state is coherent. If this is so the image is coherent too. It is true that

$$\text{Tr}_m \rho_m \Phi^{m-n}(\rho_n) = \frac{n+1}{m+1} \text{Tr}_n \Phi^{n-m}(\rho_m) \rho_n. \tag{13.88}$$

Taking ρ_n to be a pure state we see that the largest eigenvalue of $\Phi^{m-n}(\rho_n)$ is bounded from above by

$$\lambda_0 \le \frac{n+1}{m+1}. \tag{13.89}$$

Equality will hold only if $\Phi^{n-m}(\rho_m) = \rho_n = |\psi\rangle\langle\psi|$, and since the image of Φ^{n-m} is pure only if ρ_m is a coherent pure state it follows that the bound is saturated if and only if the state ρ_n is pure and coherent.

Let us now act with Φ^{m-n} on a Dicke state in $\mathcal{H}^{n+1}$. The eigenvectors of $\Phi^{m-n}(|n-r,r\rangle\langle n-r,r|)$ are again Dicke states in $\mathcal{H}^{m+1}$, namely $|m-i,i\rangle$ for $r \le i \le m-n+r$. The eigenvalues are

$$\lambda_i^{(r)} = \frac{(n+1)!\,(m-n)!}{(m+1)!\,r!\,(n-r)!} \frac{i!\,(m-i)!}{(i-r)!\,(m-n-i+r)!}. \tag{13.90}$$

It is convenient to introduce a new index $k = i - r$, so that the eigenvalues become

$$\lambda_k^{(r)} = \frac{(n+1)!\,(m-n)!}{(m+1)!\,r!\,(n-r)!} \frac{(k+r)!\,(m-r-k)!}{k!\,(m-n-k)!}, \qquad 0 \le k \le m-n, \tag{13.91}$$

with eigenvectors $|m-k-r, r+k\rangle$. The most important case is that of coherent states, for which we get

$$\lambda_k^{(0)} = (n+1)\frac{(m-n)!}{(m+1)!} \frac{(m-k)!}{(m-n-k)!}. \tag{13.92}$$

Lieb and Solovej go on to show (using induction) that the spectrum of the image of a coherent state majorises all other spectra in the image. We refer to their paper for the proof [604], but we illustrate how pure states behave under Φ^2 in Figure 13.4. These illustrations show that, in general, the states do not even come close to saturating the bounds provided by majorisation.

---

[17] The map has independent interest, and had already been used to find the best possible quantum state cloning device [341, 959].

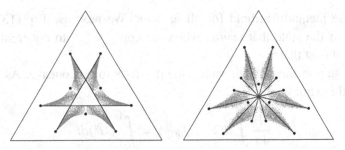

Figure 13.4 Spectra of the states obtained by applying the map Φ^2 to pure states in four (left) or five (right) dimensions. The black dots show where the Dicke states end up. The straight lines connecting some pairs of Dicke states are the images of two-dimensional subspaces containing them. To each picture the image of 3000 states picked at random have been added.

Now we can prove the inequality (13.83). The proof begins by using Eq. (13.88), together with the fact that a Bloch coherent state $|z\rangle$ in $\mathbb{C}^{n+1}$ is the image under the reversed map of a coherent state in $\mathbb{C}^{m+1}$. Hence

$$\langle z|\rho_n|z\rangle = \mathrm{Tr}_n\rho_n\Phi^{n-m}(|z\rangle\langle z|) = \frac{m+1}{n+1}\langle z|\Phi^{m-n}(\rho_n)|z\rangle. \tag{13.93}$$

We appeal to the *Berezin–Lieb inequality*, which states that for any concave function $f : [0, 1] \to \mathbb{R}$, and for any density matrix ρ, there holds

$$\frac{n+1}{4\pi}\int \mathrm{d}\Omega f(\langle z|\rho|z\rangle) \geq \mathrm{Tr}f(\rho). \tag{13.94}$$

This is an easy consequence of Eq. (6.57), which says that the Bloch coherent states form a resolution of the identity.[18] The Berezin–Lieb inequality implies that

$$\frac{n+1}{4\pi}\int \mathrm{d}\Omega f(\langle z|\rho_n|z\rangle) \geq \frac{n+1}{m+1}\mathrm{Tr}\left(\frac{m+1}{n+1}\Phi^{m-n}(\rho_n)\right). \tag{13.95}$$

The next step is the key one. We rely on Karamata's inequality, see Section 2.1, which says that if $\vec{b}$ is majorised by $\vec{a}$ and if $f : \mathbb{R} \to \mathbb{R}$ is concave then it follows that

$$\sum_i f(b_i) \geq \sum_i f(a_i). \tag{13.96}$$

We know that the spectrum of $\Phi^{m-n}(\rho_n)$ is majorised by the spectrum of the image of a coherent state, and since f is concave it follows that

$$\frac{n+1}{m+1}\mathrm{Tr}\left(\frac{m+1}{n+1}\Phi^{m-n}(\rho_n)\right) \geq \frac{n+1}{m+1}\mathrm{Tr}\left(\frac{m+1}{n+1}\Phi^{m-n}(|n, 0\rangle\langle n, 0|)\right). \tag{13.97}$$

---

[18] This inequality has been used to great effect in the study of the classical limit of spin systems [598].

Both of these inequalities hold for all $m > n$. We next use Eq. (13.92) for the eigenvalues of the state that results when we apply Φ^{m-n} to coherent states, and then we are almost there.

The final step in the proof is valid also if f fails to be concave. As long as f is continuous there holds

$$\frac{1}{4\pi} \int d\Omega\, f(Q_{|n,0\rangle}) = \int_0^1 f(t^n)\,dt \tag{13.98}$$

(obviously), as well as (not obviously)

$$\lim_{m\to\infty} \frac{1}{m+1} \sum_{i=0}^{m-n} f\left(\frac{(m-n)!\,(m-i)!}{m!\,(m-n-i)!}\right) = \int_0^1 f(t^n)\,dt. \tag{13.99}$$

To see this last one we manipulate the left-hand side according to

$$\frac{1}{m+1} \sum_{j=n}^{m} f\left(\frac{(m-n)!\,j!}{m!\,(j-n)!}\right) \sim \frac{1}{m+1} \sum_{j=n}^{m} f\left(\frac{j^n}{m^n}\right)$$

$$\sim \frac{1}{m+1} \int_n^m dx\, f\left(\frac{x^n}{m^n}\right) \sim \frac{m}{m+1} \int_0^1 f(t^n)\,dt. \tag{13.100}$$

When f is concave the limit is always approached from below (but not very quickly).

This ends the proof. A minor blemish is that uniqueness of the minimum is not shown. The theorem can be extended to symmetric representations of $SU(K)$ [605], that is to say to the coherent states discussed in Section 6.4. And perhaps further, but this we do not know.

13.7 Entropy dynamics

What we have not discussed so far is the role of entropy as an arrow of time – which is how entropy has been regarded ever since the word was coined by Clausius. If this question is turned into the question of how the von Neumann entropy of a state changes under the action of some quantum operation $\Phi : \rho \to \rho'$, it does have a clear cut answer. Because of Eq. (13.48), it follows from monotonicity of relative entropy that a CP map increases the von Neumann entropy of every state if and only if it is unital (bistochastic), that is if it transforms the maximally mixed state into itself. For quantum operations that are stochastic, but not bistochastic, this is no longer true – for such quantum channels the von Neumann entropy may decrease. Consider for instance the decaying or amplitude damping channel (Section 10.7), which describes the effect of spontaneous emission on a qubit. This sends any

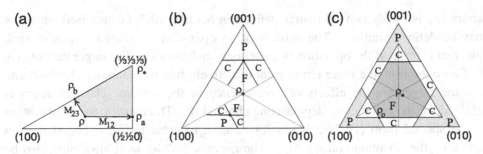

Figure 13.5 The eigenvalue simplex for $N = 3$: (a) a Weyl chamber; the shaded region is accessible from ρ with bistochastic maps. (b) The shape of the 'light cone' depends on the degeneracy of the spectrum. F denotes the future, P the past, and C the non-comparable states. (c) Splitting the simplex into Weyl chambers.

mixed state towards the pure ground state, for which the entropy is zero. But then this is not an isolated system, so this would not worry Clausius.

Even for bistochastic maps, when the von Neumann entropy does serve as an arrow of time, it does not point very accurately to the future. See Figure 13.5. Relative to any given state, the state space splits into three parts, the 'future' F that consists of all states that can be reached from the given state by bistochastic maps, the 'past' P that consists of all states from which the given state can be reached by such maps, and a set of incomparable states that we denote by C in the figure. This is reminiscent of the causal structure in special relativity, where the light cone divides Minkowski space into the future, the past, and the set of points that cannot communicate in either direction with a point sitting at the vertex of the light cone. There is also the obvious difference that the actual shape of the 'future' depends somewhat on the position of the given state, and very much so when its eigenvalues degenerate. The isoentropy curves of the von Neumann entropy do not do justice to this picture. To do better one would have to bring in a complete set of Schur concave functions such as the Rényi entropies (see Figure 2.14).

Naturally, the majorisation order may not be the last word on the future. Depending on the physics, it may well be that majorisation provides a necessary but not sufficient condition for singling it out.[19]

We turn from the future to a more modest subject, namely the *entropy of an operation* Φ. This can be conveniently defined as the von Neumann entropy of the state that corresponds to the operation via the Jamiołkowski isomorphism, namely as

$$S(\Phi) \equiv S\left(\frac{1}{N}D_{\Phi}\right) \in [0, \ln N^2]. \qquad (13.101)$$

---

[19] For a lovely example from thermodynamics, see Alberti and Uhlmann [15].

where D_Φ is the dynamical matrix defined in Section 10.3. Generalised entropies may be defined similarly. The entropy of an operation vanishes if D_Φ is of rank one, that is to say if the operation is a unitary transformation. The larger the entropy S of an operation, the more terms enter effectively into the canonical Kraus form, and the larger are the effects of decoherence in the system. The maximum is attained for the completely depolarizing channel Φ_*. The entropy for an operation of the special form (10.72), that is for random external fields, is bounded from above by the Shannon entropy $S(\vec{p})$. The norm $\sqrt{\text{Tr}\Phi\Phi^\dagger} = ||\Phi||_{\text{HS}}$ may also be used to characterise the decoherence induced by the map. This varies from unity for Φ_* (total decoherence) to N for a unitary operation (no decoherence) – see Table 10.2.

A different way to characterise a quantum operation is to compute the amount of entropy it creates when acting on an initially pure state. In Section 2.3 we defined the entropy of a stochastic matrix with respect to a fixed probability distribution. This definition, and the bound (2.39), has a quantum analogue due to Lindblad [612], and it will lead us to our present goal. Consider a CP map Φ represented in the canonical Kraus form $\rho' = \sum_{i=1}^{r} A_i \rho A_i^\dagger$. Define an operator acting on an auxiliary r dimensional space $\mathcal{H}_r$ by

$$\sigma_{ij} = \text{Tr}\rho A_j^\dagger A_i. \tag{13.102}$$

In Problem 10.3 we show that σ is a density operator in its own right. This can be considered as an output state of the complementary map $\tilde{\Phi}$ defined in (10.65), so σ represents the state of the environment after the action of the channel. The entropy $S(\sigma)$ depends on the map Φ and on the initial state and equals $S(\Phi)$, as defined above, when ρ is the maximally mixed state. Next we define a density matrix in the composite Hilbert space $\mathcal{H}_N \otimes \mathcal{H}_r$,

$$\omega = \sum_{i=1}^{r}\sum_{j=1}^{r} A_i \rho A_j^\dagger \otimes |i\rangle\langle j| = W\rho W^\dagger, \tag{13.103}$$

where $|i\rangle$ is an orthonormal basis in $\mathcal{H}_r$. The operator W maps a state $|\phi\rangle$ in $\mathcal{H}_N$ into $\sum_{j=1}^{r} A_j|\phi\rangle \otimes |j\rangle$, and the completeness of the Kraus operators implies that $W^\dagger W = \mathbb{1}_N$. It follows that $S(\omega) = S(\rho)$. Since it is easy to see that $\text{Tr}_N\omega = \sigma$ and $\text{Tr}_r\omega = \rho$, we may use the triangle inequalities (13.28) and (13.32), and some slight rearrangement, to deduce that

$$|S(\rho) - S(\sigma)| \leq S(\rho') \leq S(\sigma) + S(\rho), \tag{13.104}$$

in exact analogy to the classical bound (2.39). If the initial state is pure, that is if $S(\rho) = 0$, we find that the final state has entropy $S(\sigma)$. For this reason $S(\sigma)$ is sometimes referred to as the *entropy exchange* of the operation [809].

To get better bounds on the dynamics of entropy of an initial state ρ subjected to an operation Φ represented by its Kraus operators A_i define probabilities $p_i = \mathrm{Tr} A_i \rho A_i^\dagger$ and the output states $\rho_i = A_i \rho A_i^\dagger / p_i$. Their entropies allow one to derive another bound [776] for the entropy of the output state ρ', which involves the auxiliary state (13.102),

$$S(\rho') \leq S(\sigma) + \sum_i p_i S(\rho_i). \tag{13.105}$$

Concavity of entropy implies $\sum_i p_i S(\rho_i) \leq S(\sum_i p_i \rho_i) = S(\rho')$, while an early work of Lindblad [607] shows that

$$\sum_i p_i S(\rho_i) \leq S(\rho). \tag{13.106}$$

The above result implies that the rightmost inequality in (13.104) is weaker than (13.105). To prove the latter inequality one uses the strong subadditivity (13.34) applied to state (13.103) extended to $\Omega = \sum_i^r \sum_j^r A_i \rho A_j^\dagger \otimes |i\rangle\langle j| \otimes |i\rangle\langle j|$. Since different channels can give the same ensemble $\{p_i, \rho_i\}$, the right-hand side of (13.105) can be optimised over the channels [309]. Applying the Araki–Lieb inequality to the state Ω one obtains the Lindblad result (13.106).

Finally – and in order to give a taste of a subject that we omit – let us define the capacity of a quantum channel. To this end for any map Φ and any state ρ one introduces the *Holevo quantity* [443]

$$\chi(\Phi, \rho) \equiv \max_{\mathcal{E}} \sum_i q_i S(\Phi(\rho_i) || \Phi(\rho)) = \min_{\mathcal{E}} \left[S(\Phi(\rho)) - \sum_i q_i S(\Phi(\rho_i)) \right],$$
$$\tag{13.107}$$

which will be discussed in Section 14.1 under the name *Jensen–Shannon divergence*. The optimisation is performed over all ensembles $\mathcal{E} = \{\rho_i; q_i\}$ such that $\rho = \sum_i q_i \rho_i$. This is not an easy one to carry out. In the next step one performs maximisation over the set of all states, which yields the *Holevo information* of a channel,

$$\chi(\Phi) \equiv \max_\rho \left[\chi(\Phi, \rho) \right]. \tag{13.108}$$

There is a theorem due to Holevo [444, 446] and Schumacher–Westmoreland [812] which employs these definitions to give an upper bound on the information carrying classical capacity C of a quantum channel Φ, obtained by regularisation of the Holevo information,

$$C(\Phi) = \lim_{n\to\infty} \frac{1}{n} \chi(\Phi^{\otimes n}). \tag{13.109}$$

Together with the quantum noiseless coding theorem [511, 808], this result brings quantum information theory up to the level achieved for classical information theory in Shannon's classical work [828]. But these matters [415, 733, 969] are beyond the ken of our book. Let us just mention the famous additivity conjecture which states that

$$\chi(\Phi_1 \otimes \Phi_2) \overset{?}{=} \chi(\Phi_1) + \chi(\Phi_2). \tag{13.110}$$

In one respect its status was similar to that of strong subadditivity before the latter was proven – it was equivalent to many other outstanding conjectures.[20] There was a key difference however: the strong subadditivity conjecture was proven true, while the additivity conjecture was proven *false*, as implied by a theorem of Hasting's [410]. See Section 15.8.

Problems

13.1 Show, for any positive operator A, that

$$\ln(A + xB) - \ln A = \int_0^\infty \frac{1}{A+u} xB \frac{1}{A+xB+u} du. \tag{13.111}$$

13.2 Compute the two contour integrals

$$S(\rho) = -\frac{1}{2\pi i} \oint (\ln z) \text{Tr} \left(\mathbb{1} - \frac{\rho}{z} \right)^{-1} dz \tag{13.112}$$

and

$$S_Q(\rho) = -\frac{1}{2\pi i} \oint (\ln z) \det \left(\mathbb{1} - \frac{\rho}{z} \right)^{-1} dz, \tag{13.113}$$

with a contour that encloses all the eigenvalues of ρ. The second quantity is known as *subentropy* [510].

13.3 Prove *Donald's identity* [272]: for any mixed state $\rho = \sum_k p_k \rho_k$ and another state σ

$$\sum_k p_k S(\rho_k || \sigma) = \sum_k p_k S(\rho_k || \rho) + S(\rho || \sigma). \tag{13.114}$$

13.4 Compute the Wehrl entropy for the Husimi function (13.65) of a two-qubit pure state written in its Schmidt decomposition.

13.5 Prove that Euclidean distances between orbits can be read off from a picture of the Weyl chamber, that is, prove Eq. (8.52).

13.6 Prove that $(\det(A + B))^{1/N} \geq (\det A)^{1/N} + (\det B)^{1/N}$, where A and B are positive matrices of size N.

---

[20] See the review by Amosov, Holevo and Werner [32] and the work by Shor [837].

13.7 For any operation Φ given by its canonical Kraus form (10.55) one defines its *purity Hamiltonian*

$$\Omega \equiv \sum_{i=1}^{r} \sum_{j=1}^{r} A_j^\dagger A_i \otimes A_i^\dagger A_j, \qquad (13.115)$$

the trace of which characterises an average decoherence induced by Φ [994]. Show that $\mathrm{Tr}\,\Omega = \|\Phi\|_{\mathrm{HS}}^2 = \mathrm{Tr}D^2$, hence that it is proportional to the purity $\mathrm{Tr}\,\rho_\Phi^2$ of the state corresponding to Φ in the isomorphism (11.22).

14

Distinguishability measures

Niels Bohr supposedly said that if quantum mechanics did not make you dizzy then you did not understand it. I think that the same can be said about statistical inference.

– Robert D. Cousins[1]

In this chapter we quantify how easy it may be to distinguish probability distributions from each other (a discussion that was started in Chapter 2). The issue is a very practical one, and arises whenever one is called upon to make a decision based on imperfect data. There is no unique answer, because everything depends on the data– the l_1-distance appears if there has been just one sampling of the distributions, the relative entropy governs the approach to the 'true' distribution as the number of samplings goes to infinity, and so on.

The quantum case is even subtler. A quantum state always stands ready to produce a large variety of classical probability distributions, depending on the choice of measurement procedure. It is no longer possible to distinguish pure states from each other with certainty, unless they are orthogonal. The basic idea behind the quantum distinguishability measures is the same as that which allowed us, in Section 5.3, to relate the Fubini–Study metric to the Fisher–Rao metric. We will optimise over all possible measurements.

14.1 Classical distinguishability measures

If a distance function has an operational significance as a measure of statistical distinguishability, then we expect it to be monotone (and decreasing) under general stochastic maps. Coarse graining means that information is being discarded, and this cannot increase distinguishability. From Čencov's theorem (Section 2.5) we

---

[1] Reproduced from Robert D. Cousins. Why isn't every physicist a Bayesian? *Am. J. Phys.* 63:398, 1995.

know that the Fisher–Rao metric is the only Riemannian metric on the probability simplex that is monotone under general stochastic maps. But there is another simple distance measure that does have the desirable property, namely the l_1-distance from Eq. (1.57). The proof of monotonicity uses the observation that the difference of two probability vectors can be written in the form

$$p_i - q_i = N_i^+ - N_i^-, \tag{14.1}$$

where N^+ and N^- are two positive vectors with orthogonal support, meaning that for each component i at least one of N_i^+ and N_i^- is zero. We follow this up with the triangle inequality and condition (ii) from Eq. (2.4) that defines a stochastic matrix T:

$$||Tp - Tq||_1 = ||TN^+ - TN^-||_1 \leq ||TN^+||_1 + ||TN^-||_1$$
$$= \frac{1}{2}\sum_{i,j} T_{ij}N_j^+ + \frac{1}{2}\sum_{i,j} T_{ij}N_j^- = \frac{1}{2}\sum_j (N_j^+ + N_j^-) = ||p - q||_1.$$

$$\tag{14.2}$$

By contrast, the Euclidean l_2-distance is not monotone. To see this, consider a coarse graining stochastic matrix such as

$$T = \begin{bmatrix} 1 & 0 & 0 \\ 0 & 1 & 1 \end{bmatrix}. \tag{14.3}$$

Applying this transformation has the effect of collapsing the entire simplex onto one of its edges. If we draw a little picture of this, as in Figure 14.1, it becomes evident why $p = 1$ is the only value of p for which the l_p-distance is monotone under this map. The picture should also make it clear that it is coarse graining maps like (14.3) that may cause problems with monotonicity – monotonicity under bistochastic maps that cause a contraction of the probability simplex towards its centre, is much easier to ensure. In fact the flat l_2-distance is monotone under bistochastic maps. Incidentally, it is clear from the picture that the l_1-distance succeeds in being monotone (under general stochastic maps) only because the distance between probability distributions with orthogonal support is constant (and maximal). This is a property that the l_1-distance shares with the monotone Fisher–Rao distance – and if a distance is to quantify how easily two probability distributions can be distinguished, then it must be monotone.

The question remains to what extent, and in what sense, our various monotone notions of distance – the Bhattacharyya and Hellinger distances, and the l_1-distance – have any clear-cut operational significance. For the latter, an answer

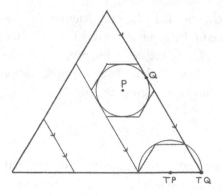

Figure 14.1 Coarse graining, according to Eq. (14.3), collapses the probability simplex to an edge. The l_1-distance never increases (the hexagon is unchanged), but the l_2 distance sometimes does (the circle grows).

is known. Consider two probability distributions P and Q over N events, and mix them, so that the probability for event i is

$$r_i = \pi_0 p_i + \pi_1 q_i. \tag{14.4}$$

A possible interpretation here is that Alice sends Bob a message in the form of an event drawn from one of two possible probability distributions. Bob is ignorant of which particular distribution Alice uses, and his ignorance is expressed by the distribution (π_0, π_1). Having sampled once, Bob is called upon to guess what distribution was used by Alice. It is clear – and this answer can be dignified with technical terms like 'Bayes' decision rule' – that his best guess, given that event i occurs, is to guess P if $p_i > q_i$ and Q if $q_i > p_i$. (If equality holds the two guesses are equally good.) Given this strategy, the probability that Bob's guess is right is

$$P_R(P, Q) = \sum_{i=1}^{N} \max\{\pi_0 p_i, \pi_1 q_i\} \tag{14.5}$$

and the probability of error is

$$P_E(P, Q) = \sum_{i=1}^{N} \min\{\pi_0 p_i, \pi_1 q_i\}. \tag{14.6}$$

Now consider the case $\pi_0 = \pi_1 = 1/2$. Then Bob has no reason to prefer any distribution in advance. In this situation it is easily shown that $P_R - P_E = D_1$, or equivalently

$$D_1(P, Q) = \frac{1}{2} \sum_{i=1}^{N} |p_i - q_i| = 1 - 2P_E(P, Q), \tag{14.7}$$

that is, the l_1-distance grows as the probability of error goes down. In this sense the l_1-distance has a precise meaning, as a measure of how reliably two probability distributions can be distinguished by means of a single sampling.

Of course it is not clear why we should restrict ourselves to one sampling; the probability of error goes down as the number of samplings $\mathcal{N}$ increases. There is a theorem that governs how it does so:

Chernoff's theorem. *Let $P_E^{(\mathcal{N})}(P, Q)$ be the probability of error after $\mathcal{N}$ samplings of two probability distributions. Then*

$$P_E^{(\mathcal{N})}(P, Q) \leq \left(\min_{\alpha \in [0,1]} \sum_{i=1}^{N} p_i^{\alpha} q_i^{1-\alpha} \right)^{\mathcal{N}}. \tag{14.8}$$

The bound is approached asymptotically when $\mathcal{N}$ goes to infinity.

Unfortunately it is not easy to obtain an analytic expression for the Chernoff bound (the one that is approached asymptotically), but we do not have to find the minimum in order to obtain useful upper bounds. The non-minimal bounds are of interest in themselves. They are related to the *relative Rényi entropy*

$$I_q(P, Q) = \frac{1}{1-q} \ln \left[\sum_{i=1}^{N} p_i^q q_i^{1-q} \right]. \tag{14.9}$$

When $q = 1/2$ the relative Rényi entropy is symmetric, and it is a monotone function of the geodesic Bhattacharyya distance D_{Bhatt} from Eq. (2.56).

In the limit $q \to 1$ the relative Rényi entropy tends to the usual relative entropy $S(P||Q)$, which figured in a different calculation of the probability of error in Section 2.3. The setting there was that we made a choice between the distributions P and Q, using a large number of samplings, in a situation where it happened to be the case that the statistics were governed by Q. The probability of erroneously concluding that the true distribution is P was shown to be

$$P_E(P, Q) \sim e^{-\mathcal{N}S(P||Q)}. \tag{14.10}$$

The asymmetry of the relative entropy reflects the asymmetry of the situation. In fact, suppose the choice is between a fair coin and a biased coin that only shows heads. Using Eq. (2.32) we find that

$$P_E(\text{fair}||\text{biased}) = e^{-\mathcal{N}\cdot\infty} = 0 \text{ and } P_E(\text{biased}||\text{fair}) = e^{-\mathcal{N}\ln 2} = \frac{1}{2^{\mathcal{N}}}. \tag{14.11}$$

This is exactly what intuition dictates; the fair coin can produce the frequencies expected from the biased coin, but not the other way around.

But sometimes we insist on true distance functions. Relative entropy cannot be turned into a true distance just by symmetrisation, because the triangle inequality will still be violated. However, there is a simple modification that does lead to a proper distance function. Given two probability distributions P and Q, let us define their mean R by

$$R = \frac{1}{2}P + \frac{1}{2}Q \quad \Leftrightarrow \quad r_i = \frac{1}{2}p_i + \frac{1}{2}q_i. \tag{14.12}$$

Then the *Jensen–Shannon divergence* is defined by

$$J(P, Q) \equiv 2S(R) - S(P) - S(Q), \tag{14.13}$$

where S is the Shannon entropy. An easy calculation shows that this is related to relative entropy:

$$J(P, Q) = \sum_{i=1}^{N}\left(p_i \ln \frac{2p_i}{p_i + q_i} + q_i \ln \frac{2q_i}{p_i + q_i}\right) = S(P||R) + S(Q||R). \tag{14.14}$$

Interestingly, the function $D(P, Q) = \sqrt{J(P, Q)}$ is not only symmetric but obeys the triangle inequality as well, and hence it qualifies as a true distance function – moreover as a distance function that is consistent with the Fisher–Rao metric.[2]

The Jensen–Shannon divergence can be generalised to a measure of the divergence between an arbitrary number of M probability distributions $P_{(m)}$, weighted by some probability distribution π over M events:

$$J(P_{(1)}, P_{(2)}, \ldots, P_{(M)}) \equiv S\left(\sum_{m=1}^{M} \pi_m P_{(m)}\right) - \sum_{m=1}^{M} \pi_m S(P_{(m)}). \tag{14.15}$$

In this form it has been used in the study of DNA sequences – and in the definition (13.107) of the capacity of a quantum channel. Its interpretation as a distinguishability measure emerges when we sample from a statistical mixture of probability distributions. Given that the Shannon entropy measures the information gained when sampling a distribution, the Jensen–Shannon divergence measures the average gain of information about how that mixture was made (that is about π), since we subtract that part of the information that concerns the sampling of each individual distribution in the mixture.

The reader may now have begun to suspect that there are many measures of distinguishability available, some of them more useful, and some of them easier to compute, than others. Fortunately there are inequalities that relate different measures of distinguishability. An example is the *Pinsker inequality* that relates the l_1-distance to the relative entropy:

---

[2] This was shown in 2003 by Endres and Schindelin [295], and the result is expected to generalise to the quantum case [175, 570]. For a survey of the Jensen–Shannon divergence and its properties, see Lin [606].

$$S(P||Q) \geq \frac{1}{2}\left(\sum_{i=1}^{N}|p_i - q_i|\right)^2 = 2D_1^2(P,Q) \tag{14.16}$$

This is a stronger bound than (2.30) since $D_1 \geq D_2$. The proof is quite interesting. First one uses brute force to establish that

$$2(p-q)^2 \leq p\ln\frac{p}{q} + (1-p)\ln\frac{1-p}{1-q} \tag{14.17}$$

wherever $0 \leq q \leq p \leq 1$. This is the Pinsker inequality for $N = 2$. We are going to reduce the general case to this. Without loss of generality we assume that $p_i \geq q_i$ for $1 \leq i \leq K$, and $p_i < q_i$ otherwise. Then we define a stochastic matrix T by

$$\begin{bmatrix} T_{1,1} & \cdots & T_{1,K} & T_{1,K+1} & \cdots & T_{1,N} \\ T_{2,1} & \cdots & T_{2,K} & T_{2,K+1} & \cdots & T_{2,N} \end{bmatrix} = \begin{bmatrix} 1 & \cdots & 1 & 0 & \cdots & 0 \\ 0 & \cdots & 0 & 1 & \cdots & 1 \end{bmatrix}. \tag{14.18}$$

We get two binomial distributions TP and TQ, and define

$$p \equiv \sum_{i=1}^{K}p_i = \sum_{i=1}^{N}T_{1i}p_i, \qquad q \equiv \sum_{i=1}^{K}q_i = \sum_{i=1}^{N}T_{1i}q_i. \tag{14.19}$$

It is easy to see that $D_1(P,Q) = p - q$. Using this, and Eq. (14.17), we get

$$2D_1^2(P,Q) \leq S(TP||TQ) \leq S(P||Q). \tag{14.20}$$

Thus monotonicity of relative entropy was used in the punchline.

The Pinsker inequality is not sharp; it has been improved to[3]

$$S(P||Q) \geq 2D_1^2 + \frac{4}{9}D_1^4 + \frac{32}{135}D_1^6 + \frac{7072}{42525}D_1^8. \tag{14.21}$$

Relative entropy is unbounded from above. But it can be shown [606] that

$$2D_1(P,Q) \geq J(P,Q). \tag{14.22}$$

Hence the l_1-distance bounds the Jensen–Shannon divergence from above.

14.2 Quantum distinguishability measures

We now turn to the quantum case. When density matrices rather than probability distributions are sampled we face new problems, since the probability distribution $P(E, \rho)$ that governs the sampling will depend, not only on the density matrix ρ, but on the POVM that represents the measurement as well. The probabilities that we actually have to play with are given by

---

[3] Inequality (14.16) is due to Pinsker (1964) [737], while (14.21) is due to Topsøe [897].

$$p_i(E, \rho) = \text{Tr } E_i \rho, \tag{14.23}$$

where the collection of effects $\{E_i\}_{i=1}^K$ forms a POVM. The quantum distinguishability measures will be defined by varying over all possible POVMs until the classical distinguishability of the resulting probability distributions is maximal. In this way any classical distinguishability measure will have a quantum counterpart – except that for some of them, notably for the Jensen–Shannon divergence, the optimisation over all POVMs is very difficult to carry out, and we will have to fall back on bounds and inequalities.[4]

Before we begin, let us define the L_p-norm of an operator A by

$$||A||_p \equiv \left(\frac{1}{2} \text{Tr} |A|^p \right)^{1/p}, \tag{14.24}$$

where the absolute value of the operator was defined in Eq. (8.12). The factor of $1/2$ is included in the definition because it is convenient when we restrict ourselves to density matrices. In obvious analogy with Section 1.4 we can now define the L_p-*distance* between two operators as

$$D_p(A, B) \equiv ||A - B||_p. \tag{14.25}$$

Like all distances based on a norm, these distances are nice because convex mixtures will appear as (segments of) metric lines. The factor $1/2$ in the definition ensures that all the L_p-distances coincide when $N = 2$. For two by two matrices, an L_p ball looks like an ordinary ball. Although the story becomes more involved when $N > 2$, it will always be true that all the L_p-distances coincide for a pair of pure states, simply because a pair of pure states taken in isolation always span a two-dimensional Hilbert space. We may also observe that, given two density matrices ρ and σ, the operator $\rho - \sigma$ can be diagonalised, and the distance $D_p(\rho, \sigma)$ becomes the l_p-distance expressed in terms of the eigenvalues of that operator.

For $p = 2$ the L_p-distance is Euclidean. It has the virtue of simplicity, and we have already used it extensively. For $p = 1$ we have the *trace distance*[5]

$$D_{tr}(A, B) = \frac{1}{2} \text{Tr} |A - B| = \frac{1}{2} D_{\text{Tr}}(A, B). \tag{14.26}$$

It will come as no surprise to learn that the trace distance will play a role similar to that of the l_1-distance in the classical case. It is interesting to get some understanding of the shape of its unit ball. All L_p-distances can be computed from the eigenvalues of the operator $\rho - \sigma$, and therefore Eq. (1.58) for the radii of its in and

---

[4] The subject of quantum decision theory, which we are entering here, was founded by Helstrom (1976) [422] and Holevo (1982) [445]. A very useful (and more modern) account is given by Fuchs [319]; see also Fuchs and van de Graaf [323]. Here we give some glimpses only.
[5] For convenience we are going to use in parallel two symbols, $D_{\text{Tr}} = 2 D_{tr}$.

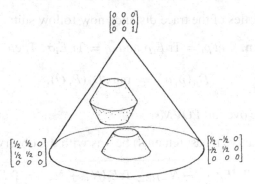

Figure 14.2 Some balls of constant radius, as measured by the trace distance, inside a three-dimensional slice of the space of density matrices (obtained by rotating the eigenvalue simplex around an axis).

outspheres can be directly taken over to the quantum case. But there is a difference between the trace and the l_1-distances, and we see it as soon as we look at a set of density matrices that cannot be diagonalised at the same time (Figure 14.2).

Thus equipped, we take up the task of quantifying the probability of error in choosing between two density matrices ρ and σ, based on a single measurement. Mathematically, the task is to maximise the l_1-distance over all possible POVMs $\{E_i\}$, given ρ and σ. Thus our quantum distinguishability measure D is defined by

$$D(\rho, \sigma) \equiv \max_E D_1\Big(P(E, \rho), P(E, \sigma)\Big). \tag{14.27}$$

As the reader may suspect already, the answer is the trace distance. We will carry out the maximisation for projective measurements only – the generalisation to arbitrary POVMs being quite easy – and start with a lemma that contains a key fact about the trace distance:

Trace distance lemma. *If P is any projector onto a subspace of Hilbert space then*

$$D_{\mathrm{tr}}(\rho, \sigma) \geq \mathrm{Tr}P(\rho - \sigma) = D_1(\rho, \sigma). \tag{14.28}$$

Equality holds if and only if P projects onto the support of N_+, where $\rho - \sigma = N_+ - N_-$, with N_+ and N_- being positive operators of orthogonal support.

To prove this, observe that by construction $\mathrm{Tr}N_+ = \mathrm{Tr}N_-$ (since their difference is a traceless matrix), so that $D_{\mathrm{tr}}(\rho, \sigma) = \mathrm{Tr}N_+$. Then

$$\mathrm{Tr}P(\rho - \sigma) = \mathrm{Tr}P(N_+ - N_-) \leq \mathrm{Tr}PN_+ \leq \mathrm{Tr}N_+ = D_{\mathrm{tr}}(\rho, \sigma). \tag{14.29}$$

Clearly, equality holds if and only if $P = P_+$, where $P_+N_- = 0$ and $P_+N_+ = N_+$.

The useful properties of the trace distance now follow suit:

Helstrom's theorem. *Let* $p_i = \mathrm{Tr}\, E_i \rho$ *and* $q_i = \mathrm{Tr}\, E_i \sigma$. *Then*

$$D_{tr}(\rho, \sigma) = \max_E D_1(P, Q), \tag{14.30}$$

where we maximise over all POVMs.

The proof (for projective measurements) begins with the observation that

$$\mathrm{Tr}|E_i(\rho - \sigma)| = \mathrm{Tr}|E_i(N_+ - N_-)| \leq \mathrm{Tr}E_i(N_+ + N_-) = \mathrm{Tr}E_i|\rho - \sigma|. \tag{14.31}$$

For every POVM, and the pair of probability distributions derived from it, this implies that

$$D_1(P, Q) = \frac{1}{2}\sum_i \mathrm{Tr}|E_i(\rho - \sigma)| \leq \frac{1}{2}\sum_i \mathrm{Tr}E_i|\rho - \sigma| = D_{tr}(\rho, \sigma). \tag{14.32}$$

The inequality is saturated when we choose a POVM that contains one projector onto the support of N_+ and one projector onto the support of N_-. The interpretation of $D_{tr}(\rho, \sigma)$ as a quantum distinguishability measure for 'one shot samplings' is thereby established.

It is important to check that the trace distance is monotone under trace preserving CP maps $\rho \rightarrow \Phi(\rho)$. This is not hard to do if we first use our lemma to find a projector P such that

$$D_{tr}(\Phi(\rho), \Phi(\sigma)) = \mathrm{Tr}P(\Phi(\rho) - \Phi(\sigma)). \tag{14.33}$$

We decompose $\rho - \sigma$ as above. Since the map is trace preserving it is true that $\mathrm{Tr}\Phi(N_+) = \mathrm{Tr}\Phi(N_-)$. Then

$$D_{tr}(\rho, \sigma) = \frac{1}{2}(N_+ + N_-) = \frac{1}{2}(\Phi(N_+) + \Phi(N_-)) = \mathrm{Tr}\Phi(N_+)$$
$$\geq \mathrm{Tr}P\Phi(N_+) \geq \mathrm{Tr}P(\Phi(N_+) - \Phi(N_-)) = \mathrm{Tr}P(\Phi(\rho) - \Phi(\sigma)) \tag{14.34}$$

and the monotonicity of the trace distance follows when we take into account how P was selected. The lemma can also be used to prove a strong convexity result for the trace distance, namely,

$$D_{tr}\left(\sum_i p_i \rho_i, \sum_i q_i \sigma_i\right) \leq D_1(P, Q) + \sum_i p_i D_{tr}(\rho_i, \sigma_i). \tag{14.35}$$

We omit the proof [692]. Joint convexity follows if we set $P = Q$.

The trace distance sets limits on how much the von Neumann entropy of a given state may change under a small perturbation. To be precise, we have Fannes' lemma:

Fannes' lemma. *Let the quantum states ρ and σ act on an N-dimensional Hilbert space, and be close enough in the sense of the trace metric so that $D_{\text{tr}}(\rho, \sigma) \leq 1/(2e)$. Then*

$$|S(\rho) - S(\sigma)| \leq 2D_{\text{tr}}(\rho, \sigma) \ln \frac{N}{2D_{\text{tr}}(\rho, \sigma)}. \qquad (14.36)$$

Again we omit the proof [308], but we take note of a rather interesting intermediate step: let the eigenvalues of ρ and σ be r_i and s_i, respectively, and assume that they have been arranged in decreasing order (e.g. $r_1 \geq r_2 \geq \cdots \geq r_N$). Then

$$D_{\text{tr}}(\rho, \sigma) \geq \frac{1}{2} \sum_i |r_i - s_i|. \qquad (14.37)$$

The closeness assumption in Fannes' lemma has to do with the fact that the function $-x \ln x$ is monotone on the interval $(0, 1/e)$. A weaker bound holds if it is not fulfilled.

The relative entropy between any two states is bounded by their trace distance by a quantum analogue of the Pinsker inequality (14.16)

$$S(\rho \| \sigma) \geq 2[D_{\text{tr}}(\rho, \sigma)]^2. \qquad (14.38)$$

The idea of the proof [432] is similar to that used in the classical case, that is, one relies on Eq. (14.17) and on the monotonicity of relative entropy.

What about relative entropy itself? The results of Hiai and Petz [433], briefly reported in Section 13.3, can be paraphrased as saying that, in certain well defined circumstances, the probability of error when performing measurements on a large number $\mathcal{N}$ of copies of a quantum system is given by

$$P_E(\rho, \sigma) = e^{-\mathcal{N}S(\rho \| \sigma)}. \qquad (14.39)$$

That is to say, this is the smallest achievable probability of erroneously concluding that the state is ρ, given that the state in fact is σ. Although our account of the story ends here, the story itself does not. Let us just mention that the step from one to many samplings turns into a giant leap in quantum mechanics, because the set of all possible measurements on density operators such as $\rho \otimes \rho \otimes \cdots \otimes \rho$ will include sophisticated measurements performed on the whole ensemble, that cannot be described as measurements on the systems separately.[6]

---

[6] For further discussion of this interesting point, see Peres and Wootters [728] and Bennett et al. [120].

14.3 Fidelity and statistical distance

Among the quantum distinguishability measures, we single out the fidelity function for special attention. It is much used and it is closely connected to the Bures geometry of quantum states. It was defined in Section 9.4 as

$$F(\rho_1, \rho_2) = \left(\mathrm{Tr}\sqrt{\sqrt{\rho_1}\rho_2\sqrt{\rho_1}} \right)^2 = \left(\mathrm{Tr}|\sqrt{\rho_1}\sqrt{\rho_2}| \right)^2. \tag{14.40}$$

Actually, in Section 9.4 we worked mostly with the *root fidelity*

$$\sqrt{F}(\rho_1, \rho_2) = \mathrm{Tr}\sqrt{\sqrt{\rho_1}\rho_2\sqrt{\rho_1}}. \tag{14.41}$$

But in some contexts fidelity is the more useful notion. If both states are pure it equals the transition probability between the states. A little more generally, suppose that one of the states is pure, $\rho_1 = |\psi\rangle\langle\psi|$. Then ρ_1 equals its own square root and in fact

$$F(\rho_1, \rho_2) = \langle\psi|\rho_2|\psi\rangle. \tag{14.42}$$

In this situation fidelity has a direct interpretation as the probability that the state ρ_2 will pass the yes/no test associated with the pure state ρ_1. It serves as a figure of merit in many statistical estimation problems.

   This still does not explain why we use the definition (14.40) of fidelity – for the quantum noiseless coding theorem we used Eq. (14.42) only, and there are many expressions that reduce to this equation when one of the states is pure (such as $\mathrm{Tr}\rho_1\rho_2$). The definition not only looks odd, it has obvious drawbacks too: in order to compute it we have to compute two square roots of positive operators – that is to say that we must go through the laborious process of diagonalizing a Hermitian matrix twice. But on further inspection the virtues of fidelity emerge. The key statement about it is Uhlmann's theorem (proved in Section 9.4). The theorem says that $F(\rho_1, \rho_2)$ equals the maximal transition probability between a pair of purifications of ρ_1 and ρ_2. It also enjoys a number of other interesting properties [509]:

   (i) $0 \leq F(\rho_1, \rho_2) \leq 1$;
   (ii) $F(\rho_1, \rho_2) = 1$ if and only if $\rho_1 = \rho_2$ and $F(\rho_1, \rho_2) = 0$ if and only if ρ_1 and ρ_2 have orthogonal supports;
   (iii) Symmetry, $F(\rho_1, \rho_2) = F(\rho_2, \rho_1)$;
   (iv) Concavity, $F(\rho, a\rho_1 + (1 - a)\rho_2) \geq aF(\rho, \rho_1) + (1 - a)F(\rho, \rho_2)$;
   (v) Multiplicativity, $F(\rho_1 \otimes \rho_2, \rho_3 \otimes \rho_4) = F(\rho_1, \rho_3)\,F(\rho_2, \rho_4)$;
   (vi) Unitary invariance, $F(\rho_1, \rho_2) = F(U\rho_1 U^\dagger, U\rho_2 U^\dagger)$;
   (vii) Monotonicity, $F(\Phi(\rho_1), \Phi(\rho_2)) \geq F(\rho_1, \rho_2)$ where Φ is a trace preserving CP map.

Root fidelity enjoys all these properties too, as well as the stronger property of joint concavity in its arguments.

It is interesting to prove property (iii) directly. To do so, observe that the trace in expression (14.41) can be written in terms of the square roots of the non-zero eigenvalues λ_n of a positive operator, as follows:

$$\sqrt{F} = \sum_n \sqrt{\lambda_n} \quad \text{where} \quad AA^\dagger|\psi_n\rangle = \lambda_n|\psi_n\rangle, \quad A \equiv \sqrt{\rho_1}\sqrt{\rho_2}. \tag{14.43}$$

But an easy argument shows that the non-zero eigenvalues of $AA^\dagger$ are the same as those of $A^\dagger A$:

$$AA^\dagger|\psi_n\rangle = \lambda_n|\psi_n\rangle \quad \Rightarrow \quad A^\dagger AA^\dagger|\psi_n\rangle = \lambda_n A^\dagger|\psi_n\rangle. \tag{14.44}$$

Unless $A^\dagger|\psi_n\rangle = 0$ this shows that any eigenvalue of $AA^\dagger$ is an eigenvalue of $A^\dagger A$. Therefore we can equivalently express the fidelity in terms of the square roots of the non-zero eigenvalues of $A^\dagger A$, in which case the roles of ρ_1 and ρ_2 are interchanged.

Property (vii) is a key entry: fidelity is a *monotone* function. The proof [94] is a simple consequence of Uhlmann's theorem (Section 9.4). We can find a purification of our density matrices, such that $F(\rho_1, \rho_2) = |\langle\psi_1|\psi_2\rangle|^2$. We can also introduce an environment – a rather 'mathematical' environment, but useful for our proof – that starts in the pure state $|0\rangle$, so that the quantum operation is described by a unitary transformation $|\psi\rangle|0\rangle \rightarrow U|\psi\rangle|0\rangle$. Then Uhlmann's theorem implies that $F(\Phi(\rho_1), \Phi(\rho_2)) \geq |\langle\psi_1|\langle0|U^\dagger U|\psi_2\rangle|0\rangle|^2 = |\langle\psi_1|\langle0|\psi_2\rangle|0\rangle|^2 = F(\rho_1, \rho_2)$. Thus the fidelity is non-decreasing with respect to any physical operation, including measurement.

Finally, we observe that the fidelity may be defined implicitly [14] by

$$F(\rho_1, \rho_2) = \inf\left[\text{Tr}(A\rho_1)\,\text{Tr}(A^{-1}\rho_2)\right], \tag{14.45}$$

where the infimum is taken over all invertible positive operators A. There is a closely related representation of the root fidelity as an infimum over the same set of operators A [17],

$$\sqrt{F}(\rho_1, \rho_2) = \frac{1}{2}\inf\left[\text{Tr}(A\rho_1) + \text{Tr}(A^{-1}\rho_2)\right], \tag{14.46}$$

since after squaring this expression only-cross terms contribute to (14.45).

In Section 9.4 we introduced the Bures distance as a function of the fidelity. This is also a monotone function, and no physical operation can increase it. It follows that the corresponding metric, the Bures metric, is a *monotone metric* under stochastic maps, and may be a candidate for a quantum version of the Fisher–Rao metric. It is a good candidate.[7] To see this, let us choose a POVM $\{E_i\}$. A given density matrix ρ will respond with the probability distribution $P(E, \rho)$. For a pair of density matrices we can define the *quantum Bhattacharyya coefficient*

$$B(\rho, \sigma) \equiv \min_E B\Big(P(E, \rho), P(E, \sigma)\Big) = \min_E \sum_i \sqrt{p_i q_i}, \tag{14.47}$$

where

$$p_i = \mathrm{Tr} E_i \rho, \qquad q_i = \mathrm{Tr} E_i \sigma, \tag{14.48}$$

and the minimisation is carried out over all possible POVMs. If we succeed in doing this, we will obtain a quantum analogue of the Fisher–Rao distance as a function of $B(\rho, \sigma)$.

We will assume that both density matrices are invertible. As a preliminary step, we rewrite p_i, using an arbitrary unitary operator U, as

$$p_i = \mathrm{Tr}\left((U\sqrt{\rho}\sqrt{E_i})(U\sqrt{\rho}\sqrt{E_i})^\dagger\right). \tag{14.49}$$

Then we use the Cauchy–Schwarz inequality (for the Hilbert–Schmidt inner product) to set a lower bound:

$$p_i q_i = \mathrm{Tr}\left((U\sqrt{\rho}\sqrt{E_i})(U\sqrt{\rho}\sqrt{E_i})^\dagger\right) \mathrm{Tr}\left((\sqrt{\sigma}\sqrt{E_i})(\sqrt{\sigma}\sqrt{E_i})^\dagger\right)$$
$$\geq \left(\mathrm{Tr}\left((U\sqrt{\rho}\sqrt{E_i})(\sqrt{\sigma}\sqrt{E_i})^\dagger\right)\right)^2. \tag{14.50}$$

Equality holds if and only if

$$\sqrt{\sigma}\sqrt{E_i} = \mu_i U\sqrt{\rho}\sqrt{E_i} \tag{14.51}$$

for some real number μ_i. Depending on the choice of U, this equation may or may not have a solution. Anyway, using the linearity of the trace, it is now easy to see that

$$\sum_i \sqrt{p_i q_i} \geq \sum_i \left|\mathrm{Tr}(U\sqrt{\rho}E_i\sqrt{\sigma})\right| \geq \left|\mathrm{Tr}\left(\sum_i U\sqrt{\rho}E_i\sqrt{\sigma}\right)\right| = \mathrm{Tr}(U\sqrt{\rho}\sqrt{\sigma}). \tag{14.52}$$

The question is: how should we choose U if we wish to obtain a sharp inequality? We have to make sure that Eq. (14.51) holds, and also that all the terms in (14.52) are positive. A somewhat tricky argument [320] shows that the answer is

$$U = \sqrt{\sqrt{\sigma}\rho\sqrt{\sigma}}\,\frac{1}{\sqrt{\sigma}}\frac{1}{\sqrt{\rho}}. \tag{14.53}$$

---

[7] The link between Bures and statistical distance was forged by Helstrom (1976) [422], Holevo (1982) [445] and Braunstein and Caves (1994) [162]. Our version of the argument follows Fuchs and Caves [320]. Actually the precise sense in which the Bures metric is the analogue of the classical Fisher–Rao metric is not quite clear [93].

This gives $\sum_i \sqrt{p_i q_i} \geq \sqrt{F(\rho, \sigma)}$, where $\sqrt{F}$ is the root fidelity. The optimal POVM turns out to be a projective measurement, associated with the Hermitian observable

$$M = \frac{1}{\sqrt{\sigma}} \sqrt{\sqrt{\sigma} \rho \sqrt{\sigma}} \frac{1}{\sqrt{\sigma}}. \qquad (14.54)$$

The end result is that

$$B(\rho, \sigma) \equiv \min_E B\Big(P(E, \rho), P(E, \sigma)\Big) = \text{Tr}\sqrt{\sqrt{\sigma} \rho \sqrt{\sigma}} \equiv \sqrt{F(\rho, \sigma)}. \qquad (14.55)$$

It follows that the Bures angle distance $D_A = \cos^{-1} \sqrt{F(\rho, \sigma)}$ is precisely the Fisher–Rao distance, maximised over all the probability distributions that one can obtain by varying the POVM.

For the case when the two states to be distinguished are pure we have already seen (in Section 5.3) that the Fubini–Study distance is the answer. These two answers are consistent. But in the pure state case the optimal measurement is not uniquely determined, while here it is: we obtained an explicit expression for the observable that gives optimal distinguishability, namely M. The operator has an ugly look, but it has a name: it is the *geometric mean* of the operators σ^{-1} and ρ. As such it was briefly discussed in Section 13.1, where we observed that the geometric mean is symmetric in its arguments. From this fact it follows that $M(\sigma, \rho) = M^{-1}(\rho, \sigma)$. Therefore $M(\sigma, \rho)$ and $M(\rho, \sigma)$ define the same measurement. The operator M also turned up in our discussion of geodesics with respect to the Bures metric, in Eq. (9.57). When $N = 2$ this fact can be used to determine M in an easy way: draw the unique geodesic that connects the two states, given that we view the Bloch ball as a round hemi-3-sphere. This geodesic will meet the boundary of the Bloch ball in two points, and these points are precisely the eigenstates of M.

We now have a firm link between statistical distance and the Bures metric, but we are not yet done with it – we will come back to it in Chapter 15. Meanwhile, let us compare the three distances that we have brought into play (Table 14.1). The first observation is that the two monotone distances, trace and Bures, have the property that the distance between states of orthogonal support is maximal:

$$\text{supp}(\rho) \perp \text{supp}(\sigma) \quad \Leftrightarrow \quad D_{\text{tr}}(\rho, \sigma) = 1 \quad \Leftrightarrow \quad D_B(\rho, \sigma) = \sqrt{2}. \qquad (14.56)$$

Table 14.1 *Metrics in the space of quantum states.*

Metric	Bures	Hilbert–Schmidt	Trace
Is it Riemannian?	Yes	Yes	No
Is it monotone?	Yes	No	Yes

This is not true for the Hilbert–Schmidt distance. The second observation concerns a bound [323] that relates fidelity (and hence the Bures distance) to the trace distance, viz.

$$1 - \sqrt{F(\rho,\sigma)} \leq D_{tr}(\rho,\sigma) \leq \sqrt{1 - F(\rho,\sigma)}. \qquad (14.57)$$

To prove that the upper bound holds, observe that it becomes an equality for a pair of pure states (which is easy to check, since we can work in the two-dimensional Hilbert space spanned by the two pure states). But Uhlmann's theorem means that we can find a purification such that $F(\rho,\sigma) = |\langle \psi | \phi \rangle|^2$. In the purifying Hilbert space the bound is saturated, and taking a partial trace can only decrease the trace distance (because of its monotonicity), while the fidelity stays constant by definition. For the lower bound, see Problem 14.3.

The Bures and trace distances are both monotone, and the close relation between them means that, for many purposes, they can be used interchangeably. There exist also relations between the Bures and Hilbert–Schmidt distances, but the latter does not have the same fundamental importance. It is evident, from the way that the Bloch sphere is deformed by an orthographic projection from the flat Hilbert–Schmidt Bloch ball to the round Bures hemi-3-sphere, that it may happen that $D_2(\rho_a, \rho_b) > D_2(\rho_c, \rho_d)$ while $D_B(\rho_a, \rho_b) < D_B(\rho_c, \rho_d)$. To find a concrete example, place $\rho_a = \rho_c$ at the north pole, ρ_b on the surface, and ρ_d on the polar axis through the Bloch ball.

For $N = 2$ we can use the explicit formula (9.48) for the Bures distance to compare it with the flat Hilbert–Schmidt distance. Since for one-qubit states the trace and Hilbert–Schmidt distances agree, we arrive in this way at strict bounds between $D_B = D_B(\rho_a, \rho_b)$ and $D_{tr} = D_{tr}(\rho_a, \rho_b)$ valid for any $N = 2$ states,

$$\sqrt{2 - 2\sqrt{1 - (D_{tr})^2}} \leq D_B \leq \sqrt{2 - 2\sqrt{1 - D_{tr}}}. \qquad (14.58)$$

The lower bound comes from pure states. The upper bound comes from the family of mixed states situated on an axis through the Bloch ball, and does not hold in higher dimensions. However, making use of the relation (9.31) between the Bures distance and fidelity we may translate the general bounds (14.57) into

$$\sqrt{2 - 2\sqrt{1 - (D_{tr})^2}} \leq D_B \leq \sqrt{2 D_{tr}}. \qquad (14.59)$$

This upper bound, valid for an arbitrary N, is not strict. Figure 14.3 presents Bures distance versus trace distance for 500 pairs consisting of a random pure state and a random mixed state distributed according to the Hilbert–Schmidt measure (see Chapter 15). The upper bound (14.58) is violated for $N > 2$.

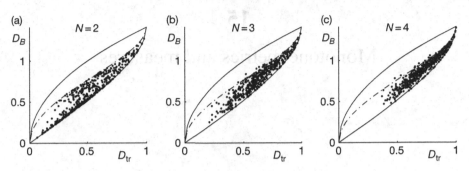

Figure 14.3 Bures distance plotted against trace distance for random density matrices of size (a) $N = 2$, (b) $N = 3$ and (c) $N = 4$. The single dots are randomly drawn density matrices, the solid lines denote the bounds (14.59) and the dotted lines the upper bound in (14.58).

Problems

14.1 Prove that the flat metric on the classical probability simplex is monotone under bistochastic maps.

14.2 What is the largest possible trace distance between two density matrices that can be connected by a unitary transformation?

14.3 Complete the proof of the inequality (14.57).

14.4 Derive the inequalities (a): $F(\sigma, \rho) \geq \left(\mathrm{Tr}\sqrt{\sigma}\sqrt{\rho}\right)^2$ and (b): $F(\sigma, \rho) \geq \mathrm{Tr}\sigma\rho$. What are the conditions for equality?

15

Monotone metrics and measures

Probability theory is a measure theory with a 'soul' provided ... by the most ancient and noble of all mathematical disciplines, namely Geometry.

– Mark Kac[1]

Section 2.6 was devoted to *classical ensembles*, that is to say ensembles defined by probability measures on the set of classical probability distributions over N events. In this chapter *quantum ensembles* are defined by choosing probability measures on the set of density matrices of size N. A warning should be issued first: there is no single, naturally distinguished measure in $\mathcal{M}^{(N)}$, so we have to analyse several measures, each of them with different physical motivations, advantages and drawbacks. This is in contrast to the set of pure quantum states, where the Fubini–Study (FS) measure is the only natural choice for a measure that defines 'random states'.

A simple way to define a probability measure goes through a metric. Hence we will start this chapter with a review of the metrics defined on the set of mixed quantum states.

15.1 Monotone metrics

In Section 2.5 we explained how the Fisher metric holds its distinguished position due to the theorem of Čencov, which states that the Fisher metric is the unique monotone metric on the probability simplex Δ_{N-1}. Now that the latter has been replaced by the space of quantum states $\mathcal{M}^{(N)}$ we must look again at the question of metrics. Since the uniqueness in the classical case came from the behaviour under stochastic maps, we turn our attention to stochastic quantum maps – the completely

---

[1] Reproduced from Mark Kac. Preface to L. A. Santalo. *Integral Geometry and Geometrical Probability.* Addison-Wesley, 1976.

positive, trace preserving maps discussed in Chapter 10. A distance D in the space of quantum states $\mathcal{M}^{(N)}$ is called *monotone* if it does not grow under the action of a stochastic map Φ,

$$D_{\text{mon}}\left(\Phi\rho, \Phi\sigma\right) \leq D_{\text{mon}}\left(\rho, \sigma\right). \tag{15.1}$$

If a monotone distance is geodesic the corresponding metric on $\mathcal{M}^{(N)}$ is called monotone. However, in contrast to the classical case it turns out that there exist infinitely many monotone Riemannian metrics on the space of quantum states.

The appropriate generalisation of Čencov's classical theorem is[2]

The Morozova–Čencov–Petz (MCP) theorem. *At a point where the density matrix is diagonal, $\rho = \text{diag}(\lambda_1, \lambda_2, \ldots, \lambda_N)$, every monotone metric on $\mathcal{M}^{(N)}$ assigns the length squared*

$$||A||^2 = \frac{1}{4}\left[C\sum_{i=1}^{N} \frac{A_{ii}^2}{\lambda_i} + 2\sum_{i<j}^{N} c(\lambda_i, \lambda_j)\,|A_{ij}|^2 \right]. \tag{15.2}$$

to any tangent vector A, where C is an arbitrary constant, the function $c(x, y)$ is symmetric,

$$c(x, y) = c(y, x), \quad \text{and obeys} \quad c(sx, sy) = s^{-1}c(x, y), \tag{15.3}$$

and the function $f(t) \equiv \frac{1}{c(t,1)}$ is operator monotone.

The tangent vector A is a traceless Hermitian matrix; if A is diagonal then the second term vanishes and the first term gives the familiar Fisher–Rao metric on the simplex. But the second term is new. And the result falls far short of providing uniqueness. Any function $f(t) : \mathbb{R}_+ \to \mathbb{R}_+$ will be called a *Morozova–Chentsov* (MC) function, if it fulfils three restrictions:

    (i) f is operator monotone,

    (ii) f is self inversive: $f(1/t) = f(t)/t$,   (15.4)

    (iii) $f(1) = 1$.

The meaning of condition (i) was discussed at some length in Section 13.1. Condition (ii) was also encountered there, as the condition for an operator mean to be symmetric. Here it ensures that the function $c(x, y) = 1/[yf(x/y)]$ satisfies (15.3). Condition (iii) is a normalisation with consequences – it forces us to set $C = 1$ in order to avoid a conical singularity at the maximally mixed state, as we will soon see explicitly in the case $N = 2$. The metric is now said to be *Fisher adjusted.*

---

[2] Here we have merged the theorem by Morozova and Chentsov (1990) [673] with a theorem by Petz (1996) [731] that completed it. See also Petz and Sudár [734], Lesniewski and Ruskai [588], and – for a guided tour through the garden of monotone metrics – Petz [732].

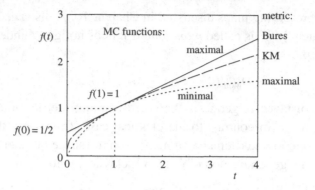

Figure 15.1 Morozova–Chentsov functions $f(t)$: minimal function for the maximal metric (dotted), Kubo–Mori metric (dashed), and maximal function for the minimal Bures metric (upper line). These metrics are Fisher adjusted, $f(1) = 1$. The Bures metric is also FS adjusted, $f(0) = 1/2$.

The MCP theorem can now be rephrased: there is a one-to-one correspondence between MC functions and monotone Riemannian metrics.

The infinite dimensional set $\mathcal{F}$ of all MC functions is convex; the set of operator monotone functions itself is convex, and an explicit calculation shows that any convex combination of two self-inversive functions is self-inversive. A monotone Riemannian metric will be called *pure* if the corresponding MC function f is an extreme point of the convex set $\mathcal{F}$. Among all operator monotone functions on $[0, +\infty)$ which are self-inversive and obey $f(1) = 1$, there exists a minimal and a maximal function [561]. In Figure 15.1 we plot three choices:

$$f_{\min}(t) = \frac{2t}{t+1}, \qquad f_{\mathrm{KM}}(t) = \frac{t-1}{\ln t}, \qquad f_{\max}(t) = \frac{1+t}{2}. \tag{15.5}$$

The maximal MC function $f_{\max}$ is a straight line and gives rise to the Bures metric, while the minimal function $f_{\min}$ is a hyperbola. The intermediate case f_{KM} leads to the Kubo–Mori metric used in quantum statistical mechanics. Our MC functions correspond to

$$c_{\min}(x, y) = \frac{x+y}{2xy}, \qquad c_{\mathrm{KM}}(x, y) = \frac{\ln x - \ln y}{x-y}, \qquad c_{\max}(x, y) = \frac{2}{x+y}, \tag{15.6}$$

so the inverse, $1/c$, is equal to the harmonic, logarithmic and arithmetic mean, respectively. For other interesting choices of $f(t)$, see Problem 15.1.

To familiarise ourselves with the frightening expression (15.2) we can take a look at the case $N = 2$, the Bloch ball. We set $\rho = \frac{1}{2}\mathrm{diag}(1 + r, 1 - r)$ and find after a minor calculation that the metric is

$$ds^2 = \frac{1}{4}\left[\frac{dr^2}{1 - r^2} + \frac{1}{f\left(\frac{1-r}{1+r}\right)} \frac{r^2}{1+r} d\Omega^2 \right], \qquad 0 < r < 1. \tag{15.7}$$

Here $d\Omega^2$ is the metric on the unit 2-sphere – the second term corresponds to the second, tangential, term in Eq. (15.2), and we used spherical symmetry to remove the restriction to diagonal density matrices. We can now see, given $C = 1$, that the condition $f(1) = 1$ means that the metric is regular at the origin. We also see that $f(0) = 0$ means that the area of the boundary is infinite. If $f(0)$ is finite the boundary is a round 2-sphere. It will have radius $1/2$ if $f(0) = 1/2$; such a metric is said to be *Fubini–Study adjusted*.

Because the MC function f appears in the denominator, the larger the (normalised) function, the smaller the area of a sphere at constant r. So – slightly confusingly – the metric that uses f_{min} will be called the *maximal metric* and the metric using f_{max} will be called the *minimal metric*.

Let us now go through the geometries corresponding to our three choices of the function $f(t)$. If $f = f_{max}$ then

$$ds_{min}^2 = \frac{1}{4}\left[\frac{dr^2}{1-r^2} + r^2 d\Omega^2\right]. \tag{15.8}$$

This is the metric on a round 3-sphere of radius $1/2$; the scalar curvature is $R_{max} = 4 \cdot 6$ and the boundary of the Bloch ball is a round 2-sphere corresponding to the equator of the 3-sphere. This is an FS adjusted metric. In fact it is the Bures metric that we have encountered in Chapters 9 and 14. Since it is given by the *maximal* function $f_{max}(t)$, it is distinguished by being the *minimal* Fisher adjusted, monotone Riemannian metric. If $f = f_{min}$ then

$$ds_{max}^2 = \frac{1}{4}\left[\frac{1}{1-r^2}(dr^2 + r^2 d\Omega^2)\right]. \tag{15.9}$$

The curvature is everywhere negative and diverges to minus infinity at $r = 1$. Indeed the area of the boundary – the space of pure states – is infinite. For the intermediate choice $f = f_{KM}$ we get

$$ds_{KM}^2 = \frac{1}{4}\left[\frac{dr^2}{1-r^2} + \frac{r}{2}\ln\left(\frac{1+r}{1-r}\right)d\Omega^2\right]. \tag{15.10}$$

The curvature is zero at the origin and decreases (very slowly at first) to minus infinity at the boundary, which again has a diverging area, although the divergence is only logarithmic.

Now where does Eq. (15.2) come from? Recall from Section 2.5 that there were two natural definitions of the (uniquely monotone) Fisher–Rao metric. We can define it as the Hessian of the relative entropy function or in terms of expectation values of score vectors (logarithmic derivatives of the probability distribution). In the classical case these definitions led to the same metric. But in the quantum case this is no longer so. The logarithmic derivative of a density matrix is an ambiguous

notion, because we have entered a non-commutative probability theory. To see this
let us assume that we are looking at a set of density matrices ρ_θ parametrized by
some parameters θ^a. For simplicity we consider an affine parametrization of $\mathcal{M}^{(N)}$
itself,

$$\rho_\theta = \frac{1}{N}\mathbb{1} + \sum_{a=1}^{N}\theta^a A_a, \tag{15.11}$$

where A_a are a set of traceless Hermitian matrices, that is to say that they are tangent
vectors of $\mathcal{M}^{(N)}$. Evidently

$$\partial_a \rho_\theta = A_a. \tag{15.12}$$

We may define the logarithmic derivative, that is a *quantum score vector* L_a, by

$$A_a = \rho_\theta L_a \quad \Rightarrow \quad L_a = \rho_\theta^{-1} A_a. \tag{15.13}$$

In analogy with the classical equation (2.65), we define the squared length of a
tangent vector as

$$||A||^2 = \frac{1}{4}\mathrm{Tr}\rho_\theta LL = \frac{1}{4}\mathrm{Tr}AL = \frac{1}{4}\mathrm{Tr}\rho_\theta^{-1} A^2. \tag{15.14}$$

This defines a metric. If we set $\rho = \mathrm{diag}(\lambda_1, \ldots, \lambda_N)$ and perform a pleasant
calculation we find that it is exactly the maximal metric, that is the metric (15.2)
for the choice $f = f_{\min}$.

But other definitions of the logarithmic derivative suggest themselves, such as
the *symmetrical logarithmic derivative* L_a occurring in the equation

$$A_a = \rho_\theta \circ L_a = \frac{1}{2}(\rho_\theta L_a + L_a \rho_\theta), \tag{15.15}$$

where we found a use for the Jordan product from Section 8.7.[3] This equation for
L_a was first encountered in Eq. (9.27), and there we claimed that there is a unique
solution if ρ_θ is invertible. To find it, we first choose a basis where ρ_θ is diagonal
with eigenvalues λ_i. The equation for the matrix elements of $L_a \equiv L$ becomes

$$A_{ij} = \frac{1}{2}\left(\lambda_i L_{ij} + L_{ij}\lambda_j\right) = \frac{1}{2}(\lambda_i + \lambda_j)L_{ij} \tag{15.16}$$

(where no summation is involved). The solution is immediate:

$$L_{ij} = \frac{2}{\lambda_i + \lambda_j}A_{ij}. \tag{15.17}$$

---

[3] The symmetric logarithmic derivative was first introduced by Helstrom [422] and Holevo [445].

The length squared of the vector A becomes

$$||A||^2 \equiv \frac{1}{4}\mathrm{Tr}AL = \frac{1}{2}\sum_{i,j}\frac{A_{ij}A_{ji}}{\lambda_i + \lambda_j} = \frac{1}{4}\sum_i \frac{A_{ii}^2}{\lambda_i} + \sum_{i<j}\frac{|A_{ij}|^2}{\lambda_i + \lambda_j}. \quad (15.18)$$

Again a pleasant calculation confirms that this agrees with Eq. 15.2, this time for the choice $f = f_{\max}$.

The Kubo–Mori metric is a different kettle of fish.[4] It arises when we try to define a metric as the Hessian of the relative entropy. It takes a little effort to do this explicitly. We need to know that, for any positive operator A,

$$\ln(A + xB) - \ln A = \int_0^\infty \frac{1}{A+u}xB\frac{1}{A+xB+u}\,du. \quad (15.19)$$

To prove this is an exercise (namely Problem 13.1). It follows that

$$\partial_x \ln(\rho + xA)_{|x=0} = \int_0^\infty \frac{1}{\rho+u}A\frac{1}{\rho+u}\,du. \quad (15.20)$$

With this result in hand it is straightforward to compute

$$-\partial_\alpha\partial_\beta S(\rho + \alpha A||\rho + \beta B)_{|\alpha=\beta=0} = \int_0^\infty \mathrm{Tr}A(\rho+u)^{-1}B(\rho+u)^{-1}\,du. \quad (15.21)$$

This is precisely the expression that defines the *Kubo–Mori* scalar product $g(A,B)(\rho)$, in analogy with the classical equation (2.59). If we evaluate this expression for a diagonal density matrix ρ we recover Eq. (15.2), with the function $c = c_{\mathrm{KM}}$ given in Eq. (15.6).

The Kubo–Mori metric does have a uniqueness property: it is the only monotone metric for which the two flat affine connections, mixture and exponential, are mutually dual [359]. Therefore it allows us to play a quantum version of the classical game played in Section 3.2. There is also a conjecture by Petz [730] that says that the Kubo–Mori metric is the only metric having the property that the scalar curvature is monotone, in the sense that if $\rho_1 \prec \rho_2$, that is if ρ_1 is majorised by ρ_2, then $R(\rho_1) \leq R(\rho_2)$ (which is not true for the Bures metric, as mentioned in Section 9.6). On the other hand the fact that the Kubo–Mori metric is the Hessian of the relative entropy is not really a uniqueness property, because it can be shown that every monotone metric is the Hessian of a certain monotone relative entropy – not Umegaki's relative entropy, but one of the larger family of monotone relative entropies[5] whose existence we hinted at in Section 13.4.

---

[4] Studies of this metric include those of Ingarden [488], Balian et al. [86], and Petz [730].
[5] Lesniewski and Ruskai [588] explored this matter in depth; see also Jenčova [501].

15.2 Product measures and flag manifolds

It will not have escaped the reader that the price of putting a Riemannian geometry on $\mathcal{M}^{(N)}$ is rather high, in the sense that the monotone metrics that appear are quite difficult to work with when $N > 2$. Fortunately the measures that come from our monotone metrics are not that difficult to handle, so that calculations of volumes and the like are quite doable. The basic trick that we will use is the same as that used in flat space, when the Euclidean measure is decomposed into a product of a measure on the orbits of the rotation group and a measure in the radial direction (in other words, when one uses spherical polar coordinates). The set of density matrices that can be written in the form $\rho = U\Lambda U^\dagger$, for a fixed diagonal matrix Λ with strictly positive eigenvalues, is a flag manifold $\mathbf{F}^{(N)} = U(N)/[U(1)]^N$ (see Section 8.5). A natural assumption concerning a probability distribution in $\mathcal{M}^{(N)}$ is to require invariance with respect to unitary rotations, $P(\varrho) = P(W\varrho W^\dagger)$. This is the case if (a) the choice of eigenvalues and eigenvectors is independent, and (b) the eigenvectors are drawn according to the Haar measure, $\mathrm{d}\nu_H(W) = \mathrm{d}\nu_H(UW)$.

Such a *product measure*, $\mathrm{d}V = \mathrm{d}\nu_H(U) \times \mathrm{d}\mu(\vec{\lambda})$, defined on the Cartesian product $\mathbf{F}^{(N)} \times \Delta_{N-1}$, leads to the probability distribution,

$$P(\rho) = P_H(\mathbf{F}^{(N)}) \times P(\vec{\lambda}) \tag{15.22}$$

in which the first factor denotes the natural, unitarily invariant distribution on the flag manifold $\mathbf{F}^{(N)}$, induced by the Haar measure on $U(N)$.

We are going to compute the volume of $\mathbf{F}^{(N)}$ with respect to this measure. Let us rewrite the complex flag manifold as a Cartesian product of complex projective spaces,

$$\mathbf{F}^{(N)} = \frac{U(N)}{[U(1)]^N} \simeq \frac{U(N)}{U(N-1) \times U(1)} \frac{U(N-1)}{U(N-2) \times U(1)} \cdots \frac{U(2)}{U(1) \times U(1)}$$
$$\simeq \mathbb{CP}^{N-1} \times \mathbb{CP}^{N-2} \times \cdots \times \mathbb{CP}^1, \tag{15.23}$$

where $\simeq$ means 'equal, if the volumes are concerned'. This ignores a number of topological complications, but our previous experience with fibre bundles, say in Section 4.7, makes it at least plausible that we can proceed like this. (And the result is correct!) Making use of Eq. (4.88), $\mathrm{Vol}(\mathbb{CP}^k) = \pi^k/k!$, we find for $N \geq 2$

$$\mathrm{Vol}(\mathbf{F}^{(N)}) = \prod_{k=1}^{N-1} \mathrm{Vol}(\mathbb{CP}^k) = \frac{\pi^{N(N-1)/2}}{\Xi_N}, \tag{15.24}$$

with

$$\Xi_N \equiv 0! \ 1! \ 2! \cdots (N-1)! = \prod_{k=1}^{N} \Gamma(k). \tag{15.25}$$

The result for $\mathrm{Vol}(\mathbf{F}^{(N)})$ still depends on a free multiplicative factor which sets the scale, just as the volume of a sphere depends on its radius. In Eq. (15.24) we have implicitly fixed the scale by the requirement that a closed geodesic on $\mathbb{CP}^k$ has circumference π. In this way we have adjusted the metric with Eq. (3.137), which corresponds to the following normalisation of the measure on the unitary group

$$ds^2 \equiv -a\,\mathrm{Tr}(U^{-1}dU)^2 \quad \text{with} \quad a = \frac{1}{2}. \tag{15.26}$$

Direct integration over the circle $U(1)$ gives $\mathrm{Vol}[U(1)] = 2\pi\sqrt{a} = \sqrt{2}\pi$. This result combined with (15.24) allows us to write the volume of the unitary group[6]

$$\mathrm{Vol}[U(N)] = \mathrm{Vol}(\mathbf{F}^{(N)})\,(\mathrm{Vol}[U(1)])^N = 2^{N/2}\frac{\pi^{N(N+1)/2}}{\Xi_N}. \tag{15.27}$$

For completeness let us state an analogous result for the special unitary group,

$$\mathrm{Vol}[SU(N)] = 2^{(N-1)/2}\sqrt{N}\,\frac{\pi^{(N+2)(N-1)/2}}{\Xi_N}. \tag{15.28}$$

This is again the quotient of the volumes of $U(N)$ and $U(1)$, but it is a different $U(1)$ than the one above, hence the presence of the stretching factor $\sqrt{N}$.

Let us return to the discussion of the product measures (15.22). The second factor $P(\vec{\lambda})$ may in principle be an arbitrary measure on the simplex Δ_{N-1}. To obtain a measure supported on the set of mixed states one may use the Dirichlet distribution (2.73), for example with the parameter s equal to 1 and $1/2$, for the

Table 15.1 *Volumes of flag manifolds and unitary groups;* $\Xi_N = \prod_{k=1}^{N}\Gamma(k)$.

Manifold	Dimension	Vol[X], $a = 1/2$	Vol'[X], $a = 1$
$\mathbb{CP}^N$	$2N$	$\dfrac{\pi^N}{N!}$	$\dfrac{(2\pi)^N}{N!}$
$\mathbf{F}^{(N)} = \dfrac{U(N)}{[U(1)]^N}$	$N(N-1)$	$\dfrac{\pi^{N(N-1)/2}}{\Xi_N}$	$\dfrac{(2\pi)^{N(N-1)/2}}{\Xi_N}$
$U(N)$	N^2	$2^{N/2}\dfrac{\pi^{N(N+1)/2}}{\Xi_N}$	$\dfrac{(2\pi)^{N(N+1)/2}}{\Xi_N}$
$SU(N)$	$N^2 - 1$	$2^{(N-1)/2}\sqrt{N}\dfrac{\pi^{(N+2)(N-1)/2}}{\Xi_N}$	$\sqrt{N}\dfrac{(2\pi)^{(N+2)(N-1)/2}}{\Xi_N}$

[6] This result was derived by Hua [474] with the normalisation $a = 1$, for which the volume of the circle becomes $\mathrm{Vol}'[U(1)] = 2\pi$. Different normalisations give proportional volumes, $\mathrm{Vol}'(X) = (\sqrt{2})^{\dim(X)}\mathrm{Vol}(X)$. For the reader's convenience we have collected the results for both normalisations in Table 15.1. For orthogonal groups, see Problem 15.2. For more details consult [153, 1017] and references therein.

flat and round measures on the simplex. These measures were called *unitary* and *orthogonal product measures* [1007], respectively, since they represent the distribution of the squared components of a random complex (real) vector and are induced by the Haar measures on the unitary (orthogonal) groups. The unitary product measure P_u is induced by the coarse graining map – see Eq. (15.81) – but as discussed in the following sections of this chapter, other product measures seem to be better motivated by physical assumptions.

15.3 Hilbert–Schmidt measure

A metric always generates a measure. For Riemannian metrics this was explained in Section 1.4, but we can also use more exotic metrics like the L_p-metrics defined in Section 14.2. All L_p-metrics, including the trace metric for $p = 1$, will generate the same *Lebesque measure*. Here we concentrate on the $p = 2$ case, the Hilbert–Schmidt (HS) metric, and when deriving the measure we treat it in the same way that we would treat any Riemannian metric. The HS metric is defined by the line element squared,

$$ds_{HS}^2 = \frac{1}{2}\text{Tr}[(d\rho)^2], \tag{15.29}$$

valid for any dimension N.[7] Making use of the diagonal form $\rho = U\Lambda U^{-1}$ and of the differentiation rule of Leibniz we write

$$d\rho = U[d\Lambda + U^{-1}dU\Lambda - \Lambda U^{-1}dU]U^{-1}. \tag{15.30}$$

Hence (15.29) can be rewritten as

$$ds_{HS}^2 = \frac{1}{2}\sum_{i=1}^{N}(d\lambda_i)^2 + \sum_{i<j}^{N}(\lambda_i - \lambda_j)^2 |(U^{-1}dU)_{ij}|^2. \tag{15.31}$$

Due to the normalisation condition $\sum_{i=1}^{N}\lambda_i = 1$, the sum of differentials vanishes, $\sum_{i=1}^{N}d\lambda_i = 0$. Thus we may consider the variation of the Nth eigenvalue as a dependent one, $d\lambda_N = -\sum_{i=1}^{N-1}d\lambda_i$, which implies

$$\frac{1}{2}\sum_{i=1}^{N}(d\lambda_i)^2 = \frac{1}{2}\sum_{i=1}^{N-1}(d\lambda_i)^2 + \frac{1}{2}\left(\sum_{i=1}^{N-1}d\lambda_i\right)^2 = \sum_{i,j=1}^{N-1}d\lambda_i\,g_{ij}\,d\lambda_j. \tag{15.32}$$

---

[7] As in Chapter 8, a factor $1/2$ is included here to ensure that the length of a closed geodesic in $\mathbb{CP}^n$ equals π.

The corresponding volume element gains a factor $\sqrt{\det g}$, where g is the metric in the $(N-1)$ dimensional simplex Δ_{N-1} of eigenvalues. From (15.32) one may read out the explicit form of the metric

$$
g = \frac{1}{2}\begin{bmatrix} 1 & & 0 \\ & \ddots & \\ 0 & & 1 \end{bmatrix} + \frac{1}{2}\begin{bmatrix} 1 & \cdots & 1 \\ \vdots & \ddots & \vdots \\ 1 & \cdots & 1 \end{bmatrix}. \tag{15.33}
$$

It is easy to check that the $N-1$ dimensional matrix g has one eigenvalue equal to $N/2$ and $N-2$ eigenvalues equal to $1/2$, so that $\det g = N/2^{N-1}$. Thus the HS volume element has the product form

$$
\mathrm{d}V_{\mathrm{HS}} = \frac{\sqrt{N}}{2^{(N-1)/2}} \prod_{j=1}^{N-1} \mathrm{d}\lambda_j \prod_{j<k}^{1\cdots N} (\lambda_j - \lambda_k)^2 \left| \prod_{j<k}^{1\cdots N} \mathrm{Re}(U^{-1}\mathrm{d}U)_{jk}\mathrm{Im}(U^{-1}\mathrm{d}U)_{jk} \right|. \tag{15.34}
$$

The first factor, depending only on the spectrum $\vec{\lambda}$, induces the *Hilbert–Schmidt probability distribution* in the eigenvalue simplex [161, 392],

$$
P_{\mathrm{HS}}(\vec{\lambda}) = C_N^{\mathrm{HS}}\, \delta\left(1 - \sum_{j=1}^{N}\lambda_j\right) \prod_{j<k}^{N} (\lambda_j - \lambda_k)^2, \tag{15.35}
$$

where the normalisation constant

$$
C_N^{\mathrm{HS}} = \frac{\Gamma(N^2)}{\prod_{k=1}^{N} \Gamma(k)\Gamma(k+1)} = \frac{\Gamma(N^2)}{\Xi_N \Xi_{N+1}} \tag{15.36}
$$

ensures that the integral over the simplex Δ_{N-1} is equal to unity [1016].

Let us now analyse the second factor in (15.34). This arises from the off-diagonal elements of the invariant metric (15.26) on the unitary group

$$
\mathrm{d}s^2 \equiv -\frac{1}{2}\mathrm{Tr}(U^{-1}\mathrm{d}U)^2 = \frac{1}{2}\sum_{j=1}^{N} |(U^{-1}\mathrm{d}U)_{jj}|^2 + \sum_{j<k=1}^{N} |(U^{-1}\mathrm{d}U)_{jk}|^2. \tag{15.37}
$$

Hence the HS measure in the space of density matrices belongs to the class of product measures (15.22).

Since the diagonal elements of (15.37) do not contribute to the volume element (15.34), integrating it over the unitary group we obtain the volume of the complex flag manifold (15.24). To compute the HS volume of the set of mixed states $\mathcal{M}^{(N)}$ we need also to integrate (15.34) over the simplex of eigenvalues Δ_{N-1}.

Normalisation (15.35) implies that the latter integral is equal to $1/C_N^{HS}$. Taking into account the prefactor $\sqrt{\det g} = \sqrt{N/2^{N-1}}$, present in (15.34), we obtain[8]

$$\text{Vol}_{HS}\left(\mathcal{M}^{(N)}\right) = \sqrt{\frac{N}{2^{N-1}}} \frac{\text{Vol}\left(\mathbf{F}^{(N)}\right)}{N! \; C_N^{HS}} = \sqrt{N} \frac{\pi^{N(N-1)/2}}{2^{(N-1)/2}} \frac{\Gamma(1)\cdots\Gamma(N)}{\Gamma(N^2)}. \qquad (15.38)$$

The factor $N!$ in the denominator compensates for the fact that we should really integrate over only one Weyl chamber in the eigenvalue simplex. In other words, different permutations of the vector $\vec{\lambda}$ of N generically different eigenvalues belong to the same unitary orbit, so we may restrict the order of the eigenvalues to, say, $\lambda_1 \geq \lambda_2 \geq \cdots \geq \lambda_N$. Substituting $N = 2$ it is satisfying to see that $V_2 = \pi/6$ – exactly the volume of the Bloch ball of radius $1/2$.

The next result $V_3 = \pi^3/(13440\sqrt{3})$ allows us to characterise the difference between the set $\mathcal{M}^{(3)} \subset \mathbb{R}^8$ and the ball $\mathbf{B}^8$. The set of mixed states is inscribed into a sphere of radius $R_3 = \sqrt{1/3} \approx 0.577$, while the maximal ball contained inside has the radius $r_3 = R_3/2 \approx 0.289$. These numbers may be compared with the *volume radius* $\bar{r}_3$ of $\mathcal{M}^{(3)}$, that is the radius of an 8-ball of the same volume. Using Eq. (1.24) we find $\bar{r}_3 \approx 0.368$. The distance from the centre of $\mathcal{M}^{(3)}$ to its boundary varies with the direction in $\mathbb{R}^8$ from r_3 to R_3.

The volume V_N tends to zero if $N \to \infty$, but there is no reason to worry about it. The same is true for the volume (1.24) of the N-balls, as this is a consequence of the choice of the units. To get some more information on the properties of $\mathcal{M}^{(N)}$, let us compute the (hyper) area of its boundary. Although the boundary $\partial \mathcal{M}^{(N)}$ is far from being trivial and contains orbits of different dimensionality, its full measure is formed of density matrices with exactly one eigenvalue equal to zero, see Section 8.5. Hence its area may be computed by integrating (15.34) over $\Delta_{N-2} \times \mathbf{F}^{(N)}$. The result is [1017]

$$\text{Vol}_{HS}\left(\partial\mathcal{M}^{(N)}\right) = \sqrt{N-1} \frac{\pi^{N(N-1)/2}}{2^{(N-2)/2}} \frac{\Gamma(1)\cdots\Gamma(N+1)}{\Gamma(N)\Gamma(N^2-1)}. \qquad (15.39)$$

For $N = 2$ we get π, just the area of the Bloch sphere of radius $R_2 = 1/2$.

Knowing the HS volume of the set of mixed states, and the area of its boundary, we may compute their ratio. As in Section 1.2 we fix the scale by multiplying the ratio with the radius of the outsphere, $R_N = \sqrt{(N-1)/2N}$, compare Eq. (8.35). The result is

$$\eta_N^{HS} \equiv R_N \frac{\text{Vol}_{HS}\left(\partial\mathcal{M}^{(N)}\right)}{\text{Vol}_{HS}\left(\mathcal{M}^{(N)}\right)} = (N-1)(N^2-1). \qquad (15.40)$$

---

[8] This differs from the result in [1017] by a factor $2^{(N^2-1)/2}$; in [1017] the length of the closed geodesic in $\mathbb{C}\mathbf{P}^n$ equals 2π.

Table 15.2 *Scaling properties of convex bodies in* $\mathbb{R}^D$ *with volume V, area A. Radii of the inspheres and outspheres are denoted by r and R.*

Body	X	dim	$\zeta = \frac{R}{r}$	$\eta = R\frac{A}{V}$	$\gamma = r\frac{A}{V}$
Round ball	$\bigcirc_D$	D	D^0	D^1	D^1
Cube	$\square_D$	D	$D^{1/2}$	$D^{3/2}$	D^1
Quantum states	$\mathcal{M}^{(N)}$	$N^2 - 1$	$D^{1/2}$	$D^{3/2}$	D^1
Cross-polytope	$\Diamond_D$	D	$D^{1/2}$	$D^{3/2}$	D^1
Simplex	Δ_D	D	D^1	D^2	D^1

This grows quickly with N. We can compare this with the results obtained for balls, simplices and cubes – if we remember that the dimension of $\mathcal{M}^{(N)}$ is $D = N^2 - 1$. Then we see that the area/volume ratio for the mixed states scales with the dimension[9] as $\eta \sim D^{3/2}$, just as for D-cubes (1.31): see Table 15.2. It is remarkable that the area to volume ratio for $\mathcal{M}^{(N)}$ behaves just like that for cubes of the same dimension.

Note that $\gamma \equiv rA/V$ is the same in each case. There is a reason for this: *a convex body X has $\gamma = D$ if and only if each of its boundary points is contained in a face tangent to its insphere* (which is then necessarily unique). To prove this let us set $r = 1$ for convenience (freeing the letter r for other purposes), regard the points ω on the unit sphere as unit vectors, and define a function $r(\omega)$ such that $r(\omega)\omega$ lies on ∂X (as shown to be possible in Figure 1.4). Using the standard measure $d\Omega$ on the unit sphere we find that

$$V(X) = \int_{S^{D-1}} d\Omega \int_0^{r(\omega)} dr \, r^{D-1} = \frac{1}{D} \int_{S^{D-1}} d\Omega \, r(\omega)^D. \tag{15.41}$$

Let $n(\omega)$ be the unit normal of X lying 'above' ω. The normal of a convex body is unique, and continuous, almost everywhere. The area of X is

$$A(X) = \int_{S^{D-1}} d\Omega \frac{r(\omega)^{D-1}}{\langle \omega, n(\omega) \rangle}. \tag{15.42}$$

If we now assume that the face containing $r(\omega)\omega$, and the support hyperplane containing it, is tangent to the unit sphere it follows that the scalar product $\langle r(\omega)\omega, n(\omega) \rangle = 1$, and from this we deduce that $A/V = D$, as advertised. It is not hard to show [878] that this is an 'if and only if' statement.

Convex bodies for which $\gamma = D$ are called *bodies of constant height*. The reason why all bodies in Table 15.2 have constant height is evidently that that their

---

[9] The same scaling also describes the set of real density matrices [1017].

boundary points are contained in faces tangent to the insphere. Another interesting characterisation of such bodies (again with unit inradius for convenience) is [878]: *a convex body X is a body of constant height if and only if there exists a closed set Y on its insphere such that the convex hull of Y contains the centre of the insphere, and such that X is the dual of Y*, in the sense that X consists of all points such that $\{\mathbf{x} \mid \mathbf{x} \cdot \mathbf{y} \leq 1\}$ for all $\mathbf{y}$ in Y. (Consult the discussion of convex duality in Section 1.1, and also note that for the self dual set $\mathcal{M}^{(N)}$ the set Y is a copy of $\mathbb{C}\mathbf{P}^n$.) It follows from this that the intersection of two bodies of constant height sharing the same insphere is again a body of constant height, a fact that will be useful in Section 16.5.

15.4 Bures measure

We can repeat the same analysis for the monotone metrics given by the MCP theorem; they may be much more complicated to deal with than is the HS metric, but they are unitarily invariant, so all of them will lead to product measures of the form (15.22). To see how this works, we first observe that in a coordinate system where ρ is diagonal Eq. (15.30) simplifies to

$$d\rho_{ij} = d\lambda_i \delta_{ij} + (\lambda_j - \lambda_i)(U^{-1}dU)_{ij}. \tag{15.43}$$

Sticking this into the expression that defines the metric, Eq. (15.2), we obtain

$$ds_f^2 = \frac{1}{4}\left[\sum_{j=1}^N \frac{d\lambda_j^2}{\lambda_j} + 2\sum_{i<j} c(\lambda_i, \lambda_j)(\lambda_i - \lambda_j)^2 |(U^{-1}dU)_{ij}|^2\right], \tag{15.44}$$

where f is an arbitrary MC function and $c(x, y) = 1/[yf(x/y)]$ enters the definition (15.2). For $c(x, y) = c_{\max}(x, y) = 2/(x+y)$ this becomes the Bures metric, our topic in this section:

$$ds_{\mathrm{B}}^2 = \frac{1}{4}\sum_{i=1}^N \frac{(d\lambda_i)^2}{\lambda_i} + \sum_{i<j} \frac{(\lambda_i - \lambda_j)^2}{\lambda_i + \lambda_j}|(U^{-1}dU)_{ij}|^2. \tag{15.45}$$

Since $\mathrm{Tr}\rho = 1$ not all $d\lambda_j$ are independent. Eliminating $d\lambda_N$ we obtain

$$\frac{1}{4}\sum_{j=1}^N \frac{(d\lambda_j)^2}{\lambda_j} = \frac{1}{4}\sum_{j=1}^{N-1} \frac{(d\lambda_j)^2}{\lambda_j} + \frac{1}{4}\left(\sum_{j=1}^{N-1} d\lambda_j\right)^2 \frac{1}{\lambda_N}. \tag{15.46}$$

The metric (15.46) in the $(N-1)$ dimensional simplex is of the form $g_{ik} = (\delta_{ik}/4\lambda_i + \mathbb{1}/4\lambda_N)$ with determinant $\det g = (\lambda_1+\lambda_2+...+\lambda_N)/(4^{N-1}\lambda_1\lambda_2...\lambda_N) = 4^{1-N}/\det \rho$

since $\text{Tr}\rho = 1$. Thus the volume element gains a factor $\sqrt{\det g} = 2^{1-N}(\det \rho)^{-1/2}$, so we obtain an expression

$$dV_{\text{B}} = \frac{1}{2^{N-1}} \frac{1}{\sqrt{\det \rho}} \prod_{j=1}^{N-1} d\lambda_j \prod_{j<k} \frac{(\lambda_j - \lambda_k)^2}{\lambda_j + \lambda_k} \left| \prod_{j<k}^{1...N} \text{Re}(U^{-1}dU)_{jk} \text{Im}(U^{-1}dU)_{jk} \right|$$

(15.47)

analogous to (15.34). This volume element gives the *Bures probability distribution* in the eigenvalue simplex [392, 195],

$$P_{\text{B}}(\lambda_1, \ldots, \lambda_N) = C_N^{\text{B}} \frac{\delta(1 - \sum_{i=1}^{N} \lambda_i)}{(\lambda_1 \cdots \lambda_N)^{1/2}} \prod_{j<k}^{N} \frac{(\lambda_j - \lambda_k)^2}{\lambda_j + \lambda_k}.$$

(15.48)

The Bures normalisation constants[10] read [849]

$$C_N^{\text{B}} = 2^{N^2-N} \frac{\Gamma(N^2/2)}{\pi^{N/2} \Gamma(1) \cdots \Gamma(N+1)} = 2^{N^2-N} \frac{\Gamma(N^2/2)}{\pi^{N/2} \Xi_{N+1}}.$$

(15.49)

Integrating (15.47) over $\mathbf{F}^{(N)} \times \Delta_{N-1}$ we get the Bures volume

$$\text{Vol}_{\text{B}}(\mathcal{M}^{(N)}) = \frac{2^{1-N}}{C_N^{\text{B}}} \frac{\text{Vol}(\mathbf{F}^{(N)})}{N!} = \frac{1}{2^{N^2-1}} \frac{\pi^{N^2/2}}{\Gamma(N^2/2)}.$$

(15.50)

Observe that the Bures volume of the set of mixed states is equal to the volume of an $(N^2 - 1)$ dimensional hemisphere of radius $R_{\text{B}} = 1/2$. In a similar way one computes the Bures volume of the boundary of the set of mixed states [849]

$$\text{Vol}_{\text{B}}(\partial \mathcal{M}^{(N)}) = \frac{N}{2^{N^2-2}} \frac{\pi^{(N^2-1)/2}}{\Gamma[(N^2-1)/2]}.$$

(15.51)

We are pleased to see that for $N = 2$ the above results describe the Uhlmann hemisphere of radius $R_{\text{B}} = 1/2$,

$$\text{Vol}_{\text{B}}(\mathcal{M}^{(2)}) = \frac{1}{2}\text{Vol}(S^3) R_{\text{B}}^3 = \frac{\pi^2}{8}, \qquad \text{Vol}_{\text{B}}(\partial \mathcal{M}^{(2)}) = \text{Vol}(S^2) R_{\text{B}}^2 = \pi.$$

(15.52)

Although for $N \geq 2$ the Bures geometry is not that of a hemisphere (see Section 9.6), we see that the ratio

$$\eta_N^{\text{B}} = R_{\text{B}} \frac{\text{Vol}_{\text{B}}(\partial \mathcal{M}^{(N)})}{\text{Vol}_{\text{B}}(\mathcal{M}^{(N)})} = \frac{N}{\sqrt{\pi}} \frac{\Gamma(N^2/2)}{\Gamma(N^2/2 - 1/2)} \sim D$$

(15.53)

asymptotically increases linearly with the dimensionality $D = N^2 - 1$, which is typical for hemispheres.

---

[10] The constant $C_2^{\text{B}} = 2/\pi$ was computed by Hall [392], while $C_3^{\text{B}} = 35/\pi$ and $C_4^{\text{B}} = 2^{11}35/\pi^2$ were found by Slater [842].

15.5 Induced measures

The measures discussed so far belong to the wide class of

S1. Metric related measures. Any metric generates a measure.

Other families of measures may be defined in an operational manner by specifying a recipe to generate a density matrix. Consider a pure state $|\psi\rangle$ of a bipartite $N \times K$ composite quantum system described in $\mathcal{H} = \mathcal{H}_N \otimes \mathcal{H}_K$. It is convenient to work in an arbitrary orthogonal product basis, $|i, k\rangle = |i\rangle \otimes |k\rangle$, where $|i\rangle \in \mathcal{H}_N$ and $|k\rangle \in \mathcal{H}_K$. The pure state $|\psi\rangle$ is then represented by a $N \times K$ rectangular complex matrix A with $A_{ik} = \langle i, k|\psi\rangle$. The normalisation condition, $\|\psi\|^2 = \mathrm{Tr}AA^\dagger = 1$, is the only constraint imposed on this matrix. The corresponding density matrix $\sigma = |\psi\rangle\langle\psi|$, acting on the composite Hilbert space $\mathcal{H}$, is represented in this basis by a matrix labelled by four indices, $\sigma_{i'k'}^{ik} = A_{ik}A_{i'k'}^*$. Partial tracing with respect to the subspace $\mathcal{H}_K$ gives the reduced density matrix of size N

$$\rho = \mathrm{Tr}_K(\sigma) = AA^\dagger, \tag{15.54}$$

while partial tracing over the first subsystem leads to the reduced density matrix $\rho' = \mathrm{Tr}_N(\sigma) = A^\dagger A$ of size K. Now we are prepared to define a family of measures in the space of mixed states $\mathcal{M}^{(N)}$ labelled by a single parameter – the size K of the ancilla [625, 161, 392].

S2. Measures $P_{N,K}^{\mathrm{trace}}(\rho)$ induced by partial trace over a K-dimensional environment (15.54) of an ensemble of pure states distributed according to the unique, unitarily invariant FS measure on the space $\mathbb{CP}^{KN-1}$ of pure states of the composite system.

There is a simple physical motivation for such measures: they can be used if we do not know anything about the density matrix, apart from the dimensionality K of the environment. When $K = 1$ we get the FS measure on the space of pure states. Since the rank of ρ is limited by K, the induced measure covers the full set of $\mathcal{M}^{(N)}$ for $K \geq N$. When $K < N$ the measure $P_{N,K}^{\mathrm{trace}}$ is supported on the subspace of density matrices of rank K belonging to the boundary $\partial\mathcal{M}^{(N)}$.

Since the pure state $|\psi\rangle$ is drawn according to the FS measure, the induced measure $P_{N,K}^{\mathrm{trace}}$ enjoys the product form (15.22). Hence the distribution of the eigenvectors of ρ is determined by the Haar measure on $U(N)$, and we need to find the joint distribution of the eigenvalues in the simplex Δ_{N-1}.

In the first step we use the relation (15.54) to write down the distribution of matrix elements

$$P(\rho) \propto \int [dA]\, \delta(\rho - AA^\dagger)\, \delta(\mathrm{Tr}AA^\dagger - 1), \tag{15.55}$$

in which the first delta function assures the (semi)positive definiteness while the second delta function provides the normalisation. Let us assume that $K \geq N$, so that $\rho = AA^\dagger$ is generically positive definite (the *Wishart case*, since $AA^\dagger$ is called a *Wishart matrix*[11]). Thus one can perform the transformation

$$A = \sqrt{\rho}\,\tilde{A} \quad \text{and} \quad [dA] = \det\rho^K\,[d\tilde{A}]. \tag{15.56}$$

Note that $[dA]$ includes alternating factors dA_{ik} and dA_{ik}^*. The matrix delta function may now be written as

$$\delta\left(\sqrt{\rho}(1 - \tilde{A}\tilde{A}^\dagger)\sqrt{\rho}\right) = \det\rho^{-N}\,\delta\left(1 - \tilde{A}\tilde{A}^\dagger\right), \tag{15.57}$$

where the first factor on the right-hand side is the inverse Jacobian of the corresponding transformation. This implies

$$P(\rho) \propto \theta(\rho)\,\delta(\text{Tr}\rho - 1)\,\det\rho^{K-N}, \tag{15.58}$$

in which the step function ensures that ρ is positive definite. The joint probability distribution of eigenvalues is given by [625, 617, 713, 392]

$$P_{N,K}^{\text{trace}}(\lambda_1, \ldots, \lambda_N) = C_{N,K}\,\delta(1 - \sum_i \lambda_i)\prod_i \lambda_i^{K-N}\prod_{i<j}(\lambda_i - \lambda_j)^2 \tag{15.59}$$

with the normalisation constant [651, 1016]

$$C_{N,K} = \frac{\Gamma(KN)}{\prod_{j=0}^{N-1}\Gamma(K-j)\Gamma(N-j+1)} = \frac{\Xi_{K-N}\Gamma(KN)}{\Xi_K\Xi_{N+1}} \tag{15.60}$$

written here for $K \geq N$ in terms of Ξ_N defined in (15.25) extended by a convention $\Xi_0 = 1$. If the size K of the ancilla equals the size N of the system, the measure $P_{N,N}^{\text{trace}}$ induced by partial tracing of the pure states in $\mathbb{CP}^{N^2-1}$ coincides with the HS measure (15.35), and $C_{N,N} = C_N^{\text{HS}}$. For instance, the partial trace of $N = 4$ random complex pure states induces the uniform measure in the Bloch ball of $N = 2$ mixed states.

Integrating out all eigenvalues but one, from the joint probability distribution (15.59), one arrives at the density of eigenvalues. This task was performed by Page [713], who derived the distribution

$$P_{N,K}(x) = \frac{\sqrt{(x - a_-)(a_+ - x)}}{2\pi x}, \quad \text{with} \quad a_\pm = 1 + \frac{K}{N} \pm 2\sqrt{\frac{K}{N}} \tag{15.61}$$

valid for $K \geq N \gg 1$. Here $x = N\lambda$ stands for the rescaled eigenvalue, so asymptotically the rescaled spectrum is supported on the interval $[a_-, a_+]$. Up to

---

[11] In the opposite, so called *anti-Wishart*, case [991, 496] the reduced density matrix $\rho = A^\dagger A$ has $N - K$ zero eigenvalues, but the reduced matrix ρ' of size K is positive definite and has the same positive eigenvalues. The formulae (15.59–15.61) hold with N and K exchanged.

an overall normalisation factor this problem is equivalent to finding the asymptotic density of the spectrum of random Wishart matrices $H = AA^\dagger$. In the symmetric case $K = N$, corresponding to the square matrices, the above expression reduces to the *Marchenko–Pastur distribution* [643]

$$MP(x) = \frac{1}{2\pi}\sqrt{\frac{4-x}{x}}, \tag{15.62}$$

valid for a large N. Note that this distribution is supported in $[0, 4]$, so asymptotically the largest eigenvalues are only four times larger than the mean value $1/N$. In the rescaled variables, $y = \sqrt{x}$, this distribution is equivalent to the *quarter-circle law*, $P_{qc}(y) = \sqrt{4 - y^2}/\pi$ for $y \in [0, 2]$.

A random mixed state may also be obtained as a convex sum of K pure states from $\mathbb{C}P^{N-1}$. The probability distribution of the weights may be arbitrary, but for simplicity we will consider the uniform distribution, $p_i = 1/K$. The number of pure states K, which governs the rank of ρ, may be treated as a free parameter labelling the measure

S3. Mixtures of random pure states obtained as a combination of K independent random pure states $|\psi_i^{\text{rand}}\rangle$ drawn according to the FS measure,

$$\rho = \frac{1}{K}\sum_{j=1}^{K} |\psi_i^{\text{rand}}\rangle\langle\psi_i^{\text{rand}}|. \tag{15.63}$$

The measure $P_{N,K}^{\text{mixt}}(\rho)$ defined in this way has the product form (15.22). For $K < N$ the measure is supported on a subspace of lower rank included in the boundary $\partial\mathcal{M}^{(N)}$. By construction $P_{N,1}^{\text{mixt}} = P_{N,1}^{\text{trace}}$ is equivalent to the FS measure on the manifold of pure states. However, for larger K both ways of generating mixed states do not coincide.

In general, one may distinguish the symmetric case, $P_{N,N}^{\text{mixt}}(\rho)$, since the number $K = N$ is the minimal one, which typically gives mixed states of the full rank. For instance, analysing the position of the barycentre of two independent random points placed on the surface of the Bloch sphere we infer that $P_{2,2}^{\text{mixt}}(\lambda_1, \lambda_2) \propto \delta(1 - \lambda_1 - \lambda_2)|\lambda_1 - \lambda_2|$. In larger dimensions the distributions $P_{N,N}^{\text{mixt}}(\vec{\lambda})$ get more complicated and differ from $P_{N,N}^{\text{trace}}(\vec{\lambda}) = P_{\text{HS}}(\vec{\lambda})$. The larger the number K of the states in the mixture, the larger the entropy of the resulting mixed state ρ. In the limit $K \to \infty$ it tends to the maximal value $\ln N$.

15.6 Random density matrices

The measures that we have discussed allow us to generate random density matrices. The picture is particularly transparent for $N = 2$, in which the product form (15.22) assures the rotational symmetry inside the Bloch ball, and the only thing to settle is

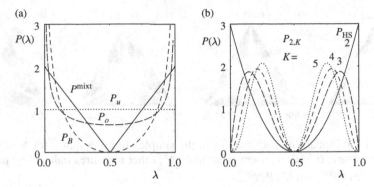

Figure 15.2 Distribution of an eigenvalue λ of density matrices of size $N = 2$: (a) P_u (dotted line), P_o 'cosine distribution' (dashed line), $P_{2,2}^{\text{mixed}}$ (solid line) and P_B Bures measure (dash-dotted line); (b) measures $P_{2,K}^{\text{trace}}$ induced by partial tracing; $K = 2$ (i.e. P_{HS}) (solid line), $K = 3$ (dashed line), $K = 4$ (dash-dotted line) and $K = 5$ (dotted line).

the radial distribution $P(r)$, where the radius $r = |\lambda_1 - 1/2|$ is equal to the length of the Bloch vector, $r = |\vec{\tau}|$.

For $N = 2$ the HS measure (15.35) gives $P_{\text{HS}}(r) = 24r^2$ (for $r \in [0, 1/2]$). This is one possible way of saying that the distribution is uniform inside the Bloch ball. The Bures measure (15.48) induces the distribution

$$P_B(r) = \frac{32\,r^2}{\pi\sqrt{1 - 4r^2}}. \tag{15.64}$$

This is the uniform distribution on the Uhlmann hemisphere. Comparing the HS and Bures measures we realise that the latter is more concentrated on states of high purity (with large r). For $N = 2$ one has $\lambda_1 + \lambda_2 = 1$, the denominator in (15.48) equals unity, and the Bures measure coincides with the induced measure $P_B(\lambda_1, \lambda_2) = P_{2,3/2}^{\text{trace}}(\lambda_1, \lambda_2)$. Even though there is no subsystem of dimension $K = 3/2$ and such an induced measure has no physical interpretation, this relation is useful for computing some averages over the Bures measure by an analytical continuation in the parameter K.

Some exemplary radial distributions inside the Bloch ball, sketched in Figure 15.2, include the $s = 1$ Dirichlet distribution (2.73), generated by the unitary product measure, $P_u(r) = 2$, and the $s = 1/2$ Dirichlet distribution, related to the orthogonal product measure, $P_o(r) = 4/(\pi\sqrt{1 - 4r^2})$. Figure 15.2 presents also a family of distributions implied by the induced measures (15.59). The larger the dimension K of the auxiliary space $\mathcal{H}_B$, the more is the induced distribution $P_{N,K}^{\text{trace}}$ concentrated in the centre of the Bloch ball [392].

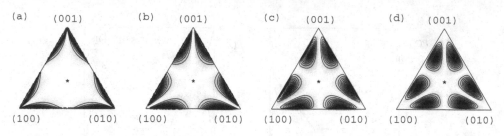

Figure 15.3 Probability distributions in the simplex of eigenvalues for $N = 3$ (a) Bures measure, (b) HS measure equal to $P_{3,3}^{\text{trace}}$; other measures induced by partial tracing: (c) $P_{3,4}^{\text{trace}}$ and (d) $P_{3,5}^{\text{trace}}$.

A similar effect is shown in Figure 15.3 for $N = 3$, in which probability distributions are plotted in the eigenvalue simplex. Again the Bures measure is more localised on states of high purity, as compared to the HS measure. Due to the factor $(\lambda_i - \lambda_j)^2$ in (15.59) the degeneracies in spectrum are avoided, which is reflected by a low probability (white colour) along all three bisectrices of the triangle. On the other hand, the distribution $P_{3,2}^{\text{trace}}(\vec{\lambda})$ is singular, and located at the edges of the triangle, which represent density matrices of rank 2. It is equal to $P_{2,3}^{\text{trace}}(\vec{\lambda})$ which is represented by a dashed line in Figure 15.2.

In order to characterise the average degree of mixing of random states we compute the mean moments $\langle \text{Tr}\rho^k \rangle$, averaged with respect to the measures introduced in this chapter. Averaging over the HS measure (15.35) gives the exact results [625, 392, 1016] for $k = 2, 3$

$$\langle \text{Tr}\rho^2 \rangle_{\text{HS}} = \frac{2N}{N^2 + 1} \quad \text{and} \quad \langle \text{Tr}\rho^3 \rangle_{\text{HS}} = \frac{5N^2 + 1}{(N^2 + 1)(N^2 + 2)} \tag{15.65}$$

and an asymptotic behaviour for large N,

$$\langle \text{Tr}\rho^k \rangle_{\text{HS}} = N^{1-k} \frac{\Gamma(1 + 2k)}{\Gamma(1 + k)\,\Gamma(2 + k)} \left(1 + O\left(\frac{1}{N}\right)\right). \tag{15.66}$$

Analogous averages with respect to the Bures measure (15.48) read [850, 706]

$$\langle \text{Tr}\rho^2 \rangle_{\text{B}} = \frac{5N^2 + 1}{2N(N^2 + 2)} \quad \text{and} \quad \langle \text{Tr}\rho^3 \rangle_{\text{B}} = \frac{8N^2 + 7}{(N^2 + 2)(N^2 + 4)}, \tag{15.67}$$

and give asymptotically

$$\langle \text{Tr}\rho^k \rangle_{\text{B}} = N^{1-k} \frac{2^k \Gamma[(3k + 1)/2]}{\Gamma[(1 + k)/2]\,\Gamma(2 + k)} \left(1 + O\left(\frac{1}{N}\right)\right). \tag{15.68}$$

Observe that $\langle \text{Tr}\rho^2 \rangle_{\text{HS}} < \langle \text{Tr}\rho^2 \rangle_{\text{B}}$, which shows that among these two measures the Bures measure is concentrated at the states of higher purity. Allowing the parameter

k to be real and performing the limit $k \to 1$ we obtain the asymptotics for the average von Neumann entropy $\langle S \rangle = -\lim_{k \to 1} \partial \langle \mathrm{Tr}\rho^k \rangle / \partial k$,

$$\langle S(\rho) \rangle_{\mathrm{HS}} = \ln N - \frac{1}{2} + O\left(\frac{\ln N}{N}\right) \qquad (15.69)$$

and

$$\langle S(\rho) \rangle_{\mathrm{B}} = \ln N - \ln 2 + O\left(\frac{\ln N}{N}\right). \qquad (15.70)$$

For comparison note that the mean entropy of the components of complex random vectors of size N, or their average Wehrl entropy (7.75), behaves as $\ln N + \gamma - 1 \approx \ln N - 0.4228$.

The above analysis may be extended for the measures (15.59) induced by partial tracing. The average purity is[12]

$$\langle \mathrm{Tr}\rho^2 \rangle_{N,K} = \frac{N+K}{NK+1}. \qquad (15.71)$$

For $K = N$ this reduces to (15.65). An exact formula for the average entropy

$$\langle S(\rho) \rangle_{N,K} = \left(\sum_{j=K+1}^{KN} \frac{1}{j}\right) - \frac{N-1}{2K} = \Psi(NK+1) - \Psi(K+1) - \frac{N-1}{2K} \qquad (15.72)$$

was conjectured by Page [713], and proved soon afterwards [315, 788, 825]. The Page formula implies that the average entropy is close to the maximal value $\ln N$, if the size K of the environment is sufficiently large. Making use of the asymptotic properties of the digamma function, $\Psi(x+1) \sim \ln x + 1/2x$ [852], we infer that

$$\langle S \rangle_{N,K} \sim \ln N - \frac{N^2-1}{2NK}, \qquad (15.73)$$

which reduces to (15.69) when $K = N$. Hence a typical pure state of an $N \times N$ system is almost maximally entangled (compare Eq. 16.35), while the probability of finding a state with the entropy S smaller by one than the average $\langle S \rangle_{N,N}$ is exponentially small [418].

To wind up the section on random density matrices let us discuss practical methods to generate them [1012]. It is rather simple to draw random mixed states with respect to the induced measures: as discussed in Chapter 7 we generate K independent complex random pure states according to the natural measure on N dimensional Hilbert space and prepare their mixture (15.63). To draw a random mixed

---

[12] This was derived by Lubkin (1978) [625], who also deemed Eq. (15.73) below to be 'plausible'. Average values of the higher moments are computed in Problem 15.5.

state according to $P_{N,K}^{\text{trace}}(\rho)$ we generate a complex random pure state on NKb dimensional Hilbert space, and then perform the partial trace (15.54) over the K dimensional subsystem [161, 392, 1016].

Alternatively, to obtain a random state according to this measure we generate a rectangular, $K \times N$ random matrix A with all entries being independent complex Gaussian numbers, and compute $\rho = AA^\dagger/(\text{Tr}AA^\dagger)$. By definition such a matrix is normalised and positive definite, while random matrix theory allows one to show that the joint probability distribution of spectrum coincides with (15.59). In particular, for $K = 1$ the density matrices represent random pure states (see Section 7.6), while for $K = N$ we deal with non-Hermitian square random matrices characteristic of the *Ginibre ensemble* [336, 651]. Fixing the trace we arrive at the HS measure [1016, 901].

To obtain random matrices with respect to unitary (orthogonal) product measures one needs to generate a random unitary (orthogonal) matrix with respect to the Haar measure on $U(N)$ (or $O(N)$, respectively) [750]. The squared moduli of its components sum to unity and provide the diagonal matrix Λ with a spectrum of the density matrix. Taking another random unitary matrix W, one obtains the random state by $\rho = W\Lambda W^\dagger$.

Random mixed states according to the Bures measure may easily be generated in the $N = 2$ case [392]. One picks a point on the Uhlmann hemisphere at random. For $N \geq 3$ one can use a general recipe involving a Haar random unitary U and a square complex Ginibre matrix G of order N,

$$\rho = \frac{(\mathbb{1} + U)GG^\dagger(\mathbb{1} + U^\dagger)}{\text{Tr}(\mathbb{1} + U)GG^\dagger(\mathbb{1} + U^\dagger)}. \tag{15.74}$$

By construction this ensemble is unitarily invariant and a direct calculation shows [706] that the eigenvalues are described by the Bures distribution (15.48). A random density matrix distributed according to the Bures distribution can be interpreted [1012] as the normalised partial trace, $\text{Tr}_B|\phi\rangle\langle\phi|$, of a symmetric superposition $|\phi\rangle = |\phi_1\rangle + |\phi_2\rangle$ of a random bipartite pure state, $|\phi_1\rangle \in \mathcal{H}_N \otimes \mathcal{H}_N$, and its locally transformed copy, $|\phi_2\rangle = (U \otimes \mathbb{1}_B)|\phi_1\rangle$.

15.7 Random operations

Analysing properties of random density matrices we should also discuss random operations, for which the Kraus operators A_i entering the Kraus form (10.53) are taken at random, provided the condition (10.54) is fulfilled. There are several reasons for doing this; for instance, due to the Jamiołkowski isomorphism (Section 11.3), measures in the sets of quantum states and quantum maps are closely related. Hence we should not expect that there exists a unique way to generate random operations. Indeed, there exist several relevant measures in the space of quantum maps, and we start our short review with:

M1. Random maps induced by random states. Any measure in the space $\mathcal{M}^{(N^2)}$ of mixed quantum states of increased dimensionality induces, by the Jamiołkowski isomorphism, a measure in the space of trace preserving, completely positive maps $\Phi : \mathcal{M}^{(N)} \rightarrow \mathcal{M}^{(N)}$. It is sufficient to take a random state ρ acting on $\mathcal{H}_A \otimes \mathcal{H}_B$, rescale it by writing $\tilde{D} = N\rho$, and compute the partial trace $\tilde{D}_B = \mathrm{Tr}_A \tilde{D}$. This positive matrix allows us to take its square root which is generically invertible. Thus defining a matrix

$$D = \left(\mathbb{1}_N \otimes \frac{1}{\sqrt{\mathrm{Tr}_A \tilde{D}}} \right) \tilde{D} \left(\mathbb{1}_N \otimes \frac{1}{\sqrt{\mathrm{Tr}_A \tilde{D}}} \right) \qquad (15.75)$$

one arrives [73, 182] at a legitimate dynamical matrix D, which is positive, satisfies the trace preserving condition, $\mathrm{Tr}_A D = \mathbb{1}$, and defines by (11.22) a random stochastic map, $\Phi = D^R$. The probability measure in the space of maps depends on the choice of the measure in the space of mixed states of size N^2. Selecting the induced measure $P^{\mathrm{trace}}_{N^2,K}$ one obtains an ensemble of random dynamical matrices governed by the probability distribution analogous to (15.58),

$$P(D) \propto \det(D^{K-N^2}) \, \delta(\mathrm{Tr}_A D - \mathbb{1}). \qquad (15.76)$$

When $K = N^2$ the first factor vanishes and the only constraint on the matrix D is the partial trace condition (10.46), so the maps are generated uniformly with respect to the flat measure in the space of quantum operations. For instance, the choice $N = 2$ and $K = 4$ implies the HS measure in the space of states $\mathcal{M}^{(4)}$ and accordingly the uniform measure supported on the entire space of one-qubit operations [182]. Note that a generic random operation Φ is not bistochastic.

Another interesting class of random operations arises from the environmental representation (10.59). Let us treat the size K of the environment as a free parameter and assume that the unitary matrix U is distributed according to the Haar measure on $U(KN)$. This natural assumption characterises eigenvectors of the initial state σ of the environment, so to specify a random operation we need to characterise its spectrum. In principle one may take for this purpose any probability distribution on the simplex Δ_{K-1}, but we shall discuss only two extremal cases.

M2. Random operations with a pure K-dimensional environment,

$$\rho' = \Phi_{\mathrm{p}}(\rho) = \mathrm{Tr}_K[U(\rho \otimes |v\rangle\langle v|)U^\dagger]. \qquad (15.77)$$

Such a random operation Φ_p is thus specified by a random unitary matrix of size KN, since the choice of the pure state $|v\rangle = |1\rangle \in \mathcal{H}_K$ does not influence the measure. Looking again at (10.61) we see that the K Kraus operators representing the map Φ_p arise as blocks of size N of the random matrix U,

$$(A_i)_{lm} \equiv U_{l\mu} \quad \text{with} \quad \mu = N(i-1) + m; \quad l, m = 1, \ldots, N; \; i = 1, \ldots, K. \quad (15.78)$$

Due to unitarity of U the set of Kraus operators $\{A_i\}_{i=1}^K$ satisfies the completeness relation (10.54) and defines a quantum operation (10.53).

To get a generic operation described by a dynamical matrix of full rank N^2 one needs to choose $K \geq N^2$. Assuming that the unitary matrix U is generated according to the Haar measure on $U(KN)$ we arrive at an ensemble of random maps, each described by a collection of K operators A_i. There are no other constraints for Kraus operators besides the trace preserving condition, so for $K = N^2$ the probability distribution for the map coincides with the particular case of (15.76). Therefore the ensemble **M2** of maps generated by a random unitary interaction with a pure state of environment of size N^2 becomes equivalent [182] to the ensemble **M1** of random maps induced with (15.75) by the HS measure in the space of extended states and produces the uniform measure in the entire set of quantum operations.

An individual Kraus operator A_i of size N might be considered as a truncation of the random unitary matrix U of order KN. For large dimensionality N the unitarity constraints become (relatively) weaker and the non-Hermitian random matrices A are described by the Ginibre ensemble [336, 651]: their spectra cover uniformly the disc of radius $1/\sqrt{N}$ in the complex plane [1015].

M3. Random operations with a mixed K-dimensional environment, $\rho' = \Phi_m(\rho) = \text{Tr}_K[U(\rho \otimes \mathbb{1}/K)U^\dagger]$, where U is a random unitary of size KN. These random operations are by construction bistochastic. The Haar measure on $U(NK)$ generates a definite measure in the space of bistochastic maps parametrized by the size K of the ancilla [684]. The case $K = N$ is the unistochastic map (10.64). Random operations Φ_m describe the situation in which only the size K of the ancilla is known; they are also called *noisy maps* [461].

M4. Random external fields (REF) defined as a convex combination of K unitary transformations [26],

$$\rho' = \Phi_{\text{REF}}(\rho) = \sum_{i=1}^K p_i V_i \rho V_i^\dagger, \quad (15.79)$$

where unitary matrices V_i are drawn according to the Haar measure on $U(N)$. Probability vector $\vec{p}$ may be drawn with an arbitrary probability distribution on Δ_{K-1}, but one assumes often that $p_i = 1/K$. As discussed in Section 10.6, for $N \geq 3$ REF's form a proper subset of the set of bistochastic maps.

M5. Maps generated by a measure in the space of stochastic matrices. Take a random stochastic matrix S of size N. Out of its rows construct N diagonal matrices,

$E_{kl}^{(i)} = \sqrt{S_{ki}}\,\delta_{kl}$, and define the set operators, $A_i = UE^{(i)}V$. Here U and V are random unitary matrices generated according to the Haar measure on $U(N)$. Since S is stochastic and satisfies (2.4), the Kraus operators A_i fulfil the completeness relation (10.54),

$$\sum_{i=1}^{N} A_i^\dagger A_i = \sum_{i=1}^{N} V^\dagger \sqrt{E^{(i)}} U^\dagger U \sqrt{E^{(i)}} V = V^\dagger \left[\sum_{i=1}^{N} E^{(i)} \right] V = \mathbb{1}, \tag{15.80}$$

hence define a quantum operation. In the same way one demonstrates that the dual condition (10.71) is fulfilled, so the random operation is bistochastic. Instead of using random stochastic matrices one may also take a concrete matrix S, for example related to some specific physically interesting map Φ, and then randomise it[13] by introducing random unitary rotations as in (15.80).

After discussing the methods to generate random maps, let us stress that any measure in the space of quantum operations $\mathcal{CP}_N$ induces a measure in the space of density matrices. Alternatively, random mixed states may be generated by any individual operation assuming that an initial pure state $|\psi\rangle$ is drawn according to the natural FS measure.

S4. Operation induced random mixed states defined by a certain operation Φ,

$$\rho' = U \left[\Phi \left(V|1\rangle\langle1|V^\dagger \right) \right] U^\dagger, \tag{15.81}$$

where U and V are independent random unitary matrices distributed according to the Haar measure on $U(N)$.

Hence any CP map Φ determines the *operation induced measure* μ_Φ. In the trivial case, the identity map, $\Phi = \mathbb{1}$, induces the unitarily invariant FS measure on the space of pure states, $\mu_{\mathbb{1}} = \mu_{\text{FS}}$.

It is instructive to study measures induced in the $N = 2$ case by planar or linear maps defined in Table 10.4 (see Problem 15.8). The latter case may be generalised for arbitrary N. As follows from the properties of the FS measure (see Eq. (7.66) and Section 7.6) the distribution of the diagonal elements of the density matrix $|\psi\rangle\langle\psi|$ of a random pure state is uniform in the simplex of eigenvalues. In this way we arrive at an important result: the coarse graining map, $\Psi_{\text{CG}}(\rho) = \text{diag}(\rho)$, induces by (15.81) the unitary measure $P_u(\vec{\lambda}) = \text{const}$, uniform in Δ_{N-1}.

Allowing operations which reduce the dimensionality of the system, we see that the measures (15.54) also belong to this class, since they are induced by the operation of partial trace, $\Phi(\rho) = \text{Tr}_K \rho$. A more general class of measures is induced via (15.81) by a family of operations Φ_a and a concrete probability distribution $P(a)$. For instance, an interesting measure supported on the subspace of degenerated

---

[13] Such random maps obtained for $S = \mathbb{1}$ were used to describe the influence of measurement on a random basis on the time evolution of quantum baker maps [27].

states arises by usage of the depolarizing channels, $\Phi_a = a\Phi_* + (1 - a)\mathbb{1}$, with a uniform distribution of the noise level, $P(a) = 1$ for $a \in [0, 1]$.

Let us conclude by emphasizing again that there is no single, naturally distinguished probability measure in the set of quantum states. Guessing at random what the state may be, we can use any available additional information. For instance, if a mixed state has arisen by the partial tracing over a K-dimensional environment, the induced measure (15.59) should be used. More generally, if a mixed state has arisen as an image of an initially pure state under the action of a known operation Φ, the operation-related measure (15.81) may be applied. Without any prior information whatsoever, it will be legitimate to use the Bures measure (15.48), related to Jeffreys' prior, statistical distance, fidelity and quantum distinguishability.

15.8 Concentration of measure

Discussing high dimensional geometry in the introductory Section 1.2 we observed that a random point from a ball $\mathbf{B}^n$ in n dimensions will likely be situated close to the boundary $\mathbf{S}^{n-1}$. Hence the skin of a high dimensional fruit contains most of its mass, so we can be pleased that we do not have to peel an orange in seven dimensions and to place the remnants in the waste basket. A closely related observation states that if one chooses any point from a high dimensional sphere $\mathbf{S}^n$ as the north pole and selects a random point according to the uniform measure on the sphere, it is likely to be located close to the equator. This fact follows directly from the integral expression (1.24) for the volume of the unit n-sphere. In geographic terms, the latitude typically takes values close to zero, so the equator can be called 'fat'.

Let us put aside these counter-intuitive geometric effects for a while and analyse a simple probabilistic problem. Consider n independent random variables ξ_i, each of them taking the values ± 1 with probability $1/2$. It is clear that the expectation value of the sum, $\zeta_n = \sum_{i=1}^{n} \xi_i$, is equal to zero, as in a long sequence of tossing a fair coin the number of the tails should be close to the number of heads. Furthermore, the central limit theorem implies that fluctuations of ζ_n are of the order of $\sqrt{n}$. Although ζ_n can take values as large as n, it typically takes values close to the mean, $E(\zeta_n) = 0$, as the *Hoeffding exponential inequality* holds [439],

$$P\left(\frac{|\zeta_n|}{n} \geq \epsilon\right) \leq 2\exp(-n\epsilon^2/2). \tag{15.82}$$

Hence the probability of observing ϵ-deviations from the mean becomes exponentially small with the number of variables.

It is instructive to compare the above two observations stemming from apparently distant fields of mathematics. Both illustrate the same *measure concentration phenomenon,*[14] which occurs for high n. A general formulation of the probabilistic effect was proposed by Talagrand [885, 886], who noticed that if a function depends on several independent random variables in a 'balanced, smooth way' the effects of randomness nearly compensate, so the function is essentially constant and fluctuates only weakly around its mean.

To describe quantitatively the 'smooth behaviour' of a function f one uses the Lipschitz property:

Definition 15.1. *A function f defined on any set $Y \subset \mathbb{R}^n$ is called Lipschitz with constant η if for any $x, y \in Y$ the following inequality is satisfied,*

$$|f(x) - f(y)| \leq \eta \, |x - y|. \tag{15.83}$$

Note that the linear function $y = ax$ is globally Lipschitz with constant a, the quadratic function $f(x) = x^2$ is locally Lipschitz in $[0, b]$ with constant $\eta = 2b$, while $\sqrt{x}$ is not Lipschitz in $[0, 1]$ as its derivative diverges.

The notion of a Lipschitz function allows us to state a crucial lemma, which bounds the probability of observing large fluctuations of a slowly varying function [662].

Lévy's Lemma. *Let $f : \mathbf{S}^{n-1} \to \mathbb{R}$ be a Lipschitz function, with the constant η. Let M_f denote the median of f, while $\langle f \rangle$ stands for its mean value. Select a random point x on the sphere $\mathbf{S}^{n-1}$ with $n > 2$ according to the uniform measure. Then the probability of observing a value of f much different from the reference value is exponentially small,*

$$P\left(|f(x) - M_f| > \epsilon\right) \leq \exp\left(-\frac{n\epsilon^2}{2\eta^2}\right), \tag{15.84}$$

$$P\left(|f(x) - \langle f \rangle| > \epsilon\right) \leq 2\exp\left(-\frac{(n-1)\epsilon^2}{2\eta^2}\right). \tag{15.85}$$

This salient lemma was derived from the isoperimetric inequality [87]: among all sets of the same volume the ball has the smallest surface area. Therefore, the area of $\mathbf{S}^{n-1}$ is sharply concentrated around any set with a measure bigger than $1/2$. Even though a Lipschitz function f varies considerably, the above fact implies that it is nearly equal to the reference value on almost all high dimensional spheres. The lemma can be generalised in various directions and formulated for other manifolds [580, 149, 69].

---

[14] Introduced by Milman in his works on the theory of Banach spaces in the early 1970s, which extended ideas from the 1919 lectures of Paul Lévy [591, 90]. For a modern account see the introduction by Ball [87], the book by Ledoux [580] and a later complementary opus [149]. A book by Aubrun and Szarek [69] covers the interface with quantum information theory.

To demonstrate Lévy's lemma in action, consider a function f of the first Carte-sian component of a point on the sphere $\mathbf{S}^{n-1}$ with $n > 2$, so that $f(x_1, \ldots, x_n) = x_1$. Then Lévy's lemma quantifies the 'fat equator' effect: the probability of finding a random point on $\mathbf{S}^{n-1}$ outside a band along the equator of width 2ϵ is bounded by $2 \exp(-n\epsilon^2/2)$, which becomes exponentially small in high dimensions.

As a pure quantum state of size M can be identified with a Hopf circle on the sphere $\mathbf{S}^{2M-1}$, one can apply Lévy's lemma to analyse typical properties of quantum states. Consider, for instance, a random pure state $|\psi\rangle$ of a bipartite $N \times K$ system, such that $K \geq N \geq 3$. As discussed is Section 15.3 its partial trace produces a mixed state, $\rho(\psi) = \mathrm{Tr}_K |\psi\rangle\langle\psi|$, distributed with respect to the induced measure μ_K, which for $K = N$ reduces to the HS measure. From (15.72) we know that the average von Neumann entropy $\langle S_{N,K} \rangle$ of a random state ρ for large N behaves as $\ln N - N/2K$.

Lévy's lemma implies that the probability of large deviations from the mean value becomes exponentially small,

$$P\left(\left|S(\rho) - \langle S_{N,K}\rangle\right| \geq \alpha\right) \leq 2 \exp\left(-\frac{(NK - 1)}{8(\pi \ln N)^2}\alpha^2\right) \qquad (15.86)$$

To arrive at this result Hayden, Leung and Winter [417] considered the function $f(\psi) = S(\rho(\psi))$, estimated its Lipschitz constant, $\eta \sim \ln N$, and used Lévy's lemma in the form in which the fluctuations are measured with respect to the mean value. For large N the typical entropy of a random state is thus very close to the mean value and the distribution of entropy converges to the Dirac delta, $P(S) \to \delta(S - \langle S \rangle)$. In the next chapter we will see that the entropy of the partial trace $\rho(\phi)$ characterises the degree of entanglement of the corresponding bipartite pure state $|\psi\rangle$, so the above result implies that a generic bipartite pure state is strongly entangled.

A similar technique can be used to bound deviations of the purity $\langle \mathrm{Tr}\rho^2 \rangle$ from the mean value (15.71) for a random mixed state [245]. Observe that this statement can be formulated in terms of moments of random matrices defined according to the prescribed measure (15.54). The phenomenon of measure concentration is directly linked to high dimensional geometry and to the theory of non-commutative probability and random matrices.

Consider a fixed trace random Wishart matrix $\rho = GG^\dagger/\mathrm{Tr}GG^\dagger$, where G is a square non-Hermitian random Ginibre matrix of size N with independent complex Gaussian entries. We have learned that the level density for this ensemble of random mixed states distributed according to the HS measure, is asymptotically described by the Marchenko–Pastur distribution (15.62). Concentration of measure allows us to arrive at a stronger statement: level density of a *single* random matrix ρ of a large size N is well described by this distribution. Take an arbitrary unitary matrix U and construct the orbit of unitarily equivalent states, $\rho' = U\rho U^\dagger$. It is easy to see by counting the parameters that such an orbit has zero measure in the set $\mathcal{M}^{(N)}$ of all mixed states. However, due to concentration of measure, for a large N an ϵ-neighbourhood along this orbit covers the entire measure of $\mathcal{M}^{(N)}$, in the same way that an ϵ-band along the equator asymptotically carries the full mass of the sphere $\mathbf{S}^n$.

Analogous reasoning can also be applied to quantum maps. Consider a random stochastic map Φ acting on density matrices of order N and distributed with respect to the flat, HS measure (15.76). The corresponding dynamical matrix D can be obtained from a random Wishart matrix W of size N^2 by normalising it according to (15.75) to assure the trace preservation condition, $D_B = \text{Tr}_A D = \mathbb{1}$. Consider now the dual partial trace, $D_A = \text{Tr}_B D$. All its off-diagonal elements are given by sums of N complex variables which for large N will be close to zero, while diagonal elements are equal to sums of N non-negative real numbers with mean $1/N$. Thus for large N the matrix D_A becomes close to identity, so the corresponding random stochastic map Φ becomes nearly bistochastic. Although the set of bistochastic maps has zero measure in the set $\mathcal{TPCP}$ of all quantum operations, due to the concentration of measure its ϵ-neighbourhood covers the entire mass of the larger set $\mathcal{TPCP}$. A generic stochastic map sends the maximally mixed state, $\rho_* = \mathbb{1}/N$, close to the input, $\Phi(\rho_*) \approx \rho_*$, so the output entropy is almost exactly close to $\ln N$.

The lemma of Lévy does not only describe amazing effects in higher dimensional spaces, it also leads to ground breaking results influencing the entire field of quantum information. In Section 13.7 we mentioned the additivity conjecture, formulated in (13.110) for the Holevo information χ. This was known to hold for some particular class of channels [32, 530, 836], so it was tempting to speculate that it is true in general. If this were the case, one would have $\chi(\Phi^{\otimes n}) = n\chi(\Phi)$, so there would be no need to use the regularised quantities and the channel capacity, $C(\Phi) = \lim_{n\to\infty} \frac{1}{n}\chi(\Phi^{\otimes n})$, would coincide with the Holevo information, $\chi(\Phi)$.

In a seminal paper [837] Shor showed that additivity of the Holevo information is equivalent to the additivity of other relevant quantities, including the minimal output entropy $S_{\min}(\Phi_1 \otimes \Phi_2) = S_{\min}(\Phi_1) + S_{\min}(\Phi_2)$, where $S_{\min}(\Phi) = \min_\rho S(\Phi(\rho))$. The latter quantity is easier to handle and it allowed Hastings to obtain the following crucial result.

Hastings' non-additivity theorem. *There exist quantum channels Φ_1 and Φ_2 for which the minimal output entropy is strictly subadditive,*

$$S_{\min}(\Phi_1 \otimes \Phi_2) < S_{\min}(\Phi_1) + S_{\min}(\Phi_2). \tag{15.87}$$

The proof [410] is non-constructive. It relies on analysing random external fields (15.79) acting on an N-dimensional space and equivalent to a mixture of K unitaries with a careful choice of random probabilities $\{p_i\}_{i=1}^{K}$, finding clever bounds for the average minimal output entropy, and looking for the dimension N and the number of unitaries K such that strict inequality (15.87) typically holds. Subadditivity implies that there exists a correlated input state of both subsystems, which has a lower output entropy, and therefore is more resistant to noise than any product state.

The ideas of Hastings, based on the earlier result of Hayden and Winter [420], in which non-additivity was established for Rényi entropies with Rényi parameter $q > 1$, were soon developed and extended [157, 68, 325, 104]. The approach of Aubrun, Szarek and Werner [68] is based on *Dvoretzky's theorem* [279, 87], which implies that a generic section of a convex body in high dimensions is close to spherical. To taste the flavour of the ingredients of the proof let us have a brief look at the simplified formulation by Brandão and M. Horodecki [157]. Consider a random operation Φ complementary to (15.77), which acts on a density matrix ρ of size N as $\Phi(\rho) = \mathrm{Tr}_N[U(\rho \otimes |v\rangle\langle v|)U^\dagger]$, where $U \in U(NK)$. Note that the original subspace $\mathcal{H}_N$ is traced out, so the output state σ has dimension K of the environment. We consider the case $N \gg K$. The dual map $\tilde\Phi$ is defined by replacing U by its complex conjugate $\bar U$. In the first step one derives the lower bound

$$S_{\min}(\Phi \otimes \tilde\Phi) \leq 2\ln K - \frac{\ln K}{K}. \tag{15.88}$$

This is achieved by bounding the operator norm of the image of the maximally entangled state with respect to the tensor map, $||(\Phi \otimes \tilde\Phi)\rho_+||_\infty \geq 1/K$, where $\rho_+ = |\phi^+\rangle\langle\phi^+|$ with $|\phi^+\rangle = \frac{1}{\sqrt N}\sum_{i=1}^{N}|i,i\rangle$, and by applying the result presented in Problem 15.10. In the second, more involved step, one selects a positive constant $c_0 > 0$ and shows that for any $c \geq c_0$ the following lower bound for the minimal output entropy of a single map almost surely holds,

$$\ln K - \frac{c}{K} \leq S_{\min}(\Phi). \tag{15.89}$$

In this place one applies Lévy's lemma and works out a precise estimation of the Lipschitz constant to show that the probability that the map Φ sends a random pure state into a low entropy state is negligible. To conclude the proof one adds inequality (15.88) and inequality (15.89) multiplied by two. Making use of the identity for dual maps, $S_{\min}(\Phi) = S_{\min}(\tilde\Phi)$, one can bring the resulting inequality into the form

$$S_{\min}(\Phi \otimes \tilde\Phi) \leq S_{\min}(\Phi) + S_{\min}(\tilde\Phi) - \frac{(\ln K - 2c)}{K}. \tag{15.90}$$

Choosing a dimension K that is large enough we ensure that the term in the bracket is strictly positive, which implies that in this case the additivity is generically violated. Note that the above reasoning is non-constructive, in analogy to the original proof by Hastings: we infer that there exist quantum channels Φ_1 and Φ_2 for which inequality (15.87) holds without specifying them explicitly.

As the minimal output entropy is strictly subadditive (15.87), the Holevo information (13.107), in which one maximises the entropy difference, is strictly superadditive [410, 969]: there exist quantum channels for which $\chi(\Phi_1 \otimes \Phi_2) > \chi(\Phi_1) + \chi(\Phi_2)$. This implies that while sending information through two quantum channels it can be advantageous to apply both channels in parallel to a correlated input state of the composed system. The problem of finding the maximum capacity of a quantum channel remains thus open, as one needs to perform an intractable optimisation over spaces of an arbitrary high dimension. On the other hand, this very feature reveals a key difference between the theory of quantum information processing and its classical analogue.

Problems

15.1 Show that the following functions $f(t)$ generate monotone Riemannian metrics: $\sqrt{t}$, $(t-1)/\ln t$, $2[(t-1)/\ln t]^2/(1+t)$ (non-informative metric), $2t^{\alpha+1/2}/(1+t^{2\alpha})$ for $\alpha \in [0, 1/2]$ [734], $t^{t/(t-1)}/e$ (quasi-Bures metric) [843], and $f_{WY} = (\sqrt{t}+1)^2/4$ (Wigner–Yanase metric). Is the latter metric pure?

15.2 Compute the volume of the orthogonal group with respect to the measure $(ds)^2 \equiv -\frac{1}{2} \mathrm{Tr}(O^{-1}dO)^2$ analogous to (15.26). Show that $\mathrm{Vol}(\mathbb{R}\mathbf{P}^N) = \frac{1}{2}\mathrm{Vol}(\mathbf{S}^N) = \pi^{(N+1)/2}/\Gamma[(N+1)/2] = \mathrm{Vol}[O(N)]/(\mathrm{Vol}[O(N-1)]\,\mathrm{Vol}[O(1)])$; $\mathrm{Vol}[O(2)] = 2$ and $\mathrm{Vol}(\mathbf{F}_{\mathbb{R}}^{(N)}) = \mathrm{Vol}[O(N)]/2^N$ [1017].

15.3 Show that the volume radius of the set $\mathcal{M}^{(N)}$ of mixed states behaves for large N as $e^{-1/4}/\sqrt{N}$ [877].

15.4 Compute the mean von Neumann entropy of $N = 2$ random mixed states averaged over HS, Bures, orthogonal and unitary measures.

15.5 Find the mean moments $\langle \mathrm{Tr}\rho^3 \rangle$ and $\langle \mathrm{Tr}\rho^4 \rangle$ averaged over the induced measures (15.59).

15.6 Derive the distribution of fidelity $P(F)$ between two random N-dimensional complex pure states.

15.7 Compute the mean fidelity between two $N = 2$ independent random states distributed according to (a) HS measure and (b) Bures measure.

15.8 Analyse operation induced measures (15.81) defined for $N = 2$ by planar or linear channels (for definitions see Table 10.4) and show that they are isotropic inside the Bloch ball with the radial distributions $P_{\mathrm{plan}}(r) = 4r/\sqrt{1 - 4r^2}$ and $P_{\mathrm{lin}}(r) = 2 = P_u(r)$, respectively, where $r \in [0, 1/2]$.

15.9 Making use of the Marchenko–Pastur distribution (15.62) find (a) average entropy $\langle S \rangle$ and (b) average purity $\langle \mathrm{Tr}\rho^2 \rangle$ for random states of a large size N generated according to the HS measure.

15.10 Assuming that the operator norm of a state ρ of size N is bounded from below, $||\rho||_\infty \geq x > 1/N$, show that the von Neumann entropy obeys $S(\rho) \leq (1-x)\ln(N-1) + S(x, 1-x)$, where the entropy function $S(x, 1-x)$ is given in Eq. (2.12).

16

Quantum entanglement

Entanglement is not *one* but rather *the* characteristic trait of quantum mechanics.

— Erwin Schrödinger[1]

16.1 Introducing entanglement

So far, when working in a Hilbert space that is a tensor product of the form $\mathcal{H} = \mathcal{H}_A \otimes \mathcal{H}_B$, we were really interested in only one of the factors; the other factor played the role of an ancilla describing an environment outside our control. Now the perspective changes: we are interested in a situation where there are two masters. The fate of both subsystems are of equal importance, although they may be sitting in two different laboratories.

The operations performed independently in the two laboratories are described using operators of the form $\Phi_A \otimes \mathbb{1}$ and $\mathbb{1} \otimes \Phi_B$ respectively, but due perhaps to past history, the global state of the system may not be a product state. In general, it may be described by an arbitrary density operator ρ acting on the composite Hilbert space $\mathcal{H}$.

The peculiarities of this situation were highlighted in 1935 by Einstein, Podolsky and Rosen [285]. Their basic observation was that if the global state of the system is chosen suitably then it is possible to change – and to some extent to choose – the state assignment in laboratory A by performing operations in laboratory B. The physicists in laboratory A will be unaware of this until they are told, but they can check in retrospect that the experiments they performed were consistent with the state assignment arrived at from afar – even though there was an element of choice in arriving at that state assignment. Einstein's considered opinion was that 'on one supposition we should ... absolutely hold fast: the real factual situation of the

---

[1] Reproduced from Erwin Schrödinger, Discussion of probability relations between separated systems. *Proc. Camb. Phil. Soc.*, 31:555, 1935.

system $\mathcal{S}_2$ is independent of what is done with the system $\mathcal{S}_1$, which is spatially separated from the former' [284]. Then we seem to be forced to the conclusion that quantum mechanics is an incomplete theory in the sense that its state assignment does not fully describe the factual situation in laboratory A.

In his reply to EPR, Schrödinger argued that in quantum mechanics 'the best possible knowledge of a *whole* does not include the best possible knowledge of all its *parts*, even though they may be entirely separated and therefore virtually capable of being "best possibly known" '.[2] Schrödinger introduced the word *Verschränkung* to describe this phenomenon, personally translated it into English as *entanglement*, and made some striking observations about it. The subject then lay dormant for many years.

To make the concept of entanglement concrete, we recall that the state of the subsystem in laboratory A is given by the partially traced density matrix $\rho_A = \mathrm{Tr}_B \rho$. This need not be a pure state, even if ρ itself is pure. In the simplest possible case, namely when both $\mathcal{H}_A$ and $\mathcal{H}_B$ are two dimensional, we find an orthogonal basis of four states that exhibit this property in an extreme form. This is the *Bell basis*, first mentioned in Table 11.1:

$$|\psi^{\pm}\rangle = \frac{1}{\sqrt{2}}\left(|0\rangle|1\rangle \pm |1\rangle|0\rangle\right) \qquad |\phi^{\pm}\rangle = \frac{1}{\sqrt{2}}\left(|0\rangle|0\rangle \pm |1\rangle|1\rangle\right). \qquad (16.1)$$

The Bell states all have the property that $\rho_A = \frac{1}{2}\mathbb{1}$, which means that we know nothing at all about the subsystems – even though we have maximal knowledge of the whole. At the opposite extreme we have product states such as $|0\rangle|0\rangle$ and so on; if the global state of the system is in a pure product state then ρ_A is a projector and the two subsystems are in pure states of their own. Pure product states are called *separable*. All other pure states are *entangled*.

Now the point is that if a projective measurement is performed in laboratory B, corresponding to an operator of the form $\mathbb{1} \otimes \Phi_B$, then the global state will collapse to a product state. Indeed, depending on what measurement that B chooses to perform, and depending on its outcome, the state in laboratory A can become any pure state in the support of ρ_A. (This conclusion was drawn by Schrödinger from his mixture theorem. He found it 'repugnant'.) Of course, if the global state was one of the Bell states to begin with, then the experimenters in laboratory A still labour under the assumption that their state is $\rho_A = \frac{1}{2}\mathbb{1}$, and it is clear that any measurement results in A will be consistent with this state assignment. Nevertheless it would seem as if the real factual situation in A has changed from afar.

In the early sixties John Bell [105] was able to show that if we hold fast to the locality assumption then there cannot exist a completion of quantum mechanics in

---

[2] Schrödinger's (1935) 'general confession' consisted of a series of three papers [803, 804, 805].

the sense of EPR; the very meaning of the expression 'real factual situation' is at stake in entangled systems.[3] The idea is that if the quantum mechanical probabilities arise as marginals of a probability distribution over some kind of a set of real factual situations, then the mere existence of the latter gives rise to inequalities for the marginal distributions that, as a matter of fact, are disobeyed by the probabilities predicted by quantum mechanics.[4]

Bell's work caused much excitement in philosophically oriented circles; it seemed to put severe limits on the world view offered by physics. In the early 1990s the emphasis began to shift. Entanglement came to be regarded as a *resource* that allows us to do otherwise impossible things. An early and influential example is that of *quantum teleportation*. The task is to send classical information that allows a distant receiver to reconstruct a quantum system whose state is unknown to the sender. But since only a single copy of the system is available the sender is unable to figure out what the state to be 'teleported' actually is. So the task appears impossible. The solution (described in Section 16.3) turns out to be to equip sender and receiver with a shared maximally entangled state. Once this is arranged the teleportation can be achieved, using only ordinary classical communication between the parties.[5] The entanglement is used up in the process. In this sense then entanglement is a resource, just as the equally abstract concept of energy is a resource. Moreover it has emerged that there are many interesting tasks for which entanglement can be used, including *quantum cryptography* and *quantum computing* [369, 692, 751].

If entanglement is a resource we naturally want to know how much of it we have. As we will see it is by no means easy to answer this question, but it is easy to take a first step in the situation when the global state is a pure one. It is clear that there is no entanglement in a product state, when the subsystems are in pure states too and the von Neumann entropy of the partially traced state vanishes. On the other hand, a pure state is maximally entangled if its reduced density matrix is a maximally mixed state. For the case of two qubits the von Neumann entropy then assumes its maximum value ln 2, and the amount of entanglement in such a state is known as an

---

[3] Opinions diverge here. Bell wrote a sympathetic review of David Bohm's viewpoint, that we do not need to hold absolutely fast to Einstein's notion of locality [107]. Followers of Everett [304] argue that the system in *A* went from being entangled with the system in *B* to being entangled with the measurement apparatus in *B*, with no change of the real factual situation in *A*.

[4] For a thorough discussion of the Bell inequalities consult Clauser and Shimony [226], or a more recent review [180]; experimental tests, notably by Aspect et al. (1982) [64], show that violation of the Bell inequalities does indeed occur in the laboratory. Various loopholes in the early experiments were carefully discussed, see Gill [333] and Larsson [574], and dealt with in a new generation of experiments. Loophole-free experiments were reported in 2015 [343, 429, 827].

[5] To send information allowing us to reconstruct a given state elsewhere is called teleportation in the science fiction literature, where it is usually assumed to be trivial for the sender to verify what the state to be sent may be. The idea of teleporting an unknown state is from 1993 [117].

e-bit. States that are neither separable nor maximally entangled will require more thought, and mixed states will require much more thought.

At this stage entanglement may appear to be such an abstract notion that the need to quantify it does not seem to be urgent but then, once upon a time, 'energy' must have seemed a very abstract notion indeed, and now there are thriving industries whose role is to deliver it in precisely quantified amounts. Perhaps our governments will eventually have special *Departments of Entanglement* to deal with these things. But that is in the far future; here we will concentrate on a geometrical description of entanglement and how it is to be quantified.

16.2 Two-qubit pure states: entanglement illustrated

Our first serious move will be to take a look (literally) at entanglement in the two qubit case.[6] Our Hilbert space has four complex dimensions, so the space of pure states is $\mathbb{CP}^3$. We can make a picture of this space along the lines of Section 4.6. So we draw the positive hyperoctant of a 3-sphere and imagine a 3-torus sitting over each point, using the coordinates

$$\left(Z^0, Z^1, Z^2, Z^3\right) = \left(n_0, \, n_1 e^{iv_1}, \, n_2 e^{iv_2}, \, n_3 e^{iv_3}\right). \tag{16.2}$$

The four non-negative real numbers n_0, etc. obey

$$n_0^2 + n_1^2 + n_2^2 + n_3^2 = 1. \tag{16.3}$$

To draw a picture of this set we use a gnomonic projection of the 3-sphere centred at

$$(n_0, n_1, n_2, n_3) = \frac{1}{2}(1, 1, 1, 1). \tag{16.4}$$

The result is a nice picture of the hyperoctant, consisting of a tetrahedron centred at the above point, with geodesics on the 3-sphere appearing as straight lines in the picture. The 3-torus sitting above each interior point can be pictured as a rhomboid that is squashed in a position dependent way.

Mathematically, all points in $\mathbb{CP}^3$ are equal. In physics, points represent states, and some states are more equal than others. In Chapter 6, this happened because we singled out a particular subgroup of the unitary group to generate coherent states. Now it is assumed that the underlying Hilbert space is presented as a product of two factors in a definite way, and this singles out the orbits of $U(N) \times U(N) \subset U(N^2)$ for special attention. More specifically there is a preferred way of using the entries

---

[6] Such a geometric approach to the problem was initiated by Brody and Hughston [177], and developed by many others [112, 565, 589, 674].

Γ_{ij} of an $N \times N$ matrix as homogeneous coordinates. Thus any (normalised) state vector can be written as

$$|\Psi\rangle = \frac{1}{\sqrt{N}} \sum_{i=0}^{n} \sum_{j=0}^{n} \Gamma_{ij} |i\rangle |j\rangle. \quad (16.5)$$

For two-qubit entanglement $N = n + 1 = 2$, and it is agreed that

$$(Z^0, Z^1, Z^2, Z^3) \equiv (\Gamma_{00}, \Gamma_{01}, \Gamma_{10}, \Gamma_{11}). \quad (16.6)$$

Let us first take a look at the separable states. For such states

$$|\Psi\rangle = \sum_{i=0}^{n} \sum_{j=0}^{n} (a_i |i\rangle)(b_j |j\rangle) \quad \Leftrightarrow \quad \Gamma_{ij} = a_i b_j. \quad (16.7)$$

In terms of coordinates a two-qubit state is separable if and only if

$$Z^0 Z^3 - Z^1 Z^2 = 0. \quad (16.8)$$

We recognise this quadric equation from Section 4.3. It defines the Segre embedding of $\mathbb{CP}^1 \times \mathbb{CP}^1$ into $\mathbb{CP}^3$. Thus the separable states form a four real-dimensional submanifold of the six real-dimensional space of all states – had we regarded $\mathbb{CP}^1$ as a classical phase space, this submanifold would have been enough to describe all the phase space of the composite system.

What we did not discuss in Chapter 4 is the fact that the Segre embedding can be nicely described in the octant picture. Eq. (16.8) splits into two real equations:

$$n_0 n_3 - n_1 n_2 = 0 \quad (16.9)$$
$$\nu_1 + \nu_2 - \nu_3 = 0. \quad (16.10)$$

Hence we can draw the space of separable states as a two-dimensional surface in the octant, with a two-dimensional surface in the torus that sits above each separable point in the octant. The surface in the octant has an interesting structure, related to Figure 4.6. In Eq. (16.7) we can keep the state of one of the subsystems fixed; say that b_0/b_1 is some fixed complex number with modulus k. Then

$$\frac{Z^0}{Z^1} = \frac{b_0}{b_1} \quad \Rightarrow \quad n_0 = k n_1 \quad (16.11)$$

$$\frac{Z^2}{Z^3} = \frac{b_0}{b_1} \quad \Rightarrow \quad n_2 = k n_3. \quad (16.12)$$

As we vary the state of the other subsystem we sweep out a curve in the octant that is in fact a geodesic in the hyperoctant (the intersection between the 3-sphere and two hyperplanes through the origin in the embedding space). In the gnomonic coordinates that we are using this curve will appear as a straight line, so what we

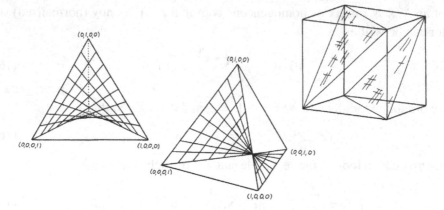

Figure 16.1 The separable states, or the Segre embedding of $\mathbb{CP}^1 \times \mathbb{CP}^1$ in $\mathbb{CP}^3$. Two different perspectives of the tetrahedron are given.

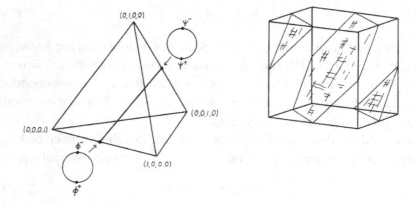

Figure 16.2 The maximally entangled states form an $\mathbb{RP}^3$, appearing as a straight line in the octant and a surface in the tori. The location of the Bell states is also shown.

see when we look at how the separable states sit in the hyperoctant is a surface that is ruled by two families of straight lines.

There is an interesting relation to the Hopf fibration (see Section 3.5) here. Each family of straight lines is also a one-parameter family of Hopf circles, and there are two such families because there are two Hopf fibrations, with different twist. We can use our hyperoctant to represent real projective space $\mathbb{RP}^3$, in analogy with Figure 4.12. The Hopf circles that rule the separable surface are precisely those that get mapped onto each other when we 'fold' the hemisphere into a hyperoctant.

We now turn to the maximally entangled states, for which the reduced density matrix is the maximally mixed state. Using composite indices we write

$$\rho_{ij \atop kl} = \frac{1}{N}\Gamma_{ij}\Gamma^*_{kl} \quad \Rightarrow \quad \rho^A_{ik} = \sum_{j=0}^{n} \rho_{ij \atop kj}, \qquad (16.13)$$

see Eq. (10.27). Thus

$$\rho_{ik}^A = \frac{1}{N}\mathbb{1} \quad \Leftrightarrow \quad \sum_{j=0}^{n} \Gamma_{ij}\Gamma_{kj}^* = \delta_{ik}. \tag{16.14}$$

Therefore the state is maximally entangled if and only if the matrix Γ is unitary. Recalling that an overall factor of this matrix is irrelevant for the state, it follows that the set of maximally entangled states of two qubits is the manifold $SU(2)/Z_2 = \mathbb{RP}^3$. To see what this space looks like in the octant picture we observe that

$$\Gamma_{ij} = \begin{bmatrix} \alpha & \beta \\ -\beta^* & \alpha^* \end{bmatrix} \quad \Rightarrow \quad Z^\alpha = (\alpha, \beta, -\beta^*, \alpha^*). \tag{16.15}$$

In our coordinates this yields three real equations; the space of maximally entangled states will appear in the picture as a straight line connecting two entangled edges and passing through the centre of the tetrahedron, while there is a two-dimensional surface in the tori. The latter is shifted relative to the separable surface in such a way that the separable and maximally entangled states manage to keep their distance in the torus also when they meet in the octant (at the centre of the tetrahedron where the torus is large). Our picture thus displays $\mathbb{RP}^3$ as a one-parameter family of two-dimensional flat tori, degenerating to circles at the ends of the interval. This is similar to our picture of the 3-sphere, except that this time the lengths of the two intersecting shortest circles on the tori stay constant while the angle between them is changing. It is amusing to convince oneself of the validity of this picture, and to verify that it is really a consequence of the way that the 3-tori are being squashed as we move around the octant.

As a further illustration we can consider the collapse of a maximally entangled state, say $|\psi^+\rangle$ for definiteness, when a measurement is performed in laboratory B. The result will be a separable state, and because the global state is maximally entangled all the possible outcomes will be equally likely. It is easily confirmed that the possible outcomes form a 2-sphere's worth of points on the separable surface, distinguished by the fact that they are all lying on the same distance $D_{FS} = \pi/4$ from the original state. This is the minimal Fubini–Study distance between a separable and a maximally entangled state. The collapse is illustrated in Figure 16.3.

We can deal with states of intermediate entanglement by writing them in their Schmidt form $|\Psi\rangle = \cos\chi\,|00\rangle + \sin\chi\,|11\rangle$, see Section 9.2. The reduced density matrix is

$$\rho_A = \text{Tr}_B|\Psi\rangle\langle\Psi| = \begin{bmatrix} \cos^2\chi & 0 \\ 0 & \sin^2\chi \end{bmatrix}. \tag{16.16}$$

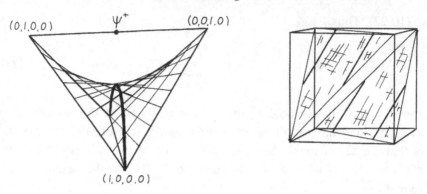

$(0,1,0,0)$ ψ^+ $(0,0,1,0)$

$(1,0,0,0)$

Figure 16.3 A complete measurement on one of the subsystems will collapse the Bell state $|\psi^+\rangle$ to a point on a sphere on the separable surface; it appears as a one-parameter family of circles in our picture. All points on this sphere are equally likely.

The *Schmidt angle* $\chi \in [0, \pi/4]$ parametrizes the amount of ignorance about the state of the subsystem, that is to say the amount of entanglement. A nice thing about it is that its value cannot be changed by local unitary transformations of the form $U(2) \times U(2)$. A set of states of intermediate entanglement, quantified by some given value of the Schmidt angle χ, is difficult to draw (although it can be done). For the extreme cases of zero or one e-bit's worth of entanglement we found the submanifolds $\mathbb{CP}^1 \times \mathbb{CP}^1$ and $SU(2)/Z_2$, respectively. There is a simple reason why these spaces turn up, namely that the amount of entanglement must be left invariant under local unitary transformations belonging to the group $SU(2) \times SU(2)$. In effect therefore we are looking for orbits of this group, and what we have found are the two obvious possibilities. More generally we will get a stratification of $\mathbb{CP}^3$ into orbits of $SU(2) \times SU(2)$; the problem is rather similar to that discussed in Section 7.2. Of the two exceptional orbits, the separable states form a Kähler manifold and can therefore serve as a classical phase space, much like the coherent states we discussed in Section 6.1. A generic orbit will be five real dimensional and the set of such orbits will be labelled by the Schmidt angle χ, which is also the minimal distance from a given orbit to the set of separable states. A generic orbit is rather difficult to describe however. Topologically it is a non-trivial fibre bundle with an $\mathbf{S}^2$ as base space and $\mathbb{R}\mathbf{P}^3$ as fibre.[7] In the octant picture it appears as a three-dimensional volume in the octant and a two-dimensional surface in the torus. And with this observation our tour of the two qubit Hilbert space is at an end.

---

[7] This can be seen in an elegant way using the Hopf fibration of $\mathbf{S}^7$ – the space of normalised state vectors – as $\mathbf{S}^4 = \mathbf{S}^7/\mathbf{S}^3$; Mosseri and Dandoloff [674] provide the details.

16.3 Maximally entangled states

The step from two qubits to two quNits is easily taken. We return to the expression (16.5) and use Γ to label both the state and the matrix of its components,

$$|\Gamma\rangle = \frac{1}{\sqrt{N}} \sum_{i=0}^{n} \sum_{j=0}^{n} \Gamma_{ij} |e_i\rangle \otimes |f_j\rangle. \tag{16.17}$$

This time we will try to avoid the cluttering of indices in Eqs. (16.13–16.14). Instead we aim for a bird's eye view. If we have two such state vectors, with components given by two different matrices Γ and Γ', their scalar product is quickly calculated to be

$$\langle \Gamma | \Gamma' \rangle = \frac{1}{N} \mathrm{Tr} \Gamma^\dagger \Gamma'. \tag{16.18}$$

In particular a state is normalised if $\mathrm{Tr}\Gamma^\dagger\Gamma = N$. The partial trace of the normalised density matrix – which summarises the information that Alice can use to predict the outcomes of her own experiments – is

$$\rho_A = \mathrm{Tr}_B |\Gamma\rangle\langle\Gamma| = \frac{1}{N} \sum_{i,j} (\Gamma\Gamma^\dagger)_{ij} |e_i\rangle\langle e_j|. \tag{16.19}$$

The eigenvalues of the Hermitian matrix $\Gamma\Gamma^\dagger$ are precisely the Schmidt coefficients of the state, and cannot be changed by local unitaries. Indeed local unitaries act according to

$$U \otimes V |\Gamma\rangle = |U\Gamma V^T\rangle. \tag{16.20}$$

Note that the transpose of V occurs on the right-hand side.

Now let us suppose that Γ is a unitary matrix, $\Gamma = U \Rightarrow \Gamma\Gamma^\dagger = \mathbb{1}$. Then all the Schmidt coefficients are equal, and it follows from Eq. (16.19) that the partial trace ρ_A of the density matrix ρ is proportional to the identity matrix. On her own, Alice can make no useful predictions at all. Moreover, anything that Alice can do on her own can be undone by Bob. An example of a maximally entangled state is obtained by choosing Γ to be the unit matrix. This is the Jamiołkowski state of size $N = n + 1$,

$$|\phi^+\rangle = \frac{1}{\sqrt{N}} \sum_{i,j} \delta_{ij} |e_i\rangle |f_j\rangle = \frac{1}{\sqrt{N}} \sum_{i=0}^{n} |e_i\rangle |f_i\rangle. \tag{16.21}$$

If Alice applies the operator $U \otimes \mathbb{1}$ to this state, thus changing it to the state $|U\rangle$, Bob can undo this action by the application of $\mathbb{1} \otimes U^*$. And whatever they do with local unitaries will leave their reduced density matrices ρ_A and ρ_B unchanged. Such a state is indeed *maximally entangled*.

Every maximally entangled state can be reached from the Jamiołkowski state by applying a local unitary transformation, and in fact by applying local unitaries of the form $U \otimes \mathbb{1}$ only. Thus

$$U \otimes \mathbb{1}|\phi^+\rangle = \mathbb{1} \otimes U^T|\phi^+\rangle = \frac{1}{\sqrt{N}}\sum_{i=0}^{n}\sum_{j=0}^{n}U_{ij}|e_i\rangle \otimes |f_j\rangle \equiv |U\rangle. \qquad (16.22)$$

This shows not only that the set of maximally entangled states is an orbit under the unitary group, but also – remembering that we ignore an overall phase – that it is identical to the group manifold $SU(N)/Z_N$.

From a geometrical point of view the set of maximally entangled states is an interesting submanifold of $\mathbb{CP}^{N^2-1}$. By definition a *minimal submanifold* is such that any deformation of a bounded piece of it leads to an increase in its volume. The equator of a sphere is minimal in this sense. The set of maximally entangled states is minimal.[8] Next we recall (from Section 3.4) that a Lagrangian submanifold of a symplectic space is a submanifold of half the dimension of the full space, such that the symplectic form vanishes when restricted to it. Using expression (4.51) for the Fubini–Study symplectic form, we find when we evaluate it for maximally entangled states that

$$\Omega = \frac{i}{N^2}\left(\mathrm{Tr}UU^\dagger\mathrm{Tr}\mathrm{d}U \wedge \mathrm{d}U^\dagger - \mathrm{Tr}\mathrm{d}UU^\dagger \wedge \mathrm{Tr}U\mathrm{d}U^\dagger\right). \qquad (16.23)$$

The wedge product affects only the ordering of the one-forms, so we can perform matrix manipulations freely. Using $\mathrm{d}U^\dagger = -U^\dagger\mathrm{d}UU^\dagger$ it is not hard to see that both terms vanish due to the anti-symmetry of the wedge product, so in fact $\Omega = 0$ when restricted to the set of maximally entangled states. Since its dimension is right we conclude that they form a minimal Lagrangian submanifold. This is as different as it can be from the set of separable states, which is in itself a Kähler space with a non-degenerate induced symplectic form.

According to an aphorism, 'in symplectic geometry every important object is a Lagrangian submanifold' [53]. Every complex projective space has some. By consulting Eq. (4.83) we see that examples include the real submanifold $\mathbb{RP}^n$, for which $\mathrm{d}v_i = 0$. It is an accident of the two-qubit case that $SU(2)/Z_2 = \mathbb{RP}^3$ – an accident that will be used to great effect in Section 16.9. Alternatively we can make the symplectic form vanish by setting $\mathrm{d}n_i = 0$. In particular we can set all the n_i equal to each other. Harking back to Chapter 12 we see that this means that the set of all vectors that are unbiased with respect to a given orthonormal basis form another Lagrangian submanifold, identical to the maximal torus shown in Figure 4.10. Still, the maximally entangled Lagrangian submanifold is arguably the most interesting example.

---

[8] And so is the set of separable states. The proof can be found in a book by Lawson [576].

But the question is: what can one do with maximally entangled states? Alice and Bob can make good use of a maximally entangled state if they cooperate. Two famous examples of this are *quantum dense coding* [123] and *quantum teleportation* [117], both of which can be understood [960] using the unitary operator bases introduced in Section 12.3. In the first protocol, Alice wants to inform Bob about which one out of N^2 alternatives she has chosen, while sending only a single quantum system of N dimensions. This sounds impossible. She can prepare the system in one out of N eigenstates, and if Bob knows beforehand what the eigenbasis is, he can perform the measurement and find out which. But Alice can do better if she and Bob have prepared two quNits in the Jamiołkowski state (16.21), and if they control one factor each. Alice can then apply one out of the N^2 local operators $U_I \otimes \mathbb{1}$, where U_I is one member of a unitary operator basis on which they have agreed beforehand. The global state is then transformed to the state $|U_I\rangle$, which is one member of a maximally entangled basis in $\mathcal{H}_A \otimes \mathcal{H}_B$. Alice then transmits her part of the physical system to Bob, who performs a measurement on the global system to see which out of the N^2 orthogonal states it is in. In effect two Nits have been sent using only one quNit.

For quantum teleportation a three-partite Hilbert space is involved. Alice is in possession of the state $|\Psi\rangle_1 \in \mathcal{H}_1$, and together with Bob she has prepared a Jamiołkowski state $|\Phi^+\rangle_{23} \in \mathcal{H}_2 \otimes \mathcal{H}_3$. Alice controls the factor $\mathcal{H}_1 \otimes \mathcal{H}_2$. Before doing anything, she rewrites the state using a unitary operator basis,

$$|\Psi\rangle_1 |\Phi^+\rangle_{23} = \left(\sum_{I=1}^{N^2} |U_I\rangle_{12} \langle U_I | \otimes \mathbb{1}_3 \right) |\Psi\rangle_1 |\Phi^+\rangle_{23}. \qquad (16.24)$$

Finding this a little difficult to see through she uses index notation, and observes that

$$\langle U_I | \Psi \rangle_1 |\Phi^+\rangle_{23} = \sum_{m,n} U_{mn}^* \langle m| \langle n| \sum_{r,i} \psi_r |r\rangle_1 |i\rangle_2 |i\rangle_3 = |U_I^\dagger \Psi\rangle_3. \qquad (16.25)$$

Putting things together, she concludes that

$$|\Psi\rangle_1 |\Phi^+\rangle_{23} = \sum_I |U_I\rangle_{12} |U_I^\dagger \Psi\rangle_3. \qquad (16.26)$$

Then she performs a von Neumann measurement with N^2 possible outcomes.[9] This causes the state to collapse to

$$|\Psi\rangle_1 |\Phi^+\rangle_{23} \ \rightarrow \ |U_I\rangle_{12} |U_I^\dagger \Psi\rangle_3. \qquad (16.27)$$

---

[9] Such measurements do pose a challenge in the laboratory, but they can be made. The present length record for quantum teleportation is 143 kilometres [630].

Finally Alice sends two classical Nits to inform Bob about the outcome. Given this information Bob performs the unitary transformation $\mathbb{1} \otimes \mathbb{1} \otimes U_I$, so that the overall result is

$$|\Psi\rangle_1 |\Omega\rangle_{23} \;\rightarrow\; |U_I\rangle_{12} |\Psi\rangle_3. \tag{16.28}$$

The still unknown quantum state has been teleported into Bob's factor of the Hilbert space, while the shared entanglement has been used up.

We leave these matters here with just the additional remark that maximally entangled states are very desirable things to have.

16.4 Pure states of a bipartite system

The time has come to look in more detail into how the Hilbert space $\mathcal{H}_{N_A} \otimes \mathcal{H}_{N_B}$ of a composite system is divided into *locally equivalent* sets of states related by local unitary transformations,

$$|\psi'\rangle \;=\; U \otimes V |\psi\rangle, \tag{16.29}$$

where $U \in SU(N_A)$ and $V \in SU(N_B)$. Sometimes one calls them *interconvertible states*, since they may be reversibly converted by local transformations one into another. It is clear that not all pure states are locally equivalent, since the product group $SU(N_A) \times SU(N_B)$ is a small subgroup of $SU(N_A N_B)$. We now ask for the dimension and topology of the orbit generated by local unitary transformations from a given reference state.[10]

We are going to rely on the Schmidt decomposition (9.8) of the normalised reference state $|\psi\rangle$. This consists of at most N terms, since without loss of generality we assume that $N_B \geq N_A = N$. The normalisation condition (9.17) implies that the Schmidt vector $\vec{\lambda}$ lives in the $(N-1)$ dimensional Schmidt simplex, and gives the spectra of the partially reduced states, $\rho_A = \mathrm{Tr}_B(|\psi\rangle\langle\psi|)$ and $\rho_B = \mathrm{Tr}_A(|\psi\rangle\langle\psi|)$, which differ only by $N_B - N_A$ zero eigenvalues. The *Schmidt rank* of a pure state is the number of non-zero Schmidt coefficients, equal to the rank of the reduced states. States with maximal Schmidt rank occupy the interior of the Schmidt simplex, while states of a lower rank live on its boundary. The separable states sit at the corners, and the maximally entangled states at the centre.

Let $\vec{\lambda} = (0, \cdots, 0, \kappa_1, \cdots, \kappa_1, \kappa_2, \cdots, \kappa_2, \ldots, \kappa_J, \cdots, \kappa_J)$, each value κ_n occurring m_n times, be an ordered Schmidt vector. The number m_0 of vanishing coefficients may equal zero. By definition $\sum_{n=0}^{J} m_n = N$. A local orbit $\mathcal{O}_{\mathrm{loc}}$ has the structure of a fibre bundle, in which two quotient spaces

$$\frac{U(N)}{U(m_0) \times U(m_1) \times \cdots \times U(m_J)} \quad \text{and} \quad \frac{U(N)}{U(m_0) \times U(1)} \tag{16.30}$$

<hr/>

[10] In the spirit of Section 16.3 we could also ask for the symplectic geometry of general orbits, but for this we refer to the literature [793].

form the base and the fibre, respectively [839]. In general such a bundle need not be trivial. The dimension of the local orbit may be computed from the dimensions of the coset spaces,

$$\dim(\mathcal{O}_{loc}) = 2N^2 - 1 - 2m_0^2 - \sum_{n=1}^{J} m_n^2. \tag{16.31}$$

Observe that the base describes the orbit of the unitarily similar mixed states of the reduced system ρ_A, and depends on the degeneracy of its spectrum $\vec{\lambda}$ (see Table 8.1). The fibre characterises the manifold of pure states which are projected by the partial trace to the same density matrix, and depends on the Schmidt rank. To understand this structure consider first a generic state of the maximal Schmidt rank, so that $m_0 = 0$. Acting on $|\psi\rangle$ with $U_N \otimes V_N$, where both unitary matrices are diagonal, we see that there exist N redundant phases. Since each pure state is determined up to an overall phase, the generic orbit has the local structure

$$\mathcal{O}_g \approx \frac{U(N)}{[U(1)]^N} \times \frac{U(N)}{U(1)} = \mathbf{F}^{(N)} \times \frac{U(N)}{U(1)}, \tag{16.32}$$

with dimension $\dim(\mathcal{O}_g) = 2N^2 - N - 1$. If some of the coefficients are equal, say $m_J > 1$, then we need to identify all states differing by a block diagonal unitary rotation with $U(m_J)$ in the right lower corner. In the same way one explains the meaning of the factor $U(m_0) \times U(m_1) \times \cdots \times U(m_J)$ which appears in the first quotient space of (16.30). If some Schmidt coefficients are equal to zero the action of the second unitary matrix U_B is trivial in the m_0-dimensional subspace – the second quotient space in (16.30) is $U(N)/[U(m_0) \times U(1)]$.

The set of all orbits foliate $\mathbb{C}\mathbf{P}^{N^2-1}$, the space of all pure states of the $N \times N$ system. This foliation is *singular*, since there exist measure-zero leaves of various dimensions and topology. In particular, for separable states there exists only one non-zero coefficient, $\lambda_1 = 1$, so $m_0 = N-1$. This gives the Segre embedding (4.16),

$$\mathcal{O}_{sep} = \frac{U(N)}{U(1) \times U(N-1)} \times \frac{U(N)}{U(1) \times U(N-1)} = \mathbb{C}\mathbf{P}^{N-1} \times \mathbb{C}\mathbf{P}^{N-1}, \tag{16.33}$$

of dimension $\dim(\mathcal{O}_{sep}) = 4(N-1)$. For maximally entangled states on the other hand one has $\lambda_1 = \lambda_N = 1/N$, hence $m_1 = N$ and $m_0 = 0$. Therefore

$$\mathcal{O}_{max} = \frac{U(N)}{U(N)} \times \frac{U(N)}{U(1)} = \frac{U(N)}{U(1)} = \frac{SU(N)}{Z_N}, \tag{16.34}$$

with $\dim(\mathcal{O}_{max}) = N^2 - 1$, half the total dimension of the space of pure states. It is in fact this drop in dimension, and their special position within the singular foliation, that guarantee that these orbits are minimal submanifolds [576]. The dimensions of all local orbits for $N = 2, 3$ are shown in Figure 16.4, and their intrinsic structure in Table 16.1.

Table 16.1 *Structure of local orbits of the $N \times N$ pure states. D_s denotes the dimension of the subspace of the Schmidt simplex Δ_{N-1}, while D_o represents the dimension of the local orbit. $\mathbf{F}^{(N)}$ is a flag manifold (Section 4.9).*

N	Schmidt coefficients	D_s	Part of the asymmetric simplex	Local structure: base × fibre	D_o
	(a, b)	1	line	$\mathbf{F}^{(2)} \times \mathbb{R}\mathrm{P}^3$	5
2	$(1, 0)$	0	left edge	$\mathbb{C}\mathrm{P}^1 \times \mathbb{C}\mathrm{P}^1$	4
	$(1/2, 1/2)$	0	right edge	$U(2)/U(1) = \mathbb{R}\mathrm{P}^3$	3
	(a, b, c)	2	interior of triangle	$\mathbf{F}^{(3)} \times \frac{U(3)}{U(1)}$	14
	$(a, b, 0)$	1	base	$\mathbf{F}^{(3)} \times \frac{U(3)}{[U(1)]^2}$	13
3	(a, b, b)	1	2 upper sides	$\frac{U(3)}{U(1)\times U(2)} \times \frac{U(3)}{U(1)}$	12
	$(1/2, 1/2, 0)$	0	right corner	$\frac{U(3)}{U(1)\times U(2)} \times \frac{U(3)}{[U(1)]^2}$	11
	$(1, 0, 0)$	0	left corner	$\mathbb{C}\mathrm{P}^2 \times \mathbb{C}\mathrm{P}^2$	8
	$(1/3, 1/3, 1/3)$	0	upper corner	$U(3)/U(1)$	8

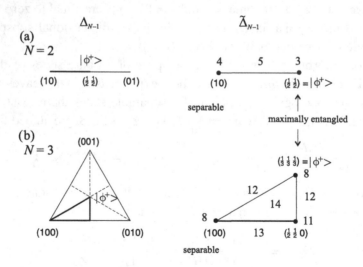

Figure 16.4 Dimensionality of local orbits generated by a given point of the Weyl chamber $\tilde{\Delta}_{N-1}$ – an asymmetric part of the Schmidt simplex Δ_{N-1} – for pure states of the $N \times N$ problem with $N = 2, 3$. Compare Table 16.1.

The local orbit defined by (16.29) contains all purifications of all mixed states acting on $\mathcal{H}_N$ isospectral with $\rho_N = \text{Tr}_B|\psi\rangle\langle\psi|$. When $N_A = N_B$ we can set $V = U$, and consider states fulfilling the *strong local equivalence* relation (SLE), $|\psi'\rangle = U\otimes U|\psi\rangle$. Pairs of states related in this way are *equal* up to selection of the reference frame used to describe the subsystems. The basis is determined by a unitary U. For a maximally entangled state there are no other states satisfying SLE, while for a separable state the orbit of SLE states forms the complex projective space $\mathbb{CP}^{N-1}$ of all pure states of a single subsystem.

The question as to whether a given pure state $|\psi\rangle \in \mathcal{H}_{N_A} \otimes \mathcal{H}_{N_B}$ is separable is easy to answer: it is enough to compute the partial trace, $\rho_A = \text{Tr}_B(|\psi\rangle\langle\psi|)$, and to check if $\text{Tr}\rho_A^2$ equals unity. If it is so the reduced state is pure, hence the initial pure state is separable. In the opposite case the pure state is entangled. The next question is: to what extent is a given state entangled?

There seems not to be a unique answer to this question. One may describe the entanglement by the entropies analysed in Chapters 2 and 13. The *entanglement entropy* is defined as the von Neumann entropy of the reduced state, which is equal to the Shannon entropy of the Schmidt vector,

$$E(|\psi\rangle) \equiv S(\rho_A) = S(\vec{\lambda}) = - \sum_{i=1}^{N} \lambda_i \ln \lambda_i. \tag{16.35}$$

The entropy $E(|\psi\rangle)$ is bounded, $0 \le E \le \ln N$. It is equal to zero for separable states and to $\ln N$ for maximally entangled states. Similarly, one can use the Rényi entropies (2.79) to define entropies $E_q \equiv S_q(\rho_A)$. We shall need a quantity related to linear entropy,

$$\tau(|\psi\rangle) \equiv 2\left(1 - \text{Tr}\rho_A^2\right) = 2\left(1 - \sum_{i=1}^{N} \lambda_i^2\right) = 2\left(1 - \exp[-E_2(|\psi\rangle)]\right), \tag{16.36}$$

which runs from 0 to $2(N-1)/N$. In the case of a two-qubit system it is known as *tangle*, while its square root $C = \sqrt{\tau}$ is called *concurrence*. It will reappear in great style in Section 16.9. The minimal entropy, $E_\infty = -\ln\lambda_{\max}$, has a nice geometric interpretation: if the Schmidt vector is ordered decreasingly and $\lambda_1 = \lambda_{\max}$ denotes its largest component then the state $|1\rangle\otimes|1\rangle$ is the separable pure state closest to $|\psi\rangle$ [96, 620]. Thus the Fubini–Study distance between $|\psi\rangle$ and the set of separable pure states reads $D_{\text{FS}}^{\min} = \arccos(\sqrt{\lambda_{\max}})$, while $1 - \lambda_{\max}$ is called the *geometric measure* of entanglement [951]. The entanglement entropy $E = E_1$ is distinguished among the Rényi entropies E_q just as the Shannon entropy is singled out by its operational meaning discussed in Section 2.2.

For the two-qubit problem the Schmidt vector has only two components, which sum to unity, so the entropy $E(|\psi\rangle) \in [0, \ln 2]$ characterises uniquely the entanglement of the pure state $|\psi\rangle$. But in general one needs a set of $N - 1$ independent quantities. Which properties should they fulfil?

Before discussing this issue we need to distinguish certain classes of quantum operations acting on bipartite systems. *Local operations* (LO) arise as the tensor product of two maps, both satisfying the trace preserving condition (10.54),

$$[\Phi_A \otimes \Phi_B](\rho) = \sum_i \sum_j (A_i \otimes B_j) \rho \left(A_i^\dagger \otimes B_j^\dagger\right). \qquad (16.37)$$

The second, important, class of maps is called *LOCC*. This name stands for *local operations and classical communication*. All quantum operations, including measurements, are allowed, provided they are performed locally in each subsystem. Classical communication allows the two parties to exchange classical information about their subsystems, and hence to introduce classical correlations between them. A third class of maps consists of operations that can be written in the form

$$\Phi_{\text{sep}}(\rho) = \sum_i (A_i \otimes B_i) \rho \left(A_i^\dagger \otimes B_i^\dagger\right). \qquad (16.38)$$

They are called *separable* (SO). Observe that this form is more general than (16.37), even though the summation goes over one index. In fact we have [119] the proper inclusion relations LO $\subset$ LOCC $\subset$ SO.

The definition of LOCC transformations looks simple but this class is not easy to characterise and describe.[11] Fortunately, if we restrict our attention to pure states, the classes LOCC and LU coincide in the sense that any two pure states of a bipartite system equivalent with respect to LOCC are also equivalent with respect to local unitaries. To show this we shall majorise Schmidt vectors corresponding to both states and rely on the following crucial result.[12]

Nielsen's majorisation theorem. [689]. *A given state $|\psi\rangle$ may be deterministically transformed into $|\phi\rangle$ by LOCC operations if and only if the corresponding Schmidt vectors satisfy the majorisation relation (2.1),*

$$|\psi\rangle \xrightarrow{\text{LOCC}} |\phi\rangle \quad \Longleftrightarrow \quad \vec{\lambda}_\psi \prec \vec{\lambda}_\phi. \qquad (16.39)$$

To prove the forward implication we follow the original proof. Assume that party A performs locally a generalised measurement, which is described by a set of k Kraus operators A_i. By classical communication the result is sent to party B, who performs a local action Φ_i, conditioned on the result i. Hence

$$\sum_{i=1}^{k} [\mathbb{1} \otimes \Phi_i] \left(A_i|\psi\rangle\langle\psi|A_i^\dagger\right) = |\phi\rangle\langle\phi|. \qquad (16.40)$$

---

[11] For a nice discussion of the subtle properties of these operations see [214].
[12] For an introduction to majorisation and its applications in quantum theory consult an unpublished text by Nielsen [688] and his review published jointly with Vidal [695].

The result is a pure state so each term in the sum needs to be proportional to the projector. Tracing out the second subsystem we get

$$A_i \rho_\psi A_i^\dagger = p_i \rho_\phi, \qquad i = 1, \ldots, k, \tag{16.41}$$

where $\sum_{i=1}^k p_i = 1$ and $\rho_\psi = \mathrm{Tr}_B(|\psi\rangle\langle\psi|)$ and $\rho_\phi = \mathrm{Tr}_B(|\phi\rangle\langle\phi|)$. Due to the polar decomposition of $A_i\sqrt{\rho_\psi}$ we may write

$$A_i \sqrt{\rho_\psi} = \sqrt{A_i \rho_\psi A_i^\dagger} \, V_i = \sqrt{p_i \rho_\phi} \, V_i \tag{16.42}$$

with unitary V_i. Making use of the completeness relation (10.54) we obtain

$$\rho_\psi = \sqrt{\rho_\psi} \, \mathbb{1} \sqrt{\rho_\psi} = \sum_{i=1}^k \sqrt{\rho_\psi} A_i^\dagger A_i \sqrt{\rho_\psi} = \sum_{i=1}^k p_i V_i^\dagger \rho_\phi V_i, \tag{16.43}$$

and the last equality follows from (16.42) and its adjoint. Hence we have arrived at an unexpected conclusion: if a local transformation $|\psi\rangle \to |\phi\rangle$ is possible, then there exists a bistochastic operation (10.72) which acts on the partially traced states with reversed time – it sends ρ_ϕ into ρ_ψ! The quantum HLP lemma (Section 13.5) implies the majorisation relation $\vec{\lambda}_\psi \prec \vec{\lambda}_\phi$. The backward implication follows from an explicit conversion protocol proposed by Nielsen, or alternative versions presented in [274, 404, 502].

The majorisation relation (16.39) introduces a partial order into the set of pure states.[13] Hence any pure state $|\psi\rangle$ allows one to split the Schmidt simplex, representing the set of all local orbits, into three regions: the set F (*future*) contains states which can be produced from $|\psi\rangle$ by LOCC, the set P (*past*) states from which $|\psi\rangle$ may be obtained, and the set C consists of *incomparable* states which cannot be joined by a local transformation in any direction.[14]

This gives a causal structure to the Schmidt simplex. See Figure 16.5, and observe the close similarity to Figure 13.5 showing paths in the simplex of eigenvalues that can be generated by bistochastic operations. But the arrow of time differs: the 'past' for the evolution in the space of density matrices corresponds to the 'future' for the local entanglement transformations and vice versa. In both cases the set C of incomparable states contains the same fragments of the simplex Δ_{N-1}. In a typical case C occupies regions close to the boundary of Δ_{N-1}, so one may expect that the larger the dimension N, the larger the relative volume of C. This is indeed the case, and in the limit $N \to \infty$ two generic pure states of the $N \times N$ system (or two generic density matrices of size N) are incomparable [228].

---

[13] A related partial order induced by LOCC into the space of mixed states is analysed in [419].
[14] For $N \geq 4$ there exists an effect of *entanglement catalysis* [244, 504, 789] that allows one to obtain certain incomparable states in the presence of additional entangled states.

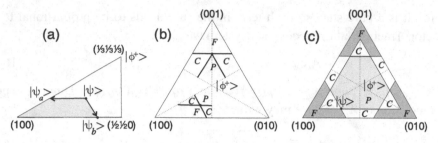

Figure 16.5 Simplex of Schmidt coefficients Δ_2 for 3×3 pure states: the corners represent separable states and the centre the maximally entangled state $|\phi^+\rangle$. Panels (a)-(c) show '*future*' and '*past*' zones with respect to LOCC and are analogous to these in Figure 13.5, but the direction of the arrow of time is reversed.

Nielsen's theorem shows that two pure states are interconvertible by means of LOCC if and only if they have the same Schmidt vectors. The point is simply that two vectors that majorise each other must be equal up to permutations. Hence pure states of a bipartite system equivalent with respect to LOCC are also equivalent with respect to local unitaries, $\psi \equiv_{\text{LOCC}} \phi \Leftrightarrow \psi \equiv_{\text{LU}} \phi$.

Consider now a single copy of an entangled state $|\psi\rangle$, majorised by the maximally entangled state $|\phi^+\rangle$. Due to the theorem of Nielsen we cannot transform $|\psi\rangle$ into $|\phi^+\rangle$ by any deterministic LOCC operation. The problem changes if we consider a broader class of operations, in which a pure state is transformed by a *probabilistic* local operation into a mixture, $|\psi\rangle\langle\psi| \to \sum_i p_i \rho_i$. Such operations are called *stochastic* LOCC, written SLOCC. A state ρ is called SLOCC convertible to σ if there exists a LOCC map Φ, not necessarily trace preserving, such that $\sigma = \Phi(\rho)/\text{Tr}\Phi(\rho)$, where $p = \text{Tr}\Phi(\rho) > 0$.

Two states of an $N_A \times N_B$ system are said to be SLOCC equivalent if they can be transformed into each other using LOCC *with some probability*, but not necessarily with certainty. Post-selection, depending on the outcomes of various measurements, is now allowed. The protocol splits probabilistically into many branches, and we keep only the outcome at the end of some branch that suits us. In each individual branch we use only one Kraus operator per system, so the SLOCC transformations act on a pure state as

$$|\psi\rangle \to A_1 \otimes A_2 |\psi\rangle. \tag{16.44}$$

Here we do not need to care about normalisation, in contrast to the formulation of the measurement postulate in Section 10.1.

For SLOCC equivalent states we must insist that both operators are invertible, but otherwise they are unrestricted. So we arrive at the conclusion that two pure states are equivalent with respect to SLOCC if and only if there exists an invertible local operation represented by two *invertible* matrices L_1 and L_2 describing the local transformation,

$$|\psi\rangle \sim_{\text{SLOCC}} |\phi\rangle = L_1 \otimes L_2 |\psi\rangle. \tag{16.45}$$

Since we do not care about normalisation here, we can rescale the operators so that they have unit determinant. Hence the group that governs SLOCC equivalence is the composite special linear group $SL(N_A, \mathbb{C}) \times SL(N_B, \mathbb{C})$.

There is mathematical order and method here, since the special linear group $SL(N, \mathbb{C})$ is the *complexification* of $SU(N)$. That is to say, if the group elements of $SU(N)$ are given in a form like (3.140), then the elements of $SL(N, \mathbb{C})$ are obtained by making the coordinates complex.

Applying a SLOCC operation to many copies of an entangled state $|\psi\rangle$ one can generate a maximally entangled state locally by proceeding in a probabilistic way: a suitable local operation produces $|\phi^+\rangle$ with probability p. It will produce unwanted states with probability $1 - p$, but we are prepared to throw them away. Consider a SLOCC probabilistic scheme to convert a pure state $|\psi\rangle$ into a target $|\phi\rangle$ with probability $p > 0$. In order to compare two vectors such that the sums of their components are not equal it is convenient to use a natural generalisation of relation (2.1) called *submajorisation*, written $x \prec^w y$. Let p_c be the maximal number such that the following submajorisation relation holds,

$$\vec{\lambda}_\psi \prec^w p_c \vec{\lambda}_\phi. \tag{16.46}$$

The theorem of Nielsen, adopted to the case of probabilistic transformations, implies that the probability p cannot be larger than p_c. The optimal conversion strategy for which $p = p_c$ was explicitly constructed by Vidal [929].

Note that the Schmidt rank cannot increase during any SLOCC conversion scheme [618], so any two SLOCC equivalent states have the same Schmidt rank. If the rank of the target state $|\phi\rangle$ is larger than the Schmidt rank of $|\psi\rangle$, then $p_c = 0$ and the probabilistic conversion cannot be performed.

This situation is illustrated in Figure 16.6, which shows the probability of accessing different regions of the Schmidt simplex for pure states of a 3×3 system for three different initial states. The shape of the black region ($p = 1$ represents deterministic transformations) is identical with the set 'future' in Figure 16.5. The more entangled the final state (closer to the maximally entangled state – in the centre of the triangle), the smaller the probability p of a successful transformation. Observe that the contour lines (at $p = 0.2, 0.4, 0.6$ and 0.8) are constructed from the isoentropy lines S_q for $q \to 0$ and $q \to \infty$ (compare Figure 2.14). The entropy S_0 is a function of the Schmidt rank r. All points inside the triangle, with $r = 3$, represent pure states equivalent with respect to SLOCC operations. Another SLOCC class

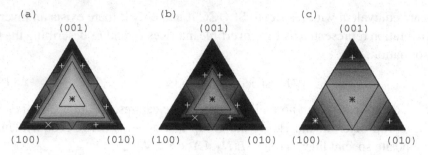

Figure 16.6 Probability of the optimal local conversion of an initial state $|\psi\rangle$ (white ×) of the 3 × 3 problem into a given pure state represented by a point in the Schmidt simplex. Initial Schmidt vector $\vec{\lambda}_\psi$ is (a) $(0.7, 0.25, 0.05)$, (b) $(0.6, 0.27, 0.03)$, and (c) $(0.8, 0.1, 0.1)$. Due to the degeneracy of λ in the latter case there exist only three interconvertible states in Δ_2, represented by (+).

corresponds to points at the edges of the Schmidt triangle, with $r = 2$. The last class with $r = 1$ corresponds to the corners of the simplex and separable states.

To describe the rules governing the changes of entanglement under local operations it is convenient to introduce a dedicated notion.

Definition 16.1 *A quantity μ which is invariant under local unitary operations and decreases, on average, under transformation belonging to the SLOCC class is called an entanglement monotone [930].*

The words 'on average' refer to the general case, in which a pure state is transformed by a probabilistic local operation into a mixture,

$$\rho \rightarrow \sum_i p_i \rho_i \quad \Rightarrow \quad \mu(\rho) \geq \sum_i p_i \mu(\rho_i). \tag{16.47}$$

We restrict our attention to the non-increasing monotones, which reflect the key paradigm of any entanglement measure: entanglement cannot increase under the action of local operations.

This property stands behind another generalisation of the theorem of Nielsen, established by Jonathan and Plenio [505], which says that a probabilistic transformation of a pure state $|\psi\rangle$ into an ensemble $\{p_j, |\phi_j\rangle\}$ by means of LOCC transformations is possible if and only if the majorisation relation holds for the averaged Schmidt vector,

$$\vec{\lambda}_\psi \prec \langle \vec{\lambda}_\phi \rangle \equiv \sum_j p_j \vec{\lambda}_{\phi_j}. \tag{16.48}$$

Nielsen's theorem implies that any Schur concave function of the Schmidt vector is an entanglement monotone. In particular, this crucial property is shared by all Rényi entropies of entanglement $E_q(\vec{\lambda})$ including the entanglement entropy (16.35). To ensure a complete description of a pure state of the $N \times N$ problem one may choose $E_1, E_2, \ldots, E_{N-1}$. Other families of entanglement monotones include partial sums of Schmidt coefficients ordered decreasingly [930], subentropy [510, 664, 247] and symmetric polynomials in Schmidt coefficients. See Problem 16.2.

Since the maximally entangled state is majorised by all pure states, it cannot be reached from other states by any deterministic local transformation. Is it at all possible to create it locally with probability one? A possible clue is hidden in the word 'average' contained in the definition of entanglement monotone.

To make use of this option we will relax the assumption that we analyse a single copy of a given state only, and treat entanglement as a kind of quantum resource. Having two glasses, each half-filled with beer, we can get one glass full and the other empty (under the sometimes realistic assumption that during such an operation we will manage not to spill a single drop of the liquid). Similarly, having at our disposal n copies of a generic pure state $|\psi\rangle$ we can locally create out of them m maximally entangled states $|\phi^+\rangle$, at the expense of the remaining $n - m$ states becoming separable.

Such protocols [115, 618] are called *entanglement concentration*. This local operation is reversible, and the reverse process of transforming m maximally entangled states and $n - m$ separable states into n entangled states is called *entanglement dilution*. The asymptotic ratio $m/n \leq 1$ obtained by an optimal concentration protocol is called *distillable entanglement* [115] of the state $|\psi\rangle$. See Problem 16.3. In agreement with (16.48), a reversible local transformation between two ensembles of states $\{q_i, |\psi_i\rangle\}$ and $\{p_j, |\phi_j\rangle\}$ exists if their averaged Schmidt vectors are equal, $\langle\vec{\lambda}_\psi\rangle = \langle\vec{\lambda}_\phi\rangle$.

In the above discussion of transformations of quantum entanglement we have deliberately stuck to the paradigm of performing local operations only, as they are easier to realise in an experiment. But to create entanglement in a composite system a physical interaction between subsystems is required, which can be described by non-local quantum gates. Just as there is a continuous spectrum of entangled states, interpolating between separable and maximally entangled states, so there are unitary gates with different degrees of non-locality. For any bipartite unitary gate U its non-local properties can be quantified by applying the operator Schmidt decomposition (10.30) – see [277, 394] – and other techniques [684, 693, 1000]. A canonical form of a two-qubit gate is known [522, 558, 640], and various measures of entangling power of U have been proposed [995, 993, 210]. We cannot analyse these issues any further here, but we emphasise that non-locality of a unitary gate is necessary to create entanglement, but not sufficient. For instance, the unitary *swap* gate (10.35), which exchanges the subsystems, is highly non-local, though its action cannot change the entanglement in the system.

The envoi of this section is: entanglement of a pure state of any bipartite system may be fully characterised by its Schmidt decomposition. All entanglement monotones are functions of the Schmidt coefficients.

16.5 A first look at entangled mixed states

This is a good time to look again at mixed states: in this section we shall analyse bipartite density matrices acting on a composite Hilbert space $\mathcal{H} = \mathcal{H}_A \otimes \mathcal{H}_B$ of finite dimension $N_A N_B$. To simplify things we will mostly consider (often without saying so explicitly) subsystems having the same dimension, and then $N_A = N_B = N$.

A state is called a *product state* if it has the form $\rho = \rho_A \otimes \rho_B$. Such states are easy to recognise. Let $\{\sigma_i\}_{i=1}^{N^2-1}$ be the traceless generators of $SU(N)$. Then we can write[15] the state in the **Fano form** [311]

$$\rho = \frac{1}{N^2}\left[\mathbb{1}_{N^2} + \sum_{i=1}^{N^2-1} \tau_i^A \sigma_i \otimes \mathbb{1}_N + \sum_{j=1}^{N^2-1} \tau_j^B \mathbb{1}_N \otimes \sigma_j + \sum_{i=1}^{N^2-1}\sum_{j=1}^{N^2-1} \beta_{ij}\sigma_i \otimes \sigma_j \right].$$

(16.49)

Here $\vec{\tau}^A$ and $\vec{\tau}^B$ are Bloch vectors of the partially reduced states. Keeping them constant and varying the matrix β (in such a way that the positivity of ρ is preserved) we obtain a large family of locally indistinguishable states. The state is a product state if $\beta_{ij} = \tau_i^A \tau_j^B$. If not, correlations are present between the two quantum subsystems.

We would like to single out the cases where these correlations are purely classical.[16] We therefore adopt the following.

Definition 16.2 *A state ρ is called separable if it can be represented as a convex sum of product states,*

$$\rho_{sep} = \sum_{j=1}^{M} q_j \rho_j^A \otimes \rho_j^B,$$

(16.50)

where ρ_j^A acts in $\mathcal{H}_A$ and ρ_j^B acts in $\mathcal{H}_B$, the weights are positive, $q_j \geq 0$, and sum to unity, $\sum_{j=1}^{M} q_j = 1$. A state which is not separable is called entangled.

Separable states, correlated or not, can be constructed locally if assisted by classical communication. Entangled states cannot. The separable states form the

---

[15] For other useful parametrizations of the set of mixed states consult [179, 185].
[16] The definition to follow was first stated clearly by Werner (1989) [958].

convex body $\mathcal{M}_S$, whose extreme points are separable pure states. However, even though we know all its extreme points, this body is many orders of magnitude more difficult to describe than is the set $\mathcal{M}$ of all states. Given a Hermitian matrix ρ, we can check if it belongs to $\mathcal{M}$ by checking if all its non-zero eigenvalues are positive. But separability is not determined by spectral properties – in Problem 16.5 we give an example of a pair of isospectral states, one of which is separable, the other not. Checking if a given density matrix ρ belongs to $\mathcal{M}_S$ can be a very hard problem.

A simple, albeit amazingly powerful criterion to decide if a state is entangled was found by Peres [731]. He noticed that the partial transpose of a separable state must be positive,

$$\rho_{\text{sep}}^{T_A} \equiv (T \otimes \mathbb{1})(\rho_{\text{sep}}) = \sum_j q_j \, (\rho_j^A)^T \otimes \rho_j^B \geq 0. \tag{16.51}$$

The same is true if we apply the map $T_B = (\mathbb{1} \otimes T)$. Thus any separable state has a positive partial transpose (is PPT), so we obtain directly:

PPT criterion. *If $\rho^{T_A} \not\geq 0$, the state ρ is entangled.*

As we shall see below the PPT criterion works in both directions only if $\dim(\mathcal{H}) \leq 6$, so there is a need for other separability criteria. However, the PPT criterion is extremely easy to use: all we need to do is to perform the partial transposition of the density matrix in question, and check if all eigenvalues are non-negative. Partial transpositions have already been defined in (10.34), but let us have a look at how these operations act on a 4×4 block matrix, with matrix elements ordered according to the prescription of Chapter 10:

$$X = \begin{bmatrix} A & B \\ C & D \end{bmatrix}, \quad X^{T_B} \equiv \begin{bmatrix} A^T & B^T \\ C^T & D^T \end{bmatrix}, \quad X^{T_A} \equiv \begin{bmatrix} A & C \\ B & D \end{bmatrix}. \tag{16.52}$$

Geometrically, partial transposition (PT) is a reflection of the set of density matrices through a plane formed by PT invariant states. In the Fano form (16.49) this means that all anti-symmetric generators σ_i acting on one of the factors switches sign. In the two-qubit case we see that the PT invariant plane has 11 dimensions.

To watch the PPT criterion in action we will introduce a one-parameter family of states for it to act on. We choose the *Werner states* [958], defined as a weighted mixture of projectors onto the symmetric and anti-symmetric subspaces. Thus

$$\rho_W(p) = (1-p)\frac{2}{N(N+1)}\Pi_{\text{sym}} + p\frac{2}{N(N-1)}\Pi_{\text{asym}}. \tag{16.53}$$

These states have a somewhat strategic position because they are invariant under *twirling*, that is to say that $\rho_W(p) = (U \otimes U)\rho_W(p)(U \otimes U)^\dagger$ for every unitary

$U \in U(N)$. Moreover they are the only states enjoying this property. It is worth noticing that there exists a linear map (a *twirling channel*) that takes an arbitrary density matrix ρ to a Werner state, namely

$$\Psi_{UU}(\rho) = \int_{U(N)} dU(U \otimes U)\rho(U \otimes U)^{\dagger}. \qquad (16.54)$$

On the right-hand side we perform an integral over the entire group manifold of $U(N)$, so that an invariant state results. We recall from Section 12.7 that the integral can be replaced by a finite sum, using a unitary 2-design.

In higher dimensions all Werner states are mixed, but for the two-qubit system they sit on a line segment going from the singlet pure state $|\psi^-\rangle\langle\psi^-|$ and through the maximally mixed state. That is to say that for two qubits we can write the family of Werner states as

$$\rho_x = x|\psi^-\rangle\langle\psi^-| + (1-x)\frac{\mathbb{1}}{4} = \frac{1}{4}\begin{bmatrix} 1-x & 0 & 0 & 0 \\ 0 & 1+x & -2x & 0 \\ 0 & -2x & 1+x & 0 \\ 0 & 0 & 0 & 1-x \end{bmatrix}. \qquad (16.55)$$

The spectrum of ρ_x is $\frac{1}{4}\{1+3x, 1-x, 1-x, 1-x\}$, so this is a density matrix if $x \in [-1/3, 1]$. Taking the partial transpose, according to the scheme in Eq. (16.52), we find that $\rho_x^{T_A} = \rho_x^{T_B}$ has the spectrum $\frac{1}{4}\{1+x, 1+x, 1+x, 1-3x\}$. This matrix is positive definite if and only if $x \leq 1/3$, hence Werner states are entangled for $x > 1/3$. It is interesting to observe that the critical state $\rho_{1/3} \in \partial\mathcal{M}_S$ sits at the distance $r_{in} = 1/\sqrt{24}$ from the maximally mixed state, so it sits on the insphere, the maximal sphere that one can inscribe into the convex set of density matrices. With a little effort one can show that $\rho_{1/3}$ is not only a PPT state, it is separable too.

The set of all bipartite states can be divided into *PPT states* (semi-positive partial transpose, $\rho^{T_A} \geq 0$) and *NPT states* (ρ^{T_A} has a negative eigenvalue). NPT states are always entangled. The question is whether PPT states are always separable. The answer is actually 'yes' for two qubits (and also for the case $N_A = 2, N_B = 3$), but beyond that the answer is 'no'. To see this we go on another detour, and introduce *unextendible product bases* (UPB).[17]

In Section 12.3, we constructed entire bases consisting of maximally entangled states only. Evidently bases consisting of only separable states also exist. In fact, in Section 9.1 we used them to define the notion of tensor product spaces. That story has a twist, however. Consider the five product states defined by the orthogonality

---

[17] Unextendible product bases were introduced by Bennett et al. [120] precisely in order to find examples of entangled PPT states. Since then UPBs have been much studied [29, 268, 740].

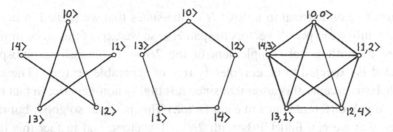

Figure 16.7 Considered as graphs, the pentagram and the pentagon are isomor-phic. The pentagon is a *cyclic* graph and has the pentagram as its *complement*, in the sense that we obtain the complete graph if we superpose them. (The cyclic graph is isomorphic to its complement only if the number of vertices equals five.) If the pentagram is realised in $\mathcal{H}_A$ and the pentagon in $\mathcal{H}_B$, the complete graph can be taken to represent five orthogonal product states in $\mathcal{H}_A \otimes \mathcal{H}_B$.

graphs in Figure 16.7. It is not difficult to realise the pentagram as an orthogonality graph in $\mathbb{R}^3$. Using the labelling given in the figure, choose

$$|j\rangle = \begin{pmatrix} \sin\theta \cos\frac{2\pi j}{5} \\ \sin\theta \sin\frac{2\pi j}{5} \\ \cos\theta \end{pmatrix}, \qquad j = 0, \dots, 4. \qquad (16.56)$$

This corresponds to drawing the pentagram on the circular base of a cone, and choosing the opening angle $\theta \approx 48°$ in order to make the five rays from the apex to the vertices orthogonal according to the pattern in the graph. This elegant solution is by no means unique – in $\mathcal{H}_3$ there is a two real parameter family of unitarily inequiv-alent realisations of the pentagram. (This counting [584] will become important as we proceed.)

Now put a pentagram and a pentagon together, in the manner of Figure 16.7. When five distinct vectors realise the pentagram one finds that any three of them span $\mathcal{H}_3$, and similarly for the pentagon. But then there cannot exist a product vector $|\psi_A\rangle \otimes |\psi_B\rangle$ orthogonal to the five product vectors corresponding to the complete graph, since at least one of its factors would have to be orthogonal to three vectors spanning $\mathcal{H}_3$. Hence our five separable states form an unextendible product basis. Moreover it follows that the four-dimensional subspace orthogonal to these five vectors cannot contain a single separable state. It is a *completely entangled subspace*.[18]

We must ask how our example generalises. Assume that we are in the Hilbert space $\mathcal{H}_N \otimes \mathcal{H}_N$, pick $2N - 1$ states at random in each factor, and form $2N - 1$ product states. If an additional product state is orthogonal to all of them, one of its

---

[18] Completely entangled subspaces appeared, in slight disguise, in Section 4.3. For discussion in depth we refer elsewhere [239, 840].

factors must be orthogonal to at least N of the states that we picked in that factor. But generically we expect N vectors to span $\mathcal{H}_N$, so we expect this to be impossible. Therefore the orthogonal complement of the $2N - 1$ dimensional subspace we constructed is expected to be completely free of separable vectors. (The standard separable basis escapes this argument since it is highly non-generic: in fact it makes do with only N different states in each factor.) This is so far so good, but does not guarantee that we can find UPBs with $2N - 1$ vectors, and in fact this does not always work. Indeed there are no orthogonal UPB in $2 \times N$ systems [120]. See also Problem 16.1.

Now we can put a UPB to use. Let us suppose it contains M orthogonal product vectors $|u_j\rangle \otimes |v_j\rangle$. Introduce a projector onto the entangled subspace that is orthogonal to all of them,

$$\rho_{\text{UPB}} = \frac{1}{N^2 - M} \left(\mathbb{1} - \sum_{j=0}^{M-1} |u_j\rangle\langle u_j| \otimes |v_j\rangle\langle v_j| \right). \tag{16.57}$$

By construction this is a density matrix which cannot lie in the convex hull of any separable states, so it is entangled. It is obviously a PPT state. So we have found an entangled PPT state. In the two-qutrit case it is a density matrix of rank four.

But its entanglement is of a peculiar kind. In Section 16.4 we discussed the distillation of entanglement: the task is to find a LOCC protocol that allows the two parties to extract a maximally entangled state out of a number of copies of a given entangled state. We found that all pure states are *distillable* in this sense. Distillation protocols are also available for mixed states [118], and indeed all two-qubit entangled states are distillable [456]. However, it was shown by Horodeccy[19] that entangled PPT states cannot be distilled [457]. This crucial observation gave rise to a new definition:

Definition 16.3 *Any entangled state of a bipartite system which cannot be distilled into a maximally entangled state is called bound entangled.*

Entanglement which can be locally transferred into singlet states and used in teleportation protocols is called *free*, while entanglement confined in any PPT entangled state is bound. PPT entangled states exist for all systems described in composite Hilbert spaces of size eight or larger. They contain a weak form of entanglement which cannot be directly used for the tasks of quantum information processing. However, bound entanglement can be activated in the sense that its presence makes possible some teleportation and concentration protocols up to an accuracy impossible to achieve without it [466].

---

[19] A person interested in languages can guess that the plural form is being used here.

Given that all entangled PPT states are bound entangled, we have to ask: do there exist bound entangled NPT states? It was shown early on that this question can be reduced to a question about Werner states [454], but despite its importance the question remains open.[20] Meanwhile this issue provides a motivation for understanding the PPT states as well as we can.

The first observation is that the PPT states form a convex body. We denote it by $\mathcal{M}_{\text{PPT}}$. It arises as an intersection of the entire body of mixed states with its reflection induced by partial transpose,

$$\mathcal{M}_{\text{PPT}} = \mathcal{M} \cap T_A(\mathcal{M}). \tag{16.58}$$

See Figure 11.2b. Since the insphere of $\mathcal{M}$ – its largest inscribed ball – goes onto itself under reflection, it is equal to the insphere of $\mathcal{M}_{\text{PPT}}$, and the dimension of the two bodies must coincide. In Section 15.3 we learned that $\mathcal{M}$ is a body of constant height, meaning that all its boundary points lie on faces that are tangent to the insphere. Also this property is conserved under reflection and intersection. For the boundaries of $\mathcal{M}$ and its reflection, one can show that the area of the intersection between them vanishes.[21]

Next one would like to find the extreme points of $\mathcal{M}_{\text{PPT}}$. This includes the separable pure states, though there must be more. It is known that there are no entangled PPT states of rank three or less [467]. This is true in general, but let us concentrate on the 3×3 system. There the UPB construction gave us an example of rank four, and it must be an extreme point (otherwise it would be a mixture of separable rank one states). In fact, because the realisation of the pentagram is not unique, we obtained a four-parameter family of unitarily inequivalent extreme points. Acting with local unitaries on any such state will give an entire orbit of extreme rank four states. But one can relax the condition that the UPB be orthonormal: convexity properties and matrix ranks are preserved by the local linear group $SL(3, \mathbb{C}) \times SL(3, \mathbb{C})$. Scalar products are not preserved, but then they are not needed in a study of convexity properties. Thus, starting from the example in Eq. (16.57), with $N = 3$, $M = 5$, we obtain an entire orbit of extreme rank four states

$$\rho_{\text{extr}} = (L_1 \otimes L_2)\rho_{\text{UPB}}(L_1 \otimes L_2)^\dagger, \tag{16.59}$$

where L_i is an $SL(3, \mathbb{C})$ matrix. (One can see by inspection of this formula that transformations belonging to the special linear group preserve positivity, though they preserve traces only if $L^\dagger = L^{-1}$.) The group $SL(3, \mathbb{C})$ has eight dimensions, and the end result is that one obtains a $2 \cdot 8 + 4 = 36$ dimensional family of

---

[20] We refer to Clarisse [225] and to Pankowski et al. [715] for overviews and complete references.

[21] This theorem is due to Szarek et al. [878]. From it one can deduce that the probability (in the Hilbert–Schmidt sense) of finding a PPT state in the interior of $\mathcal{M}$ is twice that of finding a PPT state at its boundary, as found numerically by Slater who found many relations of this type [844].

entangled rank PPT states. Remarkably, it has been proved that this includes every extreme point of rank four.[22] It is interesting that each such extreme PPT state is connected to a four-dimensional subspace of $\mathcal{H}_9$. The space of all four-dimensional subspaces of $\mathcal{H}_9$ is a Grassmannian manifold (see Section 4.9) of dimension $9^2 - 4^4 - 5^2 = 40$, so only rather exceptional subspaces appear in the construction.

Of course this is not the end of the story, even for the 3×3 system, because extreme points of higher rank will certainly occur. Lists of admissible ranks are available for several low dimensional systems [585], but our understanding of the underlying geometric structure is still incomplete [74, 208].

In Section 16.8 we will address the issue of how entanglement is to be, not only recognised, but quantified. Then we will discuss the *negativity* of a state ρ, which depends on the sum of the negative eigenvalues of ρ^{T_A}. Meanwhile, we observe that the spectrum of ρ^{T_A} of a maximally entangled state in the $N \times N$ system has $N(N-1)/2$ negative eigenvalues equal to $-1/N$, and $N(N+1)/2$ positive eigenvalues $1/N$. (It is enough to check this for $\rho = |\phi^+\rangle\langle\phi^+|$.) Actually, for a state ρ of an $N_A \times N_B$ system the matrix ρ^{T_A} can have up to $(N_A - 1)(N_B - 1)$ negative eigenvalues [208, 503], and its spectrum is contained [765] in the interval $[-1/2, 1]$. More can be said about the generic behaviour of eigenvalues for large dimensions. From our discussion in Section 15.8 we know that, for large N, the spectrum of an $N \times N$ mixed state ρ is typically contained in an interval $[0, 4/N^2]$, and described by the Marchenko–Pastur distribution, that is by Eq. (15.62), with $x = N^2\lambda$. If instead λ is an eigenvalue of ρ^{T_A}, it follows from the measure concentration effect that x is asymptotically described by the shifted semicircular law [67],

$$P_{\text{PT}}(x) = \frac{1}{2\pi}\sqrt{4 - (x-1)^2}, \tag{16.60}$$

centred at $x = 1$ and supported in $[-1, 3]$. This unexpected result allows one to estimate an average degree of entanglement for a typical mixed state [133]. See Problem 16.10.

16.6 Separability criteria

The PPT separability criterion is not the only one in the world. Other separability criteria can be divided into three classes: **(A)** sufficient and necessary, but not practically usable; **(B)** easy to use, but only necessary (or only sufficient); and **(C)** one criterion with a special status.

Before reviewing the most important of them let us introduce one more notion often used in the physical literature.

---

[22] This conclusion, supported by strong numerical evidence, was reached by Leinaas, Myrheim, and Sollid [584]. The final proof involves a fair amount of algebraic geometry, and is due to Skowronek [840] and to Chen and Đoković [209].

Definition 16.4 *An Hermitian operator* W *is called an* entanglement witness *for a given entangled state* ρ *if* $\mathrm{Tr}\rho W < 0$ *and* $\mathrm{Tr}\rho\sigma \geq 0$ *for all separable* σ *[455, 892]. For convenience the normalisation* $\mathrm{Tr}W = 1$ *is assumed.*

Horodeccy [455] proved a useful

Witness lemma. *For any entangled state* ρ *there exists an entanglement witness* W.

This is the Hahn–Banach separation theorem (Section 1.1) in slight disguise.

It is instructive to see the direct relation with the dual cones construction discussed in Chapter 11: any witness operator is proportional to a dynamical matrix, $W = D_\Phi/N$, corresponding to a non-completely positive map Φ. Since D_Φ is block positive (positive on product states), the condition $\mathrm{Tr}\,W\sigma \geq 0$ holds for all separable states for which the decomposition (16.50) exists. Conversely, a state ρ is separable if $\mathrm{Tr}\,W\rho \geq 0$ for all block positive W. This is just the definition (11.17) of a super-positive map Ψ. We arrive, therefore, at a key observation: the set $\mathcal{SP}$ of super-positive maps is isomorphic with the set $\mathcal{M}_S$ of separable states by the Jamiołkowski isomorphism, $\rho = D_\Psi/N$.

An intricate link between positive maps and the separability problem is made clear in our first criterion (which could have been taken as an alternative definition of separable states):

A1. Positive maps criterion [455]. *A state* ρ *is separable if and only if* $\rho' = (\Phi \otimes \mathbb{1})\rho$ *is positive for all positive maps* Φ.

To demonstrate that this condition is necessary, act with an extended map on the separable state (16.50),

$$\left(\Phi \otimes \mathbb{1}\right)\left(\sum_j q_j\, \rho_j^A \otimes \rho_j^B\right) = \sum_j q_j\, \Phi(\rho_j^A) \otimes \rho_j^B \geq 0. \qquad (16.61)$$

Due to positivity of Φ the above combination of positive operators is positive. To prove sufficiency, assume that $\rho' = (\Phi \otimes \mathbb{1})\rho$ is positive. Thus $\mathrm{Tr}\rho'P \geq 0$ for any projector P. Setting $P = P_+ = |\phi^+\rangle\langle\phi^+|$ and making use of the adjoint map we get $\mathrm{Tr}\rho(\Phi \otimes \mathbb{1})P_+ = \frac{1}{N}\mathrm{Tr}\rho D_\Phi \geq 0$. Since this property holds for all positive maps Φ, it implies separability of ρ due to the witness lemma $\square$.

The positive maps criterion holds also if the map acts on the second subsystem. However, this criterion is not easy to use: one needs to verify that the required inequality is satisfied for *all* positive maps. On the positive side, any quantum map which is positive but not completely positive allows us to identify some entangled states. We know an example from Section 16.5:

B1. PPT criterion. *If* $\rho^{T_A} \not\geq 0$, *the state* ρ *is entangled.*

However, something remarkable happens to this criterion for 2×2 and 2×3 systems. In this case any positive map is decomposable due to the Størmer–Woronowicz theorem and may be written as a convex combination of a CP map and a CcP map, which involves the transposition T (see Section 11.1). Hence, to apply criterion **A1** it is enough to perform a single check, working with the partial transposition $T_A = (T \otimes \mathbb{1})$. In this way we obtain

C. Peres–Horodeccy criterion [726, 455]. *A state ρ acting on the composite Hilbert spaces $\mathcal{H}_2 \otimes \mathcal{H}_2$ or $\mathcal{H}_2 \otimes \mathcal{H}_3$ is separable if and only if $\rho^{T_A} \geq 0$.*

Let us return to the discussion of a general bipartite system. A super-positive map Φ is related by the Jamiołkowski isomorphism (11.22) with a separable state. Hence $(\Phi \otimes \mathbb{1})$ acting on the maximally entangled state $|\phi^+\rangle\langle\phi^+|$ is separable. It is then not surprising that $\rho' = (\Phi \otimes \mathbb{1})\rho$ becomes separable for an arbitrary state ρ [463], which explains why SP maps are also called *entanglement breaking channels*. Furthermore, due to the positive maps criterion $(\Psi \otimes \mathbb{1})\rho' \geq 0$ for any positive map Ψ. In this way we have arrived at the first of three duality conditions equivalent to (11.16–11.18),

$$\{\Phi \in \mathcal{SP}\} \quad \Leftrightarrow \quad \Psi \cdot \Phi \in \mathcal{CP} \quad \text{for all} \quad \Psi \in \mathcal{P}, \tag{16.62}$$

$$\{\Phi \in \mathcal{CP}\} \quad \Leftrightarrow \quad \Psi \cdot \Phi \in \mathcal{CP} \quad \text{for all} \quad \Psi \in \mathcal{CP}, \tag{16.63}$$

$$\{\Phi \in \mathcal{P}\} \quad \Leftrightarrow \quad \Psi \cdot \Phi \in \mathcal{CP} \quad \text{for all} \quad \Psi \in \mathcal{SP}. \tag{16.64}$$

The second one reflects the fact that a composition of two CP maps is CP, while the third one is dual to the first.

Due to the Størmer–Woronowicz theorem and the Peres–Horodeccy criterion, all PPT states for 2×2 and 2×3 problems are separable (hence any PPT-inducing map is SP) while all NPT states are entangled. In higher dimensions there exist non-decomposable maps and consequently bound entangled states with positive partial transpose.[23]

B2. Range criterion [464]. *If a state ρ is separable, then there exists a set of pure product states such that $|\psi_i \otimes \phi_i\rangle$ span the range of ρ and $T_B(|\psi_i \otimes \phi_i\rangle)$ span the range of ρ^{T_B}.*

The action of the partial transposition on a product state gives $|\psi_i \otimes \phi_i^*\rangle$, where $*$ denotes complex conjugation in the standard basis. This criterion, proved by Paweł Horodecki [464], allowed him to identify the first PPT entangled state in the 2×4 system. Entanglement of ρ was detected by showing that none of the product states from the range of ρ, if partially conjugated, belong to the range of ρ^{T_B}. The range criterion implies that the PPT states constructed from UPB in (16.57) are entangled.

---

[23] Not completely positive maps are applied in [389, 109] to identify bound entangled states. Conversely, PPT entangled states are used in [892] to find non-decomposable positive maps.

B3. Reduction criterion [201, 454]. *If a state ρ is separable then the reduced states* $\rho_A = \mathrm{Tr}_B\rho$ *and* $\rho_B = \mathrm{Tr}_A\rho$ *satisfy*

$$\rho_A \otimes \mathbb{1} - \rho \geq 0 \quad \text{and} \quad \mathbb{1} \otimes \rho_B - \rho \geq 0. \tag{16.65}$$

This statement follows directly from the positive maps criterion with the map $\Phi_R(\sigma) = (\mathrm{Tr}\sigma)\mathbb{1} - \sigma$ applied to the first or the second subsystem. Computing the dynamical matrix for this map composed with the transposition, $\Phi' = \Phi T$, we find that $D_{\Phi'} \geq 0$, hence Φ is CcP and (trivially) decomposable. Thus the reduction criterion cannot be stronger[24] than the PPT criterion. There exists, however, a good reason to pay some attention to this criterion: the Horodeccy brothers have shown [454] that any state ρ violating (16.65) is distillable.

Any positive non-decomposable map applied into criterion **A1** allows us to identify some bound entangled states. A useful example is provided by the map studied independently by Breuer [169] and Hall [393]. This is a modification of the reduction map which acts on a state of an even dimension $2k \geq 4$,

$$\Phi_U^{BH}(\sigma) = (\mathrm{Tr}\sigma)\mathbb{1} - \sigma - U\sigma^T U^\dagger. \tag{16.66}$$

Here U stands for any anti-symmetric unitary matrix, $U^T = -U$. To see that the map is positive one can apply it to a pure state and use the fact that $\langle \psi|U|\psi^*\rangle = 0$. Acting with this on a single subsystem one arrives at a statement: if $[\Phi_U^{BH} \otimes \mathbb{1}]$ $(\rho) \not\geq 0$ then the state ρ is entangled. Since the map is non-decomposable the above criterion and its generalisations [592] can detect bound entanglement too.

B4. Majorisation criterion [694]. *If a state ρ is separable, then the reduced states* ρ_A *and* ρ_B *satisfy the majorisation relations*

$$\rho \prec \rho_A \quad \text{and} \quad \rho \prec \rho_B. \tag{16.67}$$

In brief, separable states are more disordered globally than locally. To prove this criterion one needs to find a bistochastic matrix B such that the spectra satisfy $\vec{\lambda} = B\vec{\lambda}_A$. See Problem 16.4. The majorisation relation implies that any Schur convex function satisfies the inequality (2.8). For Schur concave functions the direction of the inequality changes. In particular, this implies the following.

B5. Entropy criterion. *If a state ρ is separable, then the Rényi entropies fulfil*

$$S_q(\rho) \geq S_q(\rho_A) \quad \text{and} \quad S_q(\rho) \geq S_q(\rho_B) \quad \text{for} \quad q \geq 0. \tag{16.68}$$

The entropy criterion was originally formulated for $q = 1$ [469]. This statement may be equivalently expressed in terms of the *conditional entropy*[25] $S(A|B) = S(\rho_{AB}) - S(\rho_A)$: for any separable bipartite state $S(A|B)$ is non-negative. Thus

---

[24] This is the case for the *generalised reduction criterion* proposed in [18].
[25] The opposite quantity, $-S(A|B)$, is called *coherent quantum information* [810].

negative conditional entropy of a state ρ_{AB} confirms its entanglement [468, 810, 200]. The entropy criterion was proved for $q = 2$ in [470] and later formulated also for the Havrda–Charvat–Tsallis entropy (2.77) [3, 900, 763, 777]. Its combination with the entropic uncertainty relations of Maassen and Uffink [631] provides other separability criteria [374]. Yet another family of entropic separability criteria was obtained by applying non-decomposable positive maps [75].

A2. Contraction criterion. *A bipartite state ρ is separable if and only if any extended trace preserving positive map acts as a (weak) contraction in sense of the trace norm,*

$$||\rho'||_{\text{Tr}} = ||(\mathbb{1} \otimes \Phi)\rho||_{\text{Tr}} \leq ||\rho||_{\text{Tr}} = \text{Tr}\rho = 1. \qquad (16.69)$$

This criterion was formulated in [460] based on earlier papers [780, 206]. To prove it notice that the sufficiency follows from the positive map criterion: since $\text{Tr}\rho' = 1$, then $||\rho'||_{\text{Tr}} \leq 1$ implies that $\rho' \geq 0$. To show the converse consider a normalised product state $\rho = \rho^A \otimes \rho^B$. Any trace preserving positive map Φ acts as an isometry in the sense of the trace norm, and the same is true for the extended map,

$$||\rho'||_{\text{Tr}} = \left|\left|(\mathbb{1} \otimes \Phi)(\rho^A \otimes \rho^B)\right|\right|_{\text{Tr}} = ||\rho^A|| \cdot ||\Phi(\rho^B)||_{\text{Tr}} = 1. \qquad (16.70)$$

Since the trace norm is convex, $||A + B||_{\text{Tr}} \leq ||A||_{\text{Tr}} + ||B||_{\text{Tr}}$, any separable state fulfills

$$\left|\left|(\mathbb{1} \otimes \Phi)\left(\sum_i q_i(\rho_i^A \otimes \rho_i^B)\right)\right|\right|_{\text{Tr}} \leq \sum_i q_i ||\rho_i^A \otimes \Phi(\rho_i^B)||_{\text{Tr}} = \sum_i q_i = 1, \quad (16.71)$$

which ends the proof. □

Several particular cases of this criterion could be useful. Note that the celebrated PPT criterion **B1** follows directly, if the transposition T is selected as a trace preserving map Φ, since the norm condition, $||\rho^{T_A}||_{\text{Tr}} \leq 1$, implies positivity, $\rho^{T_A} \geq 0$. Moreover, one may formulate an analogous criterion for global maps Ψ, which act as contractions on any bipartite product states, $||\Psi(\rho_A \otimes \rho_B)||_{\text{Tr}} \leq 1$. As a representative example let us mention:

B6. Reshuffling criterion.[26] *If a bipartite state ρ is separable then reshuffling (10.33) does not increase its trace norm,*

$$||\rho^R||_{\text{Tr}} \leq ||\rho||_{\text{Tr}} = 1. \qquad (16.72)$$

---

[26] Called also *realignment* criterion [206] or *computable cross-norm criterion* [781]. It is related to the earlier *minimal cross-norm* criterion of Rudolph [780], which provides a necessary and sufficient condition for separability, but is in general not practical to use.

We shall start the proof by considering an arbitrary product state, $\sigma_A \otimes \sigma_B$. By construction its Schmidt decomposition consists of one term only. This implies

$$||(\sigma_A \otimes \sigma_B)^R||_{\text{Tr}} = 2\,||\sigma_A||_2 \cdot ||\sigma_B||_2 = \sqrt{\text{Tr}\,\sigma_A^2}\,\sqrt{\text{Tr}\,\sigma_B^2} \leq 1. \qquad (16.73)$$

Since the reshuffling transformation is linear, $(A + B)^R = A^R + B^R$, and the trace norm is convex, any separable state satisfies

$$\left|\left|\left(\sum_i q_i(\sigma_i^A \otimes \sigma_i^B)\right)^R\right|\right|_{\text{Tr}} \leq \sum_i q_i||(\sigma_i^A \otimes \sigma_i^B)^R||_{\text{Tr}} \leq \sum_i q_i = 1, \qquad (16.74)$$

which completes the reasoning. $\qquad\qquad\qquad\qquad\qquad\qquad\qquad\qquad\qquad\qquad\qquad\square$

In the simplest case of two qubits, the latter criterion is weaker than the PPT: examples of NPT states, the entanglement of which is not detected by reshuffling, were provided by Rudolph [781]. However, for some larger dimensional problems the reshuffling criterion becomes useful, since it is capable of detecting PPT entangled states, for which $||\rho^R||_{\text{Tr}} > 1$ [206]. For generalisations of this criterion see [40, 999, 832].

The problems as to which separability criterion is the strongest and what the implication chains among them are remain a subject of vivid research.[27] The question whether a given quantum state ρ of an $N \times N$ system is separable or entangled is indeed difficult. In 2003 Gurvits proved that this membership problem is *NP-hard*.[28] The abbreviation refers to 'non-deterministic polynomial' and the general belief is that there are no deterministic algorithms to solve such problems in a time that grows polynomially with the size N.

Due to the PPT criterion the problem is easy for $N = 2$, but for $N = 3$ Skowronek has shown [841] that it is not possible to reduce the separability problem to a finite number of operational criteria. To be precise, there does *not* exist a finite collection of positive maps Φ_i, $1 \leq i \leq M$, such that if all the M conditions

$$(\Phi_i \otimes \mathbb{1})\left[(L_1 \otimes L_2)\rho(L_1 \otimes L_2)^\dagger\right] \geq 0 \qquad (16.75)$$

hold, then the state ρ is separable. Note that we allow arbitrary invertible matrices L_1, L_2 in the formula, and there is still no finite solution.

Therefore, it is not surprising that all operationally feasible analytic criteria provide partial solutions only. On the other hand, one should appreciate practical

---

[27] Separability criteria not discussed here include a criterion obtained by expanding a mixed state in the Fourier basis [741], a relation between purity of a state and its maximal projection on a pure state [593], an application of uncertainty relations for non-local observables [337, 440, 370], reduction of the dimensionality of the problem [975], criteria based on the Bloch representation of density matrices [253, 833], a strong PPT condition [222, 387], separability eigenvalue equations [854], criteria based on covariance matrices [373], matrices of moments [666] and correlation tensors [81].

[28] This result is due to Gurvits [379]; see also Gharibian [329]. For a discussion of computational complexity classes in the context of quantum theory, consult Nielsen and Chuang [692] and a recent book by Aaronson [1].

methods constructed to decide separability numerically. Iterative algorithms based on an extension of the PPT criterion for higher dimensional spaces [269] or non-convex optimisation [290] are able to detect the entanglement in a finite number of steps. Another algorithm provides an explicit decomposition into pure product states [483], confirming that the given mixed state ρ is separable. A combination of these two approaches terminates after a finite time t and gives an inconclusive answer only if ρ belongs to the ϵ-neighbourhood of the boundary of the set of separable states. By increasing the computation time t one may make the width ϵ of the no man's land arbitrarily small.

16.7 Geometry of the set of separable states

Equipped with a broad spectrum of separability criteria, we may try to describe the structure of the set $\mathcal{M}_S$ of the separable states. This task becomes easier for the two-qubit system, for which positive partial transpose implies separability. Thus $\mathcal{M}_S^{(4)} = \mathcal{M}_{PPT}^{(4)}$, which certainly has a positive volume. The maximally mixed state is invariant with respect to partial transpose, $\rho_* = \rho_*^{T_B}$ and occupies the centre of the body $\mathcal{M}^{(4)}$. It is thus natural to ask: what is the radius of the separable ball centred at ρ_*? The answer is very appealing in the simplest, Euclidean geometry: the entire maximal ball inscribed in $\mathcal{M}^{(4)}$ is separable [1010].

The separable ball is sketched in two or three-dimensional cross-sections of $\mathcal{M}^{(4)}$ in Figure 16.8. To prove its separability[29] we shall invoke:

Mehta's Lemma [652]. *Let A be a Hermitian matrix of size D and let $\alpha = \mathrm{Tr}A/\sqrt{\mathrm{Tr}A^2}$. If $\alpha \geq \sqrt{D-1}$ then A is positive.*

Its proof begins with the observation that the traces are basis independent, so we may work in the eigenbasis of A. Let $(x_1, \ldots x_D)$ denote the spectrum of A. Assume first that one eigenvalue, say x_1, is negative. Making use of the right-hand side of the standard estimation between the l_1 and l_2-norms (with prefactor 1) of an N-vector, we infer

$$\mathrm{Tr}A = \sum_{i=1}^{D} x_i < \sum_{i=2}^{D} x_i \leq \sqrt{D-1}\left(\sum_{i=2}^{D} x_i^2\right)^{1/2} < \sqrt{D-1}\sqrt{\mathrm{Tr}A^2}. \quad (16.76)$$

This implies that $\alpha < \sqrt{D-1}$. Hence if the opposite is true and $\alpha \geq \sqrt{D-1}$ then none of the eigenvalues x_i could be negative, so $A \geq 0$. $\square$

The partial transpose preserves the trace and the Hilbert–Schmidt norm of any state, $||\rho^{T_B}||_2^2 = ||\rho||_2^2 = \frac{1}{2}\mathrm{Tr}\rho^2$. Taking for A a partially transposed density matrix

---

[29] An explicit separability decomposition (16.50) for any state inside the ball was provided in [163].

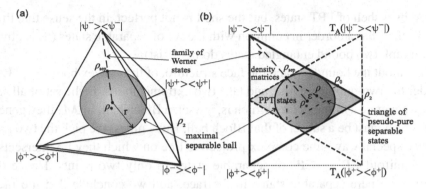

Figure 16.8 Maximal ball inscribed inside the 15-dimensional body $\mathcal{M}^{(4)}$ of mixed states is separable: (a) three-dimensional cross-section containing four Bell states, (b) two-dimensional cross-section defined by two Bell states and ρ_* with the maximal separable triangle of pseudo-pure states.

ρ^{T_B} we see that $\alpha^2 = 1/\text{Tr}\rho^2$. Let us apply the Mehta lemma to an arbitrary mixed state of an $N \times N$ bipartite system, for which the dimension $D = N^2$,

$$1/\text{Tr}\rho^2 \geq N^2 - 1 \implies \rho \text{ is PPT}. \tag{16.77}$$

Since the purity condition $\text{Tr}\rho^2 = 1/(D-1)$ characterises the insphere of $\mathcal{M}^{(D)}$, we conclude that for any bipartite[30] system the entire maximal ball inscribed inside the set of mixed states consists of PPT states only. This property implies separability for 2×2 systems. Separability of the maximal ball for higher dimensions was established by Gurvits and Barnum [380], who later estimated the radius of the separable ball for multipartite systems [381]. Useful bounds for the size of the separable ball with respect to other matrix norms are also known [831].

It is worth noticing [399] that the existence of the separable ball is in its way a non-trivial fact, connected to Eq. (9.3). If we had been working throughout with real rather than complex Hilbert spaces then every separable state would have been left invariant by partial transposition, and this would have confined all such states to a subspace of smaller dimension than that of the set of all states.

For any $N \times N$ system the Euclidean volume of the maximal separable ball may be compared with that of $\mathcal{M}^{(N^2)}$, Eq. (15.38). Their ratio decreases fast with N, which suggests that for higher dimensional systems the separable states are not typical. It is also known that for large dimensions the volume of the set $\mathcal{M}_{\text{PPT}}$ of PPT states is small with respect to the volume of the set $\mathcal{M}$ of all states, while the volume of the set of separable states is small with respect to the set $\mathcal{M}_{\text{PPT}}$ (as can be shown in a roundabout way using cones of positive maps) [879]. The set $\mathcal{M}_S$ is

---

[30] The same is true for multipartite systems [521].

covered by a shell of PPT states, but the shell is not perfect, in the sense that the set of NPT states has a border in common with the set of separable states (as is obvious because rank two bound entangled states do not exist).

What about the boundary, and the face structure, of the convex body $\mathcal{M}_S$? Let us consider the two-qubit case first, and take two extreme points. In the set of all states $\mathcal{M}$ they generate a minimal face which is, in fact, a Bloch ball. In $\mathcal{M}_S$ they generate a face which must be a subset of that Bloch ball. But – unless we pick the two points in a very special way – the complex projective line on which they lie intersects the Segre manifold $\mathbb{C}\mathbf{P}^1 \times \mathbb{C}\mathbf{P}^1$ of separable states in only two points. Hence there are only two pure separable states in the face, and we conclude that the face is identical to a line segment. In this way $\mathcal{M}_S$ acquires an edgy, 'classical' feel more reminiscent of a classical simplex than of the set of all quantum states. A similar conclusion holds for the $N_A \times N_B$ system. If we look at states that can be decomposed as a convex sum of no more than $k \leq \max(N_A, N_B)$ pure separable states we find that, exceptional cases excluded, they belong to faces that are classical simplices with k corners.[31] As is clear from the proof of Carathéodory's theorem (Problem 1.1), the face structure of a convex body is closely connected to the minimum number of terms needed in a convex decomposition of an arbitrary mixed point into extreme points. If the convex body is the body of separable states $\mathcal{M}_S$, this number is called the *cardinality*. For the two-qubit case the cardinality of a separable state is never larger than four [791]. But in higher dimensions (in fact, in all cases except possibly $2 \times N$ and 3×3) the cardinality of a random separable state in $\mathcal{M}_S$ is with probability one greater than its matrix rank [619]. Some quite basic questions about $\mathcal{M}_S$ remain open, for instance whether all its faces are exposed (Section 1.1). All faces of the self dual body $\mathcal{M}$, and also all faces of the intermediate body $\mathcal{M}_{PPT}$, are exposed.

For any finite N, the actual probability p_N to find a separable mixed state is positive, and depends on the measure used [1007, 843, 988]. Several contributions by Slater [844, 845] lead us to believe that in the simplest case of a two-qubit system the separability ratio computed with respect to the flat, Hilbert–Schmidt measure (15.34) reads $p_2 = 8/33$, but even this statement has not been rigorously proven so far. In the limit $N \to \infty$ the set of separable states is nowhere dense [227], so the probability p_N computed with respect to an arbitrary non-singular measure tends to zero.

Another method of exploring the vicinity of the maximally mixed state consists in studying *pseudo-pure states*

$$\rho_\epsilon \equiv \frac{\mathbb{1}}{N^2}(1 - \epsilon) + \epsilon|\phi\rangle\langle\phi|, \tag{16.78}$$

---

[31] This was shown by Alfsen and Schultz [19]. For more information consult Hansen et al. [399].

which are relevant for experiments with nuclear magnetic resonance (NMR), with $\epsilon \ll 1$. The set $\mathcal{M}_\epsilon$ is then defined as the convex hull of all such ϵ-pseudo-pure states. It forms a smaller copy of the entire set $\mathcal{M}$, of the same shape, and is centred at $\rho_* = \mathbb{1}/N^2$. One can show that all states in $\mathcal{M}_\epsilon$ are separable for $\epsilon \leq \epsilon_c = 1/(N^2 - 1)$. Bounds for ϵ_c in multipartite systems have been studied, partly with a view to investigating NMR quantum computing experiments [163, 381, 742, 877].

Since the cross-section of the set $\mathcal{M}^{(4)}$ shown in Figure 16.8b is a triangle, so is the set $\mathcal{M}_{\epsilon_c}$, shown as a dashed triangle located inside the dark rhombus of separable states. The rhombus is obtained as a cross-section of the separable octahedron, which arises as the common part of the tetrahedron of density matrices spanned by four Bell states and the reflected tetrahedra obtained by partial transposition [468, 50]. Together the two tetrahedra form a non-convex body called a stellated octahedron or *stella octangula*.[32] The geometry is identical to that shown in Figure 11.3.

Usually one considers states separable with respect to a fixed decomposition of the large Hilbert space, $\mathcal{H}_{N^2} = \mathcal{H}_A \otimes \mathcal{H}_B$. A mixed state ρ may be separable with respect to one decomposition and entangled with respect to another. On the other hand one may ask which states are separable with respect to all possible splittings of the composite system into subsystems A and B. This is the case if $\rho' = U\rho U^\dagger$ is separable for any global unitary U. States ρ possessing this property are called *absolutely separable* [565].

All states belonging to the maximal ball inscribed into the set of mixed states are absolutely separable. In the two-qubit case the set of absolutely separable states is larger than the maximal ball: as conjectured in [490] and proved in [922] it contains any mixed state ρ for which

$$x_1 - x_3 - 2\sqrt{x_2 x_4} \leq 0, \qquad (16.79)$$

where $\vec{x} = \{x_1 \geq x_2 \geq x_3 \geq x_4\}$ denotes the ordered spectrum of ρ. Absolutely separable states outside the maximal ball were found also for the 2×3 case [437]. In higher dimensions are only some partial results available [58].

Some insight into the geometry of the problem may be gained by studying the manifold Σ of mixed product states. To verify whether a given state ρ belongs to Σ one computes the partial traces and checks if $\rho_A \otimes \rho_B$ is equal to ρ. This is the case, for example, for the maximally mixed state, $\rho_* \in \Sigma$. All states tangent to Σ at ρ_* are separable, while the normal subspace contains the maximally entangled states.

---

[32] Ericsson has drawn a family of 3D cross-sections of $\mathcal{M}^{(4)}$ interpolating between this famous picture and the single tetrahedron seen if one chooses separable states for the basis [300]. Various 2D cross-sections [492, 925] and projections [755] (Section 8.6) provide further insight into the geometry.

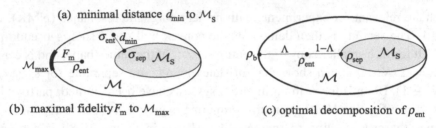

Figure 16.9 Minimal distance to a separable state (a), maximal fidelity to a maximally entangled state (b), and best separable approximation of ρ_{ent} (c).

Furthermore, for any bipartite systems the maximally mixed state ρ_* is the product state closest to any maximally entangled state (with respect to the HS distance) [621].

For a given entangled state ρ_{ent} one can ask for the separable state that lies closest to it, with respect to some given metric. Alternatively, one defines the *best separable approximation*[33]

$$\rho_{\text{ent}} = \Lambda\rho_{\text{sep}} + (1 - \Lambda)\rho_b, \tag{16.80}$$

where the separable state ρ_{sep} and the state ρ_b are chosen in such a way that the positive weight $\Lambda \in [0, 1]$ is maximal. Uniqueness of such a decomposition was proved in [595] for two qubits, and in [517] for any bipartite system. In the two-qubit problem the state ρ_b is pure and is maximally entangled for any full rank state ρ [517]. An explicit form of the decomposition (16.80) was found in [957] for a generic two-qubit state and in [12] for some particular cases in higher dimensions. A key difference is that when looking for the separable state closest to ρ we probe the boundary of the set of separable states only, while looking for its best separable approximation we must also take into account the boundary of the entire set of density matrices; see Figure 16.9.

The structure of the set of separable states may also be analysed using entanglement witnesses [743], already defined in the previous section. Any witness W, being a non-positive operator, may be represented as a point located far outside the set $\mathcal{M}$ of density matrices in its image with respect to an extended positive map $(\Phi_{\mathcal{P}} \otimes \mathbb{1})$ or (dually) as a hyperplane crossing $\mathcal{M}$. The states outside this hyperplane satisfy $\mathrm{Tr}\rho W < 0$, hence their entanglement is detected by W. A witness W_1 is called *finer* than W_0 if every entangled state detected by W_0 is also detected by W_1. A witness W_2 is called *optimal* if the corresponding map belongs to the boundary of the set of positive operators, so the hyperplane representing W_2 touches the boundary of the set $\mathcal{M}_S$ of separable states. A witness related to a generic non-CP map $\Phi_{\mathcal{P}} \in \mathcal{P}$ may

---

[33] Due to the paper [595] also called *Lewenstein–Sanpera decomposition*.

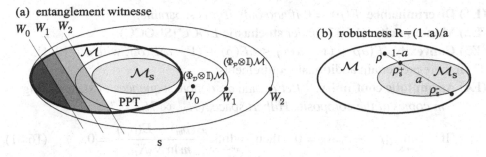

(a) entanglement witnesse

W_0 W_1 W_2

$\mathcal{M}$

$\mathcal{M}_S$

$(\Phi_P \otimes \mathbb{1})\mathcal{M}$

$(\Phi_D \otimes \mathbb{1})\mathcal{M}$

PPT

W_0 W_1 W_2

(b) robustness R=(1−a)/a

$\mathcal{M}$

ρ 1−a

ρ_s^+ $\mathcal{M}_S$

a

ρ_s^-

S

Figure 16.10 (a) The witness W_0 detects entanglement in a subset of entanglement states; W_1: optimal decomposable witness; W_2: optimal non-decomposable witness. (b) Robustness, i.e. the minimal ratio of the distance to $\mathcal{M}_S$ to the width of this set (see Section 16.8).

be *optimised* by sending it towards the boundary of $\mathcal{P}$ [594]. If a positive map Φ_D is decomposable, the corresponding witness, $W = D_{\Phi_D}/N$ is called *decomposable*. A decomposable witness (see Figure 16.10a) cannot detect PPT bound entangled states, which triggers interest in non-decomposable witnesses [388]. As the set of positive maps is convex it is advantageous to identify its extreme points and to use the corresponding extremal witnesses [223, 399].

One might argue that in general a witness $W = D_\Phi/N$ is theoretically less useful than the corresponding map Φ, since the criterion $\mathrm{Tr}\rho W < 0$ is not as powerful as $N(W^R \rho^R)^R = (\Phi \otimes \mathbb{1})\rho \geq 0$; see Eq. (11.24). However, a non-CP map Φ cannot be realised in nature, while an observable W may be measured. Suitable witness operators were actually used to detect quantum entanglement experimentally in bipartite [89, 372] and multipartite systems [151]. Furthermore, the Bell inequalities may be viewed as a kind of separability criterion, related to particular entanglement witnesses [891, 459, 486]. Experimental evidence of their violation is a detection of quantum entanglement [375].[34]

16.8 Entanglement measures

We have already learned that the degree of entanglement of any pure state of an $N_A \times N_B$ system may be characterised by the entanglement entropy (16.35) or any other Schur concave function of the Schmidt vector $\vec{\lambda}$. For mixed states the problem of quantifying entanglement becomes complicated [920, 453].

Let us first discuss the properties that any potential measure $E(\rho)$ should satisfy. Even in this respect experts seem not to share exactly the same opinions [121, 919, 471]. There are three basic axioms:

[34] The converse is not true: every pure entangled state violates some Bell inequality [339, 747], but this is not always the case for mixed entangled states [958].

(E1) Discriminance. $E(\rho) = 0$ *if and only if* ρ *is separable.*
(E2) Monotonicity. (16.47) *under* **stochastic LOCC** (SLOCC).
(E3) Convexity. $E(a\rho + (1-a)\sigma) \leq aE(\rho) + (1-a)E(\sigma)$, *with* $a \in [0, 1]$.
Then there are certain additional requirements:
(E4) Asymptotic continuity.[35] *Let* ρ_m *and* σ_m *denote sequences of states acting on m copies of the composite Hilbert space,* $(\mathcal{H}_{N_A} \otimes \mathcal{H}_{N_B})^{\otimes m}$.

$$\text{If} \quad \lim_{m \to \infty} ||\rho_m - \sigma_m||_1 = 0 \quad \text{then} \quad \lim_{m \to \infty} \frac{E(\rho_m) - E(\sigma_m)}{m \ln N_A N_B} = 0. \quad (16.81)$$

(E5) Additivity. $E(\rho \otimes \sigma) = E(\rho) + E(\sigma)$ *for any* $\rho, \sigma \in \mathcal{M}^{(N_A N_B)}$.
(E6) Normalisation. $E(|\psi\rangle\langle\psi^-|) = 1$.
(E7) Computability. *There exists an efficient method to compute E for any* ρ.
There are also alternative forms of properties **(E1–E5)**.
(E1a) Weak discriminance. *If* ρ *is separable then* $E(\rho) = 0$.
(E2a) Monotonicity *under* **deterministic LOCC.** $E(\rho) \geq E[\Phi_{\text{LOCC}}(\rho)]$.
(E3a) Pure states convexity. $E(\rho) \leq \sum_i p_i E(\phi_i)$ *where* $\rho = \sum_i p_i |\phi_i\rangle\langle\phi_i|$.
(E4a) Continuity. *If* $||\rho - \sigma||_1 \to 0$ *then* $|E(\rho) - E(\sigma)| \to 0$.
(E5a) Extensivity. $E(\rho^{\otimes n}) = nE(\rho)$.
(E5b) Subadditivity. $E(\rho \otimes \sigma) \leq E(\rho) + E(\sigma)$.
(E5c) Superadditivity. $E(\rho \otimes \sigma) \geq E(\rho) + E(\sigma)$.

The above list of postulates deserves a few comments. The rather natural 'if and only if' condition in **(E1)** is very strong: it cannot be satisfied by any measure quantifying the distillable entanglement, due to the existence of bound entangled states. Hence one often requires the weaker property **(E1a)** instead.

Monotonicity **(E2)** under probabilistic transformations is stronger than monotonicity **(E2a)** under deterministic LOCC. Since local unitary operations are reversible, the latter property implies:
(E2b) Invariance with respect to local unitary operations,

$$E(\rho) = E(U_A \otimes U_B \rho\, U_A^\dagger \otimes U_B^\dagger). \quad (16.82)$$

Convexity property **(E3)** guarantees that one cannot increase entanglement by mixing.[36] Following Vidal [930], we will call any quantity satisfying **(E2)** and **(E3)** an *entanglement monotone*.[37] These fundamental postulates reflect the key idea that quantum entanglement cannot be created locally. Or in more economical terms: it is not possible to get any entanglement for free – one needs to invest resources in global operations.

---

[35] We follow [453] here; slightly different formulations of this property are used in [458, 274].
[36] Note that entropy is a concave function of its argument: mixing of pure states increases their von Neumann entropy, but it decreases their entanglement.
[37] Some authors require also continuity **(E4a)**.

The postulate that any two neighbouring states should be characterised by similar entanglement is made precise in **(E4)**. Let us recall here Fannes' continuity lemma (14.36), which estimates the difference between von Neumann entropies of two neighbouring mixed states. Similar bounds may also be obtained for any Rényi entropy with $q > 0$, but then the bounds for S_q are weaker than for S_1. Although S_q is continuous for $q > 0$, in the asymptotic limit $n \to \infty$ only S_1 remains a continuous function of the state $\rho^{\otimes n}$. In the same way the asymptotic continuity distinguishes the entanglement entropy based on S_1 from other entropy measures related to the generalised entropies S_q [930, 274].

Additivity **(E5)** is a most welcome property of an optimal entanglement measure. For certain measures one can show sub- or super-additivity; additivity requires both of them. Unfortunately this is extremely difficult to prove for two arbitrary density matrices, so some authors suggest the requirement of extensivity **(E5a)**. Even this property is not easy to demonstrate. However, for any measure E one may consider the quantity

$$E_R(\rho) \equiv \lim_{n \to \infty} \frac{1}{n} E(\rho^{\otimes n}). \tag{16.83}$$

If such a limit exists, then the *regularised* measure E_R defined in this way satisfies **(E5a)** by construction. The normalisation property **(E6)**, useful to compare different quantities, can be achieved by a trivial rescaling.

The complete wish list **(E1–E7)** is very demanding, so it is not surprising that instead of one ideal measure of entanglement fulfilling all required properties, the literature contains a plethora of measures [181, 744, 471, 919], each of them satisfying some axioms only. The pragmatic wish **(E7)** is an especially tough one – we have learned that even the problem of deciding the separability is very hard, and quantifying entanglement cannot be easier. Instead of waiting for the discovery of a single, universal measure of entanglement, we have thus no choice but to review some approaches to the problem. In the spirit of this book we commence with

I. Geometric measures

The distance from an analysed state ρ to the set $\mathcal{M}_S$ of separable states [834, 919] satisfies **(E1)** by construction. See Figure 16.10a. However, it is not simple to find the separable state σ closest to ρ with respect to a certain metric necessary to define $D_x(\rho) \equiv D_x(\rho, \sigma)$. There are several distances to choose from, for instance:

G1. Bures distance [919] $D_B(\rho) \equiv \min_{\sigma \in \mathcal{M}_S} D_B(\rho, \sigma)$;

G2. Trace distance [287] $D_{Tr}(\rho) \equiv \min_{\sigma \in \mathcal{M}_S} D_{Tr}(\rho, \sigma)$;

G3. Hilbert–Schmidt distance [972] $D_{HS}(\rho) \equiv \min_{\sigma \in \mathcal{M}_S} D_{HS}(\rho, \sigma)$.

The Bures and the trace metrics are monotone (see Sections 14.2 and 15.1), which directly implies **(E2a)**, while D_B fulfils also the stronger property **(E2)** [919]. Since the Hilbert–Schmidt metric is not monotone [712] it is not at all clear whether the minimal HS distance is an entanglement monotone [925]. Since the diameter of the set of mixed states with respect to the above distances is finite, no distance measure can satisfy even the partial additivity **(E3a)**.

Although quantum relative entropy is not exactly a distance, but rather a contrast function, it may also be used to characterise entanglement.

G4. Relative entropy of entanglement [920] $D_R(\rho) \equiv \min_{\sigma \in \mathcal{M}_S} S(\rho||\sigma)$.

In view of the discussion in Chapter 14 this measure has an appealing interpretation as distinguishability of ρ from the closest separable state. For pure states it coincides with the entanglement entropy, $D_R(|\phi\rangle) = E_1(|\phi\rangle)$ [919]. Analytical formulae for D_R are known in certain cases only [920, 938, 489], but it may be efficiently computed numerically [769]. This measure of entanglement is convex (due to double convexity of relative entropy) and continuous [273], but not additive and not extensive [938]. It is thus useful to study the regularised quantity, $\lim_{n\to\infty} D_R(\rho^{\otimes n})/n$. This limit exists due to subadditivity of relative entropy and has been computed in some cases [70, 71].

G5. Reversed relative entropy of entanglement $D_{RR}(\rho) \equiv \min_{\sigma \in \mathcal{M}_S} S(\sigma||\rho)$.

This quantity with exchanged arguments is not so interesting per se, but its modification D'_{RR} – the minimal entropy with respect to the set $\mathcal{M}_\rho$ of separable states ρ' locally identical to ρ, $\{\rho' \in \mathcal{M}_\rho : \rho'_A = \rho_A \text{ and } \rho'_B = \rho_B\}$ – provides a distinctive example of an entanglement measure,[38] which satisfies the additivity condition **(E3)** [287].

G6. Robustness [933]. $R(\rho)$ measures the endurance of entanglement by quantifying the minimal amount of mixing with separable states needed to wipe out the entanglement,

$$R(\rho) \equiv \min_{\rho_s^- \in \mathcal{M}_S} \left(\min_{s \geq 0} s : \rho_s^+ = \frac{1}{1+s}(\rho + s\rho_s^-) \in \mathcal{M}_S \right). \tag{16.84}$$

As shown in Figure 16.10b the robustness R may be interpreted as a minimal ratio of the HS distance $1 - a = s/(1 + s)$ of ρ to the set of separable states to the width $a = 1/(1 + s)$ of this set. This construction does not depend on the boundary of the entire set $\mathcal{M}$, in contrast to the best separable approximation.[39] Robustness is known to be convex and monotone, but is not additive [933]. Robustness for

---

[38] A similar measure based on modified relative entropy was introduced by Partovi [716].
[39] In the two-qubit case the entangled state used for BSA (16.80) is pure, $\rho_b = |\phi_b\rangle\langle\phi_b|$, the weight Λ is a monotone [288], so the quantity $(1 - \Lambda)E_1(\phi_b)$ works as a measure of entanglement [595, 957].

two-qubit states diagonal in the Bell basis was found in [11]. A generalisation of this quantity was proposed in [858].

G7. Maximal fidelity F_m with respect to the set $\mathcal{M}_{\text{max}}$ of maximally entangled states [121], $F_m(\rho) \equiv \max_{\phi \in \mathcal{M}_{\text{max}}} F(\rho, |\phi\rangle\langle\phi|)$.

See Figure 16.9b. Strictly speaking the maximal fidelity cannot be considered as a measure of entanglement, since it does not satisfy even weak discriminance **(E1a)**. However, it provides a convenient way to characterise, to what extent ρ may approximate a maximally entangled state required for various tasks of quantum information processing, so in the two-qubit case it is called the *maximal singlet fraction*. Invoking (9.31) we see that F_m is a function of the minimal Bures distance from ρ to the set $\mathcal{M}_{\text{max}}$. An explicit formula for the maximal fidelity for a two-qubit state was derived in [83], while relations to other entanglement measures were analysed in [928].

II. Extensions of pure-state measures

Another class of mixed state entanglement measures can be derived from quantities characterising entanglement of pure states. There exist at least two different ways of proceeding. The idea behind the *convex roof* construction is to extend a function defined on the pure states of a convex set as 'gently' as possible to its interior.[40] If $E = E(|\phi\rangle)$ is defined on the pure states, one considers ρ as a mixture of pure states, averages E over them, and defines $E(\rho)$ as the minimal average one can obtain by considering all possible mixtures. The most important measure is induced by the entanglement entropy (16.35).

P1. Entanglement of Formation (EoF) [121]

$$E_F(\rho) \equiv \min_{\mathcal{E}_\rho} \sum_{i=1}^{M} p_i E_1(|\phi_i\rangle), \tag{16.85}$$

where the minimisation[41] is performed over all possible decompositions

$$\mathcal{E}_\rho = \{p_i, |\phi_i\rangle\}_{i=1}^{M} : \rho = \sum_{i=1}^{M} p_i |\phi_i\rangle\langle\phi_i| \quad \text{with} \quad p_i > 0, \quad \sum_{i=1}^{M} p_i = 1. \tag{16.86}$$

The ensemble $\mathcal{E}$ for which the minimum (16.85) is realised is called *optimal*. Several optimal ensembles might exist, and the minimal ensemble length M is called the *cardinality* of the state ρ. If the state is separable then $E_F(\rho) = 0$, and the

---

[40] This is particularly well explained in a review by Uhlmann [912].
[41] A dual quantity defined by maximisation over $\mathcal{E}_\rho$ is called *entanglement of assistance* [265], and both of them are related to relative entropy of entanglement of an extended system [426].

cardinality coincides with the cardinality of the separable state as defined in Section 16.7. We will study the two-qubit case in Section 16.9.

Entanglement of formation (EoF) enjoys several appealing properties: it may be interpreted as the minimal amount of pure state entanglement required to build up the mixed state. It satisfies by construction the discriminance property **(E1)** and is convex and monotone [121]. EoF is known to be continuous [690], and for pure states it is by construction equal to the entanglement entropy $E_1(|\phi\rangle)$. To be consistent with normalisation **(E6)** one often uses a rescaled quantity, $E'_F \equiv E_F/\ln 2$.

However, EoF is not an ideal entanglement measure: it satisfies subadditivity **(E5b)**, but not additivity **(E5)**.[42] This key statement follows from a work of Shor's [837], showing that additivity properties for EoF and minimal output entropy of a channel are equivalent, and from Hasting's non-additivity theorem for the latter [410] (see Section 15.8). Furthermore, EoF is not easy to evaluate.[43] Explicit analytical formulae were derived for the two-qubit system [981], and a certain class of symmetric states in higher dimensions [893, 938], while for the $2 \times N$ systems at least lower bounds are known [211, 624, 328].

P2. Generalised Entanglement of Formation (GEoF)

$$E_q(\rho) \equiv \min_{\mathcal{E}_\rho} \sum_{i=1}^{M} p_i \, E_q(|\phi_i\rangle), \tag{16.87}$$

where $E_q(|\phi\rangle) = S_q[\text{Tr}_B(|\phi\rangle\langle\phi|)]$ stands for the Rényi entropy of entanglement. Note that an optimal ensemble for a certain value of q needs not to provide the minimum for $q' \neq q$. GEoF is asymptotically continuous only in the limit $q \to 1$ for which it coincides with EoF. It is an entanglement monotone for $q \in (0, 1]$ [930].

In the very same way, the convex roof construction can be applied to extend any pure states entanglement measure for mixed states. In fact, several measures introduced so far are related to GEoF. For instance, the convex roof extended negativity [582] and *concurrence of formation* [982, 783, 663] are related to $E_{1/2}$ and E_2, respectively.

There is another way to make use of pure state entanglement measures. In analogy to the fidelity between two mixed states, equal to the maximal overlap between their purifications, one may also purify ρ by a pure state $|\psi\rangle \in (\mathcal{H}_{N_A} \otimes \mathcal{H}_{N_B})^{\otimes 2}$. Based on the entropy of entanglement (16.35) one defines:

---

[42] Additivity of EoF holds if one of the states is a product state [110], is separable [938] or supported on a specific subspace [932].

[43] EoF may be computed numerically by minimisation over the space of unitary matrices $U(M)$. A search for the optimal ensemble can be based on simulated annealing [1007], on a faster conjugate–gradient method [72], or on minimising the conditional mutual information [902].

P3. Entanglement of purification [890, 150]

$$E_P(\rho) \quad \equiv \quad \min_{|\phi\rangle:\ \rho=\text{Tr}_{KN}(|\phi\rangle\langle\phi|)} E_1(|\phi\rangle), \tag{16.88}$$

but any other measure of pure states entanglement might be used instead.

The entanglement of purification is continuous and monotone under strictly local operations (not under LOCC). It is not convex, but more importantly, it does not satisfy even weak discriminance **(E1a)**. In fact E_P measures *correlations*[44] between both subsystems, and is positive for any non-product, separable mixed state [150]. Hence entanglement of purification is not an entanglement measure, but it happens to be helpful to estimate a variant of the entanglement cost [890]. To obtain a reasonable measure one needs to allow for an arbitrary extension of the system size, as assumed by defining:

P4. Squashed entanglement [207]

$$E_S(\rho^{AB}) \quad \equiv \quad \inf_{\rho^{ABE}} \frac{1}{2}\big[S(\rho^{AE}) + S(\rho^{BE}) - S(\rho^{E}) - S(\rho^{ABE})\big], \tag{16.89}$$

where the infimum is taken over all extensions ρ^{ABE} of an unbounded size such that $\text{Tr}_E(\rho^{ABE}) = \rho^{AB}$. Here ρ^{AE} stands for $\text{Tr}_B(\rho^{ABE})$ while $\rho^{E} = \text{Tr}_{AB}(\rho^{ABE})$.

Squashed entanglement[45] is convex, monotone, vanishes for every separable state [207], and is positive for any entangled state [155]. If ρ^{AB} is pure then $\rho^{ABE} = \rho^{E} \otimes \rho^{AB}$, hence $E_S = [S(\rho^A) + S(\rho^B)]/2 = S(\rho^A)$ and the squashed entanglement reduces to the entropy of entanglement E_1. It is characterised by asymptotic continuity [25], and additivity **(E5)**, which is a consequence of the strong subadditivity of the von Neumann entropy [207]. Thus E_S would be a perfect measure of entanglement, if we only knew how to compute it.

III. Operational measures

Entanglement may also be quantified in an abstract manner by considering the minimal resources required to generate a given state or the maximal entanglement yield. These measures are defined implicitly, since one deals with an infinite set of copies of the state analysed and assumes an optimisation over all possible LOCC protocols.

O1. Entanglement cost [121, 760] $E_C(\rho) = \lim_{n\to\infty} \frac{m}{n}$ where m is the number of singlets $|\psi^-\rangle$ needed to produce locally n copies of the analysed state ρ.

Entanglement cost has been calculated, for instance for states supported on a subspace such that tracing out one of the parties forms an entanglement breaking

---

[44] To quantify them one may make use of the operator Schmidt decomposition (10.31) of ρ.
[45] Minimised quantity is proportional to quantum conditional mutual information of ρ^{ABE} [902] and its name refers to 'squashing out' the classical correlations.

channel (super-separable map) [932]. Moreover, entanglement cost was shown [421] to be equal to the regularised entanglement of formation, $E_C(\rho) = \lim_{n\to\infty} E_F(\rho^{\otimes n})/n$. Since we know that EoF is non-additive, the notions of entanglement cost and entanglement of formation are not equivalent. Entanglement cost is faithful as it satisfies [987] the discriminance property (**E1**). Asymptotic continuity of entanglement cost was shown by Winter [971].

O2. Distillable entanglement [121, 760]. $E_D(\rho) = \lim_{n\to\infty} \frac{m}{n}$ where m is the maximal number of singlets $|\psi^-\rangle$ obtained out of n copies of the state ρ by an optimal LOCC conversion protocol.

Distillable entanglement is a measure of fundamental importance, since it tells us how much entanglement one may extract out of the state analysed and use for example for cryptographic purposes. In general the entanglement payoff is not larger than the resource invested, $E_D(\rho) \le E_C(\rho)$, and the inequality becomes sharp, for example for bound entangled states. See Section 16.5.

The quantity E_D is rather difficult to compute, but there exist analytical bounds due to Rains [761, 762] and an explicit optimisation formula has been found [258]. It has been shown that E_D fails to be additive, and fails to be convex, if bound entangled PPT states exist [838]. But, as we noted in Section 16.5, this question remains open. E_D does satisfy the weaker condition (**E3a**) [274].

IV. Algebraic measures

If the partial transpose of a state ρ is not positive then ρ is entangled due to the PPT criterion **B1**. The partial transpose preserves the trace, so if $\rho^{T_A} \ge 0$ then $||\rho^{T_A}||_{\mathrm{Tr}} = \mathrm{Tr}\rho^{T_A} = 1$. Hence we can use the trace norm to characterise the degree to which the positivity of ρ^{T_A} is violated.

N1. Negativity [1010, 291], $\mathcal{N}_T(\rho) \equiv ||\rho^{T_A}||_{\mathrm{Tr}} - 1$.

Negativity is easy to compute, convex (partial transpose is linear and the trace norm is convex) and monotone [286, 934]. It is not additive, but this drawback may be cured by defining the *log-negativity*,[46] $\ln ||\rho^{T_A}||_{\mathrm{Tr}}$. However, the major deficiency of the negativity is its failure to satisfy (**E1**) – by construction $\mathcal{N}_T(\rho)$ cannot detect PPT bound entangled states. In the two-qubit case the spectrum of ρ^{T_A} contains at most a single negative eigenvalue [791], so $\mathcal{N}'_T \equiv \max\{0, -2\lambda_{\min}\} = \mathcal{N}_T$. This observation explains the name on the one hand, and on the other provides a geometric interpretation: $\mathcal{N}'_T(\rho)$ measures the minimal relative weight of the maximally mixed state ρ_* which needs to be mixed with ρ to produce a separable mixture [925]. In higher dimensions several eigenvalues of the partially transposed

---

[46] This is additive but not convex. Log-negativity was used by Rains [762] to obtain bounds on distillable entanglement.

state may be negative, so in general $\mathcal{N}_T \neq \mathcal{N}_T'$, and the latter quantity is proportional to complete co-positivity (11.13) of a map Φ associated with the state ρ.

Negativity is not the only application of the trace norm [934]. Building on the positive maps criterion **A1** for separability one might analyse analogous quantities for any (not completely) positive map, $\mathcal{N}_\Phi(\rho) \equiv ||(\Phi \otimes \mathbb{1})\rho||_{\text{Tr}} - 1$. Furthermore, one may consider another quantity related to the reshuffling criterion **B6**.

N2. Reshuffling negativity [205, 781]. $\mathcal{N}_R(\rho) \equiv ||\rho^R||_{\text{Tr}} - 1$.

This quantity is convex due to linearity of reshuffling and non-increasing under local measurements, but may increase under partial trace [781]. For certain bound entangled states $\mathcal{N}_R$ is positive; unfortunately not for all of them. A similar quantity with the minimal cross-norm $|| \cdot ||_\gamma$ was studied by Rudolph [779], who showed that $\mathcal{N}_\gamma(\rho) = ||\rho||_\gamma - 1$ is convex and monotone under local operations. However, $||\rho^R||_{\text{Tr}}$ is easily computable from the definition (10.33), in contrast to $||\rho||_\gamma$.

We end this short tour through the vast garden of entanglement measures[47] by studying how they behave for pure states. Entanglement of formation and purification coincide by construction with the entanglement entropy E_1. So also is the case for both operational measures, since conversion of n copies of an analysed pure state into m maximally entangled states is reversible. The negativities are easy to compute. For any pure state $|\psi\rangle \in \mathcal{H}_N \otimes \mathcal{H}_N$ written in the Schmidt form (9.8), the non-zero entries of the density matrix ρ_ψ of size N^2 are equal to $\sqrt{\lambda_i \lambda_j}$, $i, j = 1, \ldots, N$. The reshuffled matrix ρ_ψ^R becomes diagonal, while partially transposed matrix $\rho_\psi^{T_2}$ has a block structure: it consists of N Schmidt components λ_i at the diagonal which sum to unity and $N(N-1)/2$ off-diagonal blocks of size 2, one for each pair of different indices (i,j). Eigenvalues of each block are $\pm\sqrt{\lambda_i \lambda_j}$, so both trace norms are equal to the sum of all entries. Hence both negativities coincide for pure states,

$$\mathcal{N}_T(\rho_\psi) = \mathcal{N}_R(\rho_\psi) = \sum_{i,j=1}^{N} \sqrt{\lambda_i \lambda_j} - 1 = \left(\sum_i^N \sqrt{\lambda_i}\right)^2 - 1 = e^{E_{1/2}} - 1, \quad (16.90)$$

and vary from 0 for separable states to $N-1$ for maximally entangled states. Also maximal fidelity and robustness for pure states become related, $F = \exp(E_{1/2})/N$ and $R = \mathcal{N}_T = \exp(E_{1/2}) - 1$ [933]. On the other hand, the minimal Bures distance to the closest separable (mixed) state [919] becomes a function of the Rényi entropy of order two,

$$D_B(\rho_\phi) = \left(2 - 2\sum_{i=1}^{N} \lambda_i^2\right)^{1/2} = \sqrt{2 - 2e^{-E_2/2}}, \quad (16.91)$$

---

[47] There exist also attempts to quantify entanglement by the dynamical properties of a state and the speed of decoherence [135], or the secure key distillation rate [452, 258] and several others [744, 471].

Table 16.2 *Properties of entanglement measures: discriminance E1, monotonicity E2, convexity E3, asymptotic continuity E4, additivity E5, extensivity E5a, computability E7: explicit closed formula C, optimisation over a finite space F or an infinite space I. Here q represents the parameter of the Rényi entropy E_q, which in some cases becomes a function of a given measure specified to pure states, while Q denotes 'yes' for $q \in (0, 1]$.*

Entanglement measure	E1	E2	E3	E4	E5	E5a	E7	q
G1 Bures distance D_B	Y	Y	Y	N	N	N	F	2
G2 Trace distance D_{Tr}	Y	Y	Y	N	N	N	F	
G3 HS distance D_{HS}	Y	N	N	N	N	N	F	
G4 Relative entropy D_R	Y	Y	Y	Y	N	N	F	1
G5 Reversed RE, D'_{RR}	?	Y	Y	Y	Y	Y	F	1
G6 Robustness R	Y	Y	Y	N	N	N	F	1/2
P1 Entang. of formation E_F	Y	Y	Y	Y	N	N	C/F	1
P2 Generalised EoF, E_q	?	Q	?	N	N	N	F	q
P4 Squashed entang. E_S	Y	Y	Y	Y	Y	Y	I	1
O1 Entang. cost E_C	Y	Y	Y	Y	?	Y	I	1
O2 Distillable entang. E_D	N	Y	?	Y	?	Y	I	1
N1 Negativity $\mathcal{N}_T$	N	Y	Y	N	N	N	C	1/2
N2 Reshuffling negativity $\mathcal{N}_R$	N	N	Y	N	N	N	C	1/2

see Eq. (16.36), while the Bures distance to the closest separable pure state $D_B^{pure}(|\phi\rangle) = [2(1 - \sqrt{1 - \lambda_{max}})]^{1/2}$ is a function of $E_\infty = -\ln \lambda_{max}$. The Rényi parameters q characterising behaviour of the discussed measures of entanglement for pure states are collected in Table 16.2.

Knowing that a given state ρ can be locally transformed into ρ' implies that $E(\rho) \geq E(\rho')$ for any measure E, but the converse is not true. Two Rényi entropies of entanglement of different (positive) orders generate different orders in the space of pure states. By continuity this is also the case for mixed states, and the relation

$$E_A(\rho_1) \leq E_A(\rho_2) \iff E_B(\rho_1) \leq E_B(\rho_2) \tag{16.92}$$

does not hold. For a pair of mixed states it can happen that one state is more entangled with respect to one measure of entanglement, while another state is more entangled with respect to another measure [291]. If two measures E_A and E_B coincide for pure states they are identical or they *do not* generate the same order in the set of mixed states [936]. Hence entanglement of formation and distillable entanglement do not induce the same order. On the other hand, several entanglement measures are correlated and knowing E_A one may try to find lower and upper bounds for E_B.

The set of entanglement measures shrinks, if one imposes even some of the desired properties **(E1)–(E7)**. The asymptotic continuity **(E4)** is particularly restrictive. For instance, among the Rényi entropies it is satisfied only by the entropy of entanglement E_1 [930]. If a measure E satisfies additionally monotonicity **(E2a)** under deterministic LOCC and extensivity **(E5a)**, it is bounded by the distillable entanglement and entanglement cost [458, 453],

$$E_D(\rho) \leq E(\rho) \leq E_C(\rho). \tag{16.93}$$

Interestingly, the two first measures introduced in the pioneering paper by Bennett et al. [121] occurred to be the extreme entanglement measures. For pure states both of them coincide, and we arrive at a kind of **uniqueness theorem**: *Any monotone, extensive and asymptotically continuous entanglement measure coincides for pure states with the entanglement of formation E_F* [748, 930, 274]. This conclusion concerning pure states entanglement of bipartite systems may also be reached by an abstract, thermodynamic approach [918].

Let us try to recapitulate the similarities and differences between four classes of entanglement measures. For a geometric measure or an extension of a pure state measure it is not simple to check which of the desired properties are satisfied. Furthermore, to evaluate it for a typical mixed state one needs to perform a cumbersome optimisation scheme. Operational measures are nice, especially from the point of view of information science, and extensivity and monotonicity are direct consequence of their definitions. However, they are extremely hard to compute. In contrast, algebraic measures are easy to calculate, but they fail to detect entanglement for all non-separable states. Summarizing, many different measures of entanglement are likely to be still used in the future.

16.9 Two-qubit mixed states

Before discussing the entanglement of two-qubit mixed states let us recapitulate how the case $N = 2$ differs from $N \geq 3$.

A. Algebraic properties

  (i) $SU(N) \times SU(N)$ is homomorphic to $SO(N^2)$ for $N = 2$ only;
  (ii) $SU(N) \cong SO(N^2 - 1)$ for $N = 2$ only;
  (iii) All positive maps $\Phi : \mathcal{M}^{(N)} \to \mathcal{M}^{(N)}$ are decomposable for $N = 2$ only.

B. N-level monopartite systems

  (iv) Boundary $\partial \mathcal{M}^{(2)}$ consists of pure states only;
  (v) For any state $\rho_{\vec{\tau}} \in \mathcal{M}^{(N)}$ also its antipode $\rho_{-\vec{\tau}} = 2\rho_* - \rho_{\vec{\tau}}$ also forms a state and there exists a universal NOT operation for $N = 2$ only;
  (vi) $\mathcal{M}^{(N)} \subset \mathbb{R}^{N^2-1}$ forms a ball for $N = 2$ only.

C. $N \times N$ composite systems

(vii) For any pure state $|\psi\rangle \in \mathcal{H}_N \otimes \mathcal{H}_N$ there exist $N-1$ independent Schmidt coefficients λ_i. For $N = 2$ there exists only one independent coefficient, hence all entanglement measures are equivalent.

(viii) The maximally entangled states form the manifold $SU(N)/Z_N$, which is equivalent to the real projective space $\mathbb{RP}^{N^2-1}$ only for $N = 2$.

(ix) All PPT states of an $N \times N$ system are separable for $N = 2$ only.

(x) The optimal decomposition of a two-qubit mixed state consists of pure states of equal concurrence. Thus entanglement of formation becomes a function of concurrence of formation for 2×2 systems.

These features demonstrate why entanglement of two-qubit systems is special [937]. Several of these issues are closely related. We have already learned that decomposability (iii) is a consequence of (ii) and implies the separability (ix). We shall see now how the group-theoretic fact (i) allows us to derive a closed formula for EoF of a two-qubit system. We will follow the seminal paper of Wootters [981], building on his earlier paper with Hill [438].

Consider first a normalised two-qubit pure state $|\psi\rangle$. Its Schmidt coefficients satisfy $\mu_1 + \mu_2 = 1$. The tangle of $|\psi\rangle$, defined in (16.36), reads

$$\tau = C^2 = 2(1 - \mu_1^2 - \mu_2^2) = 4\mu_1(1 - \mu_1) = 4\mu_1\mu_2, \qquad (16.94)$$

and implies that concurrence[48] is proportional to the determinant of (16.5),

$$C = 2\,|\sqrt{\mu_1\mu_2}| = 2\,|\det\Gamma|. \qquad (16.95)$$

Inverting this relation we find the entropy of entanglement E as a function of concurrence

$$E = S(\mu_1, 1 - \mu_1) \quad \text{where} \quad \mu_1 = \frac{1}{2}\left(1 - \sqrt{1 - C^2}\right), \qquad (16.96)$$

and S stands for the Shannon entropy function, $-\sum_i \mu_i \ln \mu_i$.

Let us represent $|\psi\rangle$ in a special basis consisting of four Bell states

$$|\psi\rangle = \left[a_1|\phi^+\rangle + a_2 i|\phi^-\rangle + a_3 i|\psi^+\rangle + a_4|\psi^-\rangle\right] \qquad (16.97)$$

$$= \frac{1}{\sqrt{2}}\left[(a_1 + ia_2)|00\rangle + (ia_3 + a_4)|01\rangle + (ia_3 - a_4)|10\rangle + (a_1 - ia_2)|11\rangle\right].$$

---

[48] Concurrence was initially introduced for two qubits. Earlier we adopted the generalisation of [782, 663], but there are also other ways to generalise this notion for higher dimensions [911, 982, 82, 204].

Calculating the determinant in Eq. (16.95) we find that

$$C(|\psi\rangle) = \left|\sum_{k=1}^{4} a_k^2\right|. \tag{16.98}$$

If all coefficients of $|\psi\rangle$ in the basis (16.98) are real then $C(|\phi\rangle) = 1$, and the state is maximally entangled. This property also holds if we act on $|\psi\rangle$ with an orthogonal gate $O \in SO(4)$, and it justifies referring to (16.98) as the *magic basis* [121]. A two-qubit unitary gate represented in it by a real $SO(4)$ matrix corresponds to a local operation[49] and its action does not influence entanglement. This is how property **(i)** enters the game.

To appreciate another feature of the magic basis consider the transformation $|\psi\rangle \rightarrow |\tilde{\psi}\rangle = (\sigma_y \otimes \sigma_y)|\psi^*\rangle$, in which complex conjugation is taken in the standard basis, $\{|00\rangle, |01\rangle, |10\rangle, |11\rangle\}$. This is an anti-unitary operation, and here anti-unitaries are really at the heart of the matter.[50] It flips both spins of the system, but if the state is expressed in the magic basis the transformation is realised by complex conjugation alone. Expression (16.98) implies then

$$C(|\phi\rangle) = |\langle\psi|\tilde{\psi}\rangle|. \tag{16.99}$$

Spin flipping of mixed states is also realised by complex conjugation, if ρ is expressed in magic basis. Working in standard basis this transformation reads

$$\rho \rightarrow \tilde{\rho} = (\sigma_y \otimes \sigma_y)\rho^*(\sigma_y \otimes \sigma_y). \tag{16.100}$$

In the Fano form (16.49) flipping corresponds to reversing the signs of both Bloch vectors,

$$\tilde{\rho} = \frac{1}{4}\left[\mathbb{1}_4 - \sum_{i=1}^{3} \tau_i^A \sigma_i \otimes \mathbb{1}_2 - \sum_{j=1}^{3} \tau_j^B \mathbb{1}_2 \otimes \sigma_j + \sum_{i,j=1}^{3} \beta_{ij}\sigma_i \otimes \sigma_j\right]. \tag{16.101}$$

Root fidelity between ρ and $\tilde{\rho}$ is given by the trace of the positive matrix

$$\sqrt{F} = \sqrt{\sqrt{\rho}\tilde{\rho}\sqrt{\rho}}. \tag{16.102}$$

Let us denote by λ_i the decreasingly ordered eigenvalues of $\sqrt{F}$ (singular values of $\sqrt{\rho}\sqrt{\tilde{\rho}}$). The *concurrence* of a two-qubit mixed state is now defined by

$$C(\rho) \equiv \max\{0, \lambda_1 - \lambda_2 - \lambda_3 - \lambda_4\}. \tag{16.103}$$

The number of positive eigenvalues cannot be greater than the rank r of ρ. For a pure state the above definition is thus consistent with $C = |\langle\psi|\tilde{\psi}\rangle|$ and expression

---

[49] A gate represented by an orthogonal matrix with $\det O = -1$ is a non-local swap operation.
[50] As explicated by Uhlmann [911].

(16.98). The aim is to show that the concurrence vanishes if and only if ρ is separable.

We will work in the magical basis, so that the tilde operation is simply complex conjugation. Consider a general mixed state given by its eigendecomposition, and subnormalise its eigenvectors so that their norms are equal to the corresponding eigenvalues:

$$\rho = \sum_{i=1}^{4} |p_i\rangle\langle p_i|, \qquad \langle p_i|p_j\rangle = p_i\delta_{ij}. \qquad (16.104)$$

Next we define the matrix $W_{ij} \equiv \langle p_i|\tilde{p}_j\rangle$. Because we work in the magical basis it is a symmetric complex matrix,

$$W_{ji} = \langle p_j|\tilde{p}_i\rangle = (\langle \tilde{p}_j|p_i\rangle)^* = \langle p_i|\tilde{p}_j\rangle = W_{ij}. \qquad (16.105)$$

Using the characteristic equation we also see that the spectra of $\rho\tilde{\rho}$ and WW^* coincide. But

$$\rho\tilde{\rho} = \sqrt{\rho}\sqrt{\rho}\tilde{\rho}\sqrt{\rho}\sqrt{\rho}^{-1} = \sqrt{\rho}(\sqrt{F})^2\sqrt{\rho}^{-1}. \qquad (16.106)$$

Since similar matrices have the same spectra it follows that

$$\text{spec}(WW^\dagger) = \text{spec}(\rho\tilde{\rho}) = \text{spec}(\sqrt{F})^2 = (\lambda_1^2, \lambda_2^2, \lambda_3^2, \lambda_4^2). \qquad (16.107)$$

At this point we appeal to a quite remarkable statement known as *Takagi's factorisation*: given a symmetric complex matrix W there exists a unitary matrix U such that the matrix $D = UWU^T$ is diagonal, with real and positive eigenvalues.[51]

We take time out to prove this. Let V be a unitary that diagonalises $WW^\dagger$. A small calculation verifies that the matrix D is automatically symmetric and normal. These two properties together imply that the real and imaginary parts of $D = R + iI$ commute,

$$0 = [D, D^\dagger] = [R + iI, R - iI] = i[I, R]. \qquad (16.108)$$

Therefore we can define an orthogonal matrix O that diagonalises both R and I at the same time, and by choosing $U = OV$ we find that UWU^T is diagonal, as advertised. Phases can be adjusted to make the eigenvalues real and positive.

With Takagi's result in hand we can continue. We are going to insert the unitary matrix U we have just defined into Schrödinger's mixture theorem (8.40), which gives us all possible decompositions of the density matrix. Thus

$$|x_i\rangle \equiv \sum_j U_{ij}^*|w_j\rangle \quad \Rightarrow \quad \rho = \sum_{i=1}^{4} |w_i\rangle\langle w_i| = \sum_{i=1}^{4} |x_i\rangle\langle x_i|. \qquad (16.109)$$

---

[51] This theorem has been discovered a number of times, first by Autonne (1915) [77] and Takagi (1925) [884].

We know the spectrum of $WW^\dagger$, and when we diagonalise W it follows that

$$\langle x_i | \tilde{x}_j \rangle = (UWU^T)_{ij} = \lambda_i \delta_{ij}. \tag{16.110}$$

We are almost there. In the final step we introduce yet another set of pure states using phases η_i,

$$|z_1\rangle = \frac{1}{2}\left(e^{i\eta_1}|x_1\rangle + e^{i\eta_2}|x_2\rangle + e^{i\eta_3}|x_3\rangle + e^{i\eta_4}|x_4\rangle\right),$$

$$|z_2\rangle = \frac{1}{2}\left(e^{i\eta_1}|x_1\rangle + e^{i\eta_2}|x_2\rangle - e^{i\eta_3}|x_3\rangle - e^{i\eta_4}|x_4\rangle\right), \tag{16.111}$$

$$|z_3\rangle = \frac{1}{2}\left(e^{i\eta_1}|x_1\rangle - e^{i\eta_2}|x_2\rangle + e^{i\eta_3}|x_3\rangle - e^{i\eta_4}|x_4\rangle\right),$$

$$|z_4\rangle = \frac{1}{2}\left(e^{i\eta_1}|x_1\rangle - e^{i\eta_2}|x_2\rangle - e^{i\eta_3}|x_3\rangle + e^{i\eta_4}|x_4\rangle\right).$$

For all four states it holds that

$$\langle \tilde{z}_i | z_i \rangle = \sum_{j=1}^{4} e^{2i\eta_j} \lambda_j. \tag{16.112}$$

Note that we are in fact making use of a real Hadamard matrix at this point; see Section 12.3. We have reached our final decomposition

$$|z_i\rangle \equiv \sum_i H_{ij} e^{i\eta_j} |x_j\rangle \quad \Rightarrow \quad \rho = \sum_{i=1}^{4} |x_i\rangle\langle x_i| = \sum_{i=1}^{4} |z_i\rangle\langle z_i|. \tag{16.113}$$

The state will be separable if and only if we can choose the phase factors in Eq. (16.112) so that $\langle \tilde{z}_i | z_i \rangle = 0$, because this is the condition for a pure state to be separable.

But inspection of Figure 16.11a shows that this condition is simply the condition that a chain of four links of lengths λ_i may be closed to form a polygon. And this is possible if and only if the longest link is not longer than the sum of the three others, that is to say if the concurrence $C(\rho)$ vanishes.[52]

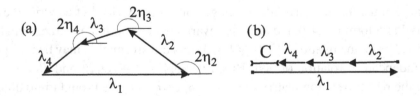

Figure 16.11 Concurrence polygon: (a) quadrangle for a separable state, (b) line for an entangled state with concurrence C.

---

[52] This reasoning holds also if the polygon reduces to, say, a triangle.

Consider now a mixed state ρ for which $C > 0$ because λ_1 is too large for the chain to close – see Figure 16.11b. Making use of the pure states $|x_i\rangle$ constructed before we introduce a set of four states

$$|y_1\rangle = |x_1\rangle, \quad |y_2\rangle = i|x_2\rangle, \quad |y_3\rangle = i|x_3\rangle, \quad |y_4\rangle = i|x_4\rangle, \tag{16.114}$$

and a symmetric matrix $Y_{ij} = \langle y_i|\tilde{y}_j\rangle$, the relative phases of which are chosen in such a way that relation (16.110) implies

$$\text{Tr } Y = \sum_{i=1}^{4} \langle y_i|\tilde{y}_i\rangle = \lambda_1 - \lambda_2 - \lambda_3 - \lambda_4 = C(\rho). \tag{16.115}$$

The states (16.114) are subnormalised, hence the above expression represents an average of a real quantity, the absolute value of which coincides with the concurrence (16.99). Using Schrödinger's theorem again one may find yet another decomposition, $|z_i\rangle \equiv \sum_i V_{ij}^*|y_j\rangle$, such that every state has *the same* concurrence, $C(|z_i\rangle) = C(\rho)$ for $i = 1, \ldots, 4$. To do so define a symmetric matrix $Z_{ij} = \langle z_i|\tilde{z}_j\rangle$ and observe that $\text{Tr}Z = \text{Tr}(VYV^T)$. This trace does not change if V is real and $V^T = V^{-1}$ follows. Hence one may find an orthogonal V such that all overlaps are equal to concurrence, $Z_{ii} = C(\rho)$, and produce the final decomposition $\rho = \sum_i |z_i\rangle\langle z_i|$.

The above decomposition is optimal for the concurrence of formation [981],

$$C_F(\rho) = \min_{\mathcal{E}_\rho} \sum_{i=1}^{M} p_i C(|\phi_i\rangle) = \min_{V} \sum_{i=1}^{m} |(VYV^T)_{ii}|, \tag{16.116}$$

where the minimum over ensembles $\mathcal{E}_\rho$ may be replaced by a minimum over $m \times n$ rectangular matrices V, containing n orthogonal vectors of size $m \le 4^2$ [910]. The function relating E_F and concurrence is convex, hence the decomposition into pure states of equal concurrence is also optimal for entropy of formation. Thus E_F of any two-qubit state ρ is given by the function (16.96) of concurrence of formation C_F equal to $C(\rho)$, and defined by (16.103).[53]

While the algebraic fact **(i)** was used to calculate concurrence, the existence of a general formula for maximal fidelity hinges on property **(ii)**. Let us write the state ρ in its Fano form (16.49) and analyse invariants of local unitary transformations $U_1 \otimes U_2$. Due to the relation $SU(2) \approx SO(3)$ this transformation may be interpreted as an independent rotation of both Bloch vectors, $\vec{\tau}^A \to O_1\vec{\tau}^A$ and $\vec{\tau}^B \to O_2\vec{\tau}^B$. Hence the real correlation matrix $\beta_{ij} = \text{Tr}(\rho\sigma_i \otimes \sigma_j)$ may be brought into diagonal

---

[53] A streamlined proof of this fact was provided in [72], while the analogous problem for two real qubits ('rebits') was solved in [196].

form $K = O_1 \beta O_2^T$. The diagonal elements may admit negative values since we require $\det O_i = +1$. Hence $|K_{ii}| = \kappa_i$ where κ_i stand for singular values of β. Let us order them decreasingly. By construction they are invariant with respect to local unitaries,[54] and govern the maximal fidelity with respect to maximally entangled states [83],

$$F_m(\rho) = \frac{1}{4}\left[1 + \kappa_1 + \kappa_2 - \text{Sign}[\det(\beta)]\,\kappa_3\right].$$ (16.117)

It is instructive to compute explicit formulae for the above entanglement measures for several families of two-qubit states. The concurrence of a Werner state (16.55) is equal to the negativity,

$$C(\rho_W(x)) = N_T(\rho_W(x)) = \begin{cases} 0 & \text{if} & x \leq 1/3 \\ (3x - 1)/2 & \text{if} & x \geq 1/3 \end{cases},$$ (16.118)

its entanglement of formation is given by (16.96), while $F_m = (3x - 1)/4$.

Another interesting family of states arises as a convex combination of a Bell state with an orthogonal separable state [459]

$$\sigma_H(a) \equiv a\,|\psi^-\rangle\langle\psi^-| + (1 - a)\,|00\rangle\langle00|.$$ (16.119)

Concurrence of such a state is by construction equal to its parameter, $C = a$, while the negativity reads $N_T = \sqrt{(1 - a)^2 + a^2} + a - 1$. The relative entanglement of entropy reads [919] $E_R = (a - 2)\ln(1 - a/2) + (1 - a)\ln(1 - a)$. We shall use also a mixture of Bell states,

$$\sigma_B(b) \equiv b\,|\psi^-\rangle\langle\psi^-| + (1 - b)\,|\psi^+\rangle\langle\psi^+|,$$ (16.120)

for which by construction $F_m = \max\{b, 1 - b\}$ and $C = N_T = 2F_m - 1$.

Entanglement measures are correlated: they vanish for separable states and coincide for maximally entangled states. For two-qubit systems several explicit bounds are known. Concurrence forms an upper bound for negativity [291, 1007]. This statement was proved in [921], where it was shown that these measures coincide if the eigenvector of ρ^{T_A} corresponding to the negative eigenvalue is maximally entangled. The lower bound

$$C \geq N_T \geq \sqrt{(1 - C)^2 + C^2} + C - 1$$ (16.121)

is achieved [921] for states (16.119).

---

[54] A two-qubit density matrix is specified by 15 parameters, the local unitaries are characterised by six variables, so there exist nine functionally independent local invariants. However, two states are locally equivalent if they share additionally nine discrete invariants which determine the signs of κ_i, τ_i^A and τ_i^B [640]. A classification of mixed states based on degeneracy and signature of K was worked out in [361, 297, 565].

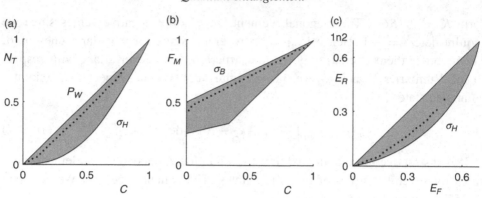

Figure 16.12 Bounds between entanglement measures for two qubits: (a) negativity versus concurrence (16.121), (b) maximal fidelity versus concurrence (16.122), (c) relative entropy of entanglement versus entanglement of formation (16.124). Labels represent families of extremal states while dots denote averages taken with respect to the HS measure in $\mathcal{M}^{(4)}$.

Analogous tight bounds between maximal fidelity and concurrence or negativity were established in [928],

$$\frac{1+C}{2} \geq F_m \geq \begin{cases} (1+C)/4 & \text{if} \quad C \leq 1/3 \\ C & \text{if} \quad C \geq 1/3 \end{cases}, \qquad (16.122)$$

$$\frac{1+\mathcal{N}_T}{2} \geq F_m \geq \begin{cases} \frac{1}{4} + \frac{1}{8}\left(\mathcal{N}_T + \sqrt{5\mathcal{N}_T^2 + 4\mathcal{N}_T}\right) & \text{if} \quad \mathcal{N}_T \leq \frac{\sqrt{5}-2}{3} \\ \sqrt{2\mathcal{N}_T(\mathcal{N}_T+1)} - \mathcal{N}_T & \text{if} \quad \mathcal{N}_T \geq \frac{\sqrt{5}-2}{3} \end{cases}. \qquad (16.123)$$

The upper bound for fidelity is realised for the family (16.120) or for any other state for which $C = \mathcal{N}_T$.

Relative entropy of entanglement is bounded from above by E_F. Numerical investigations suggest [921] that the lower bound is achieved for the family of (16.119), which implies

$$E_F \geq E_R \geq \left[(C-2)\ln(1-C/2) + (1-C)\ln(1-C)\right]. \qquad (16.124)$$

Here $C^2 = 1 - (2\mu_1 - 1)^2$ and $\mu_1 = S^{-1}(E_F)$ denotes the larger of two pre-images of the entropy function (16.96). Similar bounds between relative entropy of entanglement and concurrence or negativity were studied in [665].

Making use of the analytical formulae for entanglement measures we may try to explore the interior of the 15-dim set of mixed states. In general, the less pure a quantum state is, the less it is entangled: if $R = 1/[\text{Tr}\rho^2] \geq 3$ we enter the

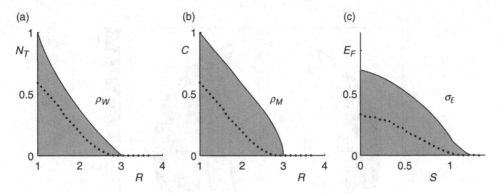

Figure 16.13 Upper bounds for measures of entanglement as a function of mixedness for two qubits: (a) negativity and (b) concurrence versus participation ratio $R = 1/(\text{Tr}\rho^2)$; (c) entanglement of formation versus von Neumann entropy. Grey shows the entire accessible region while dots denote the average taken with respect to the HS measure in $\mathcal{M}^{(4)}$.

separable ball. Movements along a global orbit $\rho \to U\rho U^\dagger$ generically changes entanglement. For a given spectrum $\vec{x}$ the largest concurrence C^*, which may be achieved on such an orbit, is given by Eq. (16.79) [490, 922]. Thus the problem of finding maximally entangled mixed states does not have a unique solution: it depends on the measure of mixedness and the measure of entanglement used. One may use for example a Rényi entropy, and for each choice of the pair of parameters q_1, q_2 one may find an extremal family of mixed states [952].

Figure 16.13 presents average entanglement plotted as a function of measures of mixedness computed with respect to the HS measure. For a fixed purity $\text{Tr}\rho^2$ the Werner states (16.55) produce the maximal negativity $\mathcal{N}_T$. On the other hand, concurrence C becomes maximal for the states [680]

$$\rho_M(y) \equiv \begin{bmatrix} a & 0 & 0 & y/2 \\ 0 & 1 - 2a & 0 & 0 \\ 0 & 0 & 0 & 0 \\ y/2 & 0 & 0 & a \end{bmatrix}, \quad \text{where} \quad \begin{cases} a = 1/3 & \text{if} \quad y \le 2/3 \\ a = y/2 & \text{if} \quad y \ge 2/3 \end{cases}$$

$$(16.125)$$

Here $y \in [0, 1]$ and $C(\rho) = y$ while $\text{Tr}\rho^2 = 1/3 + y^2/2$ in the former case and $\text{Tr}\rho^2 = 1 - 2y(1 - y)$ in the latter. A family of states σ_E providing the upper bound of E_F as a function of von Neumann entropy (see the line in Figure 16.13c) was found in [952]. Note that the HS measure restricted to pure states coincides with the Fubini–Study measure. Hence at $S = 0$ the average pure states entropy of entanglement reads $\langle E_1 \rangle_\psi = 1/3$ while for $R = 1$ we obtain $\langle C \rangle_\psi = \langle \mathcal{N}_T \rangle_\psi = 3\pi/16 \approx 0.59$ (see Section 15.6 and Problem 16.9).

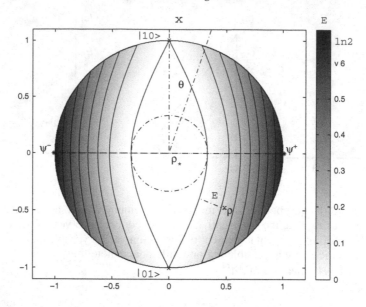

Figure 16.14 Entanglement of formation of two-qubit generalised Werner states (16.126) in polar coordinates. The white set represents separable states and the grey different degrees of entanglement.

Finally we show, in Figure 16.14, entanglement of formation for an illustrative class of two-qubit states

$$\rho(x, \vartheta) \equiv x(|\psi_\theta\rangle\langle\psi_\theta|) + (1 - x)\rho_* \quad \text{with} \quad |\psi_\theta\rangle = \frac{1}{\sqrt{2}}\left(\sin\frac{\vartheta}{2}|01\rangle + \cos\frac{\vartheta}{2}|10\rangle\right).$$
(16.126)

For $x = 1$ the pure state is separable for $\vartheta = 0, \pi$ and maximally entangled for $\vartheta = \pi/2, 3\pi/2$. The dashed horizontal line represents Werner states. The set $\mathcal{M}_S$ of separable states contains the maximal ball and touches the set of pure states in two points ($\times$). A distance E of ρ from the set $\mathcal{M}_S$ may be interpreted as a measure of entanglement.

***

Our story of bipartite entanglement has come to an end. But there is much more to learn, as one can see in reviews [375, 471]. There are also ideas here that have wider applications. The problem of dividing the set of states into orbits of a subgroup of the unitary group is familiar from the theory of coherent states, and thoughtful authors have asked [95, 546] if entanglement theory is best regarded as an instance of how quantum states are structured by a distinguished group of physical operations that are in some sense 'easy' to perform. The group can even be discrete, as is the Clifford group that is used in blueprints for the universal quantum computer. This leads to very concrete analogies to entanglement theory [166].

At the same time the idea of quantum resource theory, with entanglement as a prime example, is beginning to live a life of its own [156, 462]. So it will take some time before entanglement theory can be regarded as fully understood.

Problems

16.1 Prove that a minimal unextendible product basis, with $2N - 1$ vectors, cannot exist if N is even.

16.2 Prove that the totally symmetric polynomials of the Schmidt vector, $\mu_1 = \sum_{i\neq j} \lambda_i \lambda_j$, $\mu_2 = \sum_{i\neq j\neq k} \lambda_i \lambda_j \lambda_k$ or $\mu_{N-1} = \lambda_1 \lambda_2 \cdots \lambda_N$ are entanglement monotones [96, 839]. (See also [353] in which entanglement measures based on $(\mu_k)^{1/k}$ are introduced.)

16.3 Show that the *distillable entanglement*, the optimal asymptotic efficiency m/n of an entanglement concentration protocol transforming locally n copies of an arbitrary pure state $|\psi\rangle$ into m maximally entangled states $|\phi^+\rangle$ of the $N \times N$ system, is equal to the entanglement entropy $E(|\psi\rangle)$ [115].

16.4 Show that separability implies the majorisation relation (16.67).

16.5 Consider the following states of two qubits [694],

$$\sigma_1 = \frac{1}{3}\begin{bmatrix} 1 & 0 & 0 & 0 \\ 0 & 1 & 1 & 0 \\ 0 & 1 & 1 & 0 \\ 0 & 0 & 0 & 0 \end{bmatrix}, \quad \sigma_2 = \frac{1}{3}\begin{bmatrix} 1 & 0 & 0 & 0 \\ 0 & 0 & 0 & 0 \\ 0 & 0 & 0 & 0 \\ 0 & 0 & 0 & 2 \end{bmatrix}. \quad (16.127)$$

Show that σ_2 is separable, while σ_1 is entangled, even though both states are globally and locally isospectral (their partial traces have identical spectra).

16.6 Let us call ρ *locally diagonalizable* if it can be brought to the diagonal form $\rho = U\Lambda U^\dagger$ by a local unitary matrix $U = U_A \otimes U_B$. Are all separable states locally diagonalizable?

16.7 Show that if the smallest eigenvalue of a density matrix ρ of size N^2 is larger than $\frac{1}{N}[(N^2 - 2)/(N^2 - 1)]$, the state ρ is separable [742].

16.8 Show that the norm of the correlation matrix of the Fano form (16.49) satisfies an inequality: $\text{Tr}(\beta\beta^T) = 2\|\beta\|_2^2 \leq [N_A N_B(N_A N_B - 1) - 2N_B\|\vec{\tau}^A\|^2 - 2N_A\|\vec{\tau}^B\|^2]/4$. Is this bound sufficient to guarantee positivity of ρ?

16.9 Consider a random pure state of a 2×2 system written in its Schmidt form $|\Psi\rangle = \cos\chi|00\rangle + \sin\chi|11\rangle$ with $\chi \in [0, \pi/4]$. Show that the Fubini–Study measure on the space $\mathbb{C}P^3$ induces the probability distributions for the Schmidt angle [1016]

$$P(\chi) = 3 \cos(2\chi) \sin(4\chi) \quad \text{and} \quad P(C) = 3C\sqrt{1 - C^2} \qquad (16.128)$$

where the concurrence $C = \sin(2\chi)$. Find the mean angle and mean concurrence. For which angle χ_m is the volume of the local orbit maximal?

16.10 Estimate the average negativity $\mathcal{N}_T$ of a generic mixed state ρ of a large $N \times N$ system, distributed according to the HS measure.

17

Multipartite entanglement

The number two is ridiculous and can't exist.

– *Isaac Asimov*[1]

17.1 How much is three larger than two?

Up till now we have discussed entanglement for bipartite systems, described by states in a composite Hilbert space $\mathcal{H}_{AB}$. Any product state $|\psi_A\rangle \otimes |\psi_B\rangle$ is separable, and any other pure state is entangled. These notions can be generalised for multipartite systems in a natural way. A state of a system consisting of three subsystems and belonging to the Hilbert space $\mathcal{H}_{ABC} = \mathcal{H}_A \otimes \mathcal{H}_B \otimes \mathcal{H}_C$ is (fully) separable if it has a product form containing three factors, $|\psi_A\rangle \otimes |\psi_B\rangle \otimes |\psi_C\rangle$. All other states are entangled. This seems to be a simple and rather innocent extension, so one is tempted to pose a delicate question: Is there any huge qualitative difference between quantum entanglement in composite systems containing three or more subsystems and the known case of bipartite systems?

The answer is 'yes'. Recall that a general pure state of two subsystems with N levels each,

$$|\psi_{AB}\rangle = \sum_{i=1}^{N}\sum_{j=1}^{N} \Gamma_{ij} |i_A\rangle \otimes |j_B\rangle, \tag{17.1}$$

can *always* be written in the form

$$|\psi_{AB}\rangle = (U_A \otimes U_B) \sum_{i=1}^{N} \sqrt{\lambda_i} |i_A\rangle \otimes |i_B\rangle, \tag{17.2}$$

---

[1] Reproduced from Isaac Asimov. *The Gods Themselves*. Victor Gollancz, 1972.

and its entanglement properties are characterised by its Schmidt vector $\vec{\lambda}$. But a *general* tripartite pure state

$$|\psi_{ABC}\rangle = \sum_{i=1}^{N}\sum_{j=1}^{N}\sum_{k=1}^{N} \Gamma_{ijk} \, |i_A\rangle \otimes |j_B\rangle \otimes |k_C\rangle \qquad (17.3)$$

cannot be written in the form

$$|\psi_{ABC}\rangle = (U_A \otimes U_B \otimes U_C) \sum_{i=1}^{N} \sqrt{\lambda_i} \, |i_A\rangle \otimes |i_B\rangle \otimes |i_C\rangle. \qquad (17.4)$$

As the dimension counting argument at the end of Section 9.2 reveals, such states are really very rare (although we will see that they have interesting properties). Turning to the mathematical literature in order to standardise the tensor Γ_{ijk} in some way, we learn many interesting things [251, 552, 696], but there is no magical recipe that solves our problem. Not all algebraic notions developed for matrices work equally fine for tensors.

In short, multipartite entanglement is much more sophisticated than bipartite entanglement, and it has a rich phenomenology already for pure states. If one considers the number of parties in a quantum composite system, then three is much more than two, four is more than three, and so it goes on. The issue of entanglement in multipartite quantum systems deserves therefore a chapter of its own. We will focus our attention on multi-qubit systems and on pure states, otherwise several chapters would be needed!

17.2 Botany of states

In Section 16.2 we provided rather elaborate pictures showing where separable and entangled states can be found. If we are content with just a sampling of possibilities we can make do with less. Let the states in a product basis be represented by the corners of a square, in the fashion of Figure 17.1a. We use bitstrings to label the corners, so that 10 (say) represents the state $|10\rangle \equiv |1_A\rangle \otimes |0_B\rangle$. Occasionally we will also use notation from the multipartite Heisenberg group, so that $Z = \sigma_z$, $X = \sigma_x$, $Y = iXZ = \sigma_y$. See Section 12.2. The qubit basis is chosen so that

$$Z|0\rangle = |0\rangle, \quad Z|1\rangle = -|1\rangle, \quad X|0\rangle = |1\rangle, \quad X|1\rangle = |0\rangle. \qquad (17.5)$$

Any superposition of two states forming an edge of the square is separable. On the other hand, equal weight superpositions of states represented by two corners on a diagonal give maximally entangled Bell states, $|\phi^{\pm}\rangle = |00\rangle \pm |11\rangle$ and $|\psi^{\pm}\rangle = |01\rangle \pm |10\rangle$. (Consult Eq. (16.1) if you prefer normalised states.) Note that the local unitary transformation $\mathbb{1} \otimes Z$ interchanges $|\phi^+\rangle$ with $|\phi^-\rangle$ and $|\psi^+\rangle$ with $|\psi^-\rangle$,

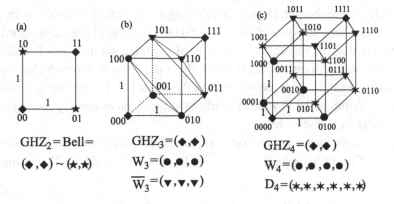

Figure 17.1 Distinguished pure states for systems of (a) two, (b) three and (c) four qubits. States represented by two corners at a diagonal of the square represent the Bell states of two qubits. Two corners on a long diagonal of a cube represent the state $|GHZ_3\rangle = |000\rangle + |111\rangle$ while two points on a long diagonal of a hypercube represent $|GHZ_4\rangle = |0000\rangle + |1111\rangle$. Two locally equivalent states W_3 and $\bar{W}_3$ are formed by two parallel triangles (case b) or tetrahedra (W_4 and $\bar{W}_4$, case c). D_4 denotes a four-qubit Dicke state (see Section 17.3).

while the equally local transformation $\mathbb{1} \otimes X$ interchanges the two diagonals of the square.

This kind of picture is easily generalised to the case of three qubits. The eight separable basis states will form the corners of a unit cube, see Figure 17.1b. To describe how far two corners are apart we count the number of edges one must traverse in order to go from the one to the other. The corners have been labelled so that this leads to the *Hamming distance* between two bitstrings of the same length, equal to the minimal number of bits which need to be flipped to transform one string into the other. Superpositions of two states at Hamming distance one, belonging to the same edge of the cube, are separable. Superpositions of states at Hamming distance two display only bipartite entanglement. For instance,

$$|\psi_{AB|C}\rangle = \frac{1}{\sqrt{2}}(|000\rangle + |110\rangle) = |\phi_{AB}^+\rangle \otimes |0_C\rangle. \qquad (17.6)$$

States that cannot be decomposed in any such way are said to exhibit *genuine multipartite entanglement*.

It is then thinkable that a balanced superposition of two states corresponding to two maximally distant corners, at Hamming distance three, is highly entangled. This is indeed the case. The *GHZ state*

$$|GHZ\rangle = \frac{1}{\sqrt{2}}(|000\rangle + |111\rangle), \qquad (17.7)$$

is named after Greenberger, Horne, and Zeilinger.[2] Its entanglement is quite fragile: if we trace out any subsystem from the GHZ state we obtain a separable state, which means that all the entanglement is of a global nature. See Problem 17.1. Interestingly, this property holds if and only if the state is *Schmidt decomposable* [894], that is to say if it can be written on the form (17.4). Among such states the GHZ state is distinguished by the property that if one traces out any two subsystems, the maximally mixed state results.

The GHZ state has many curious properties. If we rewrite it by introducing a new basis in Charlie's factor, $|+\rangle = |0\rangle + |1\rangle$ and $|-\rangle = |0\rangle - |1\rangle$, we find that it becomes

$$|GHZ\rangle = |\phi_{AB}^+\rangle \otimes |+\rangle_C + |\phi_{AB}^-\rangle \otimes |-\rangle_C, \qquad (17.8)$$

where again the maximally entangled Bell states appear. At the outset all three parties agree that the state assignment governing the measurements of Alice and Bob is $\rho_{AB} = \mathrm{Tr}_C |GHZ\rangle\langle GHZ|$. However, depending on what he chooses to do and on what the outcomes of his measurements are, Charlie can change his state assignment to one of $|0_A 0_B\rangle$, $|1_A 1_B\rangle$, $|\Phi_{AB}^+\rangle$, or $|\Phi_{AB}^-\rangle$. When Alice and Bob report the outcome of their measurement it will be consistent with that – as well as with the original state assignment ρ_{AB}. From this point of view the GHZ state describes entangled entanglement [997].

The GHZ state is an example of a stabilizer state, defined in Section 12.6 as an eigenstate of a maximal abelian subgroup of the three-partite Heisenberg group. Here this subgroup is generated by the four commuting group elements $\mathbb{1}ZZ, Z\mathbb{1}Z, ZZ\mathbb{1}, XXX$. All other equal weight superpositions related to two 'antipodal' corners of the cube separated by three Hamming units are locally equivalent to the GHZ state. There are eight of them, two on each diagonal (such as $|010\rangle \pm |101\rangle$), and all can be brought to the GHZ form by means of a local transformation (such as $\mathbb{1}X\mathbb{1}$) belonging to the Heisenberg group. And together these eight states form an orthonormal basis composed of states locally equivalent to the GHZ state.

Another genuinely multipartite entangled state is the W state[3]

$$|W\rangle = \frac{1}{\sqrt{3}} (|001\rangle + |010\rangle + |100\rangle). \qquad (17.9)$$

---

[2] These authors went beyond Bell's theorem to obtain a contradiction between quantum mechanics and local hidden variables not relying on statistics [363, 364]. The three-qubit GHZ state was created in 1999 [152]. The present experimental record is a 14-qubit GHZ state in the form of trapped ions [672].

[3] Some experts associate this name to the representation of the state in the space spanned by the energy and labels of the subsystems, which could resemble the letter W. Others tend to believe it is related to the name of one of the authors of the paper [278] in which this state was investigated. It could be called 'ZHG' or 'anti-GHZ', as it appeared in a work [997] by the authors of the GHZ paper [363]. Incidentally, one can make a case for calling GHZ the 'Svetlichny state' [874].

In Figure 17.1b it appears as a triangle. Its entanglement is more robust than that of the GHZ state, in the sense that an entangled mixed state results if we trace out any chosen subsystem, and indeed one can argue that in this respect its entanglement is maximally robust [278]. Hence the W state cannot be Schmidt decomposable. Indeed it cannot be written as a superposition of less than three separable states [278]. See Problem 17.2.

Introducing a fourth qubit into the game we need an extra dimension to construct a hypercube with 16 corners. See Figure 17.1c. Superpositions of two states at Hamming distance one are fully separable, as the tensor product consists of four factors. Corners distant by two and three Hamming units correspond to states with bi-partite and tri-partite entanglement, respectively. Superpositions of two 'antipodal' corners of the hypercube distant by four Hamming units form genuinely four-party entangled states, such as

$$|GHZ_4\rangle = \frac{1}{\sqrt{2}}\left(|0000\rangle + |1111\rangle\right). \tag{17.10}$$

This is called a four-qubit GHZ state, or sometimes a *cat state* to honour the Schrödinger cat – which existed, or so we are told, in an equal weight superposition of classical states. A four-qubit analogue of the W state corresponds to the tetrahedron obtained by four permutations of the bitstring 0001. Again the entanglement of the W state is more robust than that of the GHZ state; we can define the *persistency* of entanglement [172] as the minimum number of local measurements that need to be performed in order to ensure that the state is fully disentangled regardless of the measurement outcomes. For the GHZ_K state the persistency equals 2, for the W_K state it equals $K - 1$. But this is not to say that the one state is more entangled than the other: they are entangled in different ways.

The list of interesting states can be continued, but Figure 17.1 already makes it very clear that we are going to need an organizing principle to survey them. Botanists divide the Kingdom of Plants into classes, orders, families, genera and species. It is reasonably clear that what corresponds to a division into species of the Kingdom of Multipartite States must be a division into orbits of the local unitary group. Hence we would say that two states $|\psi\rangle$ and $|\phi\rangle$ are *LU equivalent*, written $|\psi\rangle \sim_{\text{LU}} |\phi\rangle$, if and only if there exist unitary operators U_k such that

$$|\phi\rangle = U_1 \otimes U_2 \otimes \cdots \otimes U_K |\psi\rangle. \tag{17.11}$$

Since the overall phase of each unitary can be fixed, we can choose unitary matrices with determinant set to unity and divide the set of states into orbits of the product group $G_U = SU(N)^{\otimes K}$, if we are studying K quNits.

Of more direct relevance to K parties trying to perform some quantum task would be a division into *LOCC equivalent* sets of states, consisting of states that can be transformed (with certainty) into each other by means of local operations and classical communication, abbreviated LOCC. Fortunately, for pure states LU equivalence coincides with LOCC equivalence. In the case of two qubits this result was an immediate consequence of Nielsen's majorisation theorem (Section 16.4). The argument in the multipartite case is similar but more involved.[4]

How much will this buy us? The state space of K qubits has $2 \cdot (2^K - 1)$ real dimensions, while an orbit of the local unitary group has at most $3K$ dimensions because $SU(2)$ is a three-dimensional group. Already for three qubits we can look forwards to a five-dimensional set of inequivalent orbits.

A coarser classification may therefore be useful, corresponding to the division of plants into genera. This is offered by stochastic local operations and classical communication (*SLOCC*). The definition of SLOCC operations presented in Section 16.4 for the bipartite case can be extended for multipartite systems in a natural way. For completeness let us write an analogue of Eq. (16.44) presenting a SLOCC transformation acting on a K-partite state,

$$|\psi\rangle \rightarrow A_1 \otimes A_2 \otimes \ldots A_K |\psi\rangle. \tag{17.12}$$

To formulate a relation analogous to (16.45) between two SLOCC-equivalent states all matrices have to be invertible,

$$|\psi\rangle \sim_{\text{SLOCC}} |\phi\rangle = L_1 \otimes L_2 \otimes \cdots \otimes L_K |\psi\rangle. \tag{17.13}$$

Since normalisation does not play any role here we can freely assume that the matrices L_i have unit determinant. Thus the group that governs SLOCC equivalence for a system consisting of K subsystems with N levels each is the special linear group composed with itself K times, $G_L = SL(N, \mathbb{C})^{\otimes K}$.

The group G_L is the complexification of the group G_U, and has only twice as many real dimensions as the latter: so again the number of inequivalent orbits grows quickly with the number of qubits. Like good botanists we must stand ready to introduce yet coarser classification schemes as we proceed.

17.3 Permutation symmetric states

What we really need is a way of being able to recognise a given state at a glance (so that we do not have to rely on the way it looks with respect to some perhaps arbitrary basis). We cannot quite do this, but we can if we restrict ourselves to the symmetric subspace $\mathcal{H}_{\text{sym}}^{\otimes K}$ of the full Hilbert space $\mathcal{H}^{\otimes K}$. For K qubits this subspace is in itself a Hilbert space of dimension $K + 1$, and many states of interest such

---

[4] For many qubits this result is due to Bennett et al. [122]. The criterion for SLOCC equivalence – which we are coming to – was presented by Dür, Vidal and Cirac [278].

as the GHZ and *W* states – as well as the ground states of many condensed matter systems, and more – reside in it. We will look at it through the glasses of the stellar representation[5] from our Chapter 7.

The symmetric subspace admits an orthonormal basis consisting of the states

$$
|K - k, k\rangle = \binom{K}{k}^{-\frac{1}{2}} \sum_{\text{permutations}} |0\rangle^{\otimes K - k} \otimes |1\rangle^{\otimes k}, \qquad (17.14)
$$

where $k = 0, 1, \ldots, K$ and the summation is over all $\binom{K}{k}$ permutations of K-letter strings with $K - k$ symbols $|0\rangle$ and k symbols $|1\rangle$. The basis states $|K - k, k\rangle$ can be identified with the Dicke states – which were thought of as angular momentum states in Chapter 7, but above all they were thought of as sets of stars placed at the North and South celestial poles only. The notion of composition of pure states, introduced in (7.27), is useful here. For any two symmetric states $|\psi_1\rangle$ and $|\psi_2\rangle$ of K_1 and K_2 qubits, respectively, the composite state $|\psi_1\rangle \odot |\psi_2\rangle$ of $K = K_1 + K_2$ qubits is described by superposition of all n stars. See Figure 7.5. The reason why the stellar representation turns out to be useful is that the group of local unitaries acts on the symmetric subspace through its diagonal subgroup $SU(2)$. In other words, a local unitary transformation corresponds to a rotation of the celestial sphere housing the stars. Constellations of stars that can be brought into coincidence by means of a rotation correspond to states that are equivalent under local unitaries. A SLOCC transformation can also be visualised; it is effected by some $SL(2, \mathbb{C})$ matrix. More precisely the group of Möbius transformations acts effectively on the states, and this group is famously isomorphic to the group of Lorentz transformations $SO(3, 1)$. A general Lorentz transformation will cause the constellations of stars on the sky to change. In particular, if the observer moves with constant velocity towards the north pole of the celestial sphere all the stars will be seen closer to the north pole than they would be by an observer standing still.[6] A detailed description of the action of the group is given in the caption of Figure 17.2.

Once these points are understood we can address the issue of how the symmetric subspace is divided into orbits under local unitaries and SLOCC trans-formations. In fact the former question was already discussed, in considerable detail, in Section 7.2. The set of all symmetrised states has real dimension $2K$, and a typical orbit has dimension three (equal to the dimension of the rotation group), so the set of all orbits has dimension $2K - 3$ and will quickly become unmanageable as the number n of qubits grows. For $K = 3$ things are still simple though. Any set of three stars sits

---

[5] Many nice descriptions of this idea can be found [99, 326, 475, 644, 648]. Ours is perhaps closest to that of Ribeiro and Mosseri [774].

[6] This connection is explained by Penrose and Rindler [722]. The best possible introduction to Möbius transformations is the first chapter of the book by Ford [316].

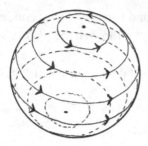
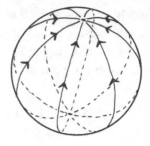

Figure 17.2 Left: a rotation of the celestial sphere, also known as an elliptic Möbius transformation [316]. Right: a Lorentz boost (a hyperbolic Möbius transformation). Combinations of the two are called four-screws [722]. A general Möbius transformation has two fixed points which can be placed anywhere on the sphere (four parameters); the fixed points are allowed to coincide. The amount of rotation and boost to include can also be chosen freely (two parameters), so the full group has six dimensions. In our illustration the fixed points sit at the north and south poles (at 0 and ∞ on the complex plane). Constellations of stars that can be related by a Möbius transformation represent SLOCC equivalent states.

on some circle on the sphere. By means of rotations this can be brought to a circle of constant latitude (one parameter), with one star at the Greenwich meridian and two stars placed at other longitudes (two parameters). Hence the set of differently entangled three-qubit states has three dimensions. By the time we get to four or more qubits a full description becomes at best very unwieldy. Still some special orbits are easily recognised in all cases. Although the physical interpretation is new, mathematically a GHZ state is identical to the noon state that we presented in Section 7.1, and its constellation of stars is a regular K-gon on some Great Circle on the sky. The set of all GHZ states form an orbit isomorphic to $SO(3)/Z_K = \mathbb{R}P^3/Z_K$. The W states are Dicke states, and their stars are confined to two antipodal points on the sphere. The set of W states is a 2-sphere.

Things simplify if we consider states equivalent under SLOCC. For symmetric pairs of qubits there are only two orbits: one entangled and one separable. See Figure 17.3. Moving on to $K = 3$ qubits, and keeping in mind the picture of the effect a Lorentz boost has on the sphere, we see that every set of three non-coinciding stars can be brought into coincidence with three stars placed at the corners of a regular triangle on the equator. An equivalent way to see this is to observe that any three distinct complex numbers can be mapped to any other three by means of a Möbius transformation. In fact there are only three orbits in this case. One orbit consists of all states with three coinciding stars. In Chapter 7 these states were called $SU(2)$ coherent states. Now they reappear as (symmetric) separable states. Then there is an orbit consisting of states for which exactly two stars coincide. This is the W class. Finally there is the GHZ class for which all stars are distinct. This gives us

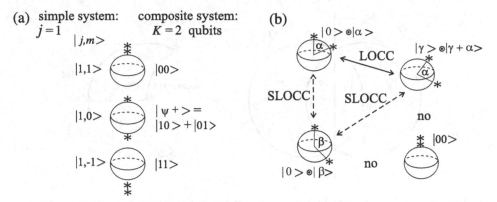

Figure 17.3 Two stars on the sphere may represent a state of a three-level system or a symmetric state of a two-qubit system. (a) Orthogonal basis $|j, m\rangle$ with $j = 1$ and $m = -j, \ldots, j$ is equivalent to $|00\rangle$, $|\psi^+\rangle$, $|11\rangle$. (b) Equivalence with respect to LOCC transformations (rotations of the sphere) and SLOCC transformations (preserving the degeneracy of the constellation).

three *degeneracy types*, denoted respectively by $\{3\}$, $\{2, 1\}$ and $\{1, 1, 1\}$. Three stars thrown randlomly on the sphere will land at different points, hence type $\{1, 1, 1\}$ is the generic one. States with two stars merged together can be well approximated with states of the latter type, while the converse statement does not hold.

When $n = 4$ the classification remains manageable, but requires some work [723, 774]. We define the *anharmonic cross ratio* of four ordered complex numbers as

$$(z_1, z_2; z_3, z_4) = \frac{(z_1 - z_3)(z_2 - z_4)}{(z_2 - z_3)(z_1 - z_4)}. \tag{17.15}$$

Recall that we are on the extended complex plane, so that ∞ is a respectable number representing the south pole of the Poincaré sphere. The point about this definition is that the cross ratio is invariant under Möbius transformations, or in our language that this function of the positions of the stars is invariant under SLOCC. We are not quite done though, because we are interested in unordered constellations of stars, and as we permute the stars the cross ratio will assume six different values

$$\left\{ \lambda, \frac{1}{\lambda}, 1 - \lambda, \frac{1}{1 - \lambda}, \frac{\lambda}{\lambda - 1}, \frac{\lambda - 1}{\lambda} \right\}. \tag{17.16}$$

The set of orbits is given by the set of values that the cross ratio can assume, provided it is understood that values related in this way represent the same orbit. This completely solves the problem of characterising the set of differently entangled states of four symmetrised qubits, in the sense of SLOCC. Figure 17.4 is a map of the set of all SLOCC orbits in the four-qubit case. Note that if we pick two states at

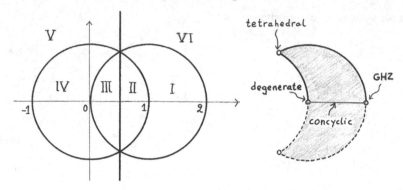

Figure 17.4 The cross ratio can take any value in the complex plane, but values related according to (17.16) are equivalent. Left: therefore the complex plane is divided into six equivalent regions – the picture would look more symmetric if drawn on the Riemann sphere. Right: using region I only we show the location of some states of special interest. Only a part of its boundary is included in the set of SLOCC equivalence classes.

random they are likely to end up in different places on the map, and then they are inequivalent under SLOCC.

Some interesting special cases deserve mention. First of all there are five different ways in which stars can coincide: $\{1, 1, 1, 1\}$, $\{2, 1, 1\}$, $\{2, 2\}$, $\{3, 1\}$ and $\{4\}$. Thus there are five degeneracy types. Type $\{1, 1, 1, 1\}$, in which all the stars sit at distinct points, is the generic one and the only one to contain a continuous family of SLOCC orbits. Within this family some special cases can be singled out. If the stars are lined up on a circle on the sky – they are then said to be *concyclic* – the cross ratio is real. If $\lambda = -1, 2$ or $\frac{1}{2}$ the state is SLOCC equivalent to a GHZ state – the four stars form a regular 4-gon on a Great Circle. If $\lambda = 0, 1, \infty$ two stars coincide. If $\lambda = e^{\pi i/3}, e^{-\pi i/3}$ the state is SLOCC equivalent to the *tetrahedral state*, in which the stars form the vertices of a regular tetrahedron.

The set of states for which some stars coincide is closed, but splits into four different orbits under SLOCC. Degeneracy type $\{2, 1, 1\}$ is the largest of these, but in itself it is an open set since states with more than two coinciding stars are not included. Its closure contains types $\{2, 2\}$ and $\{3, 1\}$, which are in themselves open orbits. Finally type $\{4\}$ consists of the separable states and forms a 2-sphere sitting in the closure of all the preceding types.[7]

As K increases beyond 4 the story becomes increasingly complicated. It is not only difficult to survey the SLOCC orbits, it even becomes hard to count the degeneracy types – this is the number p of partitions of the number n into positive

---

[7] Degeneracy types are called 'Petrov types' by Penrose and Rindler [723].

integers. This well known number theoretical problem was studied by Hardy and Ramanujan,[8] who obtained the asymptotic expression

$$p(K) \sim \frac{1}{4\sqrt{3}K} \exp\left(\pi\sqrt{2K/3}\right), \tag{17.17}$$

valid when K is large. Distinguishing between the possible degeneracy types by means of suitable invariants is then practically impossible. A typical SLOCC orbit has dimension 6 since this is the number of free parameters in the group $SL(2, \mathbb{C})$, which means that the set of orbits representing states with K non-coinciding stars always has real dimension $2K - 6$. For $K = 5$ we need two complex valued invariants – which are known functions of cross ratios, each cross ratio being a function of only four stars – but writing them down explicitly is a lengthy affair. For K qubits we would need $K - 3$ cross ratios, each computed using four stars at a time.

In Section 7.4 we mentioned the classical Thomson problem: assume the stars are charged electrons, and place them on a sphere in a way that minimises the electric potential. In a similar vein we can ask for the 'maximally entangled' constellation. We can for instance try to maximise the Fubini–Study distance to the closest separable pure state. For $n \geq 3$ it is known that the latter lies in the symmetric subspace too (although the proof is not easy [475]). The actual calculation is hard; for $n = 3$ one finds that the W-state is the winner. For $n = 4$ numerical results point to the tetrahedral state as being 'maximally entangled' in this sense, and for higher n the resulting constellations tend to be spread out in interesting patterns on the sphere [76, 85, 644, 646].

The stellar representation was used here to describe symmetric states of K qubits. The same approach works also for symmetric states of a system consisting of K subsystems with N levels each [661]. Separable states are then $SU(N)$ coherent states, see Sections 6.4 and 6.5, so a state can be represented by K unordered points in $\mathbb{CP}^{N-1}$. For symmetric states of higher dimensional systems a stellar representation works fine, but the sky in which the stars shine is just larger than ours.

17.4 Invariant theory and quantum states

In this section we make a head-on attempt to divide the set of multi-qubit states into species and genera. We would like to have one or several simple functions of the components of their state vectors, such that they take the same values only if

---

[8] The original result [402] goes back to 1918. An elementary proof of this formula was given in 1942 by Erdős [299] and simplified in 1962 by Newman [687].

the states are locally equivalent. Let us call such functions *local unitary* or *SLOCC invariants* as the case may be. We also want enough of them, so that when we have evaluated them all we can say that the states are locally equivalent if and only if the invariants agree.

In the case of bipartite entanglement we know how to do this, because the states $|\psi\rangle$ and $|\psi'\rangle$ are locally equivalent if and only if their Schmidt coefficients agree, that is if and only if the eigenvalues of the reduced density matrices $\rho_A = \text{Tr}_B|\psi\rangle\langle\psi|$ and $\rho'_A = \text{Tr}_B|\psi'\rangle\langle\psi'|$ agree. When we restrict ourselves to qubits only (with just one independent eigenvalue) this will happen if and only if their determinants agree. Letting the matrix Γ carry the components of the state vector we see from Eq. (16.19) that

$$\det \rho_A = \frac{1}{N^2}|\det \Gamma|^2. \qquad (17.18)$$

For qubits, when $N = 2$, we are in fact dealing with the *tangle*

$$\tau(|\Gamma\rangle) = |\det \Gamma|^2 = \frac{1}{4}|\epsilon_{ij}\epsilon_{i'j'}\Gamma^{ii'}\Gamma^{jj'}|^2. \qquad (17.19)$$

From now on we will be careful with tensors, just as we were in the earlier chapters. Because the transformation group is $U(2) \times U(2)$ there are two kinds of indices, unprimed and primed, and in the language of Section 1.4 $\Gamma^{ij'}$ is a contravariant tensor.[9] Indices may be summed over only if they are of the same kind and if one of them is upstairs and the other downstairs. Einstein's summation convention is in force, so repeated indices are automatically summed over. Thus when a local operation acts on a state it means that

$$\Gamma^{ii'} \rightarrow \Gamma'^{ii'} = L^i{}_j L^{i'}{}_{j'} \Gamma^{jj'}, \qquad (17.20)$$

where $L^i{}_j$ and $L^{i'}{}_{j'}$ are two independent matrices, unitary or general invertible as the case may be. As usual the anti-symmetric tensor ϵ_{ij} obeys $\epsilon_{00} = \epsilon_{11} = 0$ and $\epsilon_{01} = 1 = -\epsilon_{10}$, always. This means that it transforms not like a tensor but like a tensor density,

$$\epsilon_{ij} \rightarrow \epsilon'_{ij} = \epsilon_{kl}L^k{}_i L^l{}_j (\det L)^{-1} = \epsilon_{ij}. \qquad (17.21)$$

Hence $\det \Gamma$ changes with a determinantal factor under general linear transformations.[10] This will become important as we proceed.

For a pair of qubits the tangle is the only invariant we need. Can we do something similar for many qubits? This kind of question is addressed, in great generality, in

---

[9] In Section 4.4 we used $A, B, \ldots$ for the indices, but in this chapter A is for Alice, B is for Bob and C is (perhaps — opinions diverge on this point) for Charlie, so a change was called for. Indices always run from 0 to $N - 1$ (and usually $N = 2$).

[10] See Schrödinger [806] for a short introduction to tensor calculus, and Penrose and Rindler [722] for a long one.

the field of mathematics called *invariant theory*. Invariant theory arose in the nineteenth century like Minerva 'from Cayley's Jovian head. Her Athens over which see ruled and which she served as a tutelary and beneficient goddess was *projective geometry.*'[11] In its simplest guise the theory dealt with homogeneous polynomials in two variables, like those appearing in our Section 4.4, and this is what we need for permutation invariant states. The simplest example is the quadratic form

$$Q(u, v) = a_2 u^2 + 2a_1 uv + a_0 v^2, \tag{17.22}$$

where we renamed the independent components of the symmetric multispinor as a_2, a_1, a_0. The expression is *homogeneous* in the sense that $Q(tu, tv) = t^2 Q(u, v)$, hence it can be transformed to a second order polynomial $p(z) = Q(u, v)/u^2$ for the single variable $z = v/u$. Recall that what we really want to know is whether the polynomial $p(z)$ has multiple roots or not, and also recall that

$$p(z) = a_0 z^2 + 2a_1 z + a_2 = 0 \quad \Leftrightarrow \quad z = \frac{1}{a_0}(-a_1 \pm \sqrt{-\Delta}), \tag{17.23}$$

where we defined the *discriminant* of the polynomial,

$$\Delta \equiv a_0 a_2 - a_1^2. \tag{17.24}$$

The discriminant vanishes if and only if the roots coincide. Now suppose we make a Möbius transformation, according to the recipe in Eq. (4.34). We declare that the quadratic form Q transforms into a new quadratic form Q', according to the prescription that

$$Q'(u', v') = Q'(\alpha u + \beta v, \gamma u + \delta v) = Q(u, v). \tag{17.25}$$

This is again a homogeneous polynomial of the same degree, and its coefficients a'_i depend linearly on the original coefficients a_i. A small yet satisfying calculation confirms that

$$\Delta(a) = a_0 a_2 - a_1^2 = (\alpha\delta - \beta\gamma)^2 (a'_0 a'_2 - a'^2_1) = (\det G)^2 \Delta(a'), \tag{17.26}$$

where $\det G$ is the non-zero determinant of the invertible 2×2 matrix occurring in Eq. (4.34). The calculation shows that the number of distinct roots will be preserved by any linear transformation of the variables, and it also makes the discriminant our – and everybody's – first example of a *relative invariant* under the group $GL(2, \mathbb{C})$. It is strictly invariant under $SL(2, \mathbb{C})$.

This example can be generalised to homogeneous polynomials $Q(tu, tv) = t^n Q(u, v)$ of arbitrary order n, giving rise to nth order polynomials in one variable,

---

[11] The book by Olver [701] provides a very readable introduction to invariant theory; the book by Weyl [962] is a classic (and we hope that our sample of his prose will attract the reader).

and indeed to homogeneous polynomials in any number of variables. Under a linear transformation of the variables Eq. (17.25) will force the coefficients of Q' to become linear functions of the coefficients of Q. An invariant of *weight m* is a polynomial function of the coefficients which changes under invertible linear transformation only with a determinantal factor,

$$I(a) = (\det G)^m I(a').\tag{17.27}$$

We will also need *covariants* of weight m, which are polynomial functions of the coefficients and variables of a homogeneous polynomial $Q(u, v)$ such that

$$J(a, u, v) = (\det G)^m J(a', u', v').\tag{17.28}$$

Evidently Q itself is a covariant, but not the only one, fortunately, because we need enough of them so that they characterise the behaviour of the nth order polynomial $p(z) = Q(u, v)/t^n$.

Now a polynomial $p(z)$ has a multiple root if and only if it has a root in common with the polynomial $\partial_z p(z)$. In general two polynomials, one of order n and the other of order m, have a common root if and only if their *resultant* vanishes. By definition this is the determinant of a matrix which is constructed by writing the coefficients of the nth order polynomial followed by $m-1$ zeros in the first row. In the next $m-1$ rows one shifts the entries of the previous row cyclically one step to the right. Then comes a row which contains the coefficients of the mth order polynomial followed by $n-1$ zeros, and this is followed by $n-1$ rows where the entries are permuted as before.[12] We can now check if a cubic polynomial $p(z) = a_0 z^3 + 3a_1 z^2 + 3a_2 z + a_3$ has a multiple root by taking the resultant with its derivative. It is convenient to divide by an overall factor, and define

$$\Delta = \frac{1}{27a_0} \begin{vmatrix} a_0 & 3a_1 & 3a_2 & a_3 & 0 \\ 0 & a_0 & 3a_1 & 3a_2 & a_3 \\ 3a_0 & 6a_1 & 3a_2 & 0 & 0 \\ 0 & 3a_0 & 6a_1 & 3a_2 & 0 \\ 0 & 0 & 3a_0 & 6a_1 & 3a_2 \end{vmatrix}.\tag{17.29}$$

This invariant is called the discriminant of the cubic, and has weight 6 in the sense that $\Delta(a) = (\det G)^6 \Delta(a')$. It vanishes if and only if the corresponding quantum state has a pair of coinciding stars (in the sense of Section 17.3). But it does not allow us to distinguish between the degeneracy classes $\{2, 1, 1\}$ and $\{3\}$. To do so we need to bring in a covariant as well. For the cubic form we have the *Hessian covariant H*, which by definition is the determinant of the matrix of second derivatives of the form Q. In itself this is a quadratic form whose associated

---

[12] We present the construction as a piece of magic here, and refer to the literature for proofs [701].

discriminant is precisely equal to Δ. One can show that $H = 0$ and $Q \neq 0$ if and only if the cubic polynomial has a triple root. An additional covariant, denoted T, is often listed. It obeys

$$T^2 = 2^4 3^6 \Delta Q^2 - H^3. \qquad (17.30)$$

This is an example of a *syzygy* between invariants. (In astronomy a syzygy is said to occur when three celestial bodies line up on a great circle of the sky. The term is also used in poetry, but we omit the details.)

Note that the order of the discriminant, considered as a polynomial function of the coefficients, increases with the order of the polynomial. However, something interesting happens when we consider quartic polynomials (meaning, for us, symmetric states of four qubits). After calculating the discriminant Δ of the quartic according to the recipe, one can show that

$$\Delta = I_1^3 - 27 I_2^2, \qquad (17.31)$$

where

$$I_1 = a_0 a_4 - 4a_1 a_3 + 3a_2^2, \qquad I_2 = \begin{vmatrix} a_0 & a_1 & a_2 \\ a_1 & a_2 & a_3 \\ a_2 & a_3 & a_4 \end{vmatrix}. \qquad (17.32)$$

Both I_1 and I_2 are in themselves invariants, of weight 4 and 6 respectively. Hence we have three invariants, but the third is given as a polynomial function of the two others. The interpretation is that the quartic has a multiple root if and only if $\Delta = 0$. It has a triple or quadruple root if and only if $I_1 = I_2 = 0$. The corresponding quantum state is SLOCC equivalent to the tetrahedral state if and only if $I_1 = 0$ and $I_2 \neq 0$. There are three covariants in addition to the two independent invariants.

The difficulties increase with the order of the polynomial.[13] The question of how many independent polynomial invariants may exist has occupied numerous eminent mathematicians. An important issue was to check whether there exists a finite collection of invariants, such that any invariant can be written as their polynomial function. A deep theorem was proved by Hilbert himself:

Hilbert's basis theorem. *For any system of homogeneous polynomials, every invariant is a polynomial function of a finite number among them.*

In high brow language the invariants belong to a finitely generated ring. In a way the theorem offers more information than we really need, since we may be happy to regard a set of invariants as complete once every other invariant can be given

---

[13] Even Sylvester got himself into trouble with the septic polynomial [701], and a modern mathematician (Dixmier) says that 'la situation pour $n \geq 9$ est obscure'. We leave that as an exercise in French.

as some not necessarily polynomial function of that set, as happens in Eq. (17.30). Hilbert eventually provided a constructive proof of his theorem, but it is a moot question if his procedure can be implemented computationally. And we need a generalisation of the original theorem, since we are interested in subgroups of the full linear group. Fortunately the result holds also for the subgroups we are interested in, namely the local groups G_U and G_L [701]. It also holds for some finite subgroups: an accessible example of what Hilbert's theorem is about is the fact that every symmetric polynomial can be written as a polynomial function of the elementary symmetric functions (2.11).[14]

Leaving binary forms behind – or, in quantum mechanical terms, stepping out of the symmetric subspace – we can now try to characterise a given orbit of LU equivalent states by looking for the local unitary invariants. Each individual invariant will be a homogeneous polynomial in the components of the state vector, and we start our search with a guarantee that there exists a finite number of independent invariants, in terms of which all other invariants can be obtained by taking sums and products. If we can find such a set we are done: to verify whether two pure states are locally equivalent it will be enough to compute a finite number of invariants for both states and check whether the corresponding values coincide.

But we still have to identify a suitable set of independent invariants. In general this task requires lengthy calculations, and here we confine ourselves to describing the results in the three qubit case.[15] Any invariant under the group G_U of local unitaries can be represented as a homogeneous polynomial in the entries of the tensor (17.3) and their complex conjugates. There exists only one invariant of order two, $I_1 = \langle \psi | \psi \rangle = \Gamma^{i_1 i_2 i_3} \bar{\Gamma}_{i_1 i_2 i_3}$, equal to the norm of the state. (We now need three different kinds of indices!) There are three easily identified polynomial invariants of order four,

$$I_2 = \mathrm{Tr}\rho_A^2 = \Gamma^{i_1 i_2 i_3} \Gamma^{j_1 j_2 j_3} \bar{\Gamma}_{j_1 i_2 i_3} \bar{\Gamma}_{i_1 j_2 j_3}, \tag{17.33}$$

$$I_3 = \mathrm{Tr}\rho_B^2 = \Gamma^{i_1 i_2 i_3} \Gamma^{j_1 j_2 j_3} \bar{\Gamma}_{i_1 j_2 i_3} \bar{\Gamma}_{j_1 i_2 j_3}, \tag{17.34}$$

$$I_4 = \mathrm{Tr}\rho_C^2 = \Gamma^{i_1 i_2 i_3} \Gamma^{j_1 j_2 j_3} \bar{\Gamma}_{i_1 i_2 j_3} \bar{\Gamma}_{j_1 j_2 i_3}. \tag{17.35}$$

These are the purities of the three single-party reductions. By construction one has $1/2 \le I_i \le 1$ for $i = 2, 3, 4$. Note that the use of Einstein's summation convention saved us from writing out six sums per invariant.

At order six we have invariants such as $\mathrm{Tr}\rho_A^3$. However, because of the characteristic equation (see Section 8.1) applied to 2×2 matrices it happens that we have the syzygy $2\mathrm{Tr}\rho_A^3 = 3\mathrm{Tr}\rho_A \mathrm{Tr}\rho_A^2 - (\mathrm{Tr}\rho_A)^3$, so this is an example of an invariant

---

[14] The 14th problem on Hilbert's famous list concerns the issue of finiteness of a polynomial invariant basis for *every* subgroup of linear invertible transformations. It was settled (in the negative) in 1959 [362].
[15] We rely on Sudbery [870] and others [96, 613]. In their turn these authors relied on Weyl [962].

dependent on those we have introduced already. An independent polynomial invariant of order six, in this context called the *Kempe invariant* [519], is

$$I_5 = \Gamma^{i_1 i_2 i_3} \Gamma^{j_1 j_2 j_3} \Gamma^{k_1 k_2 k_3} \bar{\Gamma}_{i_1 j_2 k_3} \bar{\Gamma}_{j_1 k_2 i_3} \bar{\Gamma}_{k_1 i_2 j_3}. \tag{17.36}$$

It is clearly symmetric with respect to exchange of the subsystems, and one can show that $2/9 \le I_5 \le 1$. The minimum is attained by the W state. By considering various bipartite splittings of our systems one finds some further possibilities, but they are not independent invariants because

$$I_5 = 3\mathrm{Tr}[(\rho_A \otimes \rho_B) \, \rho_{AB}] - \mathrm{Tr}\rho_A^3 - \mathrm{Tr}\rho_B^3 \tag{17.37}$$

$$= 3\mathrm{Tr}[(\rho_A \otimes \rho_C) \, \rho_{AC}] - \mathrm{Tr}\rho_A^3 - \mathrm{Tr}\rho_C^3 = 3\mathrm{Tr}[(\rho_B \otimes \rho_C) \, \rho_{BC}] - \mathrm{Tr}\rho_B^3 - \mathrm{Tr}\rho_C^3.$$

Readers who try to verify these identities should be aware that doing so requires very deft handling of the characteristic equation for 2×2 matrices [870].

A sixth independent invariant is of order eight,

$$I_6 = |2\mathrm{Det}_3(\Gamma)|^2, \tag{17.38}$$

where the *hyperdeterminant* of the tensor $\Gamma^{i_1 i_2 i_3}$ is, for the moment, defined to be[16]

$$\mathrm{Det}_3(\Gamma) = \frac{1}{2}\epsilon_{i_1 j_1}\epsilon_{i_2 j_2}\epsilon_{k_1 \ell_1}\epsilon_{k_2 \ell_2}\epsilon_{i_3 k_3}\epsilon_{j_3 \ell_3} \Gamma^{i_1 i_2 i_3} \Gamma^{j_1 j_2 j_3} \Gamma^{k_1 k_2 k_3} \Gamma^{\ell_1 \ell_2 \ell_3}$$

$$= [\Gamma_{000}^2\Gamma_{111}^2 + \Gamma_{001}^2\Gamma_{110}^2 + \Gamma_{010}^2\Gamma_{101}^2 + \Gamma_{100}^2\Gamma_{011}^2]$$

$$-2[\Gamma_{000}\Gamma_{111}(\Gamma_{011}\Gamma_{100} + \Gamma_{101}\Gamma_{010} + \Gamma_{110}\Gamma_{001})$$

$$+\Gamma_{011}\Gamma_{100}(\Gamma_{101}\Gamma_{010} + \Gamma_{110}\Gamma_{001}) + \Gamma_{101}\Gamma_{010}\Gamma_{110}\Gamma_{001}]$$

$$+4[\Gamma_{000}\Gamma_{110}\Gamma_{101}\Gamma_{011} + \Gamma_{111}\Gamma_{001}\Gamma_{010}\Gamma_{100}]. \tag{17.39}$$

In the final expression indices are lowered for typographical reasons. To write out all the sums – and in fact for doing tensor algebra too – the graphical technique given in Figures 17.5–17.6 is helpful.[17] It is clear that the hyperdeterminant is somewhat analogous to the ordinary determinant, and that the invariant I_6 is analogous to the single invariant (17.19) used in the two-qubit case. We postpone the general definition of hyperdeterminants a little, and just observe that each term is a product of four components such that their 'barycentres' coincide with the centre of the cube shown in Figure 17.1(b). Examining the explicit expression for the hyperdeterminant one can convince oneself that $0 \le I_6 \le 1/4$. See also Table 17.1.

---

[16] Hyperdeterminants were introduced by Cayley in 1845 [198], who had finalised the theory of determinants a couple of years earlier. Hyperdeterminants returned in force 150 years later in the book by Gelfand, Kapranov and Zelevinsky [327], and were brought into this story by Coffman, Kundu and Wootters [230].

[17] It is due to Penrose [720, 722], who also provided the mathematical rigour behind it. Graphical notation as such goes back to Clifford [701]. Coecke and his collaborators have devised graphical techniques tailored to quantum mechanics [229].

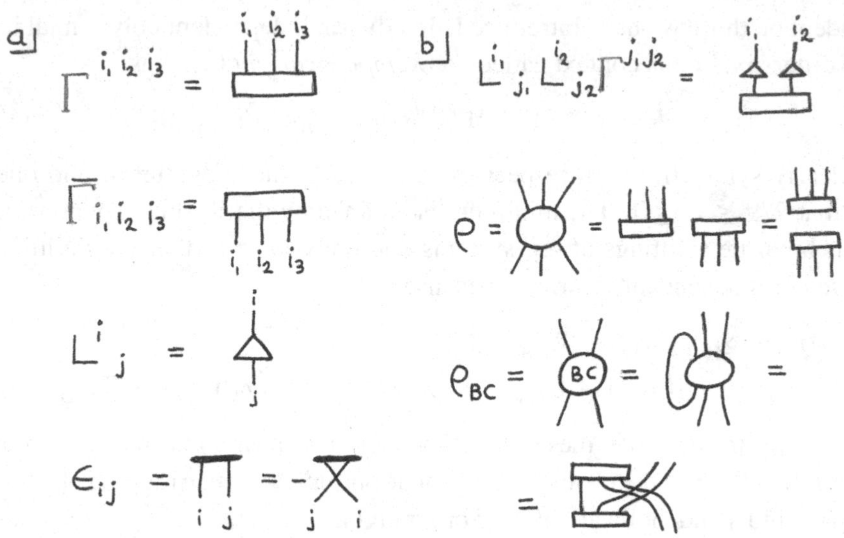

Figure 17.5 In graphical notation a tensor is represented as a shape with arms (upper indices) and legs (lower indices) attached. Open lines are labelled by the free indices, while lines connecting two tensors represent contracted indices (which need no label). A complication is that our tensors are tensors under a product group, so three different kinds of lines occur. In private calculations clarity is gained (but speed lost) if one uses three different colours to distinguish them. (a) Four examples of tensors. (b) New tensors from old. The pure state density matrix ρ equals an outer product of two tensors; it does not matter how we place the latter on the paper. We also show ρ_{BC}, obtained from a 3-party state ρ_{ABC} by means of a partial trace.

If an invariant takes different values for two states then these states cannot be locally equivalent. Reasoning in the opposite direction is more difficult, since we have to show that we have a complete set of invariants. For a normalised state $|\psi_{ABC}\rangle$ the first invariant is fixed, $I_1 = 1$, so we have only five remaining invariants to describe a local orbit. Note that one needs at least five such numbers, as the dimensionality of $\mathbb{C}P^7$ – the space of three-qubit pure states – is 14 and the number of parameters of the local unitary group $SU(2)^{\otimes 3}$ is $3 \times 3 = 9$. Unfortunately our invariants are still not enough to single out uniquely a local orbit. Clearly $I_i(|\psi\rangle) = I_i(|\psi\rangle^*)$ for all six invariants, so they cannot distinguish between a state and its complex conjugate. Grassl has found an additional invariant of order 12 which does complete the set [6]. The somewhat more modest problem of deciding when two given multi-qubit states can be connected by local unitaries has been discussed by Kraus [557].

Having described the local orbits in terms of polynomial invariants we can travel along the orbit in order to find some state distinguished by a simple *canonical*

Table 17.1 *LU invariants for some*
exemplary (normalised) states

State	I_2	I_3	I_4	I_5	I_6
separable	1	1	1	1	0
$\|\phi_A\rangle \otimes \|\psi_{BC}\rangle$	1	1/2	1/2	1/4	0
$\|\phi_B\rangle \otimes \|\psi_{AC}\rangle$	1/2	1	1/2	1/4	0
$\|\phi_C\rangle \otimes \|\psi_{AB}\rangle$	1/2	1/2	1	1/4	0
$\|W\rangle$	5/9	5/9	5/9	2/9	0
$\|GHZ\rangle$	1/2	1/2	1/2	1/4	1/4

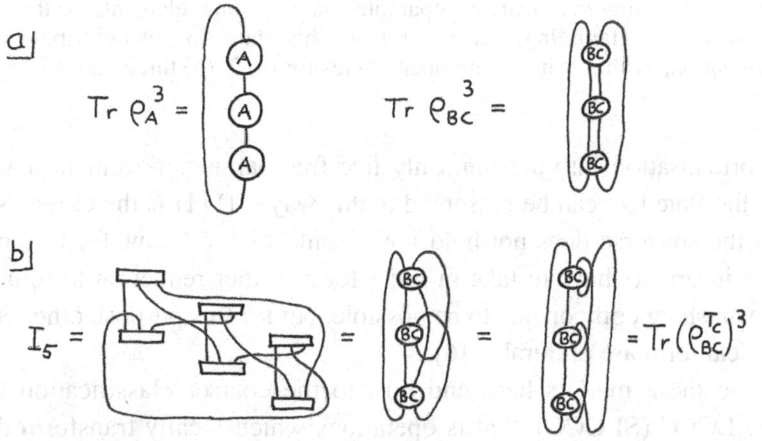

Figure 17.6 Examples of tensor calculations. (a) Here we take the trace of products of two different reduced density matrices. (b) As a non-trivial exercise we prove that $I_5 = \mathrm{Tr}(\rho_{BC}^{T_C})^3$ [133], where $\rho_{BC}^{T_C}$ is the partial transpose with respect to subsystem C of the reduced density matrix $\mathrm{Tr}_A\rho$.

form. In the two-qubit case the Schmidt decomposition provides an obvious choice. In the three-qubit case things are not so simple. By means of three local unitary transformation we can always set three out of the eight components of a given state vector to zero. See Figure 17.7. (For K quNits we can set $K \cdot N(N-1)/2$ components to zero [194].) An elegant way to achieve this is to observe that there must exist a closest separable state, in the sense of Fubini–Study distance. By changing the bases among the qubits we can assume this to be the state $|111\rangle$. An easy exercise (Problem 17.4) shows that this forces three of the components to be zero. Having come so far we can adjust the phase factors of the basis vectors so that four out of the five remaining components are real and non-negative. Thus we arrive at the canonical form

$$|\psi\rangle = r_0 e^{i\phi}|000\rangle + r_1|100\rangle + r_2|010\rangle + r_3|001\rangle + r_4|111\rangle. \qquad (17.40)$$

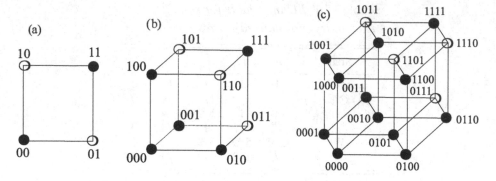

Figure 17.7 Using the nearest separable state as one element of the basis, basis vectors at Hamming distance 1 from this state do not contribute to the superposition, as shown here with open circles for (a) 2, (b) three and (c) 4 qubits.

Taking normalisation into account only five free parameters remain. (A word of warning: the state $|\psi\rangle$ can be presented in this way if $|111\rangle$ is the closest separable state, but the converse does not hold.) We would like to know, for instance, what values the invariants have to take in order for a further reduction to four, three or two non-vanishing components to be possible, but for this purpose other choices of the canonical form are preferable [6].

We leave these matters here and turn to the coarser classification based on stochastic LOCC (SLOCC), that is operations which locally transform the initial state into the target state with a non-zero probability. We will analyse the orbits arising when the group G_L acts on given pure states. In contrast to the group G_U of local unitaries, the group G_L is not compact and the orbits may not be closed – one orbit can sit in the closure of another, and hence the set of orbits will have an intricate topology.

The unitary invariants $I_1, \ldots, I_5$ will not survive as invariants under the larger group G_L – unsurprisingly, since they were constructed using complex conjugation, and G_L is the complexification of G_U. SLOCC operations can change even the norm I_1 of a state. We have one card left to play though, namely the hyperdeterminant $\mathrm{Det}_3(\Gamma)$. From its definition in terms of the tensor $\Gamma^{i_1 i_2 i_3}$ and the ϵ-tensor, in Eq. (17.39), it is clear that it changes with a determinantal factor under SLOCC. Indeed

$$\mathrm{Det}_3(\Gamma) = (\det L_1)^2 (\det L_2)^2 (\det L_3)^2 \, \mathrm{Det}_3(\Gamma'). \tag{17.41}$$

Hence the hyperdeterminant is a relative invariant, and a state for which it vanishes – such as the W state – cannot be transformed into a state for which it is non-zero – such as the GHZ state. Pure states of three qubits can be entangled in two inequivalent ways. There is some additional fine structure left, and indeed a set

Table 17.2 *SLOCC equivalence classes for three-qubit pure states* $|\psi_{ABC}\rangle$: r_A, r_B *and* r_C *denote ranks of single-partite reductions, while* $\mathrm{Det}_3(\Gamma)$ *is the hyperdeterminant of the 3-tensor representing the state.*

Class	r_A	r_B	r_C	$|\mathrm{Det}_3(\Gamma)|$	Entanglement
separable	1	1	1	$= 0$	none
$|\phi_A\rangle \otimes |\psi_{BC}\rangle$	1	2	2	$= 0$	bipartite
$|\phi_B\rangle \otimes |\psi_{AC}\rangle$	2	1	2	$= 0$	bipartite
$|\phi_C\rangle \otimes |\psi_{AB}\rangle$	2	2	1	$= 0$	bipartite
$|W\rangle$	2	2	2	$= 0$	triple bipartite
$|GHZ\rangle$	2	2	2	> 0	global tripartite

of six independent covariants exist that together provide a full classification. Rather than listing them all Table 17.2 takes the easy way out, and uses the observation that the rank of the reduced density matrices, $r_A = r(\rho_A) = r(\mathrm{Tr}_{BC}\rho_{ABC})$, cannot be changed by an invertible SLOCC transformation, so if $|\psi\rangle \equiv_{SLOCC} |\phi\rangle$ then all three local ranks have to be pairwise equal, for example $r_A(|\psi\rangle) = r_A(|\phi\rangle)$. From the table it is clear that the local ranks are enough to characterise a state as separable, or as containing bipartite entanglement only, but they cannot distinguish between the GHZ and W states – which, as we know from our tour of the symmetric subspace in Section 17.3, are SLOCC inequivalent. For this we need the hyperdeterminant.

This division into SLOCC equivalence classes is complete.[18] The GHZ class is dense in the set of all pure states while the W class is of measure zero, and any state in the latter can be well approximated by a state in the former [946]. This is an important point, and we will return to it in Section 17.6.

How do these results generalise to larger systems? For any number of quNits the local unitary group $G_U = SU(N)^{\otimes K}$ is compact, it has closed orbits when acting on the complex projective space $\mathbb{CP}^{N^K-1}$, and Hilbert's theorem (and its extensions) guarantees that these orbits can be characterised by a finite set of invariants. However, it is already clear from our overview of the situation in the symmetric subspace (Section 17.3) that the results would be so complicated that one may perhaps prefer not to know them. If instead we consider equivalence under SLOCC for modest numbers of qubits things do look better. Polynomial invariants are known for four [590, 628] and five qubits [629], although the results for the latter are partial only. In the four-qubit Hilbert space there exists infinitely many SLOCC orbits, and their classification is not straightforward [354, 707]. Still they can be organised into

---

[18] This was first shown by Dür, Vidal and Cirac [278]. Actually, in a way, the classification had already been given by Gelfand et al. [327] in their book. But they used very different words.

nine families of genuinely four-partite entangled states, six of them depending on a continuous parameter [224, 926]. We will give a slightly coarser classification of this case (into seven subcases) in Section 17.6. Osterloh and Siewert [708] have presented a way of organizing the invariants, called a *comb*, which exploits anti-unitary operators to the full. (Compare the discussion of concurrence in Section 16.9.)

To go to higher numbers of qubits we definitely need a yet coarser classification of entangled states. This is where the hyperdeterminant comes into its own – although it will have occurred to the reader that we have not really defined the hyperdeterminant, we simply gave an expression for it in the simplest case. To outline its geometrical meaning, let us return to the bipartite case, but for two N-level subsystems. The separable states are then described by the Segre embedding of $\mathbb{C}\mathbf{P}^{N-1} \times \mathbb{C}\mathbf{P}^{N-1}$ into $\mathbb{C}\mathbf{P}^{N^2-1}$. Any state can be described by a tensor Γ^{ij}, and generic entangled states can be characterised by the fact that its determinant is non-zero. For two quNits there are many intermediate cases for which the determinant vanishes even though the state is non-separable. In fact the rank of Γ gives rise to an onion-like structure, with all the non-generic cases characterised by the single equation $\det \Gamma = 0$. To unravel its meaning [669] we have to recall the notion of projective duality from Section 4.1. The idea is that a hyperplane in $\mathbb{C}\mathbf{P}^{N^2-1}$ can be regarded as a point in a dual copy of the same space. Now consider a hyperplane corresponding to the point $\bar{\Gamma}_{ij}$ in the dual space, and tangent to the separable Segre variety at the point $x^i y^j$. This hyperplane is defined by the equation

$$F = \bar{\Gamma}_{ij} x^i y^j = 0, \tag{17.42}$$

together with

$$\frac{\partial}{\partial x^i} F = \frac{\partial}{\partial y^i} F = 0 \quad \Leftrightarrow \quad \bar{\Gamma}_{ij} y^j = x^j \bar{\Gamma}_{ji} = 0. \tag{17.43}$$

The second set of equations ensures that the hyperplane has 'higher order contact' with the Segre variety, as befits a tangent hyperplane. But the condition that these equations do have a solution is precisely that $\det \bar{\Gamma} = 0$. This gives a geometrical interpretation of this very equation. It only remains to find the generalisation to the set of tangent planes of the generalised Segre variety $\mathbb{C}\mathbf{P}^{N-1} \times \mathbb{C}\mathbf{P}^{N-1} \times \cdots \times \mathbb{C}\mathbf{P}^{N-1}$.

This is not an easy task. For three qubits it turns out to be precisely the condition that the hyperdeterminant, as given above, vanishes. A similar polynomial equation (in the components of the tensor that defines the state) exists for any number of qubits. The polynomial is called the hyperdeterminant, and its degree is known. Unfortunately the latter rises quickly. For the case of four qubits the degree is 24 [327]. It is indeed useful to quantify four-qubit entanglement [628, 669], and it can be represented as a function of four other invariants of a smaller degree [271, 935], including the determinants of two-qubit reduced states. For larger number of qubits – and for quNits – it is hard to be explicit about this, but at least it is comforting to know that such a classification exists.

17.5 Monogamy relations and global multipartite entanglement

Any system consisting of three subsystems can be split into two parts in three different ways. Furthermore one can consider any two parties and investigate their entanglement. How is the entanglement between a single party A and the composite system BC, written $A|BC$, related to the pairwise entanglement $A|B$ and $A|C$?

Coffman, Kundu and Wootters analysed this question [230] using tangle as a measure of entanglement. Recall that this quantity equals the square of the concurrence, $\tau(\rho) = C^2(\rho)$, and is known analytically for any mixed state of two qubits [981]. See Eq. (16.103). They established the *monogamy relation*

$$\tau_{A|BC} \geq \tau_{A|B} + \tau_{A|C}. \tag{17.44}$$

Here $\tau_{A|B}$ denotes the tangle of the two-qubit reduced state, $\rho_{AB} = \text{Tr}_C \rho_{ABC}$, while $\tau_{A|BC}$ represents the tangle between part A and the composite system BC. Although the state of subsystems BC lives in four dimensions, its rank is not larger than two, as it is obtained by the reduction of the pure state $\rho_{ABC} = |\psi_{ABC}\rangle\langle\psi_{ABC}|$. This observation allows one to describe entanglement along the partition $A|BC$ with the two-qubit tangle (16.103) and to establish inequality (17.44).

Even though subsystem A can be simultaneously entangled with the remaining subsystems B and C, the sum of these two entanglements cannot exceed the entanglement between A and BC. This implies, for instance, that if $\tau_{A|BC} = \tau_{A|B} = 1$ then $\tau_{A|C} = 0$. Hence if the qubit A is maximally entangled with B, then it cannot be entangled with C.

To characterise entanglement in the three-qubit systems it is natural to consider the tangle averaged over all possible splittings. The average tangle of a pure state $\rho_{ABC} = |\psi_{ABC}\rangle\langle\psi_{ABC}|$ with respect to $1 + 2$ splittings reads

$$\tau_1(|\psi_{ABC}\rangle) \equiv \frac{1}{3}\left(\tau_{A|BC} + \tau_{B|AC} + \tau_{C|AB}\right). \tag{17.45}$$

Another related quantity describes the average entanglement contained in two-partite reductions,

$$\tau_2(|\psi_{ABC}\rangle) \equiv \frac{1}{3}\left(\tau_{A|B} + \tau_{B|C} + \tau_{C|A}\right). \tag{17.46}$$

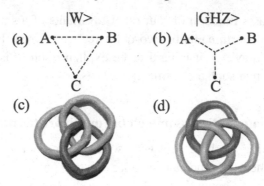

Figure 17.8 Schematic comparison between the distinguished three-qubit pure states with the help of strings and knots: if one subsystem is traced away from a system in the W state (panels a and c), the other two subsystems remain entangled. This is not the case for the GHZ state, thus it can be represented by three objects joined with a single thread or three *Borromean rings*, which enjoy an analogous property. See panels (b) and (d) and the cover of the book.

By construction, the quantities τ_1, τ_2 are non-negative and for any pure state can be computed analytically with the help of the Wootters formula (16.103). As entanglement is monotone with respect to partial trace, $\tau_{A|BC} \geq \tau_{A|B}$ (see Section 16.8) the relation $\tau_1 (|\psi_{ABC}\rangle) \geq \tau_2 (|\psi_{ABC}\rangle)$ holds for every pure state.

It is easy to check that τ_1 achieves its maximum for the GHZ state,

$$0 \leq \tau_1 (|\psi_{ABC}\rangle) \leq \tau_1 (|GHZ\rangle) = 1. \tag{17.47}$$

As discussed in Problem 17.1, after tracing out any single part of $|GHZ\rangle$, the remaining two subsystems become separable, so $\tau_2 (|GHZ\rangle) = 0$. See the cartoon sketch in Figure 17.8. The latter quantity, characterising the mean bipartite entanglement, is maximised by the W state [278],

$$0 \leq \tau_2 (|\psi_{ABC}\rangle) \leq \tau_2 (|W\rangle) = 4/9. \tag{17.48}$$

Thus taking into account the quantity τ_1 the GHZ state is 'more entangled' than the W state, while the opposite is true if we look at the measure τ_2. The fact that the maximum in (17.48) is smaller than unity is another feature of the monogamy of entanglement: an increase of the entanglement between A and B measured by tangle has to be compensated by a corresponding decrease of the entanglement between A and C or B and C.

The last, and the most important, measure, which characterises the *global entanglement*, is the *3-tangle* [230],

$$\tau_3 (|\psi_{ABC}\rangle) \equiv \tau_{A|BC} - \tau_{A|B} - \tau_{A|C}. \tag{17.49}$$

This quantity, invariant with respect to permutation of the parties, is also called *residual entanglement*, as it marks the fraction of entanglement which cannot be described by any two-body measures.

For any three-qubit pure state $|\psi_{ABC}\rangle$ written in terms of the components G^{ijk} one can use the Wootters formula (16.103) for tangle and find an analytical expression for the 3-tangle. It turns out [230, 278] that the 3-tangle is proportional to the modulus of the hyperdeterminant,

$$\tau_3 \left(|\psi_{ABC}\rangle\right) = 4|\text{Det}_3(G)|, \qquad (17.50)$$

which has already appeared in (17.39). Thus τ_3 is indeed invariant with respect to local unitary operations, as necessary for an entanglement measure and forms also an entanglement monotone [278]. Furthermore, τ_3 is a relative invariant under the action of the group $G_L = GL(2, \mathbb{C})^{\otimes 3}$, so it distinguishes the class of W states and the GHZ states with respect to SLOCC transformations. Due to the monogamy relation (17.44) the 3-tangle is non-negative, and it is equal to zero for any state separable under any cut. Its maximum is achieved for the GHZ state, for which $\tau_{A|BC} = 1$ and $\tau_{A|B} = \tau_{A|C} = 0$, so that $\tau_3 \left(|GHZ\rangle\right) = 1$. With respect to the Fubini–Study measure on the space of pure states of a three-qubit system it is possible to obtain [520] the average value $\langle\tau_3\rangle_\psi = 1/3$.

The monogamy relation (17.44), originally established for three qubits [230], was later generalised for several qubits by Osborne and Verstraete [705], while Eltschka and Siewert [293] derived monogamy equalities – exact relations between different kinds of entanglement satisfied by pure states of a system consisting of an arbitrary number of qubits.

A generalisation to quNits, with $N > 2$, is even more tricky. A monogamy relation based on tangle does not hold for several subsystems with three or more levels each [710]. Such relations can be formulated for negativity extended to mixed states by convex roof [527] and squared entanglement of formation [252]. Furthermore, for systems of an arbitrary dimension general monogamy relations hold for the squashed entanglement [528], as this measure is known to be additive.

Observe that the tangle of a pure state of two qubits can be written as $\tau(\psi) = |\langle\psi|\sigma_y \otimes \sigma_y|\psi^*\rangle|^2$, where $|\psi^*\rangle$ denotes the state after complex conjugation in the computational basis. In a similar way for a four-qubit pure state one can construct the 4-tangle $\tau_4(\psi) = |\langle\psi|\sigma_y^{\otimes 4}|\psi^*\rangle|^2$. This quantity, introduced by Wong and Christiansen [976], can be interpreted as four-party residual entanglement that cannot be shared between two-qubit bipartite cuts [354]. The 4-tangle is invariant with respect to permutations and extended for mixed states by convex roof forms an entanglement monotone [976].

17.6 Local spectra and the momentum map

The story of this section begins with an after-dinner speech by C. A. Coulson, who observed that the most interesting properties of many-electron systems tend to concern observables connecting at most two parties. To calculate them, only the reduced two-party density matrices are needed. The question he raised was: what conditions on a two-party density matrix ensure that it can have arisen as a partial trace over $K - 2$ subsystems of a pure K-partite state? This is a version of the *quantum marginal problem.*[19] In the analogous classical problem one asks for the conditions on a pair distribution p_{12}^{ij} (say) ensuring that it can arise as a marginal distribution from the joint distribution p_{123}^{ijk}. This is a question about the projection of a simplex in $\mathbb{R}^{N^3}$ to a convex body in $\mathbb{R}^{N^2} \oplus \mathbb{R}^{N^2} \oplus \mathbb{R}^{N^2}$. After finding the pure points of the projected body one has to calculate the facets of their convex hull. But this is a computationally demanding problem. Still the projection of a point is always a point, so the classical problem is trivial for pure states. Not so the quantum problem. Already when one starts from a pure quantum state one needs sophisticated tools from algebraic geometry, and from invariant theory, in order to set up an algorithm to solve it. Finding the restrictions on the resulting two-party density matrices is really hard.[20]

For bipartite pure states the story is simple. There are two reduced density matrices $\rho_A = \text{Tr}_B \rho$ and $\rho_B = \text{Tr}_A \rho$, and their spectra are the same (up to extra zeros, if the two subsystems differ in dimensionality). That spectrum can be anything however, which is dramatically different from the classical case where the subsystems are in pure states whenever the composite system is. To figure out what happens for three-qubit systems we choose a partition, say $A|BC$, and perform a Schmidt decomposition. Thus

$$|\Psi_{ABC}\rangle = \sqrt{\lambda_A} \sum_{i,j} a^{ij} |0_A i_B j_C\rangle + \sqrt{1 - \lambda_A} \sum_{i,j} b^{ij} |1_A i_B j_C\rangle. \tag{17.51}$$

By construction

$$\sum_{i,j} a^{ij} \bar{a}_{ij} = \sum_{i,j} b^{ij} \bar{b}_{ij} = 1, \qquad \sum_{i,j} a^{ij} \bar{b}_{ij} = 0. \tag{17.52}$$

As always we are free to choose the local bases, and this we do in such a way that $|0_A\rangle$ is the eigenstate of ρ_A with the smallest eigenvalue λ_A, and similarly for *all three* subsystems. We can then read off that

$$\lambda_B = \lambda_A \sum_j a^{0j} \bar{a}_{0j} + (1 - \lambda_A) \sum_j b^{0j} \bar{b}_{0j}, \tag{17.53}$$

[19] Coulson's question (1960) [236] inspired important work by A. J. Coleman [232, 231]. Their version of the problem concerned fermionic constituents. We consider distinguishable subsystems only.
[20] The problem was reduced to that of describing momentum polytopes (see below) by Klyachko [544] and by Daftuar and Hayden [243], in both cases relying on earlier work by Klyachko [545].

and similarly for λ_C. From this we find the inequalities

$$\lambda_B + \lambda_C = \lambda_A(a^{0j}\bar{a}_{0j} + a^{i0}\bar{a}_{i0}) + (1 - \lambda_A)(b^{0j}\bar{b}_{0j} + b^{i0}\bar{b}_{i0})$$

$$\geq \lambda_A \left(\sum_{i,j} a^{ij}\bar{a}_{ij} - |a_{11}|^2 \right) + (1 - \lambda_A) \left(\sum_{i,j} b^{ij}\bar{b}_{ij} - |b_{11}|^2 \right)$$

$$\geq \lambda_A \left(2 - |a_{11}|^2 - |b_{11}|^2 \right), \tag{17.54}$$

where we used $\lambda_A \leq 1 - \lambda_A$ and Eqs. (17.52) in the last step. However, a_{11} and b_{11} are the corresponding components of two orthonormal vectors, and hence they obey $|a_{11}|^2 + |b_{11}|^2 \leq 1$. (This is easy to prove using the Cauchy–Schwarz inequality, perhaps avoiding the composite indices for clarity.) Thus, repeating the exercise for all three partitions, we arrive at the triangle inequalities obeyed by the smallest eigenvalues of the three reduced density matrices, namely

$$\lambda_A \leq \lambda_B + \lambda_C ; \quad \lambda_B \leq \lambda_A + \lambda_C ; \quad \lambda_C \leq \lambda_A + \lambda_B. \tag{17.55}$$

Hence there are definite restrictions on the local spectra in the three-qubit case.

There are no other restrictions. To see this, consider the state

$$|\psi\rangle = a|001\rangle + b|010\rangle + c|100\rangle + d|111\rangle, \tag{17.56}$$

where all the components are real. Straightforward calculation verifies that

$$\begin{cases} \lambda_A = a^2 + b^2 \\ \lambda_B = a^2 + c^2 \\ \lambda_C = b^2 + c^2 \end{cases} \Rightarrow \begin{cases} 2a^2 = \lambda_A + \lambda_B - \lambda_C \\ 2b^2 = \lambda_A + \lambda_C - \lambda_B \\ 2c^2 = \lambda_B + \lambda_C - \lambda_A . \end{cases} \tag{17.57}$$

By choosing a, b, c we can realise any triple $(\lambda_A, \lambda_B, \lambda_C)$ obeying the triangle inequalities.

At the expense of some notational effort the argument can be repeated for K-qubit systems, resulting in *polygon inequalities* of the form

$$\lambda_k \leq \lambda_1 + \cdots + \lambda_{k-1} + \lambda_{k+1} + \cdots + \lambda_K. \tag{17.58}$$

Here λ_k denotes the smallest eigenvalue of the reduced density matrix for the kth subsystem. Again the inequalities are sharp: no further restrictions occur.[21]

Together with the inequalities $0 \leq \lambda_k \leq 1/2$ the polygon inequalities define a convex polytope known as the *entanglement polytope*, or sometimes as the *Kirwan polytope* (for a reason we will come to). It is the convex hull of $2^K - K$ extreme points which are easily found since the whole polytope is inscribed in a hypercube with corners whose K coordinates equal either 0 or $1/2$. Let us refer

---

[21] Full details can be found in the paper by Higuchi, Sudbery and Szulc [436], from which the whole argument is taken. Bravyi [164] provides some further results.

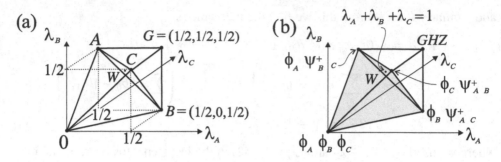

Figure 17.9 Entanglement polytope for three qubits: (a) formed by three smaller eigenvalues of the single-party reduced states, satisfying the triangle inequalities (17.55); (b) corners O and G represent fully separable and GHZ states, respectively. Three edges of the polytope connected to O denote states with bipartite entanglement only. States equivalent with respect to SLOCC to W belong to the shaded pyramid. Glueing eight such bipyramids together around the GHZ point we obtain a stellated octahedron or 'stella octangula', which is not convex. Compare Figure 11.3.

to the corner $(0, 0, \ldots, 0)$ as the separable corner, since separable states end up there. The inequalities (17.58) imply that all the K corners at Hamming distance 1 from the separable corner are missing, whereas all the other corners of the cube are there. On the long diagonal connecting the separable corner to the GHZ corner $(1/2, 1/2, \ldots, 1/2)$ – which is where the image of the GHZ state is to be found – we find the images of states in the symmetric subspace. The case $K = 3$ is an easily visualised bipyramid with five corners and six faces (Figure 17.9). When $K = 4$ there are 12 corners and 12 facets; the latter are of three different kinds and include four copies of the $K = 3$ bipyramid. When $K = 5$ there are 27 corners and 15 facets, and so it goes on.

The story becomes much more interesting once we ask where differently entangled states land in the polytope. For three qubits the story is simple. The W state lands at $(1/3, 1/3, 1/3)$, and for all states that are SLOCC equivalent to the W state one finds

$$\lambda_A + \lambda_B + \lambda_C \leq 1. \tag{17.59}$$

In fact the image of this class of states forms a polytope of its own, making up the lower pyramid in Figure 17.9. In this particular case elementary arguments suffice to prove it [395]. One can also show that states with bipartite entanglement only form the three edges emanating from the separable corner. These edges are polytopes of their own, which we can call $\mathcal{P}_{A|BC}$, $\mathcal{P}_{B|AC}$, and $\mathcal{P}_{C|AB}$. The image of the separable states is a single point $\mathcal{P}_{\text{sep}}$. Images of the generic states (SLOCC equivalent to the GHZ state) form an open set whose closure is the entire polytope.

In this way we have a hierarchy of polytopes, $\mathcal{P}_{\text{sep}} \subseteq \mathcal{P}_{A|BC} \subseteq \mathcal{P}_W \subseteq \mathcal{P}_{GHZ}$. A polytope in the hierarchy is a subpolytope of another if it corresponds to an orbit that lies in the closure of the other orbit, or in physical terms if a state from the former can be approximated by a state from the latter with arbitrary precision.

We have learned that the local spectra carry some information about the kind of entanglement: if $\lambda_A + \lambda_B + \lambda_C > 1$ the state investigated is of the GHZ type. (Perhaps we should note though that the spectra of the set of all mixed states will fill the entire cube, so if we want to use the local spectra as witnesses of GHZ-type entanglement we need a guarantee that the states we measure are close to pure.) For a state of the GHZ class there is a two-parameter family of orbits under local unitaries ending up at the same point in the entanglement polytope, while a pure state in the closure of the W class is determined uniquely up to local unitaries by its local spectra [795]. Generically this ambiguity can be resolved by considering also the spectra of all bipartite reduced density matrices, but for Schmidt decomposable states some ambiguity remains [614]: for states of the form $a|000\rangle + b|111\rangle$ the spectrum of every reduced density matrix agrees with that obtained from the mixed state $|a|^2|000\rangle\langle000| + |b|^2|111\rangle\langle111|$.

The four-dimensional entanglement polytope characterising the four-qubit system was analysed in detail by Walter et al. [945] and by Sawicki et al. [794]. Their arguments rely on invariant theory and symplectic geometry, and are no longer elementary. (The first step in one of these papers is to calculate the 170 independent covariants of the four-qubit system [171] for the nine different families found by Verstraete et al. [926].) The results are easy to describe though, and are summarised in Table 17.3. There are six kinds of four-dimensional subpolytopes, partially ordered by the inclusion relations $\mathcal{P}_5 \subseteq \mathcal{P}_3 \subseteq \mathcal{P}_2 \subseteq \mathcal{P}_1$ and $\mathcal{P}_5 \subseteq \mathcal{P}_4 \subseteq \mathcal{P}_6$. In addition to those we find lower dimensional subpolytopes describing states without genuine four-partite entanglement. This includes four facets (called F_2 in the table) that are simply copies of the three-qubit polytope. Anyway the botany has become surveyable.

Our story is not complete since we have said nothing about restrictions on the spectra when the partial trace is taken over two subsystems only. And an explicit generalisation to quNits is not easy (although the case of three qutrits is manageable [434]). But for us there is a more burning question to discuss: Why are the results as simple as they are? Why are all the conditions we encountered given by linear inequalities on the spectra of the reduced density matrices?

In the case of pure states of a system consisting of K subsystems of size N the entanglement polytope lives in $\mathbb{R}^{K(N-1)}$, so we are dealing with a map

$$\mathbb{CP}^{N^K-1} \to \mathbb{R}^{K(N-1)}. \tag{17.60}$$

Table 17.3 *The entanglement polytope for four qubits as a subpolytope of the four-dimensional cube, using notation from Section 5.7. The vertex 1111 corresponds to the spectrum* $(1/2, 1/2, 1/2, 1/2)$, *the three-partite entangled vertices are* $111\bar{1}$ *etc., the bi-partite entangled vertices are* $11\bar{1}\bar{1}$ *etc., and the separable vertex is* $\bar{1}\bar{1}\bar{1}\bar{1}$. *The extra vertex V in one of the subpolytopes corresponds to the spectrum* $(1/4, 1/2, 1/2, 1/2)$. *There are three kinds of facets* (F_i) *and six kinds of full-dimensional subpolytopes* $(\mathcal{P}_i)$. *The number of permutation equivalent copies of each is given at the bottom*

Vertex	F_1	F_2	F_3	$\mathcal{P}_1$	$\mathcal{P}_2$	$\mathcal{P}_3$	$\mathcal{P}_4$	$\mathcal{P}_5$	$\mathcal{P}_6$
1111	×						×		×
V				×					
$111\bar{1}$	×	×		×					
$11\bar{1}1$	×			×	×				
$1\bar{1}11$	×			×	×	×			×
$\bar{1}111$				×	×	×			×
$11\bar{1}\bar{1}$	×	×	×	×	×	×	×	×	×
$1\bar{1}1\bar{1}$	×		×	×	×	×	×	×	×
$1\bar{1}\bar{1}1$	×	×		×	×	×	×	×	×
$\bar{1}11\bar{1}$		×		×	×	×	×	×	×
$\bar{1}1\bar{1}1$			×	×	×	×	×	×	×
$\bar{1}\bar{1}11$			×	×	×	×	×	×	
$\bar{1}\bar{1}\bar{1}\bar{1}$		×	×	×	×	×	×	×	×
Perms	4	4	4	4	4	6	1	1	6

At this point we have to take up a thread that we left dangling at the end of Section 13.5, and explain the concepts of momentum maps and momentum polytopes. Their names suggest that they have something to do with momentum, and indeed as our motivating example (forcing us to take a detour through analytical mechanics) we choose angular momentum. On the phase space of a particle, with coordinates q^i and p_i, there exist three functions $J_i = J_i(q, p)$ such that

$$\delta q^i = \{q^i, \xi^k J_k\}, \qquad \delta p_i = \{p_i, \xi^k J_k\}, \tag{17.61}$$
$$\delta J_i = \{J_i, \xi^k J_k\} = \epsilon_{ik}{}^j \xi^k J_j \equiv \xi_i{}^j J_j. \tag{17.62}$$

Here we should think of ξ^i as a vector in the Lie algebra of the rotation group $SO(3)$, but we also rewrote it as a matrix $\xi_i{}^j$ using the natural matrix basis in the Lie algebra. For any given point x in phase space the vector J_i is a linear functional on the Lie algebra, meaning that $J_i \xi^i$ is a real number. Thus J_i sits in a vector space which is dual to the Lie algebra. The rotation group acts on its own Lie algebra by means of conjugation, and it also acts on the dual vector space by means of what we will soon refer to as the coadjoint action. In general, any Lie group G has a Lie algebra $\mathfrak{g}$. This is a vector space, and there exists a dual vector space $\mathfrak{g}^*$ such that for any element α in $\mathfrak{g}^*$ and any element ξ in $\mathfrak{g}$ there exists a real number $\langle \alpha, \xi \rangle$. The Lie group acts on $\mathfrak{g}$ by conjugation, and this gives rise to an action on $\mathfrak{g}^*$ as well. It is known as the *coadjoint action*, denoted Ad_g^* and defined by

$$\langle \text{Ad}_g^* \alpha, \xi \rangle = \langle \alpha, \text{Ad}_{g^{-1}} \xi \rangle = \langle \alpha, g^{-1} \xi g \rangle. \tag{17.63}$$

For a compact group like $SO(3)$ the distinction between $\mathfrak{g}$ and $\mathfrak{g}^*$ is slight, and the bilinear form $\langle \, , \, \rangle$ is simply given by the trace of a product of two matrices. But we have reached a point of view from which Eq. (17.62) is really quite remarkable. On the one hand, the rotation group acts on the functions J_i by means of canonical transformations of phase space. On the other hand, it acts on them through its adjoint action. And the equation says that these two actions are consistent with each other. This key observation motivates the definition of the *momentum map*, that we are now coming to.[22]

In general, suppose that we have a manifold with a symplectic form Ω defined on it (as in Section 3.4). Let a Lie group G act on it in the symplectic way. Thus to each group element g there is a map $x \rightarrow \Phi_g(x)$ from the manifold to itself, preserving the symplectic form. Let there exist a map μ from the manifold to the vector space $\mathfrak{g}^*$ dual to the Lie algebra $\mathfrak{g}$ of G. The group acts there too through the coadjoint action. Then μ is a momentum map provided it is *equivariant*,

$$\mu(\Phi_g(x)) = \text{Ad}_g^* \mu(x), \tag{17.64}$$

and provided that the vector field

$$\xi(x) = \frac{d}{dt}\bigg|_{t=0} \Phi_{e^{t\xi}}(x) \tag{17.65}$$

is the Hamiltonian vector field for the function $\mu_\xi = \langle \mu(x), \xi \rangle$. In our example the second condition is given (in different notation) by Eq. (17.61) and the first by Eq. (17.62).

Complex projective space – the space of pure quantum states – is a symplectic manifold, and in fact we have come across an important example of a momentum map already, namely

$$(Z^0, Z^1, \ldots, Z^n) \rightarrow \frac{1}{Z \cdot \bar{Z}}(|Z^0|^2, |Z^1|^2, \ldots, |Z^n|^2). \tag{17.66}$$

---

[22] A standard reference for the momentum map is the book by Guillemin and Steinberg [377]. Readers who want a gentle introduction may prefer to begin with the book by Springer [855].

The map is from $\mathbb{CP}^n$ to $\mathbb{R}^n$, and the image is a probability simplex. If we normalise the state, and write $Z^i = \sqrt{p_i}e^{i\nu_i}$, it looks even simpler. We know from Section 4.7 that p_i and ν_i are action-angle variables. The action variables p_i generate the action of an abelian Lie group which is the direct product of n copies of the circle group $U(1)$. Both the conditions for a momentum map are fulfilled. From the present point of view $\mathbb{R}^n$ is the dual space of the Lie algebra of this group. Remarkably, the image of $\mathbb{CP}^n$ is a convex subset there. So this is our first example of a momentum polytope.

A series of scintillating theorems generalise this simple observation.[23] They concern a compact Lie group G acting on a symplectic manifold M. The Lie group has a maximal abelian subgroup T ('T' for torus), with a Lie algebra $\mathfrak{t}$ (the Cartan subalgebra, see Section 6.5). We then have

First convexity theorem. *If an abelian group G admits a momentum map the image $\mu(M)$ of this momentum map is a convex polytope whose extreme points are the fixed points of the group action.*

For non-abelian groups this need not hold, but then we can divide $\mathfrak{t}$ into Weyl chambers (as in Section 8.5) and focus on the one containing vectors with positive entries in decreasing order. Call it $\mathfrak{t}_+$. Next we observe that because the momentum map is equivariant an orbit under G in M will be mapped into an orbit – called a *coadjoint orbit* – in the dual $\mathfrak{g}^*$ of the Lie algebra, and moreover each coadjoint orbit crosses the chosen Weyl chamber exactly once. (We saw this happen in Section 8.5.) We can then define a map Ψ from M to $\mathfrak{t}_+$ as the intersection of the image under the momentum map μ of an orbit in M and the positive Weyl chamber $\mathfrak{t}_+$.

Second convexity theorem. *Under the conditions stated the image $\Psi(M)$ is a convex polytope in $\mathfrak{t}_+$.*

The result holds for all symplectic manifolds, and the convex polytopes arising in this way are called *momentum polytopes*. Note that the restriction to the positive Weyl chamber is important – if we drop it we can obtain images like the non-convex stella octangula mentioned in the caption of Figure 17.9.

Let us glance back on our problem to classify orbits under the local unitary group $G_U = SU(2)^{\otimes K}$ in the many-qubit Hilbert space. It has a maximal abelian subgroup and a Cartan subalgebra $\mathfrak{t}$. The local spectra define diagonal density matrices in the dual space $\mathfrak{t}^*$, so the setting is right to apply what we have learned about the

---

[23] These theorems were motivated by the Schur–Horn theorem (Section 13.5), and are due to Atiyah [66], Guillemin and Sternberg [377], and Kirwan [533]. We remind the reader about Knutson's nice review [549]. Sawicki et al. [794] provide a good summary.

momentum map. We need a little more though, since we are mainly interested in equivalence under the SLOCC group G_L, and this group is not compact. But it is the complexification of the compact group G_U. We can then rely on the following [176, 376, 685]:

Third convexity theorem. *Let G be the complexification of a compact group, and let it act on* $\mathbb{CP}^n$. *Let* $G \cdot x$ *be an orbit through a point x, and let* $\overline{G \cdot x}$ *be its closure. Let the map* Ψ *be defined as above. Then*

  (a) *the set* $\Psi(\overline{G \cdot x})$ *is a convex polytope;*
  (b) *there is an open dense set of points for which* $\Psi(\overline{G \cdot x}) = \Psi(\mathbb{CP}^n)$;
  (c) *the number of such polytopes is finite.*

Concerning the proof we confine ourselves to the remark that invariant theory is present behind the scenes; the finiteness properties of the polytopes are related to the finitely generated ring of covariants. Anyway this proposition is the platform from which the results of this section have been derived, in the references we have cited [794, 945]. From it the problem of classifying multipartite entangled states looks at least more manageable than before.

17.7 AME states and error-correcting codes

In the bipartite case a maximally entangled state is singled out by the fact that any other state can be reached from by it means of LOCC. Moreover, a natural way to quantify bipartite entanglement is to start with a large but fixed number of copies of a state, and ask how many maximally entangled states we can distill from them by means of LOCC. That is, maximally entangled states serve as a kind of gold standard. In the multipartite case, where there are many different kinds of entanglement, we must learn to be pragmatic if we want to talk about 'maximal entanglement'. One possible way to proceed is to restrict our attention to the much simpler, bipartite entanglement that is certainly present. We can ask for the average bipartite entanglement of the individual subsystems, or perhaps for the amount of bipartite entanglement averaged over every possible bipartition of the system. Various choices of the measure of the bipartite entanglement can be made.

A very pragmatic way to proceed is to average the linear entropy (from Section 2.7) over all the reduced one-partite density matrices. Since we assume that the global state of the K qubits is pure, this is an entanglement measure. Under the name of the *Mayer–Wallach measure* it is defined by

$$Q_1(|\psi\rangle) = 2\langle S_L\rangle = 2\left(1 - \frac{1}{K}\sum_{k=1}^{K} \text{Tr}\rho_k^2\right). \tag{17.67}$$

It is normalised so that $0 \leq Q_1 \leq 1$, with $Q_1 = 0$ if and only if the state is separable and $Q_1 = 1$ if and only if each qubit taken individually is in a maximally mixed state.[24]

Now choose a subset X of qubits, with $|X| = k \leq K - k$ members, and trace out the remaining $K - k$ qubits; the assumption that $2k \leq K$ will simplify some statements. There are $K! / k! (K-k)!$ such bipartitions altogether, and we can define the entanglement measures

$$Q_k(|\psi\rangle) = \frac{2^k}{2^k - 1} \left(1 - \frac{k! (K-k)!}{K!} \sum_{|X|=k} \mathrm{Tr}\rho_X^2 \right). \qquad (17.68)$$

Again $0 \leq Q_k \leq 1$, with $Q_k = 0$ if and only if the global state $|\psi\rangle$ is separable, and $Q_k = 1$ if and only if it happens that all the reduced density matrices are maximally mixed.[25] For the W and GHZ states we find

$$Q_k(|W_K\rangle) = \frac{2^{k+1}}{2^k - 1} \frac{(K-k)k}{K^2}, \qquad Q_k(|GHZ_K\rangle) = \frac{2^{k-1}}{2^k - 1}. \qquad (17.69)$$

For the W_6 state we note that $Q_1 < Q_3 < Q_2$, which may seem odd. Moreover one can find pairs of states such that $Q_k(|\psi\rangle) > Q_k(|\phi\rangle)$ and $Q_{k'}(|\psi\rangle) < Q_{k'}(|\phi\rangle)$, for some $k' \neq k$. So there is no obvious ordering of the states into more or less entangled. Rather the measures Q_k capture different aspects of multipartite entanglement as k is varied. Moreover, if one changes the linear entropy to, say, the von Neumann entropy in the definition of Q_k, one may change the ordering of the states also when k is kept fixed. See Problem 17.7. Maximizing Q_k over the set of all states is a difficult optimisation problem, becoming computationally more expensive if we use the von Neumann entropy [148, 306]. In general, we do not obtain a convincing definition of 'maximally entangled' in this way.

States for which the upper bound $Q_k = 1$ is reached are called *k-uniform*. The GHZ state, like some other states we know (see Problem 17.3), is 1-uniform for any number of qubits. A k-uniform state (with $k \geq 2$) is always $(k - 1)$-uniform, since the partial trace of a maximally mixed state is maximally mixed. If all reduced density matrices that result when tracing over at least half of the subsystems is maximally mixed, the state is said to be *absolutely maximally entangled*, abbreviated *AME* [425]. Every measure of bipartite entanglement will have to agree that AME states – if they exist – are maximally entangled.

Let us move on from qubits to the general case of K subsystems with N levels each. In the product basis a pure state is described by

$$|\psi\rangle = \Gamma^{i_1 i_2 \dots i_K} |i_1 i_2 \dots i_K\rangle, \qquad (17.70)$$

[24] It was written in this form by Brennen [168]. Mayer and Wallach wrote it differently [656].
[25] Scott introduced these measures with the cautionary remark that they 'provide little intellectual gratification' [819].

where the indices run from 1 to N. If a bipartition is made the tensor can be described by two collective indices, μ running from 1 to N^k and ν running from 1 to N^{K-k}. Again we assume that $k \leq K - k$, and write the state as

$$|\psi\rangle = \Gamma^{\mu\nu}|\mu\nu\rangle. \tag{17.71}$$

Tracing out $K - k$ subsystems we obtain the reduced state

$$\rho_X = (\Gamma\Gamma^\dagger)^\mu_{\mu'}|\mu\rangle\langle\mu'|. \tag{17.72}$$

This state is maximally entangled (Section 16.3) if and only if the rectangular matrix $\sqrt{N^k}\Gamma^{\mu\nu}$ is a right unitary matrix (also known as an isometry), that is if and only if

$$N^K\Gamma\Gamma^\dagger = \mathbb{1}_{N^k}. \tag{17.73}$$

For a k-uniform state this has to be so for *every* bipartition of the K subsystems into $k + (K - k)$ subsystems. This is clearly putting a constraint on the tensor Γ which becomes increasingly severe as k grows. For an AME state, with an even number of subsystems, such tensors are known as *perfect* [717], while the corresponding reshaped matrices are known as *multiunitary* [356].

AME states do exist for two, three, four, five and six qubits. For three qubits they are the GHZ states. For five qubits an AME(5,2) state is[26]

$$
\begin{aligned}
|\Phi_2^5\rangle = &|00000\rangle \\
&+|11000\rangle + |01100\rangle + |00110\rangle + |00011\rangle + |10001\rangle \\
&-|11000\rangle - |01100\rangle - |00110\rangle - |00011\rangle - |10001\rangle \\
&-|11110\rangle - |01111\rangle - |10111\rangle - |11011\rangle - |11101\rangle. \tag{17.74}
\end{aligned}
$$

To see how this state arises, recall the multipartite Heisenberg group from Section 12.4, and in particular the notation used in Eq. (12.50). The cyclic properties of the state make it easy to see that it is left invariant by the operators

$$
\begin{aligned}
G_1 &= XXZ1Z, \quad G_2 = ZXXZ1, \quad G_3 = 1ZXXZ, \quad G_4 = Z1ZXX, \\
G_5 &= ZZZZZ. \tag{17.75}
\end{aligned}
$$

Moreover these five group elements generate a maximal abelian subgroup of the Heisenberg group, having 32 elements. In the terminology of Section 12.6 this means that $|\Phi_2^5\rangle$ is a stabilizer state.

---

[26] It was given in this form by Bennett et al. [121] and by Laflamme et al. [567] in a locally equivalent form with only eight terms in the superposition.

For the six-qubit AME state, see Problem 17.8. AME states do not exist for $K = 4$ [340, 435], or for more than six [476, 819], qubits. AME states built from four qutrits do exist. An example is [423]

$$\begin{aligned}
|\Phi_3^4\rangle = {} & |0000\rangle + |0112\rangle + |0221\rangle \\
& + |1011\rangle + |1120\rangle + |1202\rangle \\
& + |2022\rangle + |2101\rangle + |2210\rangle.
\end{aligned} \tag{17.76}$$

Glancing at this state we recognise the finite affine plane of order 3, constructed in Eq. (12.58). The labels from the first two factors of the basis vectors are used to define the position in the array, and then the labels from the last two factors label the remaining two sets of parallel lines.[27] It is not hard to show (Problem 17.9) that this is a stabilizer state.

The same construction, using an affine plane of order $N = 2$, gives the Svetlichny state discussed in Problem 17.5. For $N > 2$ the combinatorics of the affine plane ensures that we obtain a projector onto an N^2 dimensional subspace whenever we trace out two of the subsystems, which means that we obtain 2-uniform states of $N + 1$ quNits – provided an affine plane of order N exists, as it will if N is a prime number or a power of a prime number. Using results from classical coding theory one can show that AME states exist for any number of subsystems provided that the dimension of the subsystems is high enough [424]. For qubits, we can still ask for the highest value of k such that k-uniform states exist, and some asymptotic bounds are known for this. It is also known that if we restrict our states to the symmetric subspace, the upper bound for qubits is $k = 1$ [52]. So we cannot ask what AME states look like in the stellar representation, because it contains none (beyond GHZ$_3$).

We now turn to this section's other strand: *error-correcting codes*. We begin classically, with the problem of sending four classical bits through a noisy channel. (Perhaps the message is sent by a space probe far out in the solar system.) Assume that there is a certain probability p for any given bit to be flipped, and that these errors happen independently. In technical language, this is a *binary symmetric channel*. It may not be an accurate model of the noise. For instance, real noise has a tendency to come in bursts. But we adopt this model, and we want to be able to detect and correct the resulting errors. A simple-minded solution is to repeat the message thrice. Thus, instead of sending the message 0101 we send the message 010101010101. We can then correct any single error by means of a majority vote. But since the length of the message has increased, the probability that two errors occur has increased too. The trade-off is studied in Problem 17.10.

---

[27] There are a number of combinatorial ideas one can use to construct highly entangled states [356, 357].

It pays to adopt a geometric point of view. We regard each bitstring of length n as a vector in the discrete vector space $\mathbb{Z}_2^n$ over the finite field $\mathbb{Z}_2$ (the integers modulo 2; see Section 12.2). Our repetition code is a subspace of dimension four in $\mathbb{Z}_2^{12}$. This information is summarised by the *generator matrix* of the code, in this case

$$G_{(12,4)} = \begin{bmatrix} 1 & 0 & 0 & 0 & 1 & 0 & 0 & 0 & 1 & 0 & 0 & 0 \\ 0 & 1 & 0 & 0 & 0 & 1 & 0 & 0 & 0 & 1 & 0 & 0 \\ 0 & 0 & 1 & 0 & 0 & 0 & 1 & 0 & 0 & 0 & 1 & 0 \\ 0 & 0 & 0 & 1 & 0 & 0 & 0 & 1 & 0 & 0 & 0 & 1 \end{bmatrix}. \tag{17.77}$$

The four row vectors form a basis for the code, that is for a linear subspace containing altogether 2^4 vectors. Their first four entries carry the actual information we want to send. The reason why this code enables us to correct single errors can now be expressed geometrically. Define the *weight* of a vector as the number of its non-zero components. One can convince oneself that the minimal weight of any non-zero vector in this code is 3. Therefore the minimal Hamming distance between any pair of vectors in the code is 3, because by definition the Hamming distance between two vectors $\mathbf{u}$ and $\mathbf{v}$ is the weight of the vector $\mathbf{u} - \mathbf{v}$, which is necessarily a vector in the code since the latter is a linear subspace. Thus, what the embedding of $\mathbb{Z}_2^4$ into $\mathbb{Z}_2^{12}$ has achieved is to ensure that the 2^4 code words are well separated in terms of Hamming distance. Suppose a single error occurs during transmission, somewhere in this string. This means that we receive a vector at Hamming distance 1 from the vector $\mathbf{u}$. But the Hamming distance between two code words is never smaller than 3. If we surround each code vector with a 'ball' of vectors at Hamming distance ≤ 1 from the centre, we find that the vector received lies in one and only one such ball, and therefore we know with certainty which code word was being sent. This is why we can correct the error.

In general, a linear *classical* $[n, k, d]$ *code* is a linear subspace of dimension k in a discrete vector space of dimension n, with the minimal Hamming distance between the vectors in the subspace equal to d. The repetition code is the code $[12, 4, 3]$. Can we improve on it? Indeed, improvements are easily found by consulting the literature.[28] The generator matrix of the $[7, 4, 3]$ *Hamming code* is

$$G_{(7,4)} = \begin{bmatrix} 1 & 0 & 0 & 0 & 0 & 1 & 1 \\ 0 & 1 & 0 & 0 & 1 & 0 & 1 \\ 0 & 0 & 1 & 0 & 1 & 1 & 0 \\ 0 & 0 & 0 & 1 & 1 & 1 & 1 \end{bmatrix}. \tag{17.78}$$

---

[28] Classical error-correcting codes have been much studied since the time of Shannon's breakthrough in communication theory [828]. A standard reference is the book by MacWilliams and Sloane [635]. For a briefer account, see Pless [745].

With this code, the bitstring 0101 is encoded as the message 0101010. One can convince oneself that the minimum Hamming distance between the vectors in the code is again 3 (and indeed that these balls fill all of $\mathbb{Z}_2^7$), so that all single errors can be corrected by this code as well. Moreover, an elegant and efficient decoding procedure can be devised, so that we do not actually have to look through all the balls. To see how this works, write $G = [\mathbb{1}|A]$, and introduce the *parity check matrix* $H = [-A^{\mathrm{T}}|\mathbb{1}]$. By construction

$$HG^{\mathrm{T}} = 0. \tag{17.79}$$

In fact $H\mathbf{u} = 0$ for every vector in the code subspace. Now suppose that an error occurs during transmission, so that we receive the vector $\mathbf{u} + \mathbf{e}$. Then

$$H(\mathbf{u} + \mathbf{e}) = H\mathbf{e}. \tag{17.80}$$

It is remarkable, but easy to check, that if the error vector $\mathbf{e}$ is assumed to have weight 1 then it can be uniquely reconstructed from the vector $H\mathbf{e}$. The conclusion is that the Hamming code allows us to correct all errors of weight 1 without, in fact, ever inspecting the message itself. It is enough to inspect the *error syndrome* $H\mathbf{e}$.

The Hamming code is clearly more efficient than the simple repetition code, which required us to send three times as many bits as the number we want to send. If more than one error occur during the transmission we can no longer correct it using the [7, 4, 3] code, but there is a [23, 12, 7] code, known as the *Golay* code, that can deal with three errors. And so on. In fact, for a binary symmetric channel Shannon proved, with probabilistic methods, that the probability of errors in the transmission can be made smaller than any preassigned ϵ if we choose the dimension of the code subspace and the dimension of the space in which it is embedded suitably (with an eye on the properties of the channel) [828].

Quantum error correction, of a string of qubits rather than bits, is necessarily a subtle affair, since it has to take place without gaining any information about the quantum state that is to be corrected.[29] Suppose that the initial state is $|\Psi\rangle$, belonging to some multipartite Hilbert space $\mathcal{H}_N^{\otimes K}$. In the course of time – for concreteness assume that the transmission is through time, in the memory of a quantum computer – the state will suffer decoherence due to the unavoidable interaction with the environment. There is nothing digital about this process. But we do assume that the state is subject to a CP map of the form (10.53). We then introduce a unitary operator basis (Section 12.1) and expand all the Kraus operators in this basis. For

---

[29] The first quantum error-correcting codes took the community by surprise when they were presented by Shor [835] and Steane [856] in the mid-1990s. They were linked to the Heisenberg group soon after that [191, 352], and the theory then grew quickly.

our purposes it is best to express this in the environmental representation. Thus what has happened is that

$$|\Psi\rangle|0\rangle_{\text{env}} \rightarrow \sum_I E_I|\Psi\rangle|\psi_I\rangle_{\text{env}}. \tag{17.81}$$

The elements of the unitary operator basis are called E_I here, since this basis is soon going to deserve its alternative name 'error basis'. The environmental states $|\psi_I\rangle_{\text{env}}$ are not assumed to be orthogonal or normalised, and at this stage the introduction of the error basis is a purely formal device. If it were true that

$$\langle\Psi|E_I^\dagger E_J|\Psi\rangle = \text{Tr}|\Psi\rangle\langle\Psi|E_I^\dagger E_J = \delta_{IJ}, \tag{17.82}$$

then we would be in good shape. The elements of the error basis would then give rise to mutually exclusive alternatives, and a measurement could be devised so that the state collapses according to

$$\sum_I E_I|\Psi\rangle|\psi_I\rangle_{\text{env}} \rightarrow E_I|\Psi\rangle|\psi_I\rangle_{\text{env}}. \tag{17.83}$$

Once we knew the outcome, we could apply the operator $E_I^\dagger$ to the state, and recover the (still unknown) state $|\Psi\rangle$.

If we replace the pure state $|\Psi\rangle\langle\Psi|$ with the maximally mixed state, the second equality in Eq. (17.82) would hold, but the maximally mixed state is not worth correcting. However, this suggests that we should restrict both the noise (the set of allowed error operators E_I) and the state $|\Psi\rangle$ in such a way that the state behaves as the maximally mixed state as far as the allowed noise is concerned. Beginning with the noise, we assume that it can be expanded in terms of error operators of the form

$$E_I = E_1 \otimes E_2 \otimes \cdots \otimes E_K, \tag{17.84}$$

with at most w of the operators on the right-hand side not equal to the identity operator. This is an error operator of *weight* w. Physically we are assuming that the individual qubits are subject to independent noise, which is a reasonable assumption, and also that not too many of the qubits are affected at the same time. The operator $E_I^\dagger E_J$ will therefore have at most $2w$ factors not equal to the identity. Next, we assume that the state is a $2w$-uniform state. Coming back to Eq. (17.82), we can perform the trace over the $K - 2w$ factors where the error operators contribute with just the identity, obtaining (for quNits)

$$\langle\Psi|E_I^\dagger E_J|\Psi\rangle = \text{Tr}|\Psi\rangle\langle\Psi|E_I^\dagger E_J = \frac{1}{N^{2w}}\text{Tr}E_I'^\dagger E_J', \tag{17.85}$$

where E_I' is the non-trivial part of E_I. It follows immediately from the definition of the error basis that we obtain the desired conclusion (17.82). Hence the $2w$-uniform state can be safely sent over this noisy channel, and w errors can be corrected afterwards.

This is only a beginning of a long story. In general, a *quantum error-correcting code* of distance d, denoted $[[K, M, d]]_N$, is an N^M dimensional subspace of an N^K dimensional Hilbert space, such that errors affecting only $(d - 1)/2$ quNits can be corrected along the lines we have described. The nicest error basis of all, the Heisenberg group (Chapter 12), comes into its own when such codes are designed. We have seen that a k-uniform state of K quNits is a $[[K, 1, k + 1]]_N$ quantum error-correcting code [819], but for more information we refer elsewhere [259, 751, 857]. For us it is enough that we have pointed a moral: beautiful states have a tendency to be useful too.

17.8 Entanglement in quantum spin systems

So far we have had, at the back of our minds, the picture of a bold experimentalist able to explore all of Hilbert space. In many-body physics that picture is to be abandoned. For concreteness, imagine a cubic lattice with 'atoms' described as quNits at each lattice site. The Hamiltonian is such that only nearest neighbours interact. There are K lattice sites altogether. We will always assume that K is finite, but we may let K approach Avogadro's number, say $K = 10^{23}$. Then the total Hilbert space under consideration has $N^{10^{23}}$ dimensions. This is enormous. No experimentalist is going to explore this in full detail. Indeed, nature itself cannot have done so during the 10^{17} seconds that have passed since the creation of the universe. Still multipartite entanglement in such systems will be important.[30] Indeed the tensor product structure of Hilbert space gives rise to an intricate geography of quantum state space, and we may hope to find the ground states of physically interesting systems in very special places.

The kind of systems we will focus on are called *quantum spin systems*. Some notation: subsets of lattice sites will be denoted $X, Y, \ldots$, and the number of sites they contain will be denoted by $|X|, |Y|, \ldots$. Such a number can be regarded as the volume of the subset. The complementary subsets are denoted by $\bar{X}, \bar{Y}, \ldots$, meaning that the entire lattice is the union of X and $\bar{X}$. The boundary of a region X is denoted by ∂X. The number of edges in the lattice passing through the boundary is denoted $|\partial X|$, and is regarded as the *area* of the boundary. Each lattice site is associated with a Hilbert space of finite dimension N, and the total Hilbert space is the tensor product of all them. The Hilbert space associated to a region X is the tensor product of all the Hilbert spaces associated to sites therein.

---

[30] Many good reviews exist [31, 903], and a forthcoming book by Zeng, Chen, Zhou and Wen [998] describes how quantum information meets quantum matter. In today's laboratories it is possible to design quantum many-body systems with desirable properties [138].

With our change of perspective in mind, let us first divide the lattice into two parts (somehow). Thus the Hilbert space is

$$\mathcal{H} = \mathcal{H}_X \otimes \mathcal{H}_{\bar{X}} = (\otimes_{x \in X} \mathcal{H}_x) \otimes (\otimes_{x \in \bar{X}} \mathcal{H}_x). \tag{17.86}$$

Now consider the amount of bipartite entanglement that arises if the global state is a pure state $|\psi\rangle$ chosen at random according to the Fubini–Study measure. By tracing out the complementary set $\bar{X}$ we obtain the reduced state $\rho_X = \text{Tr}_{\bar{X}}|\psi\rangle\langle\psi|$ associated to the subsystem formed by the atoms in subset X. Assume that $|X|$ is smaller than $|\bar{X}|$. We can now apply the Page formula (15.73), keeping in mind that the size of the system is $N^{|X|}$ and the size of the environment is $N^{|\bar{X}|}$. The result is

$$\langle E(|\phi\rangle)_X \rangle = \langle S(\rho_X)\rangle \approx |X| \ln N - \frac{1}{2} N^{|X|-|\bar{X}|}. \tag{17.87}$$

The second term describes a negative correction which can be neglected if $|X| \ll |\bar{X}|$. The leading term on the other hand grows proportionally to the number $|X|$ of subsystems in the region X. In this way we arrive at the following statement:

Volume law. For a generic multipartite quantum state the entanglement between any subregion X and a larger environment $\bar{X}$ scales as the *volume* of the region, as measured by the number $|X|$ of its subsystems.

This is also how the thermodynamical entropy behaves for physical systems with short range interactions – except that the thermodynamical entropy may well vanish at zero temperature, while the entanglement entropy does not, so they are distinct. The hope is that the relevant low energy states in a many-body system are far from generic – and that they are more amenable to computer calculations than the generic states are. Indeed the number of parameters needed to describe a generic state grows exponentially with the number of atoms, and this poses a quite intractable problem in computer simulations.

At this point a sideways glance on a seemingly very different part of physics is useful. In the classical theory of general relativity black holes are assigned an entropy proportional to their area. It is known as the *Bekenstein–Hawking* entropy, and as it stands it has no microscopic origin. In an attempt to provide one it was noticed that in certain situations the entanglement entropy also grows with area.[31] This suggests that we should replace the volume law with:

Area law. For physically important states of many-body quantum systems described by local interactions the entanglement between any subregion X and a larger environment $\bar{X}$ scales as the *area* of the region, as measured by the number of interactions across its boundary ∂X.

[31] This was first seen in 1983 by Sorkin [851]; the concrete calculation performed by him and his co-workers was in the context of a non-interacting quantum field theory [145]. For a review of black hole thermodynamics see Wald's book [944].

We will give a precise statement later, but first we want to understand what kinds of states can give rise to it.

The spatial dimension of the lattice matters here. We begin with a one-dimensional lattice, an open chain with K sites. To find a suitable representation of the states relevant to such a system we are going to apply the Schmidt decomposition (see Section 9.2) stepwise to a tensor carrying K indices. First we recall Eq. (9.14), which gives the singular value decomposition of an $N_1 \times N_2$ matrix C as $C = UDV$, where D is a diagonal matrix with Schmidt rank $r \leq \min(N_1, N_2)$. The (in general) rectangular matrices U and V obey $U^\dagger U = VV^\dagger = \mathbb{1}_r$. We also define the matrix

$$\Lambda \equiv D^2. \tag{17.88}$$

If we start from a normalised state then Λ is simply the reduced density matrix in diagonal form.

Now any K-index tensor with N^K components can be viewed as an $N \times N^{K-1}$ matrix, and the singular value decomposition can be applied to it:

$$\Gamma^{i_1 i_2 \ldots i_K} = \Gamma^{i_1 | i_2 \ldots i_K} = \sum_{a=1}^{r} A^{i_1}_a (D_1 V_1)_a^{|i_2 i_3 \ldots i_K}. \tag{17.89}$$

The matrix of left eigenvectors U_1 was renamed A, because we focus on its column vectors A^{i_1}. It will be observed that

$$\sum_{i_1} (A^{i_1})^\dagger A^{i_1} = U_1^\dagger U_1 = \mathbb{1}_{r_1}, \tag{17.90}$$

$$\sum_{i_1} A^{i_1} \Lambda (A^{i_1})^\dagger = \sum_{i_1} A^{i_1} DVV^\dagger D(A^{i_1})^\dagger = \Gamma^{i_1 i_2 \ldots i_K} \bar{\Gamma}_{i_1 i_2 \ldots i_K} = 1. \tag{17.91}$$

To get the final equality we had to assume that the state is normalised.

In the next step the $r_1 \times N^{K-1}$ matrix on the right-hand side of Eq. (17.89) is reshaped into an $r_1 N \times N^{K-2}$ matrix, and the singular value decomposition again does its job:

$$\Gamma^{i_1 i_2 \ldots i_K} = \sum_{a_1=1}^{r} A^{i_1}_{a_1} (D_1 V_1)^{i_2}_{a_1}{}^{|i_3 \ldots i_K} = \sum_{a_1=1}^{r_1} \sum_{a_2=1}^{r_2} A^{i_1}_{a_1} A^{i_2}_{a_1 a_2} (D_2 V_2)_{a_2}^{|i_3 \ldots i_K}. \tag{17.92}$$

The matrix $U^{i_2}_{a_a | a_2}$ was renamed $A^{i_2}_{a_1 a_2}$, and is regarded as a collection of N matrices of size $r_1 \times r_2$. This time we have

$$\sum_{i_2} (A^{i_2})^\dagger A^{i_2} = U_2^\dagger U_2 = \mathbb{1}_{r_2} \tag{17.93}$$

$$\sum_{i_2} A^{i_2} \Lambda_2 (A^{i_2})^\dagger = \sum_{i_2} A^{i_2} D_2 V_2 V_2^\dagger D_2 (A^{i_2})^\dagger = \Lambda_1. \tag{17.94}$$

Clearly this procedure can be iterated. We have arrived at

Vidal's theorem. *Every K-partite state of K quNits can be expressed in a separable basis as*

$$\Gamma^{i_1,\ldots,i_K} = \sum_{a_1=1}^{r_1}\sum_{a_2=1}^{r_2}\cdots\sum_{a_{K-1}=1}^{r_{K-1}} A_{a_1}^{i_1} A_{a_1 a_2}^{i_2} \cdots A_{a_{K-2}a_{K-1}}^{i_{K-1}} A_{a_{K-1}}^{i_K}, \tag{17.95}$$

where, if the state is normalised,

$$\sum_{i_k=0}^{N-1} (A^{i_k})^\dagger A^{i_k} = \mathbb{1}_{r_k}, \tag{17.96}$$

$$\sum_{i_k=0}^{N-1} A^{i_k} \Lambda_k (A^{i_k})^\dagger = \Lambda_{k-1}, \tag{17.97}$$

Λ_k *is a reduced density matrix in diagonal form, and the Schmidt ranks are bounded by $r_k \leq N^{[K/2]}$.*

The largest Schmidt ranks occur in the middle of the chain; $[N/2]$ denotes the largest integer not larger than $N/2$. Conditions (17.96–17.97) fix the representation uniquely up to orderings of the Schmidt vector and possible degeneracies there.[32]

Of course, once we have a state expressed in the form (17.95) we can relax conditions (17.96–17.97) without changing the state – they simply offer a canonical form for the representation. An important variation of the theme must be mentioned. Equation (17.95) is said to use *open boundary conditions*, but one can also use *periodic boundary conditions*, which here means that one uses a collection of matrices such that

$$\Gamma^{i_1,\ldots,i_K} = \text{Tr} A^{i_1} A^{i_2} \cdots A^{i_{K-1}} A^{i_K}. \tag{17.98}$$

This time there are no vectors at the ends, and indeed there are no ends – the chain is closed. Physically this formulation is often preferred, but there is no longer an obvious way to impose a canonical form on the matrices.

The only problem is that we risk getting lost in a clutter of indices. This is where the graphical notation for tensors comes into its own, see Figure 17.10.

Having organised the N^K components of the state into products of a collection of KN matrices we must ask: was this a useful thing to do? Indeed yes, for two reasons. The first reason is that we are not interested in generic states, in fact our

---

[32] We have chosen to begin this story with a theorem by Vidal (2003) [931], which gives a canonical form for matrix product states. But the story itself is much older. It really began with Affleck, Kennedy, Lieb, and Tasaki (1987) [9], who studied the ground states of isotropic quantum antiferromagnets. This was followed up by Fannes, Nachtergaele and Werner (1992) [310]. Important precursors include works by Baxter (1968) [100] and Accardi (1981) [5].

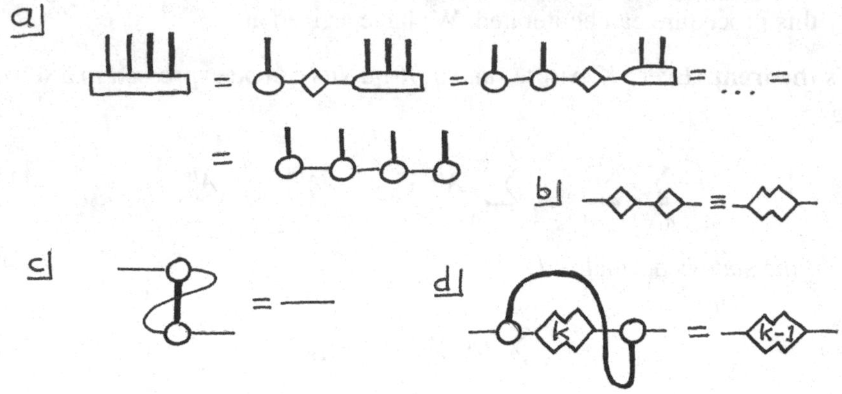

Figure 17.10 Graphical notation for matrix product states. First recall Figure 17.5. The equations that have been translated into graphical notation are: (a) (17.89), (17.92) and (17.95), (b) (17.88), (c) (17.96) and (d) (17.97). We try to make the lines proceed horizontally when the indices pertain to $\mathbb{C}^D$ (so 'arms and legs' are replaced by 'left and right arms'). Lines pertaining to $\mathbb{C}^N$ are drawn just a little bit thicker (confusion is unlikely to occur).

aim is to find special states obeying the Area Law. For this reason we now focus on states that can be written in the form (17.95), but with all the Schmidt ranks obeying $r_k \leq D$, where D is some modest integer called the *bond dimension*. We call them *matrix product states*, or *MPS* for short.[33] Splitting the chain in two and tracing out the contribution from one of the regions will result in a mixed state whose entanglement entropy behaves as

$$S(\rho_X) \sim \ln D. \tag{17.99}$$

We conclude that the strict Area Law holds for a one-dimensional chain if and only if D is independent of the number of subsystems.

There are many interesting states with low bond dimensions. Separable states have bond dimension one. The GHZ$_K$ state has bond dimension two, and moreover the matrices A^{i_k} for all sites k between 2 and $K - 1$ are the same, namely

$$A^0 = \begin{pmatrix} 1 & 0 \\ 0 & 0 \end{pmatrix}, \qquad A^1 = \begin{pmatrix} 0 & 0 \\ 0 & 1 \end{pmatrix}. \tag{17.100}$$

With a suitable choice of vectors at the ends we recover the GHZ state. The W_K state also has bond dimension two, but the latter increases for the other Dicke states. Importantly, ground states of interesting Hamiltonians (leading to the Area Law) will have modest bond dimension, in particular the AKLT ground state [9] (which

---

[33] Indispensable references, in addition to those mentioned in the previous footnote, include papers by Perez-Garcia et al. [729] and by Verstraete, Murg and Cirac [927].

started off the subject [310]) has bond dimension two. See Problem 17.12 for some examples.

The special properties of the singular value decomposition ensure that we have a reasonable approximation scheme on our hands. The *Eckart–Young theorem* says that if we want to approximate a matrix M with another matrix $\hat{M}$ of lower rank D, in such a way that the L_2-norm $\text{Tr}(\hat{M} - M)^\dagger(\hat{M} - M)$ is the smallest possible, then we can do this by performing the singular values decomposition of M and setting all but the D largest Schmidt coefficients to zero.[34] This means that we can approximate any state by setting all but the D largest Schmidt coefficients to zero (and renormalising the remaining ones so that they again sum to one), at each step of the exact expression (17.95). The number of parameters in the approximation scales as KND^2, that is to say linearly in the number of subsystems. For a given D the set of all MPS is a subset of measure zero in the set of all states, but we can make the approximation of any state better by increasing D. Truncating at some reasonable integer D in a computer calculation is not so very different from representing real numbers by rational numbers with reasonable denominators. It works fine in everyday calculations. Conversely, if an exponential growth in the bond dimensions is encountered then the calculation cannot be done efficiently on a classical computer [931].

We still have to present at least some evidence that ground states of interesting physical systems can be well approximated by MPS. The question is whether they obey the Area Law.[35] They do, as proved by Hastings [409]. The conditions imposed on the Hamiltonian are that

$$H = \sum_{k=1}^{K} H_{k,k+1}, \qquad ||H_{k,k+1}||_\infty \le J, \qquad \Delta E > 0, \qquad (17.101)$$

for some J and for some energy gap ΔE between the ground state and the first excited state. The first condition insists that only nearest neighbours are coupled, the second bounds the largest eigenvalue of each individual term. Interactions of finite range can be dealt with by grouping sites together into single sites, so the restriction to interactions between nearest neighbours is not all that severe.

Hastings' proof is beyond our scope, but one key ingredient must be mentioned because we need it to state the theorem. Consider two disjoint regions X and Y of the lattice. Take operators A and B supported in X and Y, respectively. Hence $[A, B] = 0$. Time evolve A with a Hamiltonian obeying conditions (17.101),

$$A(t) = e^{iHt}Ae^{-iHt}. \qquad (17.102)$$

Then one finds [602]

---

[34] The Eckart–Young theorem (1936) [281] was later extended to cover all the L_p-norms.
[35] Much more can be said on the topic of why and how MPS represent ground states faithfully [924]. We take a short cut here.

Lieb–Robinson theorem. *Under the conditions stated there exist constants c and a and a velocity v such that*

$$||[A(t), B]||_\infty \leq c||A||_\infty ||B||_\infty e^{-a(d(X,Y)-v|t|)}. \tag{17.103}$$

The constant a is adjustable and can be chosen to be large provided that $d(X, Y)/v|t|$ is sufficiently large. Again the proof is beyond us, but take note of the physical meaning: this is a rigorous statement saying that an effective 'light cone' appears in the system. Up to an exponentially decaying tail, influences cannot propagate outwards from the region X faster than the Lieb–Robinson velocity v, and there will be definite bounds on how fast entanglement can spread through the system under local interactions [165].

Now we can state [409]

Hastings' Area Law theorem. *Let X be the region to the left (or right) of any site on a one-dimensional chain. For the ground state of a Hamiltonian obeying the conditions stated there holds the Area Law*

$$S(\rho_X) \leq c_0 \xi \ln \xi \ln N 2^{\xi \ln N}, \tag{17.104}$$

where c_0 is a numerical constant of order unity and

$$\xi = \max\left(\frac{12v}{\Delta E}, 6a\right). \tag{17.105}$$

Here ΔE is the energy gap, v is the Lieb–Robinson velocity, and the constant a is the one that appears in the Lieb–Robinson theorem.

The point is that the upper bound on the entanglement entropy depends on the dimensionality N of the subsystems, and on the parameters of the model, but not on the number of subsystems within the region X. All the vagueness in our statement of the Area Law has disappeared, at the expense of some precise limitations on the Hamiltonians that we admit. In the case of *critical systems*, for which $\Delta E \to 0$, one observes [189, 289, 575] a logarithmic dependence of entropy on volume, $S(\rho_X) \approx c_1 \log |X| + c_2$. This is a much milder growth than one expects from a generic state obeying the Volume Law.[36]

Once it is admitted that the Area Law holds for the ground states of physically interesting Hamiltonians, the task of finding these states is greatly simplified. A standard approach is to solve the Rayleigh–Ritz problem: the ground state of the Hamiltonian H is the vector that minimises the *Rayleigh quotient*

$$\frac{\langle \psi|H|\psi \rangle}{\langle \psi|\psi \rangle}. \tag{17.106}$$

---

[36] For one-dimensional chains, area laws can also be derived assuming exponential decay of correlations in the system, without any assumptions about the energy gap [158].

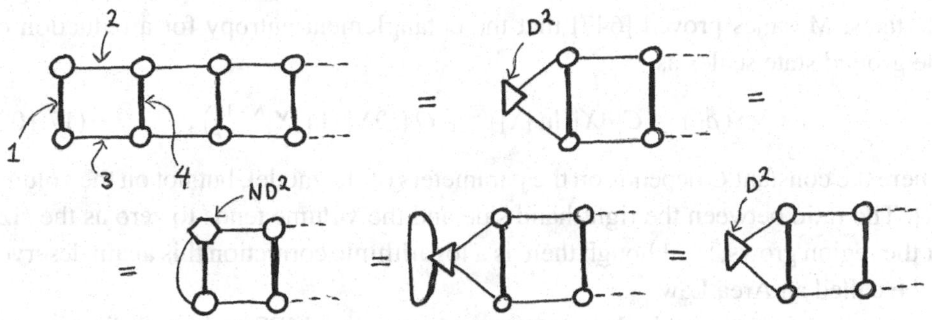

Figure 17.11 Normalising an MPS state: eventually all the sums have to be done. Doing the contractions in the order indicated ensures that the number of components that have to be stored during intermediate stages of the calculation stays reasonable (as indicated by the arrows).

But the set of all states grows exponentially with the number K of subsystems, so the best that can be done in practice is to find the minimum over a suitable set of trial states. The calculations become feasible once we assume that $|\psi\rangle$ belongs to the set of all MPS with a bond dimension growing at most polynomially in K. This idea is at the root of *the density matrix renormalization group* (DMRG) method, which has proven immensely useful in the study of strongly correlated one-dimensional systems.[37]

Note that if the computer is to be able to evaluate the Rayleigh quotient (17.106), it has to be told how. The first aim is to evaluate the single number

$$\langle\psi|\psi\rangle = \sum_{i_1,i_2,\ldots,i_K} \mathrm{Tr}A^{i_1}A^{i_2}\cdots A^{i_K}\mathrm{Tr}A^*_{i_1}A^*_{i_2}\cdots A^*_{i_K}. \tag{17.107}$$

It would clearly be a mistake to perform the traces first and the summation over the explicit indices afterwards – doing so means that the computer has to store $2N^K$ numbers at an intermediate stage of the calculation. In this case it is not difficult to propose a better strategy (and we do so in Figure 17.11), but for spin systems in spatial dimensions larger than one there are difficult issues of computational complexity to be addressed [807].[38]

So what about higher spatial dimension? Then the situation is not quite as clear-cut, but one still expects an area law to hold in suitable circumstances, even though the boundary ∂X of a region X in the lattice can have a complicated structure. In particular, with some assumptions about the decay of correlations and the density

---

[37] The DMRG is due to White (1992) [963]. The link to MPS was forged by Östlund and Rommer (1995) [709]. For a review (with useful calculational details of MPS) see Schollwöck [800].

[38] To learn how to actually perform calculations the review by Orús [703] and the paper by Huckle et al. [478] may be helpful.

of states, Masanes proved [647] that the entanglement entropy for a reduction of the ground state scales as

$$S(\rho_X) \leq C |\partial X| (\ln |X|)^N + O\left(|\partial X| (\ln |X|^{N-1})\right), \qquad (17.108)$$

where the constant C depends on the parameters of the model, but not on the volume $|X|$. The ratio between the right-hand side and the volume tends to zero as the size of the region grows, so although there is a logarithmic correction this again deserves to be called an Area Law.

So it makes sense to look for a way to generalise MPS to higher dimensions. To this end we begin by looking at the one-dimensional construction in a different way. We begin by doubling each site in the chain, so that the total Hilbert space becomes $\mathcal{H}_D^{\otimes 2K}$. The dimension of each factor is set to D, which may be larger than the original dimension N of the physical subsystems. Then we consider a quite special state there, namely a product of maximally entangled bipartite states

$$|\phi_{k,k+1}^+\rangle = \sum_{c=0}^{D-1} |c\rangle_{kR} |c\rangle_{(k+1)L}. \qquad (17.109)$$

(We worry about normalisation only at the end of the construction.) The factor Hilbert spaces that occur here are the rightmost factor from site k and the leftmost factor for site $k + 1$. In effect entanglement is being used to link the sites together. The total state of the 'virtual' (doubled) chain is taken to be the entangled pair state

$$|\Psi\rangle = |\phi_{12}^+\rangle |\phi_{23}^+\rangle \cdots |\phi_{K1}^+\rangle. \qquad (17.110)$$

Next, once for every site in the original chain, we introduce a linear map from the doubled Hilbert space $\mathbb{C}^D \otimes \mathbb{C}^D$ back to the N-dimensional Hilbert space $\mathbb{C}^N$ we started out with:

$$\mathcal{A}_k = A_{a_k b_k}^{i_k} |i_k\rangle \langle a_k b_k|, \qquad (17.111)$$

where summation over repeated indices is understood. It is now a straightforward exercise to show that

$$\mathcal{A}_1 \mathcal{A}_2 \ldots \mathcal{A}_K |\Psi\rangle = A_{a_1 a_2}^{i_1} A_{a_2 a_3}^{i_2} \ldots A_{a_K a_1}^{i_K} |i_i i_2 \ldots i_K\rangle. \qquad (17.112)$$

We have recovered a matrix product state with periodic boundary conditions, as in Eq. (17.98). When arrived at in this way it is called a *projected entangled pair state*, abbreviated *PEPS* [923]. See Figure 17.12.

The PEPS construction is easily generalised to any spatial dimension of the lattice. As shown in Figure 17.13, to describe a two-dimensional lattice we have to expand the Hilbert space $\mathbb{C}^N$ into a four-partite Hilbert space $(\mathbb{C}^N)^{\otimes 4}$, and then we introduce a perhaps site-dependent linear map $(\mathbb{C}^N)^{\otimes 4} \rightarrow \mathbb{C}^N$. Clearly any lattice,

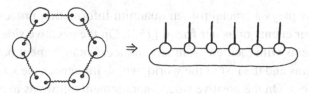

Figure 17.12 The PEPS construction. We start from a state consisting of entangled pairs in an auxiliary Hilbert space (forming a ring in this example), apply a linear map, and obtain a matrix product state (with periodic boundary conditions).

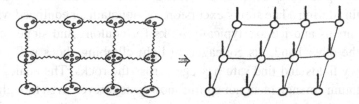

Figure 17.13 The PEPS construction works in all spatial dimensions. Applying the map we find that the tensors in the interior of the lattice have $4 + 1$ arms. In this way we obtain a tensor network, rather than just a matrix product state.

in any spatial dimension, can be handled in a similar way. But going beyond one dimension does increase every calculational difficulty, and we break off the story here. More can be found in the references we have cited.

We have now seen some simple examples of *tensor networks*. The basic idea is to view a tensor of valence m as an object with m free legs, or with both arms and legs if the distinction between upper and lower indices is important, and perhaps with several different kinds of arms and legs. Then we choose an arbitrary undirected graph – this may be a one-dimensional chain, a cubic lattice or something much more general – and assign a tensor to each vertex, in such a way that each edge in the graph corresponds to a pair of contracted indices. Tensor networks have a long history and an active present. From the beginning there was a dream to see the geometry of space emerge from the geometry of quantum states [720]. Perhaps the application of tensor networks to quantum spin systems is beginning to substantiate the dream?

***

There is much more to say. We have said nothing about graph states or toric codes, which is where the multipartite Heisenberg group comes into play. We have said nothing about the vast field of mixed multipartite states, nothing about multipartite Bell inequalities, and nothing about fermionic systems. The list of omissions can be made longer. But we have to stop somewhere.

Entanglement plays a crucial role in quantum information processing: it can be considered as our enemy or as our friend [752]. On the negative side, the inevitable interaction systems with their environment induces entanglement between the controlled subsystems and the rest of the world, which influences the state of the qubits and induces errors. On the positive side, entanglement allows us to encode a single qubit in larger quantum systems, so that the entire quantum information will not be destroyed if the environment interacts with a small number of qubits. These issues become specially significant if one considers systems consisting of several parties.

It is tempting to compare quantum entanglement with the snow found high in the mountains during a late spring excursion. A mountaineer equipped with touring skis or crampons and ice-axe typically looks for couloirs and slopes covered by snow. On the other hand, his colleague in light climbing shoes will try to avoid all the snowy fields and find safe passage across the rocks. The analogy holds as neither quantum entanglement nor spring snow lasts forever, even though the decay timescales do differ.

We wrap up this chapter with the remark that multipartite entanglement offers a lot of space for effects not present in the case of systems consisting of two subsystems only. Entanglement in many body systems is not well understood, so we are pleased to encourage the reader to contribute to this challenging field.

Problems

17.1 Trace out a subsystem from $|GHZ\rangle$, and then from $|W\rangle$. Are the resulting bipartite systems entangled?

17.2 Define the rank of a pure state as the smallest possible number of product vectors one has to superpose in order to obtain it. Show that the rank of the state $|W\rangle$ defined by (17.9) is equal to three. What changes if one adds an ϵ superposition of the state $|111\rangle$?

17.3 Compute the local spectra for the Dicke states $|K - k, k\rangle$.

17.4 Choose a separable basis such that $|111\rangle$ is the separable state closest to a given three-qubit state $|\psi\rangle$. Expand the state in that basis and show that three components vanish, $\Gamma_{011} = \Gamma_{101} = \Gamma_{110} = 0$.

17.5 (a) Classical Bernstein Paradox. Write the bitstrings (001), (010), (100) and (111) on four cards, put them in a bag and choose one at random. Let X, Y and Z be the events that the first, second and third digit on the selected card is 0. Show that $p_X = p_Y = p_Z = 1/2$. Check that the joint probabilities that two events occur read $p_{XY} = p_{XZ} = p_{YZ} = 1/4$, so that the triplet (X, Y, Z) is pairwise independent. Is it then legitimate to claim that these three events are independent, so $p_{XYZ} = p_X p_Y p_Z = 1/8$?

(b) Consider the *Svetlichny state* [874] of three qubits, $|\psi_\triangle\rangle = [|001\rangle + |010\rangle + |100\rangle + |111\rangle]/2$. To what well known state is it locally equivalent?

17.6 Construct the generalised Bell basis for three qubits making use of the state $|\psi_\triangle\rangle$ considered in Problem 17.5.

17.7 Consider the following four-qubit pure states: (a) $|GHZ_4\rangle = (|0000\rangle + |1111\rangle)/\sqrt{2}$, (b) $|B_4\rangle = \frac{1}{2}[|0000\rangle + |0111\rangle + |1001\rangle + |1110\rangle]$, (c) $|C_4\rangle = \frac{1}{2}[|0000\rangle + |1100\rangle + |0011\rangle - |1111\rangle]$, (d) a suitably enphased Dicke state [435], $|HS\rangle = [|0011\rangle + |1100\rangle + \omega(|0101\rangle + |1010\rangle) + \omega^2(|0110\rangle + |1001\rangle)]/\sqrt{6}$, where $\omega = e^{2\pi i/3}$. Find the spectra Λ of all two-qubit reduced density matrices and compute: (i) mean von Neumann entropy $\bar{S}_1(\Lambda) = [S_1(\Lambda_{AB}) + S_1(\Lambda_{AC}) + S_1(\Lambda_{AD})]/3$; (ii) mean Rényi two-entropy, $\bar{S}_2(\Lambda)$; (iii) mean Chebyshev entropy $\bar{S}_\infty(\Lambda)$; (iv) mean negativity, averaged over the three different partitions.

17.8 Making use of three-qubit entangled basis (see hints to Problem 17.6) study the six-qubit state, $|\Phi\rangle = |000\rangle|\Psi_{++}\rangle + |011\rangle|\Psi_{+-}\rangle + |101\rangle|\Psi_{-+}\rangle + |110\rangle|\Psi_{--}\rangle + |111\rangle|\Phi_{++}\rangle + |100\rangle|\Phi_{+-}\rangle + |010\rangle|\Phi_{-+}\rangle + |001\rangle|\Phi_{--}\rangle$. Analyse all its reductions. What is so special about them?

17.9 Prove that the AME state (17.76) is a stabilizer state.

17.10 For what values of the single error probability p is it advantageous to use the repetition code (17.77)? Repeat for the Hamming code (17.78).

17.11 Reshape the tensors for (a) the four-qutrit AME state (17.76), (b) the six-qubit AME state $|\Phi_6\rangle$ introduced in Problem 17.8. Write down unitary matrices U_9 and U_8 respectively. Check that these matrices are multiunitary, as all their suitable reorderings remain unitary.

17.12 Express the W_K state as a matrix product state, using (a) open boundary conditions as in Eq. (17.95) and (b) periodic boundary conditions as in Eq. (17.98). (c) What is the minimal size D of the MPS required to represent exactly the generalised GHZ state $|GHZ_K^N\rangle = 1/\sqrt{N}\left(|1\rangle^{\otimes K} + |2\rangle^{\otimes K} + \cdots + |K\rangle^{\otimes K}\right)$ of K systems with N levels each?

Epilogue

After going through the chapter on multipartite entanglement you will have reached the end of our book. As the subtitle suggests, its aim was literally to present an introduction to the subject of quantum entanglement.

In the book we have consistently used a geometric approach to highlight similarities and differences between the classical and quantum spaces of states. Interesting properties of a given quantum state, including the degree of coherence or the degree of entanglement, can be described in a geometric manner by its distance to the selected set of coherent, or separable, states. Also other problems related to quantum entanglement admit a direct geometric representation.

What is the knowledge gained by studying the book good for? We hope it will contribute to a better understanding of quantum mechanics, and will provide a solid foundation for the theory of quantum information processing. Quantum entanglement plays also an important role in several other branches of theoretical physics including the condensed matter physics and the theory of quantum fields. Furthermore, we are pleased to observe that some experts on computer science and pure mathematicians enter the field and become busy proving rigorous theorems motivated by entanglement and quantum information.

In trying to describe the intricate geometry of the space of quantum states, we have largely restricted ourselves to discussing the statics of quantum theory. We have presented the arena, in which quantum information can be processed. We have not attempted to inject any concrete dynamics into our arena, but hope that readers equipped with some knowledge of its properties may introduce into it spectacular action.

In a sense we have characterised all the peculiarities of football fields of various sizes, without even specifying the rules of the game. Having at your disposal a huge flat grassy field, you can play soccer, cricket, American football or rugby, according to your mood and wishes.

In a similar way you can play different games in the multi–dimensional arena of quantum states. It stays there right at the centre of the beautiful platonic world of quantum theory, accessible to all of us.

Its rich structure provides a real challenge especially for young researchers. We wish you a good game in fine company! Good luck!

Appendix A

Basic notions of differential geometry

This appendix explains things that are explained in every book on differential geometry.[1] It is included to make our book self contained.

A.1 Differential forms

One-forms are defined in Section 1.4. The *exterior product* of 1-forms is defined by

$$\theta^i \wedge \theta^j = -\theta^j \wedge \theta^i. \tag{A.1}$$

The result is called a *2-form*. The exterior product is assumed to be linear over the real numbers so these 2-forms can be used as a basis in which to expand any anti-symmetric covariant tensor with two indices. Continuing in the same way we can define 3-forms and so on, up to and including n-forms if the dimension of the manifold is n. We can think of the exterior product of two 1-forms as an area element spanned by the two 1-forms. Then, given an m-dimensional submanifold, we can integrate an m-form over that submanifold. A further interesting definition is the *exterior derivative* of an m-form ω, which is an $m + 1$ form defined by

$$d\omega = \partial_{i_1} \omega_{i_2 \ldots i_{m+1}} dx^{i_1} \wedge \ldots \wedge dx^{i_m}. \tag{A.2}$$

Here $\omega_{i_1 \ldots i_m}$ is an anti-symmetric tensor of the appropriate rank. The definition is a generalisation of the familiar 'curl' in vector analysis. If $d\omega = 0$ the form is said to be *closed*. If $\omega = d\theta$ (where θ is a form of rank one less than that of ω) then ω is *exact*. An exact form is closed because $d^2 = 0$; the converse holds on topologically trivial spaces such as $\mathbb{R}^n$. An example of a closed 2-form is the field

---

[1] Such as Schrödinger [806] or Murray and Rice [682].

strength tensor in classical electrodynamics. An analogue of Stokes' theorem holds; if M is a subspace and ∂M its boundary then

$$\int_M d\omega = \int_{\partial M} \omega. \tag{A.3}$$

A.2 Riemannian curvature

Given a metric tensor g_{ij} we can define the *Levi–Civita connection* by

$$\Gamma_{ij}{}^k = \frac{1}{2} g^{km} (\partial_i g_{jm} + \partial_j g_{im} - \partial_m g_{ij}). \tag{A.4}$$

It is not a tensor. In fact its transformation law contains an inhomogeneous term. Given a scalar field f its gradient $\partial_i f$ is a covariant vector, but given a contravariant (say) vector V^i the expression $\partial_j V^i$ does not transform as a tensor. Instead we use the connection to define its *covariant derivative* as

$$\nabla_j V^i = \partial_j V^i + \Gamma_{jk}{}^i V^k. \tag{A.5}$$

This does transform like a tensor although the individual terms on the right-hand side do not. The covariant derivative of a covariant vector is then defined so that $\nabla_i(U_j V^j) = \partial_i(U_j V^j)$, which transforms like a vector since $U_j V^j$ is a scalar. An analogous argument determines the covariant derivative of an arbitrary tensor. As an example,

$$\nabla_m T_{ij}{}^k = \partial_m T_{ij}{}^k + \Gamma_{mn}{}^k T_{ij}{}^n - \Gamma_{mi}{}^n T_{nj}{}^k - \Gamma_{mj}{}^n T_{in}{}^k. \tag{A.6}$$

Both the logic and the pattern should be clear. One can check that

$$\nabla_i g_{jk} = 0. \tag{A.7}$$

In fact this is how the *metric compatible affine connection* (A.4) was defined. Parallel transport of vectors using this connection conserves lengths and scalar products.

With the metric compatible affine connection in hand we can write down a differential equation for a curve $x^i(\sigma)$ whose solution, given suitable initial data $x^i(0)$ and $\dot{x}^i(0)$, is the unique geodesic starting from that point in that direction. This *geodesic equation* is

$$\dot{x}^j \nabla_j \dot{x}^i = \ddot{x}^i + \Gamma_{jk}{}^i \dot{x}^j \dot{x}^k = 0. \tag{A.8}$$

(As usual the dot signifies differentiation with respect to σ.)

The *Riemann tensor* is defined by the equation

$$(\nabla_i \nabla_j - \nabla_j \nabla_i) V^k = R_{ijl}{}^k V^l. \tag{A.9}$$

This four-index tensor plays a central role in Riemannian geometry, but we will have to refer elsewhere for its properties [806, 682]. Let us just record the slightly frightening explicit expression

$$R_{ijl}{}^k = \partial_i \Gamma_{jl}{}^k - \partial_j \Gamma_{il}{}^k + \Gamma_{im}{}^k \Gamma_{jl}{}^m - \Gamma_{jm}{}^k \Gamma_{il}{}^m. \tag{A.10}$$

Perhaps it becomes slightly less frightening if we lower one index using the metric; using square brackets to denote anti-symmetry in the indices, one can show that

$$R_{ijkl} = R_{[ij][kl]} = R_{[kl][ij]}. \tag{A.11}$$

Suppose now that we have a two-dimensional plane in the tangent space at some point, spanned by the tangent vectors m^i and n^i. Then we can define the *sectional curvature* associated to that 2-plane,

$$K = \frac{m^{[i}n^{j]}R_{ijkl}\,m^{[k}n^{l]}}{m^2\,n^2}. \tag{A.12}$$

The point is that the Riemann tensor acts like a matrix on the space of 2-planes. Using index contraction we can define the *Ricci tensor*

$$R_{ij} = R_{ikj}{}^k. \tag{A.13}$$

Because of the index symmetries of the Riemann tensor the Ricci tensor turns out to be a symmetric tensor, $R_{ij} = R_{ji}$. The *curvature scalar* is

$$R = g^{ij}R_{ij}. \tag{A.14}$$

In two-dimensional spaces the Riemann tensor can be reconstructed from the curvature scalar R. (There is only one sectional curvature to worry about.) Moreover the sign of R has a simple interpretation: if $R > 0$ nearby geodesics that start out parallel tend to attract each other, while they diverge if $R < 0$. In three dimensions the Riemann tensor can be reconstructed from the Ricci tensor, while in four dimensions and higher this is no longer possible; the full Riemann tensor is needed in higher dimensions because one can choose many independent two-dimensional tangent planes along which to measure sectional curvatures.

Space is flat if and only if the Riemann tensor vanishes. In flat space parallel transport of vectors between two points is independent of the path, and it is possible to find a coordinate system in which the affine connection vanishes.

A.3 A key fact about mappings

A key fact about tensors is that they behave nicely under (reasonable) maps. How they behave depends on whether they have their indices upstairs or downstairs. Suppose we have a map $M \to M'$ between two manifolds M and M'. We assume that the dimension of the image of M is equal to the dimension of M, but the dimension of M' may be larger, in which case we have an embedding rather than a one-to-one map. Anyway we can describe the map using coordinates as $x^i \rightarrow x^{i'} = x^{i'}(x)$. Then we have:

Theorem. *A covariant tensor on M' defines a covariant tensor on M. A contravariant tensor on M defines a contravariant tensor on the image of M in M'.*

The proof is simple, given that we know the functions $x^{i'}(x)$:

$$V_i(x) \equiv \frac{\partial x^{i'}}{\partial x^i} V_{i'}(x'(x)), \qquad V^{i'}(x') \equiv \frac{\partial x^{i'}}{\partial x^i} V^i(x(x')). \qquad (A.15)$$

If the map is not one-to-one the functions $x^i(x^{i'})$ are defined only on the image of M, so the theorem is as general as it can be.

Appendix B

Basic notions of group theory

This appendix lists a number of formulæ that are explained in every book on group theory.[1] Some conventions can be chosen at will, which is why this list is essential.

B.1 Lie groups and Lie algebras

Lie groups are groups containing a continuously infinite number of elements with the amazing property that they can to a large extent be understood through an analysis of the tangent space at the unit element of the group. This tangent space is known as the Lie algebra of the group. We will deal mostly with the *classical groups* $SU(N)$, $SO(N)$ and $Sp(N)$ and especially with the *special unitary groups* $SU(N)$. These are all, in the technical sense, simple and compact groups and have in many respects analogous properties. The unitary group $U(N)$ is not simple, but can be understood in terms of its simple subgroups $U(1)$ and $SU(N)$. The *special linear groups* $SL(N)$ are not compact.

After complexification, the Lie algebra of a compact simple group can be brought to the standard *Cartan form*

$$[H_i, H_j] = 0 , \qquad [H_i, E_\alpha] = \alpha_i E_\alpha , \tag{B.1}$$

$$[E_\alpha, E_\beta] = N_{\alpha\beta} E_{\alpha+\beta} , \qquad [E_\alpha, E_{-\alpha}] = \alpha^i H_i, \tag{B.2}$$

where α_i is a member of the set of positive *root vectors* and $N_{\alpha\beta} = 0$ if $\alpha_i + \beta_i$ is not a root vector. The H_i, $1 \leq i \leq r$ span the maximal commuting Cartan subalgebra. Their number r is the *rank* of the group, equal to $N - 1$ for $SU(N)$.

---

[1] Such as the two books by Gilmore [334, 335]. We congratulate Robert Gilmore for having rewritten a book in such a way that it became shorter.

The Lie bracket $[A, B]$ is a peculiar kind of product on a vector space. Once a matrix representation is chosen it is also meaningful to consider the usual product AB, and set $[A, B] = AB - BA$.

B.2 SU(2)

The generators of $SU(2)$ in the fundamental representation are precisely the Pauli matrices $\vec{\sigma} = \{\sigma_x, \sigma_y, \sigma_z\}$,

$$\sigma_x = \begin{bmatrix} 0 & 1 \\ 1 & 0 \end{bmatrix}, \quad \sigma_y = \begin{bmatrix} 0 & -i \\ i & 0 \end{bmatrix}, \quad \sigma_z = \begin{bmatrix} 1 & 0 \\ 0 & -1 \end{bmatrix}. \tag{B.3}$$

They form an orthonormal basis in the Lie algebra. There is one unitary representation of $SU(2)$ in every dimension $N = n + 1 = 2j + 1$. Then the generators are

$$J_x = \frac{1}{2} \begin{bmatrix} 0 & \sqrt{n} & 0 & 0 & \cdots \\ \sqrt{n} & 0 & \sqrt{2(n-1)} & 0 & \cdots \\ 0 & \sqrt{2(n-1)} & 0 & \sqrt{3(n-2)} & \cdots \\ 0 & 0 & \sqrt{3(n-2)} & 0 & \cdots \\ \cdots & \cdots & \cdots & \cdots & \cdots \end{bmatrix} \tag{B.4}$$

$$J_y = \frac{i}{2} \begin{bmatrix} 0 & -\sqrt{n} & 0 & 0 & \cdots \\ \sqrt{n} & 0 & -\sqrt{2(n-1)} & 0 & \cdots \\ 0 & \sqrt{2(n-1)} & 0 & -\sqrt{3(n-2)} & \cdots \\ 0 & 0 & \sqrt{3(n-2)} & 0 & \cdots \\ \cdots & \cdots & \cdots & \cdots & \cdots \end{bmatrix} \tag{B.5}$$

$$J_z = \frac{1}{2} \begin{bmatrix} n & 0 & 0 & \cdots \\ 0 & n-2 & 0 & \cdots \\ 0 & 0 & n-4 & \cdots \\ \cdots & \cdots & \cdots & \cdots \end{bmatrix} \equiv \begin{bmatrix} j & 0 & 0 & \cdots \\ 0 & j-1 & 0 & \cdots \\ 0 & 0 & j-2 & \cdots \\ \cdots & \cdots & \cdots & \cdots \end{bmatrix}. \tag{B.6}$$

A crucial choice here is that J_x and J_z are real while J_y is imaginary. To get to the Cartan form, set $H = J_z$ and $E_\pm = J_x \pm iJ_y$.

Schwinger's oscillator representation [816] uses two commuting pairs of creation and annihilation operators

$$[a_+, a_+^\dagger] = [a_-, a_-^\dagger] = 1. \tag{B.7}$$

There are orthonormal basis states

$$|n_+, n_-\rangle = \frac{(a_+^\dagger)^{n_+} (a_-^\dagger)^{n_-}}{\sqrt{n_+!} \sqrt{n_-!}} |0, 0\rangle. \tag{B.8}$$

In terms of these oscillators we can write the $SU(2)$ Lie algebra generators as well as a number operator $\hat{n}$:

$$J_x = \frac{1}{2}(a_+^\dagger a_- + a_-^\dagger a_+) \quad J_y = \frac{1}{2i}(a_+^\dagger a_- - a_-^\dagger a_+) \quad J_z = \frac{1}{2}(a_+^\dagger a_+ - a_-^\dagger a_-) \quad \text{(B.9)}$$

$$\hat{n} = a_+^\dagger a_+ + a_-^\dagger a_-. \quad \text{(B.10)}$$

When the representation is restricted to a subspace where $\hat{n}$ has a fixed eigenvalue n we recover the representation we gave already.

B.3 SU(N)

$SU(N)$ is an $(N^2 - 1)$ dimensional group and in the defining representation the Lie algebra consists of $N^2 - 1$ traceless Hermitian N by N matrices, labelled by the index i, $1 \le i \le N^2 - 1$. A complete orthonormal set of generators obey

$$\sigma_i \sigma_j = \frac{2}{N} \delta_{ij} + d_{ijk} \sigma_k + i f_{ijk} \sigma_k, \quad \text{(B.11)}$$

where f_{ijk} is totally anti-symmetric in its indices and d_{ijk} is totally symmetric and traceless ($d_{iik} = 0$). For the commutator and the anti-commutator respectively, this means that

$$[\sigma_i, \sigma_j] = 2i f_{ijk} \sigma_k, \qquad \{\sigma_i, \sigma_j\} = \frac{4}{N} \delta_{ij} + 2 d_{ijk} \sigma_k. \quad \text{(B.12)}$$

The generators are Hermitian matrices and obey

$$\text{Tr}\, \sigma_i \sigma_j = 2 \delta_{ij}, \qquad \text{Tr}\, \sigma_i \sigma_j \sigma_k = 2 d_{ijk} + 2i f_{ijk} \quad \text{(B.13)}$$

(the first equation here is orthonormality of the generators) as well as the completeness relation

$$(\sigma_i)_A^{\ B} (\sigma_i)_C^{\ D} = 2\delta_A^D \delta_B^C - \frac{2}{N} \delta_A^B \delta_C^D. \quad \text{(B.14)}$$

The above normalisation is implied by the case $N = 2$, since the Pauli matrices (B.3) are normalised just like this. In this case $d_{ijk} = 0$, while the group constants form the antisymmetric tensor, $f_{ijk} = \epsilon_{ijk}$.

The following identities are true for any N:

$$f_{ijm} f_{mkn} + f_{jkm} f_{min} + f_{kim} f_{mjn} = 0 \quad \text{(B.15)}$$

$$d_{ijm} f_{mkn} + d_{jkm} f_{min} + d_{kim} f_{mjn} = 0 \quad \text{(B.16)}$$

$$f_{ijm} f_{mkn} = \frac{2}{N}(\delta_{ik}\delta_{jn} - \delta_{in}\delta_{jk}) + d_{ikm}d_{mjn} - d_{jkm}d_{min} \quad \text{(B.17)}$$

$$f_{imn} f_{jmn} = N\delta_{ij} \qquad d_{imn}d_{jmn} = \frac{N^2 - 4}{N}\delta_{ij}. \quad \text{(B.18)}$$

Scanning this list one realises that some identies are 'missing'. These identities exist, but they depend on N. For further information consult the literature [633, 869].

B.4 Homomorphisms between low-dimensional groups

We sometimes make use of the following isomorphisms between Lie algebras:

$$SU(2) \cong SO(3), \tag{B.19}$$

$$SO(4) \cong SO(3) \times SO(3) \cong SU(2) \times SU(2), \tag{B.20}$$

$$SU(4) \cong SO(6). \tag{B.21}$$

Globally these are $2 \to 1$ homomorphisms between the corresponding groups; for explanations see (especially) Section 3.7.

Appendix C

Geometry: do it yourself

In this appendix we provide some additional exercises of a more practical nature.

Exercise 1: Real projective space. Cut out a disc of radius r. Prepare a narrow strip of length πr and glue it into a Möbius strip. The total length of the boundary of a strip is equal to the circumference of the disc so you may try to glue them together.[1] When you are finished, you can contemplate a fine model of a real projective space, $\mathbb{R}\mathbf{P}^2$.

Exercise 2: Hypersphere S^3 may be obtained by identifying points on the surfaces of two identical 3-balls as discussed in Section 3.1. To experience further features of the hypersphere get some playdough and prepare two cylinders of different colours with their length more than three times larger than their diameter. Form two linked tori as shown in Figure C.2.

Start gluing them together along their boundaries. After this procedure is completed[1] you will be in position to astonish your colleagues by presenting them a genuine *Heegaard decomposition* of a hypersphere.

Exercise 3: Mixed states. Make a ball out of playdough. Glue a string to its surface along the shape of the stitching of a tennis balls (see the picture). Obtain a convex hull by cutting out the redundant dough with a knife. How much of a ball is taken away?

Convince yourself that the convex hull of a one-dimensional string located on S^2 can form a considerable part of the ball B^3. In a similar way $\mathcal{M}^{(3)}$, the convex hull of the four-dimensional manifold $\mathbb{C}\mathbf{P}^2$ consisting of $N = 3$ pure states placed on S^7, contains a non-negligible part of B^8 (see Sections 8.6 and 15.3).

Exercise 4: Entangled pure states. Magnify Figure C.4 and cut out the net of the cover tetrahedron. It represents the entanglement of formation of the pure states of

---

[1] If you happen to work in three dimensions this simple task gets difficult.

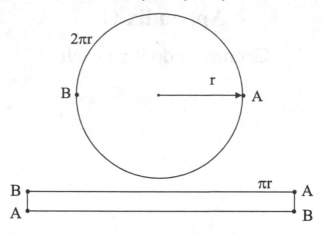

Figure C.1 A narrow Möbius strip glued with a circle produces $\mathbb{R}\mathbf{P}^2$.

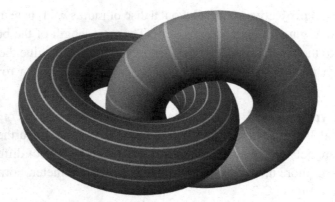

Figure C.2 Heegaard decomposition of a 3-sphere.

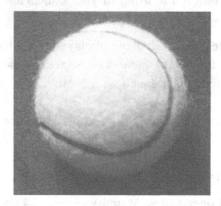

Figure C.3 Imagine a convex hull of the one-dimensional stitching of the tennis ball.

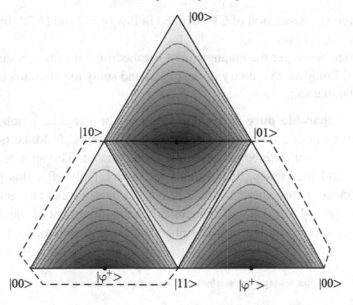

Figure C.4 Net of the tetrahedron representing entanglement for pure states of two qubits: maximally entangled states plotted in black.

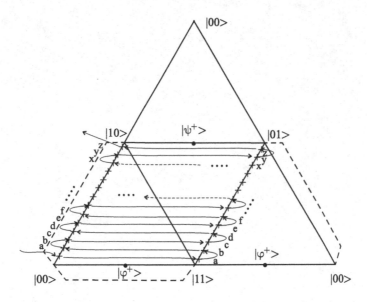

Figure C.5 Sew with a coloured thread inside a transparent tetrahedron to get the ruled surface consisting of separable pure states of two qubits.

two qubits for a cross-section of $\mathbb{C}\mathbf{P}^3$ defined in Eqs (4.71) and (4.72) by setting all phases ν_i in (4.68) to zero.

Glue it together to get the entanglement tetrahedron with four product states in four corners. Enjoy the symmetry of the object and study the contours of the states of equal entanglement.

Exercise 5: Separable pure states. Prepare a net of a regular tetrahedron from transparency according to the blueprint shown in Figure C.5. Make holes with a needle along two opposite edges as shown in the picture. Thread a needle with a (red) thread and start sewing it through your model. Only after this job is done glue the tetrahedron together.[2] If you pull out the loose thread and get the object sketched in Figure 16.1, you can contemplate, how a fragment of the subspace of separable states forms a ruled surface embedded inside the tetrahedron.

---

[2] Our experience shows that sewing after the tetrahedron is glued together is much more difficult.

Appendix D

Hints and answers to the exercises

Problem 1.1. Draw a line from a pure point $\mathbf{x}_1$ through the given point $\mathbf{x}$. It ends at a point $\mathbf{y}_1$ on some face of at most $n - 1$ dimensions. If this point is mixed, draw a line from a pure point $\mathbf{x}_2$ on that face through $\mathbf{y}_1$. It ends at a point $\mathbf{y}_2$ on some face of at most $n - 2$ dimensions. Continue until $\mathbf{x}$ is expressed as a mixture of at most $n + 1$ pure points.

Problem 1.2. One way is to construct the simplex. If we put its centre at the origin the $N = n + 1$ points can be placed at

$$
\begin{array}{ccccc}
(-r_1, & -r_2, & \ldots, & -r_{n-1}, & -r_n) \\
(R_1, & -r_2, & \ldots, & -r_{n-1}, & -r_n) \\
(0, & R_2, & \ldots, & -r_{n-1}, & -r_n). \\
\ldots & \ldots & \ldots & \ldots & \ldots \\
(0, & 0, & \ldots, & 0, & R_n)
\end{array}
$$

This helps.

Problem 2.1. Here is a plot of the structural entropy. The maximum $S_2 - S_1 \approx 0.22366$ is attained for $\vec{p} \approx (0.806, 0.097, 0.097)$ and its two other permutations. They are visible at the contour plot provided as three dark hills.

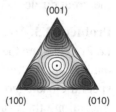

(001)

(100)   (010)

Problem 2.2. The $N = 3$ case shows the idea. We have $\vec{x} \cdot (\vec{y} - \vec{z}) = x_1(y_1 - z_1) + x_2(y_2 - z_2) + x_3(y_3 - z_3) = (x_1 - x_2)(y_1 - z_1) + (x_2 - x_3)(y_1 + y_2 - z_1 - z_2) + x_3(y_1 + y_2 + y_3 - z_1 - z_2 - z_3) \geq 0$ because of the conditions stated.

Problem 2.3. (a) Let $a, b \in [0, 1]$ and $a' = 1 - a$, $b' = 1 - b$ and define

$$B = T_1 T_2 = \begin{bmatrix} a & a' & 0 \\ a' & a & 0 \\ 0 & 0 & 1 \end{bmatrix} \begin{bmatrix} 1 & 0 & 0 \\ 0 & b & b' \\ 0 & b' & b \end{bmatrix} = \begin{bmatrix} a & a'b & a'b' \\ a' & ab & ab' \\ 0 & b' & b \end{bmatrix}. \tag{D.1}$$

B is a bistochastic matrix. It is also orthostochastic since $B_{ij} = (O_{ij})^2$, where

$$O = \begin{bmatrix} \sqrt{a} & \sqrt{a'b} & -\sqrt{a'b'} \\ \sqrt{a'} & -\sqrt{ab} & \sqrt{ab'} \\ 0 & \sqrt{b'} & \sqrt{b} \end{bmatrix}. \tag{D.2}$$

Problem 2.4. We know that $\vec{x}$ is a (non-unique) convex combination of permutation matrices acting on $\vec{y}$; this defines a bistochastic matrix according to Birkhoff's theorem.

Problem 2.5. One obtains two important cases of the Dirichlet distribution (2.73): the round measure ($s = 1/2$) for real and flat measure ($s = 1$) for complex Gaussian random numbers [1016].

Problem 2.7. For $q \leq 2$. To see this, study the second derivative close to $p = 1$.

Problem 3.1. You will obtain

$$\frac{x^i}{X^I} = \frac{2}{1 + X^0} \quad \Rightarrow \quad X^0 = \frac{4 - r^2}{4 + r^2} \quad \Rightarrow \quad ds^2 = \left(\frac{4}{4 + r^2} \right)^2 dx^i dx^i.$$

Problem 3.2. The angles are obtained by intersecting respectively the plane and the sphere with two intersecting planes. The angles will be equal if and only if both the plane and the sphere meet the line of intersection of the two planes at the same angle. But this will happen if and only if the line of intersection forms a chord of the great circle.

Problem 3.3. You can try a calculation to see whether the natural map $(x, y) \rightarrow (x, 2y)$ between the tori is analytic (it is not). Or you can observe that the tori inherit natural flat metrics from the complex plane. On each torus there will be a pair of special closed geodesics that intersect each other, namely what used to be straight lines along the x- and y-directions on the plane. Their circumferences are equal on one of the tori, and differ by a factor of two on the other. But analytic – hence conformal – transformations do not change the ratio of two lengths, and it follows that no such analytic transformation between the tori can exist.

Problem 3.4. As an intermediate step you must prove

$$\Omega^{il} \partial_l \Omega^{jk} + \Omega^{kl} \partial_l \Omega^{ji} + \Omega^{kl} \partial_l \Omega^{ij} = 0. \tag{D.3}$$

Problem 3.5. A hyperplane through the origin in embedding space meets the 3-sphere in a 2-sphere given in stereographic coordinates by

$$a_I X^I = 0 \quad \Rightarrow \quad 2a_1 x + 2a_2 y + 2a_3 z + 1 - r^2 = 0$$

(where we assumed that the fourth component of the vector equals one). A geodesic is the intersection of two such spheres; choose them to have their centres at $(a, 0, 0)$ and $(b_1, b_2, 0)$ again without loss of generality. If you also demand $r^2 = 1$ (the equator) you get three equations with the solutions $(x, y, z) = (0, 0, \pm 1)$.

Problem 3.6. The key point is that two Hopf circles with the opposite twist meet twice. Only one-half of the circumference of a Hopf circle is needed to label the members of the family of circles that twist in the other way. (Draw the torus as a flat square to see this.)

Problem 3.7. For $\tau = -\phi$ we get

$$X + iY = \cos\frac{\theta}{2} \quad \Rightarrow \quad Y = 0 \ \& \ X > 0 \quad \Rightarrow \quad y = 0 \ \& \ x > 0. \quad (\text{D.4})$$

With the exception of one point this maps $\mathbf{S}^2$ onto a half plane. The other two sections provide maps onto a hemisphere and a unit disc, respectively. In all three cases it is geometrically evident that we are selecting one point from each geodesics, except for a single one for which there is no prescription.

Problem 3.8. The group acting on the fibres is the discrete group with two elements, the unit element and the element that turns the fibre upside down. See the Figure. Left: the Möbius strip as a vector bundle, with a global section (i.e. an embedding of the circle in the bundle). Right: the principal bundle, with fibres equal to the group $\{\pm\}$.

Problem 4.1. Consider the case of three points and put one at the origin, one at $(1, 0)$. The location of the third point can be anywhere. That gives an $\mathbb{R}^2$. This coordinatisation of the space of triplets fails if the first two points coincide. That case evidently corresponds to one additional point 'at infinity', so we have a natural one-to-one correspondence between the space of triplets and a plane + the point at infinity, that is $\mathbb{CP}^1$. Consider the case of four points: if the first two points are distinct we proceed as above; the two remaining points can be coordinatised by $\mathbb{R}^2 \times \mathbb{R}^2 = \mathbb{C}^2$. If the first two points coincide we have only a triplet of points do deal with. This is a $\mathbb{CP}^1$ according to what was just shown. But $\mathbb{C}^2$ plus a $\mathbb{CP}^1$ 'at

infinity' is a $\mathbf{CP}^2$. And so on. This is useful in archaeology if we use the Fubini–Study metric to give a measure. Then we can answer questions like 'given $n + 2$ stones, what is the probability that there are k triplets of stones lying (to a given precision) on straight lines?'.

Problem 4.2. A Klein bottle; a bottle without an inside (or outside). It takes a four dimensional being to make one that does not intersect itself.

Problem 4.3. Integrate the Fubini–Study 2-form Ω over the embedded $\mathbf{CP}^1$; this gives $\int_{\mathbf{CP}^1} \Omega =$ the area $= \pi$, since Ω induces the usual Fubini–Study 2-form on $\mathbf{CP}^1$. But if $\mathbf{CP}^1$ could be shrunk to a point then this calculation could be done within a single coordinate patch, and there could be no obstruction to the calculation $\int_{\mathbf{CP}^1} \Omega = \int_{\mathbf{CP}^1} d\omega = \int_{\partial(\mathbf{CP}^1)} \omega = 0$, where we used Stokes' theorem and the fact that $\mathbf{CP}^1$ has no boundary. This is a contradiction. Alternatively one can stare at the line at infinity in the octant picture of $\mathbf{CP}^2$, and convince oneself that any attempt to move it will increase its area.

Problem 5.1. The two pure states divide a great circle into two segments. If one of the eigenstates of A lies on the shortest of these segments, the answer is $D_{\text{Bhatt}} = \theta_A + \theta/2$, otherwise it is $D_{\text{Bhatt}} = \theta/2$ (independent of A).

Problem 5.2. For the colouring problem, take a few photocopies of the graph. For the relation to the Escher print, begin by looking at the subgraph on the right. It is interesting in itself [990]. Its three outermost vertices correspond to rays going through the midpoints of the faces of a cube, the next layer of six vertices to rays going through the midpoints of its edges, and the inner four to rays going through the corners of the cube.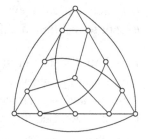

Problem 6.1. From (6.9) the Q-function of a Fock state is $Q_{|n\rangle}(z) = |z|^{2n} e^{-|z|^2}/n!$ and from (6.47) we must have $\int dz^2 Q_{|n\rangle} P_{|1\rangle} = \delta_{n1}$. The solution is the moderately singular distribution $P_{|1\rangle} = e^{|z|^2} \partial_z \partial_{\bar{z}} \delta^{(2)}(z)$.

Problem 6.2. Using spherical polars in phase space (and the integral representation of the gamma function) one finds $S_W(|n\rangle) = 1 + n + \ln n! - n\Psi(n+1)$, where Ψ is the digamma function defined in Eq. (7.54).

Problem 7.1. It is a function of the maximum of the Husimi function, so it can be read off from Eq. (7.25).

Problem 7.2. Wehrl entropy and participation number for pure states of $N = 2$–5 read

N	j	m	$S_W(|\psi\rangle)$	$R(|\psi\rangle)$
2	1/2	$\pm 1/2$	$1/2 = 0.5$	$1\frac{1}{2}$
3	1	± 1	$2/3 \approx 0.667$	$1\frac{2}{3}$
3	1	0	$5/3 - \ln 2 \approx 0.974$	$2\frac{1}{2}$
4	3/2	$\pm 3/2$	$3/4 = 0.75$	$1\frac{3}{4}$
4	3/2	$\pm 1/2$	$9/4 - \ln 3 \approx 1.151$	$2\frac{11}{12}$
4	3/2	$|\psi_\triangle\rangle$	$21/8 - \ln 4 \approx 1.239$	$3\frac{2}{11}$
5	2	± 2	$4/5 = 0.8$	$2\frac{1}{4}$
5	2	± 1	$79/30 - \ln 4 \approx 1.247$	$3\frac{3}{20}$
5	2	0	$47/15 - \ln 6 \approx 1.342$	$3\frac{1}{2}$
5	3/2	$|\psi_{\text{tetr.}}\rangle$	$165/45 - \ln 9 \approx 1.492$	$4\frac{1}{5}$

Problem 8.2. (a) yes; (b) no. Any permutation is represented by an orthogonal matrix, and multiplication by any unitary matrix does not change the singular values

Problem 8.3. (a) absolute values of real eigenvalues; (b) equal to unity; (c) absolute values of complex eigenvalues.

Problem 8.4. In fact an even stronger property is true. Directly from the definition of the singular values it follows that $sv(A) = sv(UAV)$, for arbitrary unitary U and V. However, the special case $V = U^{-1}$ is often useful in calculations.

Problem 8.5. This is the Cauchy–Schwarz inequality for the scalar product in Hilbert–Schmidt space.

Problem 8.6. We know that $\langle \psi|P|\psi\rangle \geq 0$ for all vectors. Let $|\psi\rangle$ be a basis vector.

Problem 8.7. We can bring an arbitrary vector $\tau_i\sigma_i$ into the Cartan subalgebra, $U\tau_i\sigma_i U^\dagger = \lambda_i H_i$. Generically that is the best we can do, so the number of non-zero elements will equal the dimension of the Cartan subalgebra, i.e. $N - 1$.

Problem 8.8. Definition (8.30) applied to the Pauli matrices gives

$$O = \begin{bmatrix} c^2\vartheta c(2\phi) - s^2\vartheta c(2\psi) & c^2\vartheta c(2\phi) + s^2\vartheta c(2\psi) & -s(2\vartheta)c(\phi+\psi) \\ s^2\vartheta s(2\psi) - c^2\vartheta s(2\phi) & c^2\vartheta c(2\phi) + s^2\vartheta c(2\psi) & s(2\vartheta)s(\phi+\psi) \\ s(2\vartheta)c(\psi-\phi) & -s(2\vartheta)s(\psi-\phi) & c(2\vartheta) \end{bmatrix}.$$

(D.5)

where $c \equiv \cos$, $s \equiv \sin$. This is the *Cayley parametrization* of the group $SO(3)$ and describes the rotation with respect to the axis $\vec{\Omega} = (\sin\vartheta \sin\psi, \sin\vartheta \cos\psi, \cos\vartheta \sin\phi)$ by an angle t such that $\cos(t/2) = \cos\vartheta \cos\phi$.

Problem 8.9. Diagonalise! Then $\mathrm{Tr}\rho^2 = 1$ implies that all eigenvalues obey $-1 \le \lambda_i \le 1$, and that the eigenvalue 1 can occur at most once. Also $\mathrm{Tr}\rho^2(1 - \rho) = 0$ becomes a sum of non-negative terms in which each individual term must vanish, so the eigenvalues are either 0 or 1 [507].

Problem 9.4. The spectrum consists of the MN numbers $\alpha_i\beta_j$, where $i = 1, \ldots, M$ and $j = i, \ldots, N$ (say).

Problem 9.6. Use the Schwarz inequality, $|\mathrm{Tr}(AB)|^2 \le \mathrm{Tr}(AA^\dagger) \times \mathrm{Tr}(BB^\dagger)$, and replace A by $A\rho^{1/2}$ and B by $B\rho^{1/2}$, respectively [652].

Problem 10.1. It is enough to consider a pure POVM, so that $E_i = |\phi_i\rangle\langle\phi_i|$. The vectors have components ϕ^α_i, $\alpha = 1, \ldots, N$, $i = 1, \ldots, k$. Let these components be the elements of an $N \times k$ matrix. The completeness relation $\sum_{i=1}^k (E_i)^\alpha_\beta = \delta^\alpha_\beta$ implies that the rows of this matrix are orthonormal. We can always add an additional set of $k - N$ rows to the matrix, so that it becomes unitary. The columns of the new matrix form an orthonormal basis in a k dimensional Hilbert space, and there is an obvious projection of its vectors down to the original Hilbert space.

Problem 10.2. Let $a = \mathrm{diag}(A)$ and $c = \mathrm{diag}(C)$ be diagonals of complex matrices. Show that $ABC^\dagger = (ac^\dagger) \circ B$ and use it with $A = A_i$ and $C = A_i^\dagger$ for all $i = 1, ..., k$.

Problem 10.3. To prove positivity, take an arbitrary vector V^i and define $A \equiv V^i A_i$. Then $\bar{V}^i\sigma_{ij}V^j = \mathrm{Tr}\rho AA^\dagger$ is positive because the trace of two positive operators is always positive. For the final part see (10.56).

Problem 10.5. The phase flip channel.

Problem 10.6. The dynamical matrix D_Φ is represented by $D_{mn \atop \mu\nu} = \rho_{m\mu}\delta_{n\nu}$. Writing down the elements of the image $\sigma' = \Phi_\rho(\sigma) = D^R\sigma = (\rho \otimes \mathbb{1}_N)^R\sigma$ in the standard basis we obtain the desired result, $\sigma'_{m\mu} = D_{mn \atop \mu\nu}\sigma_{n\nu} = \rho_{m\mu}\mathrm{Tr}\sigma = \rho_{m\mu}$.

Problem 11.1. Write both matrices in their eigen representations, $A = \sum_i a_i |\alpha_i\rangle \langle \alpha_i|$ and $B = \sum_i b_i |\beta_i\rangle \langle \beta_i|$. Perform decompositions $A^R = \sum_i a_i \alpha^{(i)} \otimes \bar{\alpha}^{(i)}$ and $B^R = \sum_i b_i \beta^{(i)} \otimes \bar{\beta}^{(i)}$ as in (10.58), multiply them and reshuffle again to establish positivity of $(A^R B^R)^R$. For a different setting see Havel [412].

Problem 11.2. (a) The spectrum $\vec{d}$ consists of $N(N+1)/2$ elements equal to $+1$ and $N(N-1)/2$ elements equal to -1, so its sum (the trace of D) is equal to N. (b) This canonical form contains one negative term and three positive terms.

Problem 11.3. Using the non-homogeneous form of (10.36) write down the dynamical matrix D and show that some of its eigenvalues are negative.

Problem 11.4. $\Phi_T = \frac{N-1}{N} \Phi_* + \frac{1}{N} T$

Problem 11.5. The spectrum $\vec{d}$ of D_Ψ reads $(a - 3, a, a, b, b, b, c, c, c)$. Hence $cp(\Psi) = \min(a - 3, b, c)$ and the map is CP if $a \geq 3$ and $b, c \geq 0$. The spectrum of $D_\Psi^{T_A}$ is three-fold degenerated and contains $\{a - 1, \lambda_+, \lambda_-\}$, where $\lambda_\pm = (b + c \pm \sqrt{(b - c)^2 + 4})/2$. Therefore $ccp(\Psi) = \min(a - 1, \lambda_-)$ and the map is CcP if $a \geq 1$ and $bc \geq 1$. Note that these results are consistent with (11.9) and (11.10).

Problem 11.7. As we sum contributions from the entry-wise products of a Kraus operator A_i and its complex conjugate, the entries of T are real and non-negative. If the quantum map Φ is trace preserving, $\sum_i A_i^\dagger A_i = \mathbb{1}$, then T is stochastic. If the Choi matrix becomes diagonal, due to 'decoherence' in the space of maps, $D \rightarrow \tilde{D} = \mathrm{diag}(D)$ then the map reduces to the classical map represented by a stochastic matrix, $\Phi = D^R \rightarrow \tilde{D}^R \rightarrow T$. If additionally, the map Φ is unital, $\sum_i A_i A_i^\dagger = \mathbb{1}$, then T is bistochastic.

Problem 12.1. For the matrix elements, deduce that

$$A = \sum_I U_I \frac{1}{N} \mathrm{Tr} U_I^\dagger A \quad \Rightarrow \quad \sum_I (U_I)_{m\mu} (U_I^*)_{n\nu} = N \delta_{mn} \delta_{\mu\nu}. \qquad (D.6)$$

Then contract with $A_{\mu\nu}$. In fact the operator basis does not even have to be unitary.

Problem 12.3.

$$H = \begin{bmatrix} 1 & 1 & 1 & 1 \\ 1 & e^{i\phi} & -1 & -e^{i\phi} \\ 1 & -1 & 1 & -1 \\ 1 & -e^{i\phi} & -1 & e^{i\phi} \end{bmatrix}. \qquad (D.7)$$

Repeating this exercise for $N = 6$ requires much ingenuity [516].

Problem 12.4. The field with eight elements has irreducible polynomials $p_1(x) = x^3 + x + 1$ and $p_2(x) = x^3 + x^2 + 1$; α is a root of $p_1(x)$:

Element	Polynomial	tr	trx^2	order
$0 = 0$	x	0	0	−
$\alpha^7 = 1$	$x + 1$	1	1	1
$\alpha = \alpha$	$p_1(x)$	0	0	7
$\alpha^2 = \alpha^2$	$p_1(x)$	0	0	7
$\alpha^3 = \alpha + 1$	$p_2(x)$	1	1	7
$\alpha^4 = \alpha^2 + \alpha$	$p_1(x)$	0	0	7
$\alpha^5 = \alpha^2 + \alpha + 1$	$p_2(x)$	1	1	7
$\alpha^6 = \alpha^2 + 1$	$p_2(x)$	1	1	7

For the field with nine elements there are three irreducible polynomials (of degree 2) to choose from.

Problem 12.5. Applying (12.72) to the basis vectors we obtain (ignoring phase factors)

$$|z, a\rangle \to |z', a'\rangle = \left| \frac{\alpha z + \beta}{\gamma z + \delta}, \frac{a}{\gamma z + \delta} \right\rangle, \tag{D.8}$$

except that special measures must be taken if $z = \infty$ or $\gamma z + \delta = 0$.

Problem 12.6. This is a question of completing squares in exponents.

Problem 12.7. The bases from the Mermin square are real. Using the magical basis they are maximally entangled. The concurrence, when averaged over the vectors in a 2-design, is equal to its Fubini–Study average. One can then conclude that if there are two maximally entangled bases in the set, the rest must be separable [964].

Problem 12.8. A concise answer reads: $C_{ijk} = A_i \sigma_j B_k$. The indices run $i = 0, 1$; $j = 0, 1, 2, 3$ and $k = 0, 1, 2$ and the matrices are: $A_0 = \sigma_0 = B_0 = \mathbb{1}, A_1 = V, B_1 = H$ and $B_2 = HV$. These matrices (up to a phase belonging to the Clifford group) form the octahedral group and a unitary 3-design. They represent $2 \cdot 4 \cdot 3 = 24$ proper rotations of the cube: identity, $3 \cdot 3 = 9$ rotations around axes crossing the centres of opposite faces ($\blacklozenge$), $6 \cdot 1 = 6$ rotations around axes crossing the centres of opposite edges ($\bullet$) and $4 \cdot 2 = 8$ rotations around axes passing by opposite corners ($\blacktriangledown$). The subset of 12 matrices corresponding to the tetrahedral group forms a 2-design and this number saturates the lower bound for the size K_{min} of a 2-design of unitary matrices of order two.

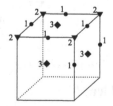

Octahedral group. Symbols represent axes crossing the centre of the cube labelled by the numbers of rotations around each axis.

Problem 12.9. A set of SIC vectors are

$$
\begin{bmatrix}
0 & 0 & 0 & e^{i\phi} & \omega e^{i\phi} & \omega^2 e^{i\phi} & 1 & 1 & 1 \\
1 & 1 & 1 & 0 & 0 & 0 & e^{i\phi} & \omega e^{i\phi} & \omega^2 e^{i\phi} \\
e^{i\phi} & \omega e^{i\phi} & \omega^2 e^{i\phi} & 1 & 1 & 1 & 0 & 0 & 0
\end{bmatrix}.
\tag{D.9}
$$

In no other (known) case is there a free parameter in a SIC. It was noted in 1844 that if we set $\phi = \pi$ there are 12 triples of linearly dependent SIC vectors, each of which is orthogonal to one of the 12 vectors in a complete set of MUB – although the language used then was that of cubic curves, not that of quantum theory [431].

Problem 13.1. Open with the observation that

$$
\ln (A + xB) = \ln (A + xB + u_0) - \int_0^{u_0} \frac{du}{A + xB + u}.
\tag{D.10}
$$

Use the fact that $A + u$ is invertible for any positive u to rewrite this as

$$
\ln (A + u_0) + \ln \left(1 + \frac{1}{A + u_0} xB \right) - \int_0^{u_0} \left[\frac{1}{A + u} - \frac{1}{A + u} xB \frac{1}{A + xB + u} \right] du.
$$

Now do the integral over the first term, collect terms, expand the remaining logarithm, and finally let $u_0 \to \infty$.

Problem 13.2. The eigenvalues of $(1 - \rho/z)^{-1}$ are $z/(z - \lambda_i)$. With ρ diagonalised, and the contour chosen suitably, the first integral equals the von Neumann entropy and the second is known as subentropy [510],

$$
S_Q(\rho) \equiv - \sum_{i=1}^{N} \left(\prod_{i \neq j} \frac{\lambda_i}{\lambda_i - \lambda_j} \right) \lambda_i \ln \lambda_i.
\tag{D.11}
$$

Problem 13.3.

$$
\sum_k p_k S(\rho_k || \sigma) = \sum_k p_k (\mathrm{Tr}\rho_k \ln \rho_k - \mathrm{Tr}\rho_k \ln \sigma) =
$$

$$
= \sum_k p_k (\mathrm{Tr}\rho_k \ln \rho_k - \mathrm{Tr}\rho_k \ln \sigma + \mathrm{Tr}\rho_k \ln \rho - \mathrm{Tr}\rho_k \ln \rho) =
$$

$$
= \sum_k p_k (\mathrm{Tr}\rho_k \ln \rho_k - \mathrm{Tr}\rho_k \ln \rho) + \mathrm{Tr}\rho \ln \rho - \mathrm{Tr}\rho \ln \sigma =
$$

$$
= \sum_k p_k S(\rho_k || \rho) + S(\rho || \sigma).
\tag{D.12}
$$

Problem 13.4. Write the Husimi function using the Schmidt decomposition $|\Psi\rangle = \sqrt{\lambda_1}|11\rangle + \sqrt{\lambda_2}|22\rangle$. Integration over the Cartesian product of two spheres gives

$$S_W(\Psi) = \frac{\lambda_1^2(1 - \ln \lambda_1)}{\lambda_1 - \lambda_2} + \frac{\lambda_2^2(1 - \ln \lambda_2)}{\lambda_2 - \lambda_1}. \tag{D.13}$$

Up to an additive constant this result is equal to the Wehrl entropy of one qubit mixed state obtained by partial trace [664] or to the subentropy (D.11) of this state.

Problem 13.5. It is enough to show that $\mathrm{Tr} d_1 d_2 \geq \mathrm{Tr} W d_1 W^\dagger d_2$, where $W = V^\dagger U$ is unitary. We can write this as an inequality for scalar products between vectors: $\vec{d}_1 \cdot \vec{d}_2 \geq (B\vec{d}_1) \cdot \vec{d}_2$ where B is unistochastic. Problem 2.2 shows that this is true.

Problem 13.6. Work in a basis where $A+B$ is diagonal. Note that $\left(\det(A+B)\right)^{1/N} = \prod_i (A_{ii} + B_{ii})^{1/N} \geq \prod_i A_{ii}^{1/N} + \prod_i B_{ii}^{1/N}$. For our purposes the second step is the more interesting: from the Schur–Horn theorem and the Schur concavity of the elementary symmetric functions it follows that $\prod_i A_{ii} \geq \det A$ (and similarly for B).

Problem 13.7. It is sufficient to compute $\mathrm{Tr} L L^\dagger$ using the representation $L = \sum_{i=1}^r A_i \otimes A_i^*$ of the superoperator.

Problem 14.1. We want to show that $||\mathbf{x}||^2 \geq ||B\mathbf{x}||^2$, so we must show that $\mathbb{1} - B^T B$ is a positive operator. This follows from the Frobenius–Perron theorem.

Problem 14.2. The probability distribution obtained from a density matrix is majorised by its eigenvalues. Helstrom's theorem implies that the trace distance between two density matrices is the maximum of the l_1-distance between the probability distributions that can be obtained from them. To maximise the trace distance the two density matrices must commute. If $\rho_2 = U\rho_1 U^\dagger$ and ρ_1 is diagonal then U must be a permutation matrix. See Figure 2.2.

Problem 14.3. We know that there is a POVM such that $\sqrt{F} = \sum_i \sqrt{p_i q_i}$, with probabilities given in Eq. (14.48). Then

$$2(1 - \sqrt{F}) = \sum_i (\sqrt{p_i} - \sqrt{q_i})^2 \leq \sum_i |\sqrt{p_i} - \sqrt{q_i}| |\sqrt{p_i} + \sqrt{q_i}| =$$

$$= \sum_i |p_i - q_i| \leq 2D_{\mathrm{Tr}}(\rho, \sigma), \tag{D.14}$$

where Helstrom's theorem was used in the last step.

Problem 14.4. (a) This follows if we set $U = \mathbb{1}$ in the argument that led to the quantum Bhattacharyya coefficient. Equality holds if $[\rho, \sigma] = 0$.
(b) Equality holds if one of the states is pure. The inequality follows from this because of concavity. Incidentally, fidelity can also be bounded [668] as $E(\sigma, \rho) \leq F(\sigma, \rho) \leq G(\sigma, \rho)$, where the lower bound, called *subfidelity*, is $E(\sigma, \rho) = \mathrm{Tr}\sigma\rho + \sqrt{2}\sqrt{(\mathrm{Tr}\sigma\rho)^2 - \mathrm{Tr}\sigma\rho\sigma\rho}$, and the upper one, called *superfidelity*, is $G(\sigma, \rho) = \mathrm{Tr}\sigma\rho + \sqrt{1 - \mathrm{Tr}\rho^2}\sqrt{1 - \mathrm{Tr}\sigma^2}$. If $N = 2$ all three quantities are equal.

Problem 15.1. No, since $f_{WY} = (f_{max} + f_{geom})/2$ where $f_{geom} = \sqrt{t}$ is related to the geometric mean, $1/c_{geom}(x, y) = \sqrt{xy}$. But it is an interesting one, because the Wigner–Yanase metric is simple, and geodesic distances can be given explicitly [332].

Problem 15.2. With $\Theta_N \equiv \prod_{k=1}^{N} \Gamma(k/2)$ we obtain:

Table D.1 *Volumes of orthogonal groups and real flag manifolds*

Manifold	Dimension	Vol[X], $a = 1/2$	Vol'[X], $a = 1$
$\mathbb{R}P^N$	N	$\dfrac{\pi^{(N+1)/2}}{\Gamma[(N+1)/2]}$	$2^{N/2}\dfrac{\pi^{(N+1)/2}}{\Gamma[(N+1)/2]}$
$F_{\mathbb{R}}^{(N)} = \dfrac{O(N)}{[O(1)]^N}$	$N(N-1)/2$	$\dfrac{\pi^{N(N+1)/4}}{\Theta_N}$	$2^{N(N-1)/4}\dfrac{\pi^{N(N+1)/4}}{\Theta_N}$
$O(N)$	$N(N-1)/2$	$2^N\dfrac{\pi^{N(N+1)/4}}{\Theta_N}$	$2^{N(N+3)/4}\dfrac{\pi^{N(N+1)/4}}{\Theta_N}$
$SO(N)$	$N(N-1)/2$	$2^{N-1}\dfrac{\pi^{N(N+1)/4}}{\Theta_N}$	$2^{(N(N+3)/4-1}\dfrac{\pi^{N(N+1)/4}}{\Theta_N}$

Problem 15.4. Integrating over respective distributions we obtain $\langle S(\rho)\rangle_{HS} = 1/3$, $\langle S(\rho)\rangle_B = 2 - 7\ln 2/6$, $\langle S(\rho)\rangle_o = 2 - \ln 2$ and $\langle S(\rho)\rangle_u = \ln 2/2$.

Problem 15.5. The averages (computed by Malacarne et al. [641]) read

$$\langle \text{Tr}\rho^3\rangle_{N,K} = \frac{(K+N)^2 + KN + 1}{(KN+1)(KN+2)}, \quad \langle \text{Tr}\rho^4\rangle_{N,K} = \frac{(K+N)[(K+N)^2 + 3KN + 5]}{(KN+1)(KN+2)(KN+3)}. \tag{D.15}$$

Problem 15.6. Fidelity between pure states is equal to the squared component of $|\phi\rangle$ expanded in a basis containing $|\psi\rangle$. One finds $P_N(F) = (N-1)(1-F)^{N-2}$ (see Section 7.6 on random pure states).

Problem 15.7. Average fidelities, $\langle F\rangle_{HS} = 1/2 + 9\pi^2/512 \approx 0.6735$ and $\langle F\rangle_B = 1/2 + 8/(9\pi^2) \approx 0.590$, exceed the average over two random pure states, $\langle F\rangle_{FS} = 1/2$ [1018].

Problem 15.9. Making use of an integral with respect to the Marchenko–Pastur distribution, $\int_0^4 [-x\ln x MP(x)]dx = -1/2$ and returning to the original variable, $\lambda = Nx$, we arrive at the asymptotic result $\langle S\rangle_{HS} \to \ln N - 1/2$. Asymptotic average purity $2/N$ is consistent with the second moment of the MP distribution, $\int_0^4 x^2 MP(x)dx = 2$. Higher traces $\langle \text{Tr}\rho^k\rangle_{HS}$ are given by higher moments of the MP distribution, equal to the *Catalan numbers*, $C_k = 1, 2, 5, 14, 42, \ldots$

Problem 15.10. Following [157] notice that since $\lambda_1 \geq x$ the spectrum λ is majorised by the vector $\nu = \{\lambda_1, (1 - \lambda_1)/(N - 1), \ldots (1 - \lambda_1)/(N - 1)\}$. The Schur concavity of the entropy function implies that $S(\rho) \leq S(\nu) = (1 - \lambda_1) \ln(N - 1) + S(\lambda_1, 1 - \lambda_1) \equiv t(\lambda_1)$. The derivative $\partial t(\lambda_1)/\partial \lambda_1$ is negative for $\lambda_1 > 1/N$, so the function $t(\lambda_1)$ is monotonously decreasing for $\lambda_1 > 1/N$. Thus $t(\lambda_1) \geq t(x)$ which completes the proof.

Problem 16.1. The complete orthogonality graph must have $(2N - 1)(2N - 2)/2$ links. Any one of Alice's vectors can supply at most $N - 1$ links, otherwise there would be linear dependencies among her vectors. The total number of links she can contribute is then the integer part of $(2N - 1)(N - 1)/2$, which is less than half of the total number if N is even. And Bob cannot do better.

Problem 16.3. The Schmidt vector of the state $|\phi_+\rangle^{\otimes m}$ consists of N^m components equal to N^{-m} each. The state consisting of n copies of the initial state $|\psi\rangle$ may be, for large n, approximated by $K = \exp[nE(|\psi\rangle)]$ terms in the Schmidt decomposition described by the vector $\vec{\lambda}$. Choosing $m \approx n[E(|\psi\rangle)]/\ln N$ we see that $\{N^{-m}, \ldots, N^{-m}\} \prec \{\lambda_1, \ldots, \lambda_K\}$. Thus Nielsen's majorisation theorem implies that such a conversion may be done [689]. Asymptotically the reverse transformation is also possible [115], so for any pure state the distillable entanglement is just equal to entanglement entropy, $E_D(|\psi\rangle) = E(|\psi\rangle)$.

Problem 16.4. Write a separable state in its eigenbasis, $\rho = \sum_j \lambda_j |\Psi_j\rangle \langle \Psi_j|$ and in its decomposition into pure product states, $\rho = \sum_i p_i |\phi_i^A\rangle \langle \phi_i^A| \otimes |\phi_i^B\rangle \langle \phi_i^B|$. Write the partial trace $\rho_A = \sum_i p_i |\phi_i^A\rangle \langle \phi_i^A|$ in its eigenbasis, $\rho_A = \sum_k \lambda_k^A |k\rangle \langle k|$. Apply Schrödinger's mixture theorem (8.39) twice, substituting $\sqrt{p_i} |\phi_i^A\rangle = \sum_k V_{ik} \sqrt{\lambda_k^A} |k\rangle$ into $\sqrt{\lambda_j} |\Psi_j\rangle = \sum_i U_{ji} \sqrt{p_i} |\phi_i^A\rangle |\phi_i^B\rangle$, where U and V are unitary. Multiply the result by its adjoint and obtain $\lambda_j = \sum_k B_{jk} \lambda_k^A$ making use of the orthonormality, $\langle k | k' \rangle = \delta_{k,k'}$. Show that B is bistochastic, what implies (16.67) due to HLP lemma.

Problem 16.6. Obviously not. A simple dimension counting will do. In the $N \times N$ problem the set of separable states has $N^4 - 1$ dimensions, while the set of locally diagonalizable states forms a $3(N^2 - 1)$ dimensional subset. (The local unitaries and the diagonalised state contribute $N^2 - 1$ parameters each.) Note that it contains the set of all product mixed states of dimension $2N^2 - 2$, equal to the dimension of the set of all pure states of the bipartite system.

Problem 16.8. This inequality follows from condition $\text{Tr}\rho^2 \leq 1$. It is not sufficient for positivity, but may be accompanied by additional inequalities involving higher traces $\text{Tr}\rho^k$, with $k = 3, 4, \ldots$ [529, 798].

Problem 16.9. Partial trace induces the HS measure (15.35) with $\lambda_1 = \cos^2 \chi$. Change of variables provide the required distributions, while integrations give the expectation values $\langle \chi \rangle_{\mathbf{CP}^3} = 1/3$ and $\langle C \rangle_{\mathbf{CP}^3} = 3\pi/16$. The distribution $P(\chi)$ achieves maximum at $\chi_m = \arccos\left[\sqrt{1/2 + 1/\sqrt{6}}\right]$, while it is most likely to find a two-qubit random pure state with concurrence $C_m = 1/\sqrt{2}$.

Problem 16.10. The spectrum of the partial transpose ρ^{T_A} of a generic $N \times N$ state is asymptotically described by the shifted semicircular law (16.60). Hence the average negativity, $\mathcal{N}_T(\rho) = \text{Tr}|\rho^{T_A}| - 1$, can be expressed by an integral, $\langle \mathcal{N}_T \rangle = \int_{-1}^0 \frac{|x|}{P} P_{\text{PT}}(x)dx = \frac{3\sqrt{3}}{4\pi} - \frac{1}{3} \simeq 0.080$. This number, compared with the negativity of the maximally entangled state, $\mathcal{N}(\rho_+) = N - 1$, shows that a generic mixed state is only weakly entangled.

Problem 17.1. For the GHZ state $\text{Tr}_A|GHZ\rangle\langle GHZ| = \frac{1}{2}(|00\rangle\langle 00| + |11\rangle\langle 11|)$, a mixture of two separable states. For the W state $\text{Tr}_A|W\rangle\langle W| = \frac{1}{3}|00\rangle\langle 00| + \frac{2}{3}|\Phi_{BC}^+\rangle\langle \Phi_{BC}^+|$, an entangled mixture of a Bell state and a separable state. Its tangle equals $4/9$, and the spectrum of $\text{Tr}_{AB}|W\rangle\langle W|$ equals $(2/3, 1/3)$.

Problem 17.2. Expression (17.9) consists of three terms, so the rank r of $|W\rangle$ cannot be larger than three. For any three-qubit state of rank two its partial trace, e.g. $\rho_{AB} = \text{Tr}_C|\psi_{ABC}\rangle\langle\psi_{ABC}|$, contains in its support at least two linearly independent product states. For $|W_3\rangle$, the reduced state ρ_{AB} analysed in Problem 17.1 contains only a single product state $|00\rangle$, so $r(|W\rangle) > 2$. The situation changes if one adds an ϵ contribution of the state $|111\rangle$. Then the reduced state ρ_{AB} supports *two* product states, $|00\rangle$ and $|11\rangle$, so the rank r *decreases* to two.

Problem 17.3. All one-party reduced density matrices are equal to

$$\rho = \begin{pmatrix} (K-k)/K & 0 \\ 0 & k/K \end{pmatrix} \tag{D.16}$$

(as can be proved by induction). For $K = 2k$ this is the maximally mixed state. The difference to the GHZ state is that entangled density matrices are encountered for bipartite reductions.

Problem 17.4. In the factors a general state is given by $|\psi_A\rangle = \cos\frac{\theta_A}{2}|0_A\rangle + \sin\frac{\theta_A}{2}e^{i\phi_A}|1_A\rangle$ etc. Write out $|\langle\psi_A|\langle\psi_B|\langle\psi_C|\psi\rangle|^2$. Take derivatives with respect to $\theta_A, \theta_B, \theta_C$. By assumption they vanish at $\theta_A = \theta_B = \theta_C = \pi$. The conclusion follows.

Problem 17.5. (a) Bernstein [126] revealed that there exist cases in which three variables X, Y, Z are pairwise independent but not independent. This is the case here. Thus the probability to register three '0' on a card is $p_{XYZ} = 0 \neq p_X p_Y p_Z$.
(b) $|\psi_\triangle\rangle$ is locally equivalent to $|GHZ\rangle$. (An easy way to check this is to use the stellar representation.) The statistics of the outcomes of measurements of the operators $\sigma_z^1, \sigma_z^2, \sigma_z^3$ coincide with that of the Bernstein cards [178].

Problem 17.6. The basis of eight orthogonal entangled states in $\mathcal{H}_{\in}^{\otimes 3}$ reads

$|\Psi_{++}\rangle = [|000\rangle + |011\rangle + |101\rangle + |110\rangle]/2, \quad |\Psi_{+-}\rangle = [|000\rangle + |011\rangle - |101\rangle - |110\rangle]/2,$

$|\Psi_{-+}\rangle = [|000\rangle - |011\rangle + |101\rangle - |110\rangle]/2, \quad |\Psi_{--}\rangle = [|000\rangle - |011\rangle - |101\rangle + |110\rangle]/2,$

$|\Phi_{++}\rangle = [|111\rangle + |100\rangle + |010\rangle + |001\rangle]/2, \quad |\Phi_{+-}\rangle = [|111\rangle + |100\rangle - |010\rangle - |001\rangle]/2,$

$|\Phi_{-+}\rangle = [|111\rangle - |100\rangle + |010\rangle - |001\rangle]/2, \quad |\Phi_{--}\rangle = [|111\rangle - |100\rangle - |010\rangle + |001\rangle]/2.$

Problem 17.7. (a) For $|GHZ_4\rangle$ all three spectra of reduced states are equal, $\Lambda_{AB} = \Lambda_{AC} = \Lambda_{AD} = \{1/2, 1/2, 0, 0\}$ so that $\bar{S}_1 = \bar{S}_2 = \bar{S}_\infty = \log 2 \approx 0.693$. (b) Two local spectra of $|B_4\rangle$ are uniform, $\{1/4, 1/4, 1/4, 1/4\}$, but the third one reads $\{1/2, 1/2, 0, 0\}$. Thus $\bar{S}_1 = \bar{S}_2 = \bar{S}_\infty = \frac{5}{3}\log 2 \approx 1.155$. (c) For the cluster state $|C_4\rangle$ the local spectra (and the mean entropies) are the same as both states are locally equivalent. (d) For the Higuchi–Sudbery state $|HS\rangle$ all three local spectra read $\{3, 1, 1, 1\}/6$, so $\bar{S}_1 = \log 2 + \frac{1}{2}\log 3 \approx 1.242$ and $\bar{S}_2 = \log 3 \approx 1.099$, while $\bar{S}_\infty = \log 2$. All the mean entropies are smaller than $\log 4 \approx 1.386$, as there are no AME states of four qubits. The notion of 'maximally entangled' then depends on the measure used: $|HS\rangle$ is conjectured to maximise von Neumann entropy $\bar{S}_1$ (and gives a local maximum [173]), while the mean Rényi entropy $\bar{S}_2$ (and the linear entropy $\bar{S}_2^{HC}$) is maximised by $|C_4\rangle$, as it leads to the minimal purity of the reduced state [354].

Problem 17.8. The state $|\Phi_3^6\rangle$ is a nice example of an absolutely maximally entangled state: for any partition of six subsystems into two triples the reduced density matrix on three subsystems is maximally mixed [148]. Hence the state is 3-uniform.

Problem 17.9. By inspection we see that the state is invariant under $\mathbb{1}XXX^2$, $X\mathbb{1}XX$, $X^2X\mathbb{1}X$, $XXX^2\mathbb{1}$ and under $\mathbb{1}ZZZ^2$, $Z\mathbb{1}ZZ$, $Z^2Z\mathbb{1}Z$, $ZZZ^2\mathbb{1}$, where X and Z are generators of the Heisenberg group $H(3)$. All elements cube to the unit element, all elements commute, and from these generators we can construct altogether 81 'words', which is the stabilizer group for this state.

Problem 17.10. The Hamming code corrects all single errors, and no double errors. The probability that the message comes through correctly is therefore $(1-p)^7 + 7p(1-p)^6$, which is larger than $(1-p)^4$ for all $p < 1/2$. The repetition code is able to correct some double and triple errors, but gives no improvement unless $p < 0.151$.

Problem 17.11. The matrix U_9 describing (17.76) is a permutation matrix obtained by taking an arbitrary single digit (e.g. **1**) from the sudoku matrix S, while the orthogonal matrix $U_8 = \frac{1}{\sqrt{8}}H_8$ is proportional to the Hadamard matrix with entries equal ± 1. Note that S enjoys a special property: apart from the standard sudoku condition that each digit appears only once in each row, each column and each block, a stronger condition holds. Each digit appears only once in each location (position inside the block), in each broken row (e.g. three upper rows of three left blocks) and each broken column (e.g. three left columns of three lower blocks) – see below. This ensures that $U_9^{T_2}$ and U_9^R are also permutation matrices, so U_9 is multiunitary [356].

$$
S = \begin{bmatrix}
8 & 1 & 6 & 2 & 4 & 9 & 5 & 7 & 3 \\
3 & 5 & 7 & 6 & 8 & 1 & 9 & 2 & 4 \\
4 & 9 & 2 & 7 & 3 & 5 & 1 & 6 & 8 \\
7 & 3 & 5 & 1 & 6 & 8 & 4 & 9 & 2 \\
2 & 4 & 9 & 5 & 7 & 3 & 8 & 1 & 6 \\
6 & 8 & 1 & 9 & 2 & 4 & 3 & 5 & 7 \\
9 & 2 & 4 & 3 & 5 & 7 & 6 & 8 & 1 \\
1 & 6 & 8 & 4 & 9 & 2 & 7 & 3 & 5 \\
5 & 7 & 3 & 8 & 1 & 6 & 2 & 4 & 9
\end{bmatrix}, \quad
H_8 = \begin{pmatrix}
- & - & - & + & - & + & + & + \\
- & - & - & + & + & - & - & - \\
- & - & + & - & - & + & - & - \\
+ & + & - & + & - & + & - & - \\
- & + & - & - & - & - & + & - \\
+ & - & + & + & - & - & + & - \\
+ & - & - & - & + & + & + & - \\
+ & - & - & - & - & - & - & +
\end{pmatrix}.
$$

$$\tag{D.17}$$

Problem 17.12. (a) Using the canonical form (17.95) one finds

$$
A^{ik} = \{A^0, A^1\} = \left\{ \begin{pmatrix} 1 & 0 \\ 0 & \sqrt{\frac{k-1}{k}} \end{pmatrix}, \begin{pmatrix} 0 & \frac{1}{\sqrt{k}} \\ 0 & 0 \end{pmatrix} \right\}. \tag{D.18}
$$

and the vectors at the end are easy to find; (b) in the periodic case (17.98) one can take

$$
A^{ik} = \{A^0, A^1\} = \left\{ \begin{pmatrix} 1 & 0 \\ 0 & 1 \end{pmatrix}, \begin{pmatrix} 0 & 1 \\ 0 & 0 \end{pmatrix} \right\} \quad \text{if } k < K, \tag{D.19}
$$

$$
A^{iK} = \{A^0, A^1\} = \left\{ \frac{1}{\sqrt{K}} \begin{pmatrix} 0 & 1 \\ 1 & 0 \end{pmatrix}, \frac{1}{\sqrt{K}} \begin{pmatrix} 1 & 0 \\ 0 & 0 \end{pmatrix} \right\}; \quad \text{c) } D = N.
$$

References

[1] S. Aaronson. *Quantum Computing since Democritus*. Cambridge: Cambridge University Press, 2013.

[2] S. Abe. A note on the q-deformation-theoretic aspect of the generalized entropy in nonextensive physics. *Phys. Lett.*, A 224:326, 1997.

[3] S. Abe and A. K. Rajagopal. Non-additive conditional entropy and its significance for local realism. *Physica*, A 289:157, 2001.

[4] L. Accardi. Non-relativistic quantum mechanics as a noncommutative Markov process. *Adv. Math.*, 20:329, 1976.

[5] L. Accardi. Topics in quantum probability. *Phys. Rep.*, 77:169, 1981.

[6] A. Acín, A. Andrianov, E. Jané, and R. Tarrach. Three-qubit pure state canonical forms. *J. Phys.*, A 34:6725, 2001.

[7] M. Adelman, J. V. Corbett, and C. A. Hurst. The geometry of state space. *Found. Phys.*, 23:211, 1993.

[8] MacAdam, *J. Opt. Soc. Am.*, 32: 247, 1942.

[9] I. Affleck, T. Kennedy, E. Lieb, and H. Tasaki. Rigorous results on valence-bond ground states in antiferromagnets. *Phys. Rev. Lett.*, 59:799, 1987.

[10] G. Agarwal. Relation between atomic coherent-state representation state multipoles and generalized phase space distributions. *Phys. Rev.*, A 24:2889, 1981.

[11] S. J. Akhtarshenas and M. A. Jafarizadeh. Robustness of entanglement for Bell decomposable states. *E. Phys. J.*, D 25:293, 2003.

[12] S. J. Akhtarshenas and M. A. Jafarizadeh. Optimal Lewenstein-Sanpera decomposition for some bipartite systems. *J. Phys.*, A 37: 2965, 2004.

[13] S. Alaca and K. S. Williams. *Introductory Algebraic Number Theory*. Cambridge UP, 2004.

[14] P. M. Alberti. A note on the transition probability over C^*-algebras. *Lett. Math. Phys.*, 7:25, 1983.

[15] P. M. Alberti and A. Uhlmann. *Dissipative Motion in State Spaces*. Leipzig: Teubner, 1981.

[16] P. M. Alberti and A. Uhlmann. *Stochasticity and Partial Order: Doubly Stochastic Maps and Unitary Mixing*. Reidel, 1982.

[17] P. M. Alberti and A. Uhlmann. On Bures distance and $*$-algebraic transition probability between inner derived positive linear forms over w^*-algebras. *Acta Appl. Math.*, 60:1, 2000.

[18] S. Albeverio, K. Chen, and S.-M. Fei. Generalized reduction criterion for separability of quantum states. *Phys. Rev.*, A 68:062313, 2003.

[19] E. Alfsen and F. Schultz. Unique decompositions, faces, and automorphisms of separable states. *J. Math. Phys.*, 51:052201, 2010.

[20] E. M. Alfsen and F. W. Shultz. *State Spaces of Operator Algebras*. Boston: Birkhäuser, 2001.

[21] E. M. Alfsen and F. W. Shultz. *Geometry of State Spaces of Operator Algebras*. Boston: Birkhäuser, 2003.

[22] S. T. Ali, J.-P. Antoine, and J.-P. Gazeau. *Coherent States, Wavelets and Their Generalizations*. Springer-Verlag, 2000.

[23] R. Alicki. Comment on 'Reduced dynamics need not be completely positive'. *Phys. Rev. Lett.*, 75:3020, 1995.

[24] R. Alicki and M. Fannes. *Quantum Dynamical Systems*. Oxford University Press, 2001.

[25] R. Alicki and M. Fannes. Continuity of quantum conditional information. *J. Phys.*, A 37:L55, 2004.

[26] R. Alicki and K. Lendi. *Quantum Dynamical Semigroups and Applications*. Springer-Verlag, 1987.

[27] R. Alicki, A. Łoziński, P. Pakoński, and K. Życzkowski. Quantum dynamical entropy and decoherence rate. *J. Phys.*, A 37:5157, 2004.

[28] W. O. Alltop. Complex sequences with low periodic correlations. *IEEE Trans. Inform. Theory.*, 26:350, 1980.

[29] N. Alon and L. Lovasz. Unextendible product bases. *J. Combinat. Theor.*, A 95:169, 2001.

[30] S. Amari. *Differential Geometrical Methods in Statistics*. Springer-Verlag, 1985.

[31] L. Amico, R. Fazio, A. Osterloh, and V. Vedral. Entanglement in many-body systems. *Rev. Mod. Phys.*, 80:517, 2008.

[32] G. G. Amosov, A. S. Holevo, and R. F. Werner. On some additivity problems in quantum information theory. *Probl. Inform. Transm.*, 36:25, 2000.

[33] J. Anandan and Y. Aharonov. Geometry of quantum evolution. *Phys. Rev. Lett.*, 65:1697, 1990.

[34] G. W. Anderson, A. Guionnet, and O. Zeitouni. *An Introduction to Random Matrices*. Cambridge: Cambridge University Press, 2010.

[35] D. Andersson, I. Bengtsson, K. Blanchfield, and H. B. Dang. States that are far from being stabilizer states. *J. Phys.*, A 48:345301, 2015.

[36] O. Andersson and I. Bengtsson. Clifford tori and unbiased vectors. *Rep. Math. Phys.*, 79:33, 2017.

[37] T. Ando. Majorization, doubly stochastic matrices, and comparison of eigenvalues. *Linear Algebra Appl.*, 118:163, 1989.

[38] T. Ando. Majorizations and inequalities in matrix theory. *Linear Algebra Appl.*, 199:17, 1994.

[39] T. Ando. Cones and norms in the tensor product of matrix spaces. *Lin. Alg. Appl.*, 379:3, 2004.

[40] P. Aniello and C. Lupo. A class of inequalities inducing new separability criteria for bipartite quantum systems. *J. Phys.*, A 41:355303, 2008.

[41] D. M. Appleby. Properties of the extended Clifford group with applications to SIC-POVMs and MUBs. preprint arXiv:0909.5233.

[42] D. M. Appleby. Symmetric informationally complete-positive operator measures and the extended Clifford group. *J. Math. Phys.*, 46:052107, 2005.

[43] M. Appleby, T. -Y. Chien, S. Flammia, and S. Waldron, Constructing exact symmetric informationally complete measurements from numerical solutions. preprint arXiv:1703.05981.

[44] D. M. Appleby, H. B. Dang, and C. A. Fuchs. Symmetric informationally-complete quantum

states as analogues to orthonormal bases and minimum-uncertainty states. *Entropy*, 16:1484, 2014.

[45] D. M. Appleby, S. Flammia, G. McConnell, and J. Yard. Generating ray class fields of real quadratic fields via complex equiangular lines. preprint arXiv:1604.06098.

[46] D. M. Appleby, H. Yadsan-Appleby, and G. Zauner. Galois automorphisms of a symmetric measurement. *Quant. Inf. Comp.*, 13:672, 2013.

[47] C. Aragone, G. Gueri, S. Salamó, and J. L. Tani. Intelligent spin states. *J. Phys.*, A 7:L149, 1974.

[48] H. Araki. On a characterization of the state space of quantum mechanics. *Commun. Math. Phys.*, 75:1, 1980.

[49] H. Araki and E. Lieb. Entropy inequalities. *Commun. Math. Phys.*, 18:160, 1970.

[50] P. Aravind. The 'twirl', stella octangula and mixed state entanglement. *Phys. Lett.*, A 233:7, 1997.

[51] P. K. Aravind. Geometry of the Schmidt decomposition and Hardy's theorem. *Am. J. Phys.*, 64:1143, 1996.

[52] L. Arnaud and N. Cerf. Exploring pure quantum states with maximally mixed reductions. *Phys. Rev.*, A 87:012319, 2013.

[53] V. I. Arnold. Symplectic geometry and topology. *J. Math. Phys.*, 41:3307, 2000.

[54] V. I. Arnold. *Dynamics, Statistics, and Projective Geometry of Galois Fields*. Cambridge University Press, 2011.

[55] P. Arrighi and C. Patricot. A note on the correspondence between qubit quantum operations and special relativity. *J. Phys.*, A 36:L287, 2003.

[56] P. Arrighi and C. Patricot. On quantum operations as quantum states. *Ann. Phys. (N.Y.)*, 311:26, 2004.

[57] E. Arthurs and J. L. Kelly, Jr. On the simultaneous measurement of a pair of conjugate variables. *Bell Sys. Tech. J.*, 44:725, 1965.

[58] S. Arunachalam, N. Johnston, and V. Russo. Is absolute separability determined by the partial transpose? *Quant. Inf. Comput.*, 15:0694, 2015.

[59] W. B. Arveson. Sub-algebras of C^*-algebras. *Acta Math.*, 123:141, 1969.

[60] Arvind, K. S. Mallesh, and N. Mukunda. A generalized Pancharatnam geometric phase formula for three-level quantum systems. *J. Phys.*, A 30:2417, 1997.

[61] M. Aschbacher, A. M. Childs, and P. Wocjan. The limitations of nice mutually unbiased bases. *J. Algebr. Comb.*, 25:111, 2007.

[62] A. Ashtekar and A. Magnon. Quantum fields in curved space-times. *Proc. Roy. Soc. A*, 346:375, 1975.

[63] A. Ashtekar and T. A. Schilling. Geometrical formulation of quantum mechanics. In A. Harvey, editor, *On Einstein's Path*, page 23. Springer, 1998.

[64] A. Aspect, J. Dalibard, and G. Roger. Experimental test of Bell's inequalities using time-varying analysers. *Phys. Rev. Lett.*, 49:1804, 1982.

[65] M. Atiyah. The non-existent complex 6-sphere. preprint arXiv: 1610.09366.

[66] M. F. Atiyah. Convexity and commuting Hamiltonians. *Bull. London Math. Soc.*, 14:1, 1982.

[67] G. Aubrun. Partial transposition of random states and non-centered semicircular distributions. *Rand. Mat. Theor. Appl.*, 1:1250001, 2012.

[68] G. Aubrun, S. Szarek, and E. Werner. Hastings' additivity counterexample via Dvoretzky's theorem. *Commun. Math. Phys.*, 305:85, 2011.

[69] G. Aubrun and S. J. Szarek. *Alice and Bob Meet Banach. The Interface of Asymptotic Geometric Analysis and Quantum Information Theory*. Amer. Math. Soc. book in preparation.

[70] K. Audenaert, J. Eisert, E. Jané, M. B. Plenio, S. Virmani, and B. D. Moor. Asymptotic relative entropy of entanglement. *Phys. Rev. Lett.*, 87:217902, 2001.

[71] K. Audenaert, B. D. Moor, K. G. H. Vollbrecht, and R. F. Werner. Asymptotic relative entropy of entanglement for orthogonally invariant states. *Phys. Rev.*, A 66:032310, 2002.

[72] K. Audenaert, F. Verstraete, and B. D. Moor. Variational characterizations of separability and entanglement of formation. *Phys. Rev.*, A 64:052304, 2001.

[73] K. M. R. Audenaert and S. Scheel. On random unitary channels. *New J. Phys.*, 10:023011, 2008.

[74] R. Augusiak, J. Grabowski, M. Kuś, and M. Lewenstein. Searching for extremal PPT entangled states. *Opt. Commun.*, 283:805, 2010.

[75] R. Augusiak, J. Stasińska, and P. Horodecki. Beyond the standard entropic inequalities: Stronger scalar separability criteria and their applications. *Phys. Rev.*, A 77:012333, 2008.

[76] M. Aulbach, D. Markham, and M. Murao. The maximally entangled symmetric state in terms of the geometric measure. *New Jour. of Phys.*, 12:073025, 2010.

[77] L. Autonne. Sur les matrices hypohermitiennes et sur les matrices unitaires. *Ann. Univ. Lyon Nouvelle Ser.*, 38:1, 1915.

[78] J. E. Avron, L. Sadun, J. Segert, and B. Simon. Chern numbers, quaternions, and Berry's phases in Fermi systems. *Commun. Math. Phys.*, 124:595, 1989.

[79] N. Ay and W. Tuschmann. Dually flat manifolds and global information geometry. *Open Sys. Inf. Dyn.*, 9:195, 2002.

[80] H. Bacry. Orbits of the rotation group on spin states. *J. Math. Phys.*, 15:1686, 1974.

[81] P. Badziąg, C. Brukner, W. Laskowski, T. Paterek, and M. Żukowski. Experimentally friendly geometrical criteria for entanglement. *Phys. Rev. Lett.*, 100:140403, 2008.

[82] P. Badziąg, P. Deaur, M. Horodecki, P. Horodecki, and R. Horodecki. Concurrence in arbitrary dimensions. *J. Mod. Optics*, 49:1289, 2002.

[83] P. Badziąg, M. Horodecki, P. Horodecki, and R. Horodecki. Local environment can enhance fidelity of quantum teleportation. *Phys. Rev.*, A 62:012311, 2000.

[84] J. Baez. Rényi entropy and free energy. preprint arXiv:1102.2098.

[85] D. Baguette, F. Damanet, O. Giraud, and J. Martin. Anticoherence of spin states with point group symmetries. *Phys. Rev.*, A 92:052333, 2015.

[86] R. Balian, Y. Alhassid, and H. Reinhardt. Dissipation in manybody systems: A geometric approach based on information theory. *Phys. Rep.*, 131:1, 1986.

[87] K. M. Ball. An elementary introduction to modern convex geometry. In S. Levy, editor, *Flavors of Geometry*, page 1. Cambridge University Press, 1997.

[88] S. Bandyopadhyay, P. O. Boykin, V. Roychowdhury, and F. Vatan. A new proof for the existence of mutually unbiased bases. *Algorithmica*, 34:512, 2002.

[89] M. Barbieri, F. D. Martini, G. D. Nepi, P. Mataloni, G. D'Ariano, and C. Macciavello. Detection of entanglement with polarized photons: Experimental realization

of an entanglement witness. *Phys. Rev. Lett.*, 91:227901, 2003.

[90] M. Barbut and L. Mazliak. Commentary on the notes for Paul Léevy's 1919 lectures on the probability calculus at the Ecole Polytechnique. *El. J. Hist. Probab. Stat.*, 4, 2008.

[91] V. Bargmann. On a Hilbert space of analytic functions and an associated analytic transform. *Commun. Pure Appl. Math.*, 14:187, 1961.

[92] V. Bargmann. Note on Wigner's theorem on symmetry operations. *J. Math. Phys.*, 5:862, 1964.

[93] O. E. Barndorff-Nielsen and R. D. Gill. Fisher information in quantum statistics. *J. Phys.*, A 33:4481, 2000.

[94] H. Barnum, C. M. Caves, C. A. Fuchs, R. Jozsa, and B. Schumacher. Non-commuting mixed states cannot be broadcast. *Phys. Rev. Lett.*, 76:2818, 1996.

[95] H. Barnum, E. Knill, G. Ortiz, and L. Viola. Generalizations of entanglement based on coherent states and convex sets. *Phys. Rev.*, A 68:032308, 2003.

[96] H. Barnum and N. Linden. Monotones and invariants for multiparticle quantum states. *J. Phys.*, A 34:6787, 2001.

[97] N. Barros e Sá. Decomposition of Hilbert space in sets of coherent states. *J. Phys.*, A 34:4831, 2001.

[98] N. Barros e Sá. Uncertainty for spin systems. *J. Math. Phys.*, 42:981, 2001.

[99] T. Bastin, S. Krins, P. Mathonet, M. Godefroid, L. Lamata, and E. Solano. Operational families of entanglement classes for symmetric n–qubit states. *Phys. Rev. Lett.*, 103:070503, 2009.

[100] R. J. Baxter. Dimers on a rectangular lattice. *J. Math. Phys.*, 9:650, 1968.

[101] C. Beck and F. Schlögl. *Thermodynamics of Chaotic Systems*. Cambridge University Press, 1993.

[102] E. F. Beckenbach and R. Bellman. *Inequalities*. Springer, 1961.

[103] B. Belavkin and P. Staszewski. c^*-algebraic generalization of relative entropy and entropy. *Ann. Inst. H. Poincaré*, A37:51, 1982.

[104] S. T. Belinschi, B. Collins, and I. Nechita. Almost one bit violation for the additivity of the minimum output entropy. *Commun. Math. Phys.*, 341:885, 2016.

[105] J. S. Bell. On the Einstein–Podolsky–Rosen paradox. *Physics*, 1:195, 1964.

[106] J. S. Bell. On the problem of hidden variables in quantum mechanics. *Rev. Mod. Phys.*, 38:447, 1966.

[107] J. S. Bell. *Speakable and Unspeakable in Quantum Mechanics*. Cambridge University Press, 1987.

[108] A. Belovs. *Welch bounds and quantum state tomography*. MSc thesis, Univ. Waterloo, 2008.

[109] F. Benatti, R. Floreanini, and M. Piani. Quantum dynamical semigroups and non-decomposable positive maps. *Phys. Lett.*, A 326:187, 2004.

[110] F. Benatti and H. Narnhofer. Additivity of the entanglement of formation. *Phys. Rev.*, A 63:042306, 2001.

[111] J. Bendat and S. Sherman. Monotone and convex operator functions. *Trans. Amer. Math. Soc.*, 79:58, 1955.

[112] I. Bengtsson, J. Brännlund, and K. Życzkowski. CP^n, or entanglement illustrated. *Int. J. Mod. Phys.*, A 17:4675, 2002.

[113] I. Bengtsson and Å. Ericsson. How to mix a density matrix. *Phys. Rev.*, A 67:012107, 2003.

[114] I. Bengtsson, S. Weis, and K. Życzkowski. Geometry of the set of mixed quantum states: An apophatic approach. In J. Kielanowski, editor, *Geometric Methods in Physics*, page 175.

Springer, Basel, 2013. Trends in Mathematics.

[115] C. H. Bennett, H. J. Bernstein, S. Popescu, and B. Schumacher. Concentrating partial entanglement by local operations. *Phys. Rev.*, A 53:2046, 1996.

[116] C. H. Bennett and G. Brassard. Quantum cryptography: public key distribution and coin tossing. In *Int. Conf. on Computers, Systems and Signal Processing, Bangalore*, page 175. IEEE, 1984.

[117] C. H. Bennett, G. Brassard, C. Crépeau, R. Jozsa, A. Peres, and W. K. Wootters. Teleporting an unknown quantum state via dual classical and Einstein–Podolsky–Rosen channels. *Phys. Rev. Lett.*, 70:1895, 1993.

[118] C. H. Bennett, G. Brassard, S. Popescu, B. Schumacher, J. A. Smolin, and W. K. Wootters. Purification of noisy entanglement and faithful teleportation via noisy channels. *Phys. Rev. Lett.*, 76:722, 1996.

[119] C. H. Bennett, D. P. DiVincenzo, C. A. Fuchs, T. Mor, E. Rains, P. W. Shor, J. A. Smolin, and W. K. Wootters. Quantum non-locality without entanglement. *Phys. Rev.*, A 59:1070, 1999.

[120] C. H. Bennett, D. P. DiVincenzo, T. Mor, P. W. Shor, J. A. Smolin, and B. M. Terhal. Unextendible product bases and bound entanglement. *Phys. Rev. Lett.*, 82:5385, 1999.

[121] C. H. Bennett, D. P. DiVincenzo, J. Smolin, and W. K. Wootters. Mixed-state entanglement and quantum error correction. *Phys. Rev.*, A 54:3824, 1996.

[122] C. H. Bennett, S. Popescu, D. Rohrlich, J. A. Smolin, and A. V. Thapliyal. Exact and asymptotic measures of multipartite pure-state entanglement. *Phys. Rev.*, A 63:012307, 2000.

[123] C. H. Bennett and S. J. Wiesner. Communication via one- and two-particle operations on Einstein-Podolsky-Rosen states. *Phys. Rev. Lett.*, 69:2881, 1992.

[124] M. K. Bennett. *Affine and Projective Geometry.* Wiley, 1995.

[125] S. Berceanu. Coherent states, phases and symplectic areas of geodesic triangles. In M. Schliechenmaier, S. T. Ali, A. Strasburger, and A. Odzijewicz, editors, *Coherent States, Quantization and Gravity*, page 129. Wydawnictwa Uniwersytetu Warszawskiego, 2001.

[126] S. N. Bernstein. *Theory of Probabilities.* Moscow: Gostechizdat, 1928.

[127] D. W. Berry and B. C. Sanders. Bounds on general entropy measures. *J. Phys.*, A 36:12255, 2003.

[128] M. V. Berry. Quantal phase factors accompanying adiabatic changes. *Proc. Roy. Soc. A*, 392:45, 1984.

[129] J. Bertrand and P. Bertrand. A tomographic approach to Wigner's function. *Found. Phys.*, 17:397, 1987.

[130] R. Bhatia. *Matrix Analysis.* Springer-Verlag, 1997.

[131] R. Bhatia and P. Rosenthal. How and why to solve the operator equation AX - XB = Y. *Bull. Lond. Math. Soc.*, 29:1, 1997.

[132] A. Bhattacharyya. On a measure of divergence between two statistical populations defined by their probability distributions. *Bull. Calcutta Math. Soc.*, 35:99, 1943.

[133] T. Bhosale, S. Tomsovic, and A. Lakshminarayan. Entanglement between two subsystems, the Wigner semicircle and extreme value statistics. *Phys. Rev.*, A 85:062331, 2012.

[134] G. Björk, A. B. Klimov, P. de la Hoz, M. Grassl, G. Leuchs, and L. L. Sanchez-Soto. Extremal quantum states and their Majorana constellations. *Phys. Rev.*, A 92: 031801, 2015.

[135] P. Blanchard, L. Jakóbczyk, and R. Olkiewicz. Measures of entanglement based on decoherence. *J. Phys.*, A 34:8501, 2001.

[136] K. Blanchfield. Orbits of mutually unbiased bases. *J. Phys.*, A 47:135303, 2014.

[137] W. Blaschke and H. Terheggen. Trigonometria Hermitiana. *Rend. Sem. Mat. Univ. Roma, Ser. 4*, 3:153, 1939.

[138] I. Bloch, J. Dalibard, and W. Zwerger. Many–body physics with ultracold gases. *Rev. Mod. Phys.*, 80:885, 2008.

[139] F. J. Bloore. Geometrical description of the convex sets of states for systems with spin−1/2 and spin−1. *J. Phys.*, A 9:2059, 1976.

[140] B. G. Bodmann. A lower bound for the Wehrl entropy of quantum spin with sharp high-spin asymptotics. *Commun. Math. Phys.*, 250:287, 2004.

[141] E. Bogomolny, O. Bohigas, and P. Lebœuf. Distribution of roots of random polynomials. *Phys. Rev. Lett.*, 68:2726, 1992.

[142] E. Bogomolny, O. Bohigas, and P. Lebœuf. Quantum chaotic dynamics and random polynomials. *J. Stat. Phys.*, 85:639, 1996.

[143] J. J. Bollinger, W. M. Itano, D. J. Wineland, and D. J. Heinzen. Optimal frequency measurements with maximally correlated states. *Phys. Rev.*, A54:R4649, 1996.

[144] B. Bolt, T. G. Room, and G. E. Wall. On the Clifford collineation, transform and similarity groups I. *J. Austral. Math. Soc.*, 2:60, 1960.

[145] L. Bombelli, R. K. Koul, J. Lee, and R. D. Sorkin. Quantum source of entropy for black holes. *Phys. Rev.*, D 34:373, 1986.

[146] A. Bondal and I. Zhdanovskiy. Orthogonal pairs and mutually unbiased bases. preprint arXiv: 1510.05317.

[147] M. Born and E. Wolf. *Principles of Optics*. New York: Pergamon, 1987.

[148] A. Borras, A. R. Plastino, J. Batle, C. Zander, M. Casas, and A. Plastino. Multi-qubit systems: highly entangled states and entanglement distribution. *J. Phys.*, A 40:13407, 2007.

[149] S. Boucheron, G. Lugosi, and P. Massart. *Concentration Inequalities. A Nonasymptotic Theory of Independence*. Oxford: Oxford University Press, 2013.

[150] J. Bouda and V. Bužek. Purification and correlated measurements of bipartite mixed states. *Phys. Rev.*, A 65:034304, 2002.

[151] M. Bourennane, M. Eibl, C. Kurtsiefer, H. Weinfurter, O. Guehne, P. Hyllus, D. Bruß, M. Lewenstein, and A. Sanpera. Witnessing multipartite entanglement. *Phys. Rev. Lett.*, 92:087902, 2004.

[152] D. Bouwmeester, J. W. Pan, M. Daniell, H. Weinfurter, and A. Zeilinger. Observation of three-photon Greenberger–Horne–Zeilinger entanglement. *Phys. Rev. Lett.*, 82:1345, 1999.

[153] L. J. Boya, E. C. G. Sudarshan, and T. Tilma. Volumes of compact manifolds. *Rep. Math. Phys.*, 52:401, 2003.

[154] L. Boyle. Perfect porcupines: ideal networks for low frequency gravitational wave astronomy. preprint arXiv:1003.4946.

[155] F. G. S. L. Brandão, M. Christandl, and J. Yard. Faithful squashed entanglement. *Commun. Math. Phys.*, 306:805, 2011.

[156] F. G. S. L. Brandão and G. Gour. The general structure of quantum resource theories. *Phys. Rev. Lett.*, 115:070503, 2015.

[157] F. G. S. L. Brandão and M. Horodecki. On Hastings' counterexamples to the minimum output entropy additivity conjecture. *Open Syst. Inf. Dyn.*, 17:31, 2010.

[158] F. G. S. L. Brandão and M. Horodecki. An area law for

entanglement from exponential decay of correlations. *Nat. Phys.*, 9:721, 2013.

[159] D. Braun, O. Giraud, I. Nechita, C. Pellegrini, and M. Žnidarič. A universal set of qubit quantum channels. *J. Phys.*, A 47:135302, 2014.

[160] D. Braun, M. Kuś, and K. Życzkowski. Time-reversal symmetry and random polynomials. *J. Phys.*, A 30:L117, 1997.

[161] S. L. Braunstein. Geometry of quantum inference. *Phys. Lett.*, A 219:169, 1996.

[162] S. L. Braunstein and C. M. Caves. Statistical distance and the geometry of quantum states. *Phys. Rev. Lett.*, 72:3439, 1994.

[163] S. L. Braunstein, C. M. Caves, R. Jozsa, N. Linden, S. Popescu, and R. Schack. Separability of very noisy mixed states and implications for NMR quantum computing. *Phys. Rev. Lett.*, 83:1054, 1999.

[164] S. Bravyi. Requirements for compatibility between local and multipartite quantum states. *Quantum Inf. Comput.*, 4:12, 2004.

[165] S. Bravyi, M. B. Hastings, and F. Verstraete. Lieb-Robinson bounds and the generation of correlations and topological quantum order. *Phys. Rev. Lett*, 97:050401, 2006.

[166] S. Bravyi and A. Kitaev. Universal quantum computation with ideal Clifford gates and noisy ancillas. *Phys. Rev.*, A71:022316, 2005.

[167] U. Brehm. The shape invariant of triangles and trigonometry in two-point homogeneous spaces. *Geometriae Dedicata*, 33:59, 1990.

[168] G. K. Brennen. An observable measure of entanglement for pure states of multi-qubit systems. *Quant. Inf. Comp.*, 3:619, 2003.

[169] H.-P. Breuer. Optimal entanglement criterion for mixed quantum states. *Phys. Rev. Lett.*, 97:080501, 2006.

[170] H.-P. Breuer and F. Petruccione. *The Theory of Open Quantum Systems*. Oxford University Press, 2002.

[171] E. Briand, J.-G. Luque, and J.-Y. Thibon. A complete set of covariants for the four qubit system. *J. Phys.*, A 38:9915, 2003.

[172] H. J. Briegel and R. Raussendorf. Persistent entanglement in arrays of interacting particles. *Phys. Rev. Lett.*, A 86:910, 2001.

[173] S. Brierley and A. Higuchi. On maximal entanglement between two pairs in four-qubit pure states. *J. Phys.*, A 40:8455, 2007.

[174] S. Brierley and S. Weigert. Maximal sets of mutually unbiased quantum states in dimension six. *Phys. Rev.*, A 78:042312, 2008.

[175] J. Briët and P. Harremoës. Properties of classical and quantum Jensen–Shannon divergence. *Phys. Rev.*, A 79:052311, 2009.

[176] M. Brion. Sur l'image de l'application moment. In M. P. Malliavin, editor, *Séminaire d'Algebre Paul Dubreil et Marie-Paule Malliavin*, page 177. Springer, 1987.

[177] D. C. Brody and L. P. Hughston. Geometric quantum mechanics. *J. Geom. Phys.*, 38:19, 2001.

[178] D. C. Brody, L. P. Hughston, and D. M. Meier. Fragile entanglement statistics. *J. Phys.*, A 48:425301, 2015.

[179] E. Bruening, H. Makela, A. Messina, and F. Petruccione. Parametrizations of density matrices. *J. Mod. Opt.*, 59:1, 2012.

[180] N. Brunner, D. Cavalcanti, S. Pironio, V. Scarani, and S. Wehner. Bell nonlocality. *Rev. Mod. Phys.*, 86:419, 2014.

[181] D. Bruß. Characterizing entanglement. *J. Math. Phys.*, 43:4237, 2002.

[182] W. Bruzda, V. Cappellini, H.-J. Sommers, and K. Życzkowski. Random quantum operations. *Phys. Lett.*, A 373:320, 2009.

[183] D. J. C. Bures. An extension of Kakutani's theorem on infinite product measures to the tensor product of semifinite w^*- algebras. *Trans. Am. Math. Soc.*, 135:199, 1969.

[184] P. Busch, P. Lahti, and P. Mittelstaedt. *The Quantum Theory of Measurement*. Springer-Verlag, 1991.

[185] P. Caban, J. Rembieliński, K. A. Smoliński, and Z. Walczak. Classification of two-qubit states. *Quant. Infor. Process.*, 14:4665, 2015.

[186] A. Cabello, J. M. Estebaranz, and G. García-Alcaine. Bell–Kochen–Specker theorem: A proof with 18 vectors. *Phys. Lett.*, A 212:183, 1996.

[187] A. Cabello, S. Severini, and A. Winter. Graph theoretic approach to quantum correlations. *Phys. Rev. Lett.*, 112:040401, 2014.

[188] K. Cahill and R. Glauber. Density operators and quasi-probability distributions. *Phys. Rev.*, 177:1882, 1969.

[189] P. Calabrese and J. Cardy. Entanglement entropy and conformal field theory. *J. Phys.*, A 42:504005, 2009.

[190] A. R. Calderbank, P. J. Cameron, W. M. Kantor, and J. J. Seidel. Z_4-Kerdock codes, orthogonal spreads, and extremal Euclidean line sets. *Proc. London Math. Soc.*, 75:436, 1997.

[191] A. R. Calderbank, E. M. Rains, P. W. Shor, and N. J. A. Sloane. Quantum error correction and orthogonal geometry. *Phys. Rev. Lett.*, 78:405, 1997.

[192] L. L. Campbell. An extended Čencov characterization of the information metric. *Proc. Amer. Math. Soc.*, 98:135, 1986.

[193] B. Carlson and J. M. Keller. Eigenvalues of density matrices. *Phys. Rev.*, 121:659, 1961.

[194] H. A. Carteret, A. Higuchi, and A. Sudbery. Multipartite generalization of the Schmidt decomposition. *J. Math. Phys.*, 41:7932, 2000.

[195] C. M. Caves. Measures and volumes for spheres, the probability simplex, projective Hilbert spaces and density operators. http://info .phys.unm.edu/caves/reports

[196] C. M. Caves, C. A. Fuchs, and P. Rungta. Entanglement of formation of an arbitrary state of two rebits. *Found. Phys. Lett.*, 14:199, 2001.

[197] C. M. Caves, C. A. Fuchs, and R. Schack. Unknown quantum states: The quantum de Finetti representation. *J. Math. Phys.*, 43:4537, 2001.

[198] A. Cayley. On the theory of linear transformations. *Camb. Math. J.*, 4:193, 1845.

[199] N. N. Čencov. *Statistical Decision Rules and Optimal Inference*. American Mathematical Society, 1982.

[200] N. J. Cerf and C. Adami. Quantum extension of conditional probability. *Phys. Rev.*, A 60:893, 1999.

[201] N. J. Cerf, C. Adami, and R. M. Gingrich. Reduction criterion for separability. *Phys. Rev.*, A 60:898, 1999.

[202] N. J. Cerf, M. Bourennane, A. Karlsson, and N. Gisin. Security of quantum key distribution using d-level systems. *Phys. Rev. Lett.*, 88:127902, 2002.

[203] H. F. Chau. Unconditionally secure key distribution in higher dimensions by depolarization. *IEEE Trans. Inf. Theory*, 51:1451, 2005.

[204] K. Chen, S. Albeverio, and S.-M. Fei. Concurrence of arbitrary dimensional bipartite quantum states. *Phys. Rev. Lett.*, 95:040504, 2005.

[205] K. Chen and L.-A. Wu. The generalized partial transposition criterion for separability of multipartite quantum states. *Phys. Lett.*, A 306:14, 2002.

[206] K. Chen and L.-A. Wu. A matrix realignment method for recognizing entanglement. *Quant. Inf. Comp.*, 3:193, 2003.

[207] K. Chen and L.-A. Wu. Test for entanglement using physically observable witness operators and positive maps. *Phys. Rev.*, A 69:022312, 2004.

[208] L. Chen and D. Ž. Djoković. Qubit-qudit states with positive partial transpose. *Phys. Rev.*, A 86:062332, 2012.

[209] L. Chen and D. Ž. Đoković. Description of rank four PPT states having positive partial transpose. *J. Math. Phys.*, 52:122203, 2011.

[210] L. Chen and L. Yu. Entanglement cost and entangling power of bipartite unitary and permutation operators. *Phys. Rev.*, A 93:042331, 2016.

[211] P.-X. Chen, L.-M. Liang, C.-Z. Li, and M.-Q. Huang. A lower bound on entanglement of formation of $2 \otimes n$ system. *Phys. Lett.*, A 295:175, 2002.

[212] S. Chern. *Complex Manifolds Without Potential Theory*. Springer-Verlag, 1979.

[213] M.-T. Chien and H. Nakazato. Flat portions on the boundary of the Davis Wielandt shell of 3-by-3 matrices. *Lin. Alg. Appl.*, 430:204, 2009.

[214] E. Chitambar, D. Leung, L. Mančinska, M. Ozols, and A. Winter. Everything you always wanted to know about LOCC (but were afraid to ask). *Commun. Math. Phys.*, 328:303, 2014.

[215] S.-J. Cho, S.-H. Kye, and S. G. Lee. Generalized Choi map in 3-dimensional matrix algebra. *Lin. Alg. Appl.*, 171:213, 1992.

[216] M.-D. Choi. Positive linear maps on C^*-algebras. *Can. J. Math.*, 3:520, 1972.

[217] M.-D. Choi. Completely positive linear maps on complex matrices. *Lin. Alg. Appl.*, 10:285, 1975.

[218] M.-D. Choi. Positive semidefinite biquadratic forms. *Lin. Alg. Appl.*, 12:95, 1975.

[219] M.-D. Choi. Some assorted inequalities for positive maps on C^*-algebras. *J. Operator Theory*, 4:271, 1980.

[220] M.-D. Choi and T. Lam. Extremal positive semidefinite forms. *Math. Ann.*, 231:1, 1977.

[221] D. Chruściński and A. Jamiołkowski. *Geometric Phases in Classical and Quantum Mechanics*. Birkhäuser, 2004.

[222] D. Chruściński, J. Jurkowski, and A. Kossakowski. Quantum states with strong positive partial transpose. *Phys. Rev.*, A 77:022113, 2008.

[223] D. Chruściński and G. Sarbicki. Entanglement witnesses: construction, analysis and classification. *J. Phys.*, A 47:483001, 2014.

[224] O. Chterental and D. Ž. Đoković. Normal forms and tensor ranks of pure states of four qubits. In G. D. Ling, editor, *Linear Algebra Research Advances*, page 133. Nova Science Publishers, 2007.

[225] L. Clarisse. The distillability problem revisited. *Quantum Inf. Comput.*, 6:539, 2006.

[226] J. F. Clauser and A. Shimony. Bell's theorem; experimental tests and implications. *Rep. Prog. Phys.*, 41:1881, 1978.

[227] R. Clifton and H. Halvorson. Bipartite mixed states of infinite-

dimensional systems are generically non-separable. *Phys. Rev.*, A 61:012108, 2000.

[228] R. Clifton, B. Hepburn, and C. Wuthrich. Generic incomparability of infinite-dimensional entangled states. *Phys. Lett.*, A 303:121, 2002.

[229] B. Coecke. Kindergarten quantum mechanics. In G. Adenier, A. Khrennikov, and T. Nieuwenhuizen, editors, *Quantum Theory— Reconsideration of Foundations– 3*, page 81. AIP Conf. Proc. 810, 2006.

[230] V. Coffman, J. Kundu, and W. K. Wootters. Distributed entanglement. *Phys. Rev.*, A 61:052306, 2000.

[231] A. Coleman and V. I. Yukalov. *Reduced Density Matrices.* Springer, 2000.

[232] A. J. Coleman. The structure of fermion density matrices. *Rev. Mod. Phys.*, 35:668, 1963.

[233] B. Collins and I. Nechita. Random matrix techniques in quantum information theory. *J. Math. Phys.*, 57:015215, 2016.

[234] T. Constantinescu and V. Ramakrishna. Parametrizing quantum states and channels. *Quant. Inf. Proc.*, 2:221, 2003.

[235] J. Cortese. Relative entropy and single qubit Holevo–Schumacher–Westmoreland channel capacity. preprint quant-ph/0207128.

[236] C. A. Coulson. Present state of molecular structure calculations. *Rev. Mod. Phys.*, 32:170, 1960.

[237] T. M. Cover and J. A. Thomas. *Elements of Information Theory.* Wiley, 1991.

[238] P. Cromwell. *Polyhedra.* Cambridge University Press, 1997.

[239] T. Cubitt, A. Montanaro, and A.Winter. On the dimension of subspaces with bounded Schmidt rank. *J. Math. Phys.*, 49:022107, 2008.

[240] D. G. Currie, T. F. Jordan, and E. C. G. Sudarshan. Relativistic invariance and Hamiltonian theories of interacting particles. *Rev. Mod. Phys.*, 35:350, 1963.

[241] T. Curtright, D. B. Fairlie, and C. Zachos. *A Concise Treatise on Quantum Mechanics in Phase Space.* World Scientific, 2014.

[242] L. Dąbrowski and A. Jadczyk. Quantum statistical holonomy. *J. Phys.*, A 22:3167, 1989.

[243] S. Daftuar and P. Hayden. Quantum state transformations and the Schubert calculus. *Ann. Phys. (N. Y.)*, 315:80, 2005.

[244] S. Daftuar and M. Klimesh. Mathematical structure of entanglement catalysis. *Phys. Rev.*, A 64:042314, 2001.

[245] O. C. O. Dahlsten, R. Oliveira, and M. B. Plenio. Emergence of typical entanglement in two-party random processes. *J. Phys.*, A 40:8081, 2007.

[246] C. Dankert. *Efficient Simulation of Random Quantum States and Operators.* MSc thesis, Univ. Waterloo, 2005.

[247] N. Datta, T. Dorlas, R. Jozsa, and F. Benatti. Properties of subentropy. *J. Math. Phys.*, 55:062203, 2014.

[248] E. B. Davies. *Quantum Theory of Open Systems.* London: Academic Press, 1976.

[249] C. Davis. The Toeplitz–Hausdorff theorem explained. *Canad. Math. Bull.*, 14:245, 1971.

[250] R. I. A. Davis, R. Delbourgo, and P. D. Jarvis. Covariance, correlation and entanglement. *J. Phys.*, A 33:1895, 2000.

[251] L. De Lathauwer, B. D. Moor, and J. Vandewalle. A multilinear singular value decomposition. *SIAM J. Matrix Anal. Appl.*, 21:1253, 2000.

[252] T. R. de Oliveira, M. F. Cornelio, and F. F. Fanchini. Monogamy of entanglement of formation. *Phys. Rev.*, A 89:034303, 2014.

[253] J. I. de Vicente. Separability criteria based on the Bloch representation of density matrices. *Quantum Inf. Comput.*, 7:624, 2007.

[254] R. Delbourgo. Minimal uncertainty states for the rotation group and allied groups. *J. Phys.*, A 10:1837, 1977.

[255] R. Delbourgo and J. R. Fox. Maximum weight vectors possess minimal uncertainty. *J. Phys.*, A 10:L233, 1977.

[256] M. Delbrück and G. Molière. Statistische Quantenmechanik und Thermodynamik. *Ab. Preuss. Akad. Wiss.*, 1:1, 1936.

[257] B. DeMarco, A. Ben-Kish, D. Leibfried, V. Meyer, M. Rowe, B. M. Jelenkovic, W. M. Itano, J. Britton, C. Langer, T. Rosenband, and D. Wineland. Experimental demonstration of a controlled-NOT wave-packet gate. *Phys. Rev. Lett.*, 89:267901, 2002.

[258] I. Devetak and A. Winter. Distillation of secret key and entanglement from quantum states. *Proc. Royal Soc. Lond. A Math.-Phys.*, 461:207, 2005.

[259] S. J. Devitt, W. J. Munro, and K. Naemoto. Quantum error correction for beginners. *Rep. Prog. Phys.*, 76:076001, 2013.

[260] P. Diaconis. What is . . . a random matrix? *Notices of the AMS*, 52:1348, 2005.

[261] R. H. Dicke. Coherence in spontaneous radiation processes. *Phys. Rev.*, 93:99, 1954.

[262] J. Dittmann. On the Riemannian metric on the space of density matrices. *Rep. Math. Phys.*, 36:309, 1995.

[263] J. Dittmann. Explicit formulae for the Bures metric. *J. Phys.*, A 32:2663, 1999.

[264] J. Dittmann. The scalar curvature of the Bures metric on the space of density matrices. *J. Geom. Phys.*, 31:16, 1999.

[265] D. DiVincenzo, C. A. Fuchs, H. Mabuchi, J. A. Smolin, A. Thapliyal, and A. Uhlmann. Entanglement of assistance. *Lecture Notes Comp. Sci.*, 1509:247, 1999.

[266] D. DiVincenzo, D. W. Leung, and B. Terhal. Quantum data hiding. *IEEE Trans. Inf. Theory*, 48:580, 2002.

[267] D. P. DiVincenzo. Two-bit gates are universal for quantum computation. *Phys. Rev.*, A 50:1015, 1995.

[268] D. P. DiVincenzo, T. Mor, P. W. Shor, J. A. Smolin, and B. M. Terhal. Unextendible product bases, uncompletable product bases and bound entanglement. *Commun. Math. Phys.*, 238:379, 2003.

[269] A. C. Doherty, P. A. Parillo, and F. M. Spedalieri. Distinguishing separable and entangled states. *Phys. Rev. Lett.*, 88:187904, 2002.

[270] A. C. Doherty, P. A. Parillo, and F. M. Spedalieri. Complete family of separability criteria. *Phys. Rev.*, A 69:022308, 2004.

[271] D. Ž. Doković and A. Osterloh. On polynomial invariants of several qubits. *J. Math. Phys.*, 50:033509, 2009.

[272] M. J. Donald. Further results on the relative entropy. *Math. Proc. Cam. Phil. Soc.*, 101:363, 1987.

[273] M. J. Donald and M. Horodecki. Continuity of relative entropy of entanglement. *Phys. Lett.*, A 264:257, 1999.

[274] M. J. Donald, M. Horodecki, and O. Rudolph. The uniqueness theorem for entanglement measures. *J. Math. Phys.*, 43:4252, 2002.

[275] J. P. Dowling, G. S. Agarwal, and W. P. Schleich. Wigner function of a general angular momentum state: Applications to a collection of two-level atoms. *Phys. Rev.*, A 49:4101, 1994.

[276] C. F. Dunkl, P. Gawron, J. A. Holbrook, J. A. Miszczak, Z. Puchała, and K. Życzkowski.

Numerical shadow and geometry of quantum states. *Phys. Rev.*, A 61:042314, 2000.

[277] W. Dür and J. I. Cirac. Equivalence classes of non-local operations. *Q. Inform. Comput.*, 2:240, 2002.

[278] W. Dür, G. Vidal, and J. I. Cirac. Three qubits can be entangled in two inequivalent ways. *Phys. Rev.*, A 62:062314, 2000.

[279] A. Dvoretzky. Some results on convex bodies and Banach spaces. In *Proc. Internat. Sympos. Linear Spaces, Jerusalem, 1960*, page 123, 1961.

[280] F. Dyson. The threefold way. Algebraic structure of symmetry groups and ensembles in quantum mechanics. *J. Math. Phys.*, 3:1199, 1962.

[281] C. Eckart and G. Young. The approximation of one matrix by one of lower rank. *Psychometrica*, 1:211, 1936.

[282] A. Edelman and E. Kostlan. How many zeros of a random polynomial are real? *Bull. Amer. Math. Soc.*, 32:1, 1995.

[283] H. G. Eggleston. *Convexity*. Cambridge University Press, 1958.

[284] A. Einstein. Autobiographical notes. In P. A. Schilpp, editor, *Albert Einstein: Philosopher – Scientist*, page 1. Cambridge University Press, 1949.

[285] A. Einstein, B. Podolsky, and N. Rosen. Can quantum mechanical description of reality be considered complete? *Phys. Rev.*, 47:777, 1935.

[286] J. Eisert. Entanglement in quantum information theory. Ph.D. Thesis, University of Potsdam, 2001.

[287] J. Eisert, K. Audenaert, and M. B. Plenio. Remarks on entanglement measures and non-local state distinguishability. *J. Phys.*, A 36:5605, 2003.

[288] J. Eisert and H. J. Briegel. Schmidt measure as a tool for quantifying multiparticle entanglement. *Phys. Rev.*, A 64:022306, 2001.

[289] J. Eisert, M. Cramer, and M. B. Plenio. Area laws for the entanglement entropy – a review. *Rev. Mod. Phys.*, 82:277, 2010.

[290] J. Eisert, P. Hyllus, O. Guhne, and M. Curty. Complete hierarchies of efficient approximations to problems in entanglement theory. *Phys. Rev.*, A 70:062317, 2004.

[291] J. Eisert and M. B. Plenio. A comparison of entanglement measures. *J. Mod. Opt.*, 46:145, 1999.

[292] A. Ekert and P. L. Knight. Entangled quantum systems and the Schmidt decomposition. *Am. J. Phys.*, 63:415, 1995.

[293] C. Eltschka and J. Siewert. Monogamy equalities for qubit entanglement from lorentz invariance. *Phys. Rev. Lett.*, 114: 140402, 2015.

[294] G. G. Emch. *Algebraic Methods in Statistical Mechanics and Quantum Field Theory*. Wiley Interscience, 1972.

[295] D. M. Endres and J. E. Schindelin. A new metric for probability distributions. *IEEE Trans. Inf. Theory*, 49:1858, 2003.

[296] B.-G. Englert and Y. Aharonov. The mean king's problem: Prime degrees of freedom. *Phys. Lett.*, 284:1, 2001.

[297] B.-G. Englert and N. Metwally. Remarks on 2-q-bit states. *Appl. Phys.*, B 72:35, 2001.

[298] T. Erber and G. M. Hockney. Complex systems: equilibrium configurations of N equal charges on a sphere ($2 \leq N \leq 112$). *Adv. Chem. Phys.*, 98:495, 1997.

[299] P. Erdös. On an elementary proof of some asymptotic formulas in the theory of partitions. *Phys. Rev. Lett.*, 43:437, 1942.

[300] Å. Ericsson. Separability and the stella octangula. *Phys. Lett.*, A 295:256, 2002.

[301] M. Ericsson, E. Sjöqvist, J. Brännlund, D. K. L. Oi, and

A. K. Pati. Generalization of the geometric phase to completely positive maps. *Phys. Rev.*, A 67:020101(R), 2003.

[302] G. Etesi. Complex structure on the six dimensional sphere from a spontaneous symmetry breaking. *J. Math. Phys.*, 56:043508, 2015.

[303] D. E. Evans. Quantum dynamical semigroups. *Acta Appl. Math.*, 2:333, 1984.

[304] H. Everett. 'Relative state' formulation of quantum mechanics. *Rev. Mod. Phys.*, 29:454, 1957.

[305] G. Ewald. *Combinatorial Convexity and Algebraic Geometry*. New York: Springer, 1996.

[306] P. Facchi, G. Florio, G. Parisi, and S. Pascazio. Maximally multipartite entangled states. *Phys. Rev.*, A 77:060304 R, 2008.

[307] D. K. Faddejew. Zum Begriff der Entropie eines endlichen Wahrscheinlichkeitsschemas. In H. Grell, editor, *Arbeiten zur Informationstheorie I*, page 86. Deutscher Verlag der Wissenschaften, 1957.

[308] M. Fannes. A continuity property of the entropy density for spin lattices. *Commun. Math. Phys.*, 31:291, 1973.

[309] M. Fannes, F. de Melo, W. Roga, and K. Życzkowski. Matrices of fidelities for ensembles of quantum states and the holevo quantity. *Quant. Inf. Comp.*, 12:472, 2012.

[310] M. Fannes, B. Nachtergaele, and R. F. Werner. Finitely correlated states on quantum spin chains. *Commun. Math. Phys.*, 144:443, 1992.

[311] U. Fano. Pairs of two-level systems. *Rev. Mod. Phys.*, 55:855, 1983.

[312] R. A. Fisher. Theory of statistical estimation. *Proc. Cambr. Phil. Soc.*, 22:700, 1925.

[313] R. A. Fisher. *The Design of Experiments*. Oliver and Boyd, 1935.

[314] J. Fiurášek. Structural physical approximations of unphysical maps and generalized quantum measurements. *Phys. Rev.*, A 66:052315, 2002.

[315] S. K. Foong and S. Kanno. Proof of Page's conjecture on the average entropy of a subsystem. *Phys. Rev. Lett.*, 72:1148, 1994.

[316] L. R. Ford. *Automorphic Functions*. Chelsea, New York, 1951.

[317] P. J. Forrester. *Log–Gases and Random matrices*. Princeton University Press, 2010.

[318] G. Fubini. Sulle metriche definite da una forma Hermitiana. *Atti Instituto Veneto*, 6:501, 1903.

[319] C. A. Fuchs. *Distinguishability and Accessible Information in Quantum Theory*. PhD Thesis, Univ. of New Mexico, Albuquerque, 1996.

[320] C. A. Fuchs and C. M. Caves. Mathematical techniques for quantum communication theory. *Open Sys. Inf. Dyn*, 3:345, 1995.

[321] C. A. Fuchs, M. C. Hoang, A. J. Scott, and B. C. Stacey. SICs: More numerical solutions. see www.physics.umb.edu/Research/Qbism/

[322] C. A. Fuchs and R. Schack. Quantum-Bayesian coherence. *Rev. Mod. Phys.*, 85:1693, 2013.

[323] C. A. Fuchs and J. van de Graaf. Cryptographic distinguishability measures for quantum mechanical states. *IEEE Trans. Inf. Th.*, 45:1216, 1999.

[324] A. Fujiwara and P. Algoet. One-to-one parametrization of quantum channels. *Phys. Rev.*, A 59:3290, 1999.

[325] M. Fukuda. Revisiting additivity violation of quantum channels. *Commun. Math. Phys.*, 332:713, 2014.

[326] W. Ganczarek, M. Kuś, and K. Życzkowski. Barycentric measure of quantum entanglement. *Phys. Rev.*, A 85:032314, 2012.

[327] I. M. Gelfand, M. M. Kapranov, and Z. V. Zelevinsky. *Discriminants, Resultants, and Multidimensional Determinants*. Birkhauser, 1994.

[328] E. Gerjuoy. Lower bound on entanglement of formation for the qubit–qudit system. *Phys. Rev.*, A 67:052308, 2003.

[329] S. Gharibian. Strong NP-Hardness of the quantum separability problem. *Quantum Infor. Comput.*, 10:343, 2010.

[330] G. W. Gibbons. Typical states and density matrices. *J. Geom. Phys.*, 8:147, 1992.

[331] G. W. Gibbons and H. Pohle. Complex numbers, quantum mechanics and the beginning of time. *Nucl. Phys.*, B 410:375, 1993.

[332] P. Gibilisco and T. Isola. Wigner–Yanase information on quantum state space: The geometric approach. *J. Math. Phys.*, 44:3752, 2003.

[333] R. Gill. Time, finite statistics, and Bell's fifth position. In A. Khrennikov, editor, *Foundations of Probability and Physics – 2*, page 179. Växjö University Press, 2003.

[334] R. Gilmore. *Lie Groups, Lie Algebras, and Some of Their Applications*. New York: Wiley, 1974.

[335] R. Gilmore. *Lie Groups, Physics, and Geometry: an Introduction for Physicists, Engineers and Chemists*. Cambridge UP, 2008.

[336] J. Ginibre. Statistical ensembles of complex quaternion and real matrices. *J. Math. Phys.*, 6:440, 1965.

[337] V. Giovannetti, S. Mancini, D. Vitali, and P. Tombesi. Characterizing the entanglement of bipartite quantum systems. *Phys. Rev.*, A 67:022320, 2003.

[338] N. Gisin. Stochastic quantum dynamics and relativity. *Helv. Phys. Acta*, 62:363, 1989.

[339] N. Gisin. Bell inequality holds for all non-product states. *Phys. Lett.*, A 154:201, 1991.

[340] N. Gisin and H. Bechmann-Pasquinucci. Bell inequality, Bell states and maximally entangled states for n qubits? *Phys. Lett.*, A 246:1, 1998.

[341] N. Gisin and S. Massar. Optimal quantum cloning machines. *Phys. Rev. Lett.*, 79:2153, 1997.

[342] D. Gitman and A. L. Shelepin. Coherent states of SU(N) groups. *J. Phys.*, A 26:313, 1993.

[343] M. Giustina et al. Significant-loophole-free test of Bell's theorem with entangled photons. *Phys. Rev. Lett.*, 115:250401, 2015.

[344] R. J. Glauber. Coherent and incoherent states of the radiation field. *Phys. Rev.*, 131:2766, 1963.

[345] A. M. Gleason. Measures on the closed subspaces of a Hilbert space. *J. Math. Mech.*, 6:885, 1957.

[346] S. Gnutzmann, F. Haake, and M. Kuś. Quantum chaos of $SU(3)$ observables. *J. Phys.*, A 33:143, 2000.

[347] S. Gnutzmann and M. Kuś. Coherent states and the classical limit on irreducible SU(3) representations. *J. Phys.*, A 31:9871, 1998.

[348] S. Gnutzmann and K. Życzkowski. Rényi-Wehrl entropies as measures of localization in phase space. *J. Phys.*, A 34:10123, 2001.

[349] C. Godsil and A. Roy. Equiangular lines, mutually unbiased bases, and spin models. *Eur. J. Comb.*, 30:246, 2009.

[350] S. Golden. Lower bounds for the Helmholtz function. *Phys. Rev.*, B 137:1127, 1965.

[351] V. Gorini, A. Kossakowski, and E. C. G. Sudarshan. Completely positive dynamical semigroups of n-level systems. *J. Math. Phys.*, 17:821, 1976.

[352] D. Gottesman. *Stabilizer Codes and Quantum Error Correction*. PhD thesis, California Institute of Technology, 1997.

[353] G. Gour. Family of concurrence monotones and its applications. *Phys. Rev.*, A 71:012318, 2004.

[354] G. Gour and N. R. Wallach. All maximally entangled four-qubit states. *J. Math. Phys.*, 51:112201, 2010.

[355] S. K. Goyal, B. N. Simon, R. Singh, and S. Simon. Geometry of the generalized Bloch sphere for qutrit. *J. Phys.*, A 49:165203, 2016.

[356] D. Goyeneche, D. Alsina, J. I. Latorre, A. Riera, and K. Życzkowski. Absolutely maximally entangled states, combinatorial designs and multi-unitary matrices. *Phys. Rev.*, A 92:032316, 2015.

[357] D. Goyeneche and K. Życzkowski. Genuinely multipartite entangled states and orthogonal arrays. *Phys. Rev.*, A 90:022316, 2014.

[358] J. Grabowski, A. Ibort, M. Kuś, and G. Marmo. Convex bodies of states and maps. *J. Phys.*, A 46:425301, 2013.

[359] M. R. Grasseli and R. F. Streater. On the uniqueness of the Chentsov metric in quantum information geometry. *Inf. Dim. Analysis, Quantum Prob.*, 4:173, 2001.

[360] M. Grassl. On SIC-POVMs and MUBs in dimension 6. preprint quant-ph/040675.

[361] M. Grassl, M. Rötteler, and T. Beth. Computing local invariants of qubit systems. *Phys. Rev.*, A 58:1833, 1998.

[362] J. Gray. *The Hilbert Challenge*. Oxford UP, 2000.

[363] D. M. Greenberger, M. Horne, and A. Zeilinger. Going beyond Bell's theorem. In M. Kafatos, editor, *Bell's theorem, quantum theory and conceptions of the Universe*, page 69. Dordrecht: Kluwer, 1989.

[364] D. M. Greenberger, M. A. Horne, A. Shimony, and A. Zeilinger. Bell's theorem without inequalities. *Am. J. Phys.*, 58:1131, 1990.

[365] J. Griffiths and J. Harris. *Principles of Algebraic Geometry*. Wiley, 1978.

[366] D. Gross. Hudson's theorem for finite-dimensional quantum systems. *J. Math. Phys.*, 47:122107, 2006.

[367] D. Gross, K. Audenaert, and J. Eisert. Evenly distributed unitaries: on the structure of unitary designs. *J. Math. Phys.*, 48:052104, 2007.

[368] B. Grünbaum. *Convex Polytopes, II ed.* New York: Springer Verlag, 2003.

[369] J. Gruska. *Quantum Computing*. New York: McGraw-Hill, 1999.

[370] O. Gühne. Characterizing entanglement via uncertainty relations. *Phys. Rev. Lett.*, 92:117903, 2004.

[371] O. Gühne, P. Hyllus, D. Bruß, A. Ekert, M. Lewenstein, C. Macchiavello, and A. Sanpera. Detection of entanglement with few local measurements. *Phys. Rev.*, A 66:062305, 2002.

[372] O. Gühne, P. Hyllus, D. Bruß, A. Ekert, M. Lewenstein, C. Macchiavello, and A. Sanpera. Experimental detection of entanglement via witness operators and local measurements. *J. Mod. Opt.*, 50:1079, 2003.

[373] O. Gühne, P. Hyllus, O. Gittsovich, and J. Eisert. Covariance matrices and the separability problem. *Phys. Rev. Lett.*, 99:130504, 2007.

[374] O. Gühne and M. Lewenstein. Entropic uncertainty relations and entanglement. *Phys. Rev.*, A 70:022316, 2004.

[375] O. Gühne and G. Tóth. Entanglement detection. *Physics Reports*, 474:1, 2009.

[376] V. Guillemin and R. Sjamaar. Convexity theorems for varieties invariant under a Borel subgroup. *Pure Appl. Math. Q.*, 2:637, 2006.

[377] V. Guillemin and S. Sternberg. *Symplectic Techniques in Physics.* Cambridge University Press, 1984.

[378] F. Gürsey and H. C. Tze. Complex and quaternionic analyticity in chiral and gauge theories, I. *Ann. of Phys.*, 128:29, 1980.

[379] L. Gurvits. Classical complexity and quantum entanglement. *J. Comput. Syst. Sciences*, 69:448, 2004.

[380] L. Gurvits and H. Barnum. Largest separable balls around the maximally mixed bipartite quantum state. *Phys. Rev.*, A 66:062311, 2002.

[381] L. Gurvits and H. Barnum. Separable balls around the maximally mixed multipartite quantum states. *Phys. Rev.*, A 68:042312, 2003.

[382] K. E. Gustafson and D. K. M. Rao. *Numerical Range: The Field of Values of Linear Operators and Matrices.* New York: Springer-Verlag, 1997.

[383] E. Gutkin. The Toeplitz–Hausdorff theorem revisited: relating linear algebra and geometry. *Math. Intelligencer*, 26:8, 2004.

[384] E. Gutkin, E. A. Jonckheere, and M. Karow. Convexity of the joint numerical range: topological and differential geometric viewpoints. *Lin. Alg. Appl.*, 376:143, 2004.

[385] E. Gutkin and K. Życzkowski. Joint numerical ranges, quantum maps, and joint numerical shadows. *Lin. Alg. Appl.*, 438:2394, 2013.

[386] K.-C. Ha. Atomic positive linear maps in matrix algebra. *Publ. RIMS Kyoto Univ.*, 34:591, 1998.

[387] K.-C. Ha. Separability of qubit-qudit quantum states with strong positive partial transposes. *Phys. Rev.*, A 87:024301, 2013.

[388] K.-C. Ha and S.-H. Kye. Optimality for indecomposable entanglement witnesses. *Phys. Rev.*, A 86:034301, 2012.

[389] K.-C. Ha, S.-H. Kye, and Y.-S. Park. Entangled states with positive partial transposes arising from indecomposable positive linear maps. *Phys. Lett.*, A 313:163, 2003.

[390] F. Haake. *Quantum Signatures of Chaos, 3nd. ed.* Springer, 2010.

[391] J. Hadamard. Résolution d'une question relative aux déterminants. *Bull. Sci. Math.*, 17:240, 1893.

[392] M. J. W. Hall. Random quantum correlations and density operator distributions. *Phys. Lett.*, A 242:123, 1998.

[393] W. Hall. A new criterion for indecomposability of positive maps. *J. Phys.*, 39:14119, 2006.

[394] K. Hammerer, G. Vidal, and J. I. Cirac. Characterization of non-local gates. *Phys. Rev.*, A 66:062321, 2002.

[395] Y.-J. Han, Y.-S. Zhang, and G.-C. Guo. Compatible conditions, entanglement, and invariants. *Phys. Rev.*, A70:042309, 2004.

[396] T. Hangan and G. Masala. A geometrical interpretation of the shape invariant geodesic triangles in complex projective spaces. *Geometriae Dedicata*, 49:129, 1994.

[397] J. H. Hannay. Exact statistics for 2*j* points on a sphere: The Majorana zeros for random spin state. *J. Phys.*, A 29:L101, 1996.

[398] J. H. Hannay. The Berry phase for spin in the Majorana representation. *J. Phys.*, A 31:L53, 1998.

[399] L. O. Hansen, A. Hauge, J. Myrheim, and P. Ø. Sollid. Extremal entanglement witnesses. *Int. J. Quantum Inform.*, 13:1550060, 2015.

[400] R. H. Hardin and N. J. A. Sloane. McLaren's improved snub cube and other new spherical designs in three dimensions. *Discrete Comput. Geom.*, 15:429, 1996.

[401] G. H. Hardy, J. E. Littlewood, and G. Pólya. Some simple inequalities satisfied by convex functions. *Messenger Math.*, 58:145, 1929.

[402] G. H. Hardy and S. Ramanujan. Asymptotic formulae in combinatory analysis. *Proc. Lond. Math. Soc.*, 17:75, 1918.

[403] L. Hardy. Quantum theory from five reasonable axioms. preprint quant-ph/0101012.

[404] L. Hardy. Method of areas for manipulating the entanglement properties of one copy of a two-particle pure entangled state. *Phys. Rev.*, A 60:1912, 1999.

[405] P. Harremoës and F. Topsøe. Inequalities between entropy and index of coincidence derived from information diagrams. *IEEE Trans. Inform. Theory*, 47:2944, 2001.

[406] J. E. Harriman. Geometry of density matrices 1. *Phys. Rev.*, A 17:1249, 1978.

[407] J. Harris. *Algebraic Geometry. A First Course*. Springer, 1992.

[408] R. V. L. Hartley. Transmission of information. *Bell Sys. Tech. J.*, 7:535, 1928.

[409] M. B. Hastings. An area law for one dimensional quantum systems. *J. Stat. Mech.*, page P08024, 2007.

[410] M. B. Hastings. Superadditivity of communication capacity using entangled inputs. *Nat. Phys.*, 5:255, 2009.

[411] F. Hausdorff. Der Wertvorrat einer Bilinearform. *Math. Z.*, 3:314, 1919.

[412] T. F. Havel. Robust procedures for converting among Lindblad, Kraus and matrix representations of quantum dynamical semigroups. *J. Math. Phys.*, 44:534, 2003.

[413] T. F. Havel, Y. Sharf, L. Viola, and D. G. Cory. Hadamard products of product operators and the design of gradient: Diffusion experiments for simulating decoherence by NMR spectroscopy. *Phys. Lett.*, A 280:282, 2001.

[414] J. Havrda and F. Charvát. Quantification methods of classification processes: Concept of structural α entropy. *Kybernetica*, 3:30, 1967.

[415] M. Hayashi, S. Ishizaka, A. Kawachi, G. Kimura, and T. Ogawa. *Introduction to Quantum Information Science*. Springer Verlag, 2015.

[416] P. Hayden, R. Jozsa, D. Petz, and A. Winter. Structure of states which satisfy strong subadditivity of qantum entropy with equality. *Commun. Math. Phys.*, 246:359, 2003.

[417] P. Hayden, D. Leung, and A. Winter. Aspects of generic entanglement. *Commun. Math. Phys.*, 265:95, 2006.

[418] P. Hayden, D. W. Leung, and A. Winter. Aspects of generic entanglement. *Commun. Math. Phys.*, 269:95, 2006.

[419] P. Hayden, B. M. Terhal, and A. Uhlmann. On the LOCC classification of bipartite density matrices. preprint quant-ph/0011095.

[420] P. Hayden and A. Winter. Counterexamples to the maximal p–norm multiplicativity conjecture for all $p > 1$. *Commun. Math. Phys.*, 284:263, 2008.

[421] P. M. Hayden, M. Horodecki, and B. M. Terhal. The asymptotic entanglement cost of preparing a quantum state. *J. Phys.*, A 34:6891, 2001.

[422] C. W. Helstrom. *Quantum Detection and Estimation Theory*. London: Academic Press, 1976. Mathematics in Science and Engineering vol. 123.

[423] W. Helwig. Absolutely maximally entangled qudit graph states. preprint arxiv:1306.2879.

[424] W. Helwig and W. Cui. Absolutely maximally entangled states: existence and applications. preprint arXiv:1306.2536.

[425] W. Helwig, W. Cui, J. I. Latorre, A. Riera, and H.-K. Lo. Absolute maximal entanglement and

quantum secret sharing. *Phys. Rev.*, A 86:052335, 2012.

[426] L. Henderson and V. Vedral. Information, relative entropy of entanglement and irreversibility. *Phys. Rev. Lett.*, 84:2263, 2000.

[427] D. Henrion. Semidefinite geometry of the numerical range. *Electr. J. Lin. Alg.*, 20:322, 2010.

[428] D. Henrion. Semidefinite representation of convex hulls of rational varieties. *Acta Appl. Math.*, 115:319, 2011.

[429] B. Hensen et al. Loophole-free violation of a Bell inequality using entangled electron spins separated by 1.3 km. *Nature*, 526:682, 2015.

[430] N. Herbert. FLASH—A superluminal communicator based upon a new kind of quantum measurement. *Found. Phys.*, 12:1171, 1982.

[431] O. Hesse.Über die Wendepunkte der Curven Dritten Ordnung. *J. Reine Angew. Math.*, 28:97, 1844.

[432] F. Hiai, M. Ohya, and M. Tsukada. Sufficiency, KMS condition and relative entropy in von Neumann algebras. *Pacific J. Math.*, 96:99, 1981.

[433] F. Hiai and D. Petz. The proper formula for relative entropy and its asymptotics in quantum probability. *Commun. Math. Phys.*, 143:99, 1991.

[434] A. Higuchi. On the one-particle reduced density matrices of a pure three-qutrit quantum state. preprint arXiv:quant-ph/0309186.

[435] A. Higuchi and A. Sudbery. How entangled can two couples get? *Phys. Lett.*, A 272:213, 2000.

[436] A. Higuchi, A. Sudbery, and J. Szulc. One-qubit reduced states of a pure many-qubit state: polygon inequalities. *Phys. Rev. Lett.*, 90:107902, 2003.

[437] R. Hildebrand. Positive partial transpose from spectra. *Phys. Rev.*, A 76:052325, 2007.

[438] S. Hill and W. K. Wootters. Entanglement of a pair of quantum bits. *Phys. Rev. Lett.*, 78:5022, 1997.

[439] W. Hoeffding. Probability inequalities for sums of bounded random variables. *J. Amer. Stat. Assoc.*, 58:13, 1963.

[440] H. F. Hofmann and S. Takeuchi. Violation of local uncertainty relations as a signature of entanglement. *Phys. Rev.*, A 68:032103, 2003.

[441] S. G. Hoggar. t-designs in projective spaces. *Europ. J. Combin.*, 3:233, 1982.

[442] S. G. Hoggar. 64 lines from a quaternionic polytope. *Geometriae Dedicata*, 69:287, 1998.

[443] A. S. Holevo. Information theoretical aspects of quantum measurements. *Prob. Inf. Transmission USSR*, 9:177, 1973.

[444] A. S. Holevo. Some estimates for information quantity transmitted by quantum communication channels. *Probl. Pered. Inf.*, 9:3, 1973. English tranl.: *Probl. Inf. Transm.* 9:177, 1973.

[445] A. S. Holevo. *Probabilistic and Statistical Aspects of Quantum Theory*. North-Holland, 1982.

[446] A. S. Holevo. The capacity of the quantum channel with general signal states. *IEEE Trans. Inf. Theor.*, 44:269, 1998.

[447] A. S. Holevo. *Statistical Structure of Quantum Theory*. Springer, 2001.

[448] K. J. Horadam. *Hadamard Matrices and Their Applications*. Princeton University Press, 2007.

[449] A. Horn. Doubly stochastic matrices and the diagonal of rotation. *Am. J. Math.*, 76:620, 1954.

[450] R. Horn and C. Johnson. *Matrix Analysis*. Cambridge University Press, 1985.

[451] R. Horn and C. Johnson. *Topics in Matrix Analysis*. Cambridge University Press, 1991.

[452] K. Horodecki, M. Horodecki, P. Horodecki, and J. Oppenheim.

Secure key from bound entanglement. *Phys. Rev. Lett.*, 94: 160502, 2005.

[453] M. Horodecki. Entanglement measures. *Quant. Inf. Comp.*, 1:3, 2001.

[454] M. Horodecki and P. Horodecki. Reduction criterion of separability and limits for a class of distillation protocols. *Phys. Rev.*, A 59:4206, 1999.

[455] M. Horodecki, P. Horodecki, and R. Horodecki. Separability of mixed states: Necessary and sufficient conditions. *Phys. Lett.*, A 223:1, 1996.

[456] M. Horodecki, P. Horodecki, and R. Horodecki. Inseparable two spin-1/2 density matrices can be distilled to a singlet form. *Phys. Rev. Lett.*, 78:574, 1997.

[457] M. Horodecki, P. Horodecki, and R. Horodecki. Mixed-state entanglement and distillation: Is there a 'bound' entanglement in nature? *Phys. Rev. Lett.*, 80:5239, 1998.

[458] M. Horodecki, P. Horodecki, and R. Horodecki. Limits for entanglement measures. *Phys. Rev. Lett.*, 84:2014, 2000.

[459] M. Horodecki, P. Horodecki, and R. Horodecki. Mixed-state entanglement and quantum communication. In G. Alber and M. Weiner, editors, *Quantum Information – Basic Concepts and Experiments*, page 1. Springer, 2000.

[460] M. Horodecki, P. Horodecki, and R. Horodecki. Separability of mixed quantum states: Linear contractions and permutations criteria. *Open Sys. & Information Dyn.*, 13:103, 2006.

[461] M. Horodecki, P. Horodecki, and J. Oppenheim. Reversible transformations from pure to mixed states and the unique measure of information. *Phys. Rev.*, A 67:062104, 2003.

[462] M. Horodecki and J. Oppenheim. (Quantumness in the context of)

resource theories. *Int. J. Mod. Phys.*, B27:1345019, 2013.

[463] M. Horodecki, P. W. Shor, and M. B. Ruskai. Entanglement breaking channels. *Rev. Math. Phys.*, 15:621, 2003.

[464] P. Horodecki. Separability criterion and inseparable mixed states with positive partial transposition. *Phys. Lett.*, A 232:333, 1997.

[465] P. Horodecki and A. Ekert. Method for direct detection of quantum entanglement. *Phys. Rev. Lett.*, 89:127902, 2002.

[466] P. Horodecki, M. Horodecki, and R. Horodecki. Bound entanglement can be activated. *Phys. Rev. Lett.*, 82:1056, 1999.

[467] P. Horodecki, M. Lewenstein, G. Vidal, and I. Cirac. Operational criterion and constructive checks for the separability of low rank density matrices. *Phys. Rev.*, A 62:032310, 2000.

[468] R. Horodecki and M. Horodecki. Information-theoretic aspects of quantum inseparability of mixed states. *Phys. Rev.*, A 54:1838, 1996.

[469] R. Horodecki and P. Horodecki. Quantum redundancies and local realism. *Phys. Lett.*, A 194:147, 1994.

[470] R. Horodecki, P. Horodecki, and M. Horodecki. Quantum α-entropy inequalities: Independent condition for local realism? *Phys. Lett.*, A 210:377, 1996.

[471] R. Horodecki, P. Horodecki, M. Horodecki, and K. Horodecki. Quantum entanglement. *Rev. Mod. Phys.*, 81:865, 2009.

[472] S. D. Howard, A. R. Calderbank, and W. Moran. The finite Heisenberg-Weyl groups in radar and communications. *EURASIP J. Appl. Sig. Process.*, 2006:85865, 2006.

[473] J. M. Howie. *Fields and Galois Theory*. Springer, 2006.

[474] L. K. Hua. *Harmonic Analysis of Functions of Several Variables in the Classical Domains.* American Mathematical Society, 1963. Chinese original 1958, Russian translation, Moskva 1959.

[475] R. Hübener, M. Kleinmann, T.-C. Wei, C. González-Guillén, and O. Gühne. The geometric measure of entanglement for symmetric states. *Phys. Rev.*, A 80:032324, 2009.

[476] F. Huber, O. Gühne, and J. Siewert. Absolutely maximally entangled states of seven qubits do not exist. Phys. Rev. Lett. 118:200502, 2017.

[477] M. Hübner. Explicit computation of the Bures distance for density matrices. *Phys. Lett.*, A 163:239, 1992.

[478] T. Huckle, K. Waldherr, and T. Schulte-Herbrüggen. Computations in quantum tensor networks. *Lin. Alg. Appl.*, 438:750, 2013.

[479] R. Hudson. When is the Wigner quasi-probability density non-negative? *Rep. Math. Phys.*, 6:249, 1974.

[480] L. P. Hughston. Geometric aspects of quantum mechanics. In S. Huggett, editor, *Twistor Theory.* S. Marcel Dekker, p. 59, 1995.

[481] L. P. Hughston, R. Jozsa, and W. K. Wootters. A complete classification of quantum ensembles having a given density matrix. *Phys. Lett.*, A 183:14, 1993.

[482] L. P. Hughston and S. M. Salamon. Surverying points on the complex projective plane. *Adv. Math.*, 286:1017, 2016.

[483] F. Hulpke and D. Bruß. A two-way algorithm for the entanglement problem. *J. Phys.*, A 38:5573, 2005.

[484] W. Hunziker. A note on symmetry operations in quantum mechanics. *Helv. Phys. Acta*, 45:233, 1972.

[485] K. Husimi. Some formal properties of density matrices. *Proc. Phys. Math. Soc. Japan*, 22:264, 1940.

[486] P. Hyllus, O. Gühne, D. Bruß, and M. Lewenstein. Relations between entanglement witnesses and Bell inequalities. *Phys. Rev.*, A 72:012321, 2005.

[487] R. Ingarden, A. Kossakowski, and M. Ohya. *Information Dynamics and Open Systems.* Dordrecht: Kluwer, 1997.

[488] R. S. Ingarden. Information geometry in functional spaces of classical and quantum finite statistical systems. *Int. J. Engng. Sci.*, 19:1609, 1981.

[489] S. Ishizaka. Analytical formula connecting entangled state and the closest disentangled state. *Phys. Rev.*, A 67:060301, 2003.

[490] S. Ishizaka and T. Hiroshima. Maximally entangled mixed states under non-local unitary operations in two qubits. *Phys. Rev.*, A 62:022310, 2000.

[491] I. D. Ivanović. Geometrical description of quantal state determination. *J. Phys.*, A 14:3241, 1981.

[492] L. Jakóbczyk and M. Siennicki. Geometry of Bloch vectors in two-qubit system. *Phys. Lett.*, A 286:383, 2001.

[493] P. Jaming, M. Matolsci, P. Móra, F. Szöllősi, and M. Weiner. A generalized Pauli problem an infinite family of MUB-triplets in dimension 6. *J. Phys.*, A 42:245305, 2009.

[494] A. Jamiołkowski. Linear transformations which preserve trace and positive semi-definiteness of operators. *Rep. Math. Phys.*, 3:275, 1972.

[495] A. Jamiołkowski. An effective method of investigation of positive maps on the set of positive definite operators. *Rep. Math. Phys.*, 5:415, 1975.

[496] R. A. Janik and M. A. Nowak. Wishart and anti-Wishart random matrices. *J.Phys.*, A 36:3629, 2003.

[497] J. M. Jauch. *Foundations of Quantum Mechanics.* Addison-Wesley, 1968.

[498] J. M. Jauch and C. Piron. Generalized localizability. *Helv. Phys. Acta*, 45:559, 1967.

[499] E. T. Jaynes. *Probability Theory. The Logic of Science*. Cambridge University Press, 2003.

[500] H. Jeffreys. *Theory of Probability*. Oxford: Clarendon Press, 1961.

[501] A. Jenčova. Generalized relative entropies as contrast functionals on density matrices. *Int. J. Theor. Phys.*, 43:1635, 2004.

[502] J. G. Jensen and R. Schack. A simple algorithm for local conversion of pure states. *Phys. Rev.*, A 63:062303, 2001.

[503] N. Johnston. Non-positive-partial-transpose subspaces can be as large as any entangled subspace. *Phys. Rev.*, A 87:064302, 2013.

[504] D. Jonathan and M. B. Plenio. Entanglement-assisted local manipulation of pure quantum states. *Phys. Rev. Lett.*, 83:3566, 1999.

[505] D. Jonathan and M. B. Plenio. Minimial conditions for local pure-state entanglement manipulation. *Phys. Rev. Lett.*, 83:1455, 1999.

[506] K. R. W. Jones. Entropy of random quantum states. *J. Phys.*, A 23:L1247, 1990.

[507] N. S. Jones and N. Linden. Parts of quantum states. *Phys. Rev.*, A 71:012324, 2005.

[508] P. Jordan, J. von Neumann, and E. P. Wigner. On an algebraic generalization of the quantum mechanical formalism. *Ann. Math.*, 56:29, 1934.

[509] R. Jozsa. Fidelity for mixed quantum states. *J. Mod. Opt.*, 41:2315, 1994.

[510] R. Jozsa, D. Robb, and W. K. Wootters. Lower bound for accessible information in quantum mechanics. *Phys. Rev.*, A 49:668, 1994.

[511] R. Jozsa and B. Schumacher. A new proof of the quantum noiseless coding theorem. *J. Mod. Opt.*, 41:2343, 1994.

[512] M. Kac. On the average number of real roots of a random algebraic equation. *Bull. Amer. Math. Soc.*, 49:314, 1943.

[513] W. K. Kantor. MUBs inequivalence and affine planes. *J. Math. Phys.*, 53:032204, 2012.

[514] L. V. Kantorovich. On translocation of masses. *Doklady Acad. Sci. URSS*, 37:199, 1942.

[515] J. Kapur. *Measures of Information and Their Applications*. New York: John Wiley & Sons, 1994.

[516] B. R. Karlsson. Three-parameter complex Hadamard matrices of order 6. *Linear Alg. Appl.*, 434:247, 2011.

[517] S. Karnas and M. Lewenstein. Separable approximations of density matrices of composite quantum systems. *J. Phys.*, A 34:6919, 2001.

[518] D. S. Keeler, L. Rodman, and I. M. Spitkovsky. The numerical range of 3×3 matrices. *Linear Alg. Appl.*, 252:115, 1997.

[519] J. Kempe. Multiparticle entanglement and its applications to cryptography. *Phys. Rev.*, A 60:910, 1999.

[520] V. Kendon, V. K. Nemoto, and W. Munro. Typical entanglement in multiple-qubit systems. *J. Mod. Opt.*, 49:1709, 2002.

[521] V. M. Kendon, K. Życzkowski, and W. J. Munro. Bounds on entanglement in qudit subsystems. *Phys. Rev.*, A 66:062310, 2002.

[522] N. Khaneja, R. Brockett, and S. J. Glaser. Time optimal control in spin systems. *Phys. Rev.*, A 63:0323308, 2001.

[523] M. Khatirinejad. On Weyl–Heisenberg orbits of equiangular lines. *J. Algebra. Comb.*, 28:333, 2008.

[524] A. I. Khinchin. *Mathematical Foundations of Information Theory*. Dover, 1957.

[525] T. W. B. Kibble. Geometrization of quantum mechanics. *Commun. Math. Phys.*, 65:189, 1979.

[526] H.-J. Kim and S.-H. Kye. Indecomposable extreme positive linear maps in matrix algebras. *Bull. London Math. Soc.*, 26:575, 1994.

[527] J. S. Kim, A. Das, and B. C. Sanders. Entanglement monogamy of multipartite higher-dimensional quantum systems using convex-roof extended negativity. *Phys. Rev.*, A 79:012329, 2009.

[528] J. S. Kim, G. Gour, and B. C. Sanders. Limitations to sharing entanglement. *Contemp. Phys.*, 53:417, 2012.

[529] G. Kimura. The Bloch vector for *n*-level systems. *Phys. Lett.*, A 314:339, 2003.

[530] C. King. Additivity for unital qubit channels. *J. Math. Phys.*, 43:4641, 2002.

[531] C. King and M. B. Ruskai. Minimal entropy of states emerging from noisy channels. *IEEE Trans. Inf. Th.*, 47:192, 2001.

[532] R. Kippenhahn. Ueber den Wertevorrat einer Matrix. *Math. Nachricht.*, 6:193, 1951.

[533] F. C. Kirwan. Convexity properties of the moment mapping. *Invent. Math.*, 77:547, 1984.

[534] A. Klappenecker and M. Rötteler. Beyond stabilizer codes I: Nice error bases. *IEEE Trans. Inform. Theory*, 48:2392, 2002.

[535] A. Klappenecker and M. Rötteler. Constructions of mutually unbiased bases. *Lect. Not. Computer Science*, 2948:137, 2004.

[536] A. Klappenecker and M. Rötteler. Mutually Unbiased Bases are complex projective 2–designs. In *Proc ISIT*, page 1740. Adelaide, 2005.

[537] A. Klappenecker and M. Rötteler. On the monomiality of nice error bases. *IEEE Trans. Inform. Theory*, 51:1084, 2005.

[538] J. R. Klauder. The action option and a Feynman quantization of spinor fields in terms of ordinary c-numbers. *Ann. Phys. (N.Y)*, 11:60, 1960.

[539] J. R. Klauder. Are coherent states the natural language of quantum theory? In V. Gorini and A. Frigerio, editors, *Fundamental Aspects of Quantum Theory*, page 1. Plenum, 1985.

[540] J. R. Klauder and B.-S. Skagerstam. *Coherent States. Applications in Physics and Mathematical Physics.* Singapore: World Scientific, 1985.

[541] J. R. Klauder and E. C. G. Sudarshan. *Fundamentals of Quantum Optics.* New York: W. A. Benjamin, 1968.

[542] O. Klein. Zur quantenmechanischen Begründung des zweiten Hauptsatzes der Wärmelehre. *Z. f. Physik*, 72:767, 1931.

[543] A. Klimov, J. L. Romero, G. Björk, and L. L. Sanchez-Soto. Discrete phase space structure of *n*-qubit mutually unbiased bases. *Ann. Phys.*, 324:53, 2009.

[544] A. Klyachko. Quantum marginal problem and representations of the symmetric group. preprint quant-ph/0409113.

[545] A. A. Klyachko. Stable bundles, representation theory and Hermitian operators. *Selecta Math.*, 4:419, 1998.

[546] A. A. Klyachko and A. S. Shumovsky. General entanglement. *J. Phys. Conf. Ser.*, 36:87, 2006.

[547] E. Knill. Group representations, error bases and quantum codes. preprint quant-ph/9608049.

[548] E. Knill and R. Laflamme. Power of one bit of quantum information. *Phys. Rev. Lett.*, 81:5672, 1998.

[549] A. Knutson. The symplectic and algebraic geometry of Horn's problem. *Linear Algebra Appl.*, 319:61, 2000.

[550] S. Kobayashi and K. Nomizu. *Foundations of Differential Geometry I.* Wiley Interscience, 1963.

[551] S. Kochen and E. Specker. The problem of hidden variables in quantum mechanics. *J. Math. Mech.*, 17: 59, 1967.

[552] T. G. Kolda and B. W. Bader. Tensor decompositions and applications. *SIAM Rev.*, 51:455, 2009.

[553] H. J. Korsch, C. Müller, and H. Wiescher. On the zeros of the Husimi distribution. *J. Phys.*, A 30:L677, 1997.

[554] A. Kossakowski. Remarks on positive maps of finite-dimensional simple Jordan algebras. *Rep. Math. Phys.*, 46:393, 2000.

[555] A. Kossakowski. A class of linear positive maps in matrix algebras. *Open Sys. & Information Dyn.*, 10:1, 2003.

[556] A. I. Kostrikin and P. I. Tiep. *Orthogonal Decompositions and Integral Lattices.* de Gruyter, 1994.

[557] B. Kraus. Local unitary equivalence and entanglement of multipartite pure states. *Phys. Rev.*, A82:032121, 2010.

[558] B. Kraus and J. I. Cirac. Optimal creation of entanglement using a two-qubit gate. *Phys. Rev.*, A 63:062309, 2001.

[559] K. Kraus. General state changes in quantum theory. *Ann. Phys.(N.Y.)*, 64:311, 1971.

[560] K. Kraus. *States, Effects and Operations: Fundamental Notions of Quantum Theory.* Springer-Verlag, 1983.

[561] F. Kubo and T. Ando. Means of positive linear operators. *Math. Ann.*, 246:205, 1980.

[562] R. Kueng and D. Gross. Qubit stabilizer states are complex projective 3-designs. preprint arXiv:1510.02767.

[563] S. Kullback and R. A. Leibler. On information and sufficiency. *Ann. Math. Stat.*, 22:79, 1951.

[564] M. Kuś, J. Mostowski, and F. Haake. Universality of eigenvector statistics of kicked tops of different symmetries. *J. Phys.*, A 21:L1073, 1988.

[565] M. Kuś and K. Życzkowski. Geometry of entangled states. *Phys. Rev.*, A 63:032307, 2001.

[566] S.-H. Kye. Facial structures for unital positive linear maps in the two-dimensional matrix algebra. *Lin. Alg. Appl.*, 362:57, 2003.

[567] R. Laflamme, C. Miquel, J. P. Paz, and W. Zurek. Perfect quantum error correcting code. *Phys. Rev. Lett.*, 77:198, 1996.

[568] A. Laing, T. Lawson, E. M. López, and J. L. O'Brien. Observation of quantum interference as a function of Berry's phase in a complex Hadamard network. *Phys. Rev. Lett.*, 108:260505, 2012.

[569] C. W. H. Lam. The search for a finite projective plane of order 10. *Amer. Math. Mon.*, 98:305, 1991.

[570] P. W. Lamberti, A. P. Majtey, A. Borras, M. Casas, and A. Plastino. Metric character of the quantum Jensen-Shannon divergence. *Phys. Rev.*, A77:052311, 2008.

[571] L. Landau. Das Dämpfungsproblem in der Wellenmechanik. *Z. Phys.*, 45:430, 1927.

[572] L. J. Landau and R. F. Streater. On Birkhoff's theorem for doubly stochastic completely positive maps of matrix algebras. *Lin. Algebra Appl.*, 193:107, 1993.

[573] O. Lanford and D. Robinson. Mean entropy of states in quantum statistical ensembles. *J. Math. Phys.*, 9:1120, 1968.

[574] J.-Å. Larsson. Loopholes in Bell inequality tests of local realism. *J. Phys.*, A 47:424003, 2014.

[575] J. I. Latorre and A. Riera. A short review on entanglement in quantum spin systems. *J. Phys.*, A 42:504002, 2009.

[576] H. B. Lawson. *Lectures on Minimal Submanifolds*. Publish and Perish, Inc, 1980.

[577] P. Lebœuf. Phase space approach to quantum dynamics. *J. Phys.*, A 24:4575, 1991.

[578] P. Lebœuf and P. Shukla. Universal fluctuations of zeros of chaotic wavefunctions. *J. Phys.*, A 29:4827, 1996.

[579] P. Lebœuf and A. Voros. Chaos-revealing multiplicative representation of quantum eigenstates. *J. Phys.*, A 23:1765, 1990.

[580] M. Ledoux. *The Concentration of Measure Phenomenon*. Providence: AMS Publishing, 2001.

[581] C. T. Lee. Wehrl's entropy of spin states and Lieb's conjecture. *J. Phys.*, A 21:3749, 1988.

[582] S. Lee, D. P. Chi, S. D. Oh, and J. Kim. Convex-roof extended negativity as an entanglement measure for bipartite quantum systems. *Phys. Rev.*, A 68:062304, 2003.

[583] J. M. Leinaas, J. Myrheim, and E. Ovrum. Geometrical aspects of entanglement. *Phys. Rev.*, A 74:012313, 2006.

[584] J. M. Leinaas, J. Myrheim, and P. Ø. Sollid. Low-rank extremal positive-partial-transpose states and unextendible product bases. *Phys. Rev.*, A 81:062330, 2010.

[585] J. M. Leinaas, J. Myrheim, and P. Ø. Sollid. Numerical studies of entangled positive-partial-transpose states in composite quantum systems. *Phys. Rev.*, A 81:062329, 2010.

[586] P. W. H. Lemmens and J. J. Seidel. Equiangular lines. *J. Algebra*, 24:494, 1973.

[587] U. Leonhardt. *Measuring the Quantum State of Light*. Cambridge University Press, 1997.

[588] A. Lesniewski and M. B. Ruskai. Monotone Riemannian metrics and relative entropy on non-commutative probability spaces. *J. Math. Phys.*, 40:5702, 1999.

[589] P. Lévay. The geometry of entanglement: Metrics, connections and the geometric phase. *J. Phys.*, A 37:1821, 2004.

[590] P. Lévay. On the geometry of four-qubit invariants. *J. Phys.*, A 39:9533, 2006.

[591] P. Lévy. *Leçons d'analyse fonctionnelle*. Paris: Gauthier-Villars, 1922.

[592] M. Lewenstein, R. Augusiak, D. Chruściński, S. Rana, and J. Samsonowicz. Sufficient separability criteria and linear maps. *Phys. Rev.*, A 93:042335, 2016.

[593] M. Lewenstein, D. Bruß, J. I. Cirac, B. Kraus, M. Kuś, J. Samsonowicz, A. Sanpera, and R. Tarrach. Separability and distillability in composite quantum systems: A primer. *J. Mod. Opt.*, 47:2841, 2000.

[594] M. Lewenstein, B. Kraus, J. I. Cirac, and P. Horodecki. Optimization of entanglement witnesses. *Phys. Rev.*, A 62:052310, 2000.

[595] M. Lewenstein and A. Sanpera. Separability and entanglement of composite quantum systems. *Phys. Rev. Lett.*, 80:2261, 1998.

[596] C. K. Li. A simple proof of the elliptical range theorem. *Proc. Am. Math. Soc.*, 124:1985, 1996.

[597] C.-K. Li and Y.-T. Poon. Convexity of the joint numerical range. *SIAM J. Matrix Anal.*, A 21:668, 2000.

[598] E. H. Lieb. The classical limit of quantum spin systems. *Commun. Math. Phys.*, 31:327, 1973.

[599] E. H. Lieb. Convex trace functions and the Wigner–Yanase–Dyson conjecture. *Adv. Math.*, 11:267, 1973.

[600] E. H. Lieb. Some convexity and subadditivity properties of entropy. *Bull. AMS*, 81:1, 1975.

[601] E. H. Lieb. Proof of an entropy conjecture of Wehrl. *Commun. Math. Phys.*, 62:35, 1978.

[602] E. H. Lieb and D. W. Robinson. The finite group velocity of quantum spin systems. *Commun. Math. Phys.*, 28:251, 1972.

[603] E. H. Lieb and M. B. Ruskai. Proof of the strong subadditivity of quantum mechanical entropy. *J. Math. Phys.*, 14:1938, 1973.

[604] E. H. Lieb and J. S. Solovej. Proof of an entropy conjecture for Bloch coherent states and its generalizations. *Acta Math.*, 212:379, 2014.

[605] E. H. Lieb and J. S. Solovej. Proof of the Wehrl-type entropy conjecture for symmetric SU(N) coherent states. *Commun. Math. Phys.*, 348:567, 2016.

[606] J. Lin. Divergence measures based on the Shannon entropy. *IEEE Trans. Inf. Theory*, 37:145, 1991.

[607] G. Lindblad. An entropy inequality for quantum measurements. *Commun. Math. Phys.*, 28:245, 1972.

[608] G. Lindblad. Entropy, information and quantum measurement. *Commun. Math. Phys.*, 33:305, 1973.

[609] G. Lindblad. Expectations and entropy inequalities for finite quantum systems. *Commun. Math. Phys.*, 39:111, 1974.

[610] G. Lindblad. Completely positive maps and entropy inequalities. *Commun. Math. Phys.*, 40:147, 1975.

[611] G. Lindblad. On the generators of quantum dynamical semigroups. *Commun. Math. Phys.*, 48:119, 1976.

[612] G. Lindblad. Quantum entropy and quantum measurements. In C. B. et al., editor, *Quantum Aspects of Optical Communication*, page 36. Springer-Verlag, Berlin, 1991. Lecture Notes in Physics 378.

[613] N. Linden, S. Popescu, and A. Sudbery. Non-local properties of multi-particle density matrices. *Phys. Rev. Lett.*, 83:243, 1999.

[614] N. Linden, S. Popescu, and W. K. Wootters. Almost every pure state of three qubits is completely determined by its two-particle reduced density matrices. *Phys. Rev. Lett.*, 89:207901, 2002.

[615] N. Linden and A. Winter. A new inequality for the von Neumann entropy. *Commun. Math. Phys.*, 259:129, 2005.

[616] S. Lloyd. Almost any quantum logic gate is universal. *Phys. Rev. Lett.*, 75:346, 1995.

[617] S. Lloyd and H. Pagels. Complexity as thermodynamic depth. *Ann. Phys. (N.Y.)*, 188:186, 1988.

[618] H.-K. Lo and S. Popescu. Concentrating entanglement by local actions: Beyond mean values. *Phys. Rev.*, A 63:022301, 1998.

[619] R. Lockhart. Optimal ensemble length of mixed separable states. *J. Math. Phys.*, 41:6766, 2000.

[620] R. B. Lockhart and M. J. Steiner. Preserving entanglement under perturbation and sandwiching all separable states. *Phys. Rev.*, A 65:022107, 2002.

[621] R. B. Lockhart, M. J. Steiner, and K. Gerlach. Geometry and product states. *Quant. Inf. Comp.*, 2:333, 2002.

[622] L. Lovász. Semidefinite programs and combinatorial optimization. In B. A. Reed and C. L. Sales, editors, *Recent Advances in Algorithms and Combinatorics*, page 137. Springer, 2003.

[623] K. Löwner. Über monotone Matrixfunktionen. *Math. Z.*, 38:177, 1934.

[624] A. Łoziński, A. Buchleitner, K. Życzkowski, and T. Wellens. Entanglement of $2 \times k$ quantum systems. *Europhys. Lett.*, A 62:168, 2003.

[625] E. Lubkin. Entropy of an n-system from its correlation with a k-reservoir. *J. Math. Phys.*, 19:1028, 1978.

[626] G. Lüders. Über die Zustandsänderung durch den Messprozeß. *Ann. Phys. (Leipzig)*, 8:322, 1951.

[627] S. Luo. A simple proof of Wehrl's conjecture on entropy. *J. Phys.*, A 33:3093, 2000.

[628] J.-G. Luque and J.-Y. Thibon. Polynomial invariants of four qubits. *Phys. Rev.*, A 67:042303, 2003.

[629] J.-G. Luque and J.-Y. Thibon. Algebraic invariants of five qubits. *J. Phys.*, A 39:371, 2006.

[630] X.-S. Ma et al. Quantum teleportation over 143 kilometres using active feed–forward. *Nature*, 489:269, 2012.

[631] H. Maassen and J. B. M. Uffink. Generalized entropic uncertainty relations. *Phys. Rev. Lett.*, 60:1103, 1988.

[632] D. L. MacAdam, Visual sensitivities to color differences in daylight. *J. Opt. Soc. Am.*, 32:247, 1942.

[633] A. J. MacFarlane, A. Sudbery, and P. H. Weisz. On Gell–Mann's λ-matrices, d- and f-tensors, octets and parametrization of $SU(3)$. *Commun. Math. Phys.*, 11:77, 1968.

[634] H. F. MacNeish. Euler squares. *Ann. Math.*, 23:221, 1921.

[635] F. J. MacWilliams and N. J. A. Sloane. *The Theory of Error–Correcting Codes*. North-Holland, 1977.

[636] G. Mahler and V. A. Weberruß. *Quantum Networks*. Springer, 1995, (2nd. ed.) 1998.

[637] W. A. Majewski. Transformations between quantum states. *Rep. Math. Phys.*, 8:295, 1975.

[638] W. A. Majewski and M. Marciniak. On characterization of positive maps. *J. Phys.*, A 34:5863, 2001.

[639] E. Majorana. Atomi orientati in campo magnetico variabile. *Nuovo Cimento*, 9:43, 1932.

[640] Y. Makhlin. Non-local properties of two-qubit gates and mixed states, and the optimization of quantum computations. *Quant. Inf. Proc.*, 1:243, 2002.

[641] L. C. Malacarne, R. S. Mandes, and E. K. Lenzi. Average entropy of a subsystem from its average Tsallis entropy. *Phys. Rev.*, E 65:046131, 2002.

[642] L. Mandel and E. Wolf. *Optical Coherence and Quantum Optics*. Cambridge University Press, 1995.

[643] V. A. Marchenko and L. A. Pastur. The distribution of eigenvalues in certain sets of random matrices. *Math. Sb.*, 72:507, 1967.

[644] D. J. H. Markham. Entanglement and symmetry in permutation symmetric states. *Phys. Rev.*, A 83:042332, 2011.

[645] A. W. Marshall and I. Olkin. *Inequalities: Theory of Majorization and its Applications*. New York: Academic Press, 1979.

[646] J. Martin, O. Giraud, P. A. Braun, D. Braun, and T. Bastin. Multiqubit symmetric states with high geometric entanglement. *Phys. Rev.*, A 81:062347, 2010.

[647] L. Masanes. An area law for the entropy of low-energy states. *Phys. Rev.*, A 80:052104, 2009.

[648] P. Mathonet, S. Krins, M. Godefroid, L. Lamata, E. Solano, and T. Bastin. Entanglement equivalence of n-qubit symmetric states. *Phys. Rev.*, A 81: 052315, 2010.

[649] M. Matolcsi. A Fourier analytic approach to the problem of unbiased bases. *Studia Sci. Math. Hungarica*, 49:482, 2012.

[650] K. McCrimmon. Jordan algebras and their applications. *Bull. AMS.*, 84:612, 1978.

[651] M. Mehta. *Random Matrices, 2nd. ed.* New York: Academic Press, 1991.

[652] M. L. Mehta. *Matrix Theory*. Delhi: Hindustan Publishing, 1989.

[653] C. B. Mendl and M. M. Wolf. Unital quantum channels convex structure and revivals of Birkhoff's theorem. *Commun. Math. Phys.*, 289:1057, 2009.

[654] N. D. Mermin. Hidden variables and the two theorems of John Bell. *Rev. Mod. Phys.*, 65:803, 1993.

[655] N. D. Mermin. The Ithaca interpretation of quantum mechanics. *Pramana*, 51:549, 1998.

[656] D. A. Meyer and N. R. Wallach. Global entanglement in multiparticle systems. *J. Math. Phys.*, 43:4273, 2002.

[657] F. Mezzadri. How to generate random matrices from the classical compact groups. *Notices of the AMS*, 54:592, 2007.

[658] B. Mielnik. Theory of filters. *Commun. Math. Phys.*, 15:1, 1969.

[659] B. Mielnik. Quantum theory without axioms. In C. J. Isham, R. Penrose, and D. W. Sciama, editors, *Quantum Gravity II. A Second Oxford Symposium*, page 638. Oxford University Press, 1981.

[660] B. Mielnik. Nonlinear quantum mechanics: A conflict with the Ptolomean structure? *Phys. Lett.*, A 289:1, 2001.

[661] P. Migdał, J. Rodriguez-Laguna, and M. Lewenstein. Entanglement classes of permutation-symmetric qudit states: symmetric operations suffice. *Phys. Rev.*, A 88:012335, 2013.

[662] V. D. Milman and G. Schechtman. *Asymptotic theory of finite dimensional normed spaces*. Springer-Verlag, 1986. Lecture Notes in Math. 1200.

[663] F. Mintert, M. Kuś, and A. Buchleitner. Concurrence of mixed bipartite quantum states in arbitrary dimensions. *Phys. Rev. Lett.*, 92:167902, 2004.

[664] F. Mintert and K. Życzkowski. Wehrl entropy, Lieb conjecture and entanglement monotones. *Phys. Rev.*, A 69:022317, 2004.

[665] A. Miranowicz and A. Grudka. A comparative study of relative entropy of entanglement, concurrence and negativity. *J. Opt. B:* *Quantum Semiclass. Opt.*, 6:542, 2004.

[666] A. Miranowicz, M. Piani, P. Horodecki, and R. Horodecki. Inseparability criteria based on matrices of moments. *Phys. Rev.*, A 80:052303, 2009.

[667] B. Mirbach and H. J. Korsch. A generalized entropy measuring quantum localization. *Ann. Phys. (N.Y.)*, 265:80, 1998.

[668] J. A. Miszczak, Z. Puchała, P. Horodecki, A. Uhlmann, and K. Życzkowski. Sub– and super–fidelity as bounds for quantum fidelity. *Quantum Inf. Comp.*, 9:103, 2009.

[669] A. Miyake. Classification of multipartite entangled states by multidimensional determinants. *Phys. Rev.*, A 67:012108, 2003.

[670] L. Molnár. Fidelity preserving maps on density operators. *Rep. Math. Phys.*, 48:299, 2001.

[671] C. Monroe, D. M. Meekhof, B. E. King, W. M. Itano, and D. J. Wineland. Demonstration of a fundamental quantum logic gate. *Phys. Rev. Lett.*, 75:4714, 1995.

[672] T. Monz, P. Schindler, J. T. Barreiro, M. Chwalla, D. Nigg, W. A. Coish, M. Harlander, W. Hänsel, M. Hennrich, and R. Blatt. 14-qubit entanglement: creation and coherence. *Phys. Rev. Lett.*, 106:130506, 20011.

[673] E. A. Morozova and N. N. Čencov. Markov invariant geometry on state manifolds (in Russian). *Itogi Nauki i Tehniki*, 36:69, 1990.

[674] R. Mosseri and R. Dandoloff. Geometry of entangled states, Bloch spheres and Hopf fibrations. *J. Phys.*, A 34:10243, 2001.

[675] N. Mukunda, Arvind, S. Chaturvedi, and R. Simon. Generalized coherent states and the diagonal representation of operators. *J. Math. Phys.*, 44:2479, 2003.

[676] N. Mukunda and R. Simon. Quantum kinematic approach to the geometric phase. 1. General formalism. *Ann. Phys. (N. Y)*, 228:205, 1993.

[677] M. E. Muller. A note on a method for generating points uniformly on N-dimensional spheres. *Comm. Assoc. Comp. Mach.*, 2:19, 1959.

[678] M. P. Müller and C. Ududec. Structure of reversible computation determines the self-duality of quantum theory. *Phys. Rev. Lett.*, 108:130401, 2012.

[679] D. Mumford. *Tata Lectures on Theta*. Birkhäuser, Boston, 1983.

[680] W. J. Munro, D. F. V. James, A. G. White, and P. G. Kwiat. Maximizing the entanglement of two mixed qubits. *Phys. Rev.*, A 64:030302, 2001.

[681] F. D. Murnaghan. On the field of values of a square matrix. *Proc. Natl. Acad. Sci.*, 18:246, 1932.

[682] M. K. Murray and J. W. Rice. *Differential Geometry and Statistics*. Chapman and Hall, 1993.

[683] B. Musto and J. Vicary. Quantum Latin squares and unitary error bases. Quantum Inf. Comp. 16: 1318, 2016.

[684] M. Musz, M. Kuś, and K. Życzkowski. Unitary quantum gates, perfect entanglers, and unistochastic maps. *Phys. Rev.*, A 87:022111, 2013.

[685] L. Ness. A stratification of the null cone via the moment map [with an appendix by D. Mumford]. *Amer. J. Math.*, 106:1281, 1984.

[686] M. Neuhauser. An explicit construction of the metaplectic representation over a finite field. *Journal of Lie Theory*, 12:15, 2002.

[687] D. Newman. A simplified proof of the partition formula. *Michigan Math. J.*, 9:283, 1962.

[688] M. A. Nielsen. An introduction to majorization and its applications to quantum mechanics. Course

notes, October 2002, http://michaelnielsen.org/blog.

[689] M. A. Nielsen. Conditions for a class of entanglement transformations. *Phys. Rev. Lett.*, 83:436, 1999.

[690] M. A. Nielsen. Continuity bounds for entanglement. *Phys. Rev.*, A 61:064301, 2000.

[691] M. A. Nielsen. A simple formula for the average gate fidelity of a quantum dynamical operation. *Phys. Lett.*, 303:249, 2002.

[692] M. A. Nielsen and I. L. Chuang. *Quantum Computation and Quantum Information*. Cambridge University Press, 2000.

[693] M. A. Nielsen, C. Dawson, J. Dodd, A. Gilchrist, D. Mortimer, T. Osborne, M. Bremner, A. Harrow, and A. Hines. Quantum dynamics as physical resource. *Phys. Rev.*, A 67:052301, 2003.

[694] M. A. Nielsen and J. Kempe. Separable states are more disordered globally than locally. *Phys. Rev. Lett.*, 86:5184, 2001.

[695] M. A. Nielsen and G. Vidal. Majorization and the interconversion of bipartite states. *Quantum Infor. Comput.*, 1:76, 2001.

[696] D. Nion and L. D. Lathauwer. An enhanced line search scheme for complex-valued tensor decompositions. *Signal Processing*, 88:749, 2008.

[697] G. Nogues, A. Rauschenbeutel, S. Osnaghi, P. Bertet, M. Brune, J. Raimond, S. Haroche, L. Lutterbach, and L. Davidovich. Measurement of the negative value for the Wigner function of radiation. *Phys. Rev.*, A 62:054101, 2000.

[698] S. Nonnenmacher. Crystal properties of eigenstates for quantum cat maps. *Nonlinearity*, 10:1569, 1989.

[699] M. Ohya and D. Petz. *Quantum Entropy and Its Use*. Springer,

1993. Second edition, Springer 2004.

[700] D. K. L. Oi. The geometry of single qubit maps. preprint quant-ph/0106035.

[701] P. J. Olver. *Classical Invariant Theory*. Cambridge U P, 1999.

[702] A. Orłowski. Classical entropy of quantum states of light. *Phys. Rev.*, A 48:727, 1993.

[703] R. Orús. A practical introduction to tensor networks: Matrix product states and projected entangled pair states. *Ann. Phys.*, 349:117, 2014.

[704] H. Osaka. Indecomposable positive maps in low-dimensional matrix algebra. *Lin. Alg. Appl.*, 153:73, 1991.

[705] T. Osborne and F. Verstraete. General monogamy inequality for bipartite qubit entanglement. *Phys. Rev. Lett.*, 96:220503, 2006.

[706] V. A. Osipov, H.-J. Sommers, and K. Życzkowski. Random Bures mixed states and the distribution of their purity. *J. Phys.*, A 43:055302, 2010.

[707] A. Osterloh. Classification of qubit entanglement: SL(2,C) versus SU(2) invariance. *Appl. Phys.*, B 98:609, 2010.

[708] A. Osterloh and J. Siewert. Entanglement monotones and maximally entangled states in multipartite qubit systems. *Int. J. Quant. Inf.*, 4:531, 2006.

[709] S. Östlund and S. Rommer. Thermodynamic limit of density matrix renormalization. *Phys. Rev. Lett.*, 75:3537, 1995.

[710] Y. Ou. Violation of monogamy inequality for higher-dimensional objects. *Phys. Rev.*, A 75:034305, 2007.

[711] C. J. Oxenrider and R. D. Hill. On the matrix reordering Γ and Ψ. *Lin. Alg. Appl.*, 69:205, 1985.

[712] M. Ozawa. Entanglement measures and the Hilbert–Schmidt distance. *Phys. Lett.*, A 268:158, 2001.

[713] D. N. Page. Average entropy of a subsystem. *Phys. Rev. Lett.*, 71:1291, 1993.

[714] R. E. A. C. Paley. On orthogonal matrices. *J. Math.and Phys.*, 12:311, 1933.

[715] Ł. Pankowski, M. Piani, M. Horodecki, and P. Horodecki. A few steps more towards NPT bound entanglement. *IEEE Trans. Inform. Theory*, 56:4085, 2010.

[716] M. H. Partovi. Universal measure of entanglement. *Phys. Rev. Lett.*, 92:077904, 2004.

[717] F. Pastawski, B. Yoshida, D. Harlow, and J. Preskill. Holographic quantum error-correcting codes: toy models for the bulk/boundary correspondence. *JHEP*, 06:149, 2015.

[718] T. Paterek, M. Pawłowski, M. Grassl, and Č. Brukner. On the connection between mutually unbiased bases and orthogonal Latin squares. *Phys. Scr.*, T140:014031, 2010.

[719] R. Penrose. A spinor approach to general relativity. *Ann. of Phys. (N. Y.)*, 10:171, 1960.

[720] R. Penrose. Applications of negative dimensional tensors. In D. J. A. Welsh, editor, *Combinatorial Mathematics and its Appplications*, page 221. Academic Press, 1971.

[721] R. Penrose. *The Road to Reality*. London: Jonathan Cape, 2004.

[722] R. Penrose and W. Rindler. *Spinors and Spacetime: Vol. 1*. Cambridge U P, 1984.

[723] R. Penrose and W. Rindler. *Spinors and Spacetime: Vol. 2*. Cambridge U P, 1986.

[724] A. M. Perelomov. Generalized coherent states and some of their applications. *Sov. Phys. Usp.*, 20:703, 1977.

[725] A. Peres. *Quantum Theory: Concepts and Methods*. Dordrecht: Kluwer, 1995.

[726] A. Peres. Separability criterion for density matrices. *Phys. Rev. Lett.*, 77:1413, 1996.

[727] A. Peres and D. R. Terno. Convex probability domain of generalized quantum measurements. *J. Phys.*, A 31:L671, 1998.

[728] A. Peres and W. K. Wootters. Optimal detection of quantum information. *Phys. Rev. Lett.*, 66:1119, 1991.

[729] D. Perez-Garcia, F. Verstraete, M. M. Wolf, and J. I. Cirac. Matrix product state representations. *Quant. Inf. Comput.*, 7:401, 2007.

[730] D. Petz. Geometry of canonical correlation on the state space of a quantum system. *J. Math. Phys.*, 35:780, 1994.

[731] D. Petz. Monotone metrics on matrix spaces. *Lin. Alg. Appl.*, 244:81, 1996.

[732] D. Petz. Information-geometry of quantum states. *Quantum Prob. Commun.*, X:135, 1998.

[733] D. Petz. *Quantum Information Theory and Quantum Statistics*. Springer Verlag, 2008.

[734] D. Petz and C. Sudár. Geometries of quantum states. *J. Math. Phys.*, 37:2662, 1996.

[735] D. Petz and R. Temesi. Means of positive numbers and matrices. *SIAM J. Matrix Anal. Appl.*, 27:712, 2005.

[736] R. F. Picken. The Duistermaat–Heckman integration formula on flag manifolds. *J. Math. Phys.*, 31:616, 1990.

[737] M. Pinsker. *Information and Information Stability of Random Variables and Processes*. San Francisco: Holden-Day, 1964.

[738] J. Pipek and I. Varga. Universal scheme for the spacial-localization properties of one-particle states in finite d-dimensional systems. *Phys. Rev.*, A 46:3148, 1992.

[739] I. Pitowsky. Infinite and finite Gleason's theorems and the logic of indeterminacy. *J. Math. Phys.*, 39:218, 1998.

[740] A. O. Pittenger. Unextendible product bases and the construction of inseparable states. *Lin. Alg. Appl.*, 359:235, 2003.

[741] A. O. Pittenger and M. H. Rubin. Separability and Fourier representations of density matrices. *Phys. Rev.*, A 62:032313, 2000.

[742] A. O. Pittenger and M. H. Rubin. Convexity and the separability problem of quantum mechanical density matrices. *Lin. Alg. Appl.*, 346:47, 2002.

[743] A. O. Pittenger and M. H. Rubin. Geometry of entanglement witnesses and local detection of entanglement. *Phys. Rev.*, A 67:012327, 2003.

[744] M. B. Plenio and S. Virmani. An introduction to entanglement measures. *Quantum Inf. Comput.*, 7:1, 2007.

[745] V. Pless. *Introduction to the Theory of Error–Correcting Codes*. Wiley, 1982.

[746] Y.-T. Poon and N.-K. Tsing. Inclusion relations between orthostochastic matrices and products of pinching matrices. *Lin. Multilin. Algebra*, 21:253, 1987.

[747] S. Popescu and D. Rohrlich. Generic quantum nonlocality. *Phys. Lett.*, A 166:293, 1992.

[748] S. Popescu and D. Rohrlich. Thermodynamics and the measure of entanglement. *Phys. Rev.*, A 56:R3319, 1997.

[749] D. Poulin, R. Blume–Kohout, R. Laflamme, and H. Olivier. Exponential speedup with a single bit of quantum information. *Phys. Rev. Lett.*, 92:177906, 2004.

[750] M. Poźniak, K. Życzkowski, and M. Kuś. Composed ensembles of random unitary matrices. *J. Phys.*, A 31:1059, 1998.

[751] J. Preskill. Lecture notes on quantum computation, 1998. See www

.theory.caltech.edu/people/preskill/ph229/.

[752] J. Preskill. Reliable quantum computers. *Proc. R. Soc. Lond.*, A 454:385, 1998.

[753] T. Prosen. Exact statistics of complex zeros for Gaussian random polynomials with real coefficients. *J. Phys.*, A 29:4417, 1996.

[754] T. Prosen. Parametric statistics of zeros of Husimi representations of quantum chaotic eigenstates and random polynomials. *J. Phys.*, A 29:5429, 1996.

[755] Z. Puchała, J. A. Miszczak, P. Gawron, C. F. Dunkl, J. A. Holbrook, and K. Życzkowski. Restricted numerical shadow and geometry of quantum entanglement. *J. Phys.*, A 45:415309, 2012.

[756] E. M. Rabei, Arvind, R. Simon, and N. Mukunda. Bargmann invariants and geometric phases – a generalized connection. *Phys. Rev.*, A 60:3397, 1990.

[757] S. T. Rachev. *Probability Metrics and the Stability of Stochastic Models*. New York: Wiley, 1991.

[758] S. T. Rachev and L. Rüschendorf. *Mass Transportation Problems. Vol. 1: Theory*. Springer, 1998.

[759] J. M. Radcliffe. Some properties of coherent spin states. *J. Phys.*, A 4:313, 1971.

[760] E. M. Rains. Bound on distillable entanglement. *Phys. Rev.*, A 60:179, 1999.

[761] E. M. Rains. Rigorous treatment of distillable entanglement. *Phys. Rev.*, A 60:173, 1999.

[762] E. M. Rains. A semidefinite program for distillable entanglement. *IEEE Trans. Inform. Theory*, 47:2921, 2001.

[763] A. K. Rajagopal and R. W. Rendell. Separability and correlations in composite states based on entropy methods. *Phys. Rev.*, A 66:022104, 2002.

[764] M. Ramana and A. J. Goldman. Some geometric results in semidefinite programming. *J. Global Optim.*, 7:33, 1995.

[765] S. Rana. Negative eigenvalues of partial transposition of arbitrary bipartite states. *Phys. Rev.*, A 87:05430, 2013.

[766] C. R. Rao. Information and accuracy attainable in the estimation of statistical parameters. *Bull. Calcutta Math. Soc.*, 37:81, 1945.

[767] P. Raynal, X. Lü, and B.-G. Englert. Mutually unbiased bases in dimension 6: The four most distant bases. *Phys. Rev.*, A 83:062303, 2011.

[768] M. Reck, A. Zeilinger, H. J. Bernstein, and P. Bertani. Experimental realization of any discrete unitary operator. *Phys. Rev. Lett.*, 73:58, 1994.

[769] J. Řeháček and Z. Hradil. Quantification of entanglement by means of convergent iterations. *Phys. Rev. Lett.*, 90:127904, 2003.

[770] C. Reid. *Hilbert*. Springer, 1970.

[771] M. Reimpell and R. F. Werner. A meaner king uses biased bases. *Phys. Rev.*, A 75:062334, 2008.

[772] J. M. Renes, R. Blume-Kohout, A. J. Scott, and C. M. Caves. Symmetric informationally complete quantum measurements. *J. Math. Phys.*, 45:2171, 2004.

[773] A. Rényi. On measures of entropy and information. In *Proc. of the Fourth Berkeley Symp. Math. Statist. Prob. 1960, Vol. I*, page 547. Berkeley: University of California Press, 1961.

[774] P. Ribeiro and R. Mosseri. Entanglement in the symmetric sector of *n* qubits. *Phys. Rev. Lett.*, 106:180502, 2011.

[775] A. G. Robertson. Automorphisms of spin factors and the decomposition of positive maps. *Quart. J. Math. Oxford*, 34:87, 1983.

[776] W. Roga, M. Fannes, and K. Życzkowski. Universal bounds

for the Holevo quantity, coherent information and the Jensen-Shannon divergence. *Phys. Rev. Lett.*, 105:040505, 2010.

[777] R. Rossignoli and N. Canosa. Generalized entropic criterion for separability. *Phys. Rev.*, A 66:042306, 2002.

[778] A. Roy and A. J. Scott. Unitary designs and codes. *Des. Codes Cryptogr.*, 53:13, 2009.

[779] O. Rudolph. A new class of entanglement measures. *J. Math. Phys.*, A 42:5306, 2001.

[780] O. Rudolph. On the cross norm criterion for separability. *J. Phys.*, A 36:5825, 2003.

[781] O. Rudolph. Some properties of the computable cross norm criterion for separability. *Physical Review*, A 67:032312, 2003.

[782] P. Rungta, V. Bužek, C. M. Caves, M. Hillery, and G. J. Milburn. Universal state inversion and concurrence in arbitrary dimensions. *Phys. Rev.*, A 64:042315, 2001.

[783] P. Rungta and C. M. Caves. Concurrence-based entanglement measures for isotropic states. *Phys. Rev.*, A 67:012307, 2003.

[784] M. B. Ruskai. Inequalities for quantum entropy: A review with conditions for equality. *J. Math. Phys.*, 43:4358, 2002.

[785] M. B. Ruskai, S. Szarek, and E. Werner. An analysis of completely-positive trace-preserving maps on 2 × 2 matrices. *Lin. Alg. Appl.*, 347:159, 2002.

[786] E. B. Saff and A. B. J. Kuijlaars. Distributing many points on a sphere. *Math. Intelligencer*, 19:5, 1997.

[787] T. Salvemini. Sul calcolo degli indici di concordanza tra due caratteri quantitativi. *Atti della VI Riunione della Soc. Ital. di Statistica*, 1943.

[788] J. Sánchez-Ruiz. Simple proof of Page's conjecture on the average entropy of a subsystem. *Phys. Rev.*, E 52:5653, 1995.

[789] Y. Sanders and G. Gour. Necessary conditions on entanglement catalysts. *Phys. Rev.*, A 79:054302, 2008.

[790] I. N. Sanov. On the probability of large deviations of random variables (in Russian). *Mat. Sbornik*, 42:11, 1957.

[791] A. Sanpera, R. Tarrach, and G. Vidal. Local description of quantum inseparability. *Phys. Rev.*, A 58:826, 1998.

[792] G. Sarbicki and I. Bengtsson. Dissecting the qutrit. *J. Phys.*, A 46:035306, 2013.

[793] A. Sawicki, A. Huckleberry, and M. Kuś. Symplectic geometry of entanglement. *Commun. Math. Phys.*, 305:421, 2011.

[794] A. Sawicki, M. Oszmaniec, and M. Kuś. Convexity of momentum map, Morse index and quantum entanglement. *Rev. Math. Phys.*, 26:1450004, 2014.

[795] A. Sawicki, M. Walter, and M. Kuś. When is a pure state of three qubits determined by its single-particle reduced density matrices? *J. Phys.*, A 46:055304, 2013.

[796] J. Scharlau and M. P. Müller. Quantum Horn lemma, finite heat baths, and the third law of thermodynamics. preprint arxiv:1605.06092.

[797] R. Schatten. *A Theory of Cross-spaces*. Princeton University Press, 1950.

[798] S. Schirmer, T. Zhang, and J. V. Leahy. Orbits of quantum states and geometry of Bloch vectors for *n*-level systems. *J. Phys.*, A 37:1389, 2004.

[799] E. Schmidt. Zur Theorie der linearen und nichtlinearen Integralgleichungen. *Math. Ann.*, 63:433, 1907.

[800] U. Schollwöck. The density-matrix renormalization group in the age

of matrix product states. *Annals of Phys.*, 326:96, 2011.

[801] E. Schrödinger. Der stetige Übergang von der Mikro- zur Makromechanik. *Naturwissenschaften*, 14:664, 1926.

[802] E. Schrödinger. Die Gesichtsempfindungen. In O. Lummer, editor, *Müller-Pouillets Lehrbuch der Physik, 11th edn, Vol. 2*, pages 456–560. Friedr. Vieweg und Sohn, 1926.

[803] E. Schrödinger. Die gegenwärtige Situation in der Quantenmechanik. *Naturwissenschaften*, 32:446, 1935.

[804] E. Schrödinger. Discussion of probability relations between separated systems. *Proc. Camb. Phil. Soc.*, 31:555, 1935.

[805] E. Schrödinger. Probability relations between separated systems. *Proc. Camb. Phil. Soc.*, 32:446, 1936.

[806] E. Schrödinger. *Space–Time Structure*. Cambridge University Press, 1950.

[807] N. Schuch, M. M. Wolf, F. Verstraete, and J. I. Cirac. Computational complexity of projected entangled pair states. *Phys. Rev. Lett.*, 98:140506, 2007.

[808] B. Schumacher. Quantum coding. *Phys. Rev.*, A 51:2738, 1995.

[809] B. Schumacher. Sending entanglement through noisy quantum channels. *Phys. Rev.*, A 54:2614, 1996.

[810] B. Schumacher and M. Nielsen. Quantum data processing and error correction. *Phys. Rev.*, A 54:2629, 1996.

[811] B. Schumacher and M. Westmoreland. Relative entropy in quantum information theory. preprint quant-ph/0004045.

[812] B. Schumacher and M. D. Westmoreland. Sending classical information via noisy quantum channels. *Phys. Rev.*, A 56:131, 1997.

[813] P. Schupp. On Lieb's conjecture for the Wehrl entropy of Bloch coherent states. *Commun. Math. Phys.*, 207:481, 1999.

[814] J. Schur. Über eine Klasse von Mittelbildungen mit Anwendungen auf die Determinantentheorie. *Sitzber. Berl. Math. Ges.*, 22:9, 1923.

[815] J. Schwinger. Unitary operator bases. *Proc. Natl. Acad. Sci.*, 46:570, 1960.

[816] J. Schwinger. On angular momentum. In L. C. Biedenharn and H. van Dam, editors, *Quantum Theory of Angular Momentum*, page 229. Academic Press, 1965.

[817] J. Schwinger. *Quantum Mechanics —Symbolism of Atomic Measurements*. Springer, 2003.

[818] A. J. Scott. SICs: Extending the list of solutions. preprint arXiv: 1703.03993.

[819] A. J. Scott. Multipartite entanglement, quantum-error-correcting codes, and entangling power of quantum evolutions. *Phys. Rev.*, A 69:052330, 2004.

[820] A. J. Scott. Tight informationally complete measurements. *J. Phys.*, A 39:13507, 2006.

[821] A. J. Scott and M. Grassl. SIC-POVMs: A new computer study. *J. Math. Phys.*, 51:042203, 2010.

[822] H. Scutaru. On Lieb's conjecture. preprint FT-180 (1979) Bucharest and math-ph/9909024.

[823] I. E. Segal. Postulates for general quantum mechanics. *Ann. Math.*, 48:930, 1947.

[824] C. Segre. Remarques sur les transformations uniformes des courbes elliptiques en elles-mèmes. *Math. Ann.* 27:296, 1886.

[825] S. Sen. Average entropy of a quantum subsystem. *Phys. Rev. Lett.*, 77:1, 1996.

[826] P. D. Seymour and T. Zaslawsky. Averaging sets: a generalization of mean values and spherical designs. *Adv. Math.*, 52:213, 1984.

[827] L. K. Shalm et al. A strong loophole-free test of local realism. *Phys. Rev. Lett.*, 115:250402, 2015.

[828] C. E. Shannon. A mathematical theory of communication. *Bell Sys. Tech. J.*, 27:379, 623, 1948.

[829] C. E. Shannon and W. Weaver. *The Mathematical Theory of Communication*. Univ. of Illinois Press, 1949.

[830] A. Shapere and F. Wilczek. *Geometric Phases in Physics*. World Scientific, 1989.

[831] S.-Q. Shen, M. Li, and L. Li. A bipartite separable ball and its applications. *J. Phys.*, A 48:095302, 2015.

[832] S.-Q. Shen, M.-Y. Wang, M. Li, and S.-M. Fei. Separability criteria based on the realignment of density matrices and reduced density matrices. *Phys. Rev.*, A 92:042332, 2015.

[833] S.-Q. Shen, J. Yu, M. Li, and S.-M. Fei. Improved separability criteria based on Bloch representation of density matrices. *Sci. Rep.*, 6:28850, 2016.

[834] A. Shimony. Degree of entanglement. *Ann. NY. Acad. Sci.*, 755:675, 1995.

[835] P. Shor. Scheme for reducing decoherence in quantum computer memory. *Phys. Rev.*, A52:R2493, 1995.

[836] P. W. Shor. Additivity of the classical capacity of entanglement-breaking quantum channels. *J. Math. Phys.*, 43:4334, 2002.

[837] P. W. Shor. Equivalence of additivity questions in quantum information theory. *Commun. Math. Phys.*, 246:453, 2004.

[838] P. W. Shor, J. A. Smolin, and B. M. Terhal. Non-additivity of bipartite distillable entanglement follows from a conjecture on bound entangled Werner states. *Phys. Rev. Lett.*, 86:2681, 2001.

[839] M. Sinołęcka, K. Życzkowski, and M. Kuś. Manifolds of equal entanglement for composite quantum systems. *Acta Phys. Pol.*, B 33:2081, 2002.

[840] Ł. Skowronek. Three-by-three bound entanglement with general unextendible product bases. *J. Math. Phys.*, 52:122202, 2011.

[841] Ł. Skowronek. There is no direct generalization of positive partial transpose criterion to the three-by-three case. *J. Math. Phys.*, 57:112201, 2016.

[842] P. B. Slater. Hall normalization constants for the Bures volumes of the n-state quantum systems. *J. Phys.*, A 32:8231, 1999.

[843] P. B. Slater. A priori probabilities of separable quantum states. *J. Phys.*, A 32:5261, 1999.

[844] P. B. Slater. Silver mean conjectures for 15-d volumes and 14-d hyperareas of the separable two-qubit systems. *J. Geom. Phys.*, 53:74, 2005.

[845] P. B. Slater. A concise formula for generalized two-qubit Hilbert–Schmidt separability probabilities. *J. Phys.*, A 46:445302, 2013.

[846] W. Słomczyński. Subadditivity of entropy for stochastic matrices. *Open Sys. & Information Dyn.*, 9:201, 2002.

[847] W. Słomczyński and K. Życzkowski. Mean dynamical entropy of quantum maps on the sphere diverges in the semiclassical limit. *Phys. Rev. Lett.*, 80:1880, 1998.

[848] D. T. Smithey, M. Beck, M. Raymer, and A. Faridani. Measurement of the Wigner distribution and the density matrix of a light mode using optical homodyne tomography: Application to squeezed states and the vacuum. *Phys. Rev. Lett.*, 70:1244, 1993.

[849] H.-J. Sommers and K. Życzkowski. Bures volume of the set of mixed quantum states. *J. Phys.*, A 36:10083, 2003.

[850] H.-J. Sommers and K. Życzkowski. Statistical properties of random

density matrices. *J. Phys.*, A 37: 8457, 2004.

[851] R. D. Sorkin. On the entropy of the vacuum outside a horizon. In B. Bertotti, F. de Felice, and A. Pascolini, editors, *10th Iinternational Conference on General Relativity and Gravitation, Contributed Papers, vol. II*, page 734. Roma, Consiglio Nazionale delle Richerche, 1983.

[852] J. Spanier and K. B. Oldham. *An Atlas of Functions*. Washington: Hemisphere Publishing Corporation, 1987.

[853] C. Spengler, M. Huber, S. Brierley, T. Adaktylos, and B. C. Hiesmayr. Entanglement detection via mutually unbiased bases. *Phys. Rev.*, A 86:022311, 2012.

[854] J. Sperling and W. Vogel. Necessary and sufficient conditions for bipartite entanglement. *Phys. Rev.*, A 79:022318, 2009.

[855] S. F. Springer. *Symmetry in Mechanics*. Birkhäuser, 2001.

[856] A. M. Steane. Error correcting codes in quantum theory. *Phys. Rev. Lett.*, 77:793, 1996.

[857] A. M. Steane. Introduction to quantum error correction. *Phil. Trans R. Soc. Lond.*, A356:1739, 1998.

[858] M. Steiner. Generalized robustness of entanglement. *Phys. Rev.*, A 67:054305, 2003.

[859] W. F. Stinespring. Positive functions on C^*–algebras. *Proc. Am. Math. Soc.*, 6:211, 1955.

[860] D. R. Stinson. *Combinatorial Designs: Constructions and Analysis*. Springer, 2004.

[861] E. Størmer. Positive linear maps of operator algebras. *Acta Math.*, 110:233, 1963.

[862] E. Størmer. Decomposable positive maps on C^*–algebras. *Proc. Amer. Math. Soc.*, 86:402, 1982.

[863] R. F. Streater. *Statistical Dynamics*. London: Imperial College Press, 1995.

[864] F. Strocchi. Complex coordinates and quantum mechanics. *Rev. Mod. Phys.*, 38:36, 1966.

[865] E. Study. Kürzeste Wege in komplexen Gebiet. *Math. Annalen*, 60:321, 1905.

[866] E. C. G. Sudarshan. Equivalence of semiclassical and quantum mechanical descriptions of light beams. *Phys. Rev. Lett.*, 10:277, 1963.

[867] E. C. G. Sudarshan, P. M. Mathews, and J. Rau. Stochastic dynamics of quantum-mechanical systems. *Phys. Rev.*, 121:920, 1961.

[868] E. C. G. Sudarshan and A. Shaji. Structure and parametrization of stochastic maps of density matrices. *J. Phys.*, A 36:5073, 2003.

[869] A. Sudbery. Computer-friendly d-tensor identities for $SU(n)$. *J. Phys.*, A 23:L705, 1990.

[870] A. Sudbery. On local invariants of pure three-qubit states. *J. Phys.*, A 34:643, 2001.

[871] A. Sugita. Proof of the generalized Lieb–Wehrl conjecture for integer indices more than one. *J. Phys.*, A 35:L621, 2002.

[872] A. Sugita. Moments of generalized Husimi distribution and complexity of many-body quantum states. *J. Phys.*, A 36:9081, 2003.

[873] M. A. Sustik, J. Tropp, I. S. Dhillon, and R. W. Heath. On the existence of equiangular tight frames. *Lin. Alg. Appl.*, 426:619, 2007.

[874] G. Svetlichny. Distinguishing three-body from two-body nonseparability by a Bell-type inequality. *Phys. Rev.*, D35:3066, 1987.

[875] J. Sylvester. Sur l'equation en matrices $px = xq$. *C. R. Acad. Sci. Paris*, 99:67, 115, 1884.

[876] J. J. Sylvester. A word on nonions. *John Hopkins Univ. Circulars.*, 1:1, 1882.

[877] S. Szarek. The volume of separable states is super-doubly-exponentially small. *Phys. Rev.*, A 72:032304, 2005.

[878] S. Szarek, I. Bengtsson, and K. Życzkowski. On the structure of the body of states with positive partial transpose. *J. Phys.*, A 39:L119, 2006.

[879] S. Szarek, E. Werner, and K. Życzkowski. Geometry of sets of quantum maps: A generic positive map acting on a high-dimensional system is not completely positive. *J. Math. Phys.*, 49:032113, 2008.

[880] F. Szöllősi. *Construction, classification and parametrization of complex Hadamard matrices.* PhD thesis, CEU, Budapest, 2011.

[881] F. Szöllősi. Complex Hadamard matrices of order 6: a four-parameter family. *J. London Math. Soc.*, 85:616, 2012.

[882] K. Szymański, S. Weis, and K. Życzkowski. Classification of joint numerical ranges of three hermitian matrices of size three. preprint arxiv:1603.06569.

[883] W. Tadej and K. Życzkowski. A concise guide to complex Hadamard matrices. *Open Sys. Inf. Dyn.*, 13:133, 2006.

[884] T. Takagi. On an algebraic problem related to an analytic theorem of Carathéodory and Fejér and on an allied theorem of Landau. *Japan J. Math.*, 1:83, 1925.

[885] M. Talagrand. New concentration inequalities in product spaces. *Invent. Math.*, 126:505, 1996.

[886] M. Talagrand. A new look at independence. *Ann. Probab.*, 24:1, 1996.

[887] K. Tanahashi and J. Tomiyama. Indecomposable positive maps in matrix algebra. *Canad. Math. Bull.*, 31:308, 1988.

[888] W. Tang. On positive linear maps between matrix algebra. *Lin. Alg. Appl.*, 79:33, 1986.

[889] B. Terhal, I. Chuang, D. DiVincenzo, M. Grassl, and J. Smolin. Simulating quantum operations with mixed environments. *Phys. Rev.*, A 60:88, 1999.

[890] B. Terhal, M. Horodecki, D. W. Leung, and D. DiVincenzo. The entanglement of purification. *J. Math. Phys.*, 43:4286, 2002.

[891] B. M. Terhal. Bell inequalities and the separability criterion. *Phys. Lett.*, A 271:319, 2000.

[892] B. M. Terhal. A family of indecomposable positive linear maps based on entangled quantum states. *Lin. Alg. Appl.*, 323:61, 2000.

[893] B. M. Terhal and K. G. H. Vollbrecht. Entanglement of formation for isotropic states. *Phys. Rev. Lett.*, 85:2625, 2000.

[894] A. V. Thapliyal. On multi-particle pure states entanglement. *Phys. Rev.*, A 59:3336, 1999.

[895] E. Thiele and J. Stone. A measure of quantum chaos. *J. Chem. Phys.*, 80:5187, 1984.

[896] O. Toeplitz. Das algebraische Analogon zu einem Satze von Fejér. *Math. Z.*, 2:187, 1918.

[897] F. Topsøe. Bounds for entropy and divergence for distributions over a two element set. *J. Ineq. P. Appl. Math.*, 2:25, 2001.

[898] M. Tribus and E. C. McIrvine. Energy and information. *Scient. Amer.*, 224:178, 1971.

[899] C. Tsallis. Entropic non-extensivity: a possible measure of complexity. *Chaos, Solitons, Fract.*, 13:371, 2002.

[900] C. Tsallis, S. Lloyd, and M. Baranger. Peres criterion for separability through non-extensive entropy. *Phys. Rev.*, A 63:042104, 2001.

[901] R. R. Tucci. All moments of the uniform ensemble of quantum density matrices. preprint quant-ph/0206193.

[902] R. R. Tucci. Relaxation method for calculating quantum entanglement. preprint quant-ph/0101123.

[903] J. Tura, A. B. Sainz, T. Grass, R. Augusiak, A. Acín, and M. Lewenstein. Entanglement and nonlocality in many-body systems: a primer. Proceedings of the International School of Physics "Enrico Fermi" 191, p. 505, Ed. by M. Inguscio et. al, IOS Press, Amsterdam 2016.

[904] A. Uhlmann. Endlich dimensionale Dichtematrizen. *Wiss. Z. Karl-Marx-Univ.*, 20:633, 1971.

[905] A. Uhlmann. Relative entropy and the Wigner–Yanase–Dyson–Lieb concavity in an interpolation theory. *Commun. Math. Phys.*, 54:21, 1977.

[906] A. Uhlmann. The metric of Bures and the geometric phase. In R. Gielerak, editor, *Quantum Groups and Related Topics*, page 267. Dordrecht: Kluwer, 1992.

[907] A. Uhlmann. Density operators as an arena for differential geometry. *Rep. Math. Phys.*, 33:253, 1993.

[908] A. Uhlmann. Geometric phases and related structures. *Rep. Math. Phys.*, 36:461, 1995.

[909] A. Uhlmann. Spheres and hemispheres as quantum state spaces. *J. Geom. Phys.*, 18:76, 1996.

[910] A. Uhlmann. Entropy and optimal decompositions of states relative to a maximal commutative subalgebra. *Open Sys. & Information Dyn.*, 5:209, 1998.

[911] A. Uhlmann. Fidelity and concurrence of conjugated states. *Phys. Rev.*, A 62:032307, 2000.

[912] A. Uhlmann. Roofs and convexity. *Entropy*, 2010:1799, 2010.

[913] A. Uhlmann. Anti- (conjugate) linearity. *Science China*, 59:630301, 2016.

[914] H. Umegaki. Conditional expectation in an operator algebra IV: entropy and information. *Kōdai Math. Sem. Rep.*, 14:59, 1962.

[915] W. van Dam and P. Hayden. Rényi-entropic bounds on quantum communication. preprint quant-ph/0204093.

[916] V. S. Varadarajan. *Geometry of Quantum Theory*. Springer, 1985.

[917] V. Vedral. The role of relative entropy in quantum information theory. *Rev. Mod. Phys.*, 74:197, 2002.

[918] V. Vedral and E. Kashefi. Uniqueness of the entanglement measure for bipartite pure states and thermodynamics. *Phys. Rev. Lett.*, 89:037903, 2002.

[919] V. Vedral and M. B. Plenio. Entanglement measures and purification procedures. *Phys. Rev.*, A 57:1619, 1998.

[920] V. Vedral, M. B. Plenio, M. A. Rippin, and P. L. Knight. Quantifying entanglement. *Phys. Rev. Lett.*, 78:2275, 1997.

[921] F. Verstraete, K. Audenaert, J. Dehaene, and B. DeMoor. A comparison of the entanglement measures negativity and concurrence. *J. Phys.*, A 34:10327, 2001.

[922] F. Verstraete, K. Audenaert, and B. DeMoor. Maximally entangled mixed states of two qubits. *Phys. Rev.*, A 64:012316, 2001.

[923] F. Verstraete and J. I. Cirac. Renormalization algorithms for quantum-many body systems in two and higher dimensions. preprint arxiv:0407066.

[924] F. Verstraete and J. I. Cirac. Matrix product states represent ground states faithfully. *Phys. Rev.*, B 73:094423, 2006.

[925] F. Verstraete, J. Dahaene, and B. DeMoor. On the geometry of entangled states. *J. Mod. Opt.*, 49:1277, 2002.

[926] F. Verstraete, J. Dahaene, B. DeMoor, and H. Verschelde. Four qubits can be entangled in nine different ways. *Phys. Rev.*, A 65:052112, 2002.

[927] F. Verstraete, V. Murg, and J. I. Cirac. Matrix product states, projected entangled pair states, and variational renormalization group methods for quantum spin systems. *Adv. Phys.*, 57:143, 2008.

[928] F. Verstraete and H. Verschelde. Fidelity of mixed states of two qubits. *Phys. Rev.*, A 66:022307, 2002.

[929] G. Vidal. Entanglement of pure states for a single copy. *Phys. Rev. Lett.*, 83:1046, 1999.

[930] G. Vidal. Entanglement monotones. *J. Mod. Opt.*, 47:355, 2000.

[931] G. Vidal. Efficient classical simulation of slightly entangled quantum computations. *Phys. Rev. Lett.*, 91:147902, 2003.

[932] G. Vidal, W. Dür, and J. I. Cirac. Entanglement cost of bipartite mixed states. *Phys. Rev. Lett.*, 89:027901, 2002.

[933] G. Vidal and R. Tarrach. Robustness of entanglement. *Phys. Rev.*, A 59:141, 1999.

[934] G. Vidal and R. F. Werner. A computable measure of entanglement. *Phys. Rev.*, A 65:032314, 2002.

[935] O. Viehmann, C. Eltschka, and J. Siewert. Polynomial invariants for discrimination and classification of four-qubit entanglement. *Phys. Rev.*, A 83:052330, 2011.

[936] S. Virmani and M. Plenio. Ordering states with entanglement measures. *Phys. Lett.*, A 268:31, 2000.

[937] K. G. H. Vollbrecht and R. F. Werner. Why two qubits are special. *J. Math. Phys.*, 41:6772, 2000.

[938] K. G. H. Vollbrecht and R. F. Werner. Entanglement measures under symmetry. *Phys. Rev.*, A 64:062307, 2001.

[939] R. von Mises. *Probability, Statistics and Truth.* New York: Dover, 1957.

[940] J. von Neumann. Thermodynamik quantummechanischer Gesamtheiten. *Gött. Nach.*, 1:273, 1927.

[941] J. von Neumann. *Mathematical Foundations of Quantum Mechanics.* Princeton University Press, 1955.

[942] A. Vourdas. Quantum systems with finite Hilbert space. *Rep. Prog. Phys.*, 67:267, 2004.

[943] M. Waegell and K. Aravind. Critical noncolorings of the 600-cell proving the Bell–Kochen–Specker theorem. *J. Phys.*, A43:105304, 2010.

[944] R. M. Wald. *Quantum Field Theory in Curved Spacetime and Black Hole Thermodynamics.* Univ. of Chicago Press, 1994.

[945] M. Walter, B. Doran, D. Gross, and M. Christandl. Entanglement polytopes: Multiparticle entanglement from single-particle information. *Science*, 340:6137, 2013.

[946] P. Walther, K. J. Resch, and A. Zeilinger. Local conversion of GHZ states to approximate W states. *Phys. Rev. Lett.*, 94:240501, 2005.

[947] J. Watrous. Mixing doubly stochastic quantum channels with the completely depolarizing channel. *Quant. Inform. Comput.*, 9:406, 2009.

[948] Z. Webb. The Clifford group forms a unitary 3-design. *Quant. Inf. Comp.*, 16:1379, 2016.

[949] A. Wehrl. General properties of entropy. *Rev. Mod. Phys.*, 50:221, 1978.

[950] A. Wehrl. On the relation between classical and quantum mechanical entropies. *Rep. Math. Phys.*, 16:353, 1979.

[951] T.-C. Wei and P. M. Goldbart. Geometric measure of entanglement and applications to bipartite and multipartite quantum states. *Phys. Rev.*, A 68:042307, 2003.

[952] T. C. Wei, K. Nemoto, P. M. Goldbart, P. G. Kwiat, W. J. Munro, and F. Verstraete. Maximal entanglement versus entropy

for mixed quantum states. *Phys. Rev.*, A 67:022110, 2003.

[953] S. Weigert and T. Durt. Affine constellations without mutually unbiased counterparts. *J. Phys.*, A43:402002, 2010.

[954] M. Weiner. A gap for the maximum number of mutually unbiased bases. *Proc. Amer. Math. Soc.*, 141:1963, 2013.

[955] S. Weis. Quantum convex support. *Linear Alg. Appl.*, 435:3168, 2011.

[956] L. R. Welch. Lower bounds on the maximum cross correlation of signals. *IEEE Trans. Inf. Theory.*, 20:397, 1974.

[957] T. Wellens and M. Kuś. Separable approximation for mixed states of composite quantum systems. *Phys. Rev.*, A 64:052302, 2001.

[958] R. F. Werner. Quantum states with Einstein–Podolski–Rosen correlations admitting a hidden-variable model. *Phys. Rev.*, A 40:4277, 1989.

[959] R. F. Werner. Optimal cloning of pure states. *Phys. Rev.*, A 58:1827, 1998.

[960] R. F. Werner. All teleportation and dense coding schemes. *J. Phys.*, A 34:7081, 2001.

[961] H. Weyl. *Group Theory and Quantum Mechanics*. E. P. Dutton, New York, 1932.

[962] H. Weyl. *The Classical Groups*. Princeton UP, New Jersey, 1939.

[963] S. R. White. Density matrix formulation for quantum renormalization groups. *Phys. Rev. Lett.*, 69:2863, 1992.

[964] M. Wieśniak, T. Paterek, and A. Zeilinger. Entanglement in mutually unbiased bases. *New J. Phys.*, 13:053047, 2011.

[965] E. P. Wigner. On the quantum correction for thermodynamic equilibrium. *Phys. Rev.*, 40:749, 1932.

[966] E. P. Wigner. *Group Theory and its Application to the Quantum Mechanics of Atomic Spectra*. Academic Press, 1959.

[967] E. P. Wigner. Normal form of antiunitary operators. *J. Math. Phys.*, 1:409, 1960.

[968] A. Wilce. Four and a half axioms for finite dimensional quantum mechanics. In Y. Ben-Menahem and M. Hemmo, editors, *Probability in Physics*, page 281. Springer, Berlin, 2012.

[969] M. M. Wilde. *Quantum Information Theory*. Cambridge University Press, Cambridge, 2013.

[970] S. J. Williamson and H. Z. Cummins. *Light and Color in Nature and Art*. Wiley, 1983.

[971] A. Winter. Tight uniform continuity bounds for quantum entropies: conditional entropy, relative entropy distance and energy constraints. *Commun. Math. Phys.*, 344:1, 2016.

[972] C. Witte and M. Trucks. A new entanglement measure induced by the Hilbert–Schmidt norm. *Phys. Lett.*, A 257:14, 1999.

[973] P. Wocjan and T. Beth. New construction of mutually unbiased bases in square dimensions. *Quantum Inf. Comp.*, 5:93, 2005.

[974] K. Wódkiewicz. Stochastic decoherence of qubits. *Optics Express*, 8:145, 2001.

[975] H. J. Woerdeman. The separability problem and normal completions. *Lin. Alg. Appl.*, 376:85, 2004.

[976] A. Wong and N. Christensen. Potential multiparticle entanglement measure. *Phys. Rev.*, A 63:044301, 2001.

[977] Y.-C. Wong. Differential geometry of Grassmann manifolds. *Proc. Nat. Acad. Sci. USA*, 47:589, 1967.

[978] W. K. Wootters. Statistical distance and Hilbert space. *Phys. Rev.*, D 23:357, 1981.

[979] W. K. Wootters. A Wigner function formulation of finite-state quantum mechanics. *Ann. Phys. (N.Y.)*, 176:1, 1987.

[980] W. K. Wootters. Random quantum states. *Found. Phys.*, 20:1365, 1990.

[981] W. K. Wootters. Entanglement of formation of an arbitrary state of two qubits. *Phys. Rev. Lett.*, 80:2245, 1998.

[982] W. K. Wootters. Entanglement of formation and concurrence. *Quant. Inf. Comp.*, 1:27, 2001.

[983] W. K. Wootters and B. F. Fields. Optimal state determination by mutually unbiased measurements. *Ann. Phys. (N.Y.)*, 191:363, 1989.

[984] S. L. Woronowicz. Non-extendible positive maps. *Commun. Math. Phys.*, 51:243, 1976.

[985] S. L. Woronowicz. Positive maps of low-dimensional matrix algebra. *Rep. Math. Phys.*, 10:165, 1976.

[986] N. Wu and R. Coppins. *Linear Programming and Extensions*. New York: McGraw-Hill, 1981.

[987] D. Yang, M. Horodecki, and Z. D. Wang. An additive and operational entanglement measure: conditional entanglement of mutual information. *Phys. Rev. Lett.*, 101:140501, 2008.

[988] D. Ye. On the Bures volume of separable quantum states. *J. Math. Phys.*, 50:083502, 2009.

[989] D. A. Yopp and R. D. Hill. On completely copositive and decomposable linear transformations. *Lin. Alg. Appl.*, 312:1, 2000.

[990] S. Yu and C. H. Oh. State-independent proof of Kochen–Specker theorem with 13 rays. *Phys. Rev. Lett*, 108:030402, 2012.

[991] Y.-K. Yu and Y.-C. Zhang. On the anti-Wishart distribution. *Physica*, A 312:1, 2002.

[992] C. Zalka and E. Rieffel. Quantum operations that cannot be implemented using a small mixed environment. *J. Math. Phys.*, 43:4376, 2002.

[993] P. Zanardi. Entanglement of quantum evolution. *Phys. Rev.*, A 63:040304(R), 2001.

[994] P. Zanardi and D. A. Lidar. Purity and state fidelity of quantum channels. *Phys. Rev.*, A 70:012315, 2004.

[995] P. Zanardi, C. Zalka, and L. Faoro. On the entangling power of quantum evolutions. *Phys. Rev.*, A 62:030301(R), 2000.

[996] G. Zauner. *Quantendesigns*. PhD thesis, Univ. Wien, 1999.

[997] A. Zeilinger, M. A. Horne, and D. M. Greenberger. Higher-order quantum entanglement. In D. Han, Y. S. Kim, and W. W. Zachary, editors, *Proc. of Squeezed States and Quantum Uncertainty*, pages 73–81, 1992. NASA Conf. Publ.

[998] B. Zeng, X. Chen, D.-L. Zhou, and X.-G. Wen. *Quantum Information Meets Quantum Matter*. book in preparation and preprint arxiv:1508.02595.

[999] C.-J. Zhang, Y.-S. Zhang, S. Zhang, and G.-C. Guo. Entanglement detection beyond the cross-norm or realignment criterion. *Phys. Rev.*, A 77:060301(R), 2008.

[1000] J. Zhang, J. Vala, S. Sastry, and K. Whaley. Geometric theory of non-local two-qubit operations. *Phys. Rev.*, A 67:042313, 2003.

[1001] W.-M. Zhang, D. H. Feng, and R. Gilmore. Coherent states: Theory and some applications. *Rev. Mod. Phys.*, 62:867, 1990.

[1002] H. Zhu. Multiqubit Clifford groups are unitary 3–designs. preprint arXiv:1510.02619.

[1003] H. Zhu. SIC-POVMs and Clifford groups in prime dimensions. *J. Phys.*, A 43:305305, 2010.

[1004] H. Zhu. Mutually unbiased bases as minimal Clifford covariant 2–designs. *Phys. Rev.*, A 91:060301, 2015.

[1005] H. Zhu. Permutation symmetry determines the discrete Wigner function. *Phys. Rev. Lett.*, 116: 040501, 2016.

[1006] K. Życzkowski. Indicators of quantum chaos based on eigenvector statistics. *J. Phys.*, A 23:4427, 1990.

[1007] K. Życzkowski. Volume of the set of separable states II. *Phys. Rev.*, A 60:3496, 1999.

[1008] K. Życzkowski. Localization of eigenstates & mean Wehrl entropy. *Physica*, E 9:583, 2001.

[1009] K. Życzkowski. Rényi extrapolation for Shannon entropy. *Open Sys. & Information Dyn.*, 10:297, 2003.

[1010] K. Życzkowski, P. Horodecki, A. Sanpera, and M. Lewenstein. Volume of the set of separable states. *Phys. Rev.*, A 58:883, 1998.

[1011] K. Życzkowski and M. Kuś. Random unitary matrices. *J. Phys.*, A 27:4235, 1994.

[1012] K. Życzkowski, K. A. Penson, I. Nechita, and B. Collins. Generating random density matrices. *J. Math. Phys.*, 52:062201, 2011.

[1013] K. Życzkowski and W. Słomczyński. The Monge distance between quantum states. *J. Phys.*, A 31:9095, 1998.

[1014] K. Życzkowski and W. Słomczyński. Monge metric on the sphere and geometry of quantum states. *J. Phys.*, A 34:6689, 2001.

[1015] K. Życzkowski and H.-J. Sommers. Truncations of random unitary matrices. *J. Phys.*, A 33:2045, 2000.

[1016] K. Życzkowski and H.-J. Sommers. Induced measures in the space of mixed quantum states. *J. Phys.*, A 34:7111, 2001.

[1017] K. Życzkowski and H.-J. Sommers. Hilbert–Schmidt volume of the set of mixed quantum states. *J. Phys.*, A 36:10115, 2003.

[1018] K. Życzkowski and H.-J. Sommers. Average fidelity between random quantum states. *Phys. Rev*, A 71:032313, 2005.

Index

Printed in the United States
By Bookmasters